VECTOR ANALYSIS

VECTOR ANALYSIS

Dr. N. CH. S. N. IYENGAR
Asstt. Professor
Department of Mathematics
Vellore Engineering College, Vellore

ANMOL PUBLICATIONS PVT. LTD.
NEW DELHI-110 002 (INDIA)

ANMOL PUBLICATIONS PVT. LTD.
4374/4B, Ansari Road, Daryaganj
New Delhi-110 002

Vector Analysis

First Edition 1997
Reprint, 2003

ISBN 81-7488-673-7

PRINTED IN INDIA

Published by J. L. Kumar for Anmol Publications Pvt. Ltd., New Delhi-110 002 and Printed at Mehra Offset Press, Delhi.

Contents

Preface

As a teacher, teaching Mathematics to college students at different levels about a decade I have been aware of the difficulties experienced by these students in understanding the diverse topics in Mathematics. If has been my long cherished desire to write books on the different areas of Mathematics in simple language using step by step progression so that the students can use them without the enpress help of a teacher.

This book entitled *Vector Analysis* is as attempt at realising the author's desire to cater to the needs of students preparing for B.A./B.Sc., (Pass & Hons.), Engineering, M.A./M.Sc. examinations of various Universities and also for competitive examinations.

In writing this book my objective is to present the necessary theory with definitions, theorems and a large number of solved problems with exercises which may facilitate the student in understanding the subject by self study. I hope so that this book will meet the requirement of students and teachers as well.

I am highly obliged to Thiru. G. Viswanathan, Chairman and Founder of Vellore Engineering College and to Thiru. Sekar Viswanathan, Vice Chairman of V.E.C. for their encouragement and for providing me with the necessary facilities in writing this book.

I acknowledge with thanks to Dr. V. N. Vedamurthy, Principal, Vellore Engineering College, for encouraging and inspiring me all along.

I am grateful to Mr. J. L. Kumar, Managing Director and Mr. Kripal D. Joshi, Production Manager, Anmol Publications Pvt.Ltd., New Delhi, for their painstaking attitude and untiring efforts for bringing out this book beautifully in record time.

Suggestions from teacher and students for improvement of the book are welcome.

N. Ch. S. N. Iyengar

1

Addition of Vectors

1.1. Introduction

In daily life we come across many physical quantities such as mass, length, area, volume, velocity, acceleration, force etc. Taking into account the various properties these physical quantities are classified as **scalars** and **vectors** as defined below.

(i) **Scalar.** A physical quantity which has magnitude but is not related to any fixed direction in space is called a *scalar quantity.*

Examples are length, mass, density, etc. Operations with Scalar quantities, follow the same rules as in elementary algebra and these are indicated by letters as in it. Hence we mean a real number for a scalar.

(ii) **Vector.** A physical quantity which has both magnitude and direction is called a *vector quantity.*

Examples are velocity, acceleration, force, momentum, couple, electric current, gravitational intensities etc.

1.2. Representation and Notion of a Vector

Analytically a vector is represented by a letter with an arrow over it as $|\vec{a}$ and its magnitude as $|\vec{a}|$. We can also notate a vector by a bold face letter **a** and its magnitude as | **a** |. *We use this notation only, in this book throughout.*

Graphically a vector is represented in by means of a *directed line segment* (line vector)

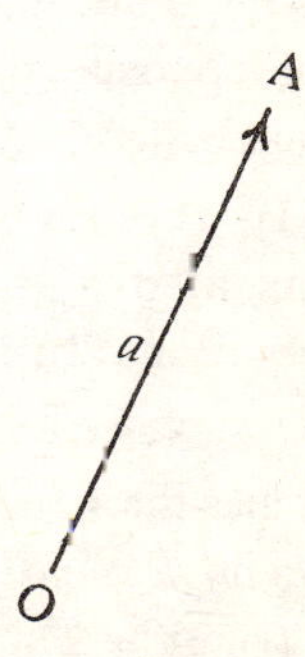

Fig. 1.1.

In the figure (1.1) the directed line segment $\overrightarrow{OA}$ represents a vector. Here the length of $\overrightarrow{OA}$ represents the magnitude $|\overrightarrow{OA}|$ and arrow defines the direction. The tail end O is called the *initial point* and the head A is called the *terminal point.* The sense of directed line segment is from its initial point to its terminal point. (which is non-negative) is called the *modulus of a vector.*

So if $\overrightarrow{OA} = \mathbf{a}$ then modulus of $\overrightarrow{OA} = |\mathbf{a}|$

1.3. Classifications of Vectors

Here under we study different kinds of vectors based on their behaviour and properties.

(i) **Null or zero Vector** A vector whose magnitude is zero and direction is indeterminate is called a *null or zero vector* and denoted by **O**. In this case the initial and terminal points coincide. Thus $\overrightarrow{AA}$, $\overrightarrow{BB}$, ... etc we are all zero vectors.

(ii) **Unit vector** A vector whose magnitude is one is called a *unit vector.* If **a** is any vector then the unit vector is the direction of **a** is represent by $\hat{\mathbf{a}}$ (read as **a** cap). If $|\mathbf{a}|$ is the magnitude of **a** then $\mathbf{a} = |\mathbf{a}|\,\hat{\mathbf{a}}$

i.e.,
$$\hat{\mathbf{a}} = \frac{\mathbf{a}}{|\mathbf{a}|} = \frac{\text{Vector } \mathbf{a}}{\text{magnitude of } \mathbf{a}}$$

(iii) **Like and unlike vectors** Vectors drawn in the same direction whatever their magnitudes may be are called *like vectors.* On the other hand if they are of opposite directions, then they are called *unlike vectors*

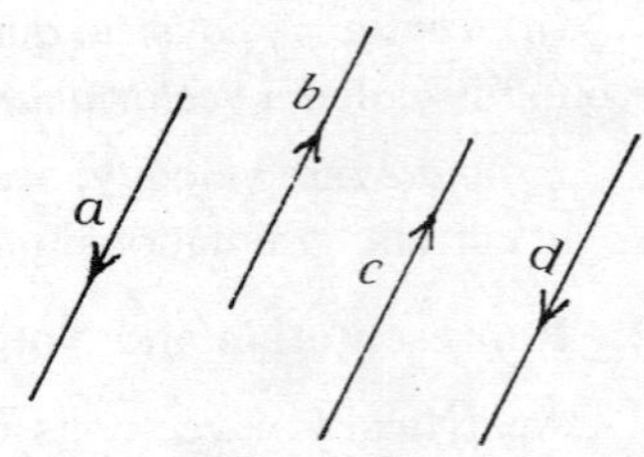

Fig. 1.2.

In the figure (1.2) the vectors **a, d** and **b, c** are like vectors. But **a, b; a, c; b, d** are unlike vectors.

(iv) **Negative vector:** Let **a** be any vector. Then the line vector which has the same magnitude as **a** but opposite in direction in called the *negative vector of* **a**.

If $\overrightarrow{AB} = \mathbf{a}$ then $\overrightarrow{BA} = -\mathbf{a}$ provided $|\overrightarrow{AB}| = |\overrightarrow{BA}|$

(v) **Collinear Vectors:** *(Parallel Vectors).* Vectors having the same line of action or having the lines of action parallel to one another are called *collinear vectors* (or parallel vectors), whatever be their magnitudes and directions.

(vi) **Equal vectors:** Two vectors are said to be *equal* iff they are parallel, have the same sense of direction and the same magnitude.

∴ If **a** and **b** are two equal vectors then we write **a** = **b**. Here initial point is unimportant. Only direction and magnitudes are important.

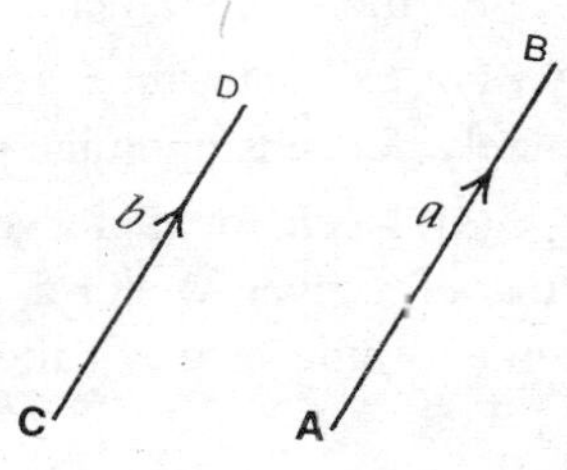

Fig. 1.3.

In the figure (1.3), vectors $\vec{AB}$ and $\vec{CD}$ are equal if $|\vec{AB}| = |\vec{CD}|$

Note: *It does not mean that if two vectors are equal they are equivalent in all aspects.*

(vii) **Coplanar vectors :** Vectors are said to be *coplanar* if they lie on the same plane or if they are paralled to the same plane.

(viii) **Co-initial vectors:** The vectors which have the same initial point are called *co–initial vectors.*

A vector remains unaltered by shifting in space parallel to itself. Hence any vector say $\vec{AB}$ = **a** may be drawn from any assigned origin O by shifting the line segment AB parallel to itself such that A coincides with O and B coincides with some other point P so that

$$\vec{AB} = \vec{OP} = \mathbf{a}$$

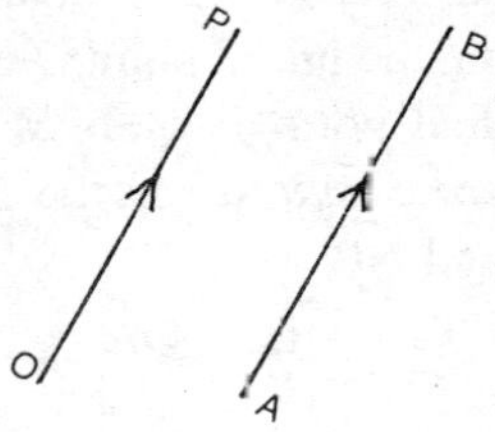

Fig. 1.4.

In the figure (1.4) $\vec{AB}$, $\vec{OP}$ are coinitial vectors. In this way all the vectors in the space can be replaced by vectors drawn from the same assigned origin O by shifting them parallel to themselves till their initial point coincides with the origin O. All such vectors having the same origin are called co-initial vectors.

(ix) **Free vectors:** Vectors whose initial points are not fixed are called *free vectors,*

(x) **Localised vectors:** If there is a restriction to choose initial point for the vectors then they are called *localised vectors.*

For example, force is a localised vector, because the effect of it purely depends upon the point where it acts.

(xi) **Reciprocal of a vector:** A vector having the same direction as that of a given vector **a**, but magnitude equal to the reciprocal of the magnitude of **a** is called reciprocal vector of **a** and is written as $\mathbf{a}^{-1}$

$$\therefore \qquad \mathbf{a}^{-1} = \frac{\mathbf{a}}{|\mathbf{a}|^2}$$

1.4. Position Vectors

The vectors which are used to specify the position of a point P with respect to some fixed point O is represented by $\overrightarrow{OP}$ and is known as *position vector* of P w.r.t O.

1.5 Addition of Vectors

If the points O, A, B are taken such that $\overrightarrow{OA} = \mathbf{a}$, $\overrightarrow{AB} = \mathbf{b}$ then the vector $\overrightarrow{OB}$ is obtained by joining the initial point O of **a** and terminal point B of **b** called the *sum or resultant* of the Vectors $\overrightarrow{AB}$ and $\overrightarrow{AB}$

$$\therefore \qquad \overrightarrow{OB} = \overrightarrow{OA} + \overrightarrow{AB}$$
$$= \mathbf{a} + \mathbf{b}$$

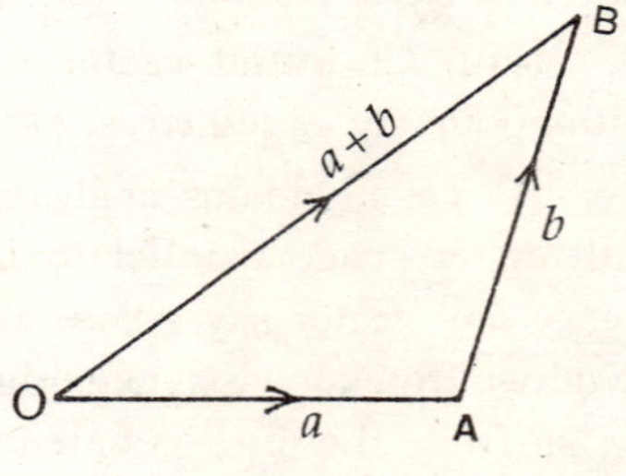

Fig. 1.5.

This law of addition is known as triangle law of addition.

1.6. Properties of Vector Addition

(i) *Vector addition is commutative*

If **a** and **b** are any two vectors then $\mathbf{a} + \mathbf{b} = \mathbf{b} + \mathbf{a}$

Proof. Let $\overrightarrow{OA} = \mathbf{a}$, $\overrightarrow{AB} = \mathbf{b}$

Then $\qquad \overrightarrow{OB} = \overrightarrow{OA} + \overrightarrow{AB}$

$$= \mathbf{a} + \mathbf{b} \qquad ...(1)$$

Complete the parallelogram OABC as shown in the figure (1.6). We have by definition of equality of vectors $\overrightarrow{OC} = \overrightarrow{AB} = \mathbf{b}$ and $\overrightarrow{CB} = \overrightarrow{OA} = \mathbf{a}$.

$$\therefore \quad \overrightarrow{OB} = \overrightarrow{OC} + \overrightarrow{CB}$$

$$= \mathbf{b} + \mathbf{a} \qquad ...(2)$$

Fig. 1.6.

$\therefore$ From (1) and (2) we have

$$\mathbf{a} + \mathbf{b} = \mathbf{b} + \mathbf{a}$$

when proves the law.

(ii) *Vector addition is associative.*

If **a, b, c** be any three vectors then

$$\mathbf{a} + (\mathbf{b} + \mathbf{c}) = (\mathbf{a} + \mathbf{b}) + \mathbf{c}$$

Proof. Let the vectors **a, b** and **c** be represented by $\overrightarrow{OA}$, $\overrightarrow{AB}$ and $\overrightarrow{BC}$ respectively.

Then $\mathbf{a} + \mathbf{b} = \overrightarrow{OA} + \overrightarrow{AB} = \overrightarrow{OB}$

and $\qquad \mathbf{b} + \mathbf{c} = \overrightarrow{AB} + \overrightarrow{BC} = \overrightarrow{AC}$

Now $\mathbf{a} + (\mathbf{b} + \mathbf{c}) = \overrightarrow{OA} + \overrightarrow{AC} = \overrightarrow{OC}$ ...(1)

and $\quad (\mathbf{a} + \mathbf{b}) + \mathbf{c} = \overrightarrow{OB} + \overrightarrow{BC} = \overrightarrow{OC}$ ...(2)

$\therefore$ From (1) and (2), we get

$$\mathbf{a} + (\mathbf{b} + \mathbf{c}) = (\mathbf{a} + \mathbf{b}) + \mathbf{c}$$

which proves the statement.

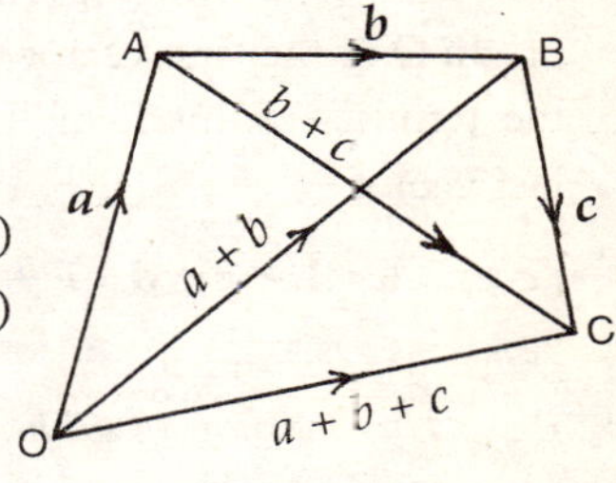

Fig. 1.7.

Note: *Sum of the three vectors* **a, b, c** *is written as* **a + b + c** *and is independent of order.*

(iii) *For every vector* **a**, *there exists a zero vector* **O** *Such that* $\mathbf{a} + \mathbf{0} = \mathbf{a} = \mathbf{0} + \mathbf{a}$

Proof. Let $\overrightarrow{OA} = \mathbf{a}$ and $\overrightarrow{AA} = \mathbf{0}$

Then $\qquad \overrightarrow{OA} + \overrightarrow{AA} = \overrightarrow{OA} \Rightarrow \mathbf{a} + \mathbf{0} = \mathbf{a}$

Similarly we can prove that $\mathbf{0} + \mathbf{a} = \mathbf{a}$.

(iv) *To every vector* **a**, *there exists the vector* **–a** *such that* $\mathbf{a} + (-\mathbf{a}) = \mathbf{0} = (-\mathbf{a}) + \mathbf{a}$, *where* **0** *is the zero vector.*

Proof. Let $\overrightarrow{OA} = \mathbf{a} \quad \therefore \quad \overrightarrow{AO} = -\mathbf{a}$

Hence $\vec{OA} + \vec{AO} = \vec{OO}$

i.e., $\mathbf{a} + (-\mathbf{a}) = \mathbf{0}$

Similarly we can prove that

$$(-\mathbf{a}) + \mathbf{a} = \mathbf{0}$$

Note. If **V** be the set of all the vectors and addition of vectors be the binary operation defined in **V** then, [**V**, +] is an abelian group in view of above properties.

1.7. Addition of any number of vectors

Let **a, b, c, d, e** be five given vectors represented by $\vec{OA}$, $\vec{AB}$, $\vec{BC}$, $\vec{CD}$, $\vec{DE}$. Then the sum (also called resultant) is represented in length and direction by $\vec{OE}$. i.e., the line joining the origin O of the first vector to the terminal point E of the last vector

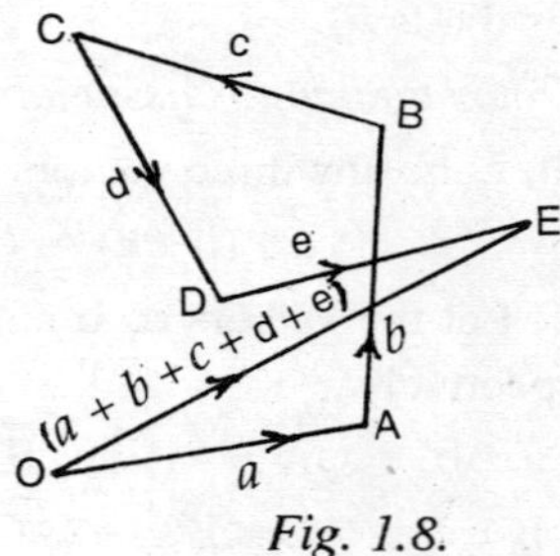

Fig. 1.8.

i.e., $\mathbf{a} + \mathbf{b} + \mathbf{c} + \mathbf{d} + \mathbf{e} = \vec{OA} + \vec{AB} + \vec{BC} + \vec{CD} + \vec{DE}$

$$= \vec{OE}$$

This method of addition is called polygon law of addition

Note 1. *The vector law of addition does not require that component vector should be coplanar ⇒OABCD need not be coplanar.*

Note 2. From above, we have

$$\vec{OA} + \vec{AB} + \vec{BC} + \vec{CD} + \vec{DE} = \vec{OE}$$

$$= -\vec{EO}$$

$$\Rightarrow \vec{OA} + \vec{AB} + \vec{BC} + \vec{CD} + \vec{DE} + \vec{EO} = O$$

⇒ *Sum of vectors determined by the sides of polyogon (true for any polygon) is zero.*

Note 3. In a triangle ABC, $\vec{AB} + \vec{BC} + \vec{CA} = \mathbf{O}$.

1.8 Subtraction of Vectors

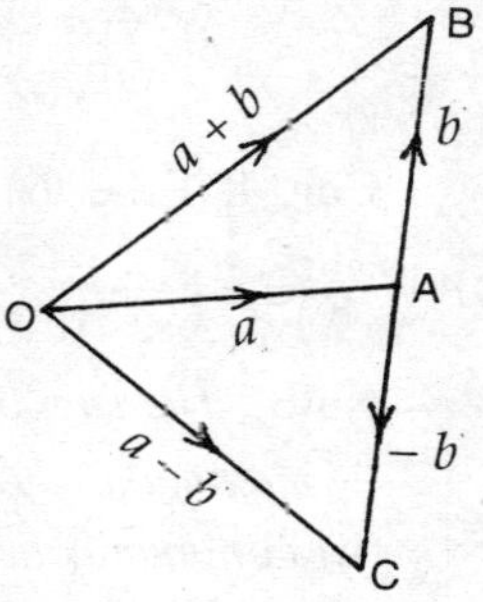

Fig. 1.9.

We have already defined that – **a** is a vector which has the same magnitude as **a** but in opposite direction.

∴ The subtraction of **b** from **a** should be taken as the addition of – **b** to **a**.

Thus we have $\mathbf{a} - \mathbf{b} = \mathbf{a} + (-\mathbf{b})$

i.e., to subtract a vector **b** from **a**, we should reverse the direction of **b** and add to **a**. In the figure $\overrightarrow{OC}$ represents **a – b**.

1.9. Expression of a vector in terms of position vectors of its end point

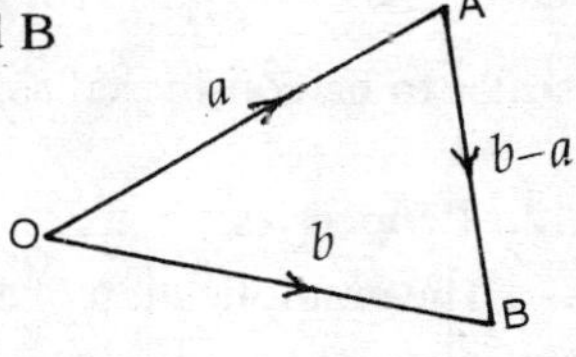

Fig. 1.10

Let O be the origin and let the position vectors of the end points A and B of a vector $\overrightarrow{AB}$ w.r.t O be **a** and **b**.

i.e., $\overrightarrow{OA} = \mathbf{a}$ and $\overrightarrow{OB} = \mathbf{b}$

These $\overrightarrow{AB} = \overrightarrow{AO} + \overrightarrow{OB}$

$= \overrightarrow{OB} - \overrightarrow{OA} = \mathbf{b} - \mathbf{a}$

= Position vector of B–
Position vector of A

1.10. Position Vector of a point P dividing the line AB in the given ratio m : n

Let O be the origin and

$\overrightarrow{OA} = \mathbf{a}, \ \overrightarrow{OB} = \mathbf{b}.$

Given that P divides AB in m : n ratio.

$\Rightarrow \quad nAP = mPB \qquad ...(1)$

But $\overrightarrow{AP} = \overrightarrow{OP} - \overrightarrow{OA}$

and $\overrightarrow{PB} = \overrightarrow{OB} - \overrightarrow{OP}$

Then from (1), we get

$n(\overrightarrow{OP} - \overrightarrow{OA}) = m(\overrightarrow{OB} - \overrightarrow{OP})$

or $(m + n)\overrightarrow{OP} = n\overrightarrow{OA} + m\overrightarrow{OB}$

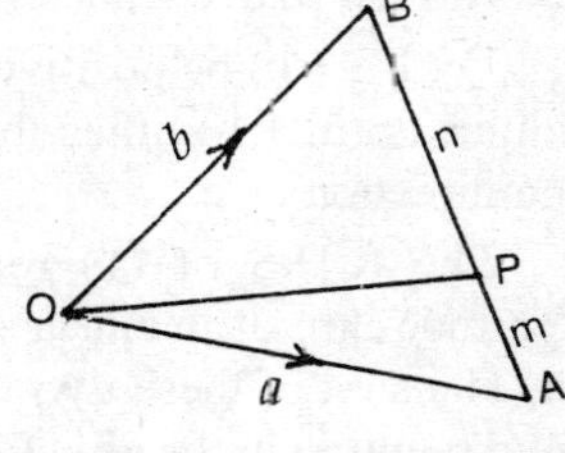

Fig. 1.11

or $$\vec{OP} = \frac{n\,\vec{OA} + m\,\vec{OB}}{m + n} = \frac{n\mathbf{a} + m\mathbf{b}}{m + n}$$

Cor. If P be the midpoint of AB then m = n and we have $\vec{OP} = \frac{1}{2}(\mathbf{a} + \mathbf{b})$

Note : *The sum of two vectors is*

(i) *maximum when the two vectors are like vectors*

(ii) *minimum when the two vectors are unlike vectors*

(iii) *zero when the two vectors are equal in magnitude but opposite in direction.*

1.11. Multiplication of a Vector by a Scalar

Let m be a scalar and **a** be a vector, then m**a** is defined as a vector whose modulus is |m| times the modulus of the vector **a** according as m is positive or negative. The division of a vector **a** by a scalar m be considered as the product of the vector **a** by $\frac{1}{m}$.

1.12. Properties

The multiplication of a vector by a scalar satisfies the following properties

(i) *Commutative law* : m**a** = **a**m

(ii) *Associative law:* : m(n**a**) = (mn)**a** = (mn)**a**

(iii) *Distributive law* : (m + n)**a** = m**a** + n**a**

and : m(**a** + **b**) = m**a** + m**b** *where* m, n *are any two scalars and* **a, b** *are any two vectors.*

Proof. The first two can be obtained direct from definition.

(iii) To prove that (m + n)**a** = m**a** + n**a** ...(1)

Let (m + n) be positive. Then L.H.S represents a vector whose modulus is (m + n) times the modulus of **a** and which points in the same direction as **a**.

The R.H.S of (1) represents the sum of two vectors of magnitudes |m| |a| and |n| |a| pointing in the direction of **a** or opposite to **a**. The sum of these two vectors is a vector of magnitude (m + n) |**a**| and pointing in the direction of **a**. If (m + n) is negative, there both the sides of (1) represent a vector of magnitude |m + n| |a| and pointing in the direction opposite to **a**. Hence the proof.

To prove that m(**a** + **b**) = m**a** + m**b**

Case I: m = 0. In this case, vectors m(**a** + **b**), m**a** and m**b** are all zero vectors and so the equality follows trivially.

Case II. m > 0.

Let **a** = OA and **b** = AB

∴ **a** + **b** = OA + AB = OB

Let A′ be on OA such that OA′ = mOA

$\vec{OA'} = m\,\vec{OA}$ = m**a**

Draw A′B′ parallel to AB and let A′B′ = m AB

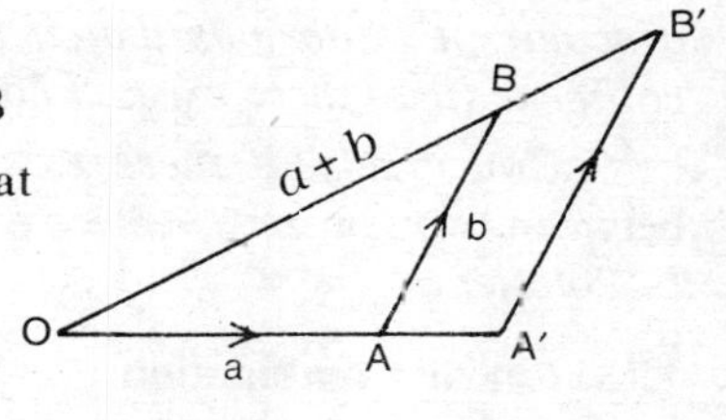

Fig. 1.12

∴ $\vec{A'B'} = m\,\vec{AB}$ = m**b**. The choice of A′ and B′

$\Rightarrow \dfrac{OA'}{OA} = \dfrac{A'B'}{AB}$

∴ Δ OAB ||| Δ OA′B′. So OB′ = mOB

$\Rightarrow \vec{OA'} + \vec{A'B'} = m\,(OA' + \vec{AB})$

⇒ m**a** + m**b** = m (**a** + **b**)

Case III. m < 0 et **a** = $\vec{OA}$ and **b** = $\vec{AB}$

∴ **a** + **b** = OA + AB = OB

Let A′ be on OA Such that *A′O* = (− *m*) *OA*

∴ $\vec{A'O} = (-m)\,\vec{OA'} \Rightarrow \vec{OA} = m\,\vec{OA}$ = m **a**

Draw A′B′ parallel to AB and let B′A′ = (− m) AB

∴ $\vec{B'A'} = (-m)\,\vec{AB}$

and so $\vec{A'B'}$ = m**b**

Now, we have

$-m = \dfrac{A'O}{OA} = \dfrac{B'A'}{AB}$

and so Δ OAB and OA′B′ are similar.

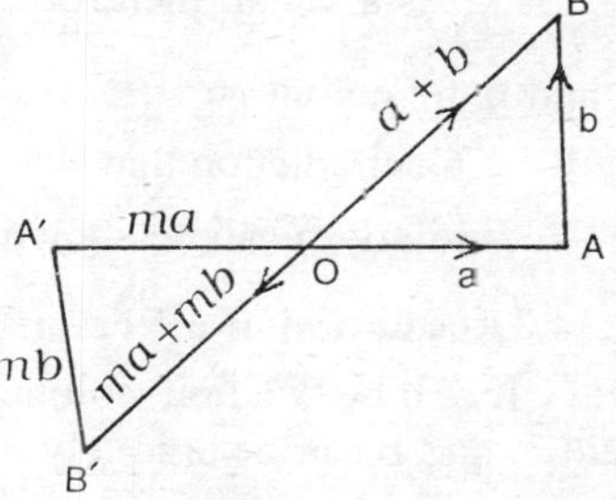

Fig. 1.13

∴ BO′ = − m OB

$\Rightarrow \vec{OB'} = m\,\vec{OB}$

$\Rightarrow \vec{OA'} + \vec{A'B'} = m\,(\vec{OA} + \vec{AB})$

$\Rightarrow \quad \mathbf{ma} + \mathbf{mb} = \mathrm{m}(\mathbf{a} + \mathbf{b})$

$\therefore$ For all possible values of m,

$$\mathrm{m}(\mathbf{a} + \mathbf{b}) = \mathbf{ma} + \mathbf{mb}$$

Imp. Note. *From the definition of multiplication of a vector by a scalar, it is obvious that if two non-zero vectors* **a** *and* **b** *are collinear (like) there exists a non-zero scalar m such that* **a** = **mb.**

Conversely, if there exists a relation of the type **a** = **mb** between two non-zero vectors **a** and **b** *then the vectors* **a** *and* **b** *must be collinear.*

1.13. Linear Combination

A vector **r** is said to be a linear combination of the given vectors **a, b, c, ...** etc, if there exists scalars x, y, z, ... such that

$$\mathbf{r} = x\,\mathbf{a} + y\mathbf{b} + z\mathbf{c} + \ldots$$

Imp Note. The necessary and sufficient condition that a linear relation $x_1\mathbf{a}_1 + x_2\,\mathbf{a}_2 + \ldots + x_n\mathbf{a}_n = \mathbf{O}$, connecting the position vectors $\mathbf{a}_1, \mathbf{a}_2, \ldots, \mathbf{a}_n$ of fixed points is independent of origin is that $x_1 + x_2 + \ldots + x_n = 0$.

Theorem. If **a**, **b** are two non-zero, non-collinear vectors and x, y are two scalars such that $x\mathbf{a} + y\mathbf{b} = \mathbf{0}$ then x = 0, y = 0.

Proof. Given that $\quad x\mathbf{a} + y\mathbf{b} = 0 \quad$ ...(1)

Let $x \neq 0$. then (1) can be written as

$$\mathbf{a} = -\frac{y}{x}\mathbf{b} \qquad \text{...(2)}$$

$\because \frac{y}{x}$ is a scalar, therefore, equation (2) implies that the vectors **a** and **b** are collinear

$\therefore$ Contradiction that $x \neq 0 \Rightarrow x = 0$

Similarly it can be shown that y = 0.

1.14. Resolution of a Vector in terns of coplanar Vectors

If a, b be two non-collinear vectors then any vector r coplanar with a and b can be uniquely expressed as $\mathbf{r} = x\mathbf{a} + y\mathbf{b}$ where x and y are scalars.

Proof. Let $\overrightarrow{OA} = \mathbf{a}$, $\overrightarrow{OB} = \mathbf{b}$ and $\overrightarrow{OP} = \mathbf{r}$.

OA, OB, OP are coplanar. Through the point P draw lines

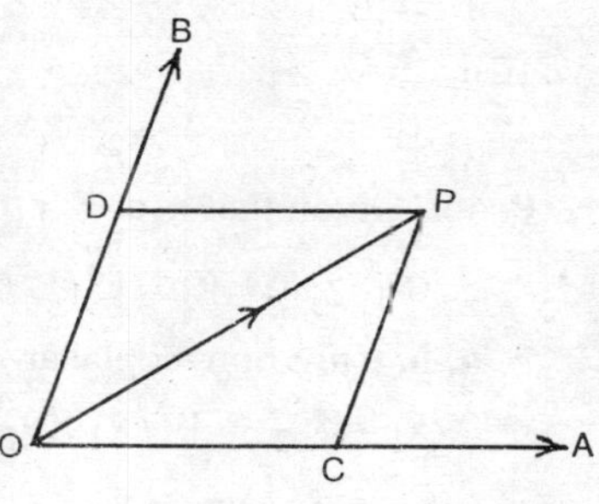

Fig. 1.14

parallel to OB and OA meeting OA and OB at C and D respectively.

Now $\overrightarrow{OC}$ and $\overrightarrow{OD}$ are collinear with $\overrightarrow{OA}$ and $\overrightarrow{OB}$ respectively.

$\therefore$ $\overrightarrow{OC} = x\,\overrightarrow{OA}$ and $\overrightarrow{OD} = y\,\overrightarrow{OB}$ where x and y are scalars

i.e., $\overrightarrow{OC} = x\mathbf{a}$ and $\overrightarrow{OD} = y\mathbf{b}$

But we have $\overrightarrow{OP} = \overrightarrow{OC} + \overrightarrow{CP} = \overrightarrow{OC} + \overrightarrow{OD} = x\mathbf{a} + y\mathbf{b}$

$\therefore$ $\mathbf{r} = x\mathbf{a} + y\mathbf{b}$

Uniqueness. If possible let $\mathbf{r} = x_1\mathbf{a} + y_1\mathbf{b}$

Then we have $x\mathbf{a} + y\mathbf{b} = x_1\mathbf{a} + y_1\mathbf{b}$

or $$(x - x_1)\,\mathbf{a} + (y - y_1)\,\mathbf{b} = \mathbf{0} \qquad ...(1)$$

$\therefore$ **a** and **b** are non-collinear vectors, therefore for the existence of the linear relation (1) between **a** and **b**, we should have

$$x - x_1 = 0 \text{ and } y - y_1 = 0$$

i.e., $x = x_1$ and $y = y_1$.

Hence the linear combination $\mathbf{r} = x\mathbf{a} + y\mathbf{b}$ is unique.

Theorem. If **a, b, c** are three non-coplanar vectors and x, y, z are scalars such that $x\mathbf{a} + y\mathbf{b} + z\mathbf{c} = \mathbf{0}$ then $x = y = z = 0$.

Proof. Given that $x\mathbf{a} + y\mathbf{b} + z\mathbf{c} = \mathbf{0}$ ...(1)

Then (1) can be written as

$$\mathbf{a} = -\frac{y}{x}\mathbf{b} - \frac{z}{x}\mathbf{c} \qquad ...(2)$$

$\therefore$ $\frac{y}{x}$ and $\frac{z}{x}$ are scalars, therefore equation (2) implies that the vector **a** is coplanar with the vectors **b** and **c** which is contrary to the hypothesis. Hence $x = 0$.

Similarly we can show that $y = 0$, and $z = 0$.

Corollary. If $x_1\mathbf{a} + y_1\mathbf{b} + z_1\mathbf{c} = x_2\mathbf{a} + y_2\mathbf{b} + z_2\mathbf{c}$ where **a, b, c** are non-coplanar vectors, x_1, y_1, z_1, x_2, y_2 and z_2 are scalars then prove that

$$x_1 = x_2,\ y_1 = y_2,\ z_1 = z_2$$

Proof. Given that $x_1\mathbf{a} + y_1\mathbf{b} + z_1\mathbf{c} = x_2\mathbf{a} + y_2\mathbf{b} + z_2\mathbf{c}$

$$\therefore \quad (x_1 - x_2)\,\mathbf{a} + (y_1 - y_2)\mathbf{b} + (z_1 - z_2)\mathbf{c} = 0$$

$\therefore$ **a, b, c** are non-coplanar

$$\Rightarrow \quad x_1 - x_2 = 0,\ y_1 - y_2 = 0,\ z_1 - z_2 = 0$$

$$\Leftrightarrow \quad x_1 = x_2,\ y_1 = y_2,\ z_1 = z_2$$

1.15. Resolution of Non-coplanar Vectors

If **a, b, c** be three given non-coplanar vectors, then any vector **r** can be uniquely expressed as a linear combination **xa + yb + zc,** where x, y, z are Scalars.

Proof. Let $\overrightarrow{OA} = \mathbf{a}$, $\overrightarrow{OB} = \mathbf{b}$ and $\overrightarrow{OC} = \mathbf{c}$. The lines OA, OB and OC are not in the same plane. Let $\overrightarrow{OP} = \mathbf{r}$. Construct a parallelopiped with OP as diagonal and other conterminous edges OV, OR and OT along OA, OB and OC respectively.

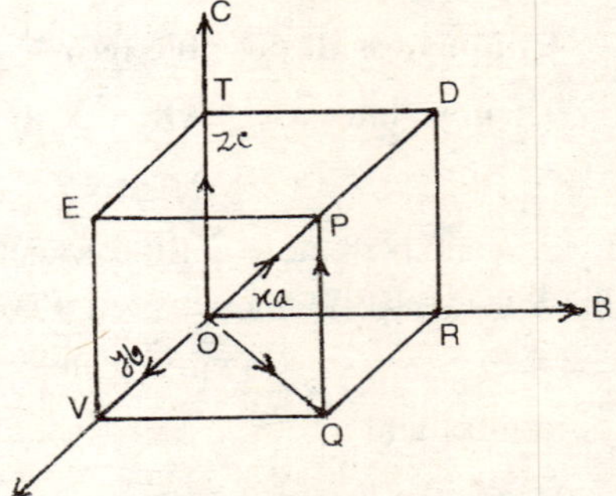

Fig. 1.15

Note. $\overrightarrow{OV} = x\,\overrightarrow{OA}$, $\overrightarrow{OR} = y\,\overrightarrow{OB}$ and $\overrightarrow{OT} = z\,\overrightarrow{OC}$.

[Since $\overrightarrow{OV}$, $\overrightarrow{OR}$ and $\overrightarrow{OT}$ are collinear with $\overrightarrow{OA}$, $\overrightarrow{OB}$ and $\overrightarrow{OC}$]

where x, y, z are scalars

$$\therefore \quad \overrightarrow{OV} = x\mathbf{a},\ \overrightarrow{OR} = y\mathbf{b} \text{ and } \overrightarrow{OT} = z\mathbf{c}$$

Note. $\overrightarrow{OP} = \overrightarrow{OQ} + \overrightarrow{QP} = \overrightarrow{OV} + \overrightarrow{VQ} + \overrightarrow{QP}$

$$= \overrightarrow{OV} + \overrightarrow{OR} + \overrightarrow{OT}$$

$$[\because \ \overrightarrow{VQ} = \overrightarrow{OR} \text{ and } \overrightarrow{QP} = \overrightarrow{OT}]$$

$$\therefore \quad \mathbf{r} = x\mathbf{a} + y\mathbf{b} + z\mathbf{c}$$

Uniqueness. If possible let $\mathbf{r} = x_1\mathbf{a} + y_1\mathbf{b} + z_1\mathbf{c}$

Then $x\mathbf{a} + y\mathbf{b} + z\mathbf{c} = x_1\mathbf{a} + y_1\mathbf{b} + z_1\mathbf{c}$

or $\quad (x - x_1)\mathbf{a} + (y - y_1)\mathbf{b} + (z - z_1)\mathbf{c} = \mathbf{0}$

$\therefore$ **a, b, c** are non–coplanar

$\Rightarrow$ $x - x_1 = 0, \ y - y_1 = 0, \ z - z_1 = 0$

or $x = x_1, \ y = y_1, \ z = z_1$

i.e., resolution is unique.

Note : In the relation $\mathbf{r} = x\mathbf{a} + y\mathbf{b} + z\mathbf{c}$, the vector **r** is called *resultant* where as x**a**, y**b** and z**c** are called *components of* **r**

1.16. Linearly Dependent and Independent System of Vectors

If the linear combination of vectors **a, b, c, ...** be zero i.e., if there is a relation

$x\mathbf{a} + y\mathbf{b} + z\mathbf{c} + \ldots = \mathbf{0}$ where x, y, z, ... are the scalars then the system of vectors **a, b, c, ...** is said to be

(i) **linearly dependent** if all the scalars are not zero i.e., all x, y, z, ... are not zero.

(ii) **linearly independent** if all the scalars are zero i.e., x = o, y = 0, z = 0, ...

SOLVED PROBLEMS

Example 1. The position vectors of the points A, B, C and D are **a, b, 2a + 3b** and **a − 2b**. Find the vectors $\vec{AC}$, $\vec{OB}$, $\vec{BC}$ and $\vec{CA}$

Solution: Let O be the origin.

Then $\vec{OA} = \mathbf{a}$, $\vec{OB} = \mathbf{b}$, $\vec{OC} = 2\mathbf{a} + 3\mathbf{b}$ and $\vec{OD} = \mathbf{a} - 2\mathbf{b}$

Now $\vec{AC}$ = P.V. of C - P.V. of A

$= \vec{OC} - \vec{OA} = \mathbf{b} - \mathbf{a}$

$\vec{DB} = \vec{OB} - \vec{OD} = \mathbf{b} - (\mathbf{a} - 2\mathbf{b}) = 3\mathbf{b} - \mathbf{a}$

$\vec{BC} = \vec{OC} - \vec{OB} = (2\mathbf{a} + 3\mathbf{b}) - \mathbf{b} = 2(\mathbf{a} + \mathbf{b})$

and $\vec{CA} = \vec{OA} - \vec{OC} = \mathbf{a} - (2\mathbf{a} + 3\mathbf{b}) = -(\mathbf{a} + 3\mathbf{b})$

Example 2. If ABCDEF is a regular hexagon, and $\vec{AB} = \mathbf{a}$, $\vec{BC} = \mathbf{b}$ then express $\vec{AC}$, $\vec{AD}$, $\vec{AE}$, $\vec{AF}$, $\vec{CD}$, $\vec{DE}$, $\vec{EF}$ and $\vec{FA}$.

Solution: Consider A as origin.

Given that $\vec{AB} = \mathbf{a}$ and $\vec{BC} = \mathbf{b}$. we know that

$$\vec{AC} = \vec{AB} + \vec{BC} = \mathbf{a} + \mathbf{b}$$

$\because$ $\vec{AD}$ is parallel to $\vec{BC}$ and double of $\vec{BC}$

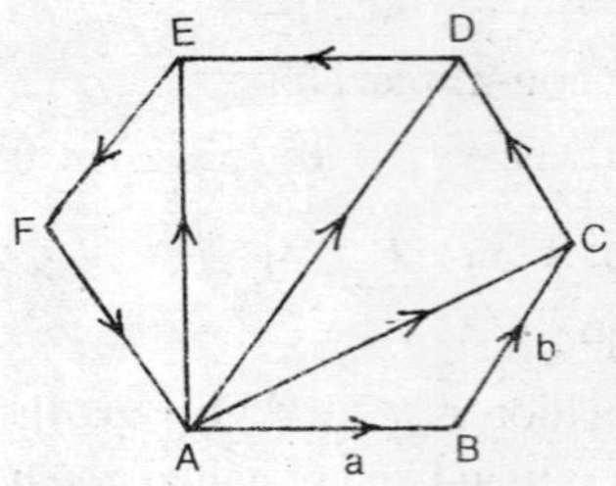

Fig. 1.16

$\therefore \quad \overrightarrow{AD} = 2\,\overrightarrow{BC} = 2\mathbf{b}$

Now $\overrightarrow{AE} + \overrightarrow{AD} + \overrightarrow{DE} = \overrightarrow{AD} + \overrightarrow{BA}$

$= \overrightarrow{AD} - \overrightarrow{AB} = 2\mathbf{b} - \mathbf{a}$

$\overrightarrow{AF} = \overrightarrow{AE} + \overrightarrow{EF} = \overrightarrow{AE} + \overrightarrow{CB}$

$= \overrightarrow{AE} - \overrightarrow{BC} = 2\mathbf{b} - \mathbf{a} - \mathbf{b} = \mathbf{b} - \mathbf{a}$

$\overrightarrow{CD} = \overrightarrow{AD} - \overrightarrow{AC} = 2\mathbf{b} - (\mathbf{a} + \mathbf{b}) = \mathbf{b} - \mathbf{a}$

$\overrightarrow{DE} = \overrightarrow{BA} = -\overrightarrow{AB} = -\mathbf{a}$

$\overrightarrow{EF} = \overrightarrow{CB} = -\overrightarrow{BC} = -\mathbf{b}$

and $\quad \overrightarrow{FA} = \overrightarrow{DC} = -\overrightarrow{CD} = -(\mathbf{b} - \mathbf{a}) = \mathbf{a} - \mathbf{b}$

Example 3. If D be the midpoint of the side BC of a ΔABC, then prove that $\overrightarrow{AB} + \overrightarrow{AC} = 2\,\overrightarrow{AD}$

Solution: From triangle law of addition in ΔABD we have

$\overrightarrow{AB} = \overrightarrow{AD} + \overrightarrow{DB}$

and from ΔACD, we have

$\overrightarrow{AC} = \overrightarrow{AD} + \overrightarrow{DC}$

Now $\overrightarrow{AB} + \overrightarrow{AC}$

$= (\overrightarrow{AD} + \overrightarrow{DB}) + (\overrightarrow{AD} + \overrightarrow{DC})$

$= 2\overrightarrow{AD} + (\overrightarrow{DB} + \overrightarrow{DC})$

$= 2\,\overrightarrow{AD} + O \qquad (\because \overrightarrow{DB} = -\overrightarrow{DC})$

$\therefore \quad \overrightarrow{AB} + \overrightarrow{AC} = 2\overrightarrow{AD}$.

Fig. 1.17

Example 4. If **a**, **b**, **c**, **d** represent the consecutive sides of a quadrilateral, show that the necessary and sufficient condition that the quadrilateral be a parallelogram is that $\mathbf{a} + \mathbf{c} = 0$ and that this implies $\mathbf{b} + \mathbf{d} = \mathbf{0}$.

Solution Let **a, b, c, d** represent the Sides $\overrightarrow{AB}$, $\overrightarrow{BC}$, $\overrightarrow{CD}$ and $\overrightarrow{DA}$ of the quadrilateral ABCD.

i.e., $\overrightarrow{AB} = \mathbf{a}$, $\overrightarrow{BC} = \mathbf{b}$, $\overrightarrow{CD} = \mathbf{c}$ and $\overrightarrow{DC} = \mathbf{d}$

$$AB + BC = AC = \mathbf{a} + \mathbf{b}$$

also $$\overrightarrow{AC} + \overrightarrow{CD} = \overrightarrow{AD}$$

$$\therefore \quad \mathbf{a} + \mathbf{b} + \mathbf{c} = -\mathbf{d}$$

or $$\mathbf{a} + \mathbf{b} + \mathbf{c} + \mathbf{d} = 0 \qquad ...(1)$$

Now if ABCD is a parallelogram then $\overrightarrow{AB}$ and $\overrightarrow{DC}$ are parallel and equal.

Fig. 1.18

$$\therefore \quad \overrightarrow{AB} = -\overrightarrow{CD}$$

i.e., $$\mathbf{a} = -\mathbf{c} \quad \text{or} \quad \mathbf{a} + \mathbf{c} = 0$$

Hence the condition is necessary. Also using (1) we get $\mathbf{b} + \mathbf{d} = 0$.

Sufficient $\because \mathbf{a} + \mathbf{c} = 0 \quad \therefore \quad \mathbf{a} = -\mathbf{c}$

i.e., $$\overrightarrow{AB} = -\overrightarrow{CD} \Rightarrow \overrightarrow{AB} = \overrightarrow{DC}$$

Thus $\overrightarrow{AB}$ and $\overrightarrow{DC}$ are parallel and equal when $\mathbf{a} + \mathbf{c} = 0$, we have from (1)

$$\mathbf{b} + \mathbf{d} = 0 \Leftrightarrow \overrightarrow{BC} = \overrightarrow{AD}.$$

$\therefore$ $\overrightarrow{BC}$ and $\overrightarrow{AD}$ are parallel and equal.

Hence ABCD is a parallelogram.

Example 5. ABCD is a parallelogram. If L, M be the middle points of $\overrightarrow{BC}$ and $\overrightarrow{CD}$, express $\overrightarrow{AL}$ and $\overrightarrow{AM}$ in terms of $\overrightarrow{AB}$ and $\overrightarrow{AD}$ and also show that $\overrightarrow{AL} + \overrightarrow{AM} = \frac{3}{2}\overrightarrow{AC}$.

Solution: Let $\overrightarrow{AB} = \mathbf{a}$ and $\overrightarrow{AD} = \mathbf{b}$, refering A as origin. Here $\overrightarrow{AB} = \overrightarrow{CD}$ and $\overrightarrow{AD} = \overrightarrow{BC}$

$$\overrightarrow{AC} = \overrightarrow{AD} + \overrightarrow{DC} = \overrightarrow{AD} + \overrightarrow{AB}$$

$$= \mathbf{b} + \mathbf{a}$$

$$\overrightarrow{AL} = \text{Position vector of L}$$

$$= \text{midpoint of } \overrightarrow{BC}$$

$$= \frac{1}{2}(\overrightarrow{AB} + \overrightarrow{AC})$$

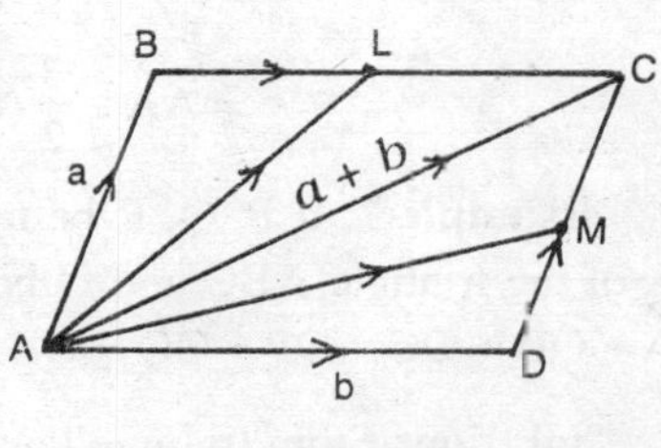

Fig. 1.19

$$= \frac{1}{2}(\mathbf{a} + \mathbf{b} + \mathbf{a}) = \mathbf{a} + \frac{1}{2}\mathbf{b} = \overrightarrow{AB} + \frac{1}{2}\overrightarrow{AD}$$

Similarly $\overrightarrow{AM}$ = Position vector of M

$$= \text{Mid point of } \overrightarrow{DC}$$

$$= \frac{1}{2}[\overrightarrow{AD} + \overrightarrow{AC}] = \frac{1}{2}[\mathbf{b} + \mathbf{b} + \mathbf{a}]$$

$$= \mathbf{b} + \frac{1}{2}\mathbf{a} = \overrightarrow{AD} + \frac{1}{2}\overrightarrow{AB}$$

$$\therefore \overrightarrow{AL} + \overrightarrow{AM} = \left(\overrightarrow{AB} + \frac{1}{2}\overrightarrow{AD}\right) + \left(\overrightarrow{AD} + \frac{1}{2}\overrightarrow{AB}\right)$$

$$= \frac{3}{2}(\overrightarrow{AB} + \overrightarrow{AD}) = \frac{3}{2}(\mathbf{b} + \mathbf{a}) = \frac{3}{2}\overrightarrow{AC}$$

Example 6. Show that the sum of three vectors determined by the medians of a triangle directed from the vertices in zero.

Solution: Let $\overrightarrow{AD}, \overrightarrow{BE}, \overrightarrow{CF}$ be the medians of the ΔABC.

There $\overrightarrow{AD} = \overrightarrow{AB} + \overrightarrow{BD}$

$$= \overrightarrow{AB} + \frac{1}{2}\overrightarrow{BC} \left(\therefore \overrightarrow{BD} = \frac{1}{2}\overrightarrow{BC}\right)$$

Similarly $\overrightarrow{BE} = \overrightarrow{BC} + \frac{1}{2}\overrightarrow{CA}$

and $\overrightarrow{CF} = \overrightarrow{CA} + \frac{1}{2}\overrightarrow{AB}$.

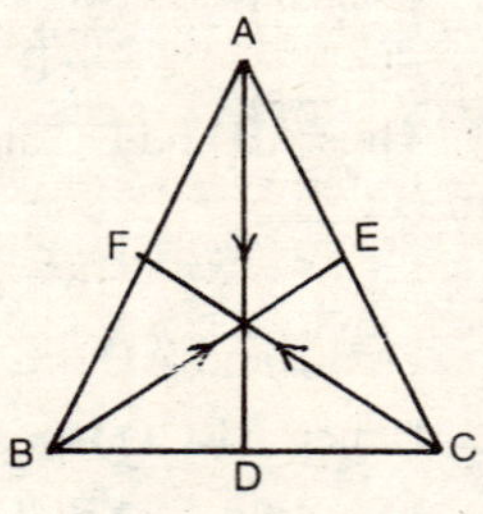

Fig. 1.20

$$\text{Now} \overrightarrow{AD} + \overrightarrow{BE} + \overrightarrow{CF} = \left(\overrightarrow{AB} + \frac{1}{2}\overrightarrow{BC}\right) + \left(\overrightarrow{BC} + \frac{1}{2}\overrightarrow{CA}\right) + \left(\overrightarrow{CA} + \frac{1}{2}\overrightarrow{AB}\right)$$

$$= \frac{3}{2}(\overrightarrow{AB} + \overrightarrow{BC} + \overrightarrow{CA})$$

$$= \frac{3}{2}\overrightarrow{AA} = \frac{3}{2}(\mathbf{0}) = \mathbf{0}$$

Example 7. If P, Q, R be midpoints of the sides AB, BC and CA of the triangle ABC and O be a point within it, then prove that $\overrightarrow{OA} + \overrightarrow{OB} + \overrightarrow{OC} = \overrightarrow{OP} + \overrightarrow{OQ} + \overrightarrow{OR}$.

Solution: From triangle law of addition,

$$\vec{OA} = \vec{OP} + \vec{PA} = \vec{OP} + \frac{1}{2}\vec{BA}$$

$$\left[\therefore\ \vec{PA} = \frac{1}{2}\vec{BA} \text{ as } \overline{PA} = \frac{1}{2}\overline{BA}\right]$$

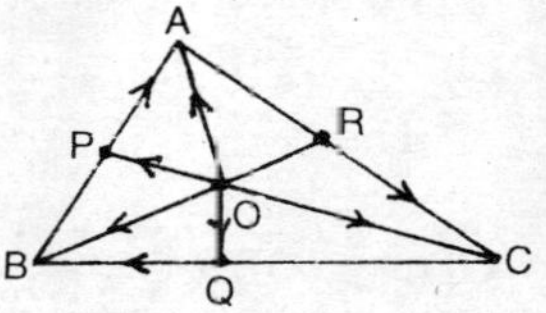

Similarly $\vec{OB} = \vec{OQ} + \frac{1}{2}\vec{CB}$

and $\vec{OC} = \vec{OR} + \frac{1}{2}\vec{AC}.$

Fig. 1.21

$$\text{Now}\ \vec{OA} + \vec{OB} + \vec{OC} = \left(\vec{OP} + \frac{1}{2}\vec{BA}\right) + \left(\vec{OQ} + \frac{1}{2}\vec{CB}\right) + \left(\vec{OR} + \frac{1}{2}\vec{AC}\right)$$

$$= \vec{OP} + \vec{OQ} + \vec{OR} + \frac{1}{2}(\vec{BA} + \vec{AC} + \vec{CB})$$

$$= \vec{OP} + \vec{OQ} + \vec{OR} + \frac{1}{2}(\vec{BB})$$

$$= \vec{OP} + \vec{OQ} + \vec{OR} \qquad (\because\ \vec{BB} = \mathbf{O})$$

Example 8. If G be the centroid of ΔABC then prove that $\vec{GA} + \vec{GB} + \vec{GC} = \mathbf{O}$.

Solution. Let AD, BE and CF be the medians of the triangle ABC and G be the centroid.

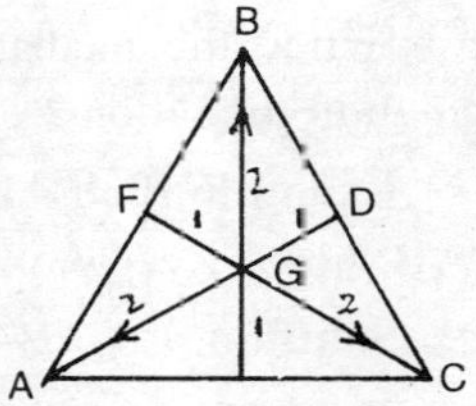

Now CG : GF = 2 : 1

More over F is the midpoint of $\vec{AB}$

$$\therefore\quad \vec{GA} + \vec{GB} = 2\,\vec{GF} \qquad ...(1)$$

Similarly CG : GF = 2 : 1

or CG = 2 GF or $\vec{CG} = 2\vec{GF}$

Fig. 1.22

Substituting in (1) we get

$$\vec{GA} + \vec{GB} = \vec{CG} = -\vec{GC}$$

or $$\vec{GA} + \vec{GB} + \vec{GC} = \mathbf{O}$$

Example 9. Prove that the medians of a triangle are concurrent.

Solution. Let **a, b, c** be the position vectors of the vertices A, B, C of ΔABC referred to some origin O.

$$\therefore\quad \vec{OA} = \mathbf{a},\ \vec{OB} = \mathbf{b},\ \vec{OC} = \mathbf{c}$$

Let D be the midpoint of the side BC of ΔABC and let G be a point on AD, such that AG : GD = 2 : 1 .

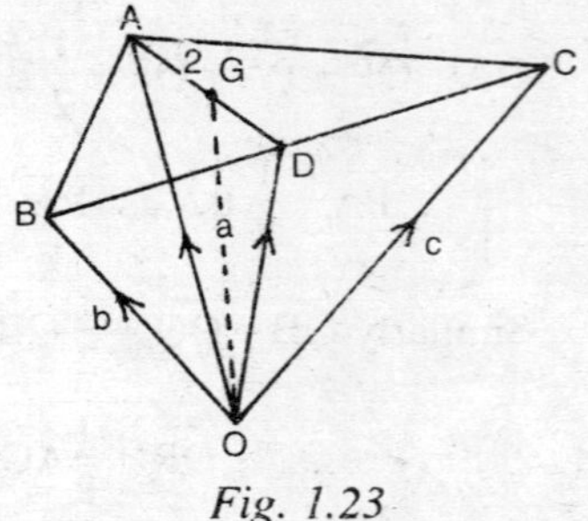

Fig. 1.23

Then

$$\vec{OD} = \frac{1}{2}(\vec{OB} + \vec{OC}) = \frac{1}{2}(b + c)$$

Also

$$\vec{OG} = \frac{1.\vec{OA} + 2.\vec{OD}}{1+2} = \frac{1.\mathbf{a} + 2.\frac{1}{2}(\mathbf{b} + \mathbf{c})}{3}$$

$$\therefore \quad \vec{OG} = \frac{\mathbf{a} + \mathbf{b} + \mathbf{c}}{3}$$

In a similar way we can show that the position vectors of the points on the other two medians which divide these medians in the ration 2 : 1 (2 from vertex and 1 from the opposite side) are also $= \frac{1}{3}(\mathbf{a} + \mathbf{b} + \mathbf{c})$

Hence the medians of ΔABC meet in G and divides each median in the ration 2 : 1.

Example 10. If ABCD is a parallelogram and E is the midpoint of AB show by vector method that DE trisects and is trisected by AC.

Solution. Let $\vec{AB} = \mathbf{a}$; $\vec{AD} = \mathbf{b}$

Then $\vec{BC} = \vec{AD} = \mathbf{b}$ and $\vec{AC} = \vec{AB} + \vec{AD} = \mathbf{a} + \mathbf{b}$

Let K be a point on AC such that AK : AC = 1 : 3 or

$$AK = \frac{1}{3}AC \Rightarrow \vec{AK} = \frac{1}{3}\vec{AC} = \frac{1}{3}(\mathbf{a} + \mathbf{b})$$

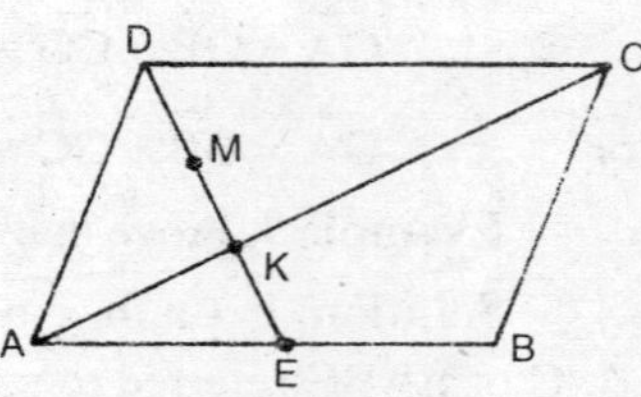

Fig. 1.24

Again E being the mid-point of AB, we have

$$\vec{AE} = \frac{1}{2}\vec{AB} = \frac{1}{2}\mathbf{a}$$

Let M be the point on DE such that DM : ME = 2 : 1.

Then $AM = \dfrac{1.\vec{AD} + 2.\vec{AE}}{1+2} = \dfrac{1.\mathbf{b} + 2\frac{1}{2}\mathbf{a}}{3} = \dfrac{\mathbf{b}+\mathbf{a}}{3}$

$\therefore \quad \vec{AK} = \frac{1}{3}(\mathbf{a}+\mathbf{b}) = \vec{AM}$

$\Rightarrow$ K and M coincides.

i.e., DE trisects AC and is trisected by AC.

Example 11. Prove that the mid point of two opposite sides of any quadrilateral and the mid-point of the diagonals are the vertices of a parallelogram.

Solution: Let ABCD be the quadrilateral and refering A as origin let $\vec{AB} = \mathbf{b}$, $\vec{AC} = \mathbf{c}$ and $\vec{AD} = \mathbf{d}$ be the position vectors respectively.

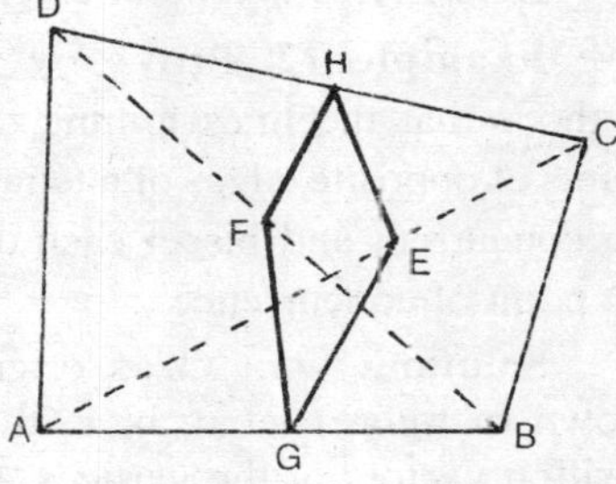

Fig. 1.25

Let E and F be the mid points of the diagonals AC and BD. Let G, H be the midpoints of the sides AB and DC. Join HF, FG, GE, and EH.

Now $\quad \vec{AG} = \frac{1}{2}\vec{AB} \quad$ ($\because$ G is mid point of AB)

$= \frac{1}{2}\mathbf{b}$

and $\quad \vec{AH} = \vec{AE} + \vec{EH}$

$= \frac{1}{2}\vec{AC} + \frac{1}{2}\vec{AD} = \frac{1}{2}(\mathbf{c}+\mathbf{d})$

Again $\quad \vec{AE} = \frac{1}{2}\vec{AC} = \frac{1}{2}\mathbf{c}$

and $\quad \vec{AF} = \frac{1}{2}\vec{AB} + \frac{1}{2}\vec{AD} = \frac{1}{2}(\mathbf{b}+\mathbf{d})$

$\therefore \quad \vec{GF} = \vec{AF} - \vec{AG} = \frac{1}{2}(\mathbf{b}+\mathbf{d}) - \frac{1}{2}\mathbf{b} = \frac{1}{2}\mathbf{d} \quad$ ...(1)

$$\overrightarrow{GE} = \overrightarrow{AE} - \overrightarrow{AG} = \frac{1}{2}\mathbf{c} - \frac{1}{2}\mathbf{b} = \frac{1}{2}(\mathbf{c} - \mathbf{b}) \qquad ...(2)$$

$$\overrightarrow{EH} = \overrightarrow{AH} - \overrightarrow{AE} = \frac{1}{2}(\mathbf{c} + \mathbf{d}) - \frac{1}{2}\mathbf{c} = \frac{1}{2}\mathbf{d} \qquad ...(3)$$

$$\overrightarrow{FH} = \overrightarrow{AH} - \overrightarrow{AF} = \frac{1}{2}(\mathbf{c} + \mathbf{d}) - \frac{1}{2}(\mathbf{b} + \mathbf{d}) = \frac{1}{2}(\mathbf{c} - \mathbf{b}) \qquad ...(4)$$

from (1) and (3), $\overrightarrow{GF} = \overrightarrow{EH}$

$\Rightarrow$ GF and EH are equal and parallel.

Again from (2) and (4) $\overrightarrow{GE} = \overrightarrow{FH}$

$\Rightarrow$ GE and FH are equal and parallel.

Here EHFG is a parallelogram.

Example 12. Prove by vector methods that the lines joining the mid points of opposite edges of a tetrahedron are concurrent and bisect each other at the point of concurrence

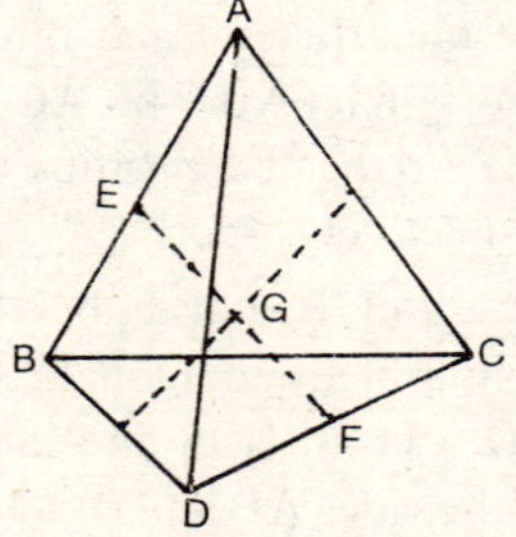

Fig. 1.26

Solution: w.r.t O as origin (not shown in figure) let **a, b, c, d** be the position vectors of the vertices A, B, C and D of a tetrahedron, Let E, F be the mid points of the opposite edges AB and CD. Then the P.V's of E and F are $\frac{1}{2}(\mathbf{a} + \mathbf{b})$ and $\frac{1}{2}(\mathbf{c} + \mathbf{d})$ respectively

If G is the mid point of EF then its P.V is

$$= \frac{1}{2}\left\{\frac{1}{2}(\mathbf{a} + \mathbf{b}) + \frac{1}{2}(\mathbf{c} + \mathbf{d})\right\}$$

$$= \frac{1}{4}(\mathbf{a} + \mathbf{b} + \mathbf{c} + \mathbf{d})$$

The symmetry of this result shows that G lies on the line joining the middle point of other pairs AC, BD and BC, AD of opposite edges of the tetrahedron.

Hence the lines joining the midpoints of opposite edges of a tetrahedron are concurrent at G and bisect each other at the point of concurrence.

Example 13. Prove by vector methods that the diagonals of a parallellogram bisect each other conversely if the diagonals of a quadrilateral bisect each other it is a parallelogram.

Solution. w.r.to O as origin let **a, b, c, d** be the P.V's of A, B, C, D of a parallelogram ABCD respectively. Since ABCD is a parallelogram the sides AB and DC are parallel

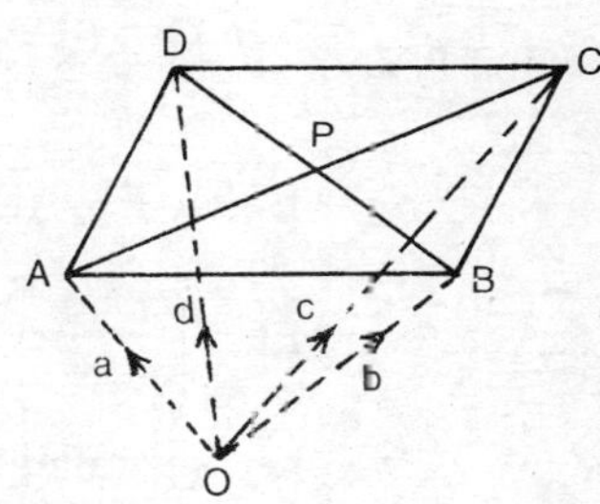

Fig. 1.27

$\therefore \quad \overrightarrow{AB} = \overrightarrow{DC}$

or $\quad \overrightarrow{OB} - \overrightarrow{OA} = \overrightarrow{OC} - \overrightarrow{OD}$

$\therefore\ \mathbf{b} - \mathbf{a} = \mathbf{c} - \mathbf{d} \Rightarrow \mathbf{b} + \mathbf{d} = \mathbf{a} + \mathbf{c}$ ...(1)

The P.V. of the mid point of AC

$$= \frac{\mathbf{a} + \mathbf{c}}{2} \quad \text{...(2)}$$

and the P.V. of the mid point of BD

$$= \frac{\mathbf{b} + \mathbf{d}}{2} \quad \text{...(3)}$$

By virtue of (1), (2) and (3) are equal

Hence diagonals of a parallelogram bisect each other.

Converse AC and BD have the same point as their midpoint.

$$\therefore \quad \frac{\mathbf{a} + \mathbf{c}}{2} = \frac{\mathbf{b} + \mathbf{d}}{2} \Rightarrow \mathbf{a} + \mathbf{c} = \mathbf{b} + \mathbf{d} \quad \text{...(4)}$$

$$\therefore \quad \mathbf{b} - \mathbf{a} = \mathbf{c} - \mathbf{d} \Rightarrow \overrightarrow{AB} = \overrightarrow{DC}$$

$\therefore$ AB and DC are parallel and equal. Similarly from (4)

$$\mathbf{d} - \mathbf{a} = \mathbf{c} - \mathbf{b} \Rightarrow \overrightarrow{AD} = \overrightarrow{BC}$$

$\therefore$ AD and BC are parallel and equal

Hence ABCD is a parallelogram.

Example 14. Show that the line joining the mid points of two non parallel sides of trapezium is parallel to the other sides and half of their sum.

Solution. Choose A as origin of vectors in the trapezium in which AB and DC are parallel sides,

Let $\overrightarrow{AB} = \mathbf{b}$ and $\overrightarrow{AD} = \mathbf{d}$

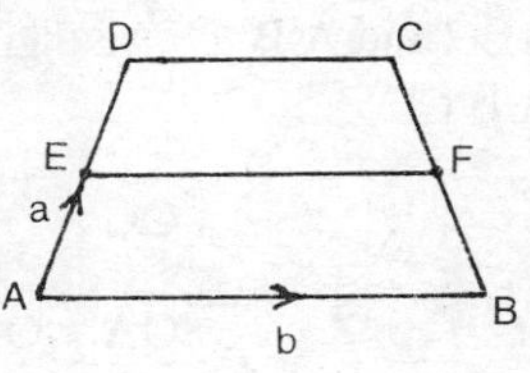

Fig. 1.28

$\therefore$ DC is parallel to AB. Let $\overrightarrow{DC} = t\,\overrightarrow{AB}$ where t is some scalar i.e., $\overrightarrow{DC} = t\mathbf{b}$

$\therefore\ \overrightarrow{AC} = \overrightarrow{AD} + \overrightarrow{DC} = \mathbf{d} + t\mathbf{b}$

E and F are the middle points of AD and BC

$\therefore$ Position vector of E $= \dfrac{\mathbf{0} + \mathbf{d}}{2} = \dfrac{\mathbf{d}}{2}$

and P.V. of $F = \dfrac{\mathbf{b} + (\mathbf{d} + t\mathbf{b})}{2} = \dfrac{\mathbf{d}}{2} + \dfrac{(t+1)\mathbf{b}}{2}$

$$\therefore \quad \overrightarrow{EF} = \overrightarrow{AF} - \overrightarrow{AE} = \frac{\mathbf{d}}{2} + \frac{(t+1)\mathbf{b}}{2} - \frac{\mathbf{d}}{2} = \frac{(t+1)\mathbf{b}}{2}$$

$$\therefore \quad \overrightarrow{EF} = \left(\frac{t+1}{2}\right)\overrightarrow{AB}$$

$\Rightarrow \quad \overrightarrow{EF}$ is parallel to $\overrightarrow{AB}$ and therefore to DC.

$$\therefore \quad \overrightarrow{EF} = \frac{t+1}{2}\,\overrightarrow{AB} \quad \therefore \quad \frac{EF}{AB} = \frac{t+1}{2}$$

$$\therefore \quad EF = \frac{t+1}{2}\,AB = \frac{1}{2}\,(AB + t\,AB)$$

or $$EF = \frac{1}{2}\,(AB + DC)$$

$$[\because \quad \overrightarrow{DC} = t\,\overrightarrow{AB} \Rightarrow DC = t\,AB]$$

Hence the line joining the mid points of two non parallel sides $= \dfrac{1}{2}$ (sum of parallel sides)

Example 15. ABC and A′B′C′ are two triangles, G, G′ are the respective points of intersection of medians. Then prove that $\overrightarrow{AA'} + \overrightarrow{BB'} + \overrightarrow{CC'} = 3\overrightarrow{GG'}$

Solution. Let O be the point of reference for both the triangles ABC and A′B′C′. Also given that G and G′ are centroids of ABC and A′B′C′

$$\therefore \quad \overrightarrow{OG} = \frac{1}{3}\,(\overrightarrow{OA} + \overrightarrow{OB} + \overrightarrow{OC})$$

or $$\overrightarrow{OA} + \overrightarrow{OB} + \overrightarrow{OC} = 3\overrightarrow{OG} \quad \text{...(1)}$$

Similarly $$\overrightarrow{OA'} + \overrightarrow{OB'} + \overrightarrow{OC'} = 2\overrightarrow{OG'} \quad \text{...(2)}$$

$$\text{Now } \overrightarrow{AA'} + \overrightarrow{BB'} + \overrightarrow{CC'} = (\overrightarrow{OA'} - \overrightarrow{OA}) + (\overrightarrow{OB'} - \overrightarrow{OB} + (\overrightarrow{OC'} - \overrightarrow{OC})$$
$$= (\overrightarrow{OA'} + \overrightarrow{OB'} + \overrightarrow{OC'}) - (\overrightarrow{OA} + \overrightarrow{OB} + \overrightarrow{OC})$$
$$= 3\overrightarrow{OG'} - 3\overrightarrow{OG} \qquad \text{[using (1) and (2)]}$$
$$= 3(\overrightarrow{OG'} - \overrightarrow{OG}) = 3\overrightarrow{GG'}$$

Hence the result.

Example 16. If **a** and **b** are consecutive vector sides of a parallelogram, express the diagonal vectors in terms of **a** and **b**. Hence show that if vectors $\pm \mathbf{a} \pm \mathbf{b}$ are drawn from a common origin then they will terminate at the vertices of a parallelogram.

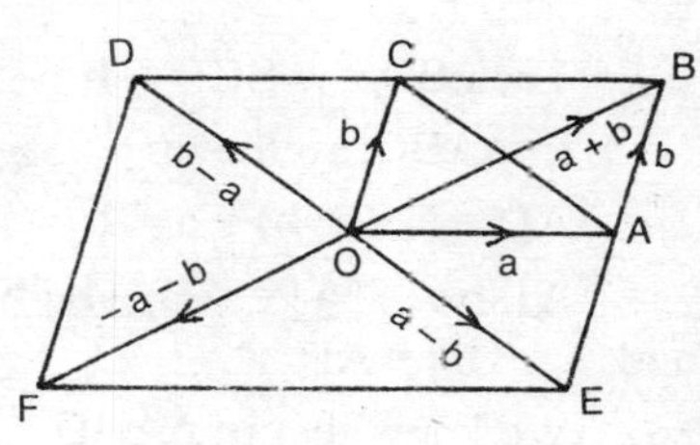

Fig. 1.29

Solution. Let OABC be a parallelogram in which $\overrightarrow{OA} = \mathbf{a}$ and $\overrightarrow{AB} = \mathbf{b} = \overrightarrow{OC}$

$\therefore \ \overrightarrow{OB} = \overrightarrow{OA} + \overrightarrow{AB} = \mathbf{a} + \mathbf{b}$ and

$\overrightarrow{AC} = \overrightarrow{OC} - \overrightarrow{OA} = \mathbf{b} - \mathbf{a}$.

Taking O as origin, we have to draw the four vectors $\pm \mathbf{a} \pm \mathbf{b}$. i.e., $\mathbf{a} + \mathbf{b}$, $\mathbf{a} - \mathbf{b}$, $-\mathbf{a} + \mathbf{b}$ and $-\mathbf{a} - \mathbf{b}$. Through O draw OD parallel and equal to AC and produce DO to E such that DO = OE. Similarly produce BO to F such that BO = OF.

Now clearly the P.V's of D, F, E and B are the four vectors $\pm \mathbf{a} \pm \mathbf{b}$. Also since O is the mid point of BF and DE and hence BDEF are the vertices of a parallelogram.

Example 17. Prove that for any two vectors, **a** and **b**,

(i) $|\mathbf{a}| - |\mathbf{b}| \le |\mathbf{a} + \mathbf{b}| \le |\mathbf{a}| + |\mathbf{b}|$

(ii) $|\mathbf{a} - \mathbf{b}| \ge |\mathbf{a}| - |\mathbf{b}|$

Solution. (i) Let $\overrightarrow{OA} = \mathbf{a}$ and $\overrightarrow{AB} = \mathbf{b}$

Then $\overrightarrow{OB} = \overrightarrow{OA} + AB = \mathbf{a} + \mathbf{b}$

Also $OA = |\mathbf{a}|$, $\overrightarrow{AB} = |\mathbf{b}|$

and $\quad OB = |\mathbf{a} + \mathbf{b}|$

We know that in a triangle ABC

$$OA + AB \ge OB$$

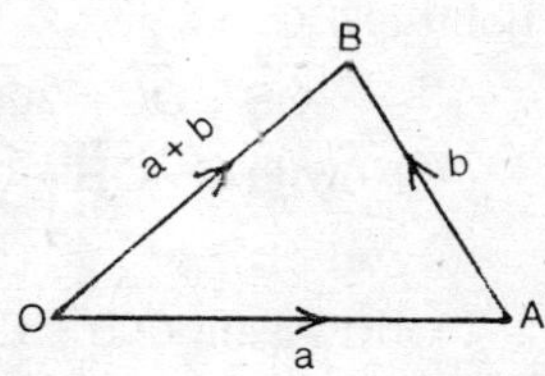

Fig. 1.30

i.e., $$|\mathbf{a}| + |\mathbf{b}| \geq |\mathbf{a} + \mathbf{b}| \quad ...(1)$$

Similarly OA − AB ≤ OB

or $$|\mathbf{a}| - |\mathbf{b}| - \leq |\mathbf{a} + \mathbf{b}| \quad ...(2)$$

∴ From (1) and (2)

$$|\mathbf{a}| - |\mathbf{b}| \leq |\mathbf{a} + \mathbf{b}| \leq |\mathbf{a}| + |\mathbf{b}|$$

(ii) Let $\vec{AB} = \mathbf{a}$ and $\vec{BC} = \mathbf{b}$ Produce CB to D, such that BD = BC

Then $\vec{BD} = -\vec{BC} = -\mathbf{b}$

In ΔABO, we have

$\vec{AD} = \mathbf{a} + (-\mathbf{b}) = \mathbf{a} - \mathbf{b}$

Also $|\mathbf{a}|$ = AB, $|\mathbf{b}|$ = BC = BD

and $|\mathbf{a} - \mathbf{b}|$ = AD

We know that in ΔABD

AB − BD ≤ AD or AD ≥ AB − BD

or $$|\mathbf{a} - \mathbf{b}| \geq |\mathbf{a}| - |\mathbf{b}|$$

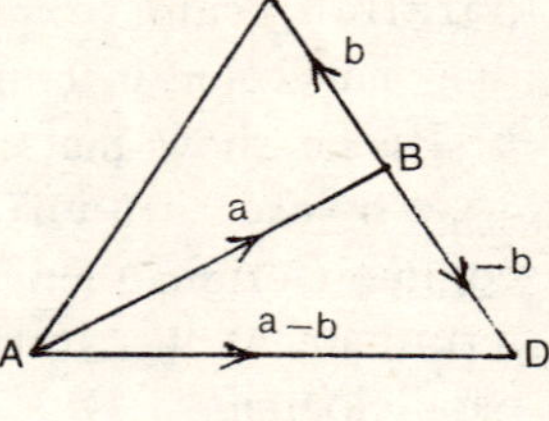

Fig. 1.31

Example 18. If O is the circumcentre and O′ the orthocentre of a triangle ABC then prove that

(i) $\vec{OA} + \vec{OB} + \vec{OC} = \vec{OO'}$

(ii) $\vec{O'A} + \vec{O'B} + \vec{O'C} = 2\vec{OO'}$

(iii) If AP is the diameter of the circum-circle the prove that $\vec{AO'} + \vec{O'B} + \vec{O'C} = \vec{AP}$.

Solution. (i) Given that O and O′ are the circumcentre and orthocentre of a triangle ABC. From O and O′ draw OE and O′D perpendiculars to BC. Then we have $\vec{AO'} = 2\vec{OE}$. Also E being the mid point of BC, we get

$$\vec{OB} + \vec{OC} = 2\vec{OE}$$

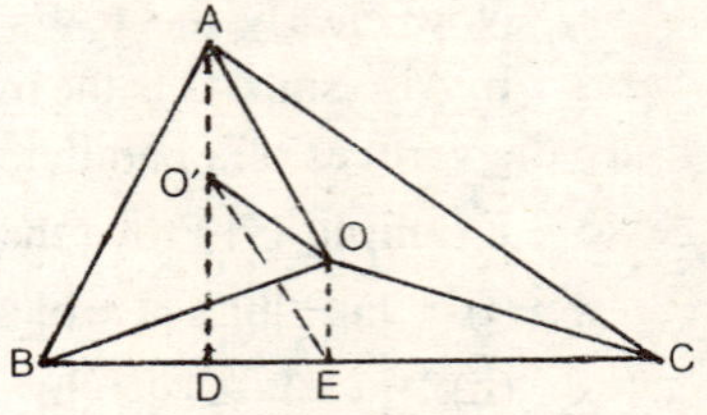

Fig. 1.32

Now $\vec{OA} + \vec{OB} + \vec{OC} = \vec{OA} + 2\vec{OE}$

$= \vec{OA} + \vec{AO'} = \vec{OO'}$

(ii) Again O′B − O′C = 20E (∴ E is the midpoint of BC)

or $\vec{O'B} + \vec{O'C} = 2(\vec{O'O} + \vec{OE}) = 2\vec{O'O} + 2\vec{OE} = 2\vec{OO'} + \vec{AO'}$

or $\overrightarrow{O'B} + \overrightarrow{O'C} = 2\overrightarrow{OO'} - \overrightarrow{O'A}$

or $\overrightarrow{O'A} + \overrightarrow{O'B} + \overrightarrow{O'C} = 2\overrightarrow{OO'}$

(iii) $\overrightarrow{A'O} + \overrightarrow{O'B} + \overrightarrow{O'C} = 2\overrightarrow{AO'} + (\overrightarrow{O'A} + \overrightarrow{O'B} + \overrightarrow{O'C})$

$= 2\overrightarrow{AO'} + 2\overrightarrow{OO'}$

$= 2\,(\overrightarrow{AO'} + \overrightarrow{O'O}) = 2\overrightarrow{AO}$

= 2 (The vector represented by the radius AO of the circumcircle)

$= \overrightarrow{AP}$ ($\because$ AP = 2AO)

Example 19. If **a, b, c** be any there non-zero, non-coplanar vectors, find the linear relation between the following system of vectors.

a + 3**b** + 4**c**, **a** − 2**b** + 3**c**, **a** + 5**b** − 2**c** and 6**a** + 14**b** + 4**c**.

Solution. Let

$$\mathbf{a} + 3\mathbf{b} + 4\mathbf{c} = x(\mathbf{a} - 2\mathbf{b} + 3\mathbf{c}) + y(\mathbf{a} + 5\mathbf{b} - 2\mathbf{c}) + z(6\mathbf{a} + 14\mathbf{b} + 4\mathbf{c}) \quad ...(1)$$

Comparing the coefficients of **a, b** and **c** on both sides, we get

$$1 = x + y + 6z$$
$$3 = -2x + 5y + 14z$$
$$4 = 3x - 2y + 4z$$

Solving these equations, we get

$$x = -2,\ y = -3 \text{ and } z = 1.$$

Hence from (1), we get

$$\mathbf{a} + 3\mathbf{b} + 4\mathbf{c} = -2(\mathbf{a} - 2b + 3\mathbf{c}) - 3(\mathbf{a} + 5\mathbf{b} - 2\mathbf{c}) + (6\mathbf{a} + 14\mathbf{b} + 4\mathbf{c})$$

Example 20. Show that the points whose position vectors are **a** + 2**b** + 5**c**, 3**a** + 2**b** + **c**, 2**a** + 2**b** + 3**c** are collinear

Solution. Let the given points be A, B, C and O be the point of reference.

Then $\overrightarrow{OA}$ = **a** + 2**b** + 5**c**, $\overrightarrow{OB}$ = 3**a** + 2**b** + **c** and $\overrightarrow{OC}$ = 2**a** + 2**b** + 3**c**. Let x, y, z, be three scalar quantities, such that

$$x\overrightarrow{OA} + y\overrightarrow{OB} + z\overrightarrow{OC} = \mathbf{O} \quad ...(1)$$

where $$x + y + z = 0 \quad ...(2)$$

Substituting $\vec{OA}$, $\vec{OB}$ and $\vec{OC}$ in (1), we get

$$x(\mathbf{a} + 2\mathbf{b} + 5\mathbf{c}) + y(3\mathbf{a} + 2\mathbf{b} + \mathbf{c}) + z(2\mathbf{a} + 2\mathbf{b} + 3\mathbf{c}) = \mathbf{0}$$

$$\Rightarrow (x + 3y + 2z)\mathbf{a} + (2x + 2y + 2z)\mathbf{b} + (5x + 2y + 3z)\mathbf{c} = \mathbf{0}$$

Comparing the co-efficients of **a, b, c** on both sides, we get,

$$x + 3y + 2z = 0 \text{ or } (x + y + z) + 2y + z = 0$$

$$2x + 2y + 2z = 0 \text{ or } x + y + z = 0$$

$$5x + 2y + 3z = 0 \text{ or } x + y + z + 2(2x + z) = 0$$

$\Leftrightarrow x = -\frac{1}{2} z, y = -\frac{1}{z}$, that satisfy (2). Hence the conditions of collinearity (1) and (2) are satisfied.

$\therefore$ The points are collinear.

Another method

$$\vec{AB} = \vec{OB} - \vec{OA} = (3\mathbf{a} + 2\mathbf{b} + \mathbf{c}) - (\mathbf{a} + 2\mathbf{b} + 5\mathbf{c})$$

$$= 2\mathbf{a} - 4\mathbf{c}$$

$$\vec{AC} = \vec{OC} - \vec{OA} = (2\mathbf{a}+ + 2\mathbf{b} + 3\mathbf{c}) - (\mathbf{a} + 2\mathbf{b} + 5\mathbf{c})$$

$$= \mathbf{a} - 2\mathbf{c}$$

$$\Rightarrow \vec{AB} = 2\vec{AC}$$

$$\Leftrightarrow \vec{AB} \text{ and } \vec{AC} \text{ are collinear.}$$

($\therefore$ Two vectors **a** and **b** are collinear if there exists a non-zero scalar λ such that $\mathbf{a} = \lambda\mathbf{b}$)

$\Leftrightarrow$ $\vec{OA}$, $\vec{OB}$, $\vec{OC}$ are collinear

i.e., the given points are collinear.

Example 21. If a and b are non-collinear vectors show that the points x_1 **a** + y_1 **b**, x_2 **a** + y_2 **b** and x_3**a** + y_3**b** are collinear if and only if

$$x_1 (y_2 - y_3) + x_2 (y_3 - y_1) + x_3 (y_1 - y_2) = 0$$

Solution. The given points are collinear if there exists scalars x, y, z such that

$$x(x_1\mathbf{a} + y_1\mathbf{b}) + y(x_2\mathbf{a} + y_2\mathbf{b}) + z(x_3\mathbf{a} + y_3\mathbf{b}) = \mathbf{0} \quad ..(1)$$

where

$$x + y + z = 0 \quad ...(2)$$

From (1)

$$(xx_1 + yx_2 + zx_3)\,\mathbf{a} + (xy_1 + yy_2 + zy_3)\mathbf{b} = \mathbf{0} \quad ...(3)$$

∵ **a** and **b** are non-collinear vectors, for existence of (3), we should have

$$xx_1 + yx_2 + zx_3 = 0 \qquad ...(4)$$

$$xy_1 + yy_2 + zy_3 = 0 \qquad ...(5)$$

Eliminating x, y, z from (4), (5) and (2), we get

$$\begin{vmatrix} x_1 & x_2 & x_3 \\ y_1 & y_2 & y_3 \\ 1 & 1 & 1 \end{vmatrix} = 0$$

or $\quad x_1 (y_2 - y_3) + x_2 (y_3 - y_1) + x_3 (y_1 - y_2) = 0$

Example 22. Show that the vectors **5a + 6b − 7c, 7a − 8b + 9c** and **3a + 2b + 5c** are coplanar where **a, b, c** being any vectors.

Solution. If three vectors are coplanar, then we can express one of them as a linear combination of the other two. Let

$$\mathbf{5a + 6b + 7c = x(7a - 8b + 9c) + y(3a + 20b + 5c)} \qquad ...(1)$$

where x and y are scalars

Comparing the coefficients of **a, b** and **c,** we get

$$1 = -2x + y \qquad ...(2)$$

$$-2 = 3x - 3y \qquad ...(3)$$

$$3 = -4x + 5y \qquad ...(4)$$

Solving (2) and (3), we get $x = -\frac{1}{3}$, $y = \frac{1}{3}$ which satisfy (4) also. Hence the given vectors are coplanar

Example 23. Show that the vectors **2a − b + 3c, a + b − 2c** and **a + b − 3c** are non-coplanar where **a, b, c** are non-coplanar vectors

Solution. Let the given vectors are coplanar. Then we can write

$$\mathbf{2a - b + 3c = x(a + b - 2c) + y\,(a + b - 3c)} \qquad ...(1)$$

where x and y are scalars.

Comparing the co-efficients of **a, b** and **c** we get

$$2 = x + y \qquad ...(2)$$

$$-1 = x + y \qquad ...(3)$$

$$3 = -2x - 3y \qquad ...(4)$$

The equations (2), (3) and (4) are inconsistent, i.e., there exists no real numbers x and y which simultaneously satisfy these equations.

Hence the given vectors are non coplanar.

Example 24. Prove that the four points **a**, $2\mathbf{b} + 3\mathbf{c}$, $-2\mathbf{a} + 3\mathbf{b} + 4\mathbf{c}$ and $-(\mathbf{b} + 2\mathbf{c})$ are coplanar.

Solution. Let $\vec{OA} = \mathbf{a}$, $\vec{OB} = 2\mathbf{b} + 3\mathbf{c}$, $\vec{OC} = -2\mathbf{a} + 3\mathbf{b} + 4\mathbf{c}$ and $\vec{OD} = -(\mathbf{b} + 2\mathbf{c})$ be the position vectors of the given four points w.r.t origin O.

Then the four points are coplanar if the vectors $\vec{AB}$, $\vec{AC}$ and $\vec{AD}$ are coplanar

$$\vec{AB} = \vec{OB} - \vec{OA} = (2\mathbf{b} + 3\mathbf{c}) - \mathbf{a} = -\mathbf{a} + 2\mathbf{b} + 3\mathbf{c}$$
$$\vec{AC} = \vec{OC} - \vec{OA} = -2\mathbf{a} + 3\mathbf{b} + 4\mathbf{c} - \mathbf{a} = -3\mathbf{a} + 3\mathbf{b} + 4\mathbf{c}$$
$$\vec{AD} = \vec{OD} - \vec{OA} = (-\mathbf{b} - 2\mathbf{c}) - \mathbf{a} = -\mathbf{a} - \mathbf{b} - 2\mathbf{c}$$

Now $\vec{AB}$, $\vec{AC}$ and $\vec{AD}$ are coplanar if one of them can be expressed as a linear combination of the other two i.e.,

$(-\mathbf{a} + 2\mathbf{b} + 3\mathbf{c}) = x(-3\mathbf{a} + 3\mathbf{b} + 4\mathbf{c}) + y(-\mathbf{a} - \mathbf{b} - 2\mathbf{c})$ where x and y are some scalars. Comparing the co-efficients of **a** and **b** and **c** we get

$$-1 = -3x - y \qquad ...(1)$$
$$2 = 3x - y \qquad ...(2)$$
$$3 = 4x - 2y \qquad ...(3)$$

Solving the first two, we get $x = \frac{1}{2}$ and $y = -\frac{1}{2}$ which satisfy equation (3). Hence the vectors $\vec{AB}$, $\vec{AC}$ and $\vec{AD}$ are coplanar.

$\Leftrightarrow$ the given four points are collinear.

Example 25. If $\mathbf{r}_1, \mathbf{r}_2, \mathbf{r}_3$ are the position vectors of three collinear points, then prove that two real constants x and y exist such that $\mathbf{r}_3 = x\mathbf{r}_1 + y\mathbf{r}_2$ where $x + y = 1$.

Solution: Let A, B, C be three points whose position vectors referred to some origin O are $\mathbf{r}_1, \mathbf{r}_2$ and $\mathbf{r}_3$ respectively i.e., $\vec{OA} = \mathbf{r}_1$, $\vec{OB} = \mathbf{r}_2$ and $\vec{OC} = \mathbf{r}_3$

Then $\qquad \vec{AB} = \vec{OB} - \vec{OA} = \mathbf{r}_2 - \mathbf{r}_1$

$$\vec{BC} = \vec{OC} - \vec{OB} = \mathbf{r}_3 - \mathbf{r}_2$$

If A, B, C are collinear points then AB and AC are in the same line and if the magnitude of $\overrightarrow{BC}$ is λ times the magnitude of $\overrightarrow{AB}$.

Then $\overrightarrow{BC} = \lambda\, \overrightarrow{AB}$

$$\mathbf{r}_3 - \mathbf{r}_2 = \lambda\,(\mathbf{r}_2 - \mathbf{r}_1) \text{ or } \mathbf{r}_3 = \lambda\, \mathbf{r}_1 + (\lambda + 1)\mathbf{r}_2$$

or $\mathbf{r}_3 = x\mathbf{r}_1 + y\mathbf{r}_2$

where $x = -\lambda,\ \ y = \lambda + 1$

or $x + y = -\lambda + \lambda + 1 = 1$

Thus $\mathbf{r}_3 = x\mathbf{r}_1 + y\mathbf{r}_2$ where $x + y = 1$.

Example 26. Examine whether the vectors **5a + 6b + 7c**, **7a − 8b + 9c** and **3a + 20b + 5c** are linearly independent or dependent where **a, b, c** are non-coplanar.

Solution. Let $\mathbf{V}_1 = 5\mathbf{a} + 6\mathbf{b} + 7\mathbf{c}$, $\mathbf{V}_2 = 7\mathbf{a} - 8\mathbf{b} + 9\mathbf{c}$ and $\mathbf{V}_3 = 3\mathbf{a} + 20\mathbf{b} + 5\mathbf{c}$.

For either independency or dependency, we have

$x\mathbf{V}_1 + y\mathbf{V}_2 + z\mathbf{V}_3 = 0$ where x, y, z are some scalars.

i.e., $x\,(5\mathbf{a} + 6\mathbf{b} + 7\mathbf{c}) + y\,(7\mathbf{a} - 8\mathbf{b} + 9\mathbf{c}) + z(3\mathbf{a} + 20\mathbf{b} + 5\mathbf{c}) = 0$

or $(5x + 7y + 3z)\mathbf{a} + (6x - 8y + 20z)\mathbf{b} + (7x + 9y + 5z)\mathbf{c} = 0$

∴ **a, b, c** are non-coplanar (linearly independent)

® $5x + 7y + 3z = 0$ ...(1)

$6x - 8y + 20z = 0$ ...(2)

$7x + 9y + 5z = 0$ ...(3)

Solving (1) and (2) by the rule of cross multiplication we get

$$\frac{x}{140 - (-24)} = \frac{y}{18 - 100} = \frac{z}{-40 - 42}$$

$$\frac{x}{2} = \frac{y}{-1} = \frac{z}{-1} = \lambda \text{ (say)}$$

∴ $x = 2\lambda,\ y = -\lambda,\ z = -\lambda$, which also satisfy (3). Hence there exist scalars x, y, z (not all zero) such that $x\mathbf{V}_1 + y\mathbf{V}_2 + z\mathbf{V}_3 = 0$ holds good.

⇒ The given vectors are linearly dependent.

Example 27. Examine whether the vectors **a − 3b + 2c**, 2a − 4**b** − **c** and 3**a** + 2**b** − **c** are linearly independent or dependent where **a, b, c** are non-coplanar.

Solution. Let $\mathbf{V_1} = \mathbf{a} - 3\mathbf{b} + 2\mathbf{c}$, $\mathbf{V_2} = 2\mathbf{a} - 4\mathbf{b} - \mathbf{c}$ and $\mathbf{V_3} = 3\mathbf{a} + 2\mathbf{b} - \mathbf{c}$. Then for either independency or dependency we have

$x\mathbf{V_1} + y\mathbf{V_2} + \mathbf{V_3} = 0$ where x, y, z are some scalars. i.e.,

$$x(\mathbf{a} - 3\mathbf{b} + 2\mathbf{c}) + y(2\mathbf{a} - 4\mathbf{b} - \mathbf{c}) + z(3\mathbf{a} + 2\mathbf{b} - \mathbf{c}) = 0$$

or $$(x + 2y + 3z)\mathbf{a} + (-3x - 4y + 2z)\mathbf{b} + (2x - y - z) = 0$$

As **a, b, c** are non-coplanar

$$x + 2y + 3z = 0 \qquad ...(1)$$

$$-3x - 4y + 2z = 0 \qquad ...(2)$$

$$2x - y - z = 0 \qquad ...(3)$$

Solving (1) and (2), we get

$$\frac{x}{16} = \frac{y}{-11} = \frac{z}{2} = \lambda \text{ (say)}$$

$$\therefore \qquad x = 16\lambda,\ y = -11\lambda,\ z = 2\lambda$$

which does not satisfy equation (3),

$\Rightarrow$ For any set of values the relation

$x\mathbf{V_1} + y\mathbf{V_2} + z\mathbf{V_3} = 0$ does not exist.

except only one case $x = y = z = 0$

$\Leftrightarrow$ the given vectors are linearly independent.

1.17. To express a vector r in terms of i, j, k

Let us take O as origin and three mutually perpendicular lines OX, OY, OZ through O as x, y and z axes respectively (right handed orthonormal frame).

Let P(x, y, z) be any point in space such that $\overrightarrow{OP} = \mathbf{r}$. Construct a rectangular parallelopiped such that OP is its diagonal and OA, OB, OC are three coterminous edges which are along OX, OY and OZ respectively. Let $\overrightarrow{OA} = x$, $\overline{OB} = y$ and $\overline{OC} = z$. and let **i, j, k** denote the unit vectors

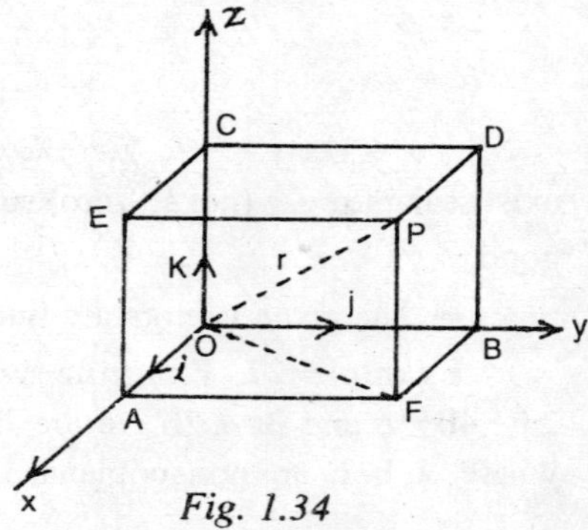

Fig. 1.34

along OX, OY and OZ

$\therefore \quad \vec{OA} = \overline{OA}\ \mathbf{i} = x\mathbf{i}; \quad \vec{OB} = \overline{OB}\ \mathbf{j} = y\mathbf{j}$

and $\quad \vec{OC} = \overline{OC}\ \mathbf{k} = z\mathbf{k}.$

Now we have $\vec{OP} = \vec{OF} + \vec{FP}$

$$= \vec{OA} + \vec{AF} + \vec{FP} = \vec{OA} + \vec{OB} + \vec{OC}$$

$$\therefore \quad \mathbf{r} = x\mathbf{i} + y\mathbf{j} + z\mathbf{k}$$

Here x**i**, y**j** and z**k** are called *resolved parts* of the vector and (x, y, z) are called *co-ordinates* of point P. Some times we write **r** (x, y, z) in place of **r** = x**i** + y**j** + z**k**.

Note 1. Modulus of **r** from the figure 1.34

$$(\overline{OP})^2 = (\overline{OF})^2 + (\overline{FP})^2$$

$$= (\overline{OA})^2 + (\overline{AF})^2 + (\overline{FP})^2 = (\overline{OA})^2 + (\overline{OB})^2 + (\overline{OC})^2$$

or $\quad r^2 = x^2 + y^2 + z^2$

or $\quad |\mathbf{r}|^2 = x^2 + y^2 + z^2$

Hence $\quad r = |\mathbf{r}| = \overline{OP} = \sqrt{x^2 + y^2 + z^2}$

i.e., the modulus of a vector **r** is the square root of the sum of the squares of the coefficients of **i, j, k** when **r** is expressed in terms of **i, j, k**

Note 2 : Unit Vector of r. If $\hat{r}$ in the unit vector of $\vec{OP}$ i.e., **r** then

$$\hat{\mathbf{r}} = \frac{\mathbf{r}}{|\mathbf{r}|}$$

1.18. Direction cosines cf r

If α, β, γ be the angles which $\vec{OP}$ makes with the positive directions of x, y and z axes, then cos α, cos β and cos γ are called the direction cosines (d. c's of r (or the line OP). They are generally denoted by l, m, n .

i.e., $\quad l = \cos\alpha,\ m = \cos\beta, \text{ and } n = \cos\gamma$

$\therefore \quad OA = OP\cos\alpha \text{ or } x = rl$

Similarly y = rm and z = rn

If $\hat{\mathbf{r}}$ is the unit vector along **r** then we can write

$$\mathbf{r} = x\mathbf{i} + y\mathbf{j} + z\mathbf{k} \text{ as}$$

$$\hat{r}r = lr\mathbf{i} + mr\mathbf{j} + nr\mathbf{k}$$

$$\hat{r}r = l\mathbf{i} + m\mathbf{j} + n\mathbf{k} = (\cos\alpha)\mathbf{i} + (\cos\beta)\mathbf{j} + (\cos\gamma)\mathbf{k}$$

$\Rightarrow$ The d.c's of vector **r** are the coefficients of **i, j, k** when the unit vector $\hat{\mathbf{r}}$ along **r** is expressed in terms of **i, j, k**

Also $x^2 + y^2 + z^2 = r^2l^2 + r^2m^2 + r^2n^2$

$$r^2 = r^2(l^2 + m^2 + n^2)$$

$$1 = l^2 + m^2 + n^2$$

or $\cos^2\alpha + \cos^2\beta + \cos^2\gamma = 1$

i.e., *The sum of the squares of direction cosines in equal to unity.*

1.19. Distance between Two Points

Let $\vec{OA}$ and $\vec{OB}$ be the position vectors of two points A and B whose rectangular cartesian co-ordinates are (x_1, y_1, z_1) and (x_2, y_2, z_2) respectively

Then $\vec{OA} = x_1\mathbf{i} + y_1\mathbf{j} + z_1\mathbf{k}$; $\vec{OB} = x_2\mathbf{i} + y_2\mathbf{j} + z_2\mathbf{k}$

Now $\vec{AB} = \vec{OB} - \vec{OA} = (x_2 - x_1)\mathbf{i} + (y_2 - y_1)\mathbf{j} + (z_2 - z_1)\mathbf{k}$

$\therefore$ $\overline{AB} = |\vec{AB}| = \sqrt{(x_2 - x_1)^2 + (y_2 - y_1)^2 + (z_2 - z_1)^2}$

Note. Two vectors $\mathbf{r_1} = a_1\mathbf{i} + b_1\mathbf{j} + c_1\mathbf{k}$ and $\mathbf{r_2} = a_2\mathbf{i} + b_2\mathbf{j} + c_2\mathbf{k}$ are collinear iff $\mathbf{r_1} = m\mathbf{r_2}$

$$\Rightarrow \quad \frac{a_1}{a_2} = \frac{b_1}{b_2} = \frac{c_1}{c_2}$$

Example 28. If the position vectors of P and Q are **i** + 2**j** – 3**k** and 2**i** + 3**j** – 5**k**, find $\vec{PQ}$, $|\vec{PQ}|$ and direction cosines of PQ.

Solution: We have

$$\vec{PQ} = \text{P. V of Q - P.V. of P}$$
$$= (2\mathbf{i} + 3\mathbf{j} - 5\mathbf{k}) - (\mathbf{i} + 2\mathbf{j} - 3\mathbf{k})$$
$$= \mathbf{j} + \mathbf{j} - 2\mathbf{k}$$

$$|\vec{PQ} = \text{modulus} = \sqrt{(1)^2 + (1)^2 + (-2)^2} = \sqrt{6}$$

Direction Cosines:

$$\text{Unit vector of } \vec{PQ} = \frac{\vec{PQ}}{|\vec{PQ}|}$$

$$= \frac{\mathbf{i} + \mathbf{j} - 2\mathbf{k}}{\sqrt{6}} = \frac{1}{\sqrt{6}}\mathbf{i} + \frac{1}{\sqrt{6}}\mathbf{j} = \frac{2}{\sqrt{6}}\mathbf{k}$$

$\therefore$ D.C's are $\left[\frac{1}{\sqrt{6}}, \frac{1}{\sqrt{6}}, \frac{-2}{\sqrt{6}}\right]$.

Example 29. Find by vector method the perimeter of the triangle whose vertices are the points (3, 1, 5), (–1, –1, 9) and (0, –5, 1).

Solution. Let P(3, 1, 5), Q (-1, -1, 9) and (0, -5, 1) be the vertices of a triangle PQR and point of reference of position vectors be O(0, 0, 0).

Then $\overrightarrow{OP} = 3\mathbf{i} + \mathbf{j} + 5\mathbf{k}, \overrightarrow{OQ} = -\mathbf{i} - \mathbf{j} + 9\mathbf{k}$

and $\overrightarrow{OR} = -5\mathbf{j} + \mathbf{k}$ **respectively.**

$\therefore$ $\overrightarrow{PQ} = \overrightarrow{OQ} - \overrightarrow{OP} = (-\mathbf{i} - \mathbf{j} + 9\mathbf{k}) - (3\mathbf{i} + \mathbf{j} + 5\mathbf{k})$

$= -4\mathbf{i} - 2\mathbf{j} + 4\mathbf{k}$

$\overrightarrow{QR} = \overrightarrow{OR} - \overrightarrow{OQ} = (-5\mathbf{j} + \mathbf{k}) - (-\mathbf{i} - \mathbf{j} + 9\mathbf{k})$

$= \mathbf{i} - 4\mathbf{j} - 8\mathbf{k}$

$\overrightarrow{RP} = \overrightarrow{OP} - \overrightarrow{OR} = (3\mathbf{i} + \mathbf{j} + 5\mathbf{k}) - (-5\mathbf{j} + \mathbf{k})$

$= 3\mathbf{i} + 6\mathbf{j} + 4\mathbf{k}$

$\therefore$ $|\overrightarrow{PQ}| = \sqrt{(-4)^2 + (-2)^2 + (4)^2} = \sqrt{36} = 6$

$|\overrightarrow{QR}| = \sqrt{(1)^2 + (-4)^2 + (-8)^2} = \sqrt{81} = 9$

$|\overrightarrow{RP}| = \sqrt{(3)^2 + (6)^2 + (4)^2} = \sqrt{61}$

$\therefore$ Perimeter $= \frac{1}{2}\{|\overrightarrow{PQ}| + |\overrightarrow{QR}| + |\overrightarrow{RP}|\}$

$= \frac{1}{2}\{6 + 9 + \sqrt{61}\} = \frac{1}{2}(15 + \sqrt{61})$

Example 30. Prove that the points $2\mathbf{i} - \mathbf{j} + \mathbf{k}$, $\mathbf{i} - 3\mathbf{j} - 5\mathbf{k}$, $3\mathbf{i} - 4\mathbf{j} - 4\mathbf{k}$ are the vertices of a right angled triangle.

Solution. Let PQR be the triangle and O be the point of reference such that

$\overrightarrow{OP} = 2\mathbf{i} - \mathbf{j} + \mathbf{k}; \overrightarrow{OQ} = \mathbf{i} - 3\mathbf{j} - 5\mathbf{k}$

and $\overrightarrow{OR} = 3\mathbf{i} - 4\mathbf{j} - 4\mathbf{k}$ respectively.

$\therefore$ $\overrightarrow{PQ} = \overrightarrow{OQ} - \overrightarrow{OP} = (\mathbf{i} - 3\mathbf{j} - 5\mathbf{k}) - (2\mathbf{i} - \mathbf{j} + \mathbf{k})$

$= -\mathbf{i} - 2\mathbf{j} - 6\mathbf{k}$

$$\overrightarrow{QR} = \overrightarrow{OR} - \overrightarrow{OQ} = (3\mathbf{i} - 4\mathbf{j} - 4\mathbf{k}) - (\mathbf{i} - 3\mathbf{j} - 5\mathbf{k})$$
$$= 2\mathbf{i} - \mathbf{j} + \mathbf{k}$$
$$\overrightarrow{RP} = \overrightarrow{OP} - \overrightarrow{OE} = (2\mathbf{i} - \mathbf{j} + \mathbf{k}) - (3\mathbf{i} - 4\mathbf{j} - 4\mathbf{k})$$
$$= -\mathbf{i} + 3\mathbf{j} + 5\mathbf{k}$$

Now $\quad |\overrightarrow{PQ}| = \sqrt{(-1)^2 + (-2)^2 + (-6)^2} = \sqrt{41}$

$\quad |\overrightarrow{QR}| = \sqrt{(2)^2 + (-1)^2 + (1)^2} = \sqrt{6}$

and $\quad |\overrightarrow{RP}| = \sqrt{(-1)^2 + (3)^2 + (5)^2} = \sqrt{35}$

Hence $\quad (\overrightarrow{PQ})^2 = (\overrightarrow{QR})^2 + (\overrightarrow{RP})^2$

$\Rightarrow$ ΔPQR is a right angled triangle.

Example 31. The position vectors of P, Q, R, S be $2\mathbf{i} + 4\mathbf{k}$, $5\mathbf{i} + 3\sqrt{3}\mathbf{j} + 4\mathbf{k}$, $-2\sqrt{3}\mathbf{j} + \mathbf{k}$, $2\mathbf{i} + \mathbf{k}$. Prove that PS is parallel to PQ and is two third of PQ.

Solution. Let O be the point of reference. Then

$$\overrightarrow{OP} = 2\mathbf{i} + 4\mathbf{k}, \quad \overrightarrow{OQ} = 5\mathbf{i} + 3\sqrt{3}\mathbf{j} + 4\mathbf{k}$$
$$\overrightarrow{OR} = -2\sqrt{3}\mathbf{j} + \mathbf{k} \text{ and } \overrightarrow{OS} = 2\mathbf{i} + \mathbf{k}$$

Now $\quad \overrightarrow{RS} = \overrightarrow{OS} - \overrightarrow{OR} = 2\mathbf{i} + \mathbf{k} - (-2\sqrt{3}\mathbf{j} + \mathbf{k})$

$$= 2\mathbf{i} + 2\sqrt{3}\mathbf{j} = 2(\mathbf{i} + \sqrt{3}\mathbf{j})$$
$$\overrightarrow{PQ} = \overrightarrow{OQ} - \overrightarrow{OP} = (5\mathbf{i} + 3\sqrt{3}\mathbf{j} + 4\mathbf{k}) - (2\mathbf{i} + 4\mathbf{k})$$
$$= 3\mathbf{i} + 3\sqrt{3}\mathbf{j} = 3(\mathbf{i} + \sqrt{3}\mathbf{j})$$

$\therefore \quad 3\overrightarrow{RS} = 2\overrightarrow{PQ} \Rightarrow \overrightarrow{PQ}$ and $\overrightarrow{RS}$ are collinear.

Also $\quad |\overrightarrow{RS}| = \frac{2}{3}|\overrightarrow{PQ}|$

$\Rightarrow$ $\overrightarrow{PQ}$ is parallel to $\overrightarrow{RS}$.

Example 32. Examine whether the vectors $\mathbf{i} - 3\mathbf{j} + 2\mathbf{k}$, $2\mathbf{i} - 4\mathbf{j} - \mathbf{k}$ and $3\mathbf{i} + 2\mathbf{j} - \mathbf{k}$ are linearly independent or dependent.

Solution. Let $\mathbf{a} = \mathbf{i} - 3\mathbf{j} + 2\mathbf{k}$, $\mathbf{b} = 2\mathbf{i} - 4\mathbf{j} - \mathbf{k}$ and $\mathbf{c} = 3\mathbf{i} + 2\mathbf{j} - \mathbf{k}$ be the vectors and α, β, γ be the scalars then either linearly independent or dependent, we have

$$\alpha\mathbf{a} + \beta\mathbf{b} + \gamma\mathbf{c} = 0 \qquad ...(1)$$

i.e., $\quad \alpha(\mathbf{i} - 3\mathbf{j} + 2\mathbf{k}) + \beta(2\mathbf{i} - 4\mathbf{j} - \mathbf{k}) + \gamma(3\mathbf{i} + 2\mathbf{j} - \mathbf{k}) = 0$

or $\quad (\alpha + 2\beta + 3\gamma)\mathbf{i} + (-3\alpha - 4\beta + 2\gamma)\mathbf{j} + (2\alpha - \beta - \gamma)\mathbf{k} = 0$

$\because$ **i, j, k** are non–coplanar

$$\Rightarrow \qquad \alpha + 2\beta + 3\gamma = 0 \qquad ...(2)$$

$$-3\alpha - 4\beta + 2\gamma = 0 \qquad ...(3)$$

$$2\alpha - \beta - \gamma = 0 \qquad ...(4)$$

Solving (2) and (3) we get

$$\frac{\alpha}{16} = \frac{\beta}{-11} + \frac{\gamma}{2} = k \text{ (say)}$$

$\therefore$ $\alpha = 16k$, $\beta = -11k$, $\gamma = 2k$, which do not satisfy (4) $\Rightarrow$ the relation (1) does not exist

$\Leftrightarrow$ $\alpha = 0$, $\beta = 0$ and $\gamma = 0$

i.e, the given vectors are linearly independent.

Example 33. Three vectors of magnitude a, 2a, 3a meet in a point and their directions are along the diagonals of the adjacent faces of a cube. Determine their resultant and its direction cosines.

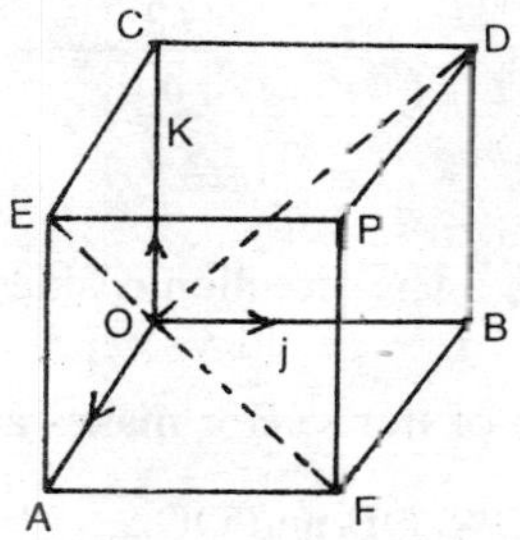

Fig. 1.35

Solution. Consider the cube as shown in the figure. Let **i, j, k** be the unit vectors along three mutually perpendicular coterminous edges OA, OB and OC respectively.

$\therefore$ $\overrightarrow{OA} = \mathbf{i}$, $\overrightarrow{OB} = \mathbf{j}$, $\overrightarrow{OC} = \mathbf{k}$, OD, OE and OF are the diagonals of three adjacent faces of the cube through O along which act the vectors of magnitudes a, 2a, 3a respectively

$$\therefore \qquad \overrightarrow{OD} = \overrightarrow{OB} + \overrightarrow{BD} = \overrightarrow{OB} + \overrightarrow{OC} = \mathbf{j} + \mathbf{k}$$

Unit vector along $\overrightarrow{OD} = \dfrac{\mathbf{j}+\mathbf{k}}{|\overrightarrow{OD}|} = \dfrac{\mathbf{j}+\mathbf{k}}{\sqrt{1+1}} = \dfrac{1}{\sqrt{2}}(\mathbf{j}+\mathbf{k})$

Similarly unit vector along $\overrightarrow{OE} = \dfrac{\mathbf{k}+\mathbf{i}}{\sqrt{2}}$

and unit vector along $\overrightarrow{OF} = \dfrac{\mathbf{i}+\mathbf{j}}{\sqrt{2}}$

A vector of magnitude a along OD $= \dfrac{a}{\sqrt{2}}(\mathbf{j}+\mathbf{k})$

A vector of magnitude 2a along OE = $\frac{2a}{\sqrt{2}}(\mathbf{k} + \mathbf{i})$

and of 3a along OF = $\frac{3a}{\sqrt{2}}(\mathbf{i} + \mathbf{j})$

Resultant **R** of all these

$$= \frac{a}{\sqrt{2}}(\mathbf{j} + \mathbf{k}) + \frac{2a}{\sqrt{2}}(\mathbf{k} + \mathbf{i}) + \frac{3a}{\sqrt{2}}(\mathbf{i} + \mathbf{j})$$

$$= \frac{a}{\sqrt{2}}(5\mathbf{i} + 4\mathbf{j} + 3\mathbf{k})$$

Unit vector in the direction of R

$$= \frac{\frac{a}{\sqrt{2}}(5\mathbf{i} + 4\mathbf{j} + 3\mathbf{k})}{\frac{a}{\sqrt{2}}\sqrt{5^2 + 4^2 + 3^2}} = \frac{1}{\sqrt{2}}\mathbf{i} + \frac{4}{5\sqrt{2}}\mathbf{j} + \frac{3}{5\sqrt{2}}\mathbf{k}$$

$\therefore$ The direction cosines of R are $\frac{1}{\sqrt{2}}, \frac{5}{5\sqrt{2}}, \frac{3}{5\sqrt{2}}$ i.e., the line of action of this vector makes angles $\cos^{-1}\frac{1}{\sqrt{2}}, \cos^{-1}\frac{4}{5\sqrt{2}}, \cos^{-1}\frac{3}{5\sqrt{2}}$ with OA, OB and OC.

EXERCISE

1. If the position vectors of three points A, B and C are 2**a**, 3**b** and **a** + 2**b** respectively find $\overrightarrow{AB}$, $\overrightarrow{AC}$ and $\overrightarrow{BC}$ in terms of **a** and **b**
2. ABCD is a parallelogram. If $\overrightarrow{AB}$ = **a**, $\overrightarrow{AD}$ = **b** then find $\overrightarrow{AC}$ and $\overrightarrow{BD}$
3. If D, E, F are the mid points of sides BC, CA and AB then show that $\overrightarrow{AB} + \overrightarrow{BE} + \overrightarrow{CF} = \mathbf{0}$
4. If ABCD is a quadrilateral prove that
$$\overrightarrow{BA} + \overrightarrow{BC} + \overrightarrow{CD} + \overrightarrow{DA} = 2\overrightarrow{BA}.$$
5. If ABCDE is a pentagon, prove that
$$\overrightarrow{AC} + \overrightarrow{AB} + \overrightarrow{AE} + \overrightarrow{CB} + \overrightarrow{DB} + \overrightarrow{EB} = 3\overrightarrow{AB}$$
6. If ABCDEF is a regular hexagon prove that

$$\vec{AB} + \vec{AC} + \vec{AD} + \vec{EA} + \vec{FA} = 4\vec{AB}$$

7. If ABCDEF is a regular hexagon and G in the centre of it then prove that

$$\vec{AB} + \vec{AC} + \vec{AD} + \vec{AE} + \vec{AF} = 6\vec{AG}$$

8. Prove that the line joining the middle point of two sides of a triangle is parallel to the third side and half of its length.

9. ABC is a triangle. The mid points of AB and AC are respectively D and E. Prove that $\vec{BE} + \vec{DC} = \frac{3}{2}\vec{BC}$.

10. If **a**, **b** are the position vectors of A, B respectively, find that point **c** in AB produced such that AC = 3AB and that of a point D in BA such that BD = 2BA. Also show that DA + CD = CB - AB.

11. Prove that the lines joining the mid points of opposite sides of a quadrilateral bisect each other.

12. If A, B, C, D be any four points in plane. Prove that the vector sum $\vec{AB} + \vec{CB} + \vec{CD} + \vec{AD} = 4\vec{EF}$ where E, F are the midpoints of AC and BD respectively.

13. ABCD is a parallelogram and P the point of intersection of its diagonals. Show that for any origin O,

$$\vec{OA} + \vec{OB} + \vec{OC} + \vec{OD} = 4\vec{OP}$$

14. ABC is a triangle and P any point in BC. IF $\vec{PQ}$ is the resultant of $\vec{AP}$, $\vec{PB}$, $\vec{PC}$ show that ABQC is a parallelogram and Q, therefore a fixed point.

15. Prove that the straight line joining the middle points of the diagonals of a trapezium is parallel to parallel sides and half of their difference.

16. Prove that the join of the middle points of the opposite edges of a tetrahedron intersect and bisect each other.

17. If the midpoints of the consecutive sides of any quadrilateral are connected by straight lines prove that the resulting in a parallelogram.

18. If M, N are the mid points of the sides AB, CD of a parallelogram ABCD, prove that DM and BN cut the diagonal

AC at its points of trisection which are also the points of trisection of DM and BN respectively.

19. Find the sum of the vectors $2\mathbf{i} + \mathbf{j} + \mathbf{k}$, $5\mathbf{i} + 4\mathbf{j} + 2\mathbf{k}$, $\mathbf{i} - 7\mathbf{j} + 2\mathbf{k}$ and $4\mathbf{i} + 2\mathbf{j} + 3\mathbf{k}$ also find the direction cosines of it.

20. The position vectors of A and B are $\mathbf{i} + 3\mathbf{j} - 4\mathbf{k}$ and $3\mathbf{i} + 2\mathbf{j} - 6\mathbf{k}$. Find $\overrightarrow{AB}$ and direction cosines.

21. For what values of x for which $x(\mathbf{i} + \mathbf{j} + \mathbf{k})$ is a unit vector.

22. Find a unit vector parallel to $3a - 2b + c$ if $\mathbf{a} = 3\mathbf{i} - \mathbf{j} - 4\mathbf{k}$, $\mathbf{b} = -2\mathbf{i} + \mathbf{j} + \mathbf{k}$ and $\mathbf{c} = 2\mathbf{i} + 3\mathbf{j} + 4\mathbf{k}$.

23. If $\mathbf{a} = 2\mathbf{i} + \mathbf{j} - \mathbf{k}$, $\mathbf{b} = \mathbf{i} - \mathbf{j}$, $\mathbf{c} = 5\mathbf{i} - \mathbf{j} + \mathbf{k}$ find a unit vector parallel to $\mathbf{a} + \mathbf{b} - c$ but opposite in direction.

24. Prove that the vectors $2\mathbf{i} + 3\mathbf{j} - 6\mathbf{k}$, $6\mathbf{i} - 2\mathbf{j} + 3\mathbf{k}$ and $3\mathbf{i} + 6\mathbf{j} - 2\mathbf{k}$ from the sides of an equilateral triangle.

25. Show the triangle ABC whose vertices A, B and C are $7\mathbf{i} + 10\mathbf{k}$, $-\mathbf{i} + 6\mathbf{j} + \mathbf{k}$, $-4\mathbf{i} + 9\mathbf{j} + 6\mathbf{k}$ is isosceles and right angled triangle.

26. The position vectors of the points A, B, C, D are $2\mathbf{i} + 4\mathbf{j} + 2\mathbf{k}$, $\mathbf{i} + 2\mathbf{j} + \mathbf{k}$, $4\mathbf{i} + 3\mathbf{j} - 2\mathbf{k}$, $3\mathbf{i} + \mathbf{j} - 3\mathbf{k}$. Show that ABCD is a parallelogram.

27. If the position vectors of A, B, C, D be $2\mathbf{i} + 4\mathbf{k}$, $5\mathbf{i} + 3\sqrt{3}\mathbf{j} + 4\mathbf{k}$, $-2\sqrt{3}\mathbf{j} + \mathbf{k}$, $2\mathbf{i} + \mathbf{k}$, prove that CD is parallel to AB and is $\frac{2}{3}$ of AB.

28. The position vectors of the points A, B, C, D are $\mathbf{i} + \mathbf{j} + \mathbf{k}$, $2\mathbf{i} + 5\mathbf{j}$, $3\mathbf{i} + 2\mathbf{j} - 3\mathbf{k}$, $\mathbf{i} - 6\mathbf{j} - \mathbf{k}$. Show that lines AB and CD are parallel on find the ratio of their lengths.

29. If the position vectors of the vertices A, B, C and D of a quadrilateral be $\mathbf{i} + \mathbf{j} + \mathbf{k}$, $\mathbf{j} + \mathbf{k} - \mathbf{i}$, $\mathbf{k} + \mathbf{i} - \mathbf{j}$ and $\mathbf{i} + \mathbf{j} - \mathbf{k}$ respectively then find the perimeter of ABCD.

30. Show that the points A (1, 1, 1), B (1, 2, 3) and C (2, -1, 1) are vertices of an isosceles triangle. Find the perimeter of it.

31. If **a, b, c** be three non-zero, non-coplanar vectors find a linear relation between the vectors

$$\mathbf{a} - \mathbf{b} + \mathbf{c},\ \mathbf{b} + \mathbf{c} - \mathbf{a},\ \mathbf{c} + \mathbf{a} + \mathbf{b},\ 2\mathbf{a} - 3\mathbf{b} + 4\mathbf{c}.$$

32. Find the value of λ and μ for which the vectors $-3\mathbf{i}+4\mathbf{j}+\lambda\mathbf{k}$ and $\mu\mathbf{i}+8\mathbf{j}+6\mathbf{k}$ are collinear.

33. If **A** and **B** are non-collinear vectors $\mathbf{A}=(x+4y)\mathbf{a}+(2x+y+1)\mathbf{b}$ and $\mathbf{B}=(y-2x+2)\mathbf{a}+(2x-3y-1)\mathbf{b}$ find x and y such that 3A = 2B.

34. Prove that the following vectors are collinear where **a**, **b**, **c** being non-coplanar vectors.

(i) $\mathbf{a}, \mathbf{b}, 3\mathbf{a}-2\mathbf{b}$

(ii) $\mathbf{a}-2\mathbf{b}+3\mathbf{c},\ 2\mathbf{a}+3\mathbf{b}-4\mathbf{c},\ -7\mathbf{a}+10\mathbf{c}$

(iii) $-2\mathbf{a}+3\mathbf{b}+5\mathbf{c},\ \mathbf{a}+2\mathbf{b}+3\mathbf{c},\ 7\mathbf{a}-\mathbf{c}$

(iv) $\mathbf{a}+\mathbf{b}+\mathbf{c},\ 4\mathbf{a}+3\mathbf{b},\ 10\mathbf{a}+7\mathbf{b}-2\mathbf{c}$

35. Show that the vectors $4\mathbf{i}+2\mathbf{j}+\mathbf{k},\ 2\mathbf{i}-\mathbf{j}+3\mathbf{k}$ and $8\mathbf{i}+7\mathbf{k}$ are coplanar.

36. Show that the following points are coplanar where **a**, **b**, **c** being non-coplanar vectors.

(i) $-\mathbf{a}+4\mathbf{b}-3\mathbf{c},\ 3\mathbf{a}+2\mathbf{b}-5\mathbf{c},\ -3\mathbf{a}+8\mathbf{b}-5\mathbf{c},\ -3\mathbf{a}+2\mathbf{b}+\mathbf{c}$

(ii) $6\mathbf{a}+2\mathbf{b}-\mathbf{c},\ -5\mathbf{a}+3\mathbf{b}-10\mathbf{c},\ 4\mathbf{a}-6\mathbf{b}-10\mathbf{c},\ 2\mathbf{b}+10\mathbf{c}$

(iii) $2\mathbf{a}+3\mathbf{b}-\mathbf{c},\ \mathbf{a}-2\mathbf{b}+3\mathbf{c},\ 3\mathbf{a}+4\mathbf{b}-2\mathbf{c},\ \mathbf{a}-6\mathbf{b}+6\mathbf{c}$

37. Show that the points $4\mathbf{i}+5\mathbf{j}+\mathbf{k},\ -\mathbf{j}-\mathbf{k},\ 5\mathbf{i}+9\mathbf{j}+4\mathbf{k},\ -\mathbf{i}+\mathbf{k}$ are coplanar

38. Show that the vectors $\mathbf{a}-2\mathbf{b}+3\mathbf{c}, -2\mathbf{a}+3\mathbf{b}-4\mathbf{c},\ -\mathbf{b}+2\mathbf{c}$ are linearly dependent where **a**, **b**, **c** are linearly independent.

39. If **a**, **b**, **c** are linearly independent show that the vectors $5\mathbf{a}-6\mathbf{b}+7\mathbf{c},\ 7\mathbf{a}-8\mathbf{b}+9\mathbf{c},\ 3\mathbf{a}+20\mathbf{b}+5\mathbf{c}$ are also linearly independent.

40. Examine whether the following vectors are linearly independent or dependent.

(i) $\mathbf{i}+2\mathbf{j}+3\mathbf{k},\ 2\mathbf{i}+\mathbf{j}+3\mathbf{k},\ \mathbf{i}+\mathbf{j}+\mathbf{k}$

(ii) $2\mathbf{i}-4\mathbf{k},\ \mathbf{i}-2\mathbf{j}-\mathbf{k},\ \mathbf{i}-4\mathbf{j}+3\mathbf{k}$

(iii) $\mathbf{i}+3\mathbf{j}+2\mathbf{k},\ \mathbf{i}-7\mathbf{j}-8\mathbf{k},\ \mathbf{i}+\mathbf{j}-\mathbf{k}$

ANSWERS

1. $\overrightarrow{AB}=3\mathbf{b}-2\mathbf{a},\ \overrightarrow{AC}=2\mathbf{b}-\mathbf{a},\ \overrightarrow{BC}=\mathbf{a}-\mathbf{b}$

2. $\overrightarrow{AC} = \mathbf{a} + \mathbf{b}$ $\overrightarrow{BD} = \mathbf{b} - \mathbf{a}$

10. $\overrightarrow{OC} = 3\mathbf{b} - 2\mathbf{a}$, $\overrightarrow{OD} = 2\mathbf{a} - \mathbf{b}$

19. $12\mathbf{i} - 2\mathbf{j} + 8\mathbf{k}$; $\frac{6}{\sqrt{53}}, \frac{-1}{\sqrt{53}}, \frac{4}{\sqrt{53}}$

20. $2\mathbf{i} - \mathbf{j} - 2\mathbf{k}$; $\frac{2}{3}, \frac{-1}{3}, \frac{-2}{3}$

21. $x = \pm \frac{1}{\sqrt{3}}$ **22.** $\frac{1}{\sqrt{329}}(15\mathbf{i} - 2\mathbf{j} - 10\mathbf{k})$

23. $\frac{1}{3}(2\mathbf{i} - \mathbf{j} + 2\mathbf{k})$ **28.** 1 : 2 **29.** $4(1 + \sqrt{2})$ 30. $\sqrt{14} + 2\sqrt{5}$

31. $2\mathbf{a} - 3\mathbf{b} + 4\mathbf{c} = \frac{7}{2}(\mathbf{a} - \mathbf{b} + \mathbf{c}) + (\mathbf{b} + \mathbf{c} - a) - \frac{1}{2}(\mathbf{c} + \mathbf{a} + \mathbf{b})$

32. $\lambda = 3$, $\mu = -6$ **33.** x = 2, y = -1

40. (i) Linearly independent (ii) and (iii) are linearly dependent

2

Product of Two Vectors

2.1. Introduction

So far we have dealt with the addition and subtraction of vectors and also the multiplication of vectors with scalars. In this chapter we shall deal with product of two vectors. There are two different ways by which vector quantities are multiplied. One being pure number is called **Scalar or dot product** while the other being a vector quantity is called **Vector or Cross product**. However in each case the product is proportional to the product of the lengths of the two vectors and they also follow distributive law.

2.2. The Scalar or Dot product of Two Vectors

The scalar product of two vectors **a** and **b** of module a and b respectively is defined as ab cos θ, where θ is the angle between the directions of the vectors **a** and **b** such that $0 \leq \theta \leq \pi$. It is denoted by **a . b** by placing a dot between the vectors and read as **a** dot **b**. Thus we have $\mathbf{a} . \mathbf{b} = |\mathbf{a}||\mathbf{b}| \cos \theta$ where $a = |\mathbf{a}|$ and $b = |\mathbf{b}|$. The product will be positive, negative or zero according as the angle θ is less than $\frac{\pi}{2}$, greater than $\frac{\pi}{2}$ or $\frac{\pi}{2}$.

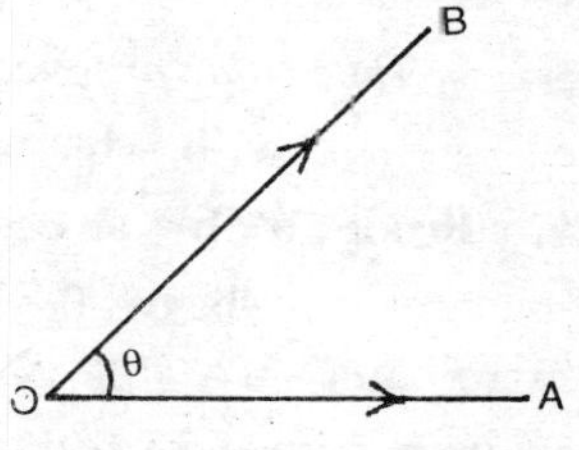

Fig. 2.1

The dot product is also called **inner product** or **direct product**.

2.3. Geometrical Interpretation

The scalar product of two vectors is the product of magnitude of one vector and the length of projection of the other in its direction.

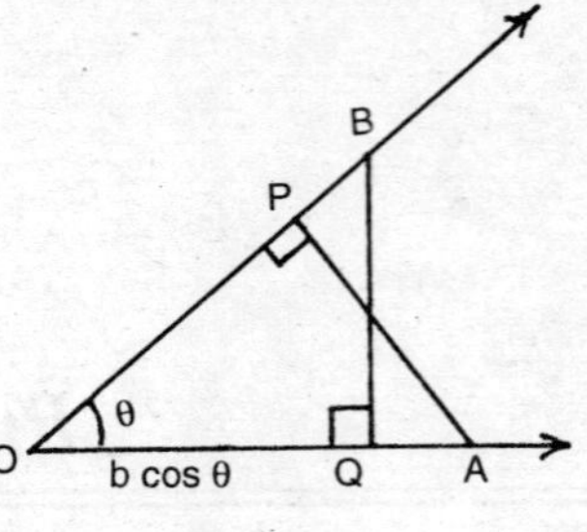

Fig. 2.2

Let $\overrightarrow{OA} = \mathbf{a}$ and $\overrightarrow{OB} = \mathbf{b}$ and BOA = θ, then $\mathbf{a} \cdot \mathbf{b} = ab \cos\theta$, where $a = |\mathbf{a}|$ and $b = |\mathbf{b}|$.

$\therefore\ \mathbf{a} \cdot \mathbf{b} = a\,[OB \cos\theta] = a[OQ]$

$= |\mathbf{a}|$. Projection of **b** in the direction of **a**

Similarly we can show that

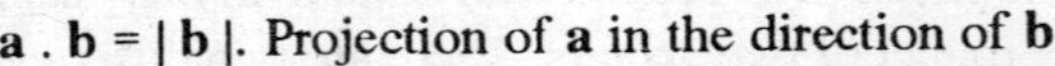

$\mathbf{a} \cdot \mathbf{b} = |\mathbf{b}|$. Projection of **a** in the direction of **b**

Note (1) Here $b \cos\theta$ is called the scalar component of **b** along **a**.

(2) If **a** and **b** are two vectors and $\boldsymbol{a} \neq 0$, then projection of **b** on **a** $= \dfrac{\mathbf{a}.\mathbf{b}}{|\mathbf{a}|} = \hat{\mathbf{a}} \cdot \mathbf{b}$.

(3) Resolved part of **b** in the direction of **a**

$$= \left(\frac{\mathbf{a}.\mathbf{b}}{|\mathbf{a}|}\right)\frac{\mathbf{a}}{|\mathbf{a}|} = (\hat{\mathbf{a}}.\mathbf{b})\,\hat{\mathbf{a}}$$

2.4. Properties of scaler product

1. *The scalar product is commutative*

i.e., $\mathbf{a} \cdot \mathbf{b} = \mathbf{b} \cdot \mathbf{a}$

Proof : $\mathbf{a} \cdot \mathbf{b} = ab \cos\theta$

$= ba \cos\theta = \mathbf{b} \cdot \mathbf{a}$

2. BOA = θ *For every pair of vectors **a**, **b** it can be easily shown that*

$\mathbf{a} \cdot (-\mathbf{b}) = -(\mathbf{a} \cdot \mathbf{b});\ (-\mathbf{a} \cdot \mathbf{b}) = -(\mathbf{a} \cdot \mathbf{b})$

and $(-\mathbf{a}) \cdot (-\mathbf{b}) = \mathbf{a} \cdot \mathbf{b}$

Proof : Let $\overrightarrow{OA} = \boldsymbol{a}$ and $\overrightarrow{OB} = \boldsymbol{b}$. Extend AO and BO to A′ and B′ such that OA′ = OA and OB′ = OB as

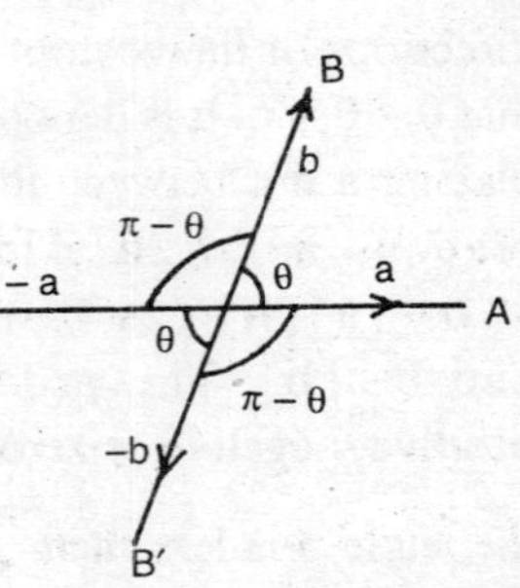

Fig. 2.3

shown in the figure. Then $\vec{OA} = -\mathbf{a}$ and $\vec{OB} = -\mathbf{b}$

Now $\quad \mathbf{a} . (-\mathbf{b}) = \vec{OA} . \vec{OB'} = |a||b| \cos(\pi - \theta)$

$$= -|a||b| \cos\theta = -(\mathbf{a} . \mathbf{b})$$

Again $\quad (-\mathbf{a}) . \mathbf{b} = \vec{OA'} . \vec{OB} = |a||b| \cos(\pi - \theta)$

$$= -|a||b| \cos\theta = (\mathbf{a} . \mathbf{b})$$

Finally $(-\mathbf{a}).(-\mathbf{b}) = \vec{OA'} . \vec{OB'}$

$$= |a||b| \cos\theta = \mathbf{a} . \mathbf{b}$$

Hence the proof.

3. *If M is any scalar and a and b are any vectors then*

$$(m\mathbf{a}) . \mathbf{b} = m(\mathbf{a} . \mathbf{b}) = \mathbf{a} . (m\mathbf{b})$$

Proof : Let θ be the angle between the vectors **a** and **b**.

Case (i) *Let m be +ve integer :*

Then the angle between m**a** and **b** will also be θ. Since the vector m**a** will be the same as that of **a**.

$$\therefore \quad (m\mathbf{a}) . \mathbf{b} = |m\mathbf{a}| . |\mathbf{b}| \cos\theta = |m||\mathbf{a}||\mathbf{b}| \cos\theta$$

$$= m|\mathbf{a}||\mathbf{b}| \cos\theta = m(\mathbf{a} . \mathbf{b}) \qquad ...(1)$$

Again $m|\mathbf{a}||\mathbf{b}| \cos\theta = |\mathbf{a}|(m|\mathbf{b}|) \cos\theta$

$$= |\mathbf{a}||m\mathbf{b}| \cos\theta = \mathbf{a} . (m\mathbf{b}) \qquad ...(2)$$

∴ From (1) and (2), we get

$$(m\mathbf{a}) . \mathbf{b} = m(\mathbf{a} . \mathbf{b} = \mathbf{a} . (m\mathbf{b}) \qquad ...(3)$$

Case (ii) *Let m be –ve integer*

Then the angle between the vectors m**a** and **b** will be π – θ because m**a** is now a vector in a direction opposite to that of **a**.

$$\therefore \quad (m\mathbf{a}) . \mathbf{b} = |m\mathbf{a}||\mathbf{b}| \cos(\pi - \theta)$$

$$= |m||\mathbf{a}||\mathbf{b}| \cos(\pi - \theta)$$

$$= (-m)|\mathbf{a}||\mathbf{b}|(-\cos\theta)$$

$$= m|\mathbf{a}||\mathbf{b}| \cos\theta = m(\mathbf{a} . \mathbf{b}) \qquad ...(4)$$

Again $m|\mathbf{a}||\mathbf{b}| \cos\theta = |\mathbf{a}|(m|\mathbf{b}|) \cos\theta$

$$= |\mathbf{a}|(-m|\mathbf{b}|)(-\cos\theta)$$

$$= |\mathbf{a}||m\mathbf{b}| \cos(\pi - \theta)$$

$$= \mathbf{a} . (m\mathbf{b}) \qquad ...(5)$$

∴ From (4) and (5), we get

$$(m\mathbf{a}) \cdot \mathbf{b} = m(\mathbf{a} \cdot \mathbf{b}) = \mathbf{a} \cdot (m\mathbf{b}) \qquad ...(6)$$

Hence from (3) and (6), for any scalar in

$$(m\mathbf{a}) \cdot \mathbf{b} = m(\mathbf{a} \cdot \mathbf{b}) = \mathbf{a} \cdot (m\mathbf{b})$$

4. *If* m, n *are any scalars and* **a, b** *be any vectors then*

$$(m\mathbf{a}) \cdot (n\mathbf{b}) = mn\,(\mathbf{a} \cdot \mathbf{b}) = mn\,\mathbf{a} \cdot \mathbf{b} = (n\mathbf{a}) \cdot (m\mathbf{b})$$

5. *If two vectors* **a** *and* **b** *have the opposite direction then*

$$\mathbf{a} \cdot \mathbf{b} = -|\mathbf{a}||\mathbf{b}|$$

$\therefore$ In such case $\theta = 0$, $\therefore$ $\cos\theta = \cos\theta = 1$

$$\mathbf{a} \cdot \mathbf{b} = |\mathbf{a}||\mathbf{b}|\cos\theta = |\mathbf{a}||\mathbf{b}|$$

6. *If two vectors* **a** *and* **b** *have opposite direction then*

$$\mathbf{a} \cdot \mathbf{b} = -|\mathbf{a}||\mathbf{b}|$$

$\therefore$ In such case $\theta = \pi$, $\therefore$ $\cos\theta = \cos\pi = -1$

$\therefore$ $\mathbf{a} \cdot \mathbf{b} = |\mathbf{a}||\mathbf{b}|\cos\pi = -|\mathbf{a}||\mathbf{b}|$.

7. *Angle between two vectors in terms of scalar product*

By definition $\mathbf{a} \cdot \mathbf{b} = |\mathbf{a}||\mathbf{b}|\cos\theta$

$$\Rightarrow \qquad \cos\theta = \frac{\mathbf{a} \cdot \mathbf{b}}{|\mathbf{a}||\mathbf{b}|}$$

or

$$\theta = \cos^{-1}\left[\frac{\mathbf{a} \cdot \mathbf{b}}{|\mathbf{a}||\mathbf{b}|}\right]$$

Note : If each of the vectors **a** and **b** is a unit vector.

i.e.,

$$\mathbf{a} = \hat{\mathbf{a}},\ \mathbf{b} = \hat{\mathbf{b}}$$

Unit Vector then $\hat{\mathbf{a}} \cdot \hat{\mathbf{b}} = \cos\theta$.

Thus scaler product of two unit vectors is equal to the cosine of the angle between their directions.

8. *If* **a** *and* **b** *are perpendicular vectors, then*

$$\mathbf{a} \cdot \mathbf{b} = 0$$

$\therefore$ If θ is the angle between the two perpendicular vectors **a** and **b** then $\theta = \pm\frac{\pi}{2}$, $\therefore$ $\cos\theta = 0$

$$\mathbf{a} \cdot \mathbf{b} = |\mathbf{a}||\mathbf{b}| \cdot 0 = 0$$

Note: If $\quad \mathbf{a} \cdot \mathbf{b} = 0 \Rightarrow |\mathbf{a}||\mathbf{b}|\cos\theta = 0$

$$|\mathbf{a}| = 0,\ |\mathbf{b}| = 0 \text{ or } \cos\theta = 0 \Rightarrow \theta = \pm\frac{\pi}{2}$$

Thus if $\mathbf{a}\cdot\mathbf{b} = 0$, *either one of the vectors must be null vector or the two vectors* **a** *and* **b** *should be perpendicular*.

Remark. The necessary and sufficient condition that two non-zero vectors should be perpendicular in that their dot product should be zero.

9. *Magnitude of a vector as a scalar product*

If **a** is any vector then

$$\mathbf{a}\cdot\mathbf{a} = |\mathbf{a}||\mathbf{a}|\cos\theta = |\mathbf{a}|^2$$

i.e.,
$$\mathbf{a}\cdot\mathbf{a} = |\mathbf{a}|^2 = a^2 = \mathbf{a}^2;\ [|\mathbf{a}| = a]$$

i.e., square of any vector is equal to square of its magnitude (modulus).

Also $\quad |\mathbf{a}| = +\sqrt{\mathbf{a}\cdot\mathbf{a}}$

Note: In particular, square of a unit vector will be equal to unity

10. *Scalar product of orthonormal vector triads* **i, j, k**

i, j, k, are unit vectors along three mutually perpendicular directions

$\therefore\ |\mathbf{i}| = |\mathbf{j}| = |\mathbf{k}| = 1$ and

$$\mathbf{i}\cdot\mathbf{i} = \mathbf{j}\cdot\mathbf{j} = \mathbf{k}\cdot\mathbf{k} = 1\,.\,1.\cos 0 = 1$$

$$\mathbf{i}\cdot\mathbf{j} = \mathbf{j}\cdot\mathbf{k} = \mathbf{k}\cdot\mathbf{i} = 1\,.\,1\cos 90^\circ = 0$$

$$\mathbf{j}\cdot\mathbf{i} = \mathbf{k}\cdot\mathbf{j} = \mathbf{i}\cdot\mathbf{k} = 1.1\cos 90^\circ = 0$$

11. Distributive law. *Dot product is distributive with respect to addition*

i.e.,
$$\mathbf{a}\cdot(\mathbf{b}+\mathbf{c}) = \mathbf{a}\cdot\mathbf{b} + \mathbf{a}\cdot\mathbf{c}$$

Proof. Take O as the origin of the vectors. Let $\overrightarrow{OA} = \mathbf{a}$, $\overrightarrow{OB} = \mathbf{b}$ and $\overrightarrow{BC} = \mathbf{c}$. Then $\overrightarrow{OC} = \overrightarrow{OB} + \overrightarrow{BC}$

$= \mathbf{b} + \mathbf{c}$

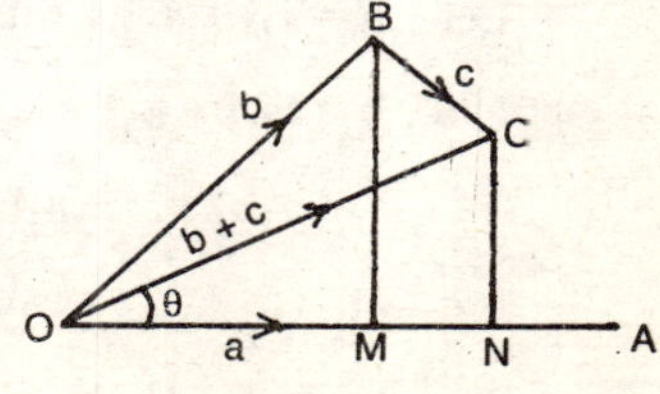

Fig. 2.4

Draw BM and CN perpendiculars from B and C on OA. Then OM and ON are the projections of OB and OC respectively and MN is the projection of BC on OA. Now let OA =a

Now $\mathbf{a} \cdot (\mathbf{b} + \mathbf{c}) = \mathbf{a} \cdot \vec{OC} = |\mathbf{a}|\, OC \cos\theta$

$= \mathbf{a}\, ON = \mathbf{a}\, OM + \mathbf{a}\, ON$

$= \mathbf{a}$ (projection of $\mathbf{b}$ on $\mathbf{a}$) + $\mathbf{a}$ (projection of $\mathbf{c}$ on $\mathbf{a}$)

$= \mathbf{a} \cdot \mathbf{b} + \mathbf{a} \cdot \mathbf{c}$

2.5. Simple Identities Based on Distributive Law

(i) $\mathbf{a} \cdot (\mathbf{b} - \mathbf{c}) = \mathbf{a} \cdot \mathbf{b} - \mathbf{a} \cdot \mathbf{c}$

L.H.S $= \mathbf{a} \cdot [\mathbf{b} + (-\mathbf{c})] = \mathbf{a} \cdot \mathbf{b} + \mathbf{a} \cdot (-\mathbf{c})$

$= \mathbf{a} \cdot \mathbf{b} - \mathbf{a} \cdot \mathbf{c} =$ R.H.S

(ii) $(\mathbf{a} + \mathbf{b}) \cdot (\mathbf{a} - \mathbf{b}) = \mathbf{a}^2 - \mathbf{b}^2$

L.H.S $= (\mathbf{a} \cdot \mathbf{a} - \mathbf{a} \cdot \mathbf{b} + \mathbf{b} \cdot \mathbf{a} - \mathbf{b} \cdot \mathbf{b}$

$= |\mathbf{a}|^2 - \mathbf{a} \cdot \mathbf{b} + \mathbf{a} \cdot \mathbf{b} - |\mathbf{b}|^2 = \mathbf{a}^2 - \mathbf{b}^2 =$ R.H.S

(iii) $(\mathbf{a} + \mathbf{b})^2 = a^2 + 2\mathbf{a} \cdot \mathbf{b} + b^2$

L.H.S $(\mathbf{a} + \mathbf{b}) \cdot (\mathbf{a} + \mathbf{b}) = \mathbf{a} \cdot \mathbf{a} + \mathbf{a} \cdot \mathbf{b} + \mathbf{b} \cdot \mathbf{a} + \mathbf{b} \cdot \mathbf{b}$

$= \mathbf{a}^2 + \mathbf{a} \cdot \mathbf{b} + \mathbf{a} \cdot \mathbf{b} + b^2$

$= \mathbf{a}^2 + 2\mathbf{a} \cdot \mathbf{b} + \mathbf{b}^2 =$ R.H.S

(iv) $(\mathbf{a} - \mathbf{b})^2 - \mathbf{a}^2 - 2\mathbf{a} \cdot \mathbf{b} + \mathbf{b}^2$

L.H.S $(\mathbf{a} - \mathbf{b}) \cdot (\mathbf{a} - \mathbf{b}) = \mathbf{a} \cdot \mathbf{a} - \mathbf{a} \cdot \mathbf{b} - \mathbf{b} \cdot \mathbf{a} + \mathbf{b} \cdot \mathbf{b}$

$= a^2 - \mathbf{a} \cdot \mathbf{b} - \mathbf{a} \cdot \mathbf{b} + b^2$

$= \mathbf{a}^2 - 2\mathbf{a} \cdot \mathbf{b} + b^2 =$ R.H.S.

(v) $(\mathbf{a} + \mathbf{b}) \cdot (\mathbf{c} + \mathbf{d}) = (\mathbf{a} + \mathbf{b}) \cdot \mathbf{c} + (\mathbf{a} + \mathbf{b}) \cdot \mathbf{d}$

$= \mathbf{a} \cdot \mathbf{c} + \mathbf{b} \cdot \mathbf{c} + \mathbf{a} \cdot \mathbf{d} + \mathbf{b} \cdot \mathbf{d}$

2.6. Scalar product of two vectors interms of their rectangular components

Let $\mathbf{a} = a_1\mathbf{i} + a_2\mathbf{j} + a_3\mathbf{k}$ and $\mathbf{b} = b_1\mathbf{i} + b_2\mathbf{j} + b_3\mathbf{k}$

$\mathbf{a} \cdot \mathbf{b} = (a_1\mathbf{i} + a_2\mathbf{j} + a_3\mathbf{k}) \cdot (b_1\mathbf{i} + b_2\mathbf{j} + b_3\mathbf{k})$

$= a_1b_1\, \mathbf{i} \cdot \mathbf{i} + a_1b_2\mathbf{i} \cdot \mathbf{j} + a_1b_3\mathbf{i} \cdot \mathbf{k} + a_2b_1\mathbf{j} \cdot \mathbf{i} + a_2b_2\mathbf{i} \cdot \mathbf{j}$

$+ a_2b_3\mathbf{k} \cdot \mathbf{k} + a_3b_1\mathbf{k} \cdot \mathbf{i} + a_3b_2\mathbf{k} \cdot \mathbf{j} + a_3b_3\mathbf{k} \cdot \mathbf{k}$

$= a_1b_1 + a_2b_2 + a_3b_3$ $\because$ [Using **property (10)** of art]

$$\therefore \qquad \mathbf{a}\cdot\mathbf{b} = a_1b_1 + a_2b_2 + a_3b_3$$

i.e the scaler product of two vectors is equal to the sum of the products of their corresponding rectangular components

Note. If $\mathbf{a} = a_1\mathbf{i} + a_2\mathbf{j} + a_3\mathbf{k}$ then

$$\mathbf{a}\cdot\mathbf{a} = a_1^2 + a_2^2 + a_3^2 = |\mathbf{a}|^2$$

$$|\mathbf{a}| = \sqrt{a_1^2 + a_2^2 + a_3^2}$$

2.7. Distance between two vectors

Let $A(a_1, a_2, a_3)$ and $B(b_1, b_2, b_3)$ be two points and O be the origin of reference. Then

$$\vec{OA} = a_1\mathbf{i} + a_2\mathbf{j} + a_3\mathbf{k} \text{ and } \vec{OB} = b_1\mathbf{i} + b_2\mathbf{j} + b_3\mathbf{k}$$

Now $\vec{AB} = \vec{OB} - \vec{OA}$

$$= (b_1\mathbf{i} + b_2\mathbf{j} + b_3\mathbf{k}) - (a_1\mathbf{i} + a_2\mathbf{j} + a_3\mathbf{k})$$

$$= (b_1 - a_1)\mathbf{i} + (b_2 - a_2)\mathbf{j} + (b_3 - a_3)\mathbf{k}$$

$$\overline{AB}^2 = [\vec{AB}]^2 = \vec{AB}\cdot\vec{AB}$$

$$= (b_1 - a_1)^2 + (b_2 - a_2)^2 + (b_3 - a_3)^2$$

or

$$\overline{AB} = \sqrt{(b_1 - a_1)^2 + (b_2 - a_2)^2 + (b_3 - a_3)^2}$$

2.8. Angle between the vectors in terms of the rectangular components of the given vectors

Let $\qquad \mathbf{a} = a_1\mathbf{i} + a_2\mathbf{j} + a_3\mathbf{k}$ and $\mathbf{b} = b_1\mathbf{i} + b_2\mathbf{j} + b_3\mathbf{k}$

$$\therefore \qquad \mathbf{a}\cdot\mathbf{b} = a_1b_1 + a_2b_2 + a_3b_3$$

$$|\mathbf{a}| = \sqrt{a_1^2 + a_2^2 + a_3^2} \quad \text{and } |\mathbf{b}| = \sqrt{b_1^2 + b_2^2 + b_3^2}$$

Let θ be the angle between the vectors **a** and **b**. Then

$$\cos\theta = \frac{\mathbf{a}\cdot\mathbf{b}}{|\mathbf{a}||\mathbf{b}|} = \frac{a_1b_1 + a_2b_2 + a_3b_3}{\sqrt{a_1^2 + a_2^2 + a_3^2}\,\sqrt{b_1^2 + b_2^2 + b_3^2}}$$

2.9. Angle between two vectors in terms of direction cosines

Let l_1, m_1, n_1 and l_2, m_2, n_2 be the direction cosines of **a** and **b**. Then the unit vectors

$$\hat{\mathbf{a}} = l_1\,\mathbf{i} + m_1\mathbf{j} + n_1\mathbf{k} \text{ and } \hat{\mathbf{b}} = l_2\mathbf{i} + m_2\mathbf{j} + n_2\mathbf{k}$$

But $\cos\theta = \hat{\mathbf{a}} \cdot \hat{\mathbf{b}} = l_1 l_2 + m_1 m_2 + n_1 n_2$

2.10. Components of a vector

To find the components of a vector **r** *which is resolved into two components*

(i) *one in the direction of* **a**

(ii) *other perpendicular to* **a** *in the plane of* **r** *and* **a**

Let **b** be a vector perpendicular to **a** in the plane of **r** and **a**. Now **r** can be expressed uniquely in terms of **a** and **b**.

Let $\mathbf{r} = \alpha\mathbf{a} + \beta\mathbf{b}$ ...(1)

Now $\mathbf{r} \cdot \mathbf{a} = \alpha(\mathbf{a} \cdot \mathbf{a}) + \beta(\mathbf{b} \cdot \mathbf{a})$

$= \alpha\mathbf{a} \cdot \mathbf{a}$ [$\because$ $\mathbf{b} \cdot \mathbf{a}$ = because $\mathbf{a} \perp \mathbf{b}$]

$= \alpha a^2$

$\therefore$ $\alpha = \dfrac{\mathbf{r} \cdot \mathbf{a}}{a^2}$

Substituting the value of α in (1) we get

$$\mathbf{r} = \frac{\mathbf{r} \cdot \mathbf{a}}{a^2}\mathbf{a} + y\mathbf{b}$$

$$y\mathbf{b} = \mathbf{r} - \frac{\mathbf{r} \cdot \mathbf{a}}{a^2}\mathbf{a}$$

Thus the resolved parts of **r** in the direction of **a** and perpendicular to **a** are $\dfrac{(\mathbf{r} \cdot \mathbf{a})\mathbf{a}}{a^2}$ and $\mathbf{r} - \dfrac{(\mathbf{r} \cdot \mathbf{a})\mathbf{a}}{a^2}$ respectively.

Note. *We can show that any vector* **r** *can be expressed as*

$$\mathbf{r} + (\mathbf{r} \cdot \mathbf{i})\mathbf{i} + (\mathbf{r} \cdot \mathbf{j})\mathbf{j} + (\mathbf{r} \cdot \mathbf{k})\mathbf{k}$$

Proof. Any vector **r** can be uniquely expressed as a linear combination of any three non-coplanar vectors.

$\therefore$ $\mathbf{r} = x\mathbf{i} + y\mathbf{j} + z\mathbf{k}$ ...(1)

Now $\mathbf{r} \cdot \mathbf{i} = x\mathbf{i} \cdot \mathbf{i} + y\mathbf{j} \cdot \mathbf{i} + z\mathbf{k} \cdot \mathbf{i} = x$

$\mathbf{r} \cdot \mathbf{j} = x\mathbf{i} \cdot \mathbf{j} + y\mathbf{j} \cdot \mathbf{j} + z\mathbf{k} \cdot \mathbf{j} = y$

and $\mathbf{r} \cdot \mathbf{k} = x\mathbf{i} \cdot \mathbf{k} + y\mathbf{j} \cdot \mathbf{k} + z\mathbf{k} \cdot \mathbf{k} = z$

Hence $(\mathbf{r} \cdot \mathbf{i})\mathbf{i} + (\mathbf{r} \cdot \mathbf{j})\mathbf{j} + (\mathbf{r} \cdot \mathbf{k})\,\mathbf{k} = x\mathbf{i} + y\mathbf{j} + z\mathbf{k} = \mathbf{r}$.

SOLVED PROBLEMS

Example 1. Show that the vectors **a** = 2**i** + 3**j** − **k** and **b** = 4**i** − 2**j** + 2**k** are prependicular to each other.

Solution: Given **a** = 2**i** + 3**j** − **k** and **b** = 4**j** − 2**j** + 2**k**

$$|\mathbf{a}| = |2\mathbf{i} + 3\mathbf{j} - \mathbf{k}| = \sqrt{4+9+1} = \sqrt{14}$$

$$|\mathbf{b}| = |4\mathbf{i} - 2\mathbf{j} + 2\mathbf{k}| = \sqrt{16+4+4} = \sqrt{24}$$

and
$$\mathbf{a}\,.\,\mathbf{b} = (2\mathbf{i} + 3\mathbf{j} - \mathbf{k})\,.\,(4\mathbf{i} - 2\mathbf{j} + 2\mathbf{k})$$
$$= (2)(4) + (3)(-2) + (-1)(2)$$
$$= 8 - 6 - 2 = 0$$

$\because$ $|\mathbf{a}| \neq |\mathbf{b}|$ and **a** . **b** = 0 the vectors **a** and **b** are prependicular to each other.

Example 2. Show that the points 2**j** − **j** + **k**, **i** − 3**j** − 5**k** and 3**i** − 4**j** − 4**k** are the vertices of a right angled triangle. Also find the angles between the sides.

Solution: Let $\overrightarrow{OA}$ = 2**i** − **j** + **k**, $\overrightarrow{OB}$ = **i** − 3**j** − 5**k** and $\overrightarrow{OC}$ = 3**i** − 4**j** − 4**k** where O is the point of reference.

Then
$$\overrightarrow{AB} = \overrightarrow{OB} - \overrightarrow{OA}$$
$$= (\mathbf{i} - 3\mathbf{j} - 5\mathbf{k}) - (2\mathbf{j} - \mathbf{j} + \mathbf{k}) = -\mathbf{i} - 2\mathbf{j} - 6\mathbf{k}$$
$$\overrightarrow{BC} = \overrightarrow{OC} - \overrightarrow{OB}$$
$$= (3\mathbf{i} - 4\mathbf{j} - 4\mathbf{k}) - (\mathbf{i} - 3\mathbf{j} - 5\mathbf{k}) = 2\mathbf{i} - \mathbf{j} + \mathbf{k}$$
$$\overrightarrow{CA} = \overrightarrow{OA} - \overrightarrow{OC}$$
$$= (2\mathbf{i} - \mathbf{j} + \mathbf{k}) - (3\mathbf{i} - 4\mathbf{j} - 4\mathbf{k}) = -\mathbf{i} + 3\mathbf{j} + 5\mathbf{k}$$

Now $\overrightarrow{BC} \perp \overrightarrow{CA} = (2\mathbf{i} - \mathbf{j} + \mathbf{k})\,.\,(-\mathbf{i} + 3\mathbf{j} + 5\mathbf{k})$
$$= 2 - 3 + 5 = 0$$

$\Rightarrow \overrightarrow{BC} \quad \overrightarrow{CA}$. Here $\angle C$ is a right angle

Now
$$|\overrightarrow{AB}| = \sqrt{1+4+36} = \sqrt{41}$$
$$|\overrightarrow{BC}| = \sqrt{4+1+1} = \sqrt{6}$$
$$|\overrightarrow{CA}| = \sqrt{1+9+25} = \sqrt{35}$$

Angle between $\overrightarrow{BC}$ and $\overrightarrow{BA}$

$$\overrightarrow{BA} = -\overrightarrow{AB} = \mathbf{i} + 2\mathbf{j} + 6\mathbf{k}$$

and
$$|\overrightarrow{BA}| = |\overrightarrow{AB}| = \sqrt{41}$$

Let θ, be the angle between $\overrightarrow{BC}$ and $\overrightarrow{BA}$

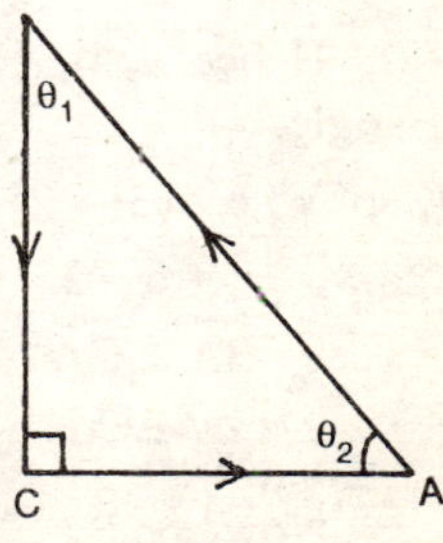

Fig. 2.5

$$\cos\theta_1 = \frac{\overrightarrow{BC}\cdot\overrightarrow{BA}}{|\overrightarrow{BC}||\overrightarrow{BA}|} = \frac{(2\mathbf{i}-\mathbf{j}+\mathbf{k})\cdot(\mathbf{i}+2\mathbf{j}+6\mathbf{k})}{\sqrt{6}\,.\sqrt{41}}$$

$$= \frac{2-2+6}{\sqrt{6}\,\sqrt{41}} = \frac{6}{\sqrt{6}\,\sqrt{41}} = \sqrt{\frac{6}{41}}$$

or $$\theta_1 = \cos^{-1}\sqrt{\frac{6}{41}}$$

Angle between $\overrightarrow{AC}$ *and* $\overrightarrow{AB}$

$\overrightarrow{AC} = -\overrightarrow{CA} = \mathbf{i}-3\mathbf{j}-5\mathbf{k}$ and $|\overrightarrow{AC}| = |\overrightarrow{CA}| = \sqrt{35}$ Let θ_2 be the angle between $\overrightarrow{AC}$ and $\overrightarrow{AB}$.

$$\therefore \quad \cos\theta_2 = \frac{\overrightarrow{AC}\cdot\overrightarrow{AB}}{|\overrightarrow{AC}|\,|\overrightarrow{AB}|} = \frac{(\mathbf{i}-3\mathbf{j}-5\mathbf{k})\cdot(-\mathbf{i}-2\mathbf{j}-6\mathbf{k})}{\sqrt{35}\,.\sqrt{41}}$$

$$= \frac{-1+6+30}{\sqrt{35}\,.\sqrt{41}} = \frac{35}{\sqrt{35}\,\sqrt{41}} = \sqrt{\frac{35}{41}}$$

$$\therefore \quad \theta_2 = \cos^{-1}\sqrt{\frac{35}{41}}$$

Example 3. Show that the vectors $\mathbf{a} = 3\mathbf{i}-2\mathbf{j}+\mathbf{k}$, $\mathbf{b} = \mathbf{i}-3\mathbf{j}+5\mathbf{k}$ and $\mathbf{c} = 2\mathbf{i}+\mathbf{j}+4\mathbf{k}$ form a right angled triangle. Also find the remaining angles of the triangle

Solution: We have

$\mathbf{b}+\mathbf{c} = (\mathbf{i}-3\mathbf{j}+5\mathbf{k}) + (2\mathbf{i}+\mathbf{j}-4\mathbf{k})$

$= 3\mathbf{i}-2\mathbf{j}+\mathbf{k} = \mathbf{a}$

$\Rightarrow \quad \mathbf{b}+\mathbf{c} = \mathbf{a}$

Thus if $\overrightarrow{BC} = \mathbf{b}$ and $\overrightarrow{CA} = \mathbf{c}$

Then $\overrightarrow{BA} = \mathbf{a}$

Hence **a**, **b**, **c** form the sides of a triangle.

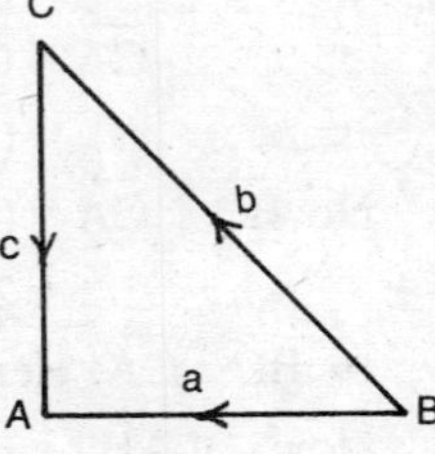

Fig. 2.6

Also $\mathbf{a}\cdot\mathbf{c} = (3\mathbf{i}-2\mathbf{j}+\mathbf{k}).(2\mathbf{i}-\mathbf{j}-4\mathbf{k})$

$= 6-2-4 = 0$

$\Rightarrow \overrightarrow{BA} \perp \overrightarrow{CA} \Rightarrow \angle A = 90°$

$\therefore$ ΔABC is a right angled triangle, the right angle at A

$$\text{Now } \cos B = \frac{\mathbf{a}\cdot\mathbf{b}}{|\mathbf{a}||\mathbf{b}|} = \frac{(3\mathbf{i}-2\mathbf{j}+\mathbf{k})\cdot(\mathbf{i}-3\mathbf{j}+4\mathbf{k})}{|3\mathbf{i}-2\mathbf{j}+\mathbf{k}||\mathbf{i}-3\mathbf{j}+5\mathbf{k}|}$$

$$= \frac{3.1 + (-2)(-3) + 1.5}{\sqrt{9+4+1}\ \sqrt{1+9+25}} = \frac{14}{\sqrt{14}\,.\sqrt{35}} = \sqrt{\frac{14}{35}}$$

$$\therefore\ B = \cos^{-1}\sqrt{\frac{14}{35}}$$

$$\text{Again } \cos C = \frac{-\mathbf{b}\,.\,\mathbf{c}}{|-\mathbf{b}|\,|\mathbf{c}|} = -\frac{\mathbf{b}\,.\,\mathbf{c}}{|\mathbf{b}|\,|\mathbf{c}|}$$

$$= \frac{-(\mathbf{i} - 3\mathbf{j} + 5\mathbf{k})\,.\,(2\mathbf{i} + \mathbf{j} - 4\mathbf{k})}{|\mathbf{i} - 3\mathbf{j} + 5\mathbf{k}|\ |2\mathbf{i} + \mathbf{j} - 4\mathbf{k}|} = \frac{-(-21)}{\sqrt{35}\,.\sqrt{21}}$$

$$C = \cos^{-1}\sqrt{\frac{21}{35}}$$

Example 4. Find the projection of the vector $\mathbf{a} = \mathbf{i} + 2\mathbf{j} + \mathbf{k}$ on the vector $\mathbf{b} = 2\mathbf{i} + \mathbf{j} - 2\mathbf{k}$

Solution. The projection of **a** on the vector **b** is

$$= \frac{\mathbf{a}\,.\,\mathbf{b}}{|\mathbf{b}|} = \frac{(\mathbf{i} + 2\mathbf{j} + \mathbf{k})\,.\,(2\mathbf{i} + \mathbf{j} - 2\mathbf{k})}{|2\mathbf{i} + \mathbf{j} - 2\mathbf{k}|} = \frac{2 + 2 - 2}{\sqrt{4+1+4}} = \frac{2}{3}$$

Example 5. A, B, C, D are the points $\mathbf{i} - \mathbf{k}$, $-\mathbf{i} + 2\mathbf{j}$, $2\mathbf{i} - 3\mathbf{k}$ and $3\mathbf{i} - 2\mathbf{j} - \mathbf{k}$ respectively. Show that the projection of $\overrightarrow{AB}$ on $\overrightarrow{CD}$ is equal to that of $\overrightarrow{CD}$ and $\overrightarrow{AB}$. Also find the cosine of their inclinations.

Solution: Given $\overrightarrow{OA} = \mathbf{i} - \mathbf{k}$, $\overrightarrow{OB} = -\mathbf{i} + 2\mathbf{j}$

$\overrightarrow{OC} = 2\mathbf{i} - 3\mathbf{k}$ and $\overrightarrow{OD} = 3\mathbf{i} - 2\mathbf{j} - \mathbf{k}$ where O is the point of reference.

$$\overrightarrow{AB} = \overrightarrow{AB} - \overrightarrow{OA} = (-\mathbf{i} + 2\mathbf{j}) - (\mathbf{i} - \mathbf{k})$$

$$= -2\mathbf{j} + 2\mathbf{j} + \mathbf{k}$$

$$|\overrightarrow{AB}| = \sqrt{4+4+1} = 3$$

$$\overrightarrow{CD} = \overrightarrow{OD} - \overrightarrow{OC} = (3\mathbf{i} - 2\mathbf{j} - \mathbf{k}) - (2\mathbf{i} - 3\mathbf{k})$$

$$= \mathbf{i} - 2\mathbf{j} + 2\mathbf{k}$$

$$|\overrightarrow{CD}| = \sqrt{1+4+4} = 3$$

Now projection of $\overrightarrow{AB}$ on $\overrightarrow{CD}$

$$= \frac{\overrightarrow{AB}\,.\,\overrightarrow{CD}}{|\overrightarrow{CD}|} = \frac{(-2\mathbf{i} + 2\mathbf{j} + \mathbf{k})\,.\,(\mathbf{i} - 2\mathbf{j} + 2\mathbf{k})}{3} = \frac{-4}{3}$$

Similarly projection of $\overrightarrow{CD}$ on $\overrightarrow{AB}$

$$= \frac{\overrightarrow{CD}\,.\,\overrightarrow{AB}}{|\overrightarrow{AB}|} = \frac{-4}{3}$$

Hence projection of $\vec{AB}$ on $\vec{CD}$ = projection of $\vec{CD}$ on $\vec{AB}$. Now angle between $\vec{AB}$ and $\vec{CD}$

$$= \cos^{-1}\left\{\frac{\vec{AB}\,.\,\vec{CD}}{|\vec{AB}|\,|\vec{CD}|}\right\} = \cos^{-1}\left\{\frac{-4}{(3)(3)}\right\} = \cos^{-1}\left(-\frac{4}{9}\right)$$

Example 6. Let P, Q, R, S be the points of (1, 1, 0), (−1, 1, 0), (1, −1, 0) and (0, −1, 1). Find the direction cosines of $\vec{PR}$ and $\vec{QS}$ and find the angle between their directions.

Solution: Let O be the point of reference.

Then $\quad \vec{OP} = \mathbf{i} + \mathbf{j}, \quad OQ = -\mathbf{i} + \mathbf{j}, \quad \vec{OR} = \mathbf{i} - \mathbf{j}$

and $\quad \vec{OS} = -\mathbf{j} + \mathbf{k}$

Now $\quad \vec{PR} = \vec{OR} - \vec{OP} = (\mathbf{i} - \mathbf{j}) - (\mathbf{i} + \mathbf{j}) = -2\mathbf{j}$

$$\vec{QS} = \vec{OS} - \vec{OQ} = (-\mathbf{j} + \mathbf{k}) - (-\mathbf{i} + \mathbf{j}) = \mathbf{i} - 2\mathbf{j} + \mathbf{k}$$

∴ Unit vector along $\vec{PR}$

$$= \frac{\vec{PR}}{|\vec{PQ}|} = \frac{-2\mathbf{j}}{|-2\mathbf{j}|} = \frac{-2\mathbf{j}}{2} = -\mathbf{j}$$

$$= (0)\mathbf{i} + (-1)\mathbf{j} + (0)\mathbf{k}$$

∴ The direction cosines of $\vec{PR}$ are (0, −1, 0)

$$\text{Unit vector along } \vec{QS} = \frac{\mathbf{i} - 2\mathbf{j} + \mathbf{k}}{|\mathbf{i} - 2\mathbf{j} + \mathbf{k}|}$$

$$= \frac{\mathbf{i} - 2\mathbf{j} + \mathbf{k}}{\sqrt{1 + 4 + 1}} = \left(\frac{1}{\sqrt{6}}\right)\mathbf{i} + \left(\frac{-2}{\sqrt{6}}\right)\mathbf{j} + \left(\frac{1}{\sqrt{6}}\right)\mathbf{k}$$

∴ The direction cosines of $\vec{QS}$ are $\left(\frac{1}{\sqrt{6}}, \frac{-2}{\sqrt{6}}, \frac{1}{\sqrt{6}}\right)$. Let θ be the angle between $\vec{PR}$ and $\vec{QS}$

$$\therefore \quad \cos\theta = l_1l_2 + m_1m_2 + n_1n_2$$

$$= (0)\left(\frac{1}{\sqrt{6}}\right) + (-1)\left(\frac{-2}{\sqrt{6}}\right) + (0)\left(\frac{1}{\sqrt{6}}\right) = \sqrt{\frac{2}{3}}$$

Example 7. If $\hat{\mathbf{a}}$ and $\hat{\mathbf{b}}$ are unit vectors and θ is the angle between them show that $\sin\frac{\theta}{2} = \frac{1}{2}(\hat{\mathbf{a}} - \hat{\mathbf{b}})$.

Solution. We know that

$$(\hat{a} - \hat{b})^2 = (\hat{a})^2 + (\hat{b})^2 - 2(\hat{a} . \hat{b})$$

or $$|\hat{a} - \hat{b}|^2 = |\hat{a}^2| + |\hat{b}|^2 - 2(\hat{a} . \hat{b})$$

$$= |\hat{a}|^2 + |\hat{b}|^2 - 2|\hat{a}||\hat{b}| \cos\theta$$

$$= 1 + 1 - 2.1.1 \cos\theta \qquad [\because |\hat{a}| = |\hat{b}| = 1]$$

$$= 2(1 - \cos\theta) = 4 \sin^2 \frac{\theta}{2}$$

or $$\sin\frac{\theta}{2} = \frac{1}{1} |\hat{a} - \hat{b}|.$$

Hence the result.

Example 8. Find the angle between the direction of the vectors **a** = **i** + **j** + **k** and **b** = 2**i** + 3**j** = **k**. Also find a unit vector perpendicular to both **a** and **b**.

Solution. Given **a** = **i** + **j** + **k** and **b** = 2**i** + 3**j** = **k**

$$\mathbf{a} . \mathbf{b} = (\mathbf{i} + \mathbf{j} + \mathbf{k}) . (2\mathbf{i} + 3\mathbf{j} - \mathbf{k}) = 2 + 3 - 1 = 4$$

$$|\mathbf{a}| = \sqrt{1 + 1 + 1} = \sqrt{3}; \ |\mathbf{b}| = \sqrt{4 + 9 + 1} = \sqrt{4}$$

Using **a** . **b** = |**a**||**b**| cos θ, we get

$$\cos\theta = \frac{\mathbf{a} . \mathbf{b}}{|\mathbf{a}||\mathbf{b}|} = \frac{4}{\sqrt{3} . \sqrt{14}} = \frac{4}{\sqrt{42}}$$

Again let tthe unit vector perpendicular to both **a** and **b** be **r** = l**i** + m**j** + n**k** where $l^2 + m^2 + n^2 = 1$ as **r** is a unit vector.

∴ **r** is perpendicular to both **a** and **b**, we have

$$\mathbf{r} . \mathbf{a} = l + m + n = 0$$

and $$\mathbf{r} . \mathbf{b} = 2l + 3m - n = 0$$

Solving by cross multiplication we get

$$\frac{l}{-4} = \frac{m}{3} = \frac{n}{1} = \frac{\sqrt{l^2 + m^2 + n^2}}{\sqrt{(-4)^2 + (3)^2 + (1)^2}}$$

$$= \frac{1}{\sqrt{26}}$$

∴ $$l = \frac{-4}{\sqrt{26}}, m = \frac{3}{\sqrt{26}}, n = \frac{1}{\sqrt{26}}$$

∴ Required unit vector is

$$\mathbf{r} = \frac{-4}{\sqrt{26}}\mathbf{i} + \frac{3}{\sqrt{26}}\mathbf{j} + \frac{1}{\sqrt{26}}\mathbf{k}$$

$$= \frac{1}{\sqrt{26}}(-4\mathbf{i} + 3\mathbf{j} + \mathbf{k})$$

Example 9. If **a** and **b** are vectors a, b are their lengths show that $\left(\frac{\mathbf{a}}{a^2} - \frac{\mathbf{b}}{b^2}\right)^2 = \left(\frac{\mathbf{a} - \mathbf{b}}{ab}\right)^2$.

Solution: Given

$$\text{LHS} = \left(\frac{\mathbf{a}}{a^2} - \frac{\mathbf{b}}{b^2}\right)^2 = \left(\frac{\mathbf{a}}{a^2} - \frac{\mathbf{b}}{b^2}\right).\left(\frac{\mathbf{a}}{a^2} - \frac{\mathbf{b}}{b^2}\right)$$

$$= \frac{\mathbf{a}.\mathbf{a}}{a^4} - \frac{\mathbf{a}.\mathbf{b}}{a^2b^2} - \frac{\mathbf{b}.\mathbf{a}}{b^2a^2} + \frac{\mathbf{b}.\mathbf{b}}{b^4}$$

$$= \frac{a^2}{a^4} - \frac{2\mathbf{a}.\mathbf{b}}{a^2b^2} + \frac{b^2}{b^4}$$

$$= \frac{1}{a^2} - \frac{2\mathbf{a}.\mathbf{b}}{a^2b^2} + \frac{1}{b^2} = \frac{b^2 - 2\mathbf{a}.\mathbf{b} + a^2}{a^2b^2}$$

$$= \frac{(\mathbf{a} - \mathbf{b}).(\mathbf{a} - \mathbf{b})}{a^2b^2} = \frac{(\mathbf{a} - \mathbf{b})^2}{a^2b^2} = \text{R.H.S.}$$

Example 10. If **a** + **b** + **c** = 0 and | **a** | = 3, | **b** | = 5, | **c** | = 7 find the angle between **a** and **b**.

Solution. Given $\mathbf{a} + \mathbf{b} + \mathbf{c} = 0 \Rightarrow \mathbf{a} + \mathbf{b} = -\mathbf{c}$

$\therefore \quad (\mathbf{a} + \mathbf{b}).(\mathbf{a} + \mathbf{b}) = (-\mathbf{c}).(-\mathbf{c})$

$\Rightarrow \quad \mathbf{a}.\mathbf{a} + \mathbf{a}.\mathbf{b} + \mathbf{b}.\mathbf{a} + \mathbf{b}.\mathbf{b} = \mathbf{c}.\mathbf{c}$

or $\quad |\mathbf{a}|^2 + 2\mathbf{a}.\mathbf{b} + |\mathbf{b}|^2 = |\mathbf{c}|^2$

or $\quad \mathbf{a}.\mathbf{b} = \frac{1}{2}\{|\mathbf{c}^2 - |\mathbf{a}|^2 - |\mathbf{b}|^2\}$

$$= \frac{1}{2}(49 - 9 - 25) = \frac{15}{2}$$

$\therefore \quad \mathbf{a}.\mathbf{b} = \frac{15}{2}$ ...(1)

Let θ be the angle between **a** and **b**.

Then $\mathbf{a} \cdot \mathbf{b} = |\mathbf{a}||\mathbf{b}| \cos\theta$

$$\frac{15}{2} + (3)(5)\cos\theta \Rightarrow \cos\theta = \frac{1}{2} \text{ or } \theta = 60°$$

Example 11. If **a, b, c** are mutually perpendicular vectors of equal magnitude show that **a + b + c** is equally inclined to **a, b, c** .

Solution. Since **a, b, c** are mutually perpendicular vectors of equal magnitude, we have

$$\mathbf{a} \cdot \mathbf{b} = \mathbf{b} \cdot \mathbf{a} = \mathbf{b} \cdot \mathbf{c} = \mathbf{c} \cdot \mathbf{b} = \mathbf{a} \cdot \mathbf{c} = \mathbf{c} \cdot \mathbf{a} = 0 \quad ...(1)$$

and $$|\mathbf{a}| = |\mathbf{b}| = |\mathbf{c}| \quad ...(2)$$

Now $|\mathbf{a} + \mathbf{b} + \mathbf{c}|^2 = (\mathbf{a} + \mathbf{b} + \mathbf{c}) \cdot (\mathbf{a} + \mathbf{b} + \mathbf{c})$

$= \mathbf{a} \cdot \mathbf{a} + \mathbf{a} \cdot \mathbf{b} + \mathbf{a} \cdot \mathbf{c} + \mathbf{b} \cdot \mathbf{a} + \mathbf{b} \cdot \mathbf{b} + \mathbf{b} \cdot \mathbf{c} + \mathbf{c} \cdot \mathbf{a} + \mathbf{c} \cdot \mathbf{b} + \mathbf{c} \cdot \mathbf{c}$

$= 3|\mathbf{a}|^2$ using (1) and (2)

$$\therefore \quad |\mathbf{a} + \mathbf{b} + \mathbf{c}| = \sqrt{3}\,|\mathbf{a}|$$

Let α, β, γ be angles at which $a + b + c$ is inclined to **a, b, c** respectively. Then we have

$$\cos\alpha = \frac{(\mathbf{a} + \mathbf{b} + \mathbf{c}) \cdot \mathbf{a}}{|\mathbf{a} + \mathbf{b} + \mathbf{c}||\mathbf{a}|} = \frac{\mathbf{a} \cdot \mathbf{a} + \mathbf{b} \cdot \mathbf{a} + \mathbf{c} \cdot \mathbf{a}}{\sqrt{3}\,|\mathbf{a}| \cdot |\mathbf{a}|}$$

$$= \frac{|\mathbf{a}|^2}{\sqrt{3}\,|\mathbf{a}|} = \frac{1}{\sqrt{3}} \text{ [using (1)]}$$

$$\therefore \quad \cos\alpha = \frac{1}{\sqrt{3}}$$

Similarly $\cos\beta = \cos\gamma = \dfrac{1}{\sqrt{3}}$

$\Rightarrow \alpha = \beta = \gamma = \dfrac{1}{\sqrt{3}}$ i.e., **a + b + c** is equally inclined to **a, b** and **c.**

Example 12. Prove that in any triangle ABC

(i) a = b cos C + C cos B

(ii) $a^2 = b^2 + c^2 - 2bc \cos A$.

Solution: Let $\overrightarrow{BC}, \overrightarrow{CA}, \overrightarrow{AB}$ represent the vectors **a, b, c** respectively. Then we have

$$\overrightarrow{BC} + \overrightarrow{CA} + \overrightarrow{AB} = 0$$

i.e., $$\mathbf{a} + \mathbf{b} + \mathbf{c} = 0$$

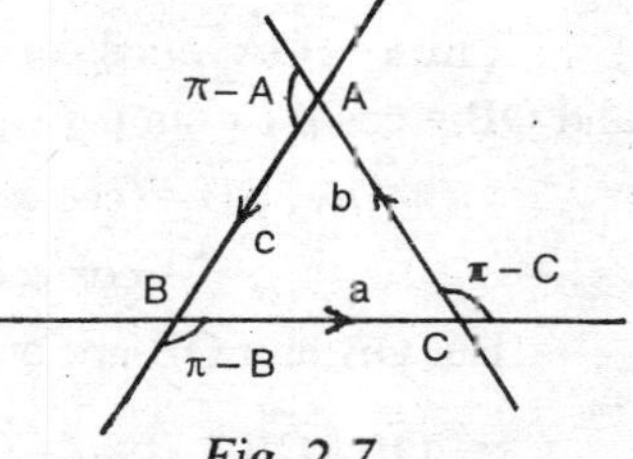

Fig. 2.7

Let $|\mathbf{a}| = a,\ |\mathbf{b}| = b$

and $|\mathbf{c}| = c$

Now we have $\mathbf{b} + \mathbf{c} - - \mathbf{a}$

or $(\mathbf{b} + \mathbf{c}) \cdot \mathbf{a} = (-\mathbf{a}) \cdot \mathbf{a}$

or $\mathbf{b} \cdot \mathbf{a} + \mathbf{c} \cdot \mathbf{a} = -\mathbf{a}^2$

or $ba \cos(\pi - c) + ca \cos(pi - B) = -a^2$ [Refer figure]

or $-ba \cos c - ca \cos B = -a_2$

or $b \cos C + c \cos B = a$

which prove the first result.

Again consider $-\mathbf{a} = \mathbf{b} + \mathbf{c}$

Now $(-\mathbf{a}) \cdot (-\mathbf{a}) = (\mathbf{b} + \mathbf{c}) \cdot (\mathbf{b} + \mathbf{c})$

$$\mathbf{a}^2 = \mathbf{b}^2 + 2\mathbf{b} \cdot \mathbf{c} + \mathbf{c}^2$$

$$a^2 = b^2 + 2\,bc \cos(\pi - A) + c^2$$

or $$a^2 = b^2 - 2\,bc \cos A + c^2$$

which proves the second result.

Example 13. Prove by vector method that

$$\cos(\alpha - \beta) = \cos\alpha \cos\beta + \sin\alpha \sin\beta$$

Solution: Let OX and OY be the two perpendicular axes, forming OXY plane O being the origin and let **i, j** be the unit vectors along OX and OY. Let OC and OD make angles α and β with OX in the same plane as shown in the figure. Then $\angle COD = \alpha - \beta$.

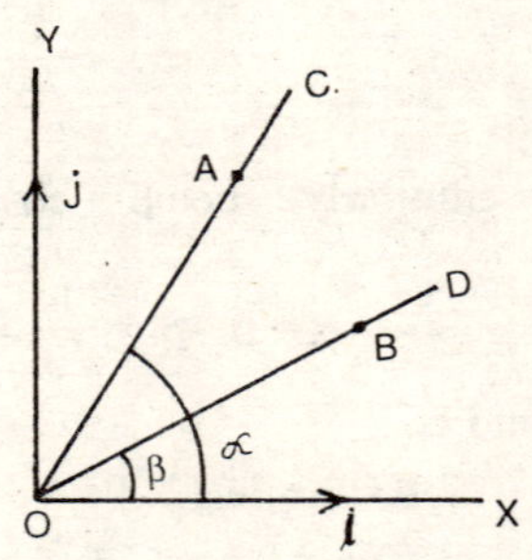

Fig. 2.8

Let $\overrightarrow{OA}$ and $\overrightarrow{OB}$ be unit vectors along OC and OD respectively.

Then $\overrightarrow{OA} = \cos\alpha\, \mathbf{i} + \sin\alpha\, \mathbf{j}$

and $\overrightarrow{OB} = \cos\beta\, \mathbf{i} + \sin\beta\, \mathbf{j}$

$$\overrightarrow{OA} \cdot \overrightarrow{OB} = (\cos\alpha\, \mathbf{i} + \sin\alpha\, \mathbf{j}) \cdot (\cos\beta\, \mathbf{i} + \sin\beta\, \mathbf{j})$$

$$= \cos\alpha \cos\beta + \sin\alpha \sin\beta \qquad ...(1)$$

But $\overrightarrow{OA}$ and $\overrightarrow{OB}$ are unit vectors.

$$\therefore\ \overrightarrow{OA} \cdot \overrightarrow{OB} = \cos\angle COD = \cos(\alpha - \beta) \qquad ...(2)$$

Hence from (1) and (2), we get

$$\cos(\alpha - \beta) = \cos\alpha\cos\beta + \sin\alpha\sin\beta$$

Example 14. Prove that the median to the base of an isosceles triangle is perpendicular to the base

Solution: Let ABC be an isosceles triangle.

Let AC = BC and D be the mid point of AB. With reference to D as origin. Let the position vectors of A, B and C be **a**, - **a** and **c**.

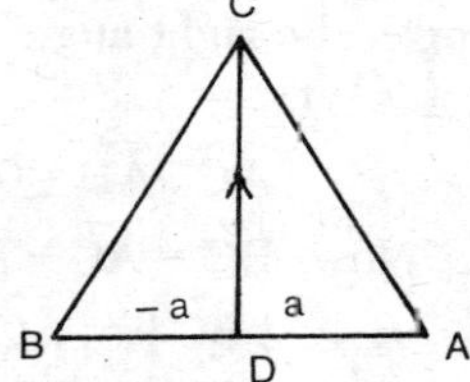

Fig. 2.9

$\therefore \quad \vec{AC} = \mathbf{c} - \mathbf{a}$ and $\vec{BC} = \mathbf{c} + \mathbf{a}$

Now $\vec{AC} = \vec{BC} \Rightarrow AC^2 = BC^2 \Leftrightarrow (\vec{AC})^2 = (\vec{BC})^2$

i.e., $\quad \vec{AC} \cdot \vec{AC} = \vec{BC} \cdot \vec{BC}$

or $\quad (\mathbf{c} - \mathbf{a}) \cdot (\mathbf{c} - \mathbf{a}) = (\mathbf{c} + \mathbf{a}) \cdot (\mathbf{c} + \mathbf{a})$

or $\quad 4\mathbf{c} \cdot \mathbf{a} = 0$ or $\mathbf{c} \cdot \mathbf{a} = 0 \Rightarrow \vec{DC} \cdot \vec{DA} = 0$

Hence $\vec{DC}$ is perpendicular to base $\vec{DA}$.

Example 15. Prove that the mid point of the hypotenuse of a right angled triangle is equidistant from its vertices.

Solution. Let ABC be a right angled triangle, the right angle being at A and D be the mid point of the hypotenuse BC. Then we have $\vec{BD} = \vec{DC}$

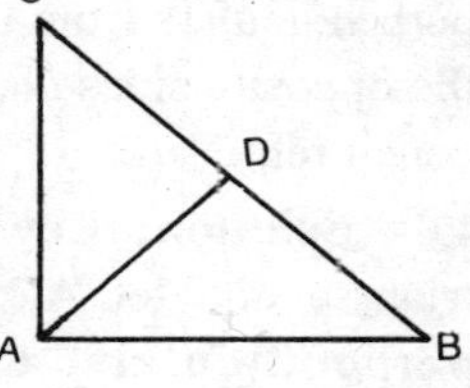

Fig. 2.10

Now $\vec{AB} = \vec{AD} + \vec{DB}$

and $\quad \vec{AC} = \vec{AD} + \vec{DC} = \vec{AD} + \vec{BD} = \vec{AD} - \vec{DB}$

But $AB \perp AC \quad \therefore \vec{AB} \cdot \vec{AC} = 0$

$\Rightarrow \quad (\vec{AD} + \vec{DB}) \cdot (\vec{AD} - \vec{DB}) = 0$

$\therefore \quad (\vec{AD})^2 - (\vec{DB})^2 = 0$, i.e., $\overline{AD}^2 - \overline{DB}^2 = 0$

$\Rightarrow \quad \overline{AD} = \overline{DB}$ ...(1)

But $\overline{DB} = \overline{DC}$. Hence from (1)

$$\overline{AD} = \overline{DB} = \overline{DC}$$

Hence mid point of the hypotenuse of a right angled triangle is equidistant from the vertices.

Example 16. Prove that in a right angled triangle the square of the hypotenuse is equal to the sum of the squares of the other two sides.

Solution. Let ABC be a right angled triangle, the right angle being at A. Now $AC \perp AB$

$\therefore \quad \overrightarrow{AC} \cdot \overrightarrow{AB} = 0$

Now $\overrightarrow{BC} = \overrightarrow{AC} - \overrightarrow{AB}$

$\therefore \quad |\overrightarrow{BC}| = |\overrightarrow{AC} - \overrightarrow{AB}|$

$\Rightarrow \quad (\overrightarrow{BC})^2 = (\overrightarrow{AC} - \overrightarrow{AB})^2$

$\Rightarrow \quad \overrightarrow{BC} \cdot \overrightarrow{BC} = (\overrightarrow{AC} - \overrightarrow{AB}) \cdot (\overrightarrow{AC} - \overrightarrow{AB})$

$\Rightarrow \quad \overrightarrow{AC} \cdot \overrightarrow{AC} - 2\overrightarrow{AC} \cdot \overrightarrow{AB} + \overrightarrow{AB} \cdot \overrightarrow{AB}$

Fig. 2.11

or $\quad (\overrightarrow{BC})^2 = (\overrightarrow{AC})^2 - 2\overrightarrow{AC} \cdot \overrightarrow{AB} + (\overline{AB})^2$

or $\quad (\overline{BC})^2 = (\overrightarrow{AC})^2 + (\overline{AB})^2 \qquad (\because \overrightarrow{AC} \cdot \overrightarrow{AB} = 0)$

Hence the result

Example 17. Prove that the perpendiculars from the vertices to the opposite sides of a triangle are concurrent.

Solution: Let ABC be a triangle and let AD, BE, be the perpendiculars to BC, AC intersecting at O. Join CO and OF. Now we have to prove that CF is perpendicular to AB .

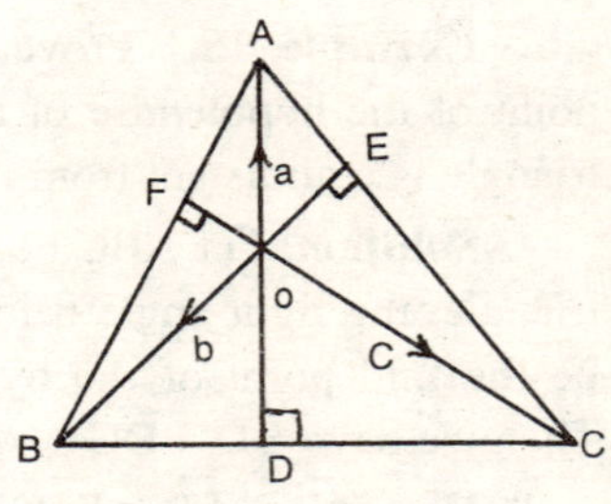

Fig. 2.12

Let **a, b, c** be the position vectors of A, B, C with reference to O as origin.

$\therefore \quad \overrightarrow{OA} = \mathbf{a}, \overrightarrow{OB} = \mathbf{b}, \overrightarrow{OC} = \mathbf{c}$

$\Rightarrow \quad \overrightarrow{AO} = -\mathbf{a}$ and $\overrightarrow{BC} = \overrightarrow{OC} - \overrightarrow{OB} = \mathbf{c} - \mathbf{b}$

$\because \quad \overrightarrow{AO} \perp \overrightarrow{BC} \Rightarrow \overrightarrow{AO} \cdot \overrightarrow{BC} = 0$

i.e., $\quad (-\mathbf{a}) \cdot (\mathbf{c} - \mathbf{b}) = 0 \Rightarrow \mathbf{a} \cdot \mathbf{c} = \mathbf{a} \cdot \mathbf{b} \qquad \ldots(1)$

Again $\quad \overrightarrow{BO} = -\mathbf{b}$ and $\overrightarrow{CA} = \overrightarrow{OA} - \overrightarrow{OC} = \mathbf{a} - \mathbf{c}$

$\overrightarrow{BO} \perp \overrightarrow{CA} \Rightarrow \overrightarrow{BO} \cdot \overrightarrow{CA} = 0$

$$(-\mathbf{b})\,.\,(\mathbf{a}-\mathbf{c}) = 0 \Rightarrow \mathbf{b}\,.\,\mathbf{a} = \mathbf{b}\,.\,\mathbf{c} = \mathbf{a}\,.\,\mathbf{b} \qquad ...(2)$$

(1) and (2) $\Rightarrow$ $\mathbf{b}\,.\,\mathbf{c} = \mathbf{a}\,.\,\mathbf{c}$

or $\qquad \mathbf{c}\,.\,\mathbf{b} = \mathbf{c}\,.\,\mathbf{a}$ $\qquad$ [using cumulative law]

or $\qquad \mathbf{c}\,.\,(b-a) = 0 \Rightarrow \overrightarrow{OC} \perp \overrightarrow{AB} \Leftrightarrow \overrightarrow{CF} \perp AB$

Hence the three perpendicular are concurrent.

Example 18. Prove that the perpendicular bisectors of the sides of a triangle are concurrent.

Solution. Let ABC be the triangle and D, E, F be the mid points of the sides BC, CA, AB and O is the point of intersection of the perpendicular bisectors of BC and CA. Join OF. Let O be the point of reference and $\overrightarrow{OA} = \mathbf{a}$, $\overrightarrow{OB} = \mathbf{b}$, $\overrightarrow{OC} = \mathbf{c}$ be the position vectors of A, B, C respectively .

$\because \qquad \overrightarrow{OD} \perp \overrightarrow{BC}$, we have

$$\frac{1}{2}(\mathbf{c}+\mathbf{b})\,.\,(\mathbf{c}-\mathbf{b}) = 0 \text{ i.e., } \mathbf{c}^2 = \mathbf{b}^2 = 0 \qquad ...(1)$$

Again since $\overrightarrow{OE} \perp \overrightarrow{CA}$, we have

$$\frac{1}{2}(\mathbf{a}+\mathbf{c})\,.\,(\mathbf{a}-\mathbf{c}) = 0 \text{ i.e., } \mathbf{a}^2 - \mathbf{b}^2 = 0 \qquad ...(2)$$

Now (1) + (2) given $\mathbf{a}^2 - \mathbf{b}^2 = 0$

or $$\frac{1}{2}(\mathbf{a}+\mathbf{b})\,.\,(\mathbf{a}-\mathbf{c}) = 0$$

$\Rightarrow \overrightarrow{OF} \perp \overrightarrow{AB}$ i.e., $\overrightarrow{OF}$ is perpendicular bisector of AB. Hence the perpendicular bisectors in a triangle are concurrent.

Example 19. If two medians of a triangle are equal the triangle is isosceles.

Solution. Let ABC be the triangle and D, E be the mid points of the sides AC and AB respectively. Choose A as point of reference.

Let $\overrightarrow{AB} = \mathbf{b}$ and $\overrightarrow{AC} = \mathbf{c}$ so that

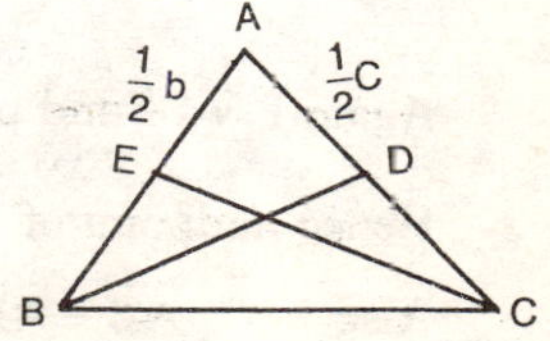

Fig. 2.13

$$\vec{AE} = \frac{1}{2}\mathbf{b} \text{ and } \vec{AD} = \frac{1}{2}\mathbf{c}$$

Now $$\vec{BD} = \vec{AD} - \vec{AB} = \frac{1}{2}\mathbf{c} - \mathbf{b}$$

and $$\vec{CE} = \vec{AE} - \vec{AC} = \frac{1}{2}\mathbf{b} - \mathbf{c}$$

Given $$\vec{BD} = \vec{CE} \Rightarrow \vec{BD} \cdot \vec{BD} = \vec{CE} \cdot \vec{CE}$$

$$\Leftrightarrow \left(\frac{1}{2}\mathbf{c} - \mathbf{b}\right) \cdot \left(\frac{1}{2}\mathbf{c} - \mathbf{b}\right) = \left(\frac{1}{2}\mathbf{b} - \mathbf{c}\right) \cdot \left(\frac{1}{2}\mathbf{b} - \mathbf{c}\right)$$

or $$\frac{1}{4}\mathbf{c}^2 + \mathbf{b}^2 - \mathbf{c} \cdot \mathbf{b} = \frac{1}{4}\mathbf{b}^2 + \mathbf{c}^2 - \mathbf{b} \cdot \mathbf{c}$$

or $$\mathbf{b}^2 = \mathbf{c}^2 \Rightarrow (\overline{AB})^2 = (\overline{AC})^2 \Leftrightarrow \overline{AB} = \overline{AC}$$

Hence ΔABC is isosceles.

Example 20. Show that the diagonals in a rhombus bisect each other.

Solution. Let ABCB be the rhombus so that the diagonals $\overline{AC}$ and $\overline{BC}$ intersect each other at O. Let A be the point of reference such that $\vec{AB} = \mathbf{b}$ and $\vec{AD} = \mathbf{d}$

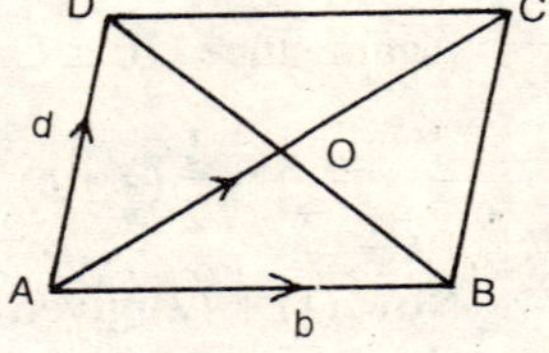

Fig. 2.14

Now AD and BC are parallel and equal.

$$\therefore \quad \vec{AD} = \vec{BC} = \mathbf{d}$$

$$\vec{AC} = \vec{AB} + \vec{BC} = \mathbf{b} + \mathbf{d};$$

$$\vec{DB} = \vec{DA} + \vec{AB} = -\mathbf{d} + \mathbf{b} = \mathbf{b} - \mathbf{d}$$

Now P. V of mid point of $\vec{AC} = \dfrac{\mathbf{0} + (\mathbf{b} + \mathbf{d})}{2} = \dfrac{\mathbf{b} + \mathbf{d}}{2}$

Again P.V of mid point of $\vec{BD} = \dfrac{\mathbf{b} + \mathbf{d}}{2}$

Hence midpoint of AC coincides with the mid point of BD

Also $\overline{AB} = \overline{AD}$ $\qquad$ ($\because$ ABCD is a rhombas)

$$\therefore \quad (\overline{AB})^2 = (\overline{AD})^2 \Rightarrow (\vec{AB})^2 = (\vec{AD})^2$$

$\therefore \quad b^2 = d^2$ or $\mathbf{b}^2 - \mathbf{d}^2 = 0$ or $(\mathbf{b} + \mathbf{d}) . (\mathbf{b} - \mathbf{d}) = 0$

or $\quad \overrightarrow{AC} . \overrightarrow{DB} = 0 \Rightarrow \overrightarrow{AC} \perp \overrightarrow{DB}$

Hence diagonals of a rhombus bisect each other at right angles.

Example 21. Prove that the angle in a semi-circle is a right angle.

Solution. Let O be the centre of the semi-circle and let AB be the diameter. Let P be any point on the circumference of the semi-circle.

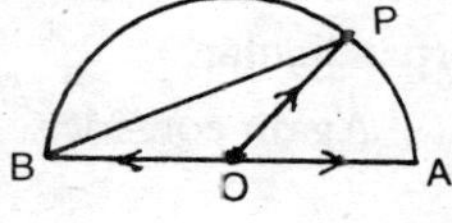

Fig. 2.15

Let $\overrightarrow{OA} = \mathbf{a}$ so that $\overrightarrow{OB} = -\mathbf{a}$ and let $\overrightarrow{OP} = \mathbf{r}$

Now $\overline{OP} = \overline{OA} \Rightarrow (\overrightarrow{OP})^2 = (\overrightarrow{OA})^2$

i.e., $\quad \mathbf{r}^2 = \mathbf{a}^2$

or $\quad \mathbf{r}^2 - \mathbf{a}^2 = 0 \quad$ or $\quad (\mathbf{r} - \mathbf{a}) . (\mathbf{r} + \mathbf{a}) = 0 \quad$ ...(1)

Now $\overrightarrow{AP} = \overrightarrow{OP} - \overrightarrow{OA} = \mathbf{r} - \mathbf{a}$

and $\quad \overrightarrow{BP} = \overrightarrow{OP} - OB = \mathbf{r} - (-\mathbf{a}) = \mathbf{r} + \mathbf{a}$

Hence from (1) we get

$\overrightarrow{AP} . \overrightarrow{BP} = 0 \Rightarrow AP \perp BP \Leftrightarrow \angle APB = 90°$

i.e., the angle in a semicircle in a right angle

Example 22. In a tetrahedron, if two pairs of opposite edges are perpendicular to each other and then prove that the third pair is also perpendicular to each other and the sum of the squares of two opposite edges in the same for each pair.

Solution. Let ABCD be the tetrahedron. Choosing D as point of reference

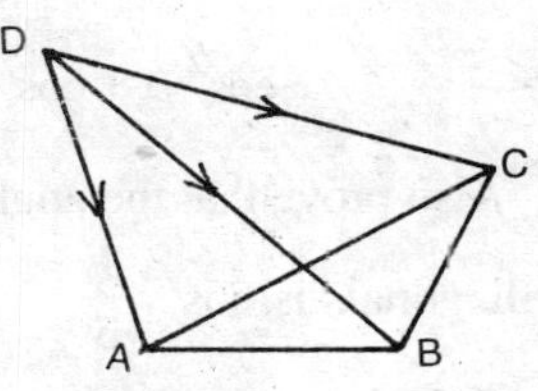

Fig. 2.16

Let $\quad \overrightarrow{DA} = \mathbf{a}, \overrightarrow{DB} = \mathbf{b}, \overrightarrow{DC} = \mathbf{c}$

$\overrightarrow{AB} = \overrightarrow{DB} - \overrightarrow{DA} = \mathbf{b} - \mathbf{a}$

$\overrightarrow{BC} = \overrightarrow{DC} - \overrightarrow{DB} = \mathbf{c} - \mathbf{b}$

$\overrightarrow{AC} = \overrightarrow{DC} - \overrightarrow{DA} = \mathbf{c} - \mathbf{a}$

Now $\vec{AB} \perp \vec{DC} \Rightarrow \vec{AB} \perp \vec{DC} = 0$

$\therefore \quad (\mathbf{b} - \mathbf{a}) \,.\, \mathbf{c} =$ or $\mathbf{b} \,.\, \mathbf{c} = \mathbf{a} \,.\, \mathbf{c}$ ...(1)

Again $\vec{AC} \perp \vec{DB} \Rightarrow \vec{AC} \,.\, \vec{DB} = 0$

$\therefore \quad (\mathbf{c} - \mathbf{a})\, \mathbf{b} = 0$ or $\mathbf{c} \,.\, \mathbf{b} = \mathbf{a} \,.\, \mathbf{b}$ ...(2)

From (1) and (2), we get $\mathbf{a} \,.\mathbf{c} = \mathbf{a} \,.\, \mathbf{b}$

or $\quad \mathbf{a} \,.\, \mathbf{c} - \mathbf{a} \,.\, \mathbf{b} = 0$ or $\mathbf{a} \,.\, (\mathbf{c} - \mathbf{b}) = 0 \Rightarrow \vec{DA} \,.\, \vec{BC} = 0$

i.e., $\vec{DA} \perp \vec{BC}$ which mean the third pair of opposite edges is also perpendicular.

Again consider

$$\mathbf{a} \,.\, \mathbf{b} = \mathbf{b} \,.\, \mathbf{c} = \mathbf{c} \,.\, \mathbf{a} \qquad ...(3)$$

Now $(\overline{AB})^2 + (\overline{DC})^2 = (\vec{AB})^2 + (\vec{DC})^2 = (\mathbf{b} - \mathbf{a})^2 + \mathbf{c}^2$

$$= a^2 + b^2 + c^2 - 2\, \mathbf{b} \,.\, \mathbf{a} \qquad ...(4)$$

$$(\overline{BC})^2 + (\overline{DA})^2 = (\vec{BC})^2 + (\vec{DA})^2 = (\mathbf{c} - \mathbf{b})^2 + \mathbf{a}^2$$

$$= a^2 + b^2 + c^2 - 2\, \mathbf{c} \,.\, \mathbf{b} \qquad ...(5)$$

and $\quad (\overline{AC})^2 + (\overline{DB})^2 = (\vec{AC})^2 + (\vec{DB})^2 = (\mathbf{c} - \mathbf{a})^2 + \mathbf{b}^2$

$$= a^2 + b^2 + c^2 - 2\, \mathbf{c} \,.\, \mathbf{a} \qquad ...(6)$$

Using (3) in (4), (5) and (6), we get

$$(\overline{AB})^2 + (\overline{DC})^2 = (\overline{BC})^2 + ((\overline{DA}))^2 = (\overline{AC})^2 + (\overline{DB})^2.$$

Example 23. A line makes angle $\alpha, \beta, \gamma, \delta$ with the diagonals of a cube. Prove that

$$\cos^2 \alpha + \cos^2 \beta + \cos^2 \gamma + \cos^2 \delta = \frac{4}{3}.$$

Also prove that the angle between two diagonals is $\cos^{-1} \frac{1}{3}$

Solution. Let a be the edge of the cube, so that

$\vec{OA} = a\mathbf{i}, \vec{OB} = a\mathbf{j}, \vec{OC} = a\mathbf{k}.$

Then

$\vec{OD} = a\,(\mathbf{i} + \mathbf{j})$

$\vec{OE} = a\,(\mathbf{j} + \mathbf{k})$

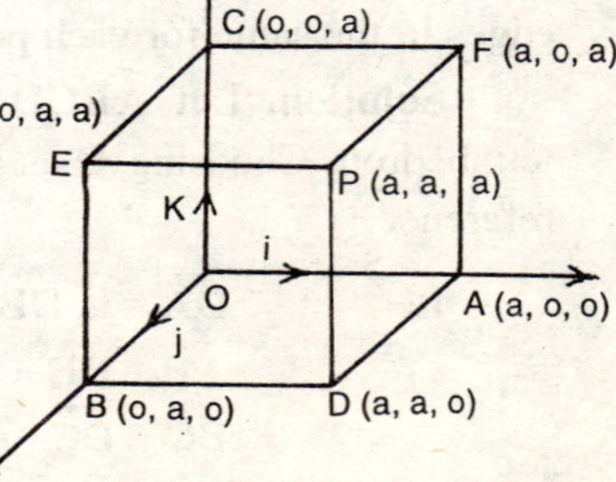

Fig. 2.17

$$\vec{OF} = a\,(\mathbf{k} + \mathbf{i})$$

$$\vec{OP} = a\,(\mathbf{i} + \mathbf{j} + \mathbf{k})$$

$$\vec{CD} = a\,(\mathbf{i} + \mathbf{j} - \mathbf{k})$$

$$\vec{AE} = a\,(\mathbf{j} + \mathbf{k} - \mathrm{i}) \text{ and } \vec{BF} = a\,(\mathbf{k} + \mathbf{i} - \mathbf{j})$$

Here $\vec{OP}$, $\vec{CD}$, $\vec{AE}$ and $\vec{BF}$ are the diagonals.

Let OL be any line given by

$$\vec{OL} = x\,\mathbf{i} + y\,\mathbf{j} + z\,\mathbf{k} \text{ so that } \overline{OL} = \sqrt{x^2 + y^2 + z^2}$$

Now α is the angle made by $\vec{OL}$ with $\vec{OP}$

$$\therefore \quad \cos\alpha = \frac{\vec{OP}\,.\,\vec{OL}}{\overline{OP}\,.\,\overline{OL}} = \frac{a\,(\mathbf{i} + \mathbf{j} + \mathbf{k})\,.\,(x\,\mathbf{i} + y\,\mathbf{j} + z\,\mathbf{k})}{a\sqrt{1 + 1 + 1}\,.\sqrt{x^2 + y^2 + z^2}}$$

$$\cos\alpha = \frac{x + y + z}{\sqrt{3\,(x^2 + y^2 + z^2)}}$$

Similarly β, γ, δ are the angles made by $\vec{OL}$ with $\vec{CD}$, $\vec{AE}$ and $\vec{BF}$. Hence

$$\cos\beta = \frac{x + y - z}{\sqrt{3\,(x^2 + y^2 + z^2)}}, \quad \cos\gamma = \frac{-x + y + z}{\sqrt{3\,(x^2 + y^2 + z^2)}}$$

and

$$\cos\delta = \frac{x - y + z}{\sqrt{3(x^2 + y^2 + z^2)}}$$

Now $\cos^2\alpha + \cos^2\beta + \cos^2\gamma + \cos^2\delta$

$$= \frac{(x + y + z)^2 + (x + y - z)^2 + (-x + y + z)^2 + (x - y + z)^2}{3(x^2 + y^2 + z^2)} = \frac{4}{3}$$

$$\therefore \quad \cos^2\alpha + \cos^2\beta + \cos^2\alpha + \cos^2\delta = \frac{4}{3}.$$

Second part

Angle between two diagonals say $\vec{OP}$ and $\vec{CD}$ is

$$\cos\theta = \frac{\vec{OP}.\vec{CD}}{\vec{OP}.\vec{CD}} = \frac{a(\mathbf{i} + \mathbf{j} + \mathbf{k}).a(\mathbf{i} + \mathbf{j} - \mathbf{k})}{(a\sqrt{3})\,(a\sqrt{3})} = \frac{1}{3}$$

or

$$\theta = \cos^{-1}\frac{1}{3}$$

Example 24. If **a, b, c** are coplanar vectors prove that

$$\begin{vmatrix} \mathbf{a} & \mathbf{b} & \mathbf{b} \\ \mathbf{a}\cdot\mathbf{a} & \mathbf{a}\cdot\mathbf{b} & \mathbf{a}\cdot\mathbf{c} \\ \mathbf{b}\cdot\mathbf{a} & \mathbf{b}\cdot\mathbf{b} & \mathbf{b}\cdot\mathbf{c} \end{vmatrix} = 0$$

Solution: Since **a, b, c** are coplanar there always exists a relation between them such that

$$x\mathbf{a} + y\mathbf{b} + z\mathbf{c} = 0 \qquad ...(1)$$

Multiplying scalarly both sides of (1) by **a** and **b** successively we get

$$x\,\mathbf{a}\cdot\mathbf{a} + y\,\mathbf{a}\cdot\mathbf{b} + z\,\mathbf{a}\cdot\mathbf{c} = 0 \qquad ...(2)$$

$$x\,\mathbf{b}\cdot\mathbf{a} + y\,\mathbf{b}\cdot\mathbf{b} + z\,\mathbf{b}\cdot\mathbf{c} = 0 \qquad ...(3)$$

Eliminating x, y, z between (1), (2) and (3) we get the required result as

$$\begin{vmatrix} \mathbf{a} & \mathbf{b} & \mathbf{c} \\ \mathbf{a}\cdot\mathbf{a} & \mathbf{a}\cdot\mathbf{b} & \mathbf{a}\cdot\mathbf{c} \\ \mathbf{b}\cdot\mathbf{a} & \mathbf{b}\cdot\mathbf{b} & \mathbf{b}\cdot\mathbf{c} \end{vmatrix} = 0$$

Note: On expansion the result can be put in the form

$$\mathbf{a}\begin{vmatrix} \mathbf{a}\cdot\mathbf{b} & \mathbf{a}\cdot\mathbf{c} \\ \mathbf{b}\cdot\mathbf{b} & \mathbf{b}\cdot\mathbf{c} \end{vmatrix} - \mathbf{b}\begin{vmatrix} \mathbf{a}\cdot\mathbf{a} & \mathbf{a}\cdot\mathbf{c} \\ \mathbf{b}\cdot\mathbf{a} & \mathbf{b}\cdot\mathbf{c} \end{vmatrix} + \mathbf{c}\begin{vmatrix} \mathbf{a}\cdot\mathbf{a} & \mathbf{a}\cdot\mathbf{b} \\ \mathbf{b}\cdot\mathbf{a} & \mathbf{b}\cdot\mathbf{b} \end{vmatrix} = 0$$

or

$$\mathbf{c} = \frac{\mathbf{a}\begin{vmatrix} \mathbf{a}\cdot\mathbf{b} & \mathbf{a}\cdot\mathbf{c} \\ \mathbf{b}\cdot\mathbf{b} & \mathbf{b}\cdot\mathbf{c} \end{vmatrix} + \mathbf{b}\begin{vmatrix} \mathbf{a}\cdot\mathbf{c} & \mathbf{a}\cdot\mathbf{a} \\ \mathbf{b}\cdot\mathbf{c} & \mathbf{b}\cdot\mathbf{a} \end{vmatrix}}{-[a^2 b^2 - (\mathbf{ab})^2]}$$

Example 25. The position vector of the foci of an ellipse are **b** and − **b** and the length of its major axis is 2a. Prove that the equation of the ellipse is

$$a^4 - a^2(\mathbf{r}^2 + \mathbf{b}^2) + (\mathbf{b}\cdot\mathbf{r})^2 = 0$$

Solution. We know that from co-ordinate geometry if P(r) be any point on the ellipse with foci S_1 and S_2 than $PS_1 + PS_2 = 2a$ (= length of major axis)

$$\therefore \qquad |\mathbf{r} - \mathbf{b}| + |\mathbf{r} + \mathbf{b}| = 2a$$

or

$$|\mathbf{r} + \mathbf{b}| = 2a - |\mathbf{r} - \mathbf{b}|$$

$$(\mathbf{r} + \mathbf{b})^2 = \{2a - |\mathbf{r} - \mathbf{b}|\}^2$$

$$= 4a^2 - 4a\,|\,\mathbf{r} - \mathbf{b}\,| + (\mathbf{r} - \mathbf{b})^2$$

$$(\mathbf{r} + \mathbf{b})^2 - (\mathbf{r} - \mathbf{b})^2 = 4a^2 - 4a\,|\mathbf{r} - \mathbf{b}|$$

$$4\,\mathbf{r}\,.\,\mathbf{b} = 4a^2 - 4a\,|\,\mathbf{r} - \mathbf{b}\,|$$

$$\mathbf{r}\,.\,\mathbf{b} - a^2 = -\,a\,|\,\mathbf{r} - \mathbf{b}\,|$$

Squaring again

$$[\mathbf{r}\,.\,\mathbf{b} - a^2]^2 = a^2\,(\mathbf{r} - \mathbf{b})^2$$

$$(\mathbf{r}\,.\,\mathbf{b})^2 - 2a^2\,(\mathbf{r}\,.\,\mathbf{b}) + a^4 = a^4\,(\mathbf{r} - \mathbf{b})^2$$

$$= a^4\,(\mathbf{r}^2 + \mathbf{b}^2 - 2\mathbf{r}\,.\,\mathbf{b})$$

or $$a^4 - a^2\,(\mathbf{r}^2 + \mathbf{b}^2) + (\mathbf{r}\,.\,\mathbf{b})^2 = 0.$$

Example 26. If **a, b, c** denote the position vectors of the vertices of a triangle, find the position vector of the orthocentre of the triangle

Solution. Let the P.V's of B and C w.r.t A as origin be **b** and **c**. Let AP be the perpendicular on to BC dividing it in $\lambda : 1$ ratio

$$\therefore \quad \text{P. V. of P} = \overrightarrow{AP} = \frac{\lambda\mathbf{c} + \mathbf{b}}{\lambda + 1} \qquad ...(1)$$

To find λ We know that AP $\perp$ BC

$$\therefore \quad \overrightarrow{AP}\,.\,\overrightarrow{BC} = 0, \text{ i.e., } \left(\frac{\lambda\mathbf{c} + \mathbf{b}}{\lambda + 1}\right).\,(\mathbf{c} - \mathbf{b}) = 0$$

or $$\lambda\mathbf{c}^2 - \lambda\,\mathbf{b}\,.\,\mathbf{c} + \mathbf{b}\,.\,\mathbf{c} - \mathbf{b}^2 = 0$$

$$\therefore\ \lambda = \frac{\mathbf{b}^2 - \mathbf{b}\,.\,\mathbf{c}}{\mathbf{c}^2 - \mathbf{c}\,.\,\mathbf{b}} = \frac{AB^2 - (AB)\,(AC)\cos A}{AC^2 - (AC)\,(AB)\cos A}$$

or $$\gamma^2 = \frac{c^2 - bc\cos A}{b^2 - bc\cos A} \qquad [\because AB = c, AC = b]$$

$$\therefore \gamma = \frac{c\,(c - b\cos A)}{b\,(b - bc\cos A} = \frac{\sin C\,(a\cos B)}{\sin B\,(a\cos C)}$$

$$[\because\ c = 2R\sin C,\ b = 2R\sin B \text{ and } a = b\cos C + c\cos B]$$

$$\lambda = \frac{\tan C}{\tan B}.$$

Substituting in (1), we get

p.v. of P = $\vec{AP} = \dfrac{\mathbf{c}\tan C + \mathbf{b}\tan B}{\tan B + \tan C}$

This is the mean (centroid) of **b** and **c** with the associated numbers tan B and tan C and it is independent of the origin of vectors.

Hence if instead of A we choose any other point as origin p.v of P remains the same. Let P.V of A w.r.t new origin **c** is **a**

∴ The p.v **r** of a point O on AP which divides it in the ratio (tan B + tan C) : tan A is

$$\mathbf{r} = \frac{(\tan B + \tan C)\dfrac{\mathbf{b}\tan B + \mathbf{c}\tan C}{\tan B + \tan C} + \mathbf{a}\tan A}{\tan A + \tan B + \tan C}$$

or $$\mathbf{r} = \frac{\mathbf{a}\tan A + \mathbf{b}\tan B + \mathbf{c}\tan C}{\tan A + \tan B + \tan C} \qquad ...(2)$$

The symmetry to the result shows that the point **r** lies on the perpendiculars from vertices on opposite sides. Hence the perpendiculars are concurrent and the point of concurrency is the orthocentre given by (2).

Example 27. If there be four coplanar straight lines and unit vectors parallel to their directions be denoted by **a, b, c** and **d** respectively and cos (**a, b**) stands for the cosine of the angle between the lines which are parallel to **a** and **b** then prove that

$$\begin{vmatrix} 1 & \cos(\mathbf{a},\mathbf{b}) & \cos(\mathbf{a},\mathbf{c}) & \cos(\mathbf{a},\mathbf{d}) \\ \cos(\mathbf{b},\mathbf{a}) & 1 & \cos(\mathbf{b},\mathbf{c}) & \cos(\mathbf{b},\mathbf{d}) \\ \cos(\mathbf{c},\mathbf{a}) & \cos(\mathbf{c},\mathbf{b}) & 1 & \cos(\mathbf{c},\mathbf{d}) \\ \cos(\mathbf{d},\mathbf{a}) & \cos(\mathbf{d},\mathbf{b}) & \cos(\mathbf{d},\mathbf{c}) & 1 \end{vmatrix} = 0$$

Solution: Since **a, b, c, d** are non-coplarnar vectors we know that there exists a linear relation such that

$$x\mathbf{a} + y\mathbf{b} + z\mathbf{c} + p\mathbf{d} = 0 \qquad ...(1)$$

where x, y, z and p are scalars

Given **a, b, c, d** are unit vectors.

Hence $|\mathbf{a}| = |\mathbf{b}| = |\mathbf{c}| = |\mathbf{d}| = 1$

and $\mathbf{a}\cdot\mathbf{b} = 1.1.\cos(\mathbf{a},\mathbf{b}) = \cos(\mathbf{a},\mathbf{b})$ etc.,

Multiplying (1) scalarly by **a, b, c** and **d** we get

$$x + y\cos(\mathbf{a},\mathbf{b}) + z\cos(\mathbf{a},\mathbf{c}) - p\cos(\mathbf{a},\mathbf{d}) = 0 \qquad ...(2)$$

$$x\cos(\mathbf{b},\mathbf{a}) + y + z\cos(\mathbf{b},\mathbf{c}) + p\cos(\mathbf{b},\mathbf{d}) = 0 \qquad ...(3)$$

$$x \cos(\mathbf{c}, \mathbf{a}) + y \cos(\mathbf{c}, \mathbf{b}) + z + p \cos(\mathbf{c}, \mathbf{d}) = 0 \qquad ...(4)$$

$$x \cos(\mathbf{d}, \mathbf{a}) + y \cos(\mathbf{d}, \mathbf{b}) + z \cos(\mathbf{d}, \mathbf{c}) + p = 0 \qquad ...(5)$$

Eliminating x, y, z and p from (2), (3), (4) and (5) we obtain the given relation.

EXERCISE

1. If $\mathbf{a} = 3\mathbf{i} - \mathbf{j} + 4\mathbf{k}$ and $\mathbf{b} = \mathbf{i} - 3\mathbf{j} + 6\mathbf{k}$ find $\mathbf{a} \cdot \mathbf{b}$.
2. If $\mathbf{a} = 3\mathbf{i} - 2\mathbf{j} + \mathbf{k}$ and $b = \mathbf{i} + \mathbf{j} - \mathbf{k}$ find $\mathbf{a} \cdot \mathbf{b}$
3. If $\mathbf{a} = \mathbf{i} - \mathbf{j} + 3\mathbf{k}$ and $\mathbf{b} = 3\mathbf{i} + 4\mathbf{j} + 2\mathbf{k}$ find $(2\mathbf{a} + \mathbf{b}) \cdot (\mathbf{a} - 3\mathbf{b})$.
4. Find the direction cosines of $3\mathbf{i} - 4\mathbf{j} + 2\mathbf{k}$.
5. Find the angles which $2\mathbf{i} - 3\mathbf{j} + 6\mathbf{k}$ makes with the co-ordinate axes.
6. Show that the vectors $3\mathbf{i} - 2\mathbf{j} + 2\mathbf{k}$ and $6\mathbf{i} + 5\mathbf{j} - 4\mathbf{k}$ are perpendicular to each other.
7. If $\mathbf{a} = \mathbf{i} + 2\mathbf{j} - 3\mathbf{k}$, $\mathbf{b} = 3\mathbf{i} - \mathbf{j} + 2\mathbf{k}$ show that $\mathbf{a} - \mathbf{b}$ and $\mathbf{a} + \mathbf{b}$ are perpendicular to each other.
8. State and prove the necessary and sufficient conditions for two non-zero vectors to be perpendicular.
9. If $3\mathbf{i} + \mathbf{j} + \mathbf{k}$ and $p\mathbf{i} - 2\mathbf{j} + 3\mathbf{k}$ are perpendicular then find p.
10. If $|\mathbf{a} + \mathbf{b}| = |\mathbf{a} - \mathbf{b}|$ then show that **a** and **b** are perpendicular.
11. Find the angle between the vectors

    (i) $2\mathbf{i} + \mathbf{j} - 3\mathbf{k}$, $3\mathbf{i} - 2\mathbf{j} - \mathbf{k}$

    (ii) $6\mathbf{i} - 3\mathbf{j} - 2\mathbf{k}$, $2\mathbf{i} + 2\mathbf{j} - \mathbf{k}$
12. If $\mathbf{a} = \mathbf{i} + 2\mathbf{j} - 3\mathbf{k}$ and $\mathbf{b} = 3\mathbf{i} - \mathbf{j} + 2\mathbf{k}$ show that

    (i) $(\mathbf{a} + \mathbf{b}) \cdot (\mathbf{a} - \mathbf{b}) = 0$

    (ii) angle between $(2\mathbf{a} + \mathbf{b})$ and $(\mathbf{a} + 2\mathbf{b})$ in $\cos^{-1}\left(\frac{31}{50}\right)$.
13. If P, Q, R, S are the points $\mathbf{i} + \mathbf{j} + \mathbf{k}$, $2\mathbf{i} + 5\mathbf{j}$, $3\mathbf{i} + 2\mathbf{j} - 3\mathbf{k}$ and $\mathbf{i} - 6\mathbf{j} - \mathbf{k}$ show that the angle between $\overrightarrow{PQ}$ and $\overrightarrow{RS}$ is π.
14. If O is the origin, A and B are the points (1, 3, −2) and (2, −3, 6) find the angle that $\overrightarrow{OA}$ makes with $\overrightarrow{OB}$
15. Prove that the triangle whose vertices are $2\mathbf{i} + 4\mathbf{j} - \mathbf{k}$, $4\mathbf{i} + 5\mathbf{j} + \mathbf{k}$ and $3\mathbf{i} + 6\mathbf{j} - 3\mathbf{k}$ is an isosceles triangle.

16. Prove that the vectors $\mathbf{a} = 7\mathbf{i} - 3\mathbf{j} - \mathbf{k}$, $\mathbf{b} = 2\mathbf{i} + \mathbf{j} + 3\mathbf{k}$, $\mathbf{c} = 5\mathbf{i} + 4\mathbf{j} - 3\mathbf{k}$ form the sides of a right angled triangle.

17. Prove that the three points whose position vectors are $\mathbf{i} + 2\mathbf{j} + 3\mathbf{k}$, $-\mathbf{i} - \mathbf{j} + 8\mathbf{k}$ and $-4\mathbf{i} + 4\mathbf{j} + 6\mathbf{k}$ form an equilateral triangle.

18. What is the unit vector perpendicular to each of the vectors $2\mathbf{i} - \mathbf{j} + \mathbf{k}$ and $3\mathbf{i} + 4\mathbf{j} - \mathbf{k}$ and calculate the sine of the angle between the two vectors.

19. Find the projection of the vector $\mathbf{a} = 2\mathbf{i} + 3\mathbf{j} + 2\mathbf{k}$ on the vector $\mathbf{b} = \mathbf{i} + 2\mathbf{j} + \mathbf{k}$.

20. Find the projection of the vector $a = \mathbf{i} - 2\mathbf{j} + \mathbf{k}$ on the vector $\mathbf{b} = 4\mathbf{i} - 4\mathbf{j} + 7\mathbf{k}$.

21. If $\mathbf{a}, \mathbf{b}, \mathbf{c}$ are three non-coplanar vectors and $\mathbf{d} \cdot \mathbf{a} = \mathbf{d} \cdot \mathbf{b} = \mathbf{d} \cdot c = 0$ then show that $\mathbf{d}$ is a null vector

22. If $\mathbf{a} = \mathbf{i} + 2\mathbf{j} + 3\mathbf{k}$, $\mathbf{b} = -\mathbf{i} + 2\mathbf{j} + \mathbf{k}$, $\mathbf{c} = 3\mathbf{i} + \mathbf{j}$ then find t such that $\mathbf{a} + t\mathbf{b}$ in perpendicular to $\mathbf{c}$.

23. If a vector $\mathbf{r}$ is perpendicular to two non-collinear vectors $\mathbf{b}$ and $\mathbf{c}$ then prove that $\mathbf{r}$ is perpendicular to every vector in the plane of $\mathbf{b}$ and $\mathbf{c}$.

24. Three vectors are $\mathbf{a} = \mathbf{i} + 2\mathbf{j} + 3\mathbf{k}$, $\mathbf{b} = -\mathbf{i} + 2\mathbf{j} + \mathbf{k}$ and $\mathbf{c} = 3\mathbf{i} + \mathbf{j}$ find a unit vector in the direction of their resultant. Also find a vector $\mathbf{d}$ which is normal to $\mathbf{a}$ and $\mathbf{b}$ both. Find also the angle between $\mathbf{c}$ and $\mathbf{d}$.

25. Prove by vector method the cauchy's inequality $(\mathbf{a} \cdot \mathbf{b})^2 \le \mathbf{a}^2\, \mathbf{b}^2$ where $\mathbf{a} = a_1\mathbf{i} + a_2\mathbf{j} + a_3\mathbf{k}$ and $\mathbf{b} = b_1\mathbf{i} + b_2\mathbf{j} + b_3\mathbf{k}$

26. Show that if $\mathbf{a}$ and $\mathbf{b}$ are two vectors inclined at an angle θ to each other then

$$|\mathbf{a} - \mathbf{b}|^2 = |\mathbf{a}|^2 + |\mathbf{b}|^2 - 2\,|\mathbf{a}|\,|\mathbf{b}| \cos\theta$$

27. If D be the mid point of the side BC of triangle ABC show that $AB^2 + AC^2 = 2\,(AD^2 + BD^2)$

28. Show that diagonals in a rhombus are at right angles

29. Prove that if the diagonals of a quadrilateral bisect each other at right angles, the quadrilateral is a rhombus

30. Prove that a parallelogram whose diagonals are equal is a rectangle

31. Prove that in any triangle if P, Q, R are the mid points of the sides BC, CA and AB then

$$2(AB^2 + BC^2 + CA^2) = 4(AP^2 + BQ^2 + CR^2)$$
$$= 9\,(AG^2 + BG^2 + CG^2)$$

where a is the centroid

32. Prove that in a parallelogram the difference of a squares on the diagonals is four times the rectangle contained by either of these sides and projection of the other upon it

33. The position vectors **a, b, c, d** of four coplanar, points A, B, C, D are such that $(\mathbf{a} - \mathbf{d}) \cdot (\mathbf{b} - \mathbf{c}) = (\mathbf{b} - \mathbf{d}) \cdot (\mathbf{c} - \mathbf{a}) = 0$ then show that the point D is the orthocentre of ΔABC.

34. Prove that in a parallopiped the sum of the squares on the diagonals is equal to sum of the squares on the edges.

35. If a straight line is equally inclined to three coplanar straight lines prove that it is perpendicular to their planes.

36. Prove that any two opposite edges of a regular tetrahedron are perpendicular.

37. Prove that sum of the squares on the edges of any tetrahedron is equal to four times the sum of the squares on the join of the midpoints of opposite edges.

38. Prove by vector methods in any triangle ABC

(i) $b = c \cos A + a \cos B$

(ii) $c = a \cos B + b \cos A$

(iii) $b^2 = a^2 + c^2 - 2ac \cos B$

(iv) $c^2 = a^2 + b^2 - 2ab \cos C$

(v) $\cos(\alpha + \beta) = \cos\alpha \cos\beta - \sin\alpha \sin\beta$

39. Find the angle between two diagonals of a cube.

40. If the edges of a rectangular parallelopiped are a, b, c show that the angles between the four diagonals are

$$\cos^{-1}\left[\frac{a^2 \pm b^2 \pm c^2}{a^2 + b^2 + c^2}\right]$$

41. Prove that the hyperbole whose foci are points b_1 and b_2 and whose transverse axis is 2a is given by

$$|\mathbf{r} - b_1| - |\mathbf{r} - b_2| = 2a.$$

42. Prove that in any triangle ABC,

$$(a + b + c)(\cos A + \cos B + \cos C) = a(1 + \cos A) + b(1 + \cos B) + c(1 + \cos C)$$

ANSWERS

(1) 30 **(2)** 0 **(3)** -90 **(4)** $\frac{3}{\sqrt{29}}, \frac{-4}{\sqrt{29}}, \frac{2}{\sqrt{29}}$

(5) $\frac{2}{7}, \frac{-3}{7}, \frac{6}{7}$ **(9)** $p = -\frac{1}{3}$ **(11)** (i) $\frac{\pi}{3}$ (ii) $\cos^{-1}\frac{4}{21}$

(14) $\cos^{-1}\left(\frac{-19}{7\sqrt{14}}\right)$ **(18)** $\frac{1}{\sqrt{155}}(-3\mathbf{i} + 5\mathbf{j} + 11\mathbf{k})$, $\sin\theta = \sqrt{\frac{155}{150}}$

(19) $\frac{5\sqrt{6}}{3}$ **(20)** $\frac{19}{9}$ **(22)** $t = 5$

(24) $\frac{1}{5\sqrt{2}}(3\mathbf{i} + 5\mathbf{j} + 4\mathbf{k})$; $\gamma(\mathbf{i} + \mathbf{j} - \mathbf{k})$, γ some scalar; $\theta = \cos^{-1}\frac{4}{\sqrt{30}}$

(39) $\cos^{-1}\frac{1}{3}$.

2.11. The vector or cross product of two vectors

The vector product of two vectors **a** and **b** of moduli a and b respectively is a vector whose modulus is ab sin θ θ being the angle between the directions of vectors **a** and **b**, and $0 \le \theta \le \pi$. The direction is that of the unit vector $\hat{\mathbf{n}}$ which is perpendicular to both **a** and **b** such that **a**, **b** and $\hat{\mathbf{n}}$ are in the right handed orientation. i.e., if we turn the vector **a** into the vector **b** then $\hat{\mathbf{n}}$ points in the direction in which a right handed screw would advance if turned in a similar manner.

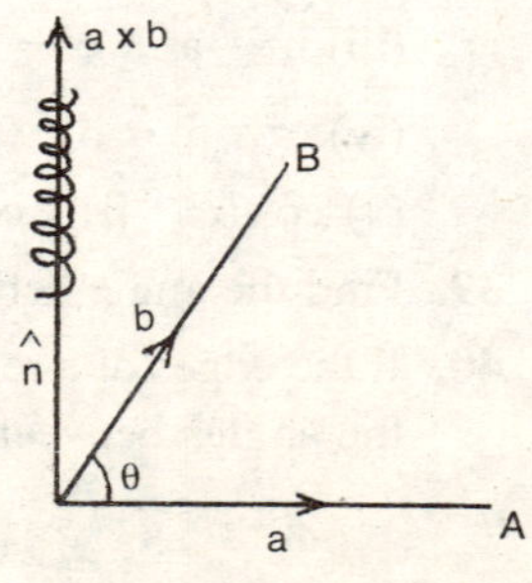

Fig. 2.18

Thus $\mathbf{a} \times \mathbf{b} = ab \sin\theta\, \hat{\mathbf{n}} = |\mathbf{a}||\mathbf{b}| \sin\theta\, \hat{\mathbf{n}}$

where $\hat{\mathbf{n}}$ is unit vector perpendicular to both **a** and **b**. We read **a** × **b** as **a** cross **b**. The vector product is also called **outer product** or **indirect product.**

Note: (1) From above definition we conclude that a × **b** is perpendicular to the plane containing **a** and **b** and a unit vector perpendicular to the plane of two given vectors **a** and **b** is

$$\hat{\mathbf{n}} = \frac{\mathbf{a} \times \mathbf{b}}{|\mathbf{a} \times \mathbf{b}|}$$

2.12. Properties of vector product

1. *Vector product is not commulative*

i.e., $\mathbf{a} \times \mathbf{b} \neq \mathbf{b} \times \mathbf{a}$

By definition **a** × **b** is a vector in which the rotation carries from **a** to **b** in anti clock wise Hence **b** × **a** is a vector in which the rotation carries from **b** to **a** is clock wise

∴ The unit vectors are $\hat{\mathbf{n}}$ and $-\hat{\mathbf{n}}$.

Hence $\mathbf{a} \times \mathbf{b} = |\mathbf{a}||\mathbf{b}| \sin\theta\, \hat{\mathbf{n}}$

$\mathbf{b} \times \mathbf{a} = |\mathbf{b}||\mathbf{a}| \sin\theta\, (-\hat{\mathbf{n}}) = -|\mathbf{b}||\mathbf{a}| \sin\theta\, \hat{\mathbf{n}}$

∴ $\mathbf{a} \times \mathbf{b} \neq \mathbf{b} \times \mathbf{a}$

But $\mathbf{b} \times \mathbf{a} = -|\mathbf{a}||\mathbf{b}| \sin\theta\, \hat{\mathbf{n}}$

$= -(\mathbf{a} \times \mathbf{b}) = -\mathbf{a} \times \mathbf{b} = \mathbf{a} \times -\mathbf{b}$

2. *Vector product is associative w.r.t scalar multiplication*

Let m, n be any two scalars and **a**, **b** vectors Then we have

$$m(\mathbf{a} \times \mathbf{b}) = (m\mathbf{a}) \times \mathbf{b} = \mathbf{a} \times (m\mathbf{b}) = (\mathbf{a} \times \mathbf{b})\, m$$

$$(m\mathbf{a} \times n\mathbf{b}) = (n\mathbf{a} \times m\mathbf{b}) = (mn\mathbf{a}) \times \mathbf{b} = \mathbf{a} \times (mn\, \mathbf{b}) = mn\, (\mathbf{a} \times \mathbf{b})$$

3. *Vector product of two parallel (or collinear) vectors*

We know that $\mathbf{a} \times \mathbf{b} = |\mathbf{a}||\mathbf{b}| \sin\theta\, \hat{\mathbf{n}}$

If the two vectors **a** and **b** are parallel (or collinear) then angle between them is O or π i.e., $\sin\theta = 0$

∴ $\mathbf{a} \times \mathbf{b} = 0$

Hence two vectors are parallel or collinear if their cross product is a null vector.

Conversely. If $\mathbf{a} \times \mathbf{b} = 0$ then $|\mathbf{a}||\mathbf{b}| \sin\theta\, \hat{\mathbf{n}} = 0$. Then either $\mathbf{a} = 0$ or $\mathbf{b} = 0$ or $\sin\theta = 0$ i.e., either any of the given vectors **a** and **b** is a zero vector or if none of them is a zero vector then $\sin\theta = 0$

Hence the necessary and sufficient condition for two non-zero vectors to be parallel or collinear is that their vector product must be equal to a zero vector.

Note: In particular $\mathbf{a} \times \mathbf{a} = 0$

4. *Vector product of two perpendicular vectors*

If the vectors **a** and **b** are perpendicular we have $\theta = 90°$

$$\therefore \quad \mathbf{a} \times \mathbf{b} = |\mathbf{a}|\,|\mathbf{b}|\,(\sin 90°)\,\hat{\mathbf{n}}$$

$$= |\mathbf{a}||\mathbf{b}|\,\hat{\mathbf{n}}$$

Thus $\mathbf{a} \times \mathbf{b}$ *is a vector whose modulus is* $|\mathbf{a}||\mathbf{b}|$ *and whose direction is such that* **a, b,** $\mathbf{a} \times \mathbf{b}$ *form a right handed system of mutually perpendicular vectors.*

5. *Vector product of two unit vectors*

If **a** and **b** are unit vectors i.e., $|\mathbf{a}| = |\mathbf{b}| = 1$ then $\mathbf{a} \times \mathbf{b} = |\mathbf{a}|\,|\mathbf{b}| \sin\theta\, \hat{\mathbf{n}} = \sin\theta\, \hat{\mathbf{n}}$.

Thus the cross product of two unit vectors is a vector whose module in equal to the sine of the angle between the directions of the given vectors.

6. *Vector product of unit vectors* **i, j, k**

We have the following important results

$$\mathbf{i} \times \mathbf{i} = \mathbf{j} \times \mathbf{j} = \mathbf{k} \times \mathbf{k} = 0$$

$$\mathbf{i} \times \mathbf{j} = \mathbf{k},\ \mathbf{j} \times \mathbf{k} = \mathbf{i},\ \mathbf{k} \times \mathbf{i} = \mathbf{j}$$

$$\mathbf{j} \times \mathbf{i} = -\mathbf{k},\ \mathbf{k} \times \mathbf{j} = -\mathbf{i},\ \mathbf{i} \times \mathbf{k} = -\mathbf{j}$$

7. Distributive property If **a, b, c** are any three vectors then

$$\mathbf{a} \times (\mathbf{b} + \mathbf{c}) = \mathbf{a} \times \mathbf{b} + \mathbf{a} \times \mathbf{c}$$

Similarly

(i) $(\mathbf{a} + \mathbf{b}) \times (\mathbf{c} + \mathbf{d}) = \mathbf{a} \times \mathbf{c} + \mathbf{a} \times \mathbf{d} + \mathbf{b} \times \mathbf{c} + \mathbf{b} \times \mathbf{d}$

(ii) $(\mathbf{a} + \mathbf{b} + \mathbf{c} + \ldots) \times (\mathbf{p} + \mathbf{q} + \mathbf{r} + \ldots)$

$$= (\mathbf{a} \times \mathbf{p} + \mathbf{a} \times \mathbf{q} + \mathbf{a} \times \mathbf{r} + \ldots) + (\mathbf{b} \times \mathbf{p} + \mathbf{b} \times \mathbf{q} + \mathbf{b} \times \mathbf{r} + \ldots) + (\mathbf{c} \times \mathbf{p} + \mathbf{c} \times \mathbf{q} + \mathbf{c} \times \mathbf{r} + \ldots)$$

(iii) $\mathbf{a} \times (\mathbf{b} - \mathbf{c}) = \mathbf{a} \times \mathbf{b} - \mathbf{a} \times \mathbf{c}$

8. *Vector product in not associative*

i.e., $\mathbf{a} \times (\mathbf{b} \times \mathbf{c}) \neq (\mathbf{a} \times \mathbf{b}) \times \mathbf{c}$

2.13. Component of a vector r perpendicular to a given vector

Let **i, j** be the unit vectors in the direction of **a** and in the direction perpendicular to **a**. Then $\mathbf{a} = a\mathbf{i}$, where a is the magnitude of **a**. Let the unit vector **k** be perpendicular to **i-j** plane. Now if **r** be any vector in **i–j** plane inclined at an angle θ to **a** then its component in the direction perpendicular to **a** i.e., **j** is

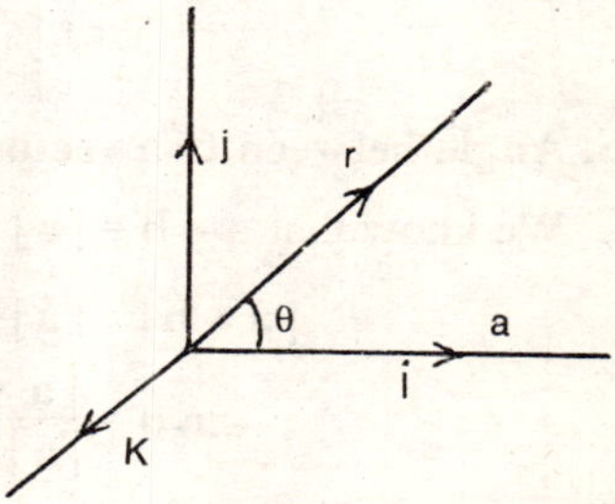

$$= (r \sin \theta)\,\mathbf{j} \qquad ...(1)$$

Fig. 2.19

Also $\mathbf{a} \times \mathbf{r} = (ar \sin \theta)\mathbf{k}$

where **k** is a unit vector perpendicular to **i-j** plane in which both **a** and **r** lie.

Now $\mathbf{k} \times \mathbf{i} = \mathbf{j}$ ...(3)

$$\therefore \quad (ar \sin \theta)\,\mathbf{j} = (ar \sin \theta)\,\mathbf{k} \times \mathbf{i} = (\mathbf{a} \times \mathbf{r}) \times \mathbf{i}$$

$$\therefore \quad (r \sin \theta)\,\mathbf{j} = \frac{(\mathbf{a} \times \mathbf{r}) \times \mathbf{i}}{a} = \frac{(\mathbf{a} \times \mathbf{r}) \times a\mathbf{i}}{a^2}$$

$$= \frac{(\mathbf{a} \times \mathbf{r}) \times \mathbf{a}}{a^2}$$

or component of **r** perpendicular to $\mathbf{a} = \dfrac{-\,\mathbf{a} \times (\mathbf{a} \times \mathbf{r})}{a^2}$.

2.14. Vector product in terms of rectangular components of the given vectors

Let $\mathbf{a} = a_1\mathbf{i} + a_2\mathbf{j} + a_3\mathbf{k}$ and $\mathbf{b} = b_1\mathbf{i} + b_2\mathbf{j} + b_3\mathbf{k}$

then $\mathbf{a} \times \mathbf{b} = (a_1\mathbf{i} + a_2\mathbf{j} + a_3\mathbf{k}) \times (b_1\mathbf{i} + b_2\mathbf{j} + b_3\mathbf{k})$

$$= a_1b_1\,\mathbf{i} \times \mathbf{i} + a_1b_2\,\mathbf{i} \times \mathbf{j} + a_1b_3\,\mathbf{i} \times \mathbf{k}$$

$$+ a_2b_1\,\mathbf{j} \times \mathbf{i} + a_2b_2\,\mathbf{j} \times \mathbf{j} + a_2b_3\mathbf{j} \times \mathbf{k}$$

$$+ a_3b_1\,\mathbf{k} \times \mathbf{i} + a_3b_2\,\mathbf{k} \times \mathbf{j} + a_3b_3\,\mathbf{k} \times \mathbf{k}$$

$$= a_1b_2\mathbf{k} - a_1b_3\mathbf{j} - a_2b_1\mathbf{k} + a_2b_3\mathbf{i} + a_3b_1\mathbf{j} - a_3b_2\mathbf{i}$$
$$= (a_2b_3 - a_3b_2)\mathbf{i} - (a_1b_3 - a_3b_1)\mathbf{j} + (a_1b_2 - a_2b_1)\mathbf{k}$$

which can be put in the determinant form as

$$\mathbf{a} \times \mathbf{b} = \begin{vmatrix} \mathbf{i} & \mathbf{j} & \mathbf{k} \\ a_1 & a_2 & a_3 \\ b_1 & b_2 & b_3 \end{vmatrix}$$

2.15. Angle between two vectors using vector product

We know that $\mathbf{a} \times \mathbf{b} = |\mathbf{a}||\mathbf{b}| \sin\theta\, \hat{\mathbf{n}}$

$$|\mathbf{a} \times \mathbf{b}| = |\mathbf{a}||\mathbf{b}| \sin\theta, \ \because\ |\hat{\mathbf{n}}| = 1$$

$$\therefore \qquad \sin\theta = \frac{|\mathbf{a} \times \mathbf{b}|}{|\mathbf{a}||\mathbf{b}|} \qquad ...(1)$$

Using **art 2.14**

$$|\mathbf{a} \times \mathbf{b}| = \sqrt{(a_2b_3 - a_3b_2)^2 + (a_1b_3 - a_3b_1)^2 + (a_1b_2 - a_2b_1)^2}$$

$$|\mathbf{a}| = \sqrt{a_1^2 + a_2^2 + a_3^2} \quad \text{and} \quad |\mathbf{b}| = \sqrt{b_1^2 + b_2^2 + b_3^2}$$

Substituting in (1), we get

$$\sin\theta = \frac{\sqrt{(a_2b_3 - a_3b_2)^2 + (a_1b_3 - a_3b_1)^2 + (a_1b_2 - a_2b_1)^2}}{\sqrt{a_1^2 + a_2^2 + a_3^2}\ \sqrt{b_1^2 + b_2^2 + b_3^2}}$$

Deduction. *If the vectors* **a** *and* **b** *are parallel*

Then $\theta = 0 \Rightarrow \sin\theta = 0$

$$\Rightarrow \quad (a_2b_3 - a_3b_2)^2 + (a_1b_3 - a_3b_1)^2 + (a_1b_2 - a_2b_1)^2 = 0$$

$$\Leftrightarrow \quad a_2b_3 - a_3b_2 = a_1b_3 - a_3b_2 = a_1b_2 - a_2b_1 = 0$$

$$\Leftrightarrow \quad \frac{a_1}{a_2} = \frac{b_1}{b_2} = \frac{c_1}{c_2}$$

Note: If l, m, n and l′, m′, n′ are direction cosines of vectors **a** and **b** then

$$l = \frac{a_1}{\sqrt{a_1^2 + a_2^2 + a_3^2}} = \frac{a_1}{|a|}; \quad m = \frac{a_2}{|a|}, \ n = \frac{a_3}{|a|}$$

$$l' = \frac{b_1}{\sqrt{b_1^2 + b_2^2 + b_3^2}} = \frac{b_1}{|b|} \ ; m' = \frac{b^2}{|b|} \ ; \ n' = \frac{b^3}{|b|} \text{ and hence}$$

$$\sin^2\theta = (mn' - m'n)^2 + (n' - nl'l)^2 + (lm' = l'm)^2$$

2.16. Geometrical interpretation of vector product

*The magnitude of the vector product of two vectors **a** and **b** in the area of a parallelogram whose adjacent sides are represented by* **a** *and* **b**.

Let $\vec{OA}$ = **a** and $\vec{OB}$ = **b** be two non-zero and non-collinear vectors and let ∠AOB = θ complete the parallelogram OACB. Then **a** × **b** = | **a** | | **b** | sin θ $\hat{\mathbf{n}}$ where $\hat{\mathbf{n}}$ is a unit vector normal to the plane of the parallelogram pointing in the direction in which a right handed screw would move if rotated in the sense OACB.

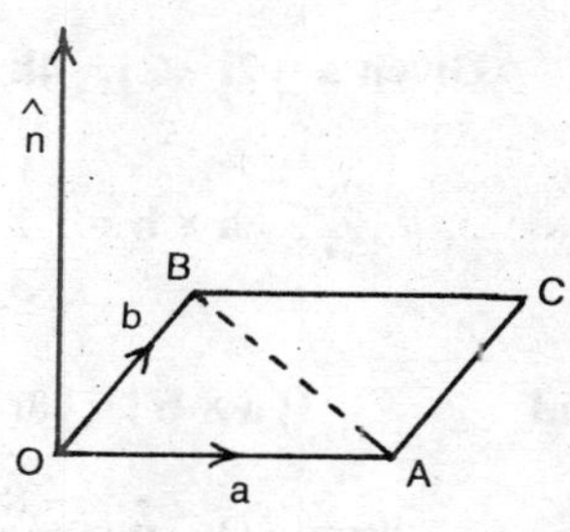

Fig. 2.20

$$\therefore \quad |\mathbf{a} \times \mathbf{b}| = |\mathbf{a}|\,|\mathbf{b}| \sin\theta$$
$$= \overline{OA}\;\overline{OB} \sin\theta = \overline{OA}\,(\overline{OB} \sin\theta)$$
$$= 2\left\{\frac{1}{2}\overline{OA}\,(\overline{OB}\sin\theta)\right\}$$
$$= 2\,\{\text{area of } \Delta\, OAB\}$$
$$= \text{area of the parallelogram OACB}$$

Note. (1) We also call **a** × **b** as vector area of a parallelogram OABC since the boundary is traced out in the anticlock wise direction.

(2) The vector area of the parallelogram OBCA will be denoted by **b** × **a**

(3) The area of ΔOAB whose adjacent sides are represented by the vectors **a** and **b** is $\frac{1}{2}$ | **a** × **b** | and vector area is $\frac{1}{2}$ **a** × **b**.

SOLVED PROBLEMS

Example 28. Find a unit vector perpendicular to each of the vectors **a** = 2**i** − 2**j** + 4**k** and **b** = 3**i** + **j** + 2**k**. Also find the sine of the angle between these vectors.

Solution: A unit vector perpendicular to each of the vectors **a** and **b** is given by

$$\hat{\mathbf{n}} = \frac{\mathbf{a} \times \mathbf{b}}{|\mathbf{a} \times \mathbf{b}|}$$

Given **a** = 2**i** − 2**j** + 4**k** and b = 3**i** + **j** + 2**k**

$$\therefore \quad \mathbf{a} \times \mathbf{b} = \begin{vmatrix} \mathbf{i} & \mathbf{j} & \mathbf{k} \\ 2 & -2 & 4 \\ 3 & 1 & 2 \end{vmatrix} = -8\mathbf{i} + 8\mathbf{j} + 8\mathbf{k}$$

and
$$|\mathbf{a} \times \mathbf{b}| = |8(-\mathbf{i} + \mathbf{j} + \mathbf{k})| = 8\sqrt{1+1+1} = 8\sqrt{3}$$

$$\therefore \text{ From (1), } \quad \hat{\mathbf{n}} = \frac{-8\mathbf{i} + 8\mathbf{j} + 8\mathbf{k}}{8\sqrt{3}} = \frac{-\mathbf{i} + \mathbf{j} + \mathbf{k}}{\sqrt{3}}$$

Now
$$\sin\theta = \frac{|\mathbf{a} \times \mathbf{b}|}{|\mathbf{a}||\mathbf{b}|} \quad \text{...(2)}$$

$$|\mathbf{a}| = |2\mathbf{i} - 2\mathbf{j} + 4\mathbf{k}| = \sqrt{4+4+16} = 2\sqrt{6}$$

$$|\mathbf{b}| = |3\mathbf{i} + \mathbf{j} + 2\mathbf{k}| = \sqrt{9+1+4} = \sqrt{14}$$

$$\therefore \quad \sin\theta = \frac{8\sqrt{3}}{2\sqrt{6}.\sqrt{14}} = \frac{2}{\sqrt{7}}$$

Example 29. If **a** = 3**i** − **j** + 2**k**, **b** = 2**i** + **j** − **k**, **c** = **i** − 2**j** + 2**k** find (**a** × **b**) × **c** and **a** × (**b** × **c**) and hence show that (**a** × **b**) × **c** ≠ **a** × (**b** × **c**). Also show that **a** × **b** is perpendicular to both **a** and **b**.

Solution: Given **a** = 3**i** − **j** + 2**k**, **b** = 2**i** + **j** − **k** and
c = **i** − 2**j** + 2**k**. Now

$$\mathbf{a} \times \mathbf{b} = \begin{vmatrix} \mathbf{i} & \mathbf{j} & \mathbf{k} \\ 3 & -1 & 2 \\ 2 & 1 & -1 \end{vmatrix} = -\mathbf{i} + 7\mathbf{j} + 5\mathbf{k}$$

$$(\mathbf{a} \times \mathbf{b}) \times \mathbf{c} = \begin{vmatrix} \mathbf{i} & \mathbf{j} & \mathbf{k} \\ -1 & 7 & 5 \\ 1 & -2 & 2 \end{vmatrix} = 24\mathbf{i} + 7\mathbf{j} - 5\mathbf{k}$$

Again
$$\mathbf{b} \times \mathbf{c} = \begin{vmatrix} \mathbf{i} & \mathbf{j} & \mathbf{k} \\ 2 & 1 & -1 \\ 1 & -2 & 2 \end{vmatrix} = -5\mathbf{j} - 5\mathbf{k}$$

and $\quad \mathbf{a} \times (\mathbf{b} \times \mathbf{c}) = \begin{vmatrix} \mathbf{i} & \mathbf{j} & \mathbf{k} \\ 3 & -1 & 2 \\ 0 & -5 & -5 \end{vmatrix} = 15\mathbf{i} + 15\mathbf{j} - 15\mathbf{k}$ ...(2)

From (1) and (2) we say that

$$(\mathbf{a} \times \mathbf{b}) \times \mathbf{c} = \mathbf{a} \times (\mathbf{b} \times \mathbf{c})$$

Now $(\mathbf{a} \times \mathbf{b}) \cdot \mathbf{a} = (-\mathbf{i} + 7\mathbf{j} + 5\mathbf{k}) \cdot (3\mathbf{i} - \mathbf{j} + 2\mathbf{k})$

$= -3 - 7 + 10 = 0$ ...(3)

and $\quad (\mathbf{a} \times \mathbf{b}) \cdot \mathbf{b} = (-\mathbf{i} + 7\mathbf{j} + 5\mathbf{k}) \cdot (2\mathbf{i} + \mathbf{j} - \mathbf{k})$

$= -2 + 7 - 5 = 0$ ...(4)

From (3) and (4), we conclude that $\mathbf{a} \times \mathbf{b}$ is perpendicular to both **a** and **b**

Example 30. If $\mathbf{a} = \mathbf{i} + \mathbf{j} + \mathbf{k}$ and $\mathbf{c} = \mathbf{j} - \mathbf{k}$ are given vectors find the vector **b** satisfying the equations $\mathbf{a} \times \mathbf{b} = \mathbf{c}$ and $\mathbf{a} \cdot \mathbf{b} = 3$.

Solution: Given $\mathbf{a} = \mathbf{i} + \mathbf{j} + \mathbf{k}$ and $\mathbf{c} = \mathbf{j} - \mathbf{k}$ let $\mathbf{b} = x\mathbf{i} + y\mathbf{j} + z\mathbf{k}$

Then $\quad \mathbf{a} \times \mathbf{b} = \begin{vmatrix} \mathbf{i} & \mathbf{j} & \mathbf{k} \\ 1 & 1 & 1 \\ x & y & z \end{vmatrix}$

$= (z - y)\mathbf{i} - (z - x)\mathbf{j} + (y - x)\mathbf{k}$

But $\quad \mathbf{a} \times \mathbf{b} = \mathbf{c}$

$\Rightarrow \quad (z - y)\mathbf{i} - (z - x)\mathbf{j} + (y - x)\mathbf{k} = \mathbf{j} - \mathbf{k}$

$\therefore \quad z - y = 0, \quad x - z = 1, \quad x - y = 1$

i.e., $\quad z = y; \quad z = x - 1; \quad y = x - 1$ ...(1)

Again $\mathbf{a} \cdot \mathbf{b} - 3 \Rightarrow (\mathbf{i} + \mathbf{j} + \mathbf{k}) \cdot (x\mathbf{i} + y\mathbf{j} + z\mathbf{k}) = 3$

or $\quad x + y + z = 3$

or $\quad x + x-1 + x-1 = 3 \quad$ or $\quad x = \dfrac{5}{3}$ [Using (1)]

$\therefore \quad y = z = x - 1 = \dfrac{5}{3} - 1 = \dfrac{2}{3}$

$\therefore \quad \mathbf{b} = \dfrac{5}{3}\mathbf{i} + \dfrac{2}{3}\mathbf{j} + \dfrac{2}{3}\mathbf{k}.$

Example 31. Find the area of the parallelogram determined by the vectors $\mathbf{i} - 2\mathbf{j} + 3\mathbf{k}$ and $-3\mathbf{j} - 2\mathbf{j} + \mathbf{k}$.

Solution: Let $\mathbf{a} = \mathbf{i} - 2\mathbf{j} + 3\mathbf{k}$ and $\mathbf{b} = -3\mathbf{i} - 2\mathbf{j} + \mathbf{k}$ be the adjacent sides of parallelogram. Then the required area = $|\mathbf{a} \times \mathbf{b}|$. Now

$$\mathbf{a} \times \mathbf{b} = \begin{vmatrix} \mathbf{i} & \mathbf{j} & \mathbf{k} \\ 1 & -2 & 3 \\ -3 & -2 & 1 \end{vmatrix} = 4\mathbf{i} - 10\mathbf{j} - 8\mathbf{k}$$

$$= 2\,(2\mathbf{i} - 5\mathbf{j} - 4\mathbf{k})$$

$$|\,\mathbf{a} \times \mathbf{b}\,| = 2\sqrt{4 + 25 + 16} = 2\sqrt{45} = 6\sqrt{5} \text{ sq. units.}$$

Example 32. Find the area of the triangle whose adjacent sides are determined by the vectors $\mathbf{i} + 4\mathbf{j} - \mathbf{k}$ and $\mathbf{i} + \mathbf{j} + 2\mathbf{k}$

Solution. Let $\mathbf{a} = \mathbf{i} + 4\mathbf{j} - \mathbf{k}$ and $\mathbf{b} = \mathbf{i} + \mathbf{j} + 2\mathbf{k}$ be the adjacent sides of triangle. Then area of the triangle = $\frac{1}{2}|\,\mathbf{a} \times \mathbf{b}\,|$. Now

$$\mathbf{a} \times \mathbf{b} = \begin{vmatrix} \mathbf{i} & \mathbf{j} & \mathbf{k} \\ 1 & 4 & -1 \\ 1 & 1 & 2 \end{vmatrix} = 9\mathbf{i} - 3\mathbf{j} - 3\mathbf{k}$$

$$= 3\,(3\mathbf{i} - \mathbf{j} - \mathbf{k})$$

$$\Delta = \frac{1}{2}|\,\mathbf{a} \times \mathbf{b}\,| = \frac{1}{2}\,.\,3\sqrt{9 + 1 + 1} = \frac{3\sqrt{11}}{2} \text{ sq. units.}$$

Example 33. Find the area of Δ ABC whose vertices are A (1, -1, 2), B (2, 1, -1) and C (3, -1, 2)

Solution: Let O be origin, then $\overrightarrow{OA} = \mathbf{i} - \mathbf{j} + 2\mathbf{k}$

$$\overrightarrow{OB} + 2\mathbf{i} + \mathbf{j} - \mathbf{k} \quad \text{and} \quad \overrightarrow{OC} = 3\mathbf{i} - \mathbf{j} + 2\mathbf{k}$$

Now $\quad \overrightarrow{AB} = \overrightarrow{OB} - \overrightarrow{OA} = (2\mathbf{i} + \mathbf{j} - \mathbf{k}) - (\mathbf{i} - \mathbf{j} + 2\mathbf{k})$

$$= \mathbf{i} + 2\mathbf{j} - 3\mathbf{k}$$

$$\overrightarrow{AC} = \overrightarrow{OC} - \overrightarrow{OA} = (3\mathbf{i} - \mathbf{j} + 2\mathbf{k}) - (\mathbf{i} - \mathbf{j} + 2\mathbf{k}) = 2\mathbf{i}$$

$$\overrightarrow{AB} \times \overrightarrow{AC} = \begin{vmatrix} \mathbf{i} & \mathbf{j} & \mathbf{k} \\ 1 & 2 & -3 \\ 2 & 0 & 0 \end{vmatrix} = -6\mathbf{j} - 4\mathbf{k}$$

$$\text{Area of } \Delta\,\text{ABC} = \frac{1}{2}|\,\overrightarrow{AB} \times \overrightarrow{AC}\,|$$

$$= \frac{1}{2}|-2(3\mathbf{j} + 2\mathbf{k})| = \frac{2}{2}\sqrt{9 + 4} = \sqrt{13} \text{ sq. units}$$

Example 34. Find the area of parallelogram ABCD having diagonals $\vec{AC}$ and $\vec{BD}$.

Calculate the area when $\vec{AC} = 3\mathbf{i} + \mathbf{j} - 2\mathbf{k}$ and $\vec{BD} = \mathbf{i} - 3\mathbf{j} + 4\mathbf{k}$

Solution. From geometry we know that area of the parallelogram ABCD = 4 times area of Δ AOB.

∴ Required area

$$= 4\frac{1}{2}|\vec{OA} \times \vec{OB}|$$

$$= 2\left|\frac{1}{2}\vec{CA} \times \frac{1}{2}\vec{DB}\right|$$

$$= \frac{1}{2}|\vec{CA} \times \vec{DB}|$$

$$= \frac{1}{2}|\vec{AC} \times \vec{BD}|$$

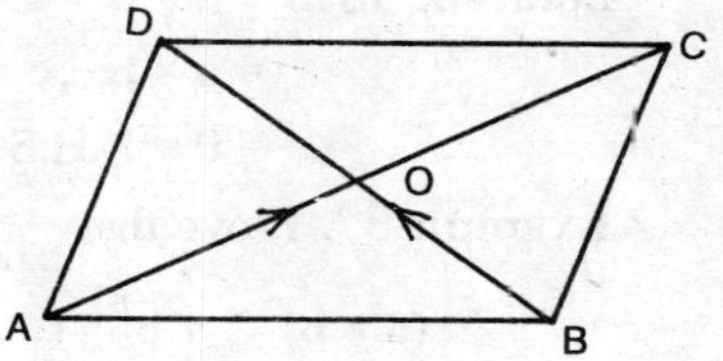

D C O A B

Fig. 2.21

$$\text{Now } \vec{AC} \times \vec{BD} = \begin{vmatrix} \mathbf{i} & \mathbf{j} & \mathbf{k} \\ 3 & 1 & -2 \\ 1 & -3 & 4 \end{vmatrix}$$

$$= -2\mathbf{i} - 14\mathbf{j} - 10\mathbf{k}$$

$$\therefore \text{ Area of parallelograms ABCD} = \frac{1}{2}|\vec{AC} \times \vec{BD}|$$

$$= \frac{1}{2}|-2\mathbf{i} - 14\mathbf{j} - 10\mathbf{k}|$$

$$= \frac{1}{2}\sqrt{4 + 196 + 100} = 5\sqrt{3} \text{ sq. units.}$$

Example 35. If $\mathbf{a} + \mathbf{b} + \mathbf{c} = 0$ prove that

$$\mathbf{a} \times \mathbf{b} = \mathbf{b} \times \mathbf{c} = \mathbf{c} \times \mathbf{a}.$$

Solution. If $\mathbf{a} + \mathbf{b} + \mathbf{c} = 0$ then $\mathbf{a} = -(\mathbf{b} + \mathbf{c})$

$$\therefore \quad \mathbf{a} \times \mathbf{b} = -(\mathbf{b} + \mathbf{c}) \times \mathbf{b}$$

$$= -\mathbf{b} \times \mathbf{b} - \mathbf{c} \times \mathbf{b}$$

$$= 0 + \mathbf{b} \times \mathbf{c} \qquad [\because\ \mathbf{b} \times \mathbf{b} = 0;\ -\mathbf{c} \times \mathbf{b} = \mathbf{b} \times \mathbf{c}]$$

$$\therefore \quad \mathbf{a} \times \mathbf{b} = \mathbf{b} \times \mathbf{c} \qquad \text{...(1)}$$

Similarly $\mathbf{c} \times \mathbf{a} = \mathbf{c} \times [-(\mathbf{b} + \mathbf{c})]$

$$= -\mathbf{c} \times (\mathbf{b} + \mathbf{c})$$

$$= -\mathbf{c} \times \mathbf{b} - \mathbf{c} \times \mathbf{c} + \mathbf{b} \times \mathbf{c}$$

$\therefore$ $\quad \mathbf{c} \times \mathbf{a} = \mathbf{b} \times \mathbf{c}$ ...(2)

Hence from (1) and (2) we get

$$\mathbf{a} \times \mathbf{b} = \mathbf{b} \times \mathbf{c} = \mathbf{c} \times \mathbf{a}.$$

Example 36. Prove that

$$\mathbf{a} \times (\mathbf{b} + \mathbf{c}) + \mathbf{b} \times (\mathbf{c} + \mathbf{a}) + \mathbf{c} \times (\mathbf{a} + \mathbf{b}) = 0$$

Solution. LHS $= \mathbf{a} \times \mathbf{b} + \mathbf{a} \times \mathbf{c} + \mathbf{b} \times \mathbf{c} + \mathbf{b} \times \mathbf{a} + \mathbf{c} \times \mathbf{a} + \mathbf{c} \times \mathbf{b}$

$= \mathbf{a} \times \mathbf{b} - \mathbf{c} \times \mathbf{a} + \mathbf{b} \times \mathbf{c} - \mathbf{a} \times \mathbf{b} + \mathbf{c} \times \mathbf{a} - \mathbf{b} \times \mathbf{c}$

$= 0 =$ R.H.S

Example 37. Prove that

$$(\mathbf{a} \times \mathbf{b})^2 = \mathbf{a}^2\mathbf{b}^2 - (\mathbf{a} \cdot \mathbf{b})^2 = \begin{vmatrix} \mathbf{a} \cdot \mathbf{a} & \mathbf{a} \cdot \mathbf{b} \\ \mathbf{a} \cdot \mathbf{b} & \mathbf{b} \cdot \mathbf{b} \end{vmatrix}$$

Solution. $(\mathbf{a} \times \mathbf{b})^2 = (\mathbf{a} \times \mathbf{b}) \cdot (\mathbf{a} \times \mathbf{b})$

$= (ab \sin \theta\, \hat{\mathbf{n}}) \cdot (ab \sin \theta\, \hat{\mathbf{n}})$

$= a^2b^2 \sin^2 \theta\, (\hat{\mathbf{n}} \cdot \hat{\mathbf{n}})$

$= a^2b^2 (1 - \cos^2 \theta).1$

$= a^2b^2 - a^2b^2 \cos^2 \theta$

$= a^2b^2 - (\mathbf{a} \cdot \mathbf{b})^2$ ...(1)

Again $\begin{vmatrix} \mathbf{a} \cdot \mathbf{a} & \mathbf{a} \cdot \mathbf{b} \\ \mathbf{a} \cdot \mathbf{b} & \mathbf{b} \cdot \mathbf{b} \end{vmatrix} = (\mathbf{a} \cdot \mathbf{a})(\mathbf{b} \cdot \mathbf{b}) - (\mathbf{a} \cdot \mathbf{b})^2$

$= a^2b^2 - (\mathbf{a}.\mathbf{b})^2$ ...(2)

Hence from (1) and (2) the proof follows.

Example 38. Show that $(\mathbf{a} - \mathbf{b}) \times (\mathbf{a} + \mathbf{b}) = 2\mathbf{a} \times \mathbf{b}$ and interpret it.

Solution. LHS $= \mathbf{a} \times \mathbf{a} + \mathbf{a} \times \mathbf{b} - \mathbf{b} \times \mathbf{a} - \mathbf{b} \times \mathbf{b}$

$= 0 + \mathbf{a} \times \mathbf{b} + \mathbf{a} \times \mathbf{b} - 0$

$= 2\mathbf{a} \times \mathbf{b} =$ R.H.S.

Interpretation Let ABCD be the parallelogram such that the diagonals

AC and BD intersect at O.

Let $\quad \overrightarrow{AO} = \mathbf{a}, \ \overrightarrow{OD} = \mathbf{b}$

$\therefore\ \overrightarrow{OB} = -\mathbf{b}$

$\overrightarrow{AD} = \overrightarrow{AO} + \overrightarrow{OD} = \mathbf{a} + \mathbf{b}$

$\overrightarrow{AB} = \overrightarrow{AO} + \overrightarrow{OB} = \mathbf{a} - \mathbf{b}$

$\therefore\ (\mathbf{a} - \mathbf{b}) \times (\mathbf{a} + \mathbf{b}) = \overrightarrow{AB} \times \overrightarrow{AD}$

= area of parallelogram ABCD

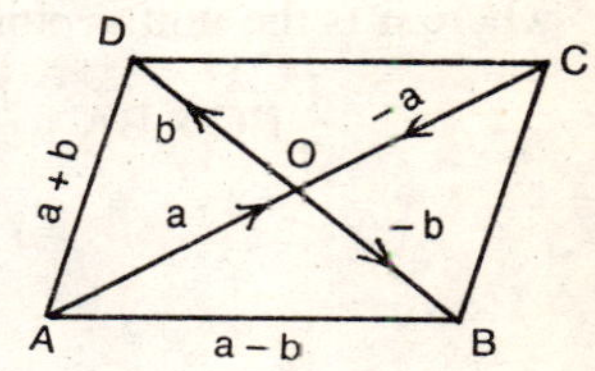

Fig. 2.22

Hence the given equation represents that area of parallelogram ABCD = 2 {area of parallelogram whose adjacent sides are the semi diagonals of the first parallelograms}.

Example 35. If **a**, **b**, **c** determine the vertices of a triangle show that $\frac{1}{2}$ [b × **c** + c × **a** + **a** × **b**] gives the vector area of the triangle. Hence deduce the condition that the three points **a**, **b**, **c** are collinear if [**b** × **c** + C × **a** + **a** × **b**] = 0. Also find the unit vector normal to the plane of the triangle.

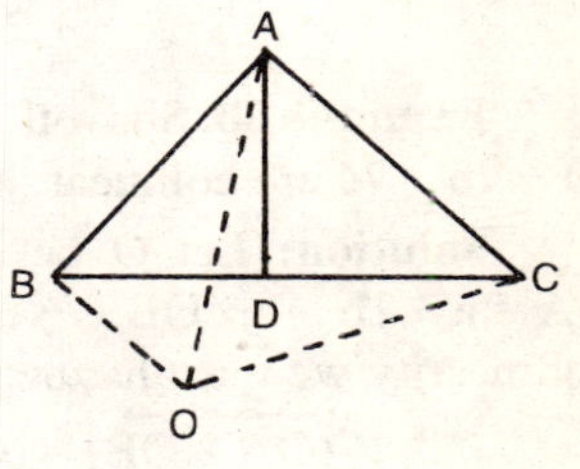

Fig. 2.23

Solution: Let O be the origin and $\overrightarrow{OA} = \mathbf{a}$, $\overrightarrow{OB} = \mathbf{b}$, $\overrightarrow{OC} = \mathbf{c}$ be the position vectors of the vertices of triangle ABC.

Then $\overrightarrow{BC} = \mathbf{c} - \mathbf{b}$ and $\overrightarrow{BA} = \mathbf{a} - \mathbf{b}$

The vector area of Δ ABC

$$= \frac{1}{2}\overrightarrow{BC} \times \overrightarrow{BA} = \frac{1}{2}(\mathbf{c} - \mathbf{b}) \times (\mathbf{a} - \mathbf{b})$$

$$= \frac{1}{2}[\mathbf{c} \times \mathbf{a} - \mathbf{c} \times \mathbf{b} - \mathbf{b} \times \mathbf{a} + \mathbf{b} \times \mathbf{b}]$$

$$= \frac{1}{2}[\mathbf{b} \times \mathbf{c} + \mathbf{c} \times \mathbf{a} + \mathbf{a} \times \mathbf{b}]$$

$$[\because\ \mathbf{b} \times \mathbf{b} = \mathbf{0},\ \mathbf{b} \times \mathbf{a} = -\mathbf{a} \times \mathbf{b}]$$

If the points A, B, C are collinear, then the vector area of Δ ABC = 0

$\Rightarrow \quad [\mathbf{b}\times\mathbf{c}+\mathbf{c}\times\mathbf{a}+\mathbf{a}\times\mathbf{b}]=0$

Again $\frac{1}{2}\vec{BC}\times\vec{BA}=\frac{1}{2}\overline{BC}\,.\,\overline{BA}\ \sin\angle ABC\,\hat{\mathbf{n}}$.

where $\hat{\mathbf{n}}$ is the unit vector perpendicular to the plane of Δ ABC

$$\therefore \quad \frac{1}{2}\vec{BC}\times\vec{BA}=\frac{1}{2}\overline{BC}\,.\,\overline{AD}\,\hat{\mathbf{n}}$$

$$=(\text{Area of }\Delta\text{ ABC})\,\hat{\mathbf{n}}$$

$$\therefore \quad \hat{\mathbf{n}}=\frac{\frac{1}{2}\vec{BC}\times\vec{BA}}{\text{Area of }\Delta\text{ ABC}}=\frac{\text{Vector area of }\Delta\text{ ABC}}{\text{Area of }\Delta\text{ ABC}}$$

$$=\frac{\frac{1}{2}(\mathbf{a}\times\mathbf{b}+\mathbf{b}\times\mathbf{c}+\mathbf{c}\times\mathbf{a})}{\left|\frac{1}{2}(\mathbf{a}\times\mathbf{b}+\mathbf{b}\times\mathbf{c}+\mathbf{c}\times\mathbf{a})\right|}$$

Example 40. Show that the points $\mathbf{a}-2\mathbf{b}-\mathbf{c}$, $-2\mathbf{a}+3\mathbf{b}-\mathbf{c}$ and $4\mathbf{a}-7\mathbf{b}+7\mathbf{c}$ are collinear, **a, b, c** being non-coplanar.

Solution: Let O be the origin and the position vectors be $\vec{DA}=\mathbf{a}-2\mathbf{b}-\mathbf{c}$, $\vec{OB}=-2\mathbf{a}+3\mathbf{b}-\mathbf{c}$ and $\vec{OC}=4\mathbf{a}-7\mathbf{b}+7\mathbf{c}$. For collinearity we must have

$$(\vec{OA}\times\vec{OB})=(\vec{OB}\times\vec{OC})+(\vec{OC}\times\vec{OA})=0 \qquad ...(1)$$

Now $(\vec{OA}\times\vec{OB})=(\mathbf{a}-2\mathbf{b}-\mathbf{c})\times(-2\mathbf{a}+3\mathbf{b}-\mathbf{c})$

$=-2(\mathbf{a}\times\mathbf{a})+3(\mathbf{a}\times\mathbf{b})-(\mathbf{a}\times\mathbf{c})+4(\mathbf{b}\times\mathbf{a})$

$-6(\mathbf{b}\times\mathbf{b})+2(\mathbf{b}\times\mathbf{c})+2(\mathbf{c}\times\mathbf{a})$

$-3(\mathbf{c}\times\mathbf{b})+(\mathbf{c}\times\mathbf{c})$

$$=-(\mathbf{a}\times\mathbf{b})-7(\mathbf{b}\times\mathbf{c})-5(\mathbf{c}\times\mathbf{a}) \qquad ...(2)$$

Similarly

$$(\vec{OB}\times\vec{OC})=2(\mathbf{a}\times\mathbf{b})+14(\mathbf{b}\times\mathbf{c})+10(\mathbf{c}\times\mathbf{a}) \qquad ...(3)$$

$$(\vec{OC}\times\vec{OA})=-(\mathbf{a}\times\mathbf{b})-7(\mathbf{b}\times\mathbf{c})-5(\mathbf{c}\times\mathbf{a}) \qquad ...(4)$$

Now adding (2), (3) and (4) we get (1). Hence the points are collinear.

Example 41. If $\mathbf{a}\times\mathbf{b}=\mathbf{a}\times\mathbf{c}$ show that either $\mathbf{a}=0$ or $\mathbf{b}=\mathbf{o}$ or and $\mathbf{b}-\mathbf{c}$ are parallel.

Solution: Given $\mathbf{a}\times\mathbf{b}=\mathbf{a}\times\mathbf{c}$

$$\Rightarrow \quad \mathbf{a}\times\mathbf{b}-\mathbf{a}\times\mathbf{c}=0 \text{ or } \mathbf{a}\times(\mathbf{b}-\mathbf{c})=0 \qquad ...(1)$$

Now (1) will hold good if either $\mathbf{a} = 0$ or $\mathbf{b} - \mathbf{c} = 0$

i.e., $\mathbf{b} = \mathbf{c}$ or $\mathbf{a}$ and $\mathbf{b} - \mathbf{c}$ are parallel.

$\Rightarrow \mathbf{b} - \mathbf{c} = \lambda\mathbf{a}$ or $\mathbf{b} = \mathbf{c} + \lambda\mathbf{a}$ where λ in any scalar.

Note. If $\mathbf{a} \times \mathbf{b} = \mathbf{a} \times \mathbf{c}$ then it does not mean that $\mathbf{b} = \mathbf{c}$. **b** differs from **c** by some vector parallel to **a** which may or may not be zero

Example 42. Show that the second vector **a** can be omitted from (i) $\mathbf{a} \times (\mathbf{a} + \mathbf{b})$ (ii) $\mathbf{a}^{-1} \times (\mathbf{a} + \mathbf{b})$

Solution. (i) $\mathbf{a} \times (\mathbf{a} + \mathbf{b}) = \mathbf{a} \times \mathbf{a} + \mathbf{a} \times \mathbf{b}$

$$= \mathbf{a} \times \mathbf{b} \qquad (\because \mathbf{a} \times \mathbf{a} = 0)$$

Hence the second vector **a** can be omitted

(ii) $\mathbf{a}^{-1}$ stands for a vector whose modulus is reciprocal of the modulus of vector **a** and is in the direction of **a**

$\therefore$ If $\mathbf{a} = a\,\hat{\mathbf{a}}$ then $\mathbf{a}^{-1} = \frac{1}{a}\hat{\mathbf{a}} = \frac{1}{a^2}a\,\hat{\mathbf{a}} = \frac{\mathbf{a}}{a^2}$

$$\text{Hence} \quad \mathbf{a}^{-1} \times (\mathbf{a} + \mathbf{b}) = \frac{\mathbf{a}}{a^2} \times (\mathbf{a} + \mathbf{b})$$

$$= \frac{1}{a^2}[\mathbf{a} \times \mathbf{a} + \mathbf{a} \times \mathbf{b}]$$

$$= \frac{1}{a^2}(\mathbf{a} \times \mathbf{b}) \quad [\because \mathbf{a} \times \mathbf{a} = 0]$$

$$= \frac{\mathbf{a}}{a^2} \times \mathbf{b} = \mathbf{a}^{-1} \times \mathbf{b}$$

Hence here also the second vector ***a*** can be omitted.

Example 43. Show that $\frac{1}{2}\overrightarrow{AC} \times \overrightarrow{BD}$ represents the vector area of the plane quadrilateral.

Solution: Let ABCD be the quadrilateral and O be the intersection of diagonals $\overrightarrow{AC}$ and $\overrightarrow{BD}$. Now

$$\overrightarrow{AC} \times \overrightarrow{BD} = (\overrightarrow{AO} + \overrightarrow{OC}) \times (\overrightarrow{BO} + \overrightarrow{OD})$$

$$= \overrightarrow{AO} \times \overrightarrow{BO} + \overrightarrow{AO} \times \overrightarrow{OD} + \overrightarrow{OC} \times \overrightarrow{BO} + \overrightarrow{OC} \times \overrightarrow{OD}$$

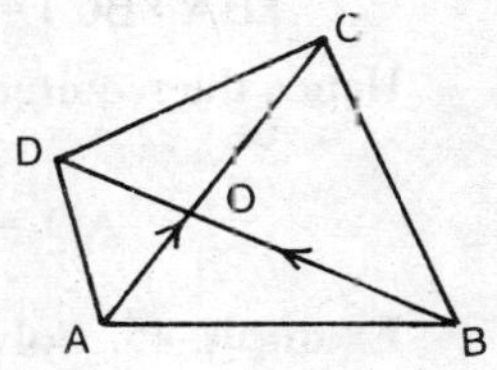

Fig. 2.24

$$= \vec{OA} \times \vec{OB} + \vec{OD} \times \vec{OA} + \vec{OB} \times \vec{OC} + \vec{OC} \times \vec{OD}$$

$$= 2\Delta\, OAB + 2\Delta\, ODA + 2\Delta OBC + 2\Delta OCD$$

$$= 2\{\text{area of quadrilateral ABCD}\}$$

or $$\frac{1}{2}\, \vec{AC} \times \vec{BD} = \text{ area of ABCD}$$

Example 44. By vector method find the perpendicular distance of A(2, 1, 3) from the line through the points B(4, 1, 2) and C(3, 2, – 4).

Solution: W.r.t O let

$\mathbf{a} = \vec{OA} = 2\mathbf{i} + \mathbf{j} + 3\mathbf{k}$

$\mathbf{b} = \vec{OB} = 4\mathbf{i} + \mathbf{j} + 2\mathbf{k}$ and

$\mathbf{c} = \vec{OC} = 3\mathbf{i} + 2\mathbf{j} - 4\mathbf{k}$

be the P.V.'s of A, B and C. Now AD is the required distance

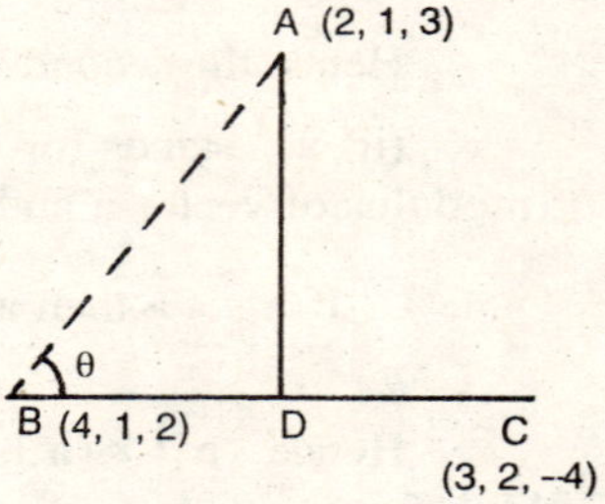

A (2, 1, 3)
θ
B (4, 1, 2) D C (3, 2, –4)

Fig. 2.25

From ΔADB we have

$$AD = BA \sin\theta = \frac{BA \times BC \sin\theta}{BC}$$

$$= \frac{|\vec{BA} \times \vec{BC}|}{|\vec{BC}|} \qquad \text{...(1)}$$

Now $\vec{BA} = \vec{OA} - \vec{OB} = -2\mathbf{i} + \mathbf{k};$

$\vec{BC} = \vec{OC} - \vec{OB} = -\mathbf{i} + \mathbf{j} - 6\mathbf{k};$

$|\vec{BC}| = \sqrt{1 + 1 + 36} = \sqrt{38}$

$$\vec{BA} \times \vec{BC} = \begin{vmatrix} \mathbf{i} & \mathbf{j} & \mathbf{k} \\ -2 & 0 & 1 \\ -1 & 1 & 6 \end{vmatrix} = -\mathbf{i} + 11\mathbf{j} - 2\mathbf{k}$$

$$|\vec{BA} \times \vec{BC}| = \sqrt{1 + 121 + 4} = \sqrt{126}$$

Hence the required perpendicular distance

$$AD = \frac{\sqrt{126}}{\sqrt{38}} = \sqrt{\frac{63}{19}} \text{ units.}$$

Example 45. Solve the simultaneous equations $\mathbf{r} \times \mathbf{a} = \mathbf{a} \times \mathbf{b}$ and $\mathbf{r.c} = 0$ where $\mathbf{b}$ and $\mathbf{c}$ are not orthogonal to each other.

Solution: Given $\mathbf{r} \times \mathbf{a} = \mathbf{a} \times \mathbf{b} \Rightarrow \mathbf{r} \times \mathbf{b} - \mathbf{a} \times \mathbf{b} = 0$

$\Rightarrow \quad (\mathbf{r} - \mathbf{a}) \times \mathbf{b} = 0$

$\Rightarrow \quad (\mathbf{r} - \mathbf{a})$ is parallel to $\mathbf{b}$

$\Rightarrow \quad \mathbf{r} - \mathbf{a} = \lambda\mathbf{b}$ where λ is some scalar

$\therefore \quad \mathbf{r} \times \mathbf{a} = \mathbf{a} \times \mathbf{b} \Rightarrow \mathbf{r} = \mathbf{a} + \lambda\mathbf{b}$ ...(1)

Now $\mathbf{r} \cdot \mathbf{c} = 0$ (given)

$\therefore \quad (\mathbf{a} + \lambda\mathbf{b}) \cdot \mathbf{c} = 0 \Rightarrow \mathbf{a} \cdot \mathbf{c} + \lambda\mathbf{b} \cdot \mathbf{c} = 0$

$\Rightarrow \quad \lambda = -\dfrac{\mathbf{a} \cdot \mathbf{c}}{\mathbf{b} \cdot \mathbf{c}}$ $[\because \mathbf{b}.\mathbf{c} \neq 0]$

Substituting in (1)

$$\mathbf{r} = \mathbf{a} - \frac{\mathbf{a} \cdot \mathbf{c}}{\mathbf{b} \cdot \mathbf{c}}\mathbf{b}$$

Example 46. Show that in any triangle ABC

$$\frac{\sin A}{a} = \frac{\sin B}{b} = \frac{\sin C}{c}$$

Solution: In the triangle ABC, let $\overrightarrow{BC} = \mathbf{a}$, $\overrightarrow{CA} = \mathbf{b}$ and $\overrightarrow{AB} = \mathbf{c}$.

Then BC = a , CA = b and AB = c

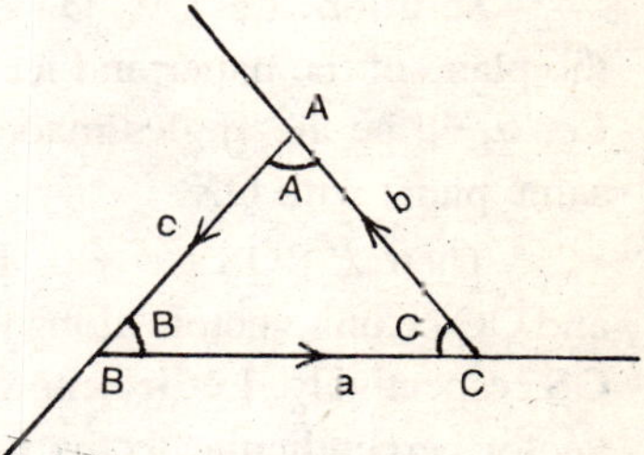

Fig. 2.26

Then, we have

$\overrightarrow{BC} + \overrightarrow{CA} + \overrightarrow{AB} = \overrightarrow{BB} = 0$

i.e., $\quad \mathbf{a} + \mathbf{b} + \mathbf{c} = \mathbf{0}$

$\therefore \quad \mathbf{a} + \mathbf{b} = -\mathbf{c}$

or $\quad \mathbf{a} \times (\mathbf{a} + \mathbf{b}) - \mathbf{a} \times (-\mathbf{c}) = -\mathbf{a} \times \mathbf{c}$

$\therefore \quad \mathbf{a} \times \mathbf{a} + \mathbf{a} \times \mathbf{b} = \mathbf{c} \times \mathbf{a}$

or $\quad \mathbf{a} \times \mathbf{b} = \mathbf{c} \times \mathbf{a}$ ...(1)

Similarly $\mathbf{b} \times (\mathbf{a} + \mathbf{b}) = \mathbf{b} \times (-\mathbf{c})$

$\Rightarrow \quad \mathbf{a} \times \mathbf{b} = \mathbf{b} \times \mathbf{c}$

$\therefore$ From (1) and (2), we have

$\mathbf{a} \times \mathbf{b} = \mathbf{b} \times \mathbf{c} = \mathbf{c} \times \mathbf{a}$ ....(3)

and also $\quad |\mathbf{a} \times \mathbf{b}| = |\mathbf{b} \times \mathbf{c}| = |\mathbf{c} \times \mathbf{a}|$

Now consider

$\mathbf{a} \times \mathbf{b} = \mathbf{b} \times \mathbf{c}$

i.e., $ab \sin(\pi - C) = bc \sin(\pi - A)$

or $\quad ab \sin C = bc \sin A$

or $$\frac{\sin C}{c} = \frac{\sin A}{a} \qquad ...(4)$$

Again consider

$$\mathbf{a} \times \mathbf{b} = \mathbf{c} \times \mathbf{a}$$

i.e., $\quad ab \sin (\pi - C) = ca \sin (\pi - B)$

or $\quad ab \sin C = ca \sin B$

or $$\frac{\sin C}{c} = \frac{\sin B}{b} \qquad ...(5)$$

Hence from (4) and (5), we get

$$\frac{\sin A}{a} = \frac{\sin B}{b} = \frac{\sin C}{c}$$

Example 47. Prove by vector methods that

$$\sin (\alpha + \beta) = \sin \alpha \cos \beta + \cos \alpha \sin \beta$$

Solution: Let OX, OY be two mutually perpendicular lines in the plane of the paper and let **i** and **j** be the unit vectors along them. Let $\alpha, -\beta$ be the angles made by two more lines OR and OS in the same plane with OX.

Then $\angle$ ROS = $\alpha + \beta$. Let $\overrightarrow{OP}$ and $\overrightarrow{OQ}$ be unit vectors along OR and OS respectively. Let **k** denote a unit vector perpendicular to the **i-j** plane such that **i, j, k** form a right handed triad. Then

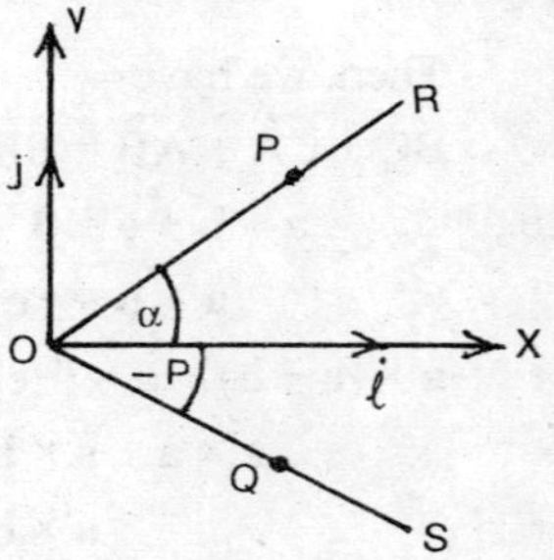

$\overrightarrow{OP} = \cos \alpha\, \mathbf{i} + \sin \alpha\, \mathbf{j}$; $\quad |\overrightarrow{OP}| = 1$

and

$\overrightarrow{OQ} = \cos \beta\, \mathbf{i} = \sin \beta\, \mathbf{j}$; $|\overrightarrow{OQ}| = 1$

Fig. 2.27

$$\text{Now } \overrightarrow{OQ} \times \overrightarrow{OP} = \begin{vmatrix} \mathbf{i} & \mathbf{j} & \mathbf{k} \\ \cos \beta & -\sin \beta & 0 \\ \cos \alpha & \sin \alpha & 0 \end{vmatrix}$$

$$= (\sin \alpha \cos \beta + \sin \beta \cos \alpha)\, \mathbf{k} \qquad ...(1)$$

$$\text{But } \overrightarrow{OQ} \times \overrightarrow{OP} = |\overrightarrow{OQ}|\,|\overrightarrow{OP}| \sin (\alpha + \beta)\, \mathbf{k}$$

$$= \sin (\alpha + \beta)\, \mathbf{k} \qquad ...(2)$$

Hence from (1) and (2), we get

$$\sin(\alpha + \beta) = \sin\alpha \cos\beta + \sin\beta \cos\alpha.$$

Example 48. Prove by vector method the parallelogram on the same base and between same parallels are equal in area.

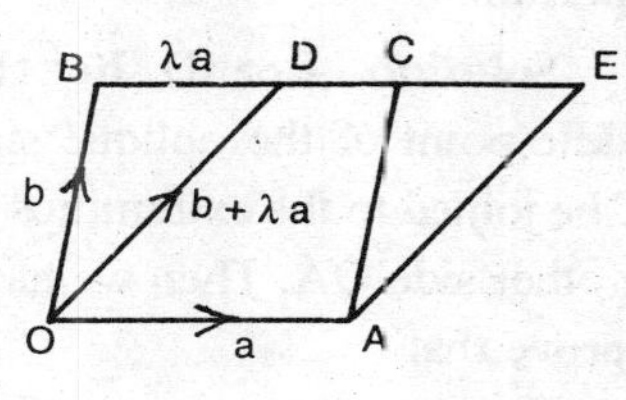

Fig. 2.28

Solution: Let OACB and OAED be the parallelograms on the same base OA, between the same parallels. Let $\overrightarrow{OA} = \mathbf{a}$ and $\overrightarrow{OB} = \mathbf{b}$.

$\therefore \overrightarrow{BC} = \mathbf{a}$ and $\overrightarrow{BD}$ is collinear with $\overrightarrow{BC}$.

$\therefore \quad \overrightarrow{BC} = \lambda\mathbf{a}$ and $\overrightarrow{OD} = \overrightarrow{OB} + \overrightarrow{BD} = \mathbf{b} + \lambda\mathbf{a}$

$\therefore$ Vector area of parallelogram OACB

$$= \overrightarrow{OA} \times \overrightarrow{OB} = \mathbf{a} \times \mathbf{b} \qquad ...(1)$$

Vector area of parallelogram OAED

$$= \overrightarrow{OA} \times \overrightarrow{OD} = \mathbf{a} \times (\mathbf{b} + \lambda\mathbf{a})$$

$$= \mathbf{a} \times \mathbf{b} + \lambda\,(\mathbf{a} \times \mathbf{a})$$

$$= \mathbf{a} \times \mathbf{b} \qquad ...(2)$$

Hence from (1) and (2) we get that parallelogram OACB is equal to parallelogram OAED.

Example 49. Find the perpendicular distance from C to the straight line through A and B where **a, b, c** are the position vectors of the points A, B, C.

Solution: Let O be the origin then $\overrightarrow{OA} = \mathbf{a}$, $\overrightarrow{OB} = \mathbf{b}$ and $\overrightarrow{OC} = \mathbf{c}$. Let CM be the perpendicular distance from C onto AB. Let θ be the angle between the lines AB and AC.

Then CM = AC sin θ

$$CM = AC\sin\theta = \frac{AB\ AC\sin\theta}{AB} = \frac{|\overrightarrow{AB} \times \overrightarrow{AC}|}{|\overrightarrow{AB}|}$$

$$= \frac{|(\mathbf{b} - \mathbf{a}) \times (\mathbf{c} - \mathbf{a})|}{|\mathbf{b} - \mathbf{a}|} = \frac{|\mathbf{b} \times \mathbf{c} + \mathbf{c} \times \mathbf{a} + \mathbf{a} \times \mathbf{b}|}{|\mathbf{b} - \mathbf{a}|}$$

which is the required perpendicular distance.

Example 50. Prove that the area of the triangle formed by joining the middle points of one of the non-parallel sides of a trapezium to the extremities of the opposite side is half that of trapezium.

Solution. Let D be the middle point of the oblique side BC be joined to the extremities of the other side OA. Then we have to prove that

$$\Delta\, \text{OAD} = \frac{1}{2}\ \text{area OABC.}$$

Let $\overrightarrow{OA} = \mathbf{a},\ \overrightarrow{AB} = \mathbf{b}$

Then $\quad \overrightarrow{OB} = \overrightarrow{OA} + \overrightarrow{AB} = \mathbf{a} + \mathbf{b}$

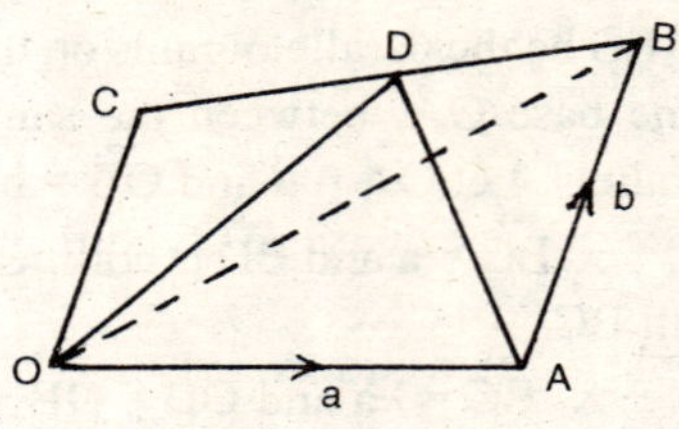

Fig. 2.29

Now $\overrightarrow{OC}$ being parallel to $\overrightarrow{AB}$

$\therefore \quad \overrightarrow{OC} = \lambda\, \overrightarrow{AB} = \lambda\mathbf{b} \qquad (\lambda$ is some scalar)

Now D is the midpoint of BC

$$= \frac{1}{2}(\mathbf{a} + \mathbf{b} + \lambda\mathbf{b})$$

$$\therefore \text{ Vector area of } \Delta\,\text{OAD} = \frac{1}{2}\overrightarrow{OA} \times \overrightarrow{OD}$$

$$= \frac{1}{2}\mathbf{a} \times \frac{1}{2}(\mathbf{a} + \mathbf{b} + \lambda\mathbf{b})$$

$$= \frac{1}{4}(\lambda + 1)\,\mathbf{a} \times \mathbf{b} \qquad \text{...(1)}$$

$$\text{Vector area of } \Delta\,\text{OAB} = \frac{1}{2}\overrightarrow{OA} \times \overrightarrow{OB}$$

$$= \frac{1}{\mathbf{a}} \times (\mathbf{a} + \mathbf{b}) = \frac{1}{2}\mathbf{a} \times \mathbf{b} \qquad \text{...(2)}$$

$$\text{Vector area of } \Delta\,\text{OBC} = \frac{1}{2}\overrightarrow{OB} \times \overrightarrow{OC}$$

$$= \frac{1}{2}(\mathbf{a} + \mathbf{b}) \times \lambda\mathbf{b} = \frac{1}{2}\lambda\mathbf{a} \times \mathbf{b} \qquad \text{...(3)}$$

$\therefore$ Vector area of the trapezium OABC

$$= \text{vector area of } \Delta\, OAB + \text{ vector area of } \Delta\, OBC$$

$$= \frac{1}{2}\mathbf{a} \times \mathbf{b} + \frac{1}{2}\lambda\mathbf{a} \times \mathbf{b} = \frac{1}{2}(\lambda + 1)\,\mathbf{a} \times \mathbf{b}$$

$$= 2\left[\frac{1}{4}(\lambda + 4)\,\mathbf{a} \times \mathbf{b}\right]$$

$$= 2 \text{ vector area of } \Delta\, OAD \qquad \text{[From (1)]}$$

Hence the proof.

EXERCISE

1. If $\mathbf{a} = 2\mathbf{j} - \mathbf{j} + \mathbf{k}$ and $\mathbf{b} = \mathbf{i} - 3\mathbf{j} - 5\mathbf{k}$ find $\mathbf{a} \times \mathbf{b}$
2. If $\mathbf{a} = 2\mathbf{i} + 3\mathbf{j} + 6\mathbf{k}$, $\mathbf{b} = 3\mathbf{i} - 6\mathbf{j} + 2\mathbf{k}$ and $c = 6\mathbf{i} + 2\mathbf{j} - 3\mathbf{k}$ show that $a \times \mathbf{b} = 7\mathbf{c}$
3. If $\mathbf{a} = 4\mathbf{i} + 3\mathbf{j} - \mathbf{k}$, $\mathbf{b} = 2\mathbf{i} + \mathbf{j} + 3\mathbf{k}$ and $\mathbf{c} = \mathbf{i} - \mathbf{j} + 2\mathbf{k}$ then find $(\mathbf{a} - \mathbf{b}) \times (\mathbf{b} + \mathbf{c})$
4. If $\mathbf{a} = 2\mathbf{i} - 3\mathbf{j} - \mathbf{k}$ and $\mathbf{b} = \mathbf{i} + 4\mathbf{j} - 2\mathbf{k}$ then find $(\mathbf{a} + \mathbf{b}) \times (\mathbf{a} - \mathbf{b})$.
5. The position vectors of the points A, B, C, D are respectively $\mathbf{a} = \mathbf{i} + \mathbf{j} + \mathbf{k}$, $\mathrm{b} = 2\mathbf{i} + 3\mathbf{j}$, $\mathrm{c} = 3\mathbf{i} + 5\mathbf{j} - 2\mathrm{k}$ and $\mathbf{d} = \mathrm{k} - \mathbf{j}$. Prove that AB is parallel to CD
6. If $\mathbf{a} = 2\mathbf{i} + 3\mathbf{j} + 4\mathbf{k}$, $\mathrm{b} = \mathbf{i} + \mathbf{j} - \mathbf{k}$ and $\mathrm{c} = \mathbf{i} - \mathbf{j} + \mathbf{k}$ then find $\mathbf{a} \times (\mathbf{b} \times \mathbf{c})$. Also show that $\mathbf{a} \times (\mathbf{b} \times \mathbf{c})$ is perpendicular to both $\mathbf{a}$ and $\mathbf{b} \times \mathbf{c}$.
7. If $\mathbf{a} = 3\mathbf{i} + 4\mathbf{j} - \mathbf{k}$, $\mathbf{b} = \mathbf{i} - 2\mathbf{j} - \mathbf{k}$ and $\mathbf{c} = (2\mathbf{i} - \mathbf{j} + 3\mathbf{k})$ find $(\mathbf{a} \times \mathbf{b}) \,.\, (\mathbf{a} \times \mathbf{c})$ and $(\mathbf{a} \times \mathbf{b}) \times (\mathbf{b} \times \mathbf{c})$
8. If $\mathbf{a} = 3\mathbf{i} - \mathbf{j} + 2\mathbf{k}$, $\mathbf{b} = 2\mathbf{i} + \mathbf{j} - \mathrm{k}$, $\mathbf{c} = \mathbf{i} - 2\mathbf{j} + 2\mathbf{k}$ show that $(\mathbf{a} \times \mathbf{b}) \times \mathbf{c} = \mathbf{a} \times (\mathbf{b} \times \mathbf{c})$
9. If $\mathbf{a} = \mathbf{i} + 2\mathbf{j} - \mathbf{k}$, $\mathbf{b} = 2\mathbf{i} + \mathbf{j} + 3\mathbf{k}$, $\mathrm{c} = \mathbf{i} - \mathbf{j} + \mathbf{k}$ and $\mathbf{d} = 3\mathbf{i} + \mathbf{j} + 2\mathbf{k}$ then evaluate $(\mathbf{a} \times \mathbf{b}) \,.\, (\mathbf{c} \times \mathbf{d})$ and $(\mathbf{a} \times \mathbf{b}) \times (\mathbf{c} \times \mathbf{d})$
10. Find a unit vector perpendicular to each of the vectors $\mathbf{i} + \mathbf{j} + \mathrm{k}$ and $2\mathbf{i} + \mathbf{j} + 3\mathbf{k}$
11. Find a unit vector perpendicular to the plane determined by $(1, -1, 2)$, $(2, 0, -1)$ and $(0, 2, 1)$

12. Find a unit vector perpendicular to each of the vectors $4\mathbf{i} + 3\mathbf{j} - \mathbf{k}$ and $2\mathbf{i} - 6\mathbf{j} - 3\mathbf{k}$. Also find the sine angle between them.

13. Find a unit vector perpendicular to each of the vectors $\mathbf{i} + 3\mathbf{j} - 2\mathbf{k}$ and $2\mathbf{i} - \mathbf{j} - \mathbf{k}$. Also find the sine angle between them

14. Find the area of the parallelogram determined by the vectors

(i) $3\mathbf{i} + \mathbf{j} - 2\mathbf{k}$, $2\mathbf{i} + 3\mathbf{j} - 4\mathbf{k}$

(ii) $\mathbf{i} - 7\mathbf{j} - 6\mathbf{k}$, $5\mathbf{i} - \mathbf{j} + 7\mathbf{k}$

15. Find the area of the triangle whose adjacent sides are determined by the vectors

(i) $\mathbf{i} - 3\mathbf{j} + 5\mathbf{k}$, $3\mathbf{i} + \mathbf{j} - \mathbf{k}$

(ii) $6\mathbf{i} - 7\mathbf{j} + 9\mathbf{k}$, $2\mathbf{i} - 8\mathbf{j} + 3\mathbf{k}$

16. Find the area of the parallelogram having $\mathbf{i} - \mathbf{j} + 3\mathbf{k}$ and $2\mathbf{i} + 7\mathbf{j} - 3\mathbf{k}$ as its diagonals

17. Find the area of triangle whose vertices are given by

(i) A (1, 2, 3), B(2, 3, 1), C(3, 1, 2)

(ii) P (1, 1, 2), Q(2, -1, 1), R(3, 2, -1)

18. Prove that the points given below are collinear

(i) $\mathbf{i} + 2\mathbf{j} + 3\mathbf{k}$, $-2\mathbf{i} + 3\mathbf{j} + 5\mathbf{k}$, $7\mathbf{i} - \mathbf{k}$

(ii) $\mathbf{a} - 2\mathbf{b} + 3\mathbf{c}$, $2\mathbf{a} + 3\mathbf{b} - 4\mathbf{c}$, $-7\mathbf{b} + 10\mathbf{c}$

(iii) $\mathbf{a} - 2\mathbf{b} + 3\mathbf{c}$, $-2\mathbf{a} + 3\mathbf{b} - \mathbf{c}$, $4\mathbf{a} - 7\mathbf{b} + 7\mathbf{c}$

19. Solve simultaneously the equations $\mathbf{b} \times \mathbf{r} = \mathbf{b} \times \mathbf{c}$ and $\mathbf{a} \cdot \mathbf{r} = \lambda$ for $\mathbf{r}$. When the solution is impossible.

20. If $\mathbf{a} = 4\mathbf{i} + 5\mathbf{j} - \mathbf{k}$, $\mathbf{b} = \mathbf{i} - 4\mathbf{j} + 5\mathbf{k}$, $\mathbf{c} = 3\mathbf{i} + \mathbf{j} - \mathbf{k}$ then find a vector $\mathbf{d}$ such that $\mathbf{a} \cdot \mathbf{d} = \mathbf{b} \cdot \mathbf{d} = 0$ and $\mathbf{d} \cdot \mathbf{c} = 21$.

21. Show that we cannot omit the second vector **a** in

$$\mathbf{a} \times (\mathbf{a} + \mathbf{b})^{-1}$$

22. Prove by vector method that

$$\sin(\alpha - \beta) = \sin\alpha\cos\beta - \cos\alpha\sin\beta$$

23. Show that the perpendicular distance from any point **a** from the line $\mathbf{r} = \mathbf{b} + t\,\mathbf{c}$ is $\dfrac{|(\mathbf{b} - \mathbf{a}) \times \mathbf{c}|}{|\mathbf{c}|}$

24. By vector method obtain the perpendicular distance of the point (5, 5, 5) from the line through the points (3, 4, –1) and (1, 3, 1).

25. If AD, BE, CF are the medians of a triangle ABC meeting at G then show that

$$\Delta\, AGB = \Delta\, BGC = \Delta\, CGA = \frac{1}{3}\Delta\, ABC$$

26. If P, Q, R are the mid points of the sides of triangle ABC then show that Δ ABC = 4Δ PQR

27. If the vector product of a constant vector $\overrightarrow{OA}$ with a variable vector $\overrightarrow{OB}$ in a fixed plane OAB be a constant vector, show that locus of B is a straight line parallel to OA.

28. In the sides BC, CA, AB of a triangle three points D, E, F are taken such that BD, CE, AF is equal to $\frac{1}{3}$ of the corresponding side prove that 3Δ DEF = Δ ABC

ANSWERS

(1) $8\mathbf{i} + 11\mathbf{j} - 5\mathbf{k}$ (3) $10\mathbf{i} - 22\mathbf{j} - 6\mathbf{k}$

(4) $-20\mathbf{i} - 6\mathbf{j} -$‘$22\mathbf{k}$ (7) $176, -22\mathbf{i} + 44\mathbf{j} - 22\mathbf{k}$

(9) $-38, -(17\mathbf{i} + 19\mathbf{j} + 8\mathbf{k})$ (10) $\frac{1}{\sqrt{6}}(2\mathbf{i} + \mathbf{j} + d\,\mathbf{k})$

(11) $\frac{1}{\sqrt{6}}(2\mathbf{i} + \mathbf{j} + \mathbf{k})$ (12) $-\frac{3}{7}\mathbf{i} + \frac{2}{7}\mathbf{i} - \frac{6}{7}\mathbf{k}, \frac{5}{\sqrt{26}}$

(13) $\frac{1}{\sqrt{83}}(5\mathbf{i} + 3\mathbf{j} + 7\mathbf{k})$ (14) (i) $\sqrt{117}$ (ii) $\sqrt{1491}$

(15) (i) $3\sqrt{10}$ (ii) $\frac{1}{2}\sqrt{2757}$ (16) $\frac{9\sqrt{6}}{2}$

(17) (i) $\frac{3\sqrt{3}}{2}$ (ii) $\frac{1}{2}\sqrt{107}$

(19) $\mathbf{r} = \mathbf{c} + \left\{\frac{\lambda - \mathbf{a}\,.\,\mathbf{c}}{\mathbf{a}\,.\,\mathbf{b}}\right\}\mathbf{b}$, **a** is perpendicular to **b**

(20) $7(\mathbf{i} - \mathbf{j} - \mathbf{k})$ (24) $\frac{8\sqrt{5}}{3}$.

3

Multiple Products

3.1. Triple products

If **a** and **b** are two vectors then we know that **a** × **b** is a vector quantity. Hence we can multiply it by another vector **c** both scalarly and vectorially i.e., as (**a** × **b**) . **c** or as (**a** × **b**) × **c**. The former (**a** × **b**) . c. is called **scalar triple product** which is purely a number where as the later (**a** × **b**) × **c** is called **vector triple product** which is again a vector quantity.

Since **a . b** is a scalar quantity, the products (**a** **b**) . **c** and (**a** . **b**) × **c** are meaning less. But (**a** . **b**)**c** represents a vector in the direction of **c** and whose modulus is (**a** .**b**) times that of **c**

3.2. Scalar triple product : Geometrical interpretation

Let OA, OB, OC be three coterminous edges of a rectangular parallelopiped representing the vectors **a**, **b** and **c** respectively both in magnitude and direction. Then the vector **a** × **b** represents a vector whose magnitude is the area of the parallelogram OABD. Let **a** × **b** = $\hat{\mathbf{n}}$, where $\hat{\mathbf{n}}$ be unit vector perpendicular to **a** and **b** such that **a**, **b** and $\hat{\mathbf{n}}$ form a right handed system as shown in the figure. If θ be the angle between the direction **a** × **b** and **c** i.e., $\hat{\mathbf{n}}$ and **c** then

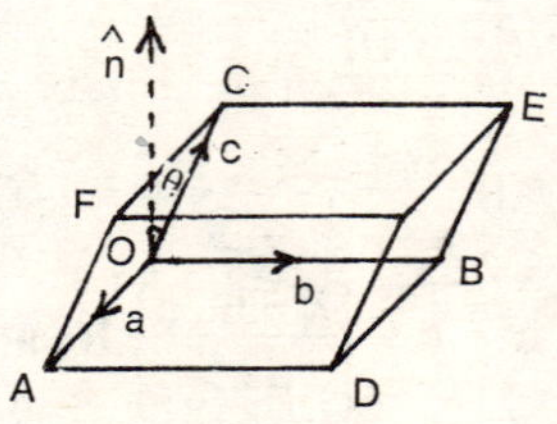

Fig. 3.1

$$(\mathbf{a} \times \mathbf{b}) \,.\, \mathbf{c} = \hat{\mathbf{n}} \,.\, \mathbf{c} = (\text{area of OABD})\ \overline{OC} \cos \theta$$

= (area of base) (height of the parallelopiped)

= Volume of the parallelopiped.

∴ *The scalar triple product* (**a** × **b**) . **c** *geometrically represents volume of a parallelopiped whose coterminous edges are the vectors* **a, b** and **c.**

One has to note that the volume will be positive if θ is acute i.e., **A**, **b** and **c** form a right handed system otherwise it will be negative i.e., if **a**, **b** and **c** form a left handed system.

Similarly (**b** × **c**) . **a** and (**c** × **a**) . **b** also represents the volume of the parallelopiped whose coterminous edges are the vectors **a, b** and **c**.

The scalar triple product (**a** × **b**) . **c** is usually written as [**a b c**] called **box product.** Similarly (**b** × **c**) . **a** as [**b c a**] and (**c** × **a**) . **b** as [**c a b**].

Note: (i) (**a** × **b**) . **c** = **c** . (**a** × **b**). . .

(ii) (**a** × **b**) . **c** = (**b** × **c**) . **a** = (**c** × **a**) . **b**

and *each of them = volume of the parallelopiped*

(iii) (**a** × **b**) . **c** = − (**b** × **a**) . **c**

(iv) (**a** × **b**) . **c** = **a** . (8**b** × **c**) i.e., *in a triple scalar product the dot and cross operations can be interchangeable*

(v) [**abc**] = − [**bac**] = [**cab**] = − [**cba**] ... i.e., the cyclic order of the factor can also be changed provided the sign in changed.

(vi) **i** . (**j** × **k**) = **i** . **i** = 1 or simply [**ijk**] = 1

3.3. Distributive law for vector product

i.e., *to prove that* **a** × (**b** + **c**) = **a** × **b** + **a** × **c**

Let **r** = **a** × (**b** + **c**) − **a** × **b** − **a** × **c** and **d** be any arbitrary vector. Then

d . **r** = **d** . [**a** × (**b** + **c**) − **a** × **b** − **a** × **c**]

= **d** . [**a** × (**b** + **c**)] − **d** . (**a** × **d**) − **d** . (**a** × **c**)

= (**d** × **a**) . (**b** + **c**) − (**d** × **a**) . **b** − (**d** × **a**) . **c**

[∴ dot and cross products can be interchangeable in triple scalar product]

= (**d** × **a**) . **b** + (**d** × **a**) . **c** − (**d** × **a**) . **b** − (**d** × **a**) . **c**

$= 0$

$\therefore$ **d** . **r** = 0 $\Rightarrow$ either **d** = 0 or **r** = 0 or **d** is perpendicular to **r**. But **d** is arbitrary $\Rightarrow$ **d** may not be a zero vector

$\Rightarrow$ **r = 0 $\Leftrightarrow$ a × (b + c) = a × b + a × c**

Hence the proof.

3.4. Properties The value of a triple scalar product is zero if two of its vectors are equal or parallel

Case (i) *If two vectors are equal.*

Then **a . (a × b) = (a × a) . b** = 0 or **[a a b]** = 0

Case (ii) *If two vectors are parallel*

Let **a** and **b** are parallel $\Rightarrow$ **a** = λ B where λ is some scalar. There

$$\mathbf{a} \,.\, (\mathbf{b} \times \mathbf{c}) = \lambda\, \mathbf{b} \,.\, (\mathbf{b} \times \mathbf{c}) = \lambda\, (\mathbf{b} \times \mathbf{b}) \,.\, \mathbf{c} = \lambda(0) = 0$$

or simply $[\mathbf{abc}] = [\lambda \mathbf{bbc}] = \lambda\, [\mathbf{bbc}] = 0$

3.5. Condition for three vectors to be coplanar

The necessary and sufficient condition that three non-parallel and non-zero vectors **a, b, c** be coplanar is that **[abc]** = 0

The condition is necessary. The **a, b, c** be three coplanar vectors. Now **a × b** is a vector perpendicular to the plane of **a** and **b**. Since **a, b, c** are coplanar therefore **a × b** also perpendicular to **c**

$\Rightarrow$ **(a × b) . c** = 0 or **[abc]** = 0

The condition is suffcient. If **[abc]** = 0 i.e., **(a × b).c** = 0 then **c** is perpendicular to **(a × b)**. But **a × b** is perpendicular to the plane containing **a** and **b** hence **c** should also lie in the plane of **a** and **c** i.e., **a, b, c** should be coplanar.

3.6. Scalar triple product in determinant form To express scalar triple product a (b × c) in the determinant form

Let $\mathbf{a} = a_1\mathbf{i} + a_2\mathbf{j} + a_3\mathbf{k}$; $\mathbf{b} = b_1\mathbf{i} + b_2\mathbf{j} + b_3\mathbf{k}$ and $\mathbf{c} = c_1\mathbf{i} + c_2\mathbf{j} + c_3\mathbf{k}$. Then

$$\mathbf{b} \times \mathbf{c} = \begin{vmatrix} \mathbf{i} & \mathbf{j} & \mathbf{k} \\ b_1 & b_2 & b_3 \\ c_1 & c_2 & c_3 \end{vmatrix}$$

$$= (b_2c_3 - b_3c_2)\mathbf{i} + (b_3c_1 - b_1c_3)\mathbf{j} + (b_1c_2 - b_2c_1)\mathbf{k}$$

Now $\mathbf{a} \cdot (\mathbf{b} \times \mathbf{c}) = (a_1\mathbf{i} + a_2\mathbf{j} + a_3\mathbf{k}) \cdot [(b_2c_3 - b_3c_2)\mathbf{i} + (b_3c_1 - b_1c_3)\mathbf{j} + (b_1c_2 - b_2c_1)\mathbf{k}]$

$$= a_1(b_2c_3 - b_3c_2) + a_2(b_3c_1 - b_1c_3) + a_3(b_1c_2 - b_2c_1)$$

or
$$\mathbf{a} \cdot (\mathbf{b} \times \mathbf{c}) = \begin{vmatrix} a_1 & a_2 & a_3 \\ b_1 & b_2 & b_3 \\ c_1 & c_2 & c_3 \end{vmatrix} = [\mathbf{abc}]$$

Note. *If OP, OQ, OR are the three concurrent edges of a parallelpiped and if* (a_1, a_2, a_3), (b_1, b_2, b_3), *and* $(c_1, c_2\ c_3)$ *be the coordinates of P, Q, R referred to O as origin then the above determinant gives the volume of the parallelopiped*

3.7. Scalar triple product in terms of non–coplanar vectors

To express scalar triple product [abc] in terms of three non-coplanr vectors l, m, n i.e., dd[lmn] ≠ 0

We know that any vector can be expressed in terms of three non-coplanar vectors as

$$\mathbf{a} = a_1\,\mathbf{l} + a_2\mathbf{m} + a_3\mathbf{n},\ \ \mathbf{b} = b_1\mathbf{l} + b_2\mathbf{m} + b_3\mathbf{n} \text{ and}$$
$$\mathbf{c} = c_1\mathbf{l} + c_2\mathbf{m} + c_3\mathbf{n}$$

$$\therefore \quad \mathbf{b} \times \mathbf{c} = (b_1\mathbf{l} + b_2\mathbf{m} + b_3\mathbf{n}) \times (c_1\mathbf{l} + c_2\mathbf{m} + c_3\mathbf{n})$$
$$= (b_2c_3 - b_3c_2)(\mathbf{m} \times \mathbf{n}) + (b_3c_1 - b_1c_3)(\mathbf{n} \times \mathbf{l}) + (b_1c_2 - b_2c_1)(\mathbf{l} \times \mathbf{m})$$

Now $\mathbf{a} \cdot (\mathbf{b} \times \mathbf{c}) = (a_1\mathbf{l} + a_2\mathbf{m} + a_3\mathbf{n}) \cdot [(b_2c_3 - b_3c_2)(\mathbf{m} \times \mathbf{n}) + (b_3c_1 - b_1c_3)(\mathbf{n} \times \mathbf{l}) + (b_1c_2 - b_2c_1)(\mathbf{l} \times \mathbf{m})]$

$$= a_1(b_2c_3 - b_3c_2)[\mathbf{l} \cdot (\mathbf{m} \times \mathbf{n})] + a_2(b_3c_1 - b_1c_3)[\mathbf{m} \cdot (\mathbf{n} \times \mathbf{l})] + a_3(b_1c_2 - b_2c_1)(\mathbf{n} \cdot (\mathbf{l} \times \mathbf{m})]$$
$$= [a_1(b_2c_3 - b_3c_2) + a_2(b_3c_1 - b_1c_3) + a_3(b_1c_2 - b_2c_1)]\,[\mathbf{lmn}]$$

$$\therefore \quad [\mathbf{abc}] = \begin{vmatrix} a_1 & a_2 & a_3 \\ b_1 & b_2 & b_3 \\ c_1 & c_2 & c_3 \end{vmatrix} [\mathbf{lmn}]$$

3.8. Vector as a linear combination of three non–coplanar vectors

To express any vector **r** *as a linear combination of three non–coplanar vectors* **a, b, c.**

Let x, y, z be three scalars. Then we can write

$$\mathbf{r} = x\mathbf{a} + y\mathbf{b} + z\mathbf{c} \qquad ...(1)$$

$$\therefore \quad \mathbf{r} \cdot (\mathbf{b} \times \mathbf{c}) = (x\mathbf{a} + y\mathbf{b} + z\mathbf{c}) \cdot (\mathbf{b} \times \mathbf{c})$$

$$[\mathbf{rbc}] = x\,[\mathbf{abc}] + y\,[\mathbf{bbc}] + z\,[\mathbf{cbc}]$$

or

$$[\mathbf{rbc}] = x\,[\mathbf{abc}] \qquad ...(2)$$

Similarly $[\mathbf{rca}] = y\,[\mathbf{bca}] = y\,[\mathbf{abc}]$ ...(3)

and

$$[\mathbf{rab}] = z\,[\mathbf{cab}] = z\,[\mathbf{abc}] \qquad ...(4)$$

Substituting (2), (3) and (4) in (1) we get

$$\mathbf{r} = \frac{[\mathbf{rbc}]}{[\mathbf{abc}]}\mathbf{a} + \frac{[\mathbf{rca}]}{[\mathbf{abc}]}\mathbf{b} + \frac{[\mathbf{rab}]}{[\mathbf{abc}]}\mathbf{c}$$

SOLVED PROBLEMS

Example 1. Evaluate $\mathbf{i} \cdot (\mathbf{j} \times \mathbf{k}) + (\mathbf{i} \times \mathbf{k}) \cdot \mathbf{j}$

Solution: $\mathbf{i} \cdot (\mathbf{j} \times \mathbf{k}) + (\mathbf{i} \times \mathbf{k}) \cdot \mathbf{j}$

$$= \mathbf{i} \cdot \mathbf{i} + (-\mathbf{j}) \cdot \mathbf{j} = \mathbf{i}^2 - \mathbf{j}^2 = 1 - 1 = 0$$

Example 2. Find the volume of the parallelopiped whose coterminous edges are represented by $\mathbf{a} = 2\mathbf{i} - \mathbf{j} + \mathbf{k}$, $\mathbf{b} = \mathbf{i} + 2\mathbf{j} - \mathbf{k}$ and $\mathbf{c} = 3\mathbf{i} + \mathbf{j} + 4\mathbf{k}$.

Solution. The required volume is given by $\mathbf{a} \cdot (\mathbf{b} \times \mathbf{c})$

$$\mathbf{a} \cdot (\mathbf{b} \times \mathbf{c}) = \begin{vmatrix} 2 & -1 & 1 \\ 1 & 2 & -1 \\ 3 & 1 & 4 \end{vmatrix}$$

$$= 2(8 + 1) + 1(4 + 3) + 1(1 - 6) = 20$$

$\therefore$ Volume of the parallelopiped whose coterminous edges are represented by the given vectors = 20 cubic units.

Example 3. Show that the vectors $\mathbf{a} = \mathbf{i} + 3\mathbf{j} + \mathbf{k}$, $\mathbf{b} = 2\mathbf{i} - \mathbf{j} - \mathbf{k}$ and $\mathbf{c} = 7\mathbf{j} + 3\mathbf{k}$ are parallel to the same plane (i.e., **a, b, c** are coplanar)

Solution. $\mathbf{a}\,.\,(\mathbf{b} \times \mathbf{c}) = \begin{vmatrix} 1 & 3 & 1 \\ 2 & -1 & -1 \\ 0 & 7 & 3 \end{vmatrix}$

$$= 1(-3+7) - 3(6-0) + 1(14-0) = 0$$

$\because$ $\mathbf{a}\,.\,(\mathbf{b} \times \mathbf{c}) = 0 \Rightarrow$ **a, b, c** are coplanar

i.e., parallel to the same plane

Example 4. Prove that the points

$$4\mathbf{i} + 8\mathbf{j} + 12\mathbf{k},\ 2\mathbf{i} + 4\mathbf{j} + 6\mathbf{k},\ 3\mathbf{i} + 5\mathbf{j} + 4\mathbf{k},\ 5\mathbf{i} + 8\mathbf{j} + 5\mathbf{k}$$

are coplanar.

Solution. Let

$\overrightarrow{OA} = 4\mathbf{i} + 8\mathbf{j} + 12\mathbf{k},\ \overrightarrow{OB} = 2\mathbf{i} + 4\mathbf{j} + 6\mathbf{k},\ \overrightarrow{OC} = 3\mathbf{i} + 5\mathbf{j} + 4\mathbf{k}$

and $\overrightarrow{OD} = 5\mathbf{i} + 8\mathbf{j} + 5\mathbf{k}$ be the position vectors of the given points w.r.t origin 0.

Then $\overrightarrow{AB}$ = P. V of B–P P.V of A

$$= (2\mathbf{i} + 4\mathbf{j} + 6\mathbf{k}) - (\mathbf{i} + 8\mathbf{j} + 12\mathbf{k}) = -2\mathbf{i} - 4\mathbf{j} - 6\mathbf{k}.$$

Similarly $\overrightarrow{BC} = \mathbf{i} + \mathbf{j} - 2\mathbf{k}$ and $\overrightarrow{CD} = 2\mathbf{i} + 3\mathbf{j} + \mathbf{k}$

Now $\overrightarrow{AB}\,.\,(\overrightarrow{BC} \times \overrightarrow{CD}) = \begin{vmatrix} -2 & -4 & -6 \\ 1 & 1 & -2 \\ 2 & 3 & 1 \end{vmatrix}$

$$= -2(1+6) + 4(1+4) - 6(3-2) = 0$$

$\Rightarrow$ The four points are coplanar

Note. Providing $\overrightarrow{AB}\,.\,(\overrightarrow{AB} \times \overrightarrow{AD}) = 0$ also implies that the points are coplanar.

Example 5. Show that $\mathbf{a} - 2\mathbf{b} + 3\mathbf{c}$, $-2\mathbf{a} + 3\mathbf{b} - 4\mathbf{c}$ and $\mathbf{a} - 3\mathbf{b} + 5\mathbf{c}$ are coplanar where **a, b, c** are non-coplanar.

Solution: Let p = $\mathbf{a} - 2\mathbf{b} + 3\mathbf{c}$, $\mathbf{q} = -2\mathbf{a} + 3\mathbf{b} - 4\mathbf{c}$ and $\mathbf{r} = \mathbf{a} - 3\mathbf{b} + 5\mathbf{c}$ then from **art (3.7)**, we have

$$[\mathbf{pqr}] = \begin{vmatrix} 1 & -2 & 3 \\ -2 & 3 & -4 \\ 1 & -3 & 5 \end{vmatrix} [\mathbf{abc}]$$

$$= \{1(15-12) + 2(-10+4) + 3(6-3)\}\ [\mathbf{abc}]$$

$$= (0)\ [\mathbf{abc}] = 0$$

$\Rightarrow$ **p, q, r** are coplanar.

Example 6. Show that the four points $\mathbf{2a + 3b - c}$, $\mathbf{a - 2b + 3c}$, $\mathbf{3a + 4b - 2c}$ and $\mathbf{a - 6b + 6c}$ are coplanar where **a, b, c** are three no-coplanar vectors.

Solution. Let

$$\overrightarrow{OA} = \mathbf{2a + 3b - c},\ \overrightarrow{OB} = \mathbf{a - 2b + 3c},\ \overrightarrow{OC} = \mathbf{3a + 4b - 2c}$$

and $\overrightarrow{OD} = \mathbf{a - 6b + 6c}$ where O is origin .

Then $\overrightarrow{AB} = \overrightarrow{OB} - \overrightarrow{OA} = \mathbf{-a - 5b + 4c}$

$\overrightarrow{BC} = \mathbf{2a + 6b - 5c}$ and $\overrightarrow{CD} = \mathbf{-2a - 10b + 8c}$.

Then using **art 3.7**

$$[\overrightarrow{AB}\ \overrightarrow{BC}\ \overrightarrow{CD}] = \begin{vmatrix} -1 & -5 & 4 \\ 2 & 6 & -5 \\ -2 & -10 & 8 \end{vmatrix} [\mathbf{abc}]$$

$$= [(-1)(48 - 50) + 5(16 - 10) + 4(-20 + 12)]\ [\mathbf{abc}]$$

$$= (0)\ [\mathbf{abc}] = 0$$

$\Rightarrow$ $\overrightarrow{AB}, \overrightarrow{BC}, \overrightarrow{CD}$ are coplanar

$\Leftrightarrow$ $\overrightarrow{OA}, \overrightarrow{OB}, \overrightarrow{OC}$ and $\overrightarrow{OD}$ are coplanar i.e. the given points are coplanar.

Example 7. If **a, b, c** are three vectors such that $\mathbf{a \times b = c}$ and $\mathbf{b \times c = a}$ show that three vectors **a, b, c** are orthogonal in pairs and $|\mathbf{b}| = 1$ and $|\mathbf{c}| = |\mathbf{a}|$.

Solution. Given $\mathbf{c = a \times b \Rightarrow a . c = a . (a \times b)} = 0$ ...(1)

Similarly $\mathbf{c = a \times b \Rightarrow b . c = b . (a \times b)} = 0$ ...(2)

and also given $\mathbf{a = b \times c \Rightarrow b . a = b . (b \times c)} = 0$ ...(3)

from (1), (2) and (3) $\Rightarrow$ $\mathbf{a . c} = 0$, $\mathbf{b . c} = 0$ and $\mathbf{c . a} = 0$

$\therefore$ **a, b, c** are pair wise orthogonal,

Now $|\mathbf{c}| = |\mathbf{a \times b}| \Rightarrow |\mathbf{c}| = |\mathbf{a}|\,|\mathbf{b}| \sin 90°$

or $|\mathbf{c}| = |\mathbf{a}|\,|\mathbf{b}|$ ...(4)

and similarly $|\mathbf{a}| = |\mathbf{b \times c}| \Rightarrow |\mathbf{a}| = |\mathbf{b}|\,|\mathbf{c}|$ ...(5)

Multiplying (4) and (5), we get

$$|\mathbf{c}|\,|\mathbf{a}| = |\mathbf{a}|\,|\mathbf{b}|^2\,|\mathbf{c}| \text{ or } 1 = |\mathbf{b}|^2 \text{ or } |\mathbf{b}| = 1$$

Using $|\mathbf{b}| = 1$ in (4), we get $|\mathbf{c}| = |\mathbf{a}|$.

Example 8. Prove that $[\mathbf{a} + \mathbf{b}, \mathbf{b} + \mathbf{c}, \mathbf{c} + \mathbf{a}] = 2[\mathbf{abc}]$

Solution. L.H.S $= (\mathbf{a} + \mathbf{b}) \cdot [\mathbf{b} + \mathbf{c}) \times (\mathbf{c} + \mathbf{a})]$

$$= (\mathbf{a} + \mathbf{b}) \cdot [\mathbf{b} \times \mathbf{c} + \mathbf{b} \times \mathbf{a} + \mathbf{c} \times \mathbf{c} + \mathbf{c} \times \mathbf{a}]$$

$$= (\mathbf{a} + \mathbf{b}) \cdot [\mathbf{b} \times \mathbf{c} + \mathbf{b} \times \mathbf{a} + \mathbf{c} \times \mathbf{a}] \quad (\because \mathbf{c} \times \mathbf{c} = 0)$$

$$= \mathbf{a} \cdot (\mathbf{b} \times \mathbf{c}) + \mathbf{a} \cdot (\mathbf{b} \times \mathbf{a}) + \mathbf{a} \cdot (\mathbf{c} \times \mathbf{a}) + \mathbf{b} \cdot (\mathbf{b} \times \mathbf{c}) + \mathbf{b} \cdot (\mathbf{b} \times \mathbf{a}) + \mathbf{b} \cdot (\mathbf{c} \times \mathbf{a})$$

$$= [\mathbf{abc}] + [\mathbf{aba}] + [\mathbf{aca}] + [\mathbf{bbc}] + [\mathbf{bba}] + [\mathbf{bca}]$$

$$= [\mathbf{abc}] + [\mathbf{abc}] = 2[\mathbf{abc}] = \text{R.H.S}$$

Example 9. Prove that

$$[\mathbf{a} \times \mathbf{b}, \mathbf{b} \times \mathbf{c}, \mathbf{c} \times \mathbf{a}] = [\mathbf{abc}]^2 = \begin{vmatrix} \mathbf{a} \cdot \mathbf{a} & \mathbf{a} \cdot \mathbf{b} & \mathbf{a} \cdot \mathbf{c} \\ \mathbf{b} \cdot \mathbf{a} & \mathbf{b} \cdot \mathbf{b} & \mathbf{b} \cdot \mathbf{c} \\ \mathbf{c} \cdot \mathbf{a} & \mathbf{c} \cdot \mathbf{b} & \mathbf{c} \cdot \mathbf{c} \end{vmatrix}$$

Solution. Let $\mathbf{a} = a_1\mathbf{i} + a_2\mathbf{j} + a_3\mathbf{k}$, $\mathbf{b} = b_1\mathbf{i} + b_2\mathbf{j} + b_3\mathbf{k}$

and $\mathbf{c} = c_1\mathbf{i} + c_2\mathbf{j} + c_3\mathbf{k}$ Then

$$[\mathbf{abc}] = \begin{vmatrix} a_1 & a_2 & a_3 \\ b_1 & b_2 & b_3 \\ c_1 & c_2 & c_3 \end{vmatrix} \quad ...(1)$$

Now $\mathbf{a} \times \mathbf{b} = \begin{vmatrix} \mathbf{i} & \mathbf{j} & \mathbf{k} \\ a_1 & a_2 & a_3 \\ b_1 & b_2 & b_3 \end{vmatrix}$

$$= (a_3b_3 - a_2b_2)\mathbf{i} + (b_1a_3 - b_3a_1)\mathbf{j} + (a_1b_2 - a_2b_1)\mathbf{k}$$

Similarly

$$\mathbf{b} \times \mathbf{c} = (b_2c_3 - c_2b_3)\mathbf{i} + (c_1b_3 - c_3b_1)\mathbf{j} + (b_1c_2 - b_2c_1)\mathbf{k}$$

$$\mathbf{c} \times \mathbf{a} = (c_2a_3 - c_3a_2)\mathbf{i} + (a_1c_3 - a_3c_1)\mathbf{j} + (c_1a_2 - c_2a_1)\mathbf{k}$$

Now $[\mathbf{a} \times \mathbf{b}, \mathbf{b} \times \mathbf{c}, \mathbf{c} \times \mathbf{a}]$

$$= \begin{vmatrix} a_2b_3 - a_3b_2 & b_1a_3 - b_3a_1 & a_1b_2 - a_2b_1 \\ b_2c_3 - c_2b_3 & c_1b_3 - c_3b_1 & b_1c_2 - b_2c_1 \\ c_2a_3 - c_3a_2 & a_1c_3 - a_3c_1 & c_1a_2 - c_2a_1 \end{vmatrix}$$

$$= \begin{vmatrix} C_1 & C_2 & C_3 \\ A_1 & A_2 & A_3 \\ B_1 & B_2 & B_3 \end{vmatrix} = \begin{vmatrix} A_1 & A_2 & A_3 \\ B_1 & B_2 & B_3 \\ C_1 & C_2 & C_3 \end{vmatrix}$$

where the capital lettus denote the cofactor of the corresponding small letters in the determinant

$$= \begin{vmatrix} a_1 & a_2 & a_3 \\ b_1 & b_2 & b_3 \\ c_1 & c_2 & c_3 \end{vmatrix}^2 = [\mathbf{abc}]^2 \qquad ...(2)$$

Now $\quad a^2 = \mathbf{a} \cdot \mathbf{a} = a_1^2 + a_2^2 + a_3^2$ etc.,

and $\quad \mathbf{a} \cdot \mathbf{b} = a_1b_1 + a_2b_2 + a_3b_3 = \mathbf{b} \cdot \mathbf{a}$ etc.,

$$\text{Again } [\mathbf{abc}]^2 = \begin{vmatrix} a_1 & a_2 & a_3 \\ b_1 & b_2 & b_3 \\ c_1 & c_2 & c_3 \end{vmatrix} \begin{vmatrix} a_1 & a_2 & a_3 \\ b_1 & b_2 & b_3 \\ c_1 & c_2 & c_3 \end{vmatrix}$$

$$= \begin{vmatrix} \Sigma a_1^2 & \Sigma a_1b_1 & \Sigma a_1c_1 \\ \Sigma b_1a_1 & \Sigma b_1^2 & \Sigma b_1c_1 \\ \Sigma c_1a_1 & \Sigma c_1b_1 & \Sigma c_1^2 \end{vmatrix}$$

$$= \begin{vmatrix} \mathbf{a} \cdot \mathbf{a} & \mathbf{a} \cdot \mathbf{b} & \mathbf{a} \cdot \mathbf{c} \\ \mathbf{b} \cdot \mathbf{a} & \mathbf{b} \cdot \mathbf{b} & \mathbf{b} \cdot \mathbf{c} \\ \mathbf{c} \cdot \mathbf{a} & \mathbf{c} \cdot \mathbf{b} & \mathbf{c} \cdot \mathbf{c} \end{vmatrix} \qquad ...(3)$$

Hence from (2) and (3) the given result is proved.

Example 10. Prove that

$$\text{(i) } [\mathbf{lmn}]\,[\mathbf{abc}] = \begin{vmatrix} \mathbf{l} \cdot \mathbf{a} & \mathbf{l} \cdot \mathbf{b} & \mathbf{l} \cdot \mathbf{c} \\ \mathbf{m} \cdot \mathbf{a} & \mathbf{m} \cdot \mathbf{b} & \mathbf{m} \cdot \mathbf{c} \\ \mathbf{n} \cdot \mathbf{a} & \mathbf{n} \cdot \mathbf{b} & \mathbf{n} \cdot \mathbf{c} \end{vmatrix} \text{ and}$$

$$\text{(ii) } [\mathbf{lmn}]\,(\mathbf{a} \times \mathbf{b}) = \begin{vmatrix} \mathbf{l} \cdot \mathbf{a} & \mathbf{l} \cdot \mathbf{b} & \mathbf{l} \\ \mathbf{m} \cdot \mathbf{a} & \mathbf{m} \cdot \mathbf{b} & \mathbf{m} \\ \mathbf{n} \cdot \mathbf{a} & \mathbf{n} \cdot \mathbf{b} & \mathbf{n} \end{vmatrix}$$

Solution. Let

$\mathbf{a} = a_1\mathbf{i} + a_2\mathbf{j} + a_3\mathbf{k}$, $\mathbf{b} = b_1\mathbf{i} + b_2\mathbf{j} + b_3\mathbf{k}$,

$\mathbf{c} = c_1\mathbf{i} + c_2\mathbf{j} + c_3\mathbf{k}$, $\mathbf{l} = l_1\mathbf{i} + l_2\mathbf{j} + l_3\mathbf{k}$, $\mathbf{m} = m_1\mathbf{i} + m_2\mathbf{j} + m_3\mathbf{k}$

and $\mathbf{n} = n_1\mathbf{i} + n_2\mathbf{j} + n_3\mathbf{k}$

$$\text{Then } [\mathbf{lmn}]\,[\mathbf{abc}] = \begin{vmatrix} l_1 & l_2 & l_3 \\ m_1 & m_2 & m_3 \\ n_1 & n_2 & n_3 \end{vmatrix} \begin{vmatrix} a_1 & a_2 & a_3 \\ b_1 & b_2 & b_3 \\ c_1 & c_2 & c_3 \end{vmatrix}$$

$$= \begin{vmatrix} l_1a_1 + l_2a_2 + l_3a_3 & l_1b_1 + l_2b_2 + l_3b_3 & l_1c_1 + l_2c_2 + l_3c_3 \\ m_1a_1 + m_2a_2 + m_3a_3 & m_1b_1 + m_2b_2 + m_3b_3 & m_1c_1 + m_2c_2 + m_3c_3 \\ n_1a_1 + n_2a_2 + n_3a_3 & n_1b_1 + n_2b_2 + n_3b_3 & n_1c_1 + n_2c_2 + n_3c_3 \end{vmatrix}$$

$$= \begin{vmatrix} \mathbf{l} \cdot \mathbf{a} & \mathbf{l} \cdot \mathbf{b} & \mathbf{l} \cdot \mathbf{c} \\ \mathbf{m} \cdot \mathbf{a} & \mathbf{m} \cdot \mathbf{b} & \mathbf{m} \cdot \mathbf{c} \\ \mathbf{n} \cdot \mathbf{a} & \mathbf{n} \cdot \mathbf{b} & \mathbf{n} \cdot \mathbf{c} \end{vmatrix}$$ which proves the first result~.

Similarly $[\mathbf{lmn}]\,[\mathbf{a} \times \mathbf{b}]$

$$= \begin{vmatrix} l_1 & l_2 & l_3 \\ m_1 & m_2 & m_3 \\ n_1 & n_2 & n_3 \end{vmatrix} \begin{vmatrix} \mathbf{i} & \mathbf{j} & \mathbf{k} \\ a_1 & a_2 & a_3 \\ b_1 & b_2 & b_3 \end{vmatrix}$$

$$= \begin{vmatrix} l_1\mathbf{i} + l_2\mathbf{j} + l_3\mathbf{k} & l_1a_1 + l_2a_2 + l_3a_3 & l_1b_1 + l_2b_2 + l_3b_3 \\ m_1\mathbf{i} + m_2\mathbf{j} + m_3\mathbf{k} & m_1a_1 + m_2a_2 + m_3a_3 & m_1b_1 + m_2b_2 + m_3b_3 \\ n_1\mathbf{i} + n_2\mathbf{j} + n_3\mathbf{k} & n_1a_1 + n_2a_2 + n_3a_3 & n_1b_1 + n_2b_2 + n_3b_3 \end{vmatrix}$$

$$= \begin{vmatrix} \mathbf{l} & \mathbf{l} \cdot \mathbf{a} & \mathbf{l} \cdot \mathbf{b} \\ \mathbf{m} & \mathbf{m} \cdot \mathbf{a} & \mathbf{m} \cdot \mathbf{b} \\ \mathbf{n} & \mathbf{n} \cdot \mathbf{a} & \mathbf{n} \cdot \mathbf{b} \end{vmatrix} = \begin{vmatrix} \mathbf{l} \cdot \mathbf{a} & \mathbf{l} \cdot \mathbf{b} & \mathbf{l} \\ \mathbf{m} \cdot \mathbf{a} & \mathbf{m} \cdot \mathbf{b} & \mathbf{m} \\ \mathbf{n} \cdot \mathbf{a} & \mathbf{n} \cdot \mathbf{b} & \mathbf{n} \end{vmatrix}$$

which proves the second result

Example 11. If **a, b, c** are three non-coplanar vectors then (i) express **a, b, c** in terms of **b × c, c × a, a × b**

(ii) Express **b × c, c × a, a × b** in terms of **a, b, c**

Solution. (i) Let $\mathbf{a} = x(\mathbf{b} \times \mathbf{c}) + y(\mathbf{c} \times \mathbf{a}) + z(\mathbf{a} \times \mathbf{b})$ ...(1)

Multiplying both sides scalarly by **a**, we get

$$\mathbf{a} \cdot \mathbf{a} = x\,\mathbf{a} \cdot (\mathbf{b} \times \mathbf{c}) + y\,\mathbf{a} \cdot (\mathbf{c} \times \mathbf{a}) + z\,\mathbf{a} \cdot (\mathbf{a} \times \mathbf{b})$$

or $\mathbf{a} \cdot \mathbf{a} = x\,[\mathbf{abc}]$ $[\because\ \mathbf{a} \cdot (\mathbf{c} \times \mathbf{a}) = \mathbf{a} \cdot (\mathbf{a} \times \mathbf{b}) = 0]$

$$\therefore \quad x = \frac{\mathbf{a} \cdot \mathbf{a}}{[\mathbf{abc}]}$$

Similarly multiplying both sides of (1) scalarly by **b** and **c**, we get

$$y = \frac{\mathbf{a} \cdot \mathbf{b}}{[\mathbf{abc}]} \text{ and } z = \frac{\mathbf{a} \cdot \mathbf{c}}{[\mathbf{abc}]}$$

Substituting the values x, y, z in (1), we get

$$\mathbf{a} = \frac{\mathbf{a} \cdot \mathbf{a}}{[\mathbf{abc}]} (\mathbf{b} \times\ ) + \frac{\mathbf{a} \cdot \mathbf{b}}{[\mathbf{abc}]} (\mathbf{c} \times \mathbf{a}) + \frac{\mathbf{a} \cdot \mathbf{c}}{[\mathbf{abc}]} (\mathbf{a} \times \mathbf{b})$$

Similarly we can write

$$\mathbf{b} = \frac{\mathbf{b} \cdot \mathbf{a}}{[\mathbf{abc}]} (\mathbf{b} \times \mathbf{c}) + \frac{\mathbf{b} \cdot \mathbf{b}}{[\mathbf{abc}]} (\mathbf{c} \times \mathbf{a}) + \frac{\mathbf{b} \cdot \mathbf{c}}{[\mathbf{abc}]} (\mathbf{a} \times \mathbf{b})$$

$$\mathbf{c} = \frac{\mathbf{c} \cdot \mathbf{a}}{[\mathbf{abc}]} (\mathbf{b} \times \mathbf{c}) + \frac{\mathbf{c} \cdot \mathbf{b}}{[\mathbf{abc}]} (\mathbf{c} \times \mathbf{a}) + \frac{\mathbf{c} \cdot \mathbf{c}}{[\mathbf{abc}]} (\mathbf{a} \times \mathbf{b})$$

(ii) Let $\mathbf{b} \times \mathbf{c} = l\mathbf{a} + m\mathbf{b} + n\mathbf{c}$ ...(1)

Multiplying scalarly by (b × c) on both sides of (1) we get,

$$(\mathbf{b} \times \mathbf{c}) \cdot (\mathbf{b} \times \mathbf{c}) = l\mathbf{a} \cdot (\mathbf{b} \times \mathbf{c}) + m\mathbf{b} \cdot (\mathbf{b} \times \mathbf{c}) + m\mathbf{c} \cdot (\mathbf{b} \times \mathbf{c})$$

or $\quad (\mathbf{b} \times \mathbf{c}) \cdot (\mathbf{b} \times \mathbf{c}) = l\, [\mathbf{abc}]$

$$\therefore \quad l = \frac{(\mathbf{b} \times \mathbf{c}) \cdot (\mathbf{b} \times \mathbf{c})}{[\mathbf{abc}]}$$

Similarly by multiplying scalarly by (**c** × **a**) and (**a** × **b**) on both sides of (1) we get

$$m = \frac{(\mathbf{b} \times \mathbf{c}) \cdot (\mathbf{c} \times \mathbf{a})}{[\mathbf{abc}]} \quad \text{and } n = \frac{(\mathbf{b} \times \mathbf{c}) \cdot (\mathbf{a} \times \mathbf{b})}{[\mathbf{abc}]}$$

Substituting the values of l, m and n in (1) we get

$$(\mathbf{b} \times \mathbf{c}) = \frac{(\mathbf{b} \times \mathbf{c}) \cdot (\mathbf{b} \times \mathbf{c})}{[\mathbf{abc}]} \mathbf{a} + \frac{(\mathbf{b} \times \mathbf{c}) \cdot (\mathbf{c} \times \mathbf{a})}{[\mathbf{abc}]} \mathbf{b} + \frac{(\mathbf{b} \times \mathbf{c}) \cdot (\mathbf{a} \times \mathbf{b})}{[\mathbf{abc}]} \mathbf{c}$$

Similarly we can write the expressions for (**c** × **a**) and (**a** × **b**).

Example 12. Show that the four points **a, b, c, d** are coplanar if [**bcd**] + [**cad**] + [**abd**] = [**abc**]

Solution. Let $\vec{OA} = \mathbf{a}$, $\vec{OB} = \mathbf{b}$, $\vec{OC} = \mathbf{c}$, $\vec{OD} = \mathbf{d}$ be the position vectors of the four points with reference to origin O.

Then $\vec{DA} = \mathbf{a} - \mathbf{d}$, $\vec{DB} = \mathbf{b} - \mathbf{d}$, $\vec{DC} = \mathbf{c} - \mathbf{d}$

The given four points are coplanr if the vectors $\vec{DA}$, $\vec{DB}$ and $\vec{DC}$ are coplanar.

$\Rightarrow \quad \vec{DA} \cdot (\vec{DB} \times \vec{DC}) = 0$

i.e., $\quad (\mathbf{a} - \mathbf{d}) \cdot [(\mathbf{b} - \mathbf{d}) \times (\mathbf{c} - \mathbf{d})] = 0$

i.e., $\quad (\mathbf{a} - \mathbf{d}) \cdot [\mathbf{b} \times \mathbf{c} - \mathbf{b} \times \mathbf{d} - \mathbf{d} \times \mathbf{c} + \mathbf{d} \times \mathbf{d}] = 0$

i.e., $\quad (\mathbf{a} - \mathbf{d}) \cdot [\mathbf{b} \times \mathbf{c} + \mathbf{d} \times \mathbf{b} + \mathbf{c} \times \mathbf{d}] = 0 \qquad (\because \mathbf{d} \times \mathbf{d} = \mathbf{0})$

i.e., $\quad \mathbf{a} \cdot (\mathbf{b} \times \mathbf{c}) + \mathbf{a} \cdot (\mathbf{d} \times \mathbf{b}) + \mathbf{a} \cdot (\mathbf{c} \times \mathbf{d}) - \mathbf{d} \cdot (\mathbf{b} \times \mathbf{c})$

$\qquad - \mathbf{d} \cdot (\mathbf{d} \times \mathbf{b}) - \mathbf{d} \cdot (\mathbf{c} \times \mathbf{d}) = 0$

i~.~e., $\quad [\mathbf{abc}] + [\mathbf{adb}] + [\mathbf{acd}] - [\mathbf{dbc}] - [\mathbf{ddb}] - [\mathbf{dcd}] = 0$

i.e., $\quad [\mathbf{abc}] + [\mathbf{adb}] + [\mathbf{acd}] - [\mathbf{dbc}] = 0$

or $\quad [\mathbf{bcd}] + [\mathbf{cad}] + [\mathbf{abd}] = [\mathbf{abc}]$

which proves the result

EXERCISE

1. Find $\mathbf{a} \cdot (\mathbf{b} \times \mathbf{c})$ if $\mathbf{a} = 3\mathbf{i} - 2\mathbf{j} + 4\mathbf{k}$, $\mathbf{b} = 2\mathbf{i} + \mathbf{j} + 3\mathbf{k}$ and $\mathbf{c} = 3\mathbf{i} + 2\mathbf{j} + 7\mathbf{k}$
2. If $\mathbf{a} = 5\mathbf{i} - 4\mathbf{j} + 3\mathbf{k}$, $\mathbf{b} = \mathbf{i} - 2\mathbf{j} - 7\mathbf{k}$ and $\mathbf{c} = 4\mathbf{i} - 3\mathbf{j} + \mathbf{k}$ show that $\mathbf{a} \cdot (\mathbf{b} \times \mathbf{c}) = \mathbf{b} \cdot (\mathbf{c} \times \mathbf{a}) = \mathbf{c} \cdot (\mathbf{a} \times \mathbf{b})$
3. Find $[\mathbf{i} + \mathbf{j}, \mathbf{j} + \mathbf{k}, \mathbf{k} + \mathbf{i}]$ and $[\mathbf{i} - \mathbf{j}, \mathbf{j} - \mathbf{k}, \mathbf{k} - \mathbf{i}]$
4. Find the volume of a parallelopiped whose coterminous edges are given by

   (i) $5\mathbf{i} - 4\mathbf{j} + \mathbf{k}$, $-4\mathbf{i} + 3\mathbf{j} - 2\mathbf{k}$, $\mathbf{i} - 2\mathbf{j} - 7\mathbf{k}$

   (ii) $2\mathbf{i} - 3\mathbf{j}$, $\mathbf{i} - \mathbf{j} - \mathbf{k}$, $3\mathbf{i} - \mathbf{k}$

   (iii) $2\mathbf{i} + \mathbf{j} + 3\mathbf{k}$, $-\mathbf{i} + 2\mathbf{j} + \mathbf{k}$, $3\mathbf{i} + \mathbf{j} + 2\mathbf{k}$

   (iv) $2\mathbf{i} - 3\mathbf{j} + \mathbf{k}$, $\mathbf{i} - \mathbf{j} + 2\mathbf{k}$, $2\mathbf{i} + \mathbf{j} - \mathbf{k}$
5. Show that the vectors $3\mathbf{i} - 4\mathbf{j}$ –'$4\mathbf{k}$, $2\mathbf{i} - \mathbf{j} + \mathbf{k}$ and $\mathbf{i} - 3\mathbf{j} - 5\mathbf{k}$ are coplanar
6. Find the value of λ so that the vectors $\lambda\mathbf{i} - \mathbf{j} + \mathbf{k}$, $\mathbf{i} + 2\mathbf{j} - 3\mathbf{k}$ and $3\mathbf{i} - 4\mathbf{j} + 5\mathbf{k}$ are coplanar
7. Prove that the vectors $5\mathbf{a} + 6\mathbf{b} - 7\mathbf{c}$, $7\mathbf{a} - 8\mathbf{b} + 9\mathbf{c}$ and $3\mathbf{a} + 20\mathbf{b} + 5\mathbf{c}$ are coplanar; $\mathbf{a}, \mathbf{b}, \mathbf{c}$ being any vectcrs.
8. Show that the four points

$-\mathbf{a} + 4\mathbf{b} - 3\mathbf{c}, 3\mathbf{a} + 2\mathbf{b} - 5\mathbf{c}, -3\mathbf{a} + 8\mathbf{b} - 5\mathbf{c}$ and $-3\mathbf{a} + 2\mathbf{b} + \mathbf{c}$ are coplanar.

9. Show that the four points (6, 2, –1), (2, –1, 3), (–1, 2, –4) and (–12, –1, –3) are coplanar.

10. Prove that $[\mathbf{a} - \mathbf{b}, \mathbf{b} - c, \mathbf{c} - \mathbf{a}] = 0$

11. If $\mathbf{a} = 2\mathbf{i} - 3\mathbf{j} + \mathbf{k}$, $\mathbf{b} = \mathbf{i} - \mathbf{j} + 2\mathbf{k}$ and $\mathbf{c} = 2\mathbf{i} + \mathbf{j} - \mathbf{k}$ find the values of $(\mathbf{a} + \mathbf{b}) . \{(\mathbf{b} + \mathbf{c}) \times (\mathbf{c} + \mathbf{a})\}$ and $(\mathbf{a} - \mathbf{b}) . \{(\mathbf{b} - \mathbf{c}) \times (\mathbf{c} - \mathbf{a})\}$

12. If $\mathbf{x} . \mathbf{a} + 0, \mathbf{x} . \mathbf{b} = 0, \mathbf{x} . \mathbf{c} = 0$ for some non–zero scalar x, show that **a, b, c** are coplanar

13. If **a, b, c** are non–coplanar then **a** + **b**, **b** + **c** and **c** + **a** are also coplanr. If this is true in case of **a** – **b**, **b** + **c**, **c** + **a**.

14. If **a, b, c** are the position vectors of A, B and C prove that $\mathbf{a} \times \mathbf{c} + \mathbf{b} \times \mathbf{c} + \mathbf{c} \times \mathbf{a}$ is a vector perpendicular to the plane of ABC.

15. If $\mathbf{a} = 2\mathbf{i} - \mathbf{j} + 3\mathbf{k}$, $\mathbf{b} = 3\mathbf{i} + 2\mathbf{j} + \mathbf{k}$ and $\mathbf{c} = 8\mathbf{i} + \mathbf{j} + 2\mathbf{k}$ find the value of $[\mathbf{a} \times \mathbf{b}, \mathbf{b} \times \mathbf{c}, \mathbf{c} \times \mathbf{a}]$

16. If **a, b, c** are linearly independent show that **a** + **b**, **b** + **c**, **c** + **a** are also linearly independent.

17. If **a, b, c** be any three non–coplanar vectors prove that

$$\mathbf{r} = \frac{\mathbf{a} . \mathbf{r}}{[\mathbf{abc}]}(\mathbf{b} \times \mathbf{c}) + \frac{\mathbf{b} . \mathbf{r}}{[\mathbf{abc}]}(\mathbf{c} \times \mathbf{a}) + \frac{\mathbf{c} . \mathbf{r}}{[\mathbf{abc}]}(\mathbf{a} \times \mathbf{b})$$

18. If $\begin{vmatrix} x & x^2 & 1 + x^3 \\ y & y^2 & 1 + y^3 \\ z & z^2 & 1 + z^3 \end{vmatrix} = 0$ where $\mathbf{a} = (1, x, x^2)$, $\mathbf{b} = (1, y, y^2)$ and $\mathbf{c} = (1, z, z^2)$ are non–coplanar then show that $xyz + 1 = 0$

19. Evaluate $\dfrac{\mathbf{a} . (\mathbf{b} \times \mathbf{c}) + \mathbf{b} . (\mathbf{a} \times \mathbf{c})}{\mathbf{c} . (\mathbf{a} \times \mathbf{b})}$ where **a, b, c** are non–coplanar

20. If **b** + **c**, **c** + **a**, **a** + **b** are coplanar show that **a, b, c** are also coplanar.

ANSWERS

(1) 17 (3) 2, 0 (4) – 224, 8, – 10, 14

(6) 2 (11) 28, 0

(13) No. If **a, b, c** are non coplanar then we can not say that **a** − **b**, **b** + **c**, **c** + **a** are necessarily coplanar.

(15) 1225 (19) 0.

3.9. Vector triple product

The vector product of two vectors one of which is itself the vector product of two vectors is a vector quantity called the **vector triple product.**

Thus if **a, b, c** are three vectors then (**a** × **b** × **c**) and (**a** × **b**) × **c** are called as *vector triple products.*

Theorem. To prove that

$$\mathbf{a} \times (\mathbf{b} \times \mathbf{c}) = (\mathbf{a} \cdot \mathbf{c})\mathbf{b} - (\mathbf{a} \cdot \mathbf{b})\mathbf{c}$$

Proof. Let $\mathbf{r} = \mathbf{a} \times (\mathbf{b} \times \mathbf{c})$ ...(1)

and $\mathbf{b} \times \mathbf{c} = \mathbf{d}$

$\because$ $\mathbf{b} \times \mathbf{c} = \mathbf{d}$

⇒ **d** is a vector perpendicular to the plane containing **b** and **c**.

Also $\mathbf{r} = \mathbf{a} \times \mathbf{d}$

⇒ **r** is a vector perpendicular to the plane containing **a** and **d**.

⇒ **r** is perpendicular to the vector **d** where **d** is perpendicular to the plane containing **b** and **c**

∴ **r** must lie in the plane containing **b** and **c**

Hence we can write

$$\mathbf{r} = x\mathbf{b} + y\mathbf{c} \qquad ...(2)$$

where x and y are scalars

∴ **r** is perpendicular to **a** ⇒ **r** . **a** = 0

$\therefore$ $(x\,\mathbf{b} + y\mathbf{c}) \cdot \mathbf{a} = 0$ or $x(\mathbf{b} \cdot \mathbf{a}) + y(\mathbf{c} \cdot \mathbf{a}) = 0$

$$\therefore \quad \frac{x}{\mathbf{c} \cdot \mathbf{a}} = \frac{-y}{\mathbf{b} \cdot \mathbf{a}} = \lambda \text{ (say)}$$

$$\therefore \quad x = \lambda(\mathbf{c} \cdot \mathbf{a});\ y = -\lambda(\mathbf{b} \cdot \mathbf{a})$$

Putting these values in (1), we get

$$\mathbf{r} = \lambda[(\mathbf{c} \cdot \mathbf{a})\,\mathbf{b} - (\mathbf{b} \cdot \mathbf{a})\,\mathbf{c}] \qquad ...(3)$$

Now we have to find the value of λ

Consider unit vectors **j** and **k**, the former parallel to **b** and the later perpendicular to it in the plane containing **b** and **c**. Then we may write

$$\mathbf{b} = b_2\mathbf{j} \text{ and } \mathbf{c} + c_2\mathbf{j} + c_3\mathbf{k}$$

Now in terms of **j** and **k** and the other unit vector **i** of the right handed system, the third vector a can be written as $\mathbf{a} = a_1\mathbf{i} + a_2\mathbf{j} + a_3\mathbf{k}$

Now $\mathbf{b} \times \mathbf{c} = (b_2\mathbf{j}) \times (c_2\mathbf{j} + c_3\mathbf{k})$

$$= b_2c_2(\mathbf{j} \times \mathbf{j}) + b_3c_3(\mathbf{j} \times \mathbf{k}) = b_2c_3\mathbf{i}$$

$\therefore$ $\mathbf{r} = \mathbf{a} \times (\mathbf{b} \times \mathbf{c}) = (a_1\mathbf{i} + a_2\mathbf{j} + a_3\mathbf{k}) \times (b_2c_3\mathbf{i})$

$$= a_1b_2c_3(\mathbf{i} \times \mathbf{i}) + a_2b_2c_3(\mathbf{j} \times \mathbf{i}) + a_3b_2c_3\,(\mathbf{k} \times \mathbf{i})$$

$$= a_2b_2c_3(-\mathbf{k}) + a_3b_2c_3\,(\mathbf{j})$$

$$= a_3b_2c_3\mathbf{j} - a_2b_2c_3\mathbf{k} \qquad ...(4)$$

Also $\mathbf{r} = \lambda\,[(\mathbf{c}\,.\,\mathbf{a})\mathbf{b} - (\mathbf{b}\,.\,\mathbf{a})\,\mathbf{c}]$

$$= \lambda\,[\{(c_1\mathbf{j} + c_2\mathbf{k})\ .\ (a_1\mathbf{i} + a_2\mathbf{j} + a_3\mathbf{k})\}b_2\mathbf{j}$$

$$- \{(b_2\mathbf{j})\,.\,(a_1\mathbf{i} + a_1\mathbf{j} + a_2\mathbf{k})\}\,(c_2\mathbf{j} + c_3\mathbf{k})]$$

$$= \lambda\,[c_2a_2b_2\mathbf{j} + c_3a_3b_2\mathbf{j} - b_2a_2c_2\mathbf{j} - b_2a_2c_3\mathbf{k}]$$

$$= \lambda\,[a_3b_2c_3\mathbf{j} - a_2b_2c_3\mathbf{k}] \qquad ...(5)$$

$\therefore$ From (4) and (5), we get $\lambda = 1$

Hence from (3)

$$\mathbf{r} = (\mathbf{c}\,.\,\mathbf{a})\mathbf{b} - (\mathbf{b}\,.\,\mathbf{a})\mathbf{c} \quad \text{or} \quad \mathbf{r} = (\mathbf{a}\,.\,\mathbf{c})\mathbf{b} - (\mathbf{a}\,.\,\mathbf{b})\mathbf{c}$$

and hence from (1), we get

$$\mathbf{a} \times (\mathbf{b} \times \mathbf{c}) = (\mathbf{a}\,.\,\mathbf{c})\,\mathbf{b} - (\mathbf{a}\,.\,\mathbf{b})\mathbf{c}$$

Note. $(\mathbf{a} \times \mathbf{b}) \times \mathbf{c} = -\,[\mathbf{c} \times (\mathbf{a} \times \mathbf{b})]$

$$= -\,[(\mathbf{c}\,.\,\mathbf{b})\mathbf{a} - (\mathbf{c}\,.\,\mathbf{a})\mathbf{b}]$$

$$= (\mathbf{c}\,.\,\mathbf{a})\mathbf{b} - (\mathbf{c}\,.\,\mathbf{b})\mathbf{a}$$

Note. Vector triple product is not associative

i.e., $\mathbf{a} \times (\mathbf{b} \times \mathbf{c}) \neq (\mathbf{a} \times \mathbf{b}) \times \mathbf{c}$

If **a, b, c** are three vectors then **a** × (**b** × **c**) is a vector lying in the plane of **b** and **c** and perpendicular to **a** where as (**a** × **b**) × **c** is a vector in the plane of **a** and **b** and perpendicular to **c**. Hence they are unequal.

SOLVED PROBLEMS

Example 13. Prove that

$$\mathbf{a} \times (\mathbf{b} \times \mathbf{c}) + \mathbf{b} \times (\mathbf{c} \times \mathbf{a}) + \mathbf{c} \times (\mathbf{a} \times \mathbf{b}) = \mathbf{0}$$

Solution. we know that

$$\mathbf{a} \times (\mathbf{b} \times \mathbf{c}) = (\mathbf{a} . \mathbf{c})\mathbf{b} - (\mathbf{a} . \mathbf{b})\mathbf{c}$$

$$\mathbf{b} \times (\mathbf{c} \times \mathbf{a}) = (\mathbf{b} . \mathbf{a})\mathbf{c} - (\mathbf{b} . \mathbf{c})\mathbf{a}$$

$$\text{and } (\mathbf{a} \times \mathbf{b}) = (\mathbf{c}\mathbf{b})\mathbf{a} - (\mathbf{c} . \mathbf{a})\mathbf{b}$$

Adding all we get

$$\mathbf{a} \times (\mathbf{b} \times \mathbf{c}) + \mathbf{b} \times (\mathbf{c} \times \mathbf{a}) + \mathbf{c} \times (\mathbf{a} \times \mathbf{b})$$

$$= (\mathbf{a} . \mathbf{c})\mathbf{b} - (\mathbf{a} . \mathbf{b})\mathbf{c} + (\mathbf{b} . \mathbf{a})\mathbf{c} - (\mathbf{b} . \mathbf{c})\mathbf{a} + (\mathbf{c} . \mathbf{b})\mathbf{a} - (\mathbf{c} . \mathbf{a})\mathbf{b}$$

$$= 0 \qquad [\because \ \mathbf{a} . \mathbf{c} + \mathbf{c} . \mathbf{a},\ \mathbf{a} . \mathbf{b} = \mathbf{b} . \mathbf{a}, \mathbf{b} . \mathbf{c} = \mathbf{c} . \mathbf{b}]$$

Example 14. Show that

$$\mathbf{i} \times (\mathbf{a} \times \mathbf{i}) + \mathbf{j} \times (\mathbf{a} \times \mathbf{j}) + \mathbf{k} \times (\mathbf{a} \times \mathbf{k}) = 2\mathbf{a}$$

Solution: $\mathbf{i} \times (\mathbf{a} \times \mathbf{i}) = (\mathbf{i} . \mathbf{i})\mathbf{a} - (\mathbf{i} . \mathbf{a})\mathbf{i}$

$$= \mathbf{a} - (\mathbf{i} . \mathbf{a})\mathbf{i} \qquad ...(1)$$

Similarly

$$\mathbf{j} \times (\mathbf{a} \times \mathbf{j}) = \mathbf{a} - (\mathbf{j} . \mathbf{a})\mathbf{j} \qquad \sim.\sim..(2)$$

and $$\mathbf{k} \times (\mathbf{a} \times \mathbf{k}) = \mathbf{a} - (\mathbf{k} . \mathbf{a})\mathbf{k} \qquad ...(3)$$

Adding (1), (2) and (3), we get

$$\mathbf{i} \times (\mathbf{a} \times \mathbf{i}) + \mathbf{j} \times (\mathbf{a} \times \mathbf{j}) + \mathbf{k} \times (\mathbf{a} \times \mathbf{k})$$

$$= 3\mathbf{a} - \{(\mathbf{i} . \mathbf{a})\mathbf{i} + (\mathbf{j} . \mathbf{a})\mathbf{j} + (\mathbf{k} . \mathbf{a})\mathbf{k}\}$$

$$= 3\mathbf{a} - \mathbf{a} = 2\mathbf{a} \qquad [\because \ (\mathbf{i} . \mathbf{a})\mathbf{i} + (\mathbf{j} . \mathbf{a})\mathbf{j} + (\mathbf{k} . \mathbf{a})\mathbf{k} = \mathbf{a}]$$

Example 15. Verify $\mathbf{a} \times (\mathbf{b} \times \mathbf{c}) = (\mathbf{a} . \mathbf{c})\,\mathbf{b} - (\mathbf{a} . \mathbf{b})\,\mathbf{c}$ given $\mathbf{a} = \mathbf{i} + 2\mathbf{j} + 3\mathbf{k}$, $\mathbf{b} = 2\mathbf{i} - \mathbf{j} + \mathbf{k}$ and $\mathbf{c} = 3\mathbf{i} + 2\mathbf{j} - 5\mathbf{k}$.

Solution. $\mathbf{a} . \mathbf{c} = (\mathbf{i} + 2\mathbf{j} + 3\mathbf{k}) . (3\mathbf{i} + 2\mathbf{j} - 5\mathbf{k})$

$$= (1)(3) + (2)(2) + (3)(-5) = -8$$

$$\mathbf{a} . \mathbf{b} = (\mathbf{i} + 2\mathbf{j} + 3\mathbf{k}) . (2\mathbf{i} - \mathbf{j} = \mathbf{k})$$

$$= (1)(2) + (2)(-1) + (3)(1) = 3$$

$$\therefore \qquad \mathbf{a} . \mathbf{b} = (\mathbf{i} + 2\mathbf{j} + \mathbf{j} + \mathbf{k}) - 3(3\mathbf{i} + 2\mathbf{j} - 5\mathbf{k})$$

$$= -25\mathbf{i} + 2\mathbf{j} + 7\mathbf{k} \qquad ...(1)$$

$$\mathbf{b} \times \mathbf{c} = \begin{vmatrix} \mathbf{i} & \mathbf{j} & \mathbf{k} \\ 2 & -1 & 1 \\ 3 & 2 & -5 \end{vmatrix} = 3\mathbf{i} + 13\mathbf{j} + 7\mathbf{k}$$

$$\mathbf{a} \times (\mathbf{b} \times \mathbf{c}) = \begin{vmatrix} \mathbf{i} & \mathbf{j} & \mathbf{k} \\ 1 & 2 & 3 \\ 3 & 13 & 7 \end{vmatrix} = -25\mathbf{i} + 2\mathbf{j} + 7\mathbf{k} \qquad ...(2)$$

From (1) and (2) $\mathbf{a} \times (\mathbf{b} \times \mathbf{c}) = (\mathbf{a} \cdot \mathbf{c})\mathbf{b} - (\mathbf{a} \cdot \mathbf{b})\mathbf{c}$

Example 16. Prove that $(\mathbf{b} \times \mathbf{c}) \times (\mathbf{c} \times \mathbf{a}) = [\mathbf{abc}]\mathbf{c}$ and hence deduce that

$$[\mathbf{b} \times \mathbf{c}, \mathbf{c} \times \mathbf{a}, \mathbf{a} \times \mathbf{b}] = [\mathbf{abc}]^2$$

Solution. Let $\mathbf{d} = \mathbf{b} \times \mathbf{c}$ Then

$$\begin{aligned}(\mathbf{b} \times \mathbf{c}) \times (\mathbf{c} \times \mathbf{a}) &= \mathbf{d} \times (\mathbf{c} \times \mathbf{a}) = (\mathbf{d} \cdot \mathbf{a})\mathbf{c} - (\mathbf{d} \cdot \mathbf{c})\mathbf{a} \\ &= \{(\mathbf{b} \times \mathbf{c}) \cdot \mathbf{a}\}\mathbf{c} - \{(\mathbf{b} \times \mathbf{c}) \cdot \mathbf{c}\}\mathbf{a} \\ &= [\mathbf{bca}]\mathbf{c} - [\mathbf{bcc}]\mathbf{a} \\ &= [\mathbf{abc}]\mathbf{c} \quad \because [\mathbf{bcc}] = 0 \text{ and } [\mathbf{bca}] = [\mathbf{abc}]\end{aligned}$$

Hence $(\mathbf{b} \times \mathbf{c}) \times (\mathbf{c} \times \mathbf{a}) = [\mathbf{abc}]\mathbf{c}$

$$\begin{aligned}\text{Now } [\mathbf{b} \times \mathbf{c}, \mathbf{c} \times \mathbf{a}, \mathbf{a} \times \mathbf{b}] &= [(\mathbf{b} \times \mathbf{c}) \times (\mathbf{c} \times \mathbf{a})] \cdot (\mathbf{a} \times \mathbf{b}) \\ &= [\mathbf{abc}]\, \mathbf{c} \cdot (\mathbf{a} \times \mathbf{b}) \\ &= [\mathbf{abc}]\,[\mathbf{cab}] = [\mathbf{abc}]^2\end{aligned}$$

Hence proved.

Example 17. Show that $(\mathbf{a} \times \mathbf{b}) \times \mathbf{c} = \mathbf{a} \times (\mathbf{b} \times \mathbf{c})$ if and only if $(\mathbf{a} \times \mathbf{c}) \times \mathbf{b} = 0$ or **a** and **c** are collinear.

Solution. Given $(\mathbf{a} \times \mathbf{b}) \times \mathbf{c} = \mathbf{a} \times (\mathbf{b} \times \mathbf{c})$ which is true if and only if

$$(\mathbf{c} \cdot \mathbf{a})\,\mathbf{b} - (\mathbf{c} \cdot \mathbf{b})\mathbf{a} = (\mathbf{a} \cdot \mathbf{c})\mathbf{b} - (\mathbf{a} \cdot \mathbf{b})\mathbf{c}$$

or $$(\mathbf{a} \cdot \mathbf{b})\mathbf{c} - (\mathbf{c} \cdot \mathbf{b})\mathbf{a} = 0$$

or $$(\mathbf{a} \times \mathbf{c}) \times \mathbf{b} = 0$$

Which concludes that either $\mathbf{b} = 0$ or $\mathbf{a} \times \mathbf{c} = 0$. But $\mathbf{b} \neq \mathbf{0}$

$\therefore \quad \mathbf{a} \times \mathbf{c} = \mathbf{0} \Rightarrow$ **b** and **c** collinear.

Example 18. If **a**, **b**, **c** be three vectors such that $\mathbf{a} \times (\mathbf{b} + \mathbf{c}) = \frac{1}{2}\mathbf{b}$ find the angles which **a** makes with **b** and **c**, **b** and **c** being non parallel.

Solution. $\mathbf{a} \times (\mathbf{b} \times \mathbf{c}) = (\mathbf{a} \cdot \mathbf{c})\mathbf{b} - (\mathbf{a} \cdot \mathbf{b})\mathbf{c} = \frac{1}{2}\mathbf{b}$ [given]

$$(\mathbf{a} \cdot \mathbf{c} - \frac{1}{2})\,\mathbf{b} - (\mathbf{a} \cdot \mathbf{b})\mathbf{c} = 0 \qquad ...(1)$$

which is of the form $\alpha\mathbf{b} - \beta\mathbf{c} = 0$ where **b** and **c** are non-parallel

$$\therefore \quad \mathbf{a} \cdot \mathbf{c} - \frac{1}{2} = 0 \text{ or } \mathbf{a} \cdot \mathbf{c} = \frac{1}{2} \text{ and } \mathbf{a} \cdot \mathbf{b} = 0$$

Let θ and ϕ be the angles which as makes with **b** and **c** respectively.

$\therefore$ **a, n, c** are unit vectors therefore

$$\mathbf{a} \cdot \mathbf{c} = \frac{1}{1} \Rightarrow \cos\theta = \frac{1}{2} \text{ or } \theta = 60°$$

and $\quad \mathbf{a} \cdot \mathbf{b} - 0 \Rightarrow \cos\phi = 0 \text{ or } \phi = 90°.$

Example 19. Show that $\mathbf{a} \times (\mathbf{b} \times \mathbf{c})$, $\mathbf{b} \times (\mathbf{c} \times \mathbf{a})$, $\mathbf{c} \times (\mathbf{a} \times \mathbf{b})$ are coplanar.

Solution. Let $\mathbf{A} = \mathbf{a} \times (\mathbf{b} \times \mathbf{c}) = (\mathbf{a} \cdot \mathbf{c})\mathbf{b} - (\mathbf{a} \cdot \mathbf{b})\mathbf{c}$

$$\mathbf{B} = \mathbf{b} \times (\mathbf{c} \times \mathbf{a}) = (\mathbf{b} \cdot \mathbf{a})\mathbf{c} - (\mathbf{b} \cdot \mathbf{c})\mathbf{a}$$

and $\quad \mathbf{C} = \mathbf{c} \times (\mathbf{a} \times \mathbf{b}) + (\mathbf{c} \cdot \mathbf{b}) = (\mathbf{c} \cdot \mathbf{b})\mathbf{a} - (\mathbf{c} \cdot \mathbf{a})\mathbf{b}$

Then **A, B, C** will be coplanar if $[\mathbf{ABC}] = 0$

i.e.,

$$[(\mathbf{a} \cdot \mathbf{c})\mathbf{b} - (\mathbf{a} \cdot \mathbf{b})\mathbf{c}] \cdot \{[(\mathbf{b} \cdot \mathbf{a})\,\mathbf{c} - (\mathbf{b} \cdot \mathbf{c})\mathbf{a}] \times [(\mathbf{c} \cdot \mathbf{b})\mathbf{a} - (\mathbf{c} \cdot \mathbf{a})\mathbf{b}]\} = 0$$

or $\quad [(\mathbf{a} \cdot \mathbf{c})\mathbf{b} - (\mathbf{a} \cdot \mathbf{b})] \cdot \{(\mathbf{b} \cdot \mathbf{a})\,(\mathbf{c} \cdot \mathbf{b})\,(\mathbf{c} \times \mathbf{a}) - (\mathbf{b} \cdot \mathbf{a})\,(\mathbf{c} \cdot \mathbf{a})\,(\times \mathbf{b})$
$\qquad - (\mathbf{b} \cdot \mathbf{c})\,(\mathbf{c} \cdot \mathbf{b})\,(\mathbf{a} \times \mathbf{a}) + (\mathbf{b} \cdot \mathbf{c})\,(\mathbf{c} \cdot \mathbf{a})\,(\mathbf{a} \times \mathbf{b})\} = 0$

or $\quad [\mathbf{a} \cdot \mathbf{c})\mathbf{b} - (\mathbf{a} \cdot \mathbf{b})\mathbf{c}] \cdot \{(\mathbf{b} \cdot \mathbf{a})\,(\mathbf{c} \cdot \mathbf{b})\,(\mathbf{c} \times \mathbf{a}) - (\mathbf{b} \cdot \mathbf{a})\,(\mathbf{c} \cdot \mathbf{a})\,(\mathbf{c} \times \mathbf{b})$
$\qquad + (\mathbf{b} \cdot \mathbf{c})\,(\mathbf{c} \cdot \mathbf{a})\,(\mathbf{a} \times \mathbf{b})\} = 0 \quad [\therefore \ \mathbf{a} \times \mathbf{a} = 0]$

or $\quad (\mathbf{a} \cdot \mathbf{c})\,(\mathbf{b} \cdot \mathbf{a})\,(\mathbf{c} \cdot \mathbf{b})\,[\mathbf{bca}] - (\mathbf{a} \cdot \mathbf{c})\,(\mathbf{b} \cdot \mathbf{a})\,(\mathbf{c} \cdot \mathbf{a})\,[\mathbf{bcb}]$
$\qquad + (\mathbf{a} \cdot \mathbf{c})\,(\mathbf{b} \cdot \mathbf{c})\,(\mathbf{c} \cdot \mathbf{a})\,[\mathbf{bab}] - (\mathbf{a} \cdot \mathbf{b})\,(\mathbf{b} \cdot \mathbf{a})\,(\mathbf{c} \cdot \mathbf{b})\,[\mathbf{cca}]$
$\qquad + (\mathbf{a} \cdot \mathbf{b})\,(\mathbf{b} \cdot \mathbf{a})\,(\mathbf{c} \cdot \mathbf{a})\,[\mathbf{ccb}] - (\mathbf{a} \cdot \mathbf{b})\,(\mathbf{b} \cdot \mathbf{c})\,(\mathbf{c} \cdot \mathbf{a})\,[\mathbf{cab}] = 0$

or $\quad (\mathbf{a} \cdot \mathbf{c})\,(\mathbf{b} \cdot \mathbf{a})\,(\mathbf{c} \cdot \mathbf{b})\,\{[\mathbf{bca}] - [\mathbf{cab}]\} = 0$

$[\because \ [\mathbf{bcb}] = [\mathbf{bab}] = [\mathbf{cca}] = [\mathbf{ccb}] = 0$
and $\mathbf{a} \cdot \mathbf{b} = \mathbf{b} \cdot \mathbf{a}, \ \mathbf{b} \cdot \mathbf{c} = \mathbf{c} \cdot \mathbf{b}, \ \mathbf{c} \cdot \mathbf{a} = \mathbf{a} \cdot \mathbf{c}]$

or $\quad (\mathbf{a} \cdot \mathbf{c})\,(\mathbf{b} \cdot \mathbf{a})\,(\mathbf{c} \cdot \mathbf{b})\,\{[\mathbf{bca}] - [\mathbf{bca}]\} = 0 \qquad [\because \ [\mathbf{bca}] = [\mathbf{cab}]]$

Hence the vectors **A, B, C** are coplanar

i.e.,**a** . (**b** × **c**), **b**(**c** × **a**), **c** . (**a** × **b**) are coplanar.

Example 20. Express a vector **r** as a linear combination of a vector **a** and another perpendicular to **a** and coplanar with **r** and **a.**

Solution. We know that the vector **a** × (**a** × **r**) is coplanar with **r** and **a** and is perpendicular to **a**. Now any vector **r** can be expressed as a linear combination of a vector **a** and another vector perpendicular to **a** and coplanar with **r** and **a**

let $$\mathbf{r} = \alpha\mathbf{a} + \beta\mathbf{a} \times (\mathbf{a} \times \mathbf{r}) \qquad ...(1)$$

where α, β are some scalars

Multiplying both sides of (1) scalarly with **a**, we get

$$\mathbf{r} \cdot \mathbf{a} = \alpha(\mathbf{a} \cdot \mathbf{a}) + \beta\, [\mathbf{a} \times (\mathbf{a} \times \mathbf{r})] \cdot \mathbf{a}$$

$$= \alpha(\mathbf{a} \cdot \mathbf{a}) + \beta\, [\mathbf{a}, \mathbf{a} \times \mathbf{r}, \mathbf{a}]$$

$$= \alpha\, (\mathbf{a} \cdot \mathbf{a}) \qquad \because [\mathbf{a}, \mathbf{a} \times \mathbf{r}, \mathbf{a}] = 0$$

$$\therefore \qquad \alpha = \frac{\mathbf{r} \cdot \mathbf{a}}{\mathbf{a} \cdot \mathbf{a}}$$

Again multiplying both sides of (1) vectorially with **a**, we get

$$\mathbf{r} \times \mathbf{a} = \alpha(\mathbf{a} \times \mathbf{a}) + \beta[\mathbf{a} \times (\mathbf{a} \times \mathbf{r})] \times \mathrm{a}$$

$$= \beta\, [(\mathbf{a} \cdot \mathbf{r})\mathbf{a} - (\mathbf{a} \cdot \mathbf{a}) \times \mathbf{a} \qquad [\because \mathbf{a} \times \mathbf{a} = 0]$$

$$= \beta\, (\mathbf{a} \cdot \mathbf{r})\, (\mathbf{a} \times \mathbf{a}) - \beta(\mathbf{a} \cdot \mathbf{a})\, (\mathbf{r} \times \mathbf{a})$$

$$= -\, \beta\, (\mathbf{a} \cdot \mathbf{a})\, (\mathbf{r} \times \mathbf{a})$$

$$\therefore \qquad \beta = -\frac{1}{\mathbf{a} \cdot \mathbf{a}}$$

Now substituting the values of α and β in (1) we get

$$\mathbf{r} = \frac{\mathbf{r} \cdot \mathbf{a}}{\mathbf{a} \cdot \mathbf{a}}\mathbf{a} - \frac{1}{\mathbf{a} \cdot \mathbf{a}}\mathbf{a} \times (\mathbf{a} \times \mathbf{r})$$

which shows that the components of **r** along and perpendicular to the direction of a are $\frac{\mathbf{r,a}}{\mathbf{a} \cdot \mathbf{a}}$ a and $-\frac{\mathbf{a} \times (\mathbf{a} \times \mathbf{r})}{\mathbf{a} \cdot \mathbf{a}}$.

EXERCISE

1. Find the value of **a** × (**b** × **c**) if

(i) **a** = **i** − 2**j** + **k**, **b** = 2**i** + **j** + **k**; **c** = **i** + 2**j** − **k**

(ii) $\mathbf{a} = 2\mathbf{i} - 10\mathbf{j} + 2\mathbf{k}$, $\mathbf{b} = 3\mathbf{i} + \mathbf{j} + 2\mathbf{k}$; $\mathbf{c} = 2\mathbf{i} + \mathbf{j} + 3\mathbf{k}$

2. If $\mathbf{a} = 2\mathbf{i} - \mathbf{j} + 3\mathbf{k}$, $\mathbf{b} = \mathbf{i} + \mathbf{j} - 3\mathbf{k}$, $\mathbf{c} = 3\mathbf{i} + 3\mathbf{j} + 2\mathbf{k}$ verify

(i) $\mathbf{a} \times (\mathbf{b} \times \mathbf{c}) = (\mathbf{a} \cdot \mathbf{c}) - (\mathbf{a} \cdot \mathbf{b})\mathbf{c}$

(ii) $\mathbf{b} \times (\mathbf{a} \times \mathbf{c}) = (\mathbf{b} \cdot \mathbf{c})\mathbf{a} - (\mathbf{b} \cdot \mathbf{a})\mathbf{c}$

3. If $\mathbf{a} = 2\mathbf{i} + 3\mathbf{j} + 4\mathbf{k}$, $\mathbf{b} = \mathbf{i} + \mathbf{j} - \mathbf{k}$, $\mathbf{c} = \mathbf{i} - \mathbf{j} + \mathbf{k}$ find $\mathbf{a} \times (\mathbf{b} \times \mathbf{c})$ and verify that $\mathbf{a} \times (\mathbf{b} \times \mathbf{c})$ is perpendicular to both $\mathbf{a}$ and $\mathbf{b} \times \mathbf{c}$.

4. If $\mathbf{a} = 3\mathbf{i} + 2\mathbf{j} - \mathbf{k}$ find $\mathbf{i} \times (\mathbf{a} \times \mathbf{i}) + \mathbf{j} \times (\mathbf{a} \times \mathbf{j}) + \mathbf{k} \times (\mathbf{a} \times \mathbf{k})$

5. Prove that $(\mathbf{a} \times \mathbf{b}) \times (\mathbf{a} \times \mathbf{c}) \cdot \mathbf{d} = (\mathbf{a} \cdot \mathbf{d})\,[\mathbf{abc}]$

6. Prove that $\mathbf{a} \times [\mathbf{b} \times (\mathbf{c} \times \mathbf{d})]$

$= (\mathbf{b} \cdot \mathbf{d})(\mathbf{a} \times \mathbf{c}) - (\mathbf{b} \cdot \mathbf{c})(\mathbf{a} \times \mathbf{d}) = [\mathbf{acd}]\mathbf{b} - (\mathbf{a} \cdot \mathbf{b})(\mathbf{c} \times \mathbf{d})$

and hence $\mathbf{a} \times [\mathbf{b} \times \{\mathbf{c} \times (\mathbf{d} \times \mathbf{e})\}]$

$= [(\mathbf{a} \cdot \mathbf{d})(\mathbf{c} \cdot \mathbf{e}) - (\mathbf{e} \cdot \mathbf{d})(\mathbf{a} \cdot \mathbf{e})]\,\mathbf{b} + (\mathbf{a} \cdot \mathbf{b})\,[(\mathbf{c} \cdot \mathbf{d})\mathbf{e} - (\mathbf{c} \cdot \mathbf{e})\mathbf{d}]$

ANSWERS

(1) (i) $-(9\mathbf{i} + 6\mathbf{j} + 3\mathbf{k})$, (ii) 0

(4) $6\mathbf{i} + 4\mathbf{j} - 2\mathbf{k}$

3.10. Scalar product of four vectors

If **a, b, c, d** be four vectors, then $\mathbf{a} \times \mathbf{b}$ and $\mathbf{c} \times \mathbf{d}$ are vector quantities and hence $(\mathbf{a} \times \mathbf{b}) \cdot (\mathbf{c} \times \mathbf{d})$ is a scalar quantity known as *scalar product of four vectors*

Theorem. To prove that

$$(\mathbf{a} \times \mathbf{b}) \cdot (\mathbf{c} \times \mathbf{d}) = \begin{vmatrix} \mathbf{a} \cdot \mathbf{c} & \mathbf{b} \cdot \mathbf{c} \\ \mathbf{a} \cdot \mathbf{d} & \mathbf{b} \cdot \mathbf{d} \end{vmatrix}$$

Proof. Let $\mathbf{c} \times \mathbf{d} = \mathbf{e}$

$\therefore \quad (\mathbf{a} \times \mathbf{b}) \cdot \mathbf{e} = \mathbf{a} \cdot (\mathbf{b} \times \mathbf{e})$ [$\because$ The dot and cross products can be interchangeable]

$$= \mathbf{a} \cdot [\mathbf{b} \times (\mathbf{c} \times \mathbf{d})]$$
$$= \mathbf{a} \cdot [(\mathbf{b} \cdot \mathbf{d})\mathbf{c} - (\mathbf{b} \cdot \mathbf{c})\mathbf{d}]$$
$$= (\mathbf{a} \cdot \mathbf{c})(\mathbf{b} \cdot \mathbf{d}) - (\mathbf{a} \cdot \mathbf{d})(\mathbf{b} \cdot \mathbf{c})$$
$$= \begin{vmatrix} \mathbf{a} \cdot \mathbf{c} & \mathbf{b} \cdot \mathbf{c} \\ \mathbf{a} \cdot \mathbf{d} & \mathbf{b} \cdot \mathbf{d} \end{vmatrix}$$

This relation is called *Lagrange's identity*

3.11. Vector product of four vectors

Again let **a, b, c, d** be four vectors. Since **a × b, c × d** be vector quantities (**a × b**) × (**c × d**) is also a vector quantity known as *vector product of four vectors*

Theorem. To prove that

$$(\mathbf{a} \times \mathbf{b}) \times (\mathbf{c} \times \mathbf{d}) = [\mathbf{abd}]\,\mathbf{c} - [\mathbf{abc}]\mathbf{d}$$
$$= [\mathbf{acd}]\,\mathbf{b} - [\mathbf{bcd}]\mathbf{a}$$

Proof. Let $\mathbf{a} \times \mathbf{b} = \mathbf{c}$

Then
$$(\mathbf{a} \times \mathbf{b}) \times (\mathbf{c} \times \mathbf{d}) = \mathbf{e} \times (\mathbf{c} \times \mathbf{d})$$
$$= (\mathbf{e} \,.\, \mathbf{d})\,\mathbf{c} - (\mathbf{e} \,.\, \mathbf{c})\,\mathbf{d}$$
$$= [(\mathbf{a} \times \mathbf{b}) \,.\, \mathbf{d}]\,\mathbf{c} - [(\mathbf{a} \times \mathbf{b})\,\mathbf{c}]\mathbf{d}$$
$$= [\mathbf{abd}]\mathbf{c} - [\mathbf{abc}]\mathbf{d} \qquad ...(1)$$

Again let $\mathbf{c} \times \mathbf{d} = \mathbf{f}$.

Then
$$(\mathbf{a} \times \mathbf{b}) \times (\mathbf{c} \times \mathbf{d}) = (\mathbf{a} \times \mathbf{b}) \times \mathbf{f}$$
$$= (\mathbf{a} \,.\, \mathbf{f})\mathbf{b} - (\mathbf{b} \,.\, \mathbf{f})\mathbf{a}$$
$$= [\mathbf{a} \,.\, (\mathbf{c} \times \mathbf{d})]\,\mathbf{b} - [\mathbf{b} \,.\, (\mathbf{c} \times \mathbf{d})]\mathbf{a}$$
$$= [\mathbf{acd}]\mathbf{b} - [\mathbf{bcd}]\mathbf{a} \qquad ...(2)$$

Hence (1) and (2) proves the theorem.

Deductions (i) From (1) and (2) of the above theorem we get

$$[\mathbf{abd}]\mathbf{c} - [\mathbf{abc}]\mathbf{d} = [\mathbf{acd}]\mathbf{b} - [\mathbf{bcd}]\mathbf{a}$$

or
$$[\mathbf{bcd}]\mathbf{a} - [\mathbf{acd}]\mathbf{b} + [\mathbf{abd}]\mathbf{c} - [\mathbf{abc}]\mathbf{d} = \mathbf{0}$$

which is a linear relation connecting the four vectors **a, b, c** and **d**

(ii) replacing **d** by **r** in above deduction we get

$$[\mathbf{bcr}]\mathbf{a} - [\mathbf{acr}]\mathbf{b} + [\mathbf{abr}]\mathbf{c} - [\mathbf{abc}]\mathbf{r} = \mathbf{0}$$

or
$$\mathbf{r} = \frac{[\mathbf{rbc}]\mathbf{a}}{[\mathbf{abc}]} + \frac{[\mathbf{rca}]\mathbf{b}}{[\mathbf{abc}]} + \frac{[\mathbf{rab}]\mathbf{c}}{[\mathbf{abc}]}$$

i.e., expression for **r** in terms of three non-coplanar vectors **a, b, c**

SOLVED PROBLEMS

Example 21. If $\mathbf{a} = \mathbf{i} + 2\mathbf{j} - \mathbf{k}$, $\mathbf{b} = 2\mathbf{i} + \mathbf{j} + 3\mathbf{k}$, $\mathbf{c} = \mathbf{i} - \mathbf{j} + \mathbf{k}$ and $\mathbf{d} = 3\mathbf{i} + \mathbf{j} + 2\mathbf{k}$ evaluate

(i) $(\mathbf{a} \times \mathbf{b}) \,.\, (\mathbf{c} \times \mathbf{d})$ and

(ii) $(\mathbf{a} \times \mathbf{b}) \times (\mathbf{c} \times \mathbf{d})$

Solution: $\mathbf{a} \times \mathbf{b} = \begin{vmatrix} \mathbf{i} & \mathbf{j} & \mathbf{k} \\ 1 & 2 & -1 \\ 2 & 1 & 3 \end{vmatrix}$

$= 7\mathbf{i} - 5\mathbf{j} - 3\mathbf{k}$

$\mathbf{c} \times \mathbf{d} = \begin{vmatrix} \mathbf{i} & \mathbf{j} & \mathbf{k} \\ 1 & -1 & 1 \\ 3 & 1 & 2 \end{vmatrix}$

$= -3\mathbf{i} + \mathbf{j} + 4\mathbf{k}$

Now (i) $(\mathbf{a} \times \mathbf{b}) \cdot (\mathbf{c} \times \mathbf{d}) = (7\mathbf{i} - 5\mathbf{j} - 3\mathbf{k}) \cdot (-3\mathbf{i} + \mathbf{j} + 4\mathbf{k})$

$= (7)(-3) + (-5)(1) + (-3)(4) = -38$

(ii) $(\mathbf{a} \times \mathbf{b}) \times (\mathbf{c} \times \mathbf{d}) = \begin{vmatrix} \mathbf{i} & \mathbf{j} & \mathbf{k} \\ 7 & -5 & -3 \\ -3 & 1 & 4 \end{vmatrix}$

$= -17\mathbf{i} - 19\mathbf{j} - 8\mathbf{k}$

Example 22. Prove that

$$(\mathbf{a} \times \mathbf{b}) \times (\mathbf{c} \times \mathbf{d}) + (\mathbf{a} \times \mathbf{c}) \times (\mathbf{d} \times \mathbf{b}) + (\mathbf{a} \times \mathbf{d}) \times (\mathbf{b} \times \mathbf{c}) = -2[\mathbf{bcd}]\mathbf{a}$$

Solution: Consider $(\mathbf{a} \times \mathbf{b}) \times (\mathbf{c} \times \mathbf{d})$ and $\mathbf{l} = \mathbf{a} \times \mathbf{b}$

Then $(\mathbf{a} \times \mathbf{b}) \times (\mathbf{c} \times \mathbf{d}) = \mathbf{l} \times (\mathbf{c} \times \mathbf{d})$

$= (\mathbf{l} \cdot \mathbf{d})\mathbf{c} - (\mathbf{l} \cdot \mathbf{c})\mathbf{d}$

$= [\mathbf{a} \times \mathbf{b}) \cdot \mathbf{d}]\, \mathbf{c} - [(\mathbf{a} \times \mathbf{b}) \cdot \mathbf{c}]\mathbf{d}$

$= [\mathbf{abd}]\mathbf{c} - [\mathbf{abc}]\mathbf{d}$ ...(1)

Similarly

$(\mathbf{a} \times \mathbf{c}) \times (\mathbf{d} \times \mathbf{b}) = [\mathbf{dba}]\mathbf{c} - [\mathbf{dbc}]\mathbf{a}$ ...(2)

and $(\mathbf{a} \times \mathbf{d}) \times (\mathbf{b} \times \mathbf{c}) = [\mathbf{abc}]\mathbf{d} - [\mathbf{bcd}]\mathbf{a}$ ...(3)

Adding (1), (2) and (3), we get

$(\mathbf{a} \times \mathbf{b}) \times (\mathbf{c} \times \mathbf{d}) + (\mathbf{a} \times \mathbf{c}) \times (\mathbf{d} \times \mathbf{b}) + (\mathbf{a} \times \mathbf{d}) \times (\mathbf{b} \times \mathbf{c})$

$= [\mathbf{abd}]\mathbf{c} - [\mathbf{abc}]\mathbf{d} + [\mathbf{dba}]\mathbf{c} - [\mathbf{dbc}]\mathbf{a} + [\mathbf{abc}]\mathbf{d} - [\mathbf{bcd}]\mathbf{a}$

$= [\mathbf{abd}]\mathbf{c} - [\mathbf{abd}]\mathbf{c} - [\mathbf{bcd}]\mathbf{a} - [\mathbf{bcd}]\mathbf{a}$

$= -2[\mathbf{bcd}]\mathbf{a}$

Hence the proof.

Example 23. Prove that

$$[\mathbf{a} \times \mathbf{p}, \mathbf{b} \times \mathbf{q}, \mathbf{c} \times \mathbf{r}] + [\mathbf{a} \times \mathbf{q}, \mathbf{b} \times \mathbf{r}, \mathbf{c} \times \mathbf{p}] + [\mathbf{a} \times \mathbf{r}, \mathbf{b} \times \mathbf{p}, \mathbf{c} \times \mathbf{q}] = 0$$

Solution. $[\mathbf{a} \times \mathbf{p}, \mathbf{b} \times \mathrm{q}, \mathbf{c} \times \mathbf{r}]$

$$= (\mathbf{a} \times \mathbf{p}) \,.\, [(\mathbf{b} \times \mathbf{q}) \times (\mathbf{c} \times \mathbf{r})]$$

$$= (\mathbf{a} \times \mathbf{p}) \,.\, \{[\mathbf{bqr}]\mathbf{c} - [\mathbf{bqc}]\mathbf{r}\}$$

$$= [\mathbf{apc}]\,[\mathbf{bqr}] - [\mathbf{apr}]\,[\mathbf{bqc}] \quad ...(1)$$

Similarly

$$[\mathbf{a} \times \mathbf{q}, \mathbf{b} \times \mathbf{r}, \mathbf{c} \times \mathbf{p}] = [\mathbf{bra}]\,[\mathbf{cpq}] - [\mathbf{bqr}]\,[\mathbf{apc}] \quad ...(2)$$

$$[\mathbf{a} \times \mathbf{r}, \mathbf{b} \times \mathbf{p}, \mathbf{c} \times \mathbf{q}] = [\mathbf{bqc}]\,[\mathbf{apr}] - [\mathbf{cpq}]\,[\mathbf{bra}] \quad ...(3)$$

Adding (1), (2) and (3), we get

$$[\mathbf{a} \times \mathbf{p}, \mathbf{b} \times \mathbf{q}, \mathbf{c} \times \mathbf{r}] + [\mathbf{a} \times \mathbf{q}, \mathbf{b} \times \mathbf{r}, \mathbf{c} \times \mathbf{p}] + [\mathbf{a} \times \mathbf{r}, \mathbf{b} \times \mathbf{p}, \mathbf{c} \times \mathbf{q}]$$

$$= [\mathbf{apc}]\,[\mathbf{bqr}] - [\mathbf{apr}]\,[\mathbf{bqc}] + [\mathbf{bra}]\,[\mathbf{cpq}]$$

$$- [\mathbf{bqr}]\,[\mathbf{apc}] + [\mathbf{bqc}]\,[\mathbf{apr}] - [\mathbf{cpq}]\,[\mathbf{bra}]$$

= 0. Hence the proof

Example 24. Prove that

$$[\mathbf{a} \times \mathbf{b}, \mathbf{c} \times \mathbf{d}, \mathbf{e} \times \mathbf{f}] = [\mathbf{abd}]\,[\mathbf{cef}] - [\mathbf{abc}]\,[\mathbf{def}]$$

$$= [\mathbf{abe}]\,[\mathbf{fcd}] - [\mathbf{abf}]\,[\mathbf{ecd}]$$

$$= [\mathbf{cda}]\,[\mathbf{bef}] - [\mathbf{cdb}]\,[\mathbf{aef}]$$

Solution. We have $[\mathbf{a} \times \mathbf{b}, \mathbf{c} \times \mathbf{d}, \mathbf{e} \times \mathbf{f}]$

$$= (\mathbf{a} \times \mathbf{b}) \,.\, \{(\mathbf{c} \times \mathbf{d}) \times (\mathbf{e} \times \mathbf{f})]$$

$$= (\mathbf{a} \times \mathbf{b}) \,.\, [\{(\mathbf{c} \times \mathbf{d}) \,.\, \mathbf{e}\}\mathbf{f} - \{(\mathbf{c} \times \mathbf{d}) \,.\, \mathbf{f}\}\mathbf{e}]$$

$$= (\mathbf{a} \times \mathbf{b}) \,.\, \{(\mathbf{cde})\mathbf{f} - [\mathbf{cdf}]\mathbf{e}\}$$

$$= [\mathbf{abf}]\,[\mathbf{cde}] - [\mathbf{cdf}]\,[\mathbf{abe}]$$

$$= -[\mathbf{abf}]\,[\mathbf{ecd}] + [\mathbf{fcd}]\,[\mathbf{abe}]$$

$$= [\mathbf{abe}]\,[\mathbf{fcd}] - [\mathbf{abf}]\,[\mathbf{ecd}] \quad ...(1)$$

Again $[\mathbf{a} \times \mathbf{b}, \mathbf{c} \times \mathbf{d}, \mathbf{e} \times \mathbf{f}] = [\mathbf{c} \times \mathbf{d}, \mathbf{e} \times \mathbf{f}, \mathbf{a} \times \mathbf{b}]$

$$= (\mathbf{c} \times \mathbf{d}) \,.\, [(\mathbf{e} \times \mathbf{f}) \times (\mathbf{a} \times \mathbf{b})]$$

$$= (\mathbf{c} \times \mathbf{d}) \,.\, [\{(\mathbf{e} \times \mathbf{f}) \,.\, \mathbf{b}\}\mathbf{a} - \{(\mathbf{e} \times \mathbf{f}) \,.\, \mathbf{a}\}\, \mathbf{b}]$$

$$= [\mathbf{cda}]\,[\mathbf{efb}] - [\mathbf{cdb}]\,[\mathbf{efa}]$$

$$= [\mathbf{cda}][\mathbf{bef}] - [\mathbf{cdb}]\,[\mathbf{aef}] \quad ...(2)$$

Again $[\mathbf{a} \times \mathbf{b}, \mathbf{c} \times \mathbf{d}, \mathbf{e} \times \mathbf{f}] = [\mathbf{e} \times \mathbf{f}, \mathbf{a} \times \mathbf{b}, \mathbf{c} \times \mathbf{d}]$

$$= (\mathbf{e} \times \mathbf{f}) \cdot [(\mathbf{a} \times \mathbf{b}) \times (\mathbf{c} \times \mathbf{d})]$$
$$= (\mathbf{e} \times \mathbf{f}) \cdot \{[\mathbf{abd}]\mathbf{c} - [\mathbf{abc}]\mathbf{d}\}$$
$$= [\mathbf{efc}]\,[\mathbf{abd}] - [\mathbf{efd}]\,[\mathbf{abc}]$$
$$= [\mathbf{abd}]\,[\mathbf{cef}] - [\mathbf{abc}]\,[\mathbf{def}] \quad ...(3)$$

Hence from (1), (2) and (3) the required results follow.

Example 25. Prove that

$$(\mathbf{b} \times \mathbf{c}) \cdot (\mathbf{a} \times \mathbf{d}) + (\mathbf{c} \times \mathbf{a}) \cdot (\mathbf{b} \times \mathbf{d}) + (\mathbf{a} \times \mathbf{b}) \cdot (\mathbf{c} \times \mathbf{d}) = 0$$

and deduce that

$$\sin (A + B) \sin (A - B) = \sin^2 A - \sin^2 B$$
$$= \cos^2 A - \sin^2 B$$

Solution: $(\mathbf{b} \times \mathbf{c}) \cdot (\mathbf{a} \times \mathbf{d}) = \begin{vmatrix} \mathbf{b.a} & \mathbf{b} \cdot \mathbf{d} \\ \mathbf{c} \cdot \mathbf{a} & \mathbf{c} \cdot \mathbf{d} \end{vmatrix}$

$$= (\mathbf{b} \cdot \mathbf{a})(\mathbf{c} \cdot \mathbf{d}) - (\mathbf{c} \cdot \mathbf{a})(\mathbf{b} \cdot \mathbf{d}) \quad ...(1)$$

Similarly

$$(\mathbf{c} \times \mathbf{a}) \cdot (\mathbf{b} \times \mathbf{d}) = (\mathbf{c} \cdot \mathbf{b})(\mathbf{a} \cdot \mathbf{d}) - (\mathbf{a} \cdot \mathbf{b})(\mathbf{c} \cdot \mathbf{d}) \quad ...(2)$$
$$(\mathbf{a} \times \mathbf{b}) \cdot (\mathbf{c} \times \mathbf{d}) = (\mathbf{a} \cdot \mathbf{c})(\mathbf{b} \cdot \mathbf{d}) - (\mathbf{b} \cdot \mathbf{c})(\mathbf{a} \cdot \mathbf{d}) \quad ...(3)$$

Adding (1), (2) and (3), we get

$$(\mathbf{b} \times \mathbf{c}) \cdot (\mathbf{a} \times \mathbf{d}) + (\mathbf{c} \times \mathbf{a}) \cdot (\mathbf{b} \times \mathbf{d}) + (\mathbf{a} \times \mathbf{b}) \cdot (\mathbf{c} \times \mathbf{d}) = 0$$

i.e. the first result is proved.

Now let $\angle AOC = A, \angle AOB = B$

$\therefore \quad \angle BOC = A - B$

Let $\angle BOD = A$

$\therefore \quad \angle COD = \angle BOD - \angle BOC$
$= A - (A - B) = B$

$\therefore \quad \angle AOD = \angle AOC + \angle COD$
$= A + B$

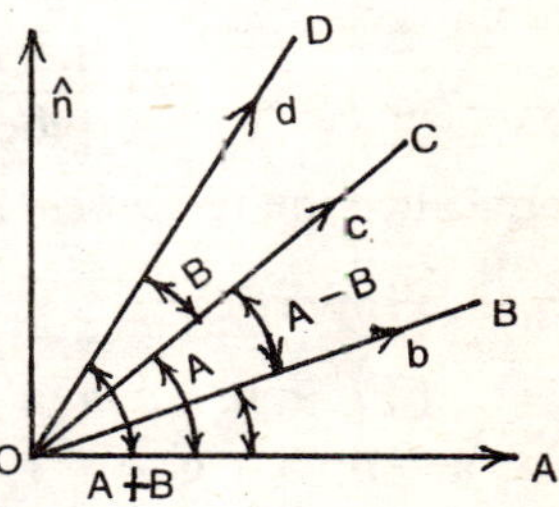

Fig. 3.2

Let $\hat{\mathbf{n}}$ be a unit vector perpendicular to the plane of **a, b, c, d** which are assumed to be coplanar.

Now we know that $\mathbf{a} \times \mathbf{b} = ab \sin \theta\, \hat{\mathbf{n}}$ where θ is the angle between the direction of **a** and **b** measured froms **a** in anticlockwise direction towards **b**. Moreever we proved that

$$(\mathbf{b} \times \mathbf{c}) \cdot (\mathbf{a} \times \mathbf{d}) + (\mathbf{c} \times \mathbf{a}) \cdot (\mathbf{b} \times \mathbf{d}) + (\mathbf{a} \times \mathbf{b}) \cdot (\mathbf{c} \times \mathbf{d}) = 0$$

or $$(\mathbf{b} \times \mathbf{c}) \cdot (\mathbf{a} \times \mathbf{d}) - (\mathbf{a} \times \mathbf{c}) \cdot (\mathbf{b} \times \mathbf{d}) + (\mathbf{a} \times \mathbf{b}) \cdot (\mathbf{c} \times \mathbf{d}) = 0$$

or $$bc \sin\angle BOC\, \hat{\mathbf{n}} \cdot ad \sin\angle AOD\, \hat{\mathbf{n}} - ac \sin\angle AOC\, \hat{\mathbf{n}} \cdot bd \sin\angle BOD\, \hat{\mathbf{n}} + ab \sin \angle AOB\, \hat{\mathbf{n}} \cdot cd \sin\angle COD\, \hat{\mathbf{n}} = 0$$

or $$abcd \sin (A - B) \sin (A + B) - abcd \sin A \sin A + abcd \sin B \sin B = 0$$

or $$\sin (A - B) \sin (A + B) = \sin^2 A - \sin^2 B \qquad ...(4)$$

$$[\because\ abcd \neq 0,\ \hat{\mathbf{n}} \cdot \hat{\mathbf{n}} = 1]$$

Again $$(\mathbf{a} \times \mathbf{b}) \cdot (\mathbf{c} \times \mathbf{d}) = \begin{vmatrix} \mathbf{a} \cdot \mathbf{c} & \mathbf{a} \cdot \mathbf{d} \\ \mathbf{b} \cdot \mathbf{c} & \mathbf{b} \cdot \mathbf{d} \end{vmatrix}$$

$$= (\mathbf{a} \cdot \mathbf{c}) (\mathbf{b} \cdot \mathbf{d}) - (\mathbf{b} \cdot \mathbf{c}) (\mathbf{a} \cdot \mathbf{d})$$

or $$(ab \sin B\, \hat{\mathbf{n}}) \cdot (cd \sin B\, \hat{\mathbf{n}}) = (ac \cos A) (bd \cos A) - [bc \cos(A - B)] [ad \cos(A + B)]$$

or $$\sin^2 B = \cos^2 A - \cos (A - B) \cos (A + B)$$

or $$\cos (A + B) \cos (A - B) = \cos^2 A - \sin^2 B \qquad ...(5)$$

Hence from (4) and (5) the second result is proved.

3.12. Reciprocal system of vectors

If **a, b, c** are any three non-coplanar vectors so that [abc] ≠ 0 then the three vectors **a′, b′, c′** defined by the equations

$$\mathbf{a}' = \frac{\mathbf{b} \times \mathbf{c}}{[\mathbf{abc}]}, \quad \mathbf{b}' = \frac{\mathbf{c} \times \mathbf{a}}{[\mathbf{abc}]}, \quad \mathbf{c}' = \frac{\mathbf{a} \times \mathbf{b}}{[\mathbf{abc}]}$$

are called the *reciprocal system of vectors* to **a, b, c**

3.13. Properties

(i) *If* **a, b, c** *and* **a′, b′, c′** *be reciprocal system of vectors then* $\mathbf{a} \cdot \mathbf{a}' = \mathbf{b} \cdot \mathbf{b}' = \mathbf{c} \cdot \mathbf{c}' = 1$

$$\mathbf{a} \cdot \mathbf{a}' = \mathbf{a} \cdot \frac{\mathbf{b} \times \mathbf{c}}{[\mathbf{abc}]} = \frac{\mathbf{a} \cdot (\mathbf{b} \times \mathbf{c})}{\mathbf{a} \cdot (\mathbf{b} \times \mathbf{c})} = 1$$

Similarly $\mathbf{b} \cdot \mathbf{b}' = \mathbf{c} \cdot \mathbf{c}' = 1$

(ii) *The product of any vector of one system with a vector of* *reciprocal system which does not correspond to it is zero. i.e.,* $\mathbf{a} \cdot \mathbf{b}' = 0$

$$\mathbf{a} \cdot \mathbf{b}' = \mathbf{a} \cdot \frac{\mathbf{c} \times \mathbf{a}}{[\mathbf{abc}]} = \frac{[\mathbf{aca}]}{[\mathbf{abc}]} = \frac{0}{[\mathbf{abc}]} = 0$$

Similarly $\mathbf{a} \cdot \mathbf{c}' = \mathbf{b} \cdot \mathbf{a}' = \mathbf{b} \cdot \mathbf{c}' = \mathbf{c} \cdot \mathbf{a}' = \mathbf{c} \cdot \mathbf{b}' = 0$

Note : Thus from above two properties we conclude that if $\mathbf{a}', \mathbf{b}', \mathbf{c}'$ be reciprocal system to **a, b, c** then **a, b, c** is a reciprocal system to $\mathbf{a}', \mathbf{b}', \mathbf{c}'$

(iii) *The scalar triple product* [**abc**] *of any three non–coplanar vectors* **a, b** *and* **c** *is reciprocal to the corresponding scalar triple product* [**a′b′c′**] *formed out of the reciprocal system of vectors* $\mathbf{a}', \mathbf{b}', \mathbf{c}'$

i.e., $$[\mathbf{a}'\,\mathbf{b}'\,\mathbf{c}']\,[\mathbf{abc}] = 1$$

Now $$[\mathbf{a}'\,\mathbf{b}'\,\mathbf{c}'] = \mathbf{a}' \cdot (\mathbf{b}' \times \mathbf{c}')$$

Substituting the values of $\mathbf{a}'$, $\mathbf{b}'$ and $\mathbf{c}'$ in terms of **a, b** and **c** we get

$$[\mathbf{a}'\,\mathbf{b}'\,\mathbf{c}'] = \frac{\mathbf{b} \times \mathbf{c}}{[\mathbf{abc}]} \cdot \left[\frac{\mathbf{c} \times \mathbf{a}}{[\mathbf{abc}]} \times \frac{\mathbf{a} \times \mathbf{b}}{[\mathbf{abc}]}\right]$$

$$= \frac{1}{[\mathbf{abc}]^3}\,[(\mathbf{b} \times \mathbf{c}) \cdot \{(\mathbf{c} \times \mathbf{a}) \times (\mathbf{a} \times \mathbf{b})\}]$$

$$= \frac{[\mathbf{abc}]^2}{[\mathbf{abc}]^3} = \frac{1}{[\mathbf{abc}]}$$

$$\therefore \quad [\mathbf{a}'\mathbf{b}'\mathbf{c}'] = \frac{1}{[\mathbf{abc}]} \quad \text{or} \quad [\mathbf{a}'\mathbf{b}'\mathbf{c}']\,[\mathbf{abc}] = 1$$

Note: The system of vectors **a, b, c** is right handed or left handed according as [**abc**] is positive or negative. By the property (iii) [**a′b′c′**] has the same sign as that of [**abc**] i.e., the system of vectors are either both right handed or left handed.

(iv) *The orthonormal triad of vectors* **i, j, k** *is self–reciprocal*

Let $\mathbf{i}', \mathbf{j}', \mathbf{k}'$ be the reciprocal system to **i, j, k** then

$$\mathbf{i}' = \frac{\mathbf{j} \times \mathbf{k}}{[\mathbf{ijk}]} = \frac{\mathbf{i}}{1} = \mathbf{i} \qquad [\because [\mathbf{ijk}] = 1]$$

Similarly $\mathbf{j}' = \mathbf{j}$ and $\mathbf{k}' = \mathbf{k}$

(v) If **a, b, c** be three non-coplanar vectors for which $[\mathbf{abc}] \neq 0$ and $\mathbf{a'}, \mathbf{b'}, \mathbf{c'}$ constitute the reciprocal system of vectors then to show that any vector **r** can be expressed as

$$\mathbf{r} = (\mathbf{r} \cdot \mathbf{a'})\mathbf{a} + (\mathbf{r} \cdot \mathbf{b'})\mathbf{b} + (\mathbf{r} \cdot \mathbf{c'})\mathbf{c}$$

Let **r** be expressed as a linear combination of the non-coplanar vectors **a, b, c** in the form

$$\mathbf{r} = x\mathbf{a} + y\mathbf{b} + z\mathbf{c} \qquad ...(1)$$

where x, y, z are some scalars. Multiplying both sides of (1) by $\mathbf{b} \times \mathbf{c}$, we get

$$\mathbf{r} \cdot (\mathbf{b} \times \mathbf{c}) = x\, \mathbf{a} \cdot (\mathbf{b} \times \mathbf{c}) + y\mathbf{b} \cdot (\mathbf{b} \times \mathbf{c}) + z\mathbf{c} \cdot (\mathbf{b} \times \mathbf{c})$$

or $$\mathbf{r} \cdot (\mathbf{b} \times \mathbf{c}) = x\, [\mathbf{abc}] + y[\mathbf{bbc}] + z[\mathbf{cbc}]$$

or $$x = \frac{\mathbf{r} \cdot (\mathbf{b} \times \mathbf{c})}{[\mathbf{abc}]} \qquad [\because \ [\mathbf{bbc}] = [\mathbf{cbc}] = 0$$

$$= \mathbf{r} \cdot \frac{\mathbf{b} \times \mathbf{c}}{[\mathbf{abc}]} = \mathbf{r} \cdot \mathbf{a'}$$

Similarly $y = \mathbf{r} \cdot \mathbf{b'}$ and $z = \mathbf{r} \cdot \mathbf{c'}$

Substituting the values of x, y, z in (1), we get

$$\mathbf{r} = (\mathbf{r} \cdot \mathbf{a'})\mathbf{a} + (\mathbf{r} \cdot \mathbf{b'})\mathbf{b} + (\mathbf{r} \cdot \mathbf{c'})\,\mathbf{c}$$

Note: (i) Similarly we can prove that

$$\mathbf{r} = (\mathbf{r} \cdot \mathbf{a})\mathbf{a'} + (\mathbf{r} \cdot \mathbf{b})\mathbf{b'} + (\mathbf{r} \cdot \mathbf{c})\,\mathbf{c'}$$

(ii) $$\mathbf{r} = (\mathbf{r} \cdot \mathbf{i})\mathbf{i} + (\mathbf{r} \cdot \mathbf{j})\mathbf{j} + (\mathbf{r} \cdot \mathbf{k})\mathbf{k}$$

and $$\mathbf{r} = (\mathbf{r} \cdot \mathbf{i'})\mathbf{i'} + (\mathbf{r} \cdot \mathbf{j'})\mathbf{j'} + (\mathbf{r} \cdot \mathbf{k'})\mathbf{k'}$$

SOLVED PROBLEMS

Example 26. Find a set of vectors reciprocal to the set

$$-\mathbf{i} + \mathbf{j} + \mathbf{k}, \mathbf{i} + \mathbf{j} + \mathbf{k}, \mathbf{i} + \mathbf{j} - \mathbf{k}$$

Solution. Let $\mathbf{a} = -\mathbf{i} + \mathbf{j} + \mathbf{k}$, $\mathbf{b} = \mathbf{i} + \mathbf{j} + \mathbf{k}$ and $\mathbf{c} = \mathbf{i} + \mathbf{j} - \mathbf{k}$, Then

$$[\mathbf{abc}] = \begin{vmatrix} -1 & 1 & 1 \\ 1 & 1 & 1 \\ 1 & 1 & -1 \end{vmatrix} = 4$$

$$\mathbf{b} \times \mathbf{c} = \begin{vmatrix} \mathbf{i} & \mathbf{j} & \mathbf{k} \\ 1 & 1 & 1 \\ 1 & 1 & -1 \end{vmatrix} = -2\mathbf{i} + 2\mathbf{j}$$

$$\mathbf{c} \times \mathbf{a} = \begin{vmatrix} \mathbf{i} & \mathbf{j} & \mathbf{k} \\ 1 & 1 & -1 \\ -1 & 1 & 1 \end{vmatrix} = 2\mathbf{i} + 2\mathbf{k}$$

$$\mathbf{a} \times \mathbf{b} = \begin{vmatrix} \mathbf{i} & \mathbf{j} & \mathbf{k} \\ -1 & 1 & 1 \\ 1 & 1 & 1 \end{vmatrix} = 2\mathbf{j} - 2\mathbf{k}$$

Now $\mathbf{a}' = \dfrac{\mathbf{b} \times \mathbf{c}}{[\mathbf{abc}]} = \dfrac{1}{4}(-2\mathbf{i} + 2\mathbf{j}) = \dfrac{1}{2}(-\mathbf{i} + \mathbf{j})$

$$\mathbf{b}' = \frac{\mathbf{c} \times \mathbf{a}}{[\mathbf{abc}]} = \frac{1}{2}(\mathbf{i} + \mathbf{j})$$

and $\mathbf{c}' = \dfrac{\mathbf{a} \times \mathbf{b}}{[\mathbf{abc}]} = \dfrac{1}{2}(\mathbf{j} - \mathbf{k})$

Example 27. Find the set of vectors reciprocal to the vectors **a, b** and **a × b.**

Solution: Let **c = a × b**. Then the vectors are a, b and c. Let **a′, b′, c′** the reciprocal system.

Now $[\mathbf{abc}] = (\mathbf{a} \times \mathbf{b}).\mathbf{c} = (\mathbf{a} \times \mathbf{b}) . (\mathbf{a} \times \mathbf{b}) = |\mathbf{a} \times \mathbf{b}|^2$

$\therefore \quad \mathbf{a}' = \dfrac{\mathbf{b} \times \mathbf{c}}{[\mathbf{abc}]} = \dfrac{\mathbf{b} \times (\mathbf{a} \times \mathbf{b})}{|\mathbf{a} \times \mathbf{b}|^2}$

$$\mathbf{b}' = \frac{\mathbf{c} \times \mathbf{a}}{[\mathbf{abc}]} = \frac{(\mathbf{a} \times \mathbf{b}) \times \mathbf{a}}{|\mathbf{a} \times \mathbf{b}|^2}$$

and $\mathbf{c}' = \dfrac{\mathbf{a} \times \mathbf{b}}{[\mathbf{abc}]} = \dfrac{\mathbf{a} \times \mathbf{b}}{|\mathbf{a} \times \mathbf{b}|^2}$

Example 28. If **a, b, c** be a set of non-coplanar vectors and

$$\mathbf{a}' = \frac{\mathbf{b} \times \mathbf{c}}{[\mathbf{abc}]},\ \mathbf{b}' = \frac{\mathbf{c} \times \mathbf{a}}{[\mathbf{abc}]},\ \mathbf{c}' = \frac{\mathbf{a} \times \mathbf{b}}{[\mathbf{abc}]}$$

then prove that

$$\mathbf{a} = \frac{\mathbf{b}' \times \mathbf{c}'}{[\mathbf{a}'\,\mathbf{b}'\,\mathbf{c}']},\ \mathbf{b} = \frac{\mathbf{c}' \times \mathbf{a}'}{[\mathbf{a}'\,\mathbf{b}'\,\mathbf{c}']},\ \mathbf{c} = \frac{\mathbf{a}' \times \mathbf{b}'}{[\mathbf{a}'\mathbf{b}'\mathbf{c}']}$$

Solution. $\mathbf{b}' \times \mathbf{c}' = \dfrac{\mathbf{c} \times \mathbf{a}}{[\mathbf{abc}]} \times \dfrac{\mathbf{a} \times \mathbf{b}}{[\mathbf{abc}]}$

$$= \frac{(\mathbf{c} \times \mathbf{a}) \times (\mathbf{a} \times \mathbf{b})}{[\mathbf{abc}]^2}$$

$$= \frac{[(\mathbf{c} \times \mathbf{a}) \,.\, \mathbf{b}]\mathbf{a} - [(\mathbf{c} \times \mathbf{a}) \,.\, \mathbf{a}]\mathbf{b}}{[\mathbf{abc}]^2}$$

$$= \frac{[\mathbf{cab}]\mathbf{a} - [\mathbf{caa}]\mathbf{b}}{[\mathbf{abc}]^2}$$

$$= \frac{[\mathbf{abc}]\mathbf{a}}{[\mathbf{abc}]^2} \qquad [\because\ [\mathbf{caa}] = 0]$$

$$\therefore \qquad \mathbf{b}' \times \mathbf{c}' = \frac{\mathbf{a}}{[\mathbf{abc}]}$$

Also $\quad [\mathbf{a}'\mathbf{b}'\mathbf{c}'] = \dfrac{1}{[\mathbf{abc}]}$ [From property (iii), **art 3.13**]

$$\therefore \qquad \mathbf{b}' \times \mathbf{c}' = \mathbf{a}\,[\mathbf{a}'\mathbf{b}'\mathbf{c}']$$

or
$$\mathbf{a} = \frac{\mathbf{b}' \times \mathbf{c}'}{[\mathbf{a}'\mathbf{b}'\mathbf{c}']}$$

Similarly $\quad \mathbf{b} = \dfrac{\mathbf{c}' \times \mathbf{a}'}{[\mathbf{a}'\mathbf{b}'\mathbf{c}']}$ and $\mathbf{c} = \dfrac{\mathbf{a}' \times \mathbf{b}'}{[\mathbf{a}'\mathbf{b}'\mathbf{c}']}$.

Example 29. If **a, b, c** and **a′, b′, c′** are reciprocal system of vectors prove that

(i) $\mathbf{a} \times \mathbf{a}' + \mathbf{b} \times \mathbf{b}' + \mathbf{c} \times \mathbf{c}' = \mathbf{0}$

(ii) $\mathbf{a}' \times \mathbf{b}' + \mathbf{b}' \times \mathbf{c}' + \mathbf{c}' \times \mathbf{a}' = \dfrac{\mathbf{a} + \mathbf{b} + \mathbf{c}}{[\mathbf{abc}]}$

(iii) $\mathbf{a} \,.\, \mathbf{a}' + \mathbf{b} \,.\, \mathbf{b}' + \mathbf{c} \,.\, \mathbf{c}' = 3$

Solution: If **a, b, c** and **a′, b′,** c are reciprocal system of vectors then

$$\mathbf{a}' = \frac{\mathbf{b} \times \mathbf{c}}{[\mathbf{abc}]},\quad \mathbf{b}' = \frac{\mathbf{c} \times \mathbf{a}}{[\mathbf{abc}]},\quad \mathbf{c}' = \frac{\mathbf{a} \times \mathbf{b}}{[\mathbf{abc}]}$$

(i) $\mathbf{a} \times \mathbf{a}' + \mathbf{b} \times \mathbf{b}' + \mathbf{c} \times \mathbf{c}'$

$$= \mathbf{a} \times \frac{\mathbf{b} \times \mathbf{c}}{[\mathbf{abc}]} + \mathbf{b} \times \frac{\mathbf{c} \times \mathbf{a}}{[\mathbf{abc}]} + \mathbf{c} \times \frac{\mathbf{a} \times \mathbf{b}}{[\mathbf{abc}]}$$

$$= \frac{1}{[\mathbf{abc}]}\,[\mathbf{a} \times (\mathbf{b} \times \mathbf{c}) + \mathbf{b} \times (\mathbf{c} \times \mathbf{a}) + \mathbf{c} \times (\mathbf{a} \times \mathbf{b})]$$

$$= \frac{1}{[\mathbf{abc}]}\,(\mathbf{0}) = 0 \qquad [\textit{Refer example } \mathbf{(13)}]$$

(ii) $\mathbf{a}' \times \mathbf{b}' + \mathbf{b}' \times \mathbf{c}' + \mathbf{c}' \times \mathbf{a}'$

$$= \frac{\mathbf{c}}{[\mathbf{abc}]} + \frac{\mathbf{a}}{[\mathbf{abc}]} + \frac{\mathbf{b}}{[\mathbf{abc}]}$$ [*Refer example* **(28)**]

$$= \frac{\mathbf{a} + \mathbf{b} + \mathbf{c}}{[\mathbf{abc}]}$$

(iii) $\mathbf{a} \,.\, \mathbf{a}' + \mathbf{b} \,.\, \mathbf{b}' + \mathbf{c} \,.\, \mathbf{c}'$

$$= \mathbf{a} \,.\, \frac{\mathbf{b} \times \mathbf{c}}{[\mathbf{abc}]} + \mathbf{b} \,.\, \frac{\mathbf{c} \times \mathbf{a}}{[\mathbf{abc}]} + \mathbf{c} \,.\, \frac{\mathbf{a} \times \mathbf{b}}{[\mathbf{abc}]}$$

$$= \frac{\mathbf{a} \,.\, (\mathbf{b} \times \mathbf{c})}{[\mathbf{abc}]} + \frac{\mathbf{b} \,.\, (\mathbf{c} \times \mathbf{a})}{[\mathbf{abc}]} + \frac{\mathbf{c} \,.\, (\mathbf{a} \times \mathbf{b})}{[\mathbf{abc}]}$$

$$= \frac{[\mathbf{abc}]}{[\mathbf{abc}]} + \frac{[\mathbf{bca}]}{[\mathbf{bca}]} + \frac{[\mathbf{cab}]}{[\mathbf{cab}]}$$

$$= 1 + 1 + 1 = 3$$

EXERCISE

1. If $\mathbf{a} = 2\mathbf{i} + \mathbf{j} - \mathbf{k}$, $\mathbf{b} = -\mathbf{i} + 2\mathbf{j} - 4\mathbf{k}$, $\mathbf{c} = \mathbf{i} + \mathbf{j} + \mathbf{k}$, find $(\mathbf{a} \times \mathbf{b}) \,.\, (\mathbf{a} \times \mathbf{c})$
2. If $\mathbf{a} = \mathbf{i} + 2\mathbf{j} - \mathbf{k}$, $\mathbf{b} = 3\mathbf{i} - 4\mathbf{k}$, $\mathbf{c} = -\mathbf{i} + \mathbf{j}$ and $\mathbf{d} = 2\mathbf{i} - \mathbf{j} - 3\mathbf{k}$ then find $(\mathbf{a} \times \mathbf{b}) \,.\, (\mathbf{c} \times \mathbf{d})$ and $(\mathbf{a} \times \mathbf{b}) \times (\mathbf{c} \times \mathbf{d})$.
3. If **a** and **b** lie in a plane normal to the plane containing **c** and **d** show that $(\mathbf{a} \times \mathbf{b}) \,.\, (\mathbf{c} \times \mathbf{d}) = 0$
4. If **a**,**b**, **c**, **d** are coplanar show that $(\mathbf{a} \times \mathbf{b}) \times (\mathbf{c} \times \mathbf{d}) = 0$
5. If $\mathbf{a} = \mathbf{i} + \mathbf{j} - \mathbf{k}$, $\mathbf{b} = 2\mathbf{i} + \mathbf{j} - 3\mathbf{k}$, $\mathbf{c} = \mathbf{i} - \mathbf{j} + 3\mathbf{k}$ and $\mathbf{d} = 3\mathbf{i} + 4\mathbf{j} - 2\mathbf{k}$ then find

   $$(\mathbf{a} \times \mathbf{b}) \,.\, (\mathbf{c} \times \mathbf{d}) + (\mathbf{c} \times \mathbf{a}) \,.\, (\mathbf{b} \times \mathbf{d}) + (\mathbf{d} \times \mathbf{a}) \,.\, (\mathbf{c} \times \mathbf{b})$$

6. If $\mathbf{a} = a_1\mathbf{i} + a_2\mathbf{j} + a_3\mathbf{k}$, $\mathbf{b} = b_1\mathbf{i} + b_2\mathbf{j} + b_3\mathbf{k}$, $\mathbf{c} = c_1\mathbf{i} + c_2\mathbf{j} + c_3\mathbf{k}$ and $\mathbf{d} = d_1\mathbf{i} + d_2\mathbf{j} + d_3\mathbf{k}$ then show that

   $$2(\mathbf{a} \times \mathbf{b}) \times (\mathbf{c} \times \mathbf{d}) = \begin{vmatrix} -\mathbf{a} & -\mathbf{b} & \mathbf{c} & \mathbf{d} \\ a_1 & b_1 & c_1 & d_1 \\ a_2 & b_2 & c_2 & d_2 \\ a_3 & b_3 & c_3 & d_3 \end{vmatrix}$$

7. Find a set of vectors reciprocal to the set of vectors

$$2\mathbf{i} + 3\mathbf{j} - \mathbf{k},\ \mathbf{i} - \mathbf{j} - 2\mathbf{k},\ -\mathbf{i} + 2\mathbf{j} + 2\mathbf{k}$$

8. If **a, b, c** are non coplanar then show that

$$\mathbf{r} = \frac{(\mathbf{r} \cdot \mathbf{a})\mathbf{b} \times \mathbf{c}}{[\mathbf{abc}]} + \frac{(\mathbf{r} \cdot \mathbf{b}) \times \mathbf{a}}{[\mathbf{abc}]} + \frac{(\mathbf{r} \cdot \mathbf{c})\, \mathbf{a} \times \mathbf{b}}{[\mathbf{abc}]}$$

9. Show that the fundamental vectors $\{\mathbf{i}, \mathbf{j}, \mathbf{k}\}$ in itself are reciprocal.

10. If the system of vectors $\mathbf{e}^1, \mathbf{e}^2, \mathbf{e}^3$ is reciprocal to the system $\mathbf{e}_1, \mathbf{e}_2, \mathbf{e}_3$ show that any vector **a** satisfies the relations

$$\mathbf{a} = (\mathbf{a} \cdot \mathbf{e}^1)\mathbf{e}_1 + (\mathbf{a} \cdot \mathbf{e}^2)\mathbf{e}_2 + (\mathbf{a} \cdot \mathbf{e}^3)\mathbf{e}_3$$

and $\mathbf{a} = (\mathbf{a} \cdot \mathbf{e}_1)\mathbf{e}^1 + (\mathbf{a} \cdot \mathbf{e}_2)\mathbf{e}^2 + (\mathbf{a} \cdot \mathbf{e}_3)\mathbf{e}^3$

ANSWERS

(1) -26 (2) $-15,\ 17\mathbf{i} - 26\mathbf{j} - 27\mathbf{k}$ (5) 0

(6) $\mathbf{a}' = \dfrac{2\mathbf{i} + \mathbf{k}}{3},\ \mathrm{b}^1 = \dfrac{-8\mathbf{i} + 7\mathbf{j} - 7\mathbf{k}}{3},\ \mathrm{c}' = \dfrac{-7\mathbf{i} + 3\mathbf{j}\ 5\mathbf{k}}{3}$

4

Centroid

(Vector Method)

4.1. Centroid

Definition If there be n points whose position vectors relative to any origin O be given by **a, b, c, d, ...** then the point G whose position vector is

$$\vec{OG} = \frac{\mathbf{a} + \mathbf{b} + \mathbf{c} + \mathbf{d} + \ldots\ldots}{n}$$

is called the **centroid** or **centre of the mean position** of the given points

Again if there be *n* Scalars $m_1, m_2, m_3, \ldots., m_n$ then the point G whose position vector is

$$\vec{OG} = \frac{m_1\mathbf{a} + m_2\mathbf{b} + m_3\mathbf{c} + \ldots\ldots.}{m_1 + m_2 + m_3 + \ldots\ldots}$$

is called the *centroid* of the given points with *associated numbers* $m_1, m_2, m_3, \ldots., m_n$.

Note : *The word associated number is to cover up all forms of centroids.* **For example :** *they may stand for the masses* $m_1, m_2, m_3, \ldots..$ *of a system of particles placed at given set of points, then the centroid is called the* **centre of mass.** Similarly **centre of gravity, centre of parallel forces** etc.

4.2. Properties

(1) *The centroid is independent of the origin of vectors*

Let O be the origin and O′ be the new origin whose position

vector referred to O is $\mathbf{r}$. Let G be the centroid of points whose position vectors referred to O be $\mathbf{a}, \mathbf{b}, \mathbf{c}$ ...., with associated numbers m_1, m_2, m_3, ..., respectively

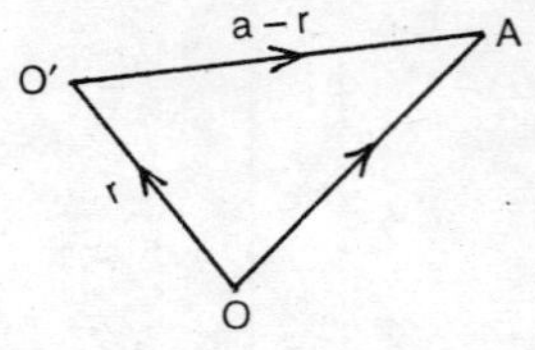

Fig. 4.1

Then $\overrightarrow{OG} = \dfrac{m_1\mathbf{a} + m_2\mathbf{b} + m_3\mathbf{c} + ...}{m_1 + m_2 + m_3 + ...}$ ...(1)

Consider the point A with associate number m_1.

If $\overrightarrow{OA} = \mathbf{a}$ then $\overrightarrow{O'A} = \mathbf{a} - \mathbf{r}$. Similarly

$\overrightarrow{OB} = \mathbf{b}$, $\overrightarrow{OC} = \mathbf{c}$, ... then $\overrightarrow{O'B} = \mathbf{b} - \mathbf{r}$, $\overrightarrow{O'C} = \mathbf{c} - \mathbf{r}$, ...

If G′ be the new centroid then

$$\overrightarrow{O'G'} = \frac{m_1(\mathbf{a} - \mathbf{r}) + m_2(\mathbf{b} - \mathbf{r}) + m_3(\mathbf{c} - \mathbf{r}) + ...}{m_1 + m_2 + m_3 + ...}$$

$$= \frac{(m_1\mathbf{a} + m_2\mathbf{b} + m_3\mathbf{c} + ...)}{m_1 + m_2 + m_3 + ...} - \mathbf{r}$$

$$\therefore \quad \overrightarrow{O'G'} = \overrightarrow{OG} - \mathbf{r} \qquad \text{[Using (1)]}$$

$$= \overrightarrow{OG} - \overrightarrow{OO'}, = \overrightarrow{O'G} \qquad [\because \overrightarrow{OO'} = \mathbf{r}]$$

Thus the points G and G′ coincide ⇒ the centroid is independent of the origin of vectors

(2) *The vector relation* $p\mathbf{a} + q\mathbf{b} + r\mathbf{c} + ... = \mathbf{0}$ *will be independent of the origin iff* $p + q + r + ... = 0$, where p, q, r ..... *are scalars*.

Let O be the origin with reference to which we have the relation

$$p\mathbf{a} + q\mathbf{b} + r\mathbf{c} + ... = \mathbf{0} \qquad ...(1)$$

Let O′ be the new origin whose P.V relative to O is $\mathbf{i}$, then w.r.t the new origin O′ the relation (1) becomes

$$p(\mathbf{a} - \mathbf{i}) + q(\mathbf{b} - \mathbf{i}) + r(\mathbf{c} - \mathbf{i}) + ... = \mathbf{0}$$

or $$(p\mathbf{a} + q\mathbf{b} + r\mathbf{c} + ...) - \mathbf{i}(p + q + r + ...) = \mathbf{0} \qquad ...(2)$$

Now if the relation (1) is independent of the origin then (1) and (2) should represent the same relation which is possible only if

$$p + q + r + ... = 0$$

(3) *If* G *be the centroid of a system of points* A, B, C, ... *whose position vectors are* **a, b, c**,....*with associated numbers* p, q, r .... *and* G′ *that of another system of points* A′, B′, C′, ... *whose position vectors are* **a′, b′, c′**, ... *with associated numbers* p′, q′, r′, ... *then the centroid of all the points taken together is the centroid of the points* G *and* G′ *with associated numbers* (p + q + r + ...) *and* (p′ + q′ + r′ + ...) *respectively*

By definition

$$\vec{OG} = \frac{p\mathbf{a} + q\mathbf{b} + r\mathbf{c} + ...}{p + q + r + ...} \quad ....(1)$$

and

$$\vec{OG'} = \frac{p'\mathbf{a'} + q'\mathbf{b'} + r'\mathbf{c'} + ...}{p' + q' + r' + ...} \quad ...(2)$$

Let H be the centroid of the points G and G′ with associated numbers (p + q + r + ...) and (p′ + q′ + r′ + ...) respectively. Then

$$\vec{OH} = \frac{(p + q + r + ...)\vec{OG} + (p' + q' + r' + ...)\vec{OG'}}{(p + q + r + ...) + (p' + q' + r' + ...)}$$

$$= \frac{(p\mathbf{a} + q\mathbf{b} + r\mathbf{c} + ...) + (p'\mathbf{a'} + q'\mathbf{b'} + r'\mathbf{c'} + ...)}{(p + q + r + ...) + (p' + q' + r' + ...)}$$

[From (1) and (2)]

showing that H is the centroid of all the points taken together.

SOLVED PROBLEMS

Example 1. Find the centroid of 3n points

i, 8i, 27i, ..., n^3i, j, 8j, 27j, ..., n^3j and k, 8k, 27k, ..., n^3k

Solution: Let G_1, G_2 and G_3 be the centroids of the three systems of points with reference to O as origin. Then

$$\vec{OG_1} = \frac{\mathbf{i} + 8\mathbf{i} + 27\mathbf{i} + ... + n^3\mathbf{i}}{n}$$

$$= \frac{1^3 + 2^3 + 3^3 + ... + n^3)\,\mathbf{i}}{n} = \frac{\frac{1}{4}n^2(n+1)^2\,\mathbf{i}}{n}$$

$$= \frac{1}{4}n(n+1)^2\,\mathbf{i} \quad \left[\because\ 1^3 + 2^3 + 3^3 + ... + n^3 = \frac{n^2(n+1)^2}{4}\right]$$

Similarly $\vec{OG}_2 = \frac{1}{4} n (n + 1)^2 \mathbf{j}$ and $\vec{OG}_3 = \frac{1}{4} n (n + 1)^2 \mathbf{k}$

If G is the required centroid then

$$\vec{OG} = \frac{1}{3} [\vec{OG}_1 + \vec{OG}_2 + \vec{OG}_3]$$

$$= \frac{1}{3} \left[\frac{1}{4} n (n + 1)^2 \mathbf{i} + \frac{1}{4} n (n + 1)^2 \mathbf{j} + \frac{1}{4} n (n + 1)^2 \mathbf{k} \right]$$

$$= \frac{n (n + 1)^2}{12} (\mathrm{i} + \mathrm{j} + \mathrm{k}).$$

Example 2. Find the centroid of the points P (2, 1, 3), Q (- 1, 1, 5), R (3, 1, - 7) with associated numbers 2, 3, 4.

Solution: Let O be the origin and a, b, c be the position vectors of P, Q, R relative to O. Then

$$\mathbf{a} = 2\mathbf{i} + \mathbf{j} + 3\mathbf{k},\ \mathbf{b} = -\mathbf{i} + \mathbf{j} + 5\mathbf{k},\ \mathbf{c} = 3\mathbf{i} + \mathbf{j} - 7\mathbf{k}$$

Let G be the centroid.

$$\therefore \quad \vec{OG} = \frac{2\mathbf{a} + 3\mathbf{b} + 4\mathbf{c}}{2 + 3 + 4} = \frac{1}{9} (2\mathbf{a} + 3\mathbf{b} + 4\mathbf{c})$$

$$= \frac{1}{9} [2(2\mathbf{i} + \mathbf{j} + 3\mathbf{k}) + 3(-\mathbf{i} + \mathbf{j} + 5\mathbf{k}) + 4(3\mathbf{i} + \mathbf{j} - 7\mathbf{k})]$$

$$= \frac{1}{9} (13\mathbf{i} + 9\mathbf{j} - 7\mathbf{k}) = \frac{13}{9} \mathbf{i} + \mathbf{j} - \frac{7}{9} \mathbf{k}$$

$\therefore$ The centroid $G = \left(\frac{13}{9}, 1, -\frac{7}{9} \right)$.

Example 3. If G be the centroid of ΔABC, show that

$$\vec{GA} + \vec{GB} + \vec{GC} = \mathbf{0}$$

Solution: If the P.V's of vertices be **a, b, c** w.r.t origin O i.e., $\vec{OA} = \mathbf{a}$, $\vec{OB} = \mathbf{b}$ and $\vec{OC} = \mathbf{c}$ then

$$\vec{OG} = \frac{\mathbf{a} + \mathbf{b} + \mathbf{c}}{3}.$$

$$\vec{GA} = \vec{OA} - \vec{OG} = \mathbf{a} - \frac{\mathbf{a} + \mathbf{b} + \mathbf{c}}{3} = \frac{2\mathbf{a} - \mathbf{b} - \mathbf{c}}{3}$$

Similarly

$$\vec{GB} = \frac{2\mathbf{b} - \mathbf{c} - \mathbf{a}}{3} \text{ and } \vec{GC} = \frac{2\mathbf{c} - \mathbf{a} - \mathbf{b}}{3}$$

Hence $\vec{GA} + \vec{GB} + \vec{GC} = \mathbf{0}$

Example 4. If G_1 is the mean centre of A_1, B_1, C_1 and G_2 that of A_2, B_2, C_2 then show that

$$\vec{A_1A_2} + \vec{B_1B_2} + \vec{C_1C_2} = 3\vec{G_1G_2}$$

Solution. Let the P.V.'s of A_1, B_1, C_1 and A_2, B_2, C_2 w.r.t O as origin be $\mathbf{a_1}, \mathbf{b_1}, \mathbf{c_1}$ and $\mathbf{a_2}, \mathbf{b_2}, \mathbf{c_2}$ i.e.,

$$\vec{OA_1} = \mathbf{a_1}, \vec{OB_1} = \mathbf{b_1}, \vec{OC_1} = \mathbf{c_1}, \vec{OA_2} = \mathbf{a_2}$$

$\vec{OB_2} = \mathbf{b_2}$ and $\vec{OC_2} = \mathbf{c_2}$. Hence

$\vec{A_1A_2} = \vec{OA_2} - \vec{OA_1} = \mathbf{a_2} - \mathbf{a_1}$. Similarly

$\vec{B_1B_2} = \mathbf{b_2} - \mathbf{b_1}$ and $\vec{C_1C_2} = \mathbf{c_2} - \mathbf{c_1}$. Now

$\vec{OG_1} = \frac{1}{3}(\mathbf{a_1} + \mathbf{b_1} + \mathbf{c_1})$ and $\vec{OG_2} = \frac{1}{3}(\mathbf{a_2} + \mathbf{b_2} + \mathbf{c_3})$

Now $3\vec{G_1G_2} = 3[\vec{OG_2} - \vec{OG_1}]$

$$= 3\left[\frac{1}{3}(\mathbf{a_2} + \mathbf{b_2} - \mathbf{c_3}) - \frac{1}{3}(\mathbf{a_1} + \mathbf{b_1} + \mathbf{c_1})\right]$$

$$= [(\mathbf{a_2} - \mathbf{a_1}) + (\mathbf{b_2} - \mathbf{b_1}) + (\mathbf{c_2} - \mathbf{c_1})]$$

$$= \vec{A_1A_2} + \vec{B_1B_2} + \vec{C_1C_2}$$

Hence the result

Example 5. Find the centre of mass of the particles 3, 5, 7, 9, 10, 11, 13, 15 gms respectively placed at the corners of a unit cube, the first four at the corners A, B, C, D of one face and the last four at their projections A′, B′, C′, D′ respectively on the opposite face.

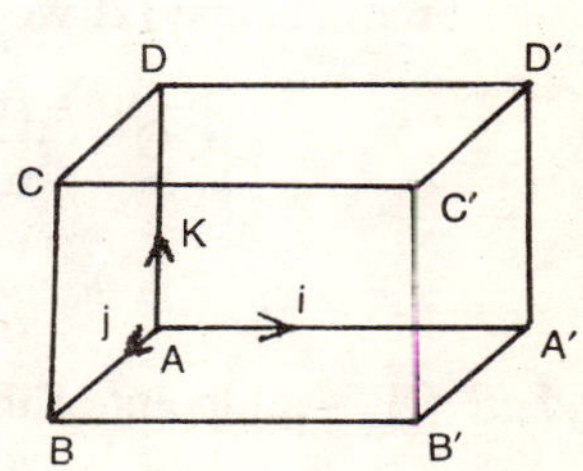

Fig. 4.2

Solution. Let A be the origin and AA′, AB and AD represent the unit vectors **i, j** and **k** respectively. Then the position vectors of the eight corners A, B, C, D and A′, B′, C′ and D′ where the masses are 3, 5, 7, 9, 10, 11, 13, 14 gms are 0, **j**, **j** + **k**, **k** and **i**, **i** + **j**, **i** + **j** + **k**, **i** + **k** respectively.

$\therefore$ If G is the required centre of mass then $\overrightarrow{AG}$

$$= \frac{3(0) + 5(\mathbf{j}) + 7(\mathbf{j} + \mathbf{k}) + 9(\mathbf{k}) - 10(\mathbf{i}) + 11(\mathbf{i} + \mathbf{j}) + 13(\mathbf{i} + \mathbf{j} + \mathbf{k}) + 15(\mathbf{i} + \mathbf{k})}{3 + 5 + 7 + 9 + 10 + 11 + 13 + 15}$$

$$= \frac{1}{73}(49\mathbf{i} + 36\mathbf{j} + 44\mathbf{k})$$

$$\therefore \qquad G = \left(\frac{49}{73}, \frac{36}{73}, \frac{44}{73}\right).$$

Example 6. Find the centre of mass of particles of equal mass placed at (n - 2) of the corners of a regular polygon of n sides.

Solution: Choose the centre O of the polygon as origin. Let **a, b, c, d,** ... be the P.V.'s of the vertices of the polygon w.r.t O.

Let **a, b** be the vertices of the polygon without particles and let the remaining (n - 2) vertices **c, d,** ... be with particles of equal masses. If G be the centre of mass of these (n - 2) equal masses then

$$\overrightarrow{OG} = \frac{\mathbf{c} + \mathbf{d} + \dots}{n - 2} \qquad \dots(1)$$

If equal masses be placed at all the vertical of the polygon then their centre of mass will be at O, whose P.V referred to O is zero

i.e., $$\mathbf{0} = \frac{\mathbf{a} + \mathbf{b} + \mathbf{c} + \mathbf{d} + \dots}{n}$$

or $$\mathbf{a} + \mathbf{b} + \mathbf{c} + \mathbf{d} + \dots = \mathbf{0} \qquad \dots(2)$$

From (1) and (2) we have

$$\overrightarrow{OG} = \frac{-(\mathbf{a} + \mathbf{b})}{n - 2} = -\frac{2}{(n-1)}\left[\frac{1}{2}(\mathbf{a} + \mathbf{b})\right]$$

$$= -\frac{2}{n - 2}\overrightarrow{OP}$$

where $\overrightarrow{OP}$ = midpoint of the join of vertices a and b

or $$\overrightarrow{OG} = \frac{2}{n - 2}\overrightarrow{PO} \qquad [\because \overrightarrow{OP} = -\overrightarrow{PO}]$$

which can also be written as

$$\therefore \quad \overrightarrow{PO} : \overrightarrow{OG} = n - 2 : 2$$

Example 7. G is the centroid of tetrahedron ABCD A'B'C'D' is another tetrahedron such that AA', BB', CC' and DD' are all bisected at G. Prove that G is also the centroid of the second tetrahedron.

Solution. Let the P.V's of A, B, C, D be **a, b, c** and **d** and let the P.V's of A′, B′, C′, D′ be **a′, b′, c′** and **d′** referred to origin O. Again let G be the centroid of ABCD and G′ be the centroid of A′B′C′D′. Then

$$\vec{OG} = \frac{1}{4}(\mathbf{a} + \mathbf{b} + \mathbf{c} + \mathbf{d}) \text{ and } \vec{OG'} = \frac{1}{4}(\mathbf{a'} + \mathbf{b'} + \mathbf{c'} + \mathbf{d'})$$

Now G is the mid point of AA′, BB′, CC′ and DD′

$$\therefore \quad = \frac{1}{4}(\mathbf{a} + \mathbf{b} + \mathbf{c} + \mathbf{d}) = \frac{1}{2}(\mathbf{a} + \mathbf{a'}) = \frac{1}{2}(\mathbf{b} + \mathbf{b'})$$

$$= \frac{1}{2}(\mathbf{c} + \mathbf{c'}) = \frac{1}{2}(\mathbf{d} + \mathbf{d'})$$

$$\Rightarrow \left.\begin{aligned} &\mathbf{a'} = (\mathbf{b} + \mathbf{c} + \mathbf{d} - \mathbf{a})\,; \quad \mathbf{b'} = \frac{1}{2}(\mathbf{c} + \mathbf{d} + \mathbf{a} - \mathbf{b}) \\ &\mathbf{c'} = \frac{1}{2}(\mathbf{d} + \mathbf{a} + \mathbf{b} - \mathbf{c})\,; \quad \mathbf{d'} = \frac{1}{2}(\mathbf{a} + \mathbf{b} + \mathbf{c} - \mathbf{d}) \end{aligned}\right\} \quad ...(1)$$

$$\text{Now} \quad \vec{OG'} = \frac{\mathbf{a'} + \mathbf{b'} + \mathbf{c'} + \mathbf{d'}}{4} = \frac{1}{4}\left[\frac{2\mathbf{a} + 2\mathbf{b} + 2\mathbf{c} + 2\mathbf{d}}{2}\right]$$

[From (1)]

$$= \frac{\mathbf{a} + \mathbf{b} + \mathbf{c} + \mathbf{d}}{4} = \vec{OG}$$

⇒ G is also the centroid of the second tetrahedron.

Example 8. A line AB is bisected in P_1, P_1B in P_2, P_2B in P_3 and so on ad-infinitum. Particles of masses $m, \frac{m}{2}, \frac{m}{2^2}, \ldots$ etc., are placed at the points $P_1, P_2, P_3, \ldots$ etc. Show that the distance of their centre of mass from B is equal to one third of the distance from A to B.

Solution: Choose B as origin and let $\vec{BA} = \mathbf{a}$ then the P.V's of $P_1, P_2, P_3, \ldots$ etc are

$$\frac{1}{2}\mathbf{a}, \frac{1}{2}\left(\frac{1}{2}\mathbf{a}\right), \frac{1}{2}\left[\frac{1}{2}\left(\frac{1}{2}\mathbf{a}\right)\right], \ldots$$

i.e., $\frac{1}{2}\mathbf{a}, \frac{1}{2^2}\mathbf{a}, \frac{1}{2^3}\mathbf{a}, \ldots$

Particles of masses $m, \frac{m}{2}, \frac{m}{2^2}, \ldots$ are placed at $P_1, P_2, P_3, \ldots$ If G is the centroid of the masses then

$$\overrightarrow{BG} = \frac{m\left(\frac{1}{2}\mathbf{a}\right) + \left(\frac{1}{2}m\right)\left(\frac{1}{2^2}\mathbf{a}\right) + \left(\frac{1}{2^2}m\right)\left(\frac{1}{2^3}\mathbf{a}\right) + \ldots \infty}{m + \frac{1}{2}m + \frac{1}{2^2}m + \ldots \infty}$$

$$= \frac{\frac{1}{2}\left[1 + \frac{1}{2^2} + \frac{1}{2^4} + \ldots \infty\right]\mathbf{a}}{\left[1 + \frac{1}{2} + \frac{1}{2^2} + \ldots\ldots \infty\right]}$$

$$= \left[\frac{1}{2}\left\{\frac{1}{1 - \frac{1}{2^2}}\right\} \div \left\{\frac{1}{1 - \frac{1}{2}}\right\}\right]\mathbf{a} = \frac{1}{3}\mathbf{a}$$

$$\left[\because \text{Sum of G. P} = \frac{a}{1 - r}\right]$$

$$\therefore \quad \overrightarrow{BG} = \frac{1}{3}\overrightarrow{BA}.$$

Example 9. Particles of masses $m_1, m_2, m_3, \ldots$ are placed at the points A, B, C ... respectively and G is their centre of mass. Prove that for any point P

$$m_1AP^2 + m_2BP^2 + m_3CP^2 + \ldots = m_1\,AG^2 + m_2BG^2 + \ldots + (\Sigma m)PG^2$$

Solution. Referring to G as origin let the P.V's of A, B, C, ... be **a, b, c, ...** let **r** be the P.V of P.

$$\overrightarrow{AP} = \text{p.v of P} - \text{P.V of A} = \mathbf{r} - \mathbf{a}$$

$$\overrightarrow{BP} = \text{P.V of P} - \text{P.V of B} = \mathbf{r} - \mathbf{b}, \ldots \text{etc}$$

Also G, the centre of mass being the origin we get

$$m_1\mathbf{a} + m_2\mathbf{b} + m_3\mathbf{c} + \ldots = \Sigma m_1\mathbf{a} = \mathbf{0}$$

$$\text{Again } AP^2 = \overrightarrow{AP}^2 = (\mathbf{r} - \mathbf{a})^2 = (\mathbf{r} - \mathbf{a}).(\mathbf{r} - \mathbf{a})$$
$$= \mathbf{r}^2 - 2\mathbf{r}.\mathbf{a} + \mathbf{a}^2 \text{ and so on}$$
$$\therefore \quad m_1AP^2 + m_2BP^2 + m_3CP^2 + ...$$
$$= m_1 [\mathbf{r}^2 - 2\mathbf{r}.\mathbf{a} + \mathbf{a}^2] + m_2 [\mathbf{r}_2 - 2\mathbf{r}.\mathbf{b} + \mathbf{b}^2]$$
$$+ m_3 [\mathbf{r}^2 - 2\mathbf{r}.\mathbf{c} + \mathbf{c}^2] + \ldots$$
$$= \mathbf{r}^2 (m_1 + m_2 + m_3 + \ldots)\, 2\mathbf{r}.\, [m_1\mathbf{a} + m_2\mathbf{b} + m_3\mathbf{c} + \ldots]$$
$$+ (m_1\, a^2 + m_2b^2 + m_3c^2 + ...)$$
$$= (\Sigma m)\, PG^2 - 2\mathbf{r}.(\mathbf{0}) + m_1AG^2 + m_2BG^2 + m_3CG^2 + \ldots$$
$$= m_1AG^2 + m_2BG^2 + m_3CG^2 + ... + (\Sigma m)PG^2$$

where $\Sigma m = m_1 + m_2 + m_3 + \ldots$

and $\quad \mathbf{a} = \overrightarrow{AG}, \mathbf{b} = \overrightarrow{BG}, \mathbf{c} = \overrightarrow{CG}, \mathbf{r} = \overrightarrow{PG}$

Hence the result

EXERCISE

1. Find the centroid of the 3n points

   (i) **i**, 2**i**, 3**i**, ..., n**i**; **j**, 2**j**, 3**j**, ..., n**j**; **k**, 2**k**, 3**k**, ..., n**k**

   (ii) **i**, 4**i**, 9**i**, ..., n^2**i**; **j**, 4**j**, 9**j**, ..., n^2**j**; **k**, 4**k**, 9**k**, ..., n^2**k**

2. Vectors are drawn from the centre of a regular polygon to its vertices. Show that their vector sum is zero.

3. If A, B, C, D, E are the vertices of a regular pentagon and O is the centre show that
$$\overrightarrow{OA} + \overrightarrow{OB} + \overrightarrow{OC} + \overrightarrow{OD} + \overrightarrow{OE} = \mathbf{0}$$

4. If G be the mean centre of A, B, C and G′ that of A′, B′, C′, prove that
$$\overrightarrow{AA'} + \overrightarrow{BB'} + \overrightarrow{CC'} = 3\overrightarrow{GG'}$$

5. Show that the centroid of the points A(2, –3, 3), B(5, –3, –4), C(2, –3, –1) with associated number 3, 4, 5 respectively is the point (3, –3, –1).

6. Find the centre of mass of particles of masses 1, 2, 3, 4 gms placed at vertices A, B, C, D respectively of the tetrahedron

ABCD, the position vectors of the vertices are given by A(3, 1, 2), B(3, 3, 2), C (–1, 3, 4) and D (1, 1, 3)

7. Show that the mean centre of the vertices of a tetrahedron bisects each of the lines joining the middle points of opposite edges and also divided in ratio 3 : 1 each of the lines joining one vertex to the mean centre of the other three.

8. Find the centre of mass of the particles 1, 2, 3, 4, 5, 6, 7, 8 gms respectively at the corners of a unit cube, the first four at the corners A, B, C, D on one face and the last four at their projections A′, B′, C′, D′ respectively on the opposite face.

9. If the points P, Q, R divide the sides BC, CA, AB of the triangle ABC in the same ratio show that A, B, C and P, Q, R have the same centroid.

10. The points D, E, F divide the sides BC, CA, AB of a triangle in 1 : 4, 3 : 2 and 3 : 7 ratio respectively. Show that sum of the vectors $\vec{AD}$, $\vec{BE}$, $\vec{CF}$ is a vector parallel to $\vec{CK}$ where K divides AB in the ratio 1 : 3

11. At each (n – 1) of the angular points of a regular polygon of n sides a particle in placed, the particles being equal, show that the distance of centre of gravity from the centre of the circle circumscribing the polygon is $\frac{\mathbf{r}}{n-1}$ where **r** is the radius of the circle.

ANSWERS

(1) (i) $\frac{n+1}{6}(\mathbf{i}+\mathbf{j}+\mathbf{k})$ (ii) $\frac{(n+1)(2n+1)}{18}(\mathbf{i}+\mathbf{j}+\mathbf{k})$

(6) $\mathbf{i} + 2\mathbf{j} + 3\mathbf{k}$ (8) $\frac{13}{18}, \frac{9}{18}, \frac{11}{18}$.

5

Vector Equation of a Straight Line

5.1. Vector equation of a straight line (Parametric form)

To find the vector equation of a straight line passing through a given point **a** *and parallel to a given vector* **b**

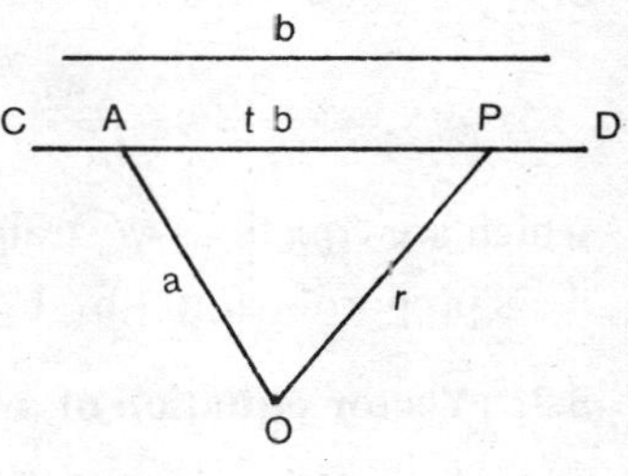

Fig. 5.1

Let O be the origin and let **a** be the P.V of A on the line CD which is parallel to the vector **b**. Let **r** be the P.V of any point P on the line CD. Since $\overrightarrow{AP}$ is parallel to **b** we have $\overrightarrow{AP} = t\mathbf{b}$ where t is some scalar positive for some points on one side of A and negative for points on another side of A.

Then $\qquad \mathbf{r} = \overrightarrow{OP} = \overrightarrow{OA} + \overrightarrow{AP} = \mathbf{a} + t\mathbf{b}$

or $\qquad \mathbf{r} = \mathbf{a} + t\mathbf{b} \qquad$ ...(1)

$\because$ Every point P on the line CD is obtained for some values of t and for every value of t, there is some point on the line. Therefore equation (1) represents the vector equation of the line in *parametric form.*

Cor. 1. *To find the vector equation of a straight line passing through the origin and parallel to given vector* **b**.

In this case $\mathbf{a} = \mathbf{0}$ $\quad$ ($\because$ the line is passing through the origin)

$\therefore$ From equation (1), we get $\mathbf{r} = t\mathbf{b}$ which is the required equation

5.2. Cestesian form

To find the castesian equations of a straight line passing through a given points (a_1, a_2, a_3) *and parallel to a given line whose direction cosines are proportional* (b_1, b_2, b_3)

Let $P = (x, y, z)$ be any point on the line

Then $\mathbf{r} = x\mathbf{i} + y\mathbf{j} + z\mathbf{k}$, also $\mathbf{a} = a_1\mathbf{i} + a_2\mathbf{j} + a_3\mathbf{k}$

Now the vector **b** parallel to the line whose d.c's are proportional to (b_1, b_2, b_3) is $b_1\mathbf{i} + b_2\mathbf{j} + b_3\mathbf{k}$

$\therefore$ The vector equation of the straight line is

$$\mathbf{r} = \mathbf{a} + t\mathbf{b}$$

i.e.,
$$x\mathbf{i} + y\mathbf{j} + z\mathbf{k} = (a_1\mathbf{i} + a_2\mathbf{j} + a_3\mathbf{k}) + t(\mathbf{b_1 i} + \mathbf{b_2 j} + b_3\mathbf{k})$$

or
$$(x - a_1)\mathbf{i} + (y - a_2)\mathbf{j} + (z - a_3)\mathbf{k} = t(b_1\mathbf{i} + b_2\mathbf{j} + b_3\mathbf{k})$$

$$\therefore \quad \frac{x - a_1}{b_1} = \frac{y - a_2}{b_2} = \frac{z - a_3}{b_3} = t$$

which are equations of straight line passing through (a_1, a_2, a_3) with d.c's proportional to (b_1, b_2, b_3) in co-ordinate geometry.

5.3. Vector equation of a straight line through two given points

To find the vector equation of a straight line passing through two given points whose position vectors are **a** and **b**

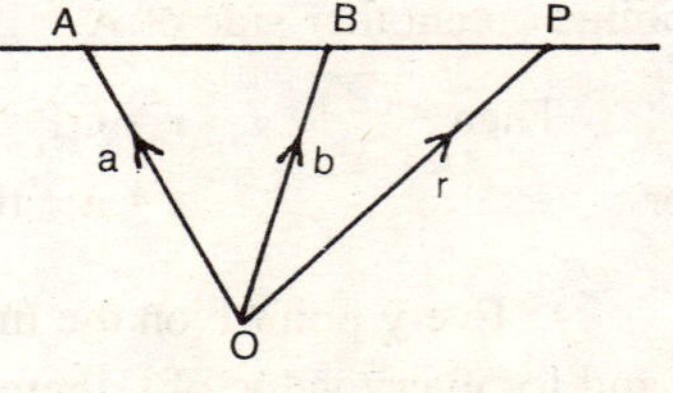

Fig. 5.2

Let **a** and **b** be the P.V's of the points A and B w.r.t the origin O. Let **r** be the position vector of a variable point P on the line

Now $\overrightarrow{AB} = \overrightarrow{OB} - \overrightarrow{OA} = \mathbf{b} - \mathbf{a}$

But $\overrightarrow{AP}$ is parallel to $\overrightarrow{AB}$

$\therefore \quad \overrightarrow{AP} = t\,\overrightarrow{AB}$ or $\overrightarrow{AP} = t(\mathbf{b} - \mathbf{a})$ where t is any scalar.

$\therefore \quad \overrightarrow{OP} = \overrightarrow{OA} + \overrightarrow{AP}$ or $\mathbf{r} = \mathbf{a} + t(\mathbf{b} - \mathbf{a})$

or
$$\mathbf{r} = (1 - t)\,\mathbf{a} + t\mathbf{b}$$

which is the required vector equation of the straight line passing through the points **a** and **b**

5.4. Condition for three points to be collinear

The necessary and sufficient condition for three points in space to be collinear is that there exists a linear relation connecting their position vectors such that the algebraic sum of the scalar coefficients in it is zero, provided the scalars are not all zero.

The condition is necessary. Let **a, b** and **c** be the P.V.'s of the points A, B and C respectively w.r.t origin O. Then vector equation of the straight line AB is

$$\mathbf{r} = (1 - t)\,\mathbf{a} + t\mathbf{b} \qquad ...(1)$$

where **r** is the P.V of any variable point on the line.

But the point **c** lies on the line

$$\therefore \qquad \mathbf{c} = (1 - t)\,\mathbf{a} + t\mathbf{b}$$

or $$(1 - t)\mathbf{a} + t\mathbf{b} - \mathbf{c} = \mathbf{0} \qquad ...(2)$$

which is a linear relation in **a, b, c** and the algebraic sum of the co-efficients of **a, b** and **c** is zero.

Hence the condition is necessary.

The condition is sufficient. Suppose that any three vectors **a, b** and **c** be connected by the linear relation $l\mathbf{a} + m\mathbf{b} + n\mathbf{c} = \mathbf{0}$ ...(3)

where $$l + m + n = 0 \qquad ...(4)$$

and l, m, n are scalars, not all zero. Let $n \neq 0$. then dividing (3) and (4) by n, we get

$$\mathbf{c} = -\frac{l}{n}\mathbf{a} - \frac{m}{n}\mathbf{b} \qquad ...(5)$$

and $$-\frac{l}{n} = \left(1 + \frac{m}{n}\right) \qquad ...(6)$$

Substituting (6) in (5), we get

$$\mathbf{c} = \left(1 + \frac{m}{n}\right)\mathbf{a} - \frac{m}{n}\mathbf{b}$$

or $$\mathbf{c} = (1 - t)\,\mathbf{a} + t\mathbf{b} \text{ where } t = -\frac{m}{n}$$

⇒ point **c** lies on the straight line joining **a** and **b** Hence the three points **a, b** and **c** are collinear.

5.5. Bisectos of angles between two straight lines

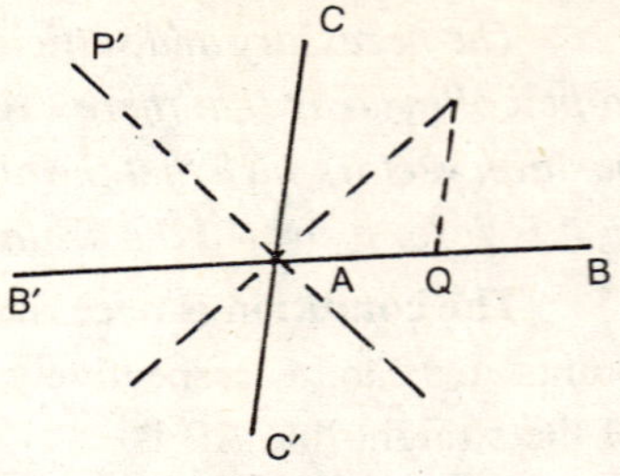

Fig. 5.3

Let BAB′ and CAC′ be two given straight lines intersecting at A. Let **a** be the P.V of A w.r.t some origin O and let $\hat{\mathbf{b}}$ and $\hat{\mathbf{c}}$ be the unit vectors along $\overrightarrow{AB}$ and $\overrightarrow{AC}$ respectively. Let P be any point on the bisector of ∠CAB. Let $\overrightarrow{OP} = \mathbf{r}$. Through the point P draw a line parallel to CA to meet AB at Q

Then $\qquad \angle CAP = \angle PAQ = \angle APQ$

$\therefore \qquad AQ = QP = t$ (Say)

$\therefore \qquad \overrightarrow{AQ} = t\hat{\mathbf{b}}$ and $\overrightarrow{QP} = t\hat{\mathbf{c}}$

Now $\qquad \overrightarrow{AP} = \overrightarrow{AQ} + \overrightarrow{QP} = t\hat{\mathbf{b}} + t\hat{\mathbf{c}} = t(\hat{\mathbf{b}} + \hat{\mathbf{c}})$

But $\qquad \overrightarrow{OP} = \overrightarrow{OA} + \overrightarrow{AP}, \ \therefore \ \mathbf{r} = \mathbf{a} + t(\hat{\mathbf{b}} + \hat{\mathbf{c}}) \qquad ...(1)$

which is the required vector equation of the internal bisector $\overrightarrow{AP}$ of ∠CAB, t being a parameter. Similarly the vector equation of the external bisector $\overrightarrow{AP'}$ of ∠B′AC whose directions are those of − **b** and **c** is given by

$$\mathbf{r} = \mathbf{a} + s(\hat{\mathbf{c}} - \hat{\mathbf{b}}) \qquad ...(2)$$

where s is a parameter

Note: (1) When $\overrightarrow{AB} = \mathbf{b}$ and $\overrightarrow{AC} = \mathbf{c}$ then

$$\hat{\mathbf{b}} = \frac{\mathbf{b}}{|\mathbf{b}|} \text{ and } \hat{\mathbf{c}} = \frac{\mathbf{c}}{|\mathbf{c}|}$$

the equations of internal and external bisectors

are $\qquad \mathbf{r} = \mathbf{a} + t\left[\frac{\mathbf{b}}{|\mathbf{b}|} + \frac{\mathbf{c}}{|\mathbf{c}|}\right]$ and

$$\mathbf{r} = \mathbf{a} + s\left[\frac{\mathbf{c}}{|\mathbf{c}|} - \frac{\mathbf{b}}{|\mathbf{b}|}\right]$$

(2) When A is origin then $\mathbf{a} = \mathbf{0}$

$$\mathbf{r} = t\,[\mathbf{b} + \mathbf{c}] \text{ and } \mathbf{r} = s\,[\mathbf{c} - \mathbf{b}]$$

are the vector equations of internal and external bisectors, and if $\overrightarrow{AB} = \mathrm{b}$ and $\overrightarrow{AC} = \mathrm{c}$ then

$$\mathbf{r} = t\left\{\frac{\mathbf{b}}{|\mathbf{b}|} + \frac{\mathbf{c}}{|\mathbf{c}|}\right\} \text{ and } \mathbf{r} = s\left\{\frac{\mathbf{c}}{|\mathbf{c}|} - \frac{\mathbf{b}}{|\mathbf{b}|}\right\}$$

5.6. Perpendicular distance

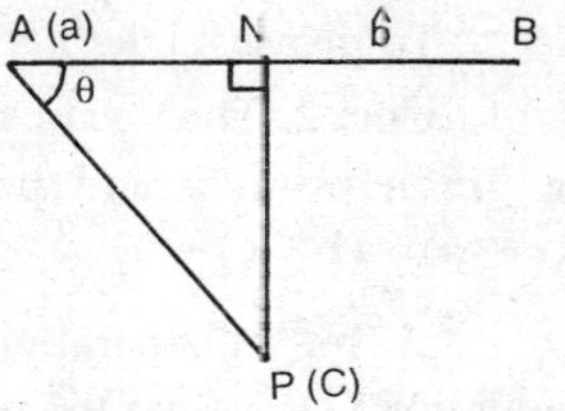

Let AB be the line through A whose P.V is **a** and let it be parallel to unit vector $\hat{\mathbf{b}}$. Then equation to AB is

$$\mathbf{r} = \mathbf{a} + t\hat{\mathbf{b}} \qquad ...(1)$$

where t is a parameter

Let PN be the perpendicular from P on AB and **c** be the p.v. of P.

Fig. 5.4

$\therefore$ AN = Projection of AP on AB

$$= \overrightarrow{AP}.\hat{\mathbf{b}} = (\mathbf{c} - \mathbf{a}).\hat{\mathbf{b}}$$

$$PN^2 = PA^2 - AN^2 = \overrightarrow{PA}^2 - AN^2$$

$$= (\mathbf{a} - \mathbf{c})^2 - [(\mathbf{c} - \mathbf{a}) \,.\, \hat{\mathbf{b}}]^2$$

which gives the perpendicular distance PN

Aliter. If AP makes an angle θ with AB.

Then $\quad PN = AP \sin\theta$ [from the fig 5.4]

$$= AP.1.\sin\theta$$

$$= |\overrightarrow{AP}| \,.\, |\hat{\mathbf{b}}|.\sin\theta$$

or $\quad \overrightarrow{PN} = \overrightarrow{AP} \times \hat{\mathbf{b}} = (\mathbf{c} - \mathbf{a}) \times \hat{\mathbf{b}}$

$\therefore \quad PN = |(\mathbf{c} - \mathbf{a}) \times \hat{\mathbf{b}}|$

5.7. Vector equation of a straight line in the non-parametric form

To show that the vector equation of a straight line passing through a point **a** *and parallel to a given vector* **b** *is* $(\mathbf{r} - \mathbf{a}) \times \mathbf{b} = \mathbf{0}$.

Let **r** be the position vector of any point P on the straight line CD and let A be the point whose position vector is **a** [Refer fig. 5.1]

Then $\overrightarrow{AP} = (\mathbf{r} - \mathbf{a})$ is always parallel to the vector **b** for every position of P on CD.

$$\therefore \quad (\mathbf{r} - \mathbf{a}) \times \mathbf{b} = \mathbf{0} \qquad ...(1)$$

Now equation (1) is satisfied by all points lying on the straight line and any point **r** that satisfies (1) must lie on the given line. Hence equation (1) i.e., $(\mathbf{r} - \mathbf{a}) \times \mathbf{b} = \mathbf{0}$ is the required equation of the straight line.

Cor. 1. The vector equation of a straight line passing through origin and parallel to vector **b** is $(\mathbf{r} \times \mathbf{b}) = \mathbf{0}$

To prove it take $\mathbf{a} = \mathbf{0}$ in (1)

Cor. 2. The vector equation of a straight line passing through a given point **a** and perpendicular to the vectors **b** and **c** is $(\mathbf{r} - \mathbf{a}) \times (\mathbf{b} - \mathbf{c}) = 0$

$\because$ $(\mathbf{b} \times \mathbf{c})$ is parallel to the given line, replacing **b** by $\mathbf{b} \times \mathbf{c}$ in equation (1), we get the required result.

SOLVED PROBLEMS

Example 1. Find the vector equation of the straight line passing through the points $4\mathbf{i} + 2\mathbf{j} - 3\mathbf{k}$ and parallel to the vector $-2\mathbf{j} + \mathbf{j} + 6\mathbf{k}$. Also express it in cantesian form.

Solution: Vector equation of the straight line passing through the point **a** and parallel to the vector **b** is $\mathbf{r} = \mathbf{a} + t\mathbf{b}$. Let O be the origin of the vectors and A, B be the given points. Then their P.V.'s are

$$\overrightarrow{OA} = \mathbf{a} = 4\mathbf{i} + 2\mathbf{j} - 3\mathbf{k} \text{ and } \overrightarrow{OB} = \mathbf{b} - 2\mathbf{i} + \mathbf{j} + 6\mathbf{k}$$

$$\therefore \quad \mathbf{r} = (4\mathbf{i} + 2\mathbf{j} - 3\mathbf{k}) + t(-2\mathbf{i} + \mathbf{j} + 6\mathbf{k})$$

t is being a parameter

Cartesian form.

$$(x\mathbf{i} + y\mathbf{j} + z\mathbf{k}) = (4\mathbf{i} + 2\mathbf{j} - 3\mathbf{k}) + t(-2\mathbf{i} + \mathbf{j} + 6\mathbf{k})$$

or $$(x - 4)\mathbf{i} + (y - 2)\mathbf{j} + (z + 3)\mathbf{k} = t(-2\mathbf{i} + \mathbf{j} + 6\mathbf{k})$$

or $$\frac{x-4}{-2} = \frac{y-2}{1} = \frac{z+3}{6} = t$$

Example 2. Find the vector equation of the straight line through the points $\mathbf{i} - \mathbf{j} - \mathbf{k}$ and $3\mathbf{i} + 4\mathbf{j} + 2\mathbf{k}$.

Solution. Let O be the origin of vectors and let A and B be the points $\mathbf{i} - \mathbf{j} - \mathbf{k}$ and $3\mathbf{i} + 4\mathbf{j} + 2\mathbf{k}$

Then $\vec{OA} = \mathbf{i} - \mathbf{j} - \mathbf{k}$, $\vec{OB} = 3\mathbf{i} + 4\mathbf{j} + 2\mathbf{k}$

$\therefore$ $\vec{AB} = \vec{OB} - \vec{OA} = (3\mathbf{i} + 4\mathbf{j} + 2\mathbf{k}) - (\mathbf{i} - \mathbf{j} - \mathbf{k})$

$= \mathbf{i} + 5\mathbf{j} + 3\mathbf{k}.$

$\therefore$ The straight line is parallel to the vector

$$\vec{AB} = \mathbf{i} + 5\mathbf{j} + 3\mathbf{k}.$$

It passes through $A(\mathbf{i} - \mathbf{j} - \mathbf{k})$

Hence its equation is

$$\mathbf{r} = (\mathbf{i} - \mathbf{j} - \mathbf{k}) + t(\mathbf{i} + 5\mathbf{j} + 3\mathbf{k}),$$

t being a parameter.

Examples 3. Prove that the line joining the points $6\mathbf{a} - 4\mathbf{b} + 4\mathbf{c}$, $-4\mathbf{c}$ and the line joining $-\mathbf{a} - 2\mathbf{b} - 3\mathbf{c}$, $\mathbf{a} + 2\mathbf{b} - 5\mathbf{c}$ intersect at $-4\mathbf{c}$.

Solution: The equation of the line joining the points $6\mathbf{a} - 4\mathbf{b} + 4\mathbf{c}$ and $-4\mathbf{c}$ is

$$\mathbf{r} = (6\mathbf{a} - 4\mathbf{b} + 4\mathbf{c}) + t[(-4\mathbf{c}) - (6\mathbf{a} - 4\mathbf{b} + 4\mathbf{c})]$$

$$= (6 - 6t)\,\mathbf{a} + (-4 + 4t)\mathbf{b} + (4 - 8t)\,\mathbf{c} \qquad \text{...(1)}$$

Similarly the line joining $-\mathbf{a} - 2\mathbf{b} - 3\mathbf{c}$ and $\mathbf{a} + 2\mathbf{b} - 5\mathbf{c}$ is

$$\mathbf{r} = (-\mathbf{a} - 2\mathbf{b} - 3\mathbf{c}) + s[(\mathbf{a} + 2\mathbf{b} - 5\mathbf{c}) - (-\mathbf{a} - 2\mathbf{b} - 3\mathbf{c})]$$

$$= (-1 + 2s)\,\mathbf{a} + (-2 + 4s)\,\mathbf{b} + (-3 - 2s)\,\mathbf{c} \qquad \text{...(2)}$$

where t and s are parameters. Now at the point of intersection of (1) and (2) the two values of **r** of (1) and (2) the two values of **r** of (1) and (2) must be the same.

Hence equating the co-efficients of **a, b** and **c** of (1) and (2), we get

$$6 - 6t = -1 + 2s \text{ or } 6t + 2s = 7 \qquad \text{...(3)}$$

$$-4 + 4t = -2 + 4s \text{ or } 2t - 2s = 1 \qquad \text{...(2)}$$

$$4 - st = -3 - 2s \text{ or } 8t - 2s = 7 \qquad \text{....(5)}$$

Solving (3) and (4) we get $t = 1$ an d $s = \frac{1}{2}$ which also satisfy (5). Putting the value of t in (1) [or value of s in (2)], we get $\mathbf{r} = -4\mathbf{c}$ which is the point of intersection.

Example 4. Find k so that the lines

$\frac{x-1}{2} = \frac{y-3}{1} = \frac{z+1}{p}$ and $\frac{x+1}{-p} = \frac{y+1}{2} = \frac{z-2}{3}$ are perpendicular.

Solution: The given lines are obviously parallel to the vectors $\mathbf{a} = 2\mathbf{i} + \mathbf{j} + p\mathbf{k}$ and $\mathbf{b} = -p\mathbf{i} + 2\mathbf{j} + 3\mathbf{k}$. Now these lines are perpendicular if $\mathbf{a}.\mathbf{b} = 0$

i.e., $(2\mathbf{i} + \mathbf{j} + p\mathbf{k}).(-p\mathbf{i} + 2\mathbf{j} + 3\mathbf{k}) = 0$

or $-2p + 2 + 3p = 0 \Rightarrow p = -2.$

Example 5. Find the vector equation of a line parallel to the vector $2\mathbf{i} - \mathbf{j} + 2\mathbf{k}$ and which passes through the point A whose position vector is $3\mathbf{i} + \mathbf{j} - \mathbf{k}$. If P is a point on this line such that AP = 15 find the position vector of P.

Solution: Let $\overrightarrow{OA} = \mathbf{a} = 3\mathbf{i} + \mathbf{j} - \mathbf{k}$ and $\mathbf{b} = 2\mathbf{i} - \mathbf{j} + 2\mathbf{k}$ where O is the origin. Now the vector equation of a line passing through a given point $\mathbf{a}$ and parallel to a given vector $\mathbf{b}$ is

$$\overrightarrow{OP} = \mathbf{r} = \mathbf{a} + t\mathbf{b} = (3\mathbf{i} + \mathbf{j} - \mathbf{k}) + t(2\mathbf{i} - \mathbf{j} + 2\mathbf{k}) \quad ...(1)$$

where t is some scalar [Ref **art 5.1**]

$\because \quad \overrightarrow{AP} = t\mathbf{b} = t(2\mathbf{i} - \mathbf{j} - 2\mathbf{k})$

$\therefore \quad AP = |t\mathbf{b}| = t\,|\mathbf{b}| = t\sqrt{4 + 1 + 4} = \pm 3t$

But given that AP = 15 i.e., $\pm 3t = 15$ or $t = \pm 5$ for $t = +5$ we get from (1)

$$\overrightarrow{OP} = \mathbf{r} = (3\mathbf{i} + \mathbf{j} - \mathbf{k}) + 5\,(2\mathbf{i} - \mathbf{j} + 2\mathbf{k})$$
$$= 13\mathbf{i} - 4\mathbf{j} + 3\mathbf{k}$$

and for $t = -5$ (denotes that P is on the opposite side of A) we get from 91)

$$\overrightarrow{OP} = \mathbf{r} = (3\mathbf{i} + \mathbf{j} - \mathbf{k}) - 5\,(2\mathbf{i} - \mathbf{j} + 2\mathbf{k})$$
$$= -7\mathbf{i} + 6\mathbf{j} - 11\mathbf{k}.$$

Example 6. Establish by vector method the equation of the straight line in the intercept form is $\frac{x}{a} + \frac{y}{b} = 1.$

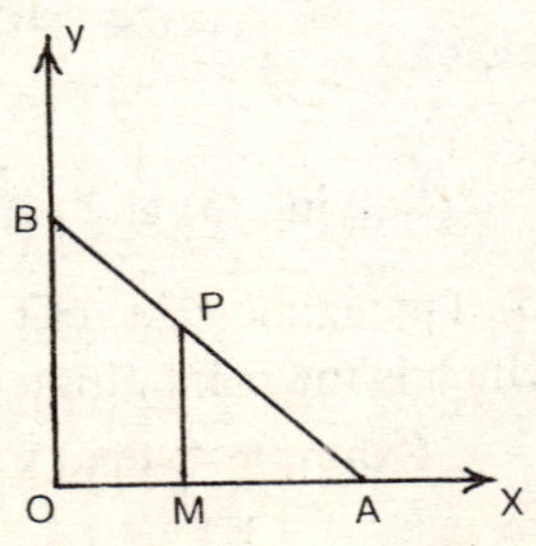

Fig. 5.5

Solution: Let the line meet the co-ordinate axes Ox and Oy in A and B such that OA = a , OB = b let **i** and **j** be the unit vectors along Ox and Oy. Then $\overrightarrow{OA} = a\mathbf{i}$ and $\overrightarrow{OB} = b\mathbf{j}$. Let P(x, y) be any point on AB Join OP and from

P draw perpendicular PM to Ox. Then OM = x**i** and PM = y**j**

$$\therefore \quad \overrightarrow{OP} = \overrightarrow{OM} + \overrightarrow{MP} = x\mathbf{i} + y\mathbf{i} \qquad ...(1)$$

Also P lies on AB, So we have

$$\overrightarrow{OP} = (1 - t)\,\overrightarrow{OA} + t\,\overrightarrow{OB}$$

$$= (1 - t)a\mathbf{i} + tb\mathbf{j} \qquad ...(2)$$

By comparing coefficients of **i** and **j** from (1) and (2) we get

$$x = (1 - t)\,a \quad \text{and} \quad y = tb$$

Eliminating t from above, we get

$$x = \left(1 - \frac{y}{b}\right) a \text{ or } \frac{x}{a} + \frac{y}{b} = 1$$

Hence the result.

Example 7. Prove that the points **a** – 2**b** + 3**c**, 2**a** + 3**b** – 4**c**, – 7**b** + 10**c** are collinear.

Solution. The equation of the line joining the first two points is

$$\mathbf{r} = (1 - t)(\mathbf{a} - 2\mathbf{b} + 3\mathbf{c}) + t(2\mathbf{a} + 3\mathbf{b} - 4\mathbf{c})$$

If the third point (–7**b** + 10**c**) lies on it then –7**b** + 10**c** = (1 – t) (**a** – 2**b** + 3**c**) + t(z**a** + 3**b** – 4**c**). Comparing the co-efficients of **a**, **b** and **c**, we get

$$0 = 1 - t + 2t \quad \Rightarrow t = -1$$

$$-7 = -2\,(1 - t) + 3t \quad \text{satisfied for } t = -1$$

$$10 = 3(1 - t) - 4t \quad \text{which also satisfied for } t = -1$$

Hence the points are collinear.

Example 8. If the vector product of a constant vector $\overrightarrow{OA}$ with a variable $\overrightarrow{OB}$ in a fixed plane AOB be a constant vector, show that the locus of B is a straight line parallel to OA.

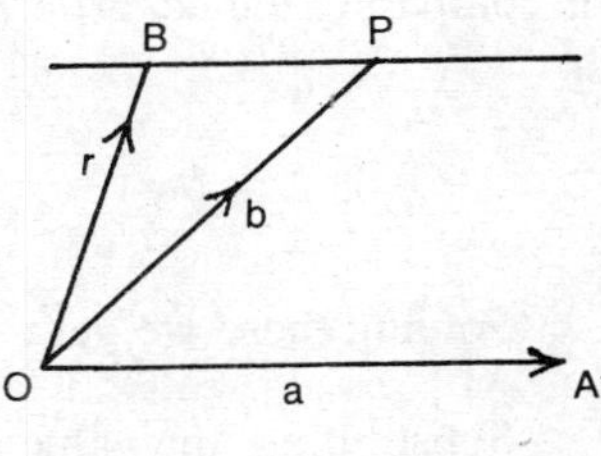

Fig. 5.6

Solution. Let $\overrightarrow{OA}$ = **a** and

$$\overrightarrow{OB} = \mathbf{r}$$

Then $\overrightarrow{OA} \times \overrightarrow{OB}$ = constant (given)

i.e., $$\mathbf{a} \times \mathbf{r} = \lambda \text{ (say)}$$

$$= a \times \overrightarrow{OP} \quad \text{(say)}$$

$$= \mathbf{a} \times \mathbf{b} \text{ where } \mathbf{b} = \overrightarrow{OP}$$

or $\quad \mathbf{a} \times (\mathbf{r} - \mathbf{b}) = \mathbf{0} \Rightarrow \mathbf{r} - \mathbf{b}$ is parallel to **a**

i.e., $\quad \mathbf{r} - \mathbf{b} = t\mathbf{a}$ where t is some scalar

or $\quad \mathbf{r} = \mathbf{b} + t\mathbf{a}$, which represents a straight line through p and parallel to **a** i.e., $\overrightarrow{OA}$

Hence locus of B is a straight line parallel to $\overrightarrow{OA}$.

Example 9. Prove that medians of a triable are concurrent and find the point of concurrency

Solution. Let ABC be the triangle and $\overrightarrow{OA} = \mathbf{a}$, $\overrightarrow{OB} = \mathbf{b}$, $\overrightarrow{OC} = \mathbf{c}$ be the P.V's of vertices A, B and C w.r.t origin O. Let D, E, F be the mid points of the sides BC, CA and AB respectively. Then the P.V's of D, E, F are $\frac{\mathbf{b}+\mathbf{c}}{2}, \frac{\mathbf{c}+\mathbf{a}}{2}, \frac{\mathbf{a}+\mathbf{b}}{2}$ respectively.

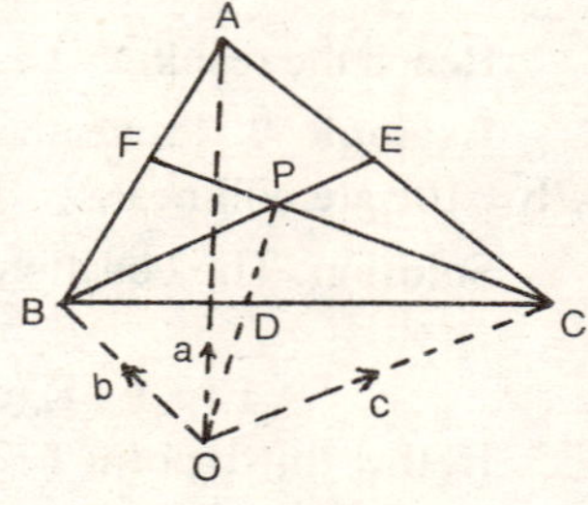

Fig. 5.7

Now the vector equation of the medians AD and BE are

$$\mathbf{r} = (1 - t)\mathbf{a} + t\left(\frac{\mathbf{b}+\mathbf{c}}{2}\right) \quad \text{...(1)}$$

and

$$\mathbf{r} = (1 - t)\,\mathbf{b} + s\left(\frac{\mathbf{c}+\mathbf{a}}{2}\right) \quad \text{...(2)}$$

If the two straight lines intersect, we should be able to find some suitable values of s and t which should give same values of **r**. For this comparing the co-efficients of like vectors in expressions **r**, we get

$$1 - t = \frac{s}{2}; \frac{t}{2} = 1 - s, \frac{t}{2} = \frac{s}{2}$$

Solving these we get $t = s = \frac{2}{3}$.

Substituting the value of t in (1) [or value of s in (2)] we get

$$\mathbf{r} = \frac{\mathbf{a}+\mathbf{b}+\mathbf{c}}{3} \quad \text{...(3)}$$

Which is the position vector of the point of intersection of the medians AD and BE.

Similarly we can show that (3) is the position vector of the point of intersection of the medians AD, CF or CF, BE. Hence the medians are concurrent and P.V of point of concurrency is given by (3)

Example 10. If a straight line is drawn parallel to the base of a triangle, the line joining the vertex to the intersection of the diagonals of the trapezium so formed bisects the base of the triangle.

Solution. Let FE be the straight line drawn parallel to the base BC of a triangle ABC. Let **b** and **c** be the P.V.'s of B and C w.r.t A i.e., $\overrightarrow{AB} = \mathbf{b}$ and $\overrightarrow{AC} = \mathbf{c}$

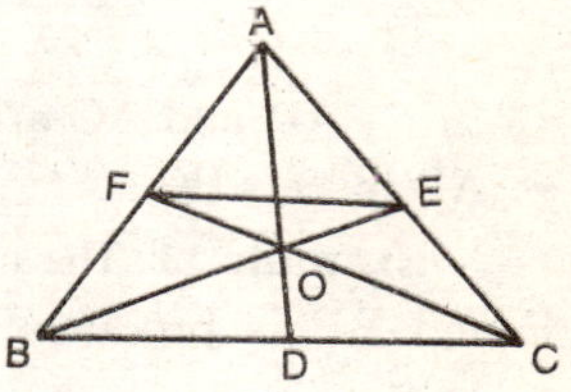

Fig. 5.8

$\because$ FE is parallel to BC

$\therefore \quad \frac{AF}{AB} = \frac{AE}{AC} = \lambda$ say

$\therefore \quad \overrightarrow{AF} = \lambda\overrightarrow{AB} = \lambda\mathbf{b}$ and

$\overrightarrow{AE} = \lambda\,\overrightarrow{AC} = \lambda\mathbf{c}$

Now equation of the line BE is

$$\mathbf{r} = (1-t)\,\mathbf{b} + t\,(\lambda\mathbf{c}) \qquad ...(1)$$

and equation to CF is

$$\mathbf{r} = (1-r)\,\mathbf{c} + s\,(\lambda\mathbf{b}) \qquad ...(2)$$

For point of intersection O of (1) and (2), we have

$$(1-t)\,\mathbf{b} + t(\lambda\mathbf{c}) = (1-s)\,\mathbf{c} + s\,(\lambda\mathbf{b})$$

Comparing the co-efficients of like vectors, we get

$$(1-t)\,\mathbf{b} + t\,(\lambda\mathbf{c}) = (1-s)\,\mathbf{c} + s\,(\lambda\mathbf{b})$$

$$1 - t = s\lambda \text{ and } t\lambda = 1 - s$$

$$\therefore \quad 1 - s = \lambda(1 - s\lambda) \text{ or } s = \frac{1-\lambda}{1-\lambda^2} = \frac{1}{1+\lambda}$$

Similarly we get $t = \frac{1}{1+\lambda}$

Substituting the value of t in (1) [or value of s in (2)] we get p.v. of the point O as

$$\vec{AO} = \left[1 - \frac{1}{1+\lambda}\right]\mathbf{b} + \frac{1}{1+\lambda}(\lambda\mathbf{c});\ \lambda \neq -1$$

$$= \frac{\lambda}{1+\lambda}(\mathbf{b}+\mathbf{c})$$

Now if D is the mid point of $\overline{BC}$ then

$$\vec{AD} = \frac{\mathbf{b}+\mathbf{c}}{2} = \frac{1+\lambda}{2\lambda}\cdot\frac{\lambda}{1+\lambda}(\mathbf{b}+\mathbf{c})$$

$$= \frac{1+\lambda}{2\lambda}\vec{AO}$$

$\therefore$ $\vec{AD}$ and $\vec{AO}$ are collinear. But D is the mid point of BC $\Rightarrow$ AD bisects BC.

Example 11. The median AD of a triangle ABC is bisected at E and BE is produced to meet the side AC in F. Prove that $AF = \frac{1}{3}AC$ and $EF = \frac{1}{4}BF$

Solution: Let **b** and **c** be the P.V's of B and C w.r.t A in a triangle ABC. If D is the mid point of BC then P.V of D is $\frac{\mathbf{b}+\mathbf{c}}{2}$ and that of E, the mid point of median AD = $\frac{\mathbf{b}+\mathbf{c}}{4}$

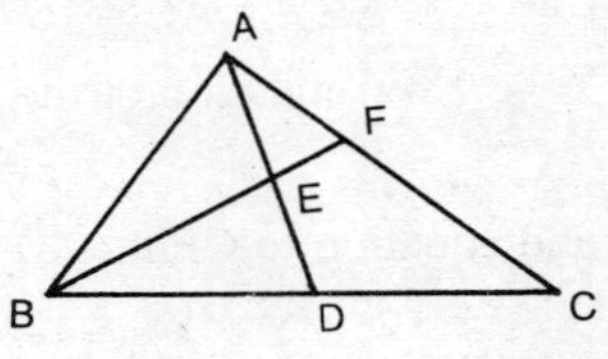

Fig. 5.9

Equation to AC is $\mathbf{r} = \lambda\mathbf{c}$ ...(1)

Equation to BE is $\mathbf{r} = (1-t)\,\mathbf{b} + t\left(\frac{\mathbf{b}+\mathbf{c}}{4}\right)$ ...(2)

Now if (1) and (2) intersect at F then

$$\lambda\mathbf{c} = (1-t)\,\mathbf{b} + t\left(\frac{\mathbf{b}+\mathbf{c}}{4}\right)$$

Comparing the coefficients of like vectors

$$1 - t + \frac{t}{4} = 0 \text{ and } \lambda = \frac{t}{4}$$

or $$t = \frac{4}{3} \text{ and } \lambda = \frac{1}{3}$$

$$\therefore \text{ P.V of F} = \overrightarrow{AF} = \frac{1}{3}\mathbf{c} = \frac{1}{3}\overrightarrow{AC} \qquad \text{[Using (1)]}$$

$$\Leftrightarrow \qquad AF = \frac{1}{3} AC$$

Again let E divide BF in the ratio K : 1 so that $\frac{BE}{EF} = K$

$$\therefore \text{ P.V of E} = \frac{K(\text{P.V. of F}) + 1(\text{P.V of B})}{K+1}$$

$$\frac{\mathbf{b}+\mathbf{c}}{4} = \frac{K\left(\frac{1}{3}\mathbf{c}\right) + 1(\mathbf{b})}{K+1} = \left\{\frac{k}{3(k+1)}\right\}\mathbf{c} + \left\{\frac{1}{k+1}\right\}\mathbf{b}$$

Comparing the like coefficients

$$\frac{1}{4} = \frac{k}{3(k+1)} \text{ and } \frac{1}{4} = \frac{1}{k+1} \text{ which gives } k = 3$$

$$\therefore \qquad \frac{BE}{EF} = 3 \text{ or } BE = 3EF$$

or $$(BF - EF) = 3EF \Rightarrow BF = 4EF \text{ or } EF = \frac{1}{4}BF.$$

Example 12. If M, N are mid points of the sides AB, CD of a parallelogram ABCD show that DM and BN cut he diagonal AC at its points of trisection. which are also points of trisection of DM and BN respectively.

Solution: W.r.t A as origin let

$\overrightarrow{AB} = \mathbf{a}$ and $\overrightarrow{AD} = \mathbf{b}$. Then

$\overrightarrow{BC} = \overrightarrow{AD} = \mathbf{b}$ and

$\overrightarrow{AC} = \overrightarrow{AB} + \overrightarrow{BC} = \mathbf{a} + \mathbf{b}$

$\therefore$ The vector equation of AC is

$\mathbf{r} = \lambda(\mathbf{a} + \mathbf{b})$ ...(1)

$\therefore$ M is the mid point of AB, the

P.V of M is $\frac{\mathbf{a}}{2}$

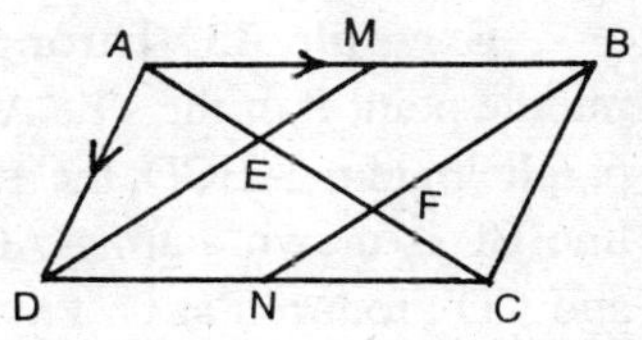

Fig. 5.10

$\therefore$ The vector equation of the line DM is

$$\mathbf{r} = \mathbf{b} + t\left(\mathbf{b} - \frac{\mathbf{a}}{2}\right) \qquad ...(2)$$

For the point of intersection E, we have from (1) and (2)

$$\lambda(\mathbf{a} + \mathbf{b}) = \mathbf{b} + \left(\mathbf{b} - \frac{\mathbf{a}}{2}\right)$$

Comparing the co-efficients of like vectors we get

$$\lambda = -\frac{t}{2} \text{ and } \lambda = 1 + t \Rightarrow \lambda = \frac{1}{3} \text{ and } t = -\frac{2}{3}$$

Putting the value of λ in (1) or value of t in (2), we get the P.V of E as

$$\vec{AE} = \frac{1}{3}(\mathbf{a} + \mathbf{b}) = \frac{1}{3}\vec{AC}$$

i.e., E trisects the diagonal $\vec{AC}$.

Now $$\vec{ME} = \vec{AE} - \vec{AM} = \frac{\mathbf{a} + \mathbf{b}}{2} - \frac{\mathbf{a}}{2} = \frac{2\mathbf{b} - \mathbf{a}}{6}$$

and $$\vec{MD} = \vec{AD} - \vec{AM} = \mathbf{b} - \frac{\mathbf{a}}{2} = \frac{2\mathbf{b} - \mathbf{a}}{2}$$

$\therefore$ $$\vec{MD} = \frac{1}{3}\vec{ME} \Rightarrow \text{E trisects MD.}$$

Similarly we can show that $AF = \frac{2}{3}AC$ and $BF = \frac{2}{3}BN$ which mean that F trisects both AC and BN.

Note: *For any line there will be two points of trisections. For example if AB is a line then there exists two points* P *and* Q *such that* AP : PB = 1 : 3 *and* BQ : QA = 1 : 3 *which are points of trisection.*

Example 13. Through the middle point P of the side AD of a parallelogram ABCD the straight line BP is drawn cutting AC at R and CD produced at Q. Prove that QR = 2RB.

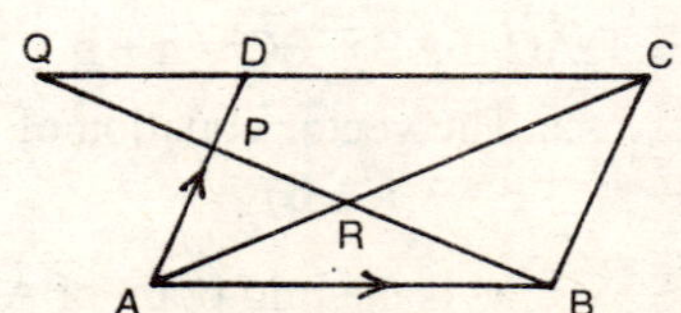

Fig. 5.11

Solution: w.r.t A as origin let the P.V of B and D be and **b** and **d**.

$\therefore \quad \overrightarrow{AB} = \mathbf{b}$ and $\overrightarrow{AD} = \mathbf{d}$

Now $\quad \overrightarrow{AC} = \overrightarrow{AB} + \overrightarrow{AD} = \mathbf{b} + \mathbf{d}$

and $\quad \overrightarrow{AP} = \frac{1}{2} AD = \frac{1}{2}\mathbf{d}$ $\quad$ ($\because$ P is the mid point of AD)

$\therefore$ Vector equation to BP is

$$\mathbf{r} = (1 - t)\,\mathbf{b} + t\left(\frac{1}{2}\mathbf{d}\right) \qquad ...(1)$$

Vector equation to CD is

$$\mathbf{r} = (1 - s)\,(\mathbf{b} + \mathbf{d}) + s\mathbf{d} \qquad ...(2)$$

and to AC is $\quad \mathbf{r} = \lambda\,(\mathbf{b} + \mathbf{d}) \qquad ...(3)$

Now Q is the point of intersection of BP and CD. Hence from (1) and (2)

$$(1 - t)\,\mathbf{b} + t\left(\frac{1}{2}\mathbf{d}\right) = (1 - s)\,(\mathbf{b} + \mathbf{d}) + s\mathbf{d}$$

which gives

$$1 - t = 1 - s \text{ and } \frac{1}{2}t = 1 - s + s.$$

On solving $t = s = 2$. Substituting the value of t in (1) or s in (2) we get the P.V of Q as

$$(1 - 2)\,\mathbf{b} + 2\left(\frac{1}{2}\mathbf{d}\right) = \mathbf{d} - \mathbf{b}$$

Similarly, at the point of intersection of R of (1) and (3) we get $1 - t = \lambda$ and $\frac{1}{2}t = \lambda$ so that $t = \frac{2}{3}, \lambda = \frac{1}{3}$

$\therefore$ From (3), the P.V of $R = \frac{1}{3}(\mathbf{b} + \mathbf{d})$

Now $\overrightarrow{QR}$ = p.v. of R − p.v. of Q

$$= \frac{1}{3}(\mathbf{b} + \mathbf{d}) - (\mathbf{d} - \mathbf{b}) = \frac{2}{3}(2\mathbf{b} - \mathbf{d}) \qquad ...(4)$$

and $\overrightarrow{RB}$ = P.V of B − P.V of R

$$= \mathbf{b} - \frac{1}{3}(\mathbf{b} + \mathbf{d}) = \frac{1}{3}(2\mathbf{b} - \mathbf{d}) \qquad ...(5)$$

From (4) and (5) $\overrightarrow{QR} = 2\overrightarrow{RB} \Rightarrow QR = 2RB$

Example 14. Prove that the diagonals of a parallelogram bisect each other and conversely if the diagonals of a quadrilateral bisect each other it is a parallelogram.

Solution: W.r.t A as origin Let the P.V's of the vertices B, C, D of the parallelogram be

$\vec{AB} = \mathbf{b}$, $\vec{AC} = \mathbf{c}$ and

$\vec{AD} = \mathbf{d}$

Then $\vec{AC} = \vec{AB} + \vec{BC} = \vec{AB} + \vec{AD}$

$\therefore$ $\mathbf{c} = \mathbf{b} + \mathbf{d}$ and the vector equation of AC is

$$\mathbf{r} = t\mathbf{c} = t(\mathbf{b} + \mathbf{d}) \qquad ...(1)$$

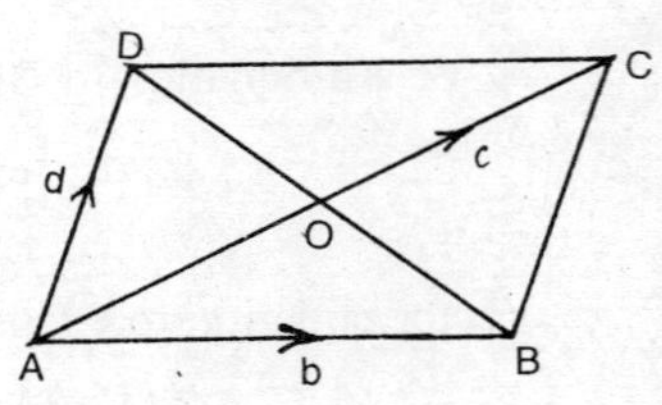

Fig. 5.12

Again the vector equation of BD is

$$\mathbf{r} = (1 - s)\mathbf{b} + s\mathbf{d} \qquad ...(2)$$

For the point of intersection of (1) and (2), we should have

$$t\,(\mathbf{b} + \mathbf{d}) = (1 = s)\,\mathbf{b} + s\mathbf{d}$$

$$\Rightarrow \quad t = 1 - s \text{ and } t = s \Leftrightarrow t = s = \frac{1}{2}$$

Putting the value of t in (1) or s in (2) we get the P.V of O as

$$\vec{AO} = \frac{1}{2}\mathbf{c} \text{ or } \frac{1}{2}(\mathbf{b} + \mathbf{d})$$

$\therefore$ $\vec{AO} = \frac{1}{2}\mathbf{c} = \frac{1}{2}\vec{AC}$ i.e., O is the midpoint diagonals AC and BD. Hence the diagonals in a parallelogram bisects each other at O.

Converse. If the diagonals AC and BD of a quadrilateral ABCD bisect at O then

$$\vec{AO} = \frac{1}{2}\vec{AC} = \frac{1}{2}\mathbf{c}$$

and $$\vec{BO} = \frac{1}{\vec{BD}} = \frac{1}{2}(\vec{AD} - \vec{AB}) \qquad (\because \text{ A is origin})$$

$$\vec{BO} = \frac{1}{2}(\mathbf{d} - \mathbf{b})$$

But $\overrightarrow{AO} = \overrightarrow{AB} + \overrightarrow{BO}$ i.e., $\frac{1}{2}\mathbf{c} = \mathbf{b} + \frac{1}{2}(\mathbf{d} - \mathbf{b})$

or $\mathbf{c} = \mathbf{b} + \mathbf{d}$ or $\overrightarrow{AC} = \overrightarrow{AB} + \overrightarrow{AD}$ or $\overrightarrow{AC} = \overrightarrow{AB} + \overrightarrow{BC}$

i.e., quadrilateral ABCD is a parallelogram.

Example 15. The line joining the vertices of a tetrahedron to the centroids of areas of opposite forces are concurrent

Solution: w.r.t some origin O let the P.V.'s of vertices A, B, C and D of the tetrahedron be **a, b, c** and **d** respectively Let G_1 and G_2 be the centroids of the triangles BCD and CDA respectively.

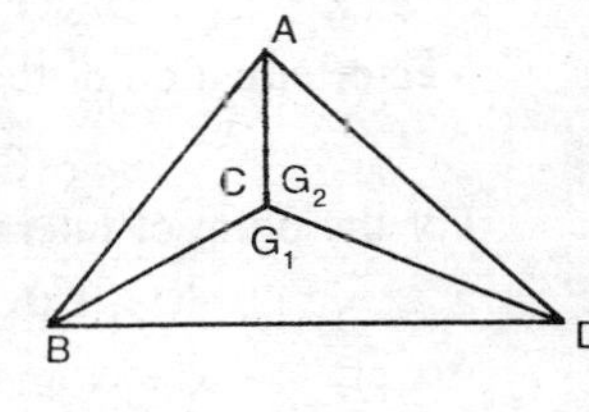

Fig. 5.13

The P.V of G_1

$$= \frac{1}{3}(\mathbf{b} + \mathbf{c} + \mathbf{d})$$

and the p.v. of $G_2 = \frac{1}{3}(\mathbf{c} + \mathbf{d} + \mathbf{a})$

Now the vector equation of AG_1 is

$$\mathbf{r} = (1 - t)\,\mathbf{a} + t\frac{1}{3}(\mathbf{b} + \mathbf{c} + \mathbf{d}) \quad ...(1)$$

and the vector equation of BG_2 is

$$\mathbf{r} = (1 - s)\,\mathbf{b} + s\frac{1}{3}(\mathbf{c} + \mathbf{d} + \mathbf{a}) \quad ...(2)$$

For the intersecting point of (1) and (2) say G, we should have

$$(1 - t)\,\mathbf{a} + t\frac{1}{3}(\mathbf{b} + \mathbf{c} + \mathbf{d}) = (1 - s)\,\mathbf{b} + s\frac{1}{3}(\mathbf{c} + \mathbf{d} + \mathbf{a})$$

$$\Leftrightarrow 1 - t = \frac{s}{3}; \frac{t}{3} = 1 - s; \frac{t}{3} = \frac{s}{3}; \frac{t}{3} = \frac{s}{3} \text{ resulting } t = s = \frac{3}{4}.$$

Substituting the value of t in (1) or sin (2), we get the P.V of intersecting point as

$$\overrightarrow{OG} = \frac{1}{4}(\mathbf{a} + \mathbf{b} + \mathbf{c} + \mathbf{d}).$$

By symmetry the result shows that this point lies on other lines joining C and D with the centroids of opposite faces.

Example 16. Three concurrent straight lines OA, OB, OC are produced to D, E, F respectively. Prove that the points of intersection of AB and DE, BC and EF, CA and FD are collinear

Solution: w.r.t O as origin let the p.v's of A, B, C be $\vec{OA} = \mathbf{a}$, $\vec{OB} = \mathbf{b}$ and $\vec{OC} = \mathbf{c}$ respectively. The p.v's of D, E, F may be taken as x**a**, y**b** and z**c** respectively where x, y, z are scalars. Let the lines BA and ED, BC and EF, CA and FD intersect at the points P, Q, R respectively.

Now the vector equation of the line BA is

$$\mathbf{r} = (1 - t)\,\mathbf{a} + t\mathbf{b} \qquad ...(1)$$

Vector equation of the line ED is

$$\mathbf{r} = (1 - s)x\mathbf{a} + s(y\mathbf{b}) \qquad ...(2)$$

For the point of intersection of (1) and (2)

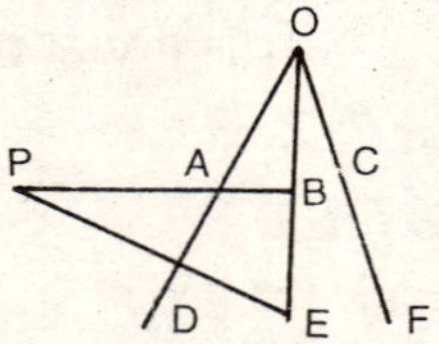

$$(1 - t)\,\mathbf{a} + t\mathbf{b} = (1 - s)\,x\mathbf{a} + s\,(y\mathbf{b})$$

$$\Rightarrow \quad 1 - t = x - sx \text{ and } t = sy$$

$$\Rightarrow \quad s = \frac{x - 1}{x - y} \text{ and } t = \frac{y\,(x - 1)}{x - y}$$

Putting the value of t in (1) [or s in (2)], we get the p.v of P(= $\mathbf{r_1}$ say) as

$$\mathbf{r_1} = \left\{1 - \frac{y\,(x - 1)}{x - y}\right\}\mathbf{a} + \left\{\frac{y\,(x - 1)}{x - y}\right\}\mathbf{b}$$

$$= \left\{\frac{x(1 - y)}{x - y}\right\}\mathbf{a} - \left\{\frac{y\,(1 - x)}{x - y}\right\}\mathbf{b}$$

or

$$\left\{\frac{x - y}{(1 - x)(1 - y)}\right\}\mathbf{r_1} = \left\{\frac{x}{1 - x}\right\}\mathbf{a} - \frac{y}{1 - y}\mathbf{b} \qquad ...(3)$$

Similarly if $\mathbf{r_2}$ and $\mathbf{r_3}$ are the P.V.'s of Q and R then, we get

$$\left\{\frac{y - z}{(1 - y)\,(1 - z)}\right\}\mathbf{r_2} = \left\{\frac{y}{1 - y}\right\}\mathbf{b} - \left\{\frac{z}{1 - z}\right\}\mathbf{c} \qquad ...(4)$$

$$\left\{\frac{z - x}{(1 - z)(1 - x)}\right\}\mathbf{r_3} = \left\{\frac{z}{1 - z}\right\}\mathbf{c} - \left\{\frac{x}{1 - x}\right\}\mathbf{a} \qquad ....(5)$$

Now adding (3), (4) and (5) we get

$$\left\{\frac{x-y}{(1-y)(1-z)}\right\}\mathbf{r}_1+\left\{\frac{y-z}{(1-y)(1-z)}\right\}\mathbf{r}_2+\left\{\frac{z-x}{(1-z)(1-x)}\right\}.\ \mathbf{r}_3=\mathbf{0} \qquad ...(6)$$

which is a linear relation in the P.V.'s of the points P, Q, R

Now the sum of coefficients of $\mathbf{r}_1$, $\mathbf{r}_2$ and $\mathbf{r}_3$ on L.H.S of (6)

$$=\frac{x-y}{(1-x)(1-y)}+\frac{y-z}{(1-y)(1-z)}+\frac{z-x}{(1-z)(1-x)}$$

$$=\frac{(1-z)(x-y)+(1-x)(y-z)+(1-y)(z-x)}{(1-x)(1-y)(1-z)}=0$$

⇒ the p.v's $\mathbf{r}_1$, $\mathbf{r}_2$, $\mathbf{r}_3$ are collinear i.e., the points P,Q and R are collinear.

Example 17. Prove that internal bisector of any angle of a triangle divides the opposite side internally in the ration of the sides containing the angle.

Solution. w.r.t A as origin Le the P.V's of B and C be $\overrightarrow{AB}=\mathbf{b}$ and $\overrightarrow{AC}=\mathbf{c}$ respectively let $\overline{AB}=c$, $\overline{BC}=a$ and $\overline{AC}=b$ respectively.

∴ Unit vector along AB

$$=\hat{\mathbf{b}}=\frac{\mathbf{b}}{AB}=\frac{\mathbf{b}}{c}$$

and unit vector along AC

$$=\hat{\mathbf{c}}=\frac{\mathbf{c}}{AC}=\frac{\mathbf{c}}{b}$$

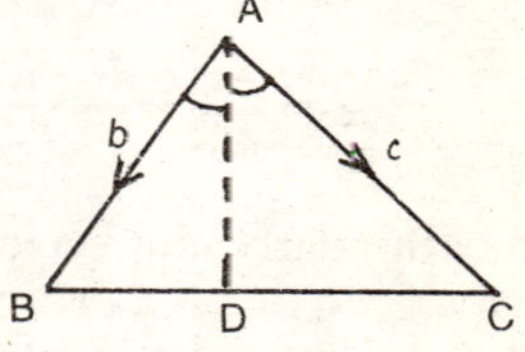

Fig. 5.15

Let AD be the internal bisector of the angle A then its equation is

$$\mathbf{r}=t(\hat{\mathbf{b}}+\hat{\mathbf{c}}) \text{ or } \mathbf{r}=t\left(\frac{\mathbf{b}}{c}+\frac{\mathbf{c}}{b}\right) \qquad ...(1)$$

Equation to BC is

$$\mathbf{r}=(1-s)\,\mathbf{b}+s\mathbf{c} \qquad ...(2)$$

The point of intersection of (1) and (2) is given by

$$t\left(\frac{\mathbf{b}}{c}+\frac{\mathbf{c}}{b}\right)=(1-s)\mathbf{b}+s\mathbf{c}$$

$\Rightarrow\ \frac{t}{c}=1-s$ and $\frac{t}{b}=s$ which gives on solving

$$t = \frac{bc}{b+c} \text{ and } s = \frac{c}{b+c}$$

Putting the value of t in (1) [or value of s in (2)] we get the P.V of D as

$$\vec{AD} = \frac{bc}{b+c}\left(\frac{\mathbf{b}}{c} + \frac{\mathbf{c}}{b}\right) = \frac{b\mathbf{b} + c\mathbf{c}}{b+c}$$

$\Rightarrow$ D divides BC internally in c : b ratio i.e., AB : AC ratio.

Example 18. Prove that internal bisectors of the angles of a triangle are concurrent.

Solution. w.r.t O as origin, let the P.V's of the vertices A,B and C of a triangle ABC be $\vec{OA} = \mathbf{a}, \vec{OB} = \mathbf{b}$ and $\vec{OC} = \mathbf{c}$ respectively. Let $\vec{AB} = c, \vec{BC} = a$ and $\vec{CA} = b$.

Then $\vec{AB} = \vec{OB} - \vec{OA} = \mathbf{b} - \mathbf{a}$ and unit vector along $\vec{AB}$

$$= \frac{\vec{AB}}{AB} = \frac{\mathbf{b} - \mathbf{a}}{c}$$

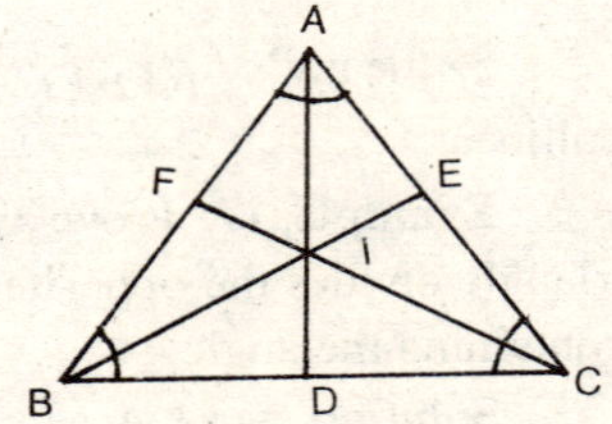

Fig. 5.16

Similarly unit vector along $\vec{AC} = \dfrac{\mathbf{c} - \mathbf{a}}{b}$

$\therefore$ The equation of the internal bisector $\vec{AD}$ is

$$\mathbf{r} = \mathbf{a} + t\left(\frac{\mathbf{b} - \mathbf{a}}{c} + \frac{\mathbf{c} - \mathbf{a}}{b}\right)$$

or

$$\mathbf{r} = \left(1 - \frac{t}{c} - \frac{t}{b}\right)\mathbf{a} + \frac{t}{c}\mathbf{b} + \frac{t}{b}\mathbf{c} \quad ...(1)$$

Again the equation of the internal bisector $\vec{BE}$ is

$$\mathbf{r} = \mathbf{b} + s\left(\frac{\mathbf{c} - \mathbf{b}}{a} + \frac{\mathbf{a} - \mathbf{b}}{c}\right)$$

or

$$\mathbf{r} = \frac{s}{c}\mathbf{a} + \left(1 - \frac{s}{a} - \frac{s}{c}\right)\mathbf{b} + \frac{s}{a}\mathbf{c} \quad ...(2)$$

For the intersecting point I (in centre) of (1) and (2), we should have

$$\left(1-\frac{t}{c}-\frac{t}{b}\right)\mathbf{a}+\frac{t}{c}\mathbf{b}+\frac{t}{b}\mathbf{c}=\frac{s}{c}\mathbf{a}+\left(1-\frac{s}{a}-\frac{s}{c}\right)\mathbf{b}+\frac{s}{a}\mathbf{c}$$

$$\Rightarrow \quad 1-\frac{t}{c}-\frac{t}{b}=\frac{s}{c};\ \frac{t}{c}=1-\frac{s}{a}-\frac{s}{c};\ \frac{t}{b}=\frac{s}{a}$$

last two equations gives $t=\dfrac{sb}{a}$

Substituting it in first equation, we get

$$1-\left(\frac{1}{c}+\frac{1}{b}\right)\frac{sb}{a}=\frac{s}{c} \quad \text{or} \quad 1=\frac{s}{c}\left[\frac{a+b+c}{a}\right]$$

$$\Rightarrow \quad s=\frac{ac}{a+b+c} \text{ and hence } t=\frac{bc}{a+b+c}$$

Putting the value of t in (1) [or value of s in (2)] we get the P.V of I as

$$\overrightarrow{OI}=\left[1-\frac{b}{a+b+c}-\frac{c}{a+b+c}\right]\mathbf{a}+\left[\frac{b}{a+b+c}\right]\mathbf{b}+\left[\frac{c}{a+b+c}\right]\mathbf{c}$$

$$=\frac{a\mathbf{a}+b\mathbf{b}+c\mathbf{c}}{a+b+c} \qquad \text{...(3)}$$

From symmetry of the result (3), it is obvious that, It is the point of intersection of any two internal bisectors of angles. Hence the internal bisectors of angles in a triangle are concurrent

Example 19. Prove that the internal bisector of an angle of a triangle and the external bisectors of the other two are concurrent.

Solution. Let ABC be any triangle w.r.t the vertex C as origin let the p.v.'s of the vertices A and B be $\overrightarrow{CA}=\mathbf{a}$ and $\overrightarrow{CB}=\mathbf{b}$. Let $\overline{AB}=c$, $\overline{BC}=a$ and $\overline{CA}=b$

Let CP be the internal bisector of angle C, AP be the external bisector of the angle between AB and CA extended and

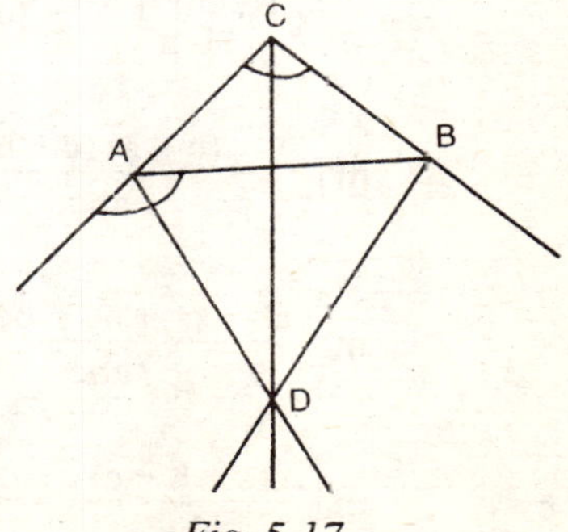

Fig. 5.17

BP is the external bisector of the angle between BA and CB extended as shown in figure. Now vector equation of CP is

$$\mathbf{r} = t\left(\frac{\overrightarrow{CA}}{CA} + \frac{\overrightarrow{CB}}{CB}\right) = t\left(\frac{\mathbf{a}}{b} + \frac{\mathbf{b}}{a}\right)$$

or $$\mathbf{r} = \frac{t}{ab}(a\mathbf{a} + b\mathbf{b}) \qquad ...(1)$$

Now the external bisector AP is

$$\mathbf{r} = \mathbf{a} + s\left(\frac{\overrightarrow{AB}}{AB} + \frac{\overrightarrow{CA}}{CA}\right) = \mathbf{a} + s\left(\frac{\mathbf{b} - \mathbf{a}}{c} + \frac{\mathbf{a}}{b}\right) \qquad ...(2)$$

and the external bisector BP is

$$\mathbf{r} = \mathbf{b} + m\left(\frac{\overrightarrow{BA}}{BA} + \frac{\overrightarrow{CB}}{CB}\right) = \mathbf{b} + m\left(\frac{\mathbf{a} - \mathbf{b}}{c} + \frac{\mathbf{b}}{a}\right) \qquad ...(3)$$

If (2) and (3) intersect then we should have

$$\mathbf{a} + s\left(\frac{\mathbf{b} - \mathbf{a}}{c} + \frac{\mathbf{a}}{b}\right) = \mathbf{b} + m\left(\frac{\mathbf{a} - \mathbf{b}}{c} + \frac{\mathbf{b}}{a}\right)$$

$$\Rightarrow \quad 1 - \frac{s}{c} + \frac{s}{b} = \frac{m}{c} \quad \text{and} \quad \frac{s}{c} = 1 - \frac{m}{c} + \frac{m}{a}$$

or $$\frac{m}{c} = 1 + \frac{s}{bc}(c - b) \quad \text{and} \quad \frac{s}{c} = 1 + \frac{m}{ac}(c - a)$$

Putting the value of $\frac{s}{c}$ from Second equation in the first equation, we get

$$\frac{m}{c} = 1 + \frac{1}{b}\left[1 + \frac{m}{ac}(c - a)\right](c - b)$$

$$m\left[\frac{1}{c} - \frac{(c - a)(c - b)}{abc}\right] = 1 + \frac{c - b}{b} = \frac{c}{b}$$

or $$m\left[\frac{ab - c^2 + ac + bc - ab}{abc}\right] = \frac{c}{b}$$

or $$m\left[\frac{c(a + b - c)}{abc}\right] = \frac{c}{b} \quad \text{or} \quad m = \frac{ac}{b + a - c}$$

Putting this value of m in second equation we get $s = \dfrac{bc}{b + a - c}$.

Substituting the value of s in (2) or m in (3), we get the P.V. of P as

$$\vec{CP} = \mathbf{a} + \left[\frac{bc}{b + a - c}\right]\left[\frac{b\,(\mathbf{b} - \mathbf{a}) + c\mathbf{a}}{bc}\right]$$

$$= \frac{(b + a - c)\,\mathbf{a} + b\mathbf{b} - b\mathbf{a} + c\mathbf{a}}{b + a - c}$$

$$\vec{CP} = \frac{a\mathbf{a} + b\mathbf{b}}{a + b - c}$$

Similarly intersection of (1) and (2), the intersection of (1) and (3) results the P.V's of P as

$$\vec{CP} = \frac{a\mathbf{a} + b\mathbf{b}}{a + b - c} \qquad \left[t = \frac{ab}{a + b - c}\right]$$

$\Rightarrow$ the internal bisector of angle C and the external bisectors of angles A and B are concurrent.

Example 20. Find the perpendicular distance of a corner of a unit cube from a diagonal not passing through it

Solution. Let O be the origin $\vec{OA} = \mathbf{i}$, $\vec{OB} = \mathbf{j}$ and $\vec{OC} = \mathbf{k}$ respectively. Let CQ be the perpendicular on to the diagonal OP which does not pass through C.

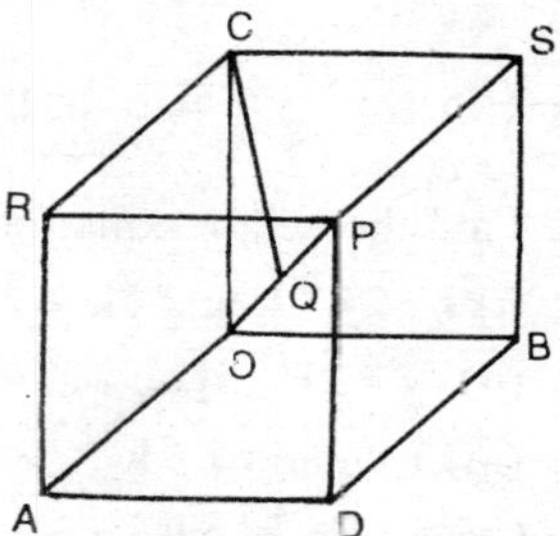

Fig. 5.18

$$\vec{OP} = \vec{OD} + \vec{DP}$$
$$= (\vec{OA} + \vec{OB}) + \vec{OC}$$
$$= \mathbf{i} + \mathbf{j} + \mathbf{k}$$

$\therefore$ Unit vector along $\vec{OP}$

$$= \frac{\vec{OP}}{|\vec{OP}|} = \frac{\mathbf{i} + \mathbf{j} + \mathbf{k}}{\sqrt{3}}$$

$\overline{OQ}$ = projection of $\overline{OC}$ on $\overline{OP} = \mathbf{k} \cdot \dfrac{\overline{OP}}{|\,\overline{OP}\,|} = \dfrac{1}{\sqrt{3}}$

From the right angled triangle OCQ, we have

$$CQ = \sqrt{\overline{OC}^2 - \overline{OQ}^2} = \sqrt{1 - \left(\frac{1}{\sqrt{3}}\right)^2} = \sqrt{\frac{2}{3}}.$$

Example 21. IF **a, b, c** be the P.V's of the points A, B and C then show that the perpendicular distance of C from a line through A and B is

$$\frac{|\mathbf{a} \times \mathbf{b} + \mathbf{b} \times \mathbf{c} + \mathbf{c} \times \mathbf{a}|}{|\mathbf{b} - \mathbf{a}|}$$

Solution: $\vec{AB} = \mathbf{b} - \mathbf{a}$

$\therefore$ Unit vector along $\vec{AB} = \dfrac{\mathbf{b} - \mathbf{a}}{AB}$

Now required perpendicular distance of C from the line AB is

$p = AC \sin\theta$, where θ is the angle between AB and AC

$= \vec{AC} \times$ unit vector along $\vec{AB}$

$$= \left| \vec{AC} \times \frac{\mathbf{b} - \mathbf{a}}{AB} \right| = \frac{1}{AB} | (\mathbf{c} - \mathbf{a}) \times (\mathbf{b} - \mathbf{a}) |$$

$$= \frac{1}{AB} | \mathbf{c} \times \mathbf{b} - \mathbf{c} \times \mathbf{a} - \mathbf{a} \times \mathbf{b} + \mathbf{a} \times \mathbf{a} |$$

$$= \frac{|\mathbf{b} \times \mathbf{c} + \mathbf{c} \times \mathbf{a} + \mathbf{a} \times \mathbf{b}|}{|\mathbf{b} - \mathbf{a}|} \quad [\because\ \mathbf{a} \times \mathbf{a} = \mathbf{0}\ ;\ \mathbf{a} \times \mathbf{c} = -\mathbf{c} \times \mathbf{a}]$$

EXERCISE

1. Find the vector equation of the joining the points
   (i) **i** − 2**j** + **k** and 3**k** − 2**j**
   (ii) 3**i** + 2**j** − 5**j** and 2**i** − 5**j**
   (iii) **i** + **j** and **j** + **k**
   (iv) **i** + **j** + **k** and **i** − **j** + **k**
2. Find the vector equation of a line through the point with the position vector 2**i** + **j** − **k** and parallel to the vector 2**i** + 3**j** + 3**k**.
3. Find the vector equation of the line joining the points P(2, −3, −1) and A(8, −1, 2) and express it in cartesian form. Find the two points on the line whose distance from A is 14.
4. Find the distance from the point P(−1, 2, 6) to the straight line through the point (2, 3, −4) and parallel to the vector (6, 3, −4).
5. Find K so that the lines

$$\frac{x+1}{1}=\frac{y+2}{k}=\frac{z-1}{-1} \text{ and } \frac{x-1}{-k}=\frac{y-1}{2}=\frac{z+2}{1}$$

are perpendicular.

6. Show that the three points whose position vectors are **a**, **b** and 3**a** − 2**b** are collinear.

7. Show that the points
(i) − 2**a** + 3**b** + 5**c**, **a** + 2**b** + 3**c**, 7**a** − **c** and
(ii) **a** + **b** + **c**, 4**a** + 3**b**, 10**a** + 7**b** − 2**c** are collinear, a, b, c being non-coplanar vectors.

8. If **a**, **b**, **c** be non-coplanar, show that the lines whose equations are

$$\mathbf{r} = (-10\mathbf{a} + \mathbf{b} - \mathbf{c}) + t(8\mathbf{a} + 2\mathbf{b} + 3\mathbf{c}) \text{ and}$$

$$\mathbf{r} = (-\mathbf{a} + 4\mathbf{b} - 3\mathbf{c}) + s(7\mathbf{a} + \mathbf{b} - 4\mathbf{c})$$

are coplanar and find their point of intersection.

9. Show that the lines **r** = **a** + t(**b** + **c**) and **r** = **b** + s(**c** + **a**) intersect and also find the point of intersection.

10. Show that the lines

$$\mathbf{r} = 8\mathbf{a} = 9\mathbf{b} + 10\mathbf{c} + t(3\mathbf{a} - 16\mathbf{b} + 7\mathbf{c}) \text{ and}$$

$$\mathbf{r} = 15\mathbf{a} + 29\mathbf{b} + 5\mathbf{c} + s(3\mathbf{a} + 8\mathbf{b} - 5\mathbf{c})$$

are non-coplanar **a**, **b**, **c** being non-coplanar vectors.

11. Show that the lines

$$\mathbf{r} = -3\mathbf{a} + 6\mathbf{b} + t(-4\mathbf{a} + 3\mathbf{b} + 2\mathbf{c}) \text{ and}$$

$$\mathbf{r} = -2\mathbf{a} + 7\mathbf{c} + t(-4\mathbf{a} + \mathbf{b} + \mathbf{c}) \text{ do not intersect.}$$

2. If **a**, **b** are two non-collinear vectors show that points $l_1\mathbf{a} + m_1\mathbf{b}$, $l_2\mathbf{a} + m_2\mathbf{b}$, $l_3\mathbf{a} + m_3\mathbf{b}$ are collinear if and only if

$$\begin{vmatrix} l_1 & l_2 & l_3 \\ m_1 & m_2 & m_3 \\ 1 & 1 & 1 \end{vmatrix} = 0$$

13. If $\mathbf{r}_1$, $\mathbf{r}_2$, $\mathbf{r}_3$ are the position vectors of three collinear points then prove that two real constants P and Q exist such that

$$\mathbf{r}_3 = p\mathbf{r}_1 + q\mathbf{r}_2 \text{ where } p + q = 1.$$

14. Prove that the necessary and sufficient conditions for three

points with position vectors **a, b, c** to be collinear is that there exists scalars x, y, z not all zero that $x\mathbf{a} + y\mathbf{b} + z\mathbf{c} = \mathbf{0}$ and $x + y + z = 0$

15. Show that the external bisectors of any angle of a triangle divides the opposite sides externally in the ratio of the sides containing the angle.

16. Show that the vector equation of a line perpendicular to the two vectors **a** and **c** is $\mathbf{r} = \mathbf{a} + t(\mathbf{b} \times \mathbf{c})$ where t is a scalar

17. Find the equation of the straight line which passes through d and makes equal angles (or equally inclined to) with the vectors **a, b, c**

18. Prove that the join of mid points of the opposite edges of a tetrahedron intersect and bisect each other.

19. If L, M are the mid points of sides BC, CD respectively of parallelogram ABCD prove that $\vec{AL} + \vec{AM} = \frac{3}{2}\vec{AC}$ and $2\vec{ML} + \vec{BD} = 0$.

20. ABCD is a parallelogram and O is the point of intersection of its diagonals. Show that for any origin the sum of the position vectors of the vertices is equal to four times that of O.

21. If a parallelogram ABCD a point P is taken on the side AD such that n.AP = AD. The line BP cuts the diagonal AC in the point Q. Prove that (n + 1)AQ = AC

22. If through any point with in a triangle, lines be drawn parallel to the sides show that the sum of the ratios of these lines to the corresponding sides is 2

23. Prove that the sides about the equal angles of equiangular triangles are proportional.

24. In the triangle ABC, points Q and R are taken in the sides CA and AB respectively such that CQ = QA and AR = 2 RB. BQ and CR intersect at 0. Prove that CO = 30 R and AO divides BC in 1 : 2 ratio.

25. If the internal and external bisectors of the angle A of a triangle ABC meet the base BC at D and E prove that BD, BC and BE are in harmonic progression.

26. ABC is a triangle in which the internal and external bisectors of angle A meet the opposite side BC in D and D′, A′ is the middle point of DD′. Similarly B′ and C′ are the middle points of EE′ and FF′ respectively show that the points A′, B′, C′ are collinear.

ANSWERS

(1) (i) $\mathbf{r} = (\mathbf{i} - 2\mathbf{j} + \mathbf{k}) + t\,(2\mathbf{k} - \mathbf{i})$

(ii) $\mathbf{r} = 3\mathbf{i} + 2\mathbf{j} - 5\mathbf{k} + t\,(-\mathbf{i} - 7\mathbf{j} + 5\mathbf{k})$

(iii) $\mathbf{r} = \mathbf{i} + \mathbf{j} + t\,(\mathbf{k} - \mathbf{i})$

(iv) $\mathbf{r} = \mathbf{i} + \mathbf{j} + \mathbf{k} - 2t\mathbf{j}$

(2) $\mathbf{r} = 2\mathbf{i} + \mathbf{j} + \mathbf{k} + t(\mathbf{i} - 3\mathbf{j} + 3\mathbf{k})$

(3) $\frac{x-2}{6} = \frac{y+3}{2} = \frac{z+1}{3}$; (14, 1, 5), (−10, − 7, − 7)

(4) 7 **(5)** k = 1

(8) $6\mathbf{a} + 5\mathbf{b} - 7\mathbf{c}$

(9) $\mathbf{a} + \mathbf{b} + \mathbf{c}$

(17) $\mathbf{r} = \mathbf{d} + t\,\{|\mathbf{a}|(\mathbf{b} \times \mathbf{c}) + |\mathbf{b}|(\mathbf{c} \times \mathbf{a}) + |\mathbf{c}|(\mathbf{a} \times \mathbf{b})\}$.

6

Vector Equation of a Plane

6.1. Vector equation of the plane (Parametric form)

To find the vector equation of a plane passing through a given point **a** *and parallel to two given vectors* **b** *and* **c**

w.r.t O as origin let the P.V of point A through which the plane passes $\overrightarrow{OA} = \mathbf{a}$. Since the plane is parallel to the vectors **b** and **c** through A draw two lines AB and AC lying in the plane such that $\overrightarrow{AB} = \mathbf{b}$ and $\overrightarrow{AC} = \mathbf{c}$. Let P be any variable point on the same plane such that $\overrightarrow{OP} = \mathbf{r}$. Join AP Now $\overrightarrow{AP}$, $\overrightarrow{AB}$ and $\overrightarrow{AC}$ are coplanar

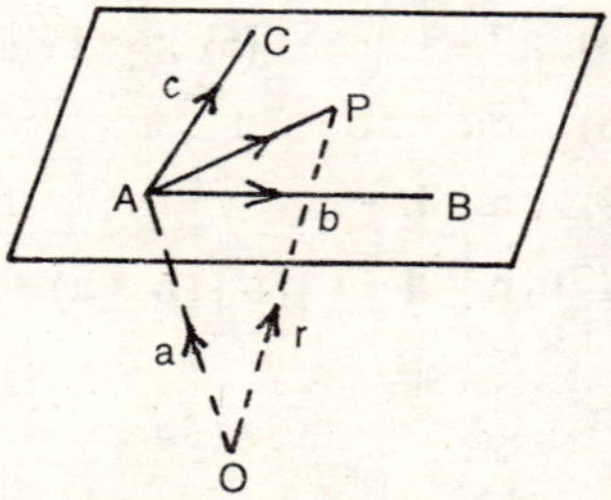

$$\therefore \quad \overrightarrow{AP} = s\,\overrightarrow{AB} + t\,\overrightarrow{AC}$$

where s and t are some scalars

Fig. 6.1

or $$(\overrightarrow{OP} - \overrightarrow{OA}) = s\,\overrightarrow{AB} + t\,\overrightarrow{AC}$$

or $$\mathbf{r} - \mathbf{a} = s\mathbf{b} + t\mathbf{c}$$

or $$\mathbf{r} = \mathbf{a} + s\mathbf{b} + t\mathbf{c}$$

which is the required equation of the given plane

Corollary. The vector equation of the plane passing through the origin and parallel to to given vectors **b** and **c** is

$$\mathbf{r} = s\mathbf{b} + t\mathbf{c} \qquad [\because\ \mathbf{a} = \mathbf{0}]$$

6.2. Vector equation of a plane through three points (Parametric form)

To find the vector equation of the plane passing through three given non–collinear points A, B, C *whose position vectors are* **a, b, c.**

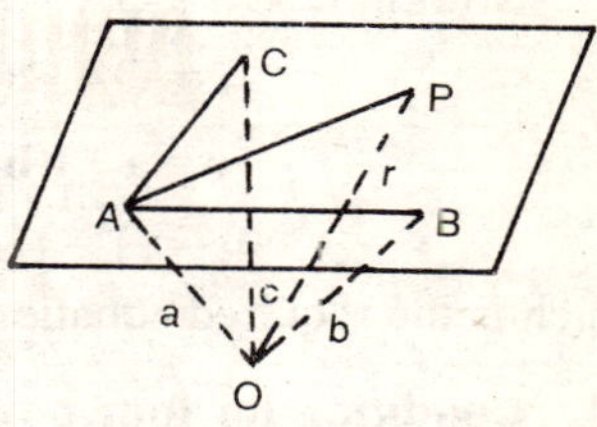

Fig. 6.2

Let O be the origin. Then the P.V's of A,B and C are $\vec{OA} = \mathbf{a}$, $\vec{OB} = \mathbf{b}$ and $\vec{OC} = \mathbf{c}$. Then $\vec{AC} = \mathbf{c} - \mathbf{a}$ and $\vec{AB} = \mathbf{b} - \mathbf{a}$. Let **r** be the p.v of any point P on the plane passing through A, B and C i.e., $\vec{OP} = \mathbf{r}$

Join AP. Now $\vec{AP}, \vec{AB}, \vec{AC}$ are coplanar

$\therefore\ \vec{AP} = s\,\vec{AB} + t\,\vec{AC}$ where s and t are some scalars.

or $$((\vec{OP} - \vec{OA}) = s\,\vec{AB} + t\,\vec{AC}$$

or $$\vec{OP} = \vec{OA} + s\,\vec{AB} + t\,\vec{AC}$$

or $$\mathbf{r} = \mathbf{a} + s(\mathbf{b} - \mathbf{a}) + t(\mathbf{c} - \mathbf{a})$$

or $$\mathbf{r} = (1 - s - t)\,\mathbf{a} + s\mathbf{b} + t\mathbf{c}$$

which is the required equation.

6.3. Vector equation of a plane through two points and parallel to a given vector (parametric form)

To find the vector equation of the plane through two given points A, B *whose position vectors are* **a, b** *and parallel to the vector* **c.**

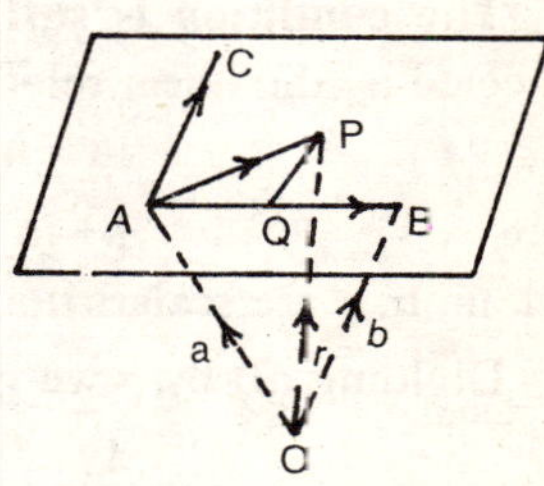

Fig. 6.3

w.r.t O as origin the P.V.'s of A and B are $\vec{OA} = a$ and $\vec{OB} = b$. Through A draw a line AC parallel to **c** lying in the same plane such that $\vec{AC} = \mathbf{c}$

Let P be any point on the plane whose p.v is $\vec{OP} = \mathbf{r}$. Draw PQ parallel to AC meeting AB in Q. Then

$$\vec{AQ} = t\,\vec{AB}$$
$$= t(\vec{OB} - \vec{OA}) = t(\mathbf{b} - \mathbf{a})$$

and $\vec{QP} = s\,\vec{AC} = s\mathbf{c}$ where t and s are some scalars

Now $\vec{OP} = \vec{OA} + \vec{AP}$

or $\mathbf{r} = \vec{OA} + (\vec{AQ} + \vec{QP})$

or $\mathbf{r} = \mathbf{a} + t(\mathbf{b} - \mathbf{a}) + s\mathbf{c}$

or $\mathbf{r} = (1 - t)\mathbf{a} + t\mathbf{b} + s\mathbf{c}$

which is the required equation

6.4. Condition for four points to the coplanar

The necessary and sufficient condition for any four points in the three diamensional space to be coplanar is that there exists a linear relation between their position vectors such that the algebraic sum of the scalar co–efficients in it is zero, provided the scalars are not all zero.

The condition is necessary W.r.t origin O let **a, b, c, d** be the P.V's of four points A, B, C and D respectively. Then the vector equation of the plane passing through the points **a, b** and **c** is

$$\mathbf{r} = (1 - s - t)\,\mathbf{a} + s\mathbf{b} + r\mathbf{c} \qquad \text{...(1)}$$

If the fourth point O whose P.V is **d** lies on (1) then

$$\mathbf{d} = (1 - s - t)\,\mathbf{a} + s\mathbf{b} + t\mathbf{c}$$

or
$$(1 - s - t)\,\mathbf{a} + s\mathbf{b} + t\mathbf{c} - \mathbf{d} = 0 \qquad \text{...(2)}$$

which is a linear relation connecting the four vectors **a, b, c** and **d**.

Sum of scalar co-efficients of **a, b, c** and **d** is

$$(1 - s - t) + s + t - 1 = 0$$

Hence the condition is necessary

The condition is sufficient. It is given that **a, b, c, d** are connected by the linear relation

$$l\mathbf{a} + m\mathbf{b} + n\mathbf{c} + p\mathbf{d} = 0 \qquad \text{...(3)}$$

where
$$l + m + n + p = 0 \qquad \text{...(4)}$$

and l, m, n, p are scalars not all zero. Let $p \neq 0$.

Dividing (3) by p we get

$$\frac{l}{p}\mathbf{a} + \frac{m}{p}\mathbf{b} + \frac{n}{p}\mathbf{c} + \mathbf{d} = 0$$

or
$$\mathbf{d} = -\frac{l}{p}\mathbf{a} - \frac{m}{p}\mathbf{b} - \frac{n}{p}\mathbf{c} \qquad \text{...(5)}$$

Again dividing (4) by p, we get

$$\frac{l}{p}+\frac{m}{p}+\frac{n}{p}+1=0 \quad \text{or} \quad -\frac{l}{p}=1+\frac{m}{p}+\frac{n}{p} \qquad ...(6)$$

Substituting (6) is (5) we get

$$\mathbf{d}=\left(1+\frac{m}{p}+\frac{n}{p}\right)\mathbf{a}-\frac{m}{p}\mathbf{b}-\frac{n}{p}\mathbf{c}$$

or $$\mathbf{d}=(1-s-t)\mathbf{a}+s\mathbf{b}+t\mathbf{c} \qquad ...(7)$$

where $$s=-\frac{m}{p} \text{ and } t=-\frac{n}{p}$$

Equation (7) shows that the point **d** lies on the plane through the points **a, b** and **c**. Hence the four points **a, b, c** and **d** are coplanar.

SOLVED PROBLEMS

Example 1. Find the vector equation of the plane through the points (–1, 1, 2), (1, –2, 1) and (2, 2, 4).

Solution. Let A = (–1, 1, 2), B = (1, –2, 1) and C= (2, 2, 4). w.r.t O as origin let the p.v's of A, B and C be

$$\vec{OA}=-\mathbf{i}+\mathbf{j}+2\mathbf{k};\ \vec{OB}=\mathbf{i}-2\mathbf{j}+\mathbf{k}$$

and $$\vec{OC}=2\mathbf{i}+2\mathbf{j}+4\mathbf{k}$$

$\therefore$ $$\vec{AB}=\vec{OB}-\vec{OA}=(\mathbf{i}-2\mathbf{j}+\mathbf{k})-(-\mathbf{i}+\mathbf{j}+2\mathbf{k})$$
$$=2\mathbf{i}-3\mathbf{j}-\mathbf{k}$$

and $$\vec{AC}=\vec{OC}-\vec{OA}=(2\mathbf{i}+2\mathbf{j}+4\mathbf{k})-(-\mathbf{i}+\mathbf{j}+2\mathbf{k})$$
$$=3\mathbf{i}+\mathbf{j}+2\mathbf{k}$$

Now equation of the plane through A and parallel to $\vec{AB}$ and $\vec{AC}$ is

$$\mathbf{r}=\vec{OA}+s\,\vec{AB}+t\,\vec{AC} \qquad \textbf{[art 6.1]}$$
$$=(-\mathbf{i}+\mathbf{j}+2\mathbf{k})+s(2\mathbf{i}-3\mathbf{j}-\mathbf{k})+t(3\mathbf{i}+\mathbf{j}+2\mathbf{k})$$

or $$\mathbf{r}=(2r+3t-1)\mathbf{i}+(1-3s+t)\mathbf{j}+(2-s+2t)\mathbf{k}$$

Example 2. Find the equation of the plane through the origin and the points 4**j** and 2**i** + **k**. Find also the point in which their plane is cut by the line joining the points **i** – 2**j** + **k** and 3**k** – 2**j**

Solution: Vector equation of the plane through the points whose p.v's are **a, b, c** is

$$\mathbf{r} = (1 - s - t)\mathbf{a} + s\mathbf{b} + t\mathbf{c}$$

If one point is origin i.e., a = 0 then it becomes

$$\mathbf{r} = s\mathbf{b} + t\mathbf{c}$$

If $\mathbf{b} = 4\mathbf{j}$ and $\mathbf{c} = 2\mathbf{i} + \mathbf{k}$ then

$$\mathbf{r} = s(4\mathbf{j}) + t(2\mathbf{i} + \mathbf{k}) \qquad ...(1)$$

Now equation of the line joining the points $\mathbf{i} - 2\mathbf{j} + \mathbf{k}$ and $3\mathbf{k} - 2\mathbf{j}$ is

$$\mathbf{r} = (1 - p)(\mathbf{i} - 2\mathbf{j} + \mathbf{k}) + p(3\mathbf{k} - 2\mathbf{j}) \qquad ...(2)$$

For the point of intersection we must have

$$s(4\mathbf{j}) + t(2\mathbf{i} + \mathbf{k}) = (1 - p)(\mathbf{i} - 2\mathbf{j} + \mathbf{k}) + p(3\mathbf{k} - 2\mathbf{j})$$

Comparing like components we get

$$2t = 1 - p\ ;\ 4s = -2(1 - p) - 2p;\ t = 1 - p + 3p$$

on solving we get

$$t = \frac{3}{4},\ s = -\frac{1}{2},\ p = -\frac{1}{5}$$

Substituting s, t in (1) or p in (2) we get the point of intersection as $\frac{1}{5}(6\mathbf{i} - 10\mathbf{j} + 3\mathbf{k})$.

Example 3. Find the point of intersection of the line joining the points $\mathbf{i} - 2\mathbf{j} - \mathbf{k}$ and $2\mathbf{i} + 3\mathbf{j} + \mathbf{k}$ with the plane through the points $2\mathbf{i} + \mathbf{j} - 3\mathbf{k}$, $4\mathbf{i} - \mathbf{j} + 2\mathbf{k}$ and $3\mathbf{i} + \mathbf{k}$

Solution: The vector equation of the line joining the points $\mathbf{i} - 2\mathbf{j} - \mathbf{k}$ and $2\mathbf{i} + 3\mathbf{j} + \mathbf{k}$ is given by

$$\mathbf{r} = (1 - p)(\mathbf{i} - 2\mathbf{j} - \mathbf{k}) + p(2\mathbf{i} + 3\mathbf{j} + \mathbf{k}) \qquad ...(1)$$

The vector equation of the plane through the three points $2\mathbf{i} + \mathbf{j} - 3\mathbf{k}$, $4\mathbf{i} - \mathbf{j} + 2\mathbf{k}$ and $3\mathbf{i} + \mathbf{k}$ is given by

$$\mathbf{r} = 2\mathbf{i} + \mathbf{j} - 3\mathbf{k} + s\{(4\mathbf{i} - \mathbf{j} + 2\mathbf{k}) - (2\mathbf{i} + \mathbf{j} - 3\mathbf{k})\} + t\{(3\mathbf{i} + \mathbf{k}) - (2\mathbf{i} + \mathbf{j} - 3\mathbf{k})\}$$

or

$$\mathbf{r} = (2\mathbf{i} + \mathbf{j} - 3\mathbf{k}) + s(2\mathbf{i} - 2\mathbf{j} + 5\mathbf{k}) + t(\mathbf{i} - \mathbf{j} + 4\mathbf{k}) \qquad ...(2)$$

For the point of intersection of (1) and (2), we must have, **R.H.S of (1) = R. H. S of (2)**

Comparing the coefficients of **i, j, k** we get

$$(1 - p) + 2p = 2 + 2s + t$$

$$-2(1 - p) + 3p = 1 - 2s - t$$

$$-(1 - p) + p = -3 + 5s + 4t$$

Solving these equations we get

$$p = \frac{2}{3}, \ s = \frac{-14}{9} \text{ and } t = \frac{25}{9}$$

Substituting p in (1) or s, t is (2), we get the point of intersection as $\frac{1}{3}(5\mathbf{i} + 8\mathbf{j} + \mathbf{k})$

Example 4. Find the point in which the plane

$$\mathbf{r} = \mathbf{a} - \mathbf{b} + a(\mathbf{a} + \mathbf{b} - \mathbf{c}) + t(\mathbf{a} + \mathbf{c} - \mathbf{b})$$

is cut by the line through the point $2\mathbf{a} + 3\mathbf{b}$ and parallel to $\mathbf{c}$

Solution: Equation of the line through $2\mathbf{a} + 3\mathbf{b}$ and parallel to $\mathbf{c}$ is

$$\mathbf{r} = 2\mathbf{a} + 3\mathbf{b} + p\mathbf{c} \qquad \text{...(1)}$$

Equation of the plane given is

$$\mathbf{r} = \mathbf{a} - \mathbf{b} + s(\mathbf{a} + \mathbf{b} - \mathbf{c}) + t\,(\mathbf{a} + \mathbf{c} - \mathbf{b}) \qquad \text{...(2)}$$

For the point of intersection of (1) and (2), we must have **R.H.S of (1) = R.H.S of (2)**

Comparing the coefficients of **a, b, c** we get

$$2 = 1 + s + t, \ 3 = -1 + s - t \text{ and } p = -s + t$$

On solving these, we get

$$s = \frac{5}{2}, t = -\frac{3}{2}, \ p = -4$$

Substituting the value of P in (1) or values of s and t in (2), we get the point of intersection as $2\mathbf{a} + 3\mathbf{b} - 4\mathbf{c}$.

Example 5. Prove that the points $2\mathbf{a} + 3\mathbf{b} - \mathbf{c}$, $\mathbf{a} - 2\mathbf{b} + 3\mathbf{c}$, $3\mathbf{b} + 4\mathbf{b} - 2\mathbf{c}$ and $\mathbf{a} - 6\mathbf{b} + 6\mathbf{c}$ are coplanar, also find the vector equation of the plane containing these points.

Solution: Vector equation of the plane through first three points is

$$\mathbf{r} = (1 - s - t)\,(2\mathbf{a} + 3\mathbf{b} - \mathbf{c}) + s(\mathbf{a} - 2\mathbf{b} + 3\mathbf{c}) + t(3\mathbf{a} + 4\mathbf{b} - 2\mathbf{c}) \qquad \text{...(1)}$$

Of fourth point also lies on it then

$$\mathbf{a} - 6\mathbf{b} + 6\mathbf{c} = (1 - s - t)\,(2\mathbf{a} + 3\mathbf{b} - \mathbf{c}) + s\,(\mathbf{a} - 2\mathbf{b} + 3\mathbf{c}) + t(3\mathbf{a} + 4\mathbf{b} - 2\mathbf{c})$$

Comparing the coefficients of **a, b** and **c** on both sides we get

$$1 = 2(1 - s - t) + s + 3t \quad \text{or} \quad s - t = 1$$
$$-6 = 3(1 - s - t) - 2s + 4t \quad \text{or} \quad 5s - t = 9$$
$$6 = -(1 - s - t) + 3s - 2t \quad \text{or} \quad 4s - t = 7$$

Solving first two we get $s = 2, t = 1$ which also satisfies the third equation. Hence the four points are coplanar and (1) represents the required vector equation of the plane.

Example 6. If **a, b, c** be three non-coplanar vectors then prove that the points $l_1\mathbf{a} + m_1\mathbf{b} + n_1\mathbf{c}$, $l_2\mathbf{a} + m_2\mathbf{b} + n_2\mathbf{c}$, $l_3\mathbf{a} + m_3\mathbf{b} + n_3\mathbf{c}$ and $l_4\mathbf{a} + m_4\mathbf{b} + n_4\mathbf{c}$ are coplanar if and only if

$$\begin{vmatrix} l_1 & l_2 & l_3 & l_4 \\ m_1 & m_2 & m_3 & m_4 \\ n_1 & n_2 & n_3 & n_4 \\ 1 & 1 & 1 & 1 \end{vmatrix} = 0$$

Solution: If the given points are co-planar, then there must exist a relation of the form

$$p(l_1\mathbf{a} + m_1\mathbf{b} + n_1\mathbf{c}) + q(l_2\mathbf{a} + m_2\mathbf{b} + n_2\mathbf{c})$$
$$+ r(l_3\mathbf{a} + m_3\mathbf{b} + n_3\mathbf{c}) + s(l_4\mathbf{a} + m_4\mathbf{b} + n_4\mathbf{c}) = \mathbf{0} \quad ...(1)$$

where $$p + q + r + s = 0 \quad ...(2)$$

The relation (1) can be rewritten as

$$(pl_1 + ql_2 + rl_3 + sl_4)\mathbf{a} + (pm_1 + qm_2 + rm_3 + sn_4)\mathbf{b}$$
$$+ (pn_1 + qn_2 + rn_3 + sn_4)\mathbf{c} = 0 \quad ...(3)$$

$\therefore$ **a, b, c** are non-coplanar, equation (3) can exist if and only if

$$pl_1 + ql_2 + rl_3 + sl_4 = 0 \quad ...(4)$$
$$pm_1 + qm_2 + rm_3 + sm_4 = 0 \quad ...(5)$$
$$pn_1 + qn_2 + rn_3 + sn_4 = 0 \quad ...(6)$$

Eliminating p, q, r and s from (4), (5), (6) and (2) we get the required result in the determinant form as given.

Example 7. If any point O with in a tetrahedron ABCD is joined to the vertices and AO, BO, CO, DO are produced to cut the planes of the opposite faces in P, Q, R and S respectively then show that

$$\frac{OP}{AP}+\frac{OQ}{BQ}+\frac{OR}{CR}+\frac{OS}{DS}=1$$

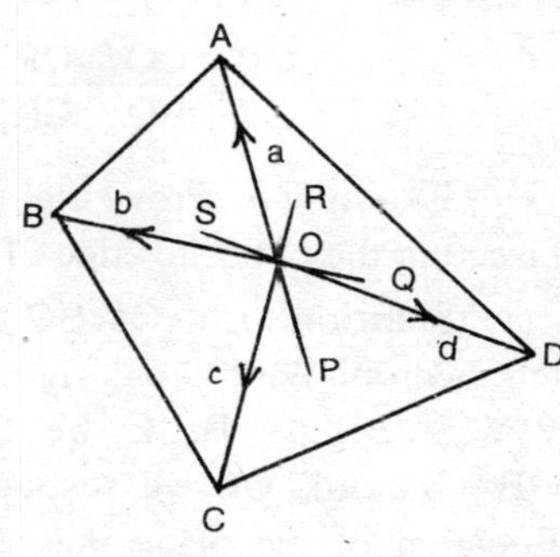

Fig. 6.4

Solution. W.r.t the origin O let the P.V.'s of the vertices A, B, C, D be $\vec{OA}=\mathbf{a}$, $\vec{OB}=\mathbf{b}$, $\vec{OC}=\mathbf{c}$ and $\vec{OD}=\mathbf{d}$ respectively. AO, BO, CO and DO are produced to meet the opposite planes in P, Q, R and S respectively. Now any one of the vectors **a, b, c** and **d** may be expressed linearly in terms of the other three. Hence there must exist a linear relation between them say

$$l\mathbf{a}+m\mathbf{b}+n\mathbf{c}+p\mathbf{d}=0$$

$$\therefore \qquad m\mathbf{b}+n\mathbf{c}+p\mathbf{d}=-l\mathbf{a}$$

or
$$\frac{m\mathbf{b}+n\mathbf{c}+p\mathbf{d}}{m+n+p}=-\frac{l\mathbf{a}}{m+n+p} \qquad ...(1)$$

Now $\dfrac{m\mathbf{b}+n\mathbf{c}+p\mathbf{d}}{m+n+p}$ is a point on the plane BCD

Let $\vec{OP}=\mathbf{r}$, then equation to OA is $\mathbf{r}=s\mathbf{a}$ i.e., to AO is $\mathbf{r}=-s\mathbf{a}$

$\therefore \dfrac{l\mathbf{a}}{m+n+p}$ is a point on AO and hence from (1) both are equal and each of them denotes the point of intersection of AO and the plane BCD. If this point be P then

$$\vec{OP}=-\frac{l\mathbf{a}}{m+n+p} \qquad ...(2)$$

Now
$$\vec{AP}=\vec{AO}+\vec{OP}=-\mathbf{a}-\frac{l\mathbf{a}}{m+n+p}=\frac{(l+m+n+p)\mathbf{a}}{m+n+p}$$

$$\frac{OP}{AP}=\frac{|\vec{OP}|}{|\vec{AP}|}=\frac{l}{m+n+p}\cdot\frac{(m+n+p)}{l+m+n+p}$$

$$=\frac{l}{l+m+n+p}$$

Similarly
$$\frac{OQ}{BQ}=\frac{m}{l+m+n+p};\quad \frac{OR}{CR}=\frac{n}{l+m+n+p}$$

and $\frac{OS}{DS} = \frac{p}{l + m + n + p}$. Adding all these, we get

$$\frac{OP}{AP} + \frac{OQ}{BQ} + \frac{OR}{CR} + \frac{OS}{DS} = \frac{l + m + n + p}{l + m + n + p} = 1$$

Example 8. Prove that the six planes containing one edge and bisecting the opposite edge of a tetrahedron bisect each other

Solution: Let OABC be the tetrahedron. W.r.t O as origin let the P.V.'s of A, B, C be $\vec{OA} = \mathbf{a}$, $\vec{OB} = \mathbf{b}$ and $\vec{OC} = \mathbf{c}$ respectively. Equation of the plane that contains the edge OA and passes through the midpoint D of opposite edge BC in

$$\mathbf{r} = s_1\mathbf{a} + t_1\left(\frac{\mathbf{b} + \mathbf{c}}{2}\right) \quad ...(1)$$

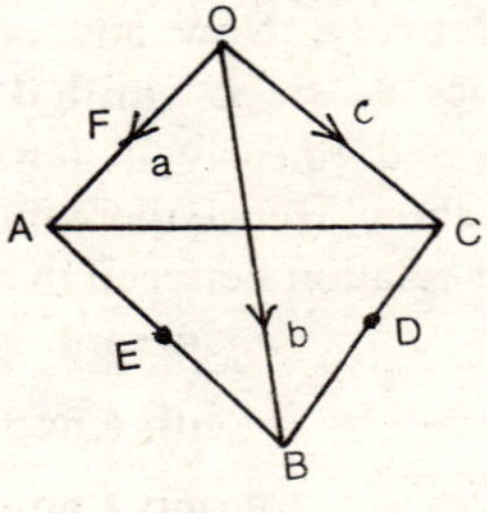

Fig. 6.5

Equation of the plane with the edge OC and passes through the mid point E of the opposite side AB is

$$\mathbf{r} = s_2 c + t_2\left(\frac{\mathbf{a} + \mathbf{b}}{2}\right) \quad ...(2)$$

Similarly equation of the plane containing the edge BC and passing through the mid point F of the opposite edge AO is

$$\mathbf{r} = (1 - s_3 - t_3)\mathbf{b} + t_3\mathbf{c} + s_3\left(\frac{\mathbf{a}}{2}\right) \quad ...(3)$$

For the point of intersection of (1), (2) and (3) the three values of **r** must be same. Equating the coefficients of **a, b, c** we get

$$s_1 = \frac{t_2}{2} = \frac{s_3}{2}; \ \frac{t_1}{2} = \frac{t_2}{2} = 1 - s_3 - t_3; \ \frac{t_1}{2} = s_2 = t_3$$

i.e., $\quad t_1 = t_2 = s_3 = 2s_2 = 2s_1 = 2t_3$

and $\quad \frac{t_1}{2} = 1 - s_3 - t_3 \Rightarrow \frac{1}{2}t_1 = 1 - t_1 - \frac{1}{2}t_1 \Rightarrow t_1 = \frac{1}{2}$

$\therefore \quad t_1 = t_2 = s_3 = \frac{1}{2}; s_2 = s_1 = t_3 = \frac{1}{4}$

Putting the values of s_1 and t_1 in (1) or s_2 and t_2 in (2) or s_3 and t_3 in (3) we get the p.v of point of intersection of the three planes (1), (2) and (3) as $\frac{1}{4}(\mathbf{a} + \mathbf{b} + \mathbf{c})$.

The symmetry of this result shows that all the six planes intersect at the point $\frac{1}{4}(\mathbf{a} + \mathbf{b} + \mathbf{c})$.

Example 9. Prove that the middle points of the six edges of a cube which do not meet a particular diagonal are coplanar.

Solution. Choose A as origin and AB, AD and AP as axes along which **i, j, k** be the unit vectors. Let the edge of the cube be of unit length. Let L, M, N, L′, M′ and N′ be the mid points of AB, AD, DS, SR, RQ and QB .

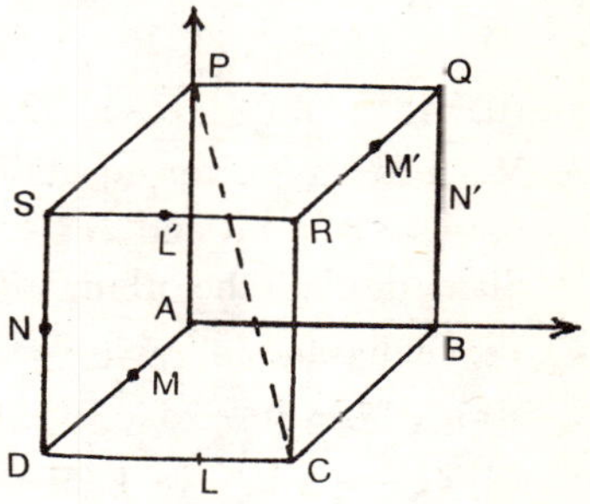

The p.v's of these points are

Fig. 6.6

$$\vec{AL} = \frac{\mathbf{i}}{2}, \ \vec{AM} = \frac{\mathbf{j}}{2}; \ \vec{AN} = \frac{\mathbf{j} + \mathbf{i} + \mathbf{k}}{2} = \mathbf{j} + \frac{\mathbf{k}}{2}$$

$$\vec{AL'} = \frac{(\mathbf{i} + \mathbf{j} + \mathbf{k}) + (\mathbf{j} + \mathbf{k})}{2} = \frac{\mathbf{i}}{2} + \mathbf{j} + \mathbf{k}$$

$$\vec{AM'} = \mathbf{i} + \frac{\mathbf{j}}{2} + \mathbf{k}; \ \vec{AN'} = \frac{\mathbf{i} + \mathbf{i} + \mathbf{k}}{2} = \mathbf{i} + \frac{\mathbf{k}}{2}$$

Now equation of the plane LMN is

$$\mathbf{r} = (1 - s - t)\left(\frac{\mathbf{i}}{2}\right) + s\left(\frac{\mathbf{j}}{2}\right) + t\left(\mathbf{j} + \frac{\mathbf{k}}{2}\right) \quad \text{...(1)}$$

If the point L′ lie on it

$$\therefore \quad \frac{\mathbf{i}}{2} + \mathbf{j} + \mathbf{k} = (1 - s - t)\left(\frac{\mathbf{i}}{2}\right) + s\left(\frac{\mathbf{j}}{2}\right) + t\left(\mathbf{j} + \frac{\mathbf{k}}{2}\right)$$

$$\Rightarrow \quad 1 = 1 - s - t; \quad 1 = \frac{s}{2} + t; \quad 1 = \frac{t}{2}$$

which are consistant since t = 2, s = − 2 satisfy all these three equations. Therefore point L′ lies on (1).

Similarly we can show that the points M′ and N′ also lie on (1). Hence the six points L, M, N, L′, M′ and N′ are coplanar

EXERCISE

1. Find the vector equation of the plane passing through the points (1, −2, 5), (0, −5, −1), (−3, 5, 0)
2. Show that the following points are coplanar

   (i) (−1 , 4, −3), (3, 2, −5), (−3, 8, −5) , (−3, 2, 1)

   (ii) (4, 5, 1), (0, −1, −1), (3, 9, 4), (−4, 4, 4)
3. What is the vector equation of the straight line through the points **i** + 2**j** + **k** and 2**i** + **j** − 3**k**. Find where this line cuts the plane through the origin and the points 2**i** − **j** + **k** and **i** − **j** − **k**
4. Prove that the following vectors are coplanar

   (i) **a** − 2**b** + 3**c**, −2**a** + 3**b** − 4**c**, **a** − 3**b** + 5**c**

   (ii) 4**i** + 5**j** + **k**, − **j** − **k**, 5**i** + 9**j** + 4**k**
5. Prove that the following points are coplanar

   (i) 6**a** + 2**b** − **c**, 2**a** − **b** + 3**c**, −**a** + 2**b** − 4**c**, −12**a** − **b** − 3**c**

   (ii) −**a** + 4**b** − 3**c**, 3**a** + 2**b** − 5**c**, −3**a** + 8**b** − 5**c**, −3**a** + 2**b** + **c**

   (iii) 4**i** + 8**j** + 12**k**, 2**i** + 4**j** + 6**k**, 3**i** + 5**j** + 4**k**, 5**i** + 8**j** + 5**k**

   (iv) 2**i** − 3**k**, **i** + 4**j**, 2**i** + 6**j** + 3**k**, 4**i** + 9**j** + 7**k**
6. If there exists a relation l**a** + m**b** + n**c** = 0, where l, m, n are numbers not all zero, then show that the vectors **a**, **b**, **c** are parallel to a plane
7. Show that the vectors **a** − **b** + **c**, 2**a** − 3**b** and a + 3**c** are parallel to a plane
8. Prove that the necessary and sufficient condition for four points with P.V's **a**, **b**, **c**, **d** to be coplanar is that there exist scalars l, m, n, p not all zero such that l**a** + m**b** + n**c** + p**d** = 0 and l + m + n + p = 0

9. Let OABC be a tetrahedron and let $\overrightarrow{OA} = \mathbf{a}$, $\overrightarrow{OB} = \mathbf{b}$ and $\overrightarrow{OC} = \mathbf{c}$. Obtain the equations of its faces

10. Find all the values of λ such that $(x, y, z) \neq (0, 0, 0)$ and $(\mathbf{i} + \mathbf{j} + 3\mathbf{k})x + (3\mathbf{i} - 3\mathbf{j} + \mathbf{k})y + (-4\mathbf{i} + 5\mathbf{j})z = \lambda(x\mathbf{i} + y\mathbf{j} + z\mathbf{k})$

11. Let OABCDEFG be a parallelopiped with O as origin. If $\overrightarrow{OA} + \mathbf{a}$, $\overrightarrow{OB} = \mathbf{b}$ and $\overrightarrow{OC} = \mathbf{c}$ then obtain the equations of its six faces.

12. If ABC is any triangle and O any point in the plane of the same. AO, BO, CO meet the sides BC, CA and AB in D, E, F respectively. Prove that

$$\frac{OD}{AD} + \frac{OE}{BE} + \frac{OF}{CF} = 1$$

13. If a tetrahedron is cut by a plane parallel to two opposite edges, show that the section is a parallelogram.

ANSWERS

(1) $\mathbf{r} = (1 - s - 4t)\mathbf{i} + (-2 - 3s + 7t)\mathbf{j} + (5 - 6s - 5t)\mathbf{k}$

(3) $\frac{1}{3}(-4\mathbf{i} + 13\mathbf{j} + 31\mathbf{k})$

(9) $\mathbf{r} = t_1\mathbf{a} + s_1\mathbf{b}$; $\mathbf{r} = t_2\mathbf{b} + s_2\mathbf{c}$; $\mathbf{r} = t_3\mathbf{a} + s_3\mathbf{c}$

$\mathbf{r} = \mathbf{a} + t_4(\mathbf{b} - \mathbf{a}) + s_4(\mathbf{c} - \mathbf{a})$

(10) $\lambda = 0, -1, -1$

(11) $\mathbf{r} = t_1\mathbf{a} + s_1\mathbf{b}$; $\mathbf{r} = t_2\mathbf{b} + s_2\mathbf{c}$; $\mathbf{r} = t_3\mathbf{a} + s_3\mathbf{c}$

$\mathbf{r} = \mathbf{c} + t_4\mathbf{a} + s_4\mathbf{b}$; $\mathbf{r} = \mathbf{a} + t_5\mathbf{b} + s_5\mathbf{c}$

$\mathbf{r} = \mathbf{b} + t_6\mathbf{a} + s_6\mathbf{c}$.

6.5. Vector equation of a plane in normal form (Non-parametric)

Let O be the origin of and ON = p be the length of perpendicular from O to the given plane. Let $\hat{\mathbf{n}}$ be the unit vector normal to the plane having positive direction from O to N.

$\therefore \quad \overrightarrow{ON} = p\hat{\mathbf{n}}$

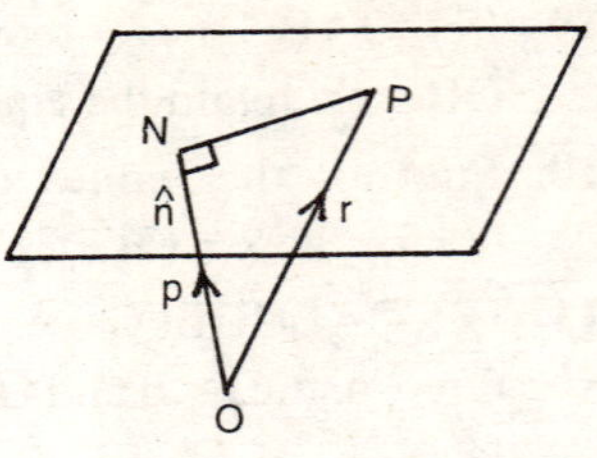

Fig. 5.7

Let P be any point on the plane with position vector $\vec{OP} = \mathbf{r}$. Join NP. Now $\vec{NP}$ is always parallel to the plane i.e., $\vec{NP}$ is perpendicular to $\vec{ON}$

$\therefore \quad \vec{NP} \cdot \vec{ON} = 0$

or $\quad (\vec{OP} - \vec{ON}) \cdot \vec{ON} = 0$

or $\quad (\mathbf{r} - p\hat{\mathbf{n}}) \cdot p\hat{\mathbf{n}} = 0$

or $\quad (\mathbf{r} - p\hat{\mathbf{n}}) \cdot \hat{\mathbf{n}} = 0$

or $\quad \mathbf{r}.\hat{\mathbf{n}} - p\hat{\mathbf{n}}.\hat{\mathbf{n}} = 0$

or $\quad \mathbf{r}.\hat{\mathbf{n}} - p = 0 \qquad (\because \hat{\mathbf{n}}.\hat{\mathbf{n}} = 1)$

or $\quad \mathbf{r}.\hat{\mathbf{n}} = p \qquad ...(1)$

Equation (1) is satisfied by the P.V's of all points lying on the plane. Hence (1) is the non-parametric form of the required plane.

Cartesian form. If P = (x, y, z) and l, m, n are d.c's of $\hat{n}$ w.r.t a rectangular axes through O then $\vec{OP} = x\mathbf{i} + y\mathbf{j} + z\mathbf{k}$ and $\hat{\mathbf{n}} = l\mathbf{i} + m\mathbf{j} + n\mathbf{k}$ and hence from (1)

$$(x\mathbf{i} + y\mathbf{j} + z\mathbf{k}) \cdot (l\mathbf{i} + m\mathbf{j} + n\mathbf{k}) = p$$

or

$$lx + my + nz = p$$

which is equation of a plane in co-ordinate geometry

Corollary. Let **n** be any vector of magnitude n and having the same direction as that of $\hat{\mathbf{n}}$. Then $\mathbf{n} = n\hat{\mathbf{n}}$.

Then from equation (1) we have

$$\mathbf{r} \cdot \frac{\mathbf{n}}{n} = p \quad \text{or} \quad \mathbf{r}.\mathbf{n} = np = q \text{ (say)} \qquad ...(2)$$

$\therefore$ (2) is the vector equation of a plane perpendicular to **n** and the perpendicular distance

$$p = \frac{q}{n} = \frac{q}{\text{modulus of } \mathbf{n}}$$

Note: If origin lies on the plane then p = 0. In such case from equation (1) we get $\mathbf{r}.\mathbf{n} = 0$

6.6. Converse

To prove that every equation of the form $\mathbf{r}.\mathbf{n} = q$ *represents a plane*

Let $\mathbf{r_1}$ and $\mathbf{r_2}$ be the p.v's of any two points on the surface represented by

$$\mathbf{r.n} = q \quad \text{...(1)}$$

Then $$\mathbf{r_1.n} = q \quad \text{...(2)}$$

and $$\mathbf{r_2.n} = q \quad \text{...(3)}$$

Multiplying (2) by x and (3) by y and adding we get

$$x\mathbf{r_1.n} + y\mathbf{r_2.n} = xq + yq$$

or $$(x\mathbf{r_1} + y\mathbf{r_2})\mathbf{.n} = (x + y)q$$

or $$\left[\frac{x\mathbf{r_1} + y\mathbf{r_2}}{x + y}\right]\mathbf{.n} = q \quad \text{...(4)}$$

Showing that $\dfrac{x\mathbf{r_1} + y\mathbf{r_2}}{x + y}$ also lies on (1).

Now $\dfrac{x\mathbf{r_1} + y\mathbf{r_2}}{x + y}$ is a point which divides the line joining $\mathbf{r_1}$ and $\mathbf{r_2}$ in the ratio y:x. But x, y are arbitrary scalars

$\therefore$ Every point on this line must lie on the surface given by (1).

Hence (1) represents the equation of a plane

6.7. Vector equation of a plane through a given point a and perpendicular to a given vector n

Let A be the given point whose p.v is $\overrightarrow{OA} = \mathbf{a}$ w.r.t origin O. and let $\overrightarrow{ON} = \mathbf{n}$. Let r be the p.v. of any point P which lies on the plane passing through A and perpendicular to ON such that $\overrightarrow{ON}$ is perpendicular to $\overrightarrow{AP}$.

$\therefore$ $\overrightarrow{AP} \cdot \overrightarrow{ON} = 0$ or

$(\overrightarrow{OP} - \overrightarrow{OA}) \cdot \overrightarrow{ON} = 0$

or $(\mathbf{r} - \mathbf{a}) \cdot \mathbf{n} = 0$ or $\mathbf{r.n} - \mathbf{a.n} = 0$

or $\mathbf{r.n} = \mathbf{a.\ n}$ which is the required equation of the plane.

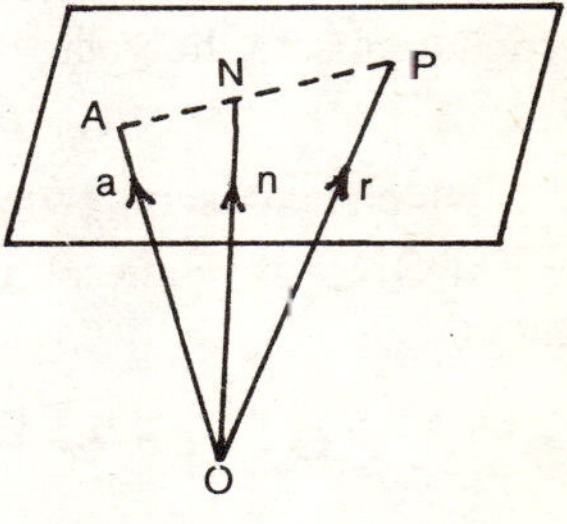

Fig. 6.8

6.8. Vector equation of the plane passing through three non–collinear points whose p.v's are a, b and c

Let **a, b, c** be the p.v.'s of three given points say A, B and C w.r.t origin O. Let **r** be the P.V of any points P on the plane passing through the point A, B, and C.

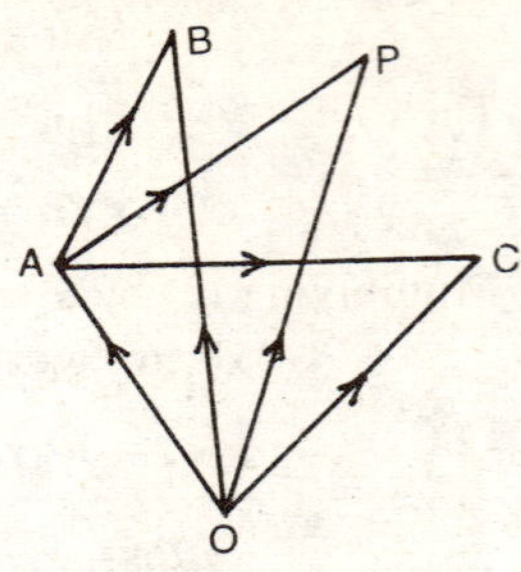

Fig. 6.9

Then $\overrightarrow{AP} = \overrightarrow{OP} - \overrightarrow{OA} = \mathbf{r} - \mathbf{a}$

Similarly $\overrightarrow{AB} = \mathbf{b} - \mathbf{a}$

and $\overrightarrow{AC} = \mathbf{c} - \mathbf{a}$

Now A, B, C and P lie on the plane

$\Rightarrow$ $\overrightarrow{AP}$, $\overrightarrow{AB}$ and $\overrightarrow{AC}$ are coplanar

$\therefore$ $\overrightarrow{AP} \cdot \{\overrightarrow{AB} \times \overrightarrow{AC}\} = 0$

i.e., $(\mathbf{r} - \mathbf{a}) \cdot \{(\mathbf{b} - \mathbf{a}) \times (\mathbf{c} - \mathbf{a})\} = 0$

or $(\mathbf{r} - \mathbf{a}) \cdot \{\mathbf{b} \times \mathbf{c} - \mathbf{b} \times \mathbf{a} - \mathbf{a} \times \mathbf{c} + \mathbf{a} \times \mathbf{a}\} = 0$

or $(\mathbf{r} - \mathbf{a}) \cdot \{\mathbf{a} \times \mathbf{b} + \mathbf{b} \times \mathbf{c} + \mathbf{c} \times \mathbf{a}\} = 0$

$[\because \ \mathbf{a} \times \mathbf{a} = \mathbf{0}, \mathbf{a} \times \mathbf{b} = -\,\mathbf{b} \times \mathbf{a}]$

or $\mathbf{r}\,(\mathbf{a} \times \mathbf{b} + \mathbf{b} \times \mathbf{c} + \mathbf{c} \times \mathbf{a}) = \mathbf{a} \cdot (\mathbf{a} \times \mathbf{b} + \mathbf{b} \times \mathbf{c} + \mathbf{c} \times \mathbf{a})$

or $\mathbf{r} \cdot (\mathbf{a} \times \mathbf{b} + \mathbf{b} \times \mathbf{c} + \mathbf{c} \times \mathbf{a}) = [\mathbf{aab}] + [\mathbf{abc}] + [\mathbf{aca}]$

or $\mathbf{r} \cdot (\mathbf{a} \times \mathbf{b} + \mathbf{b} \times \mathbf{c} + \mathbf{c} \times \mathbf{a}) = [\mathbf{abc}]$ ...(1)

$[\because \ [\mathbf{aab}] = [\mathbf{aca}] = 0]$

which is an equation in the form **r.n** = q.

Hence (1) is the required equation of the plane which is perpendicular to the vector

$$\mathbf{n} = \mathbf{a} \times \mathbf{b} + \mathbf{b} \times \mathbf{c} + \mathbf{c} \times \mathbf{a}$$

which represents twice the area of the triangle ABC

If ON = p be the perpendicular distance from origin O to this plane then

$$ON = p = \frac{q}{n} = \frac{[\mathbf{abc}]}{|\mathbf{n}|} = \frac{[\mathbf{abc}]}{|\mathbf{a} \times \mathbf{b} + \mathbf{b} \times \mathbf{c} + \mathbf{c} \times \mathbf{a}|}$$

6.9. Vector equation of the plane passing through a given point a and parallel to the vectors b and c

Let $\overrightarrow{OA} = \mathbf{a}$ be the p.v of the point A w.r.t origin O and let

$\overrightarrow{OP} = \mathbf{r}$ be the p.v of any point P on the plane which passes through A and is parallel to the vectors **b** and **c**.

Then $\overrightarrow{AP} = \overrightarrow{OP} - \overrightarrow{OA} = \mathbf{r} - \mathbf{a}$.

$\because$ $\overrightarrow{AP}$, **b** and **c** are parallel to the same plane (coplanar) we have

$$\overrightarrow{AP} \,.\, (\mathbf{b} \times \mathbf{c}) = 0 \text{ or } (\mathbf{r} - \mathbf{a}) \,.\, (\mathbf{b} \times \mathbf{c}) = 0$$

or $$\mathbf{r} \,.\, (\mathbf{b} \times \mathbf{c}) - \mathbf{a}.(\mathbf{b} \times \mathbf{c}) = 0$$

or $$\mathbf{r} \,.\, (\mathbf{b} \times \mathbf{c}) = [\mathbf{abc}]$$

which is of the form $\mathbf{r.n} = q$ and is the required equation.

Here the plane is perpendicular to $\mathbf{n} = \mathbf{b} \times \mathbf{c}$

$\therefore$ length of perpendicular $p = \dfrac{q}{|\mathbf{n}|} = \dfrac{[\mathbf{abc}]}{\mathbf{b} \times \mathbf{c}}$

6.10. Vector equation of the plane containing a given line and parallel to another line (parallel to another vector)

Let the plane contain the line $\mathbf{r} = \mathbf{a} + t\mathbf{b}$ (The line is passing through **a** and is parallel to **b**) Let the point **a** be A and any point P on the required plane be **r**. Since the plane is parallel to the vectors **b** and **c**, **b** × **c** is normal to the plane. Hence equation of the plane is

$$\overrightarrow{AP} \,.\, (\mathbf{b} \times \mathbf{c}) = 0 \text{ or } (\mathbf{r} - \mathbf{a}) \,.\, (\mathbf{b} \times \mathbf{c}) = 0$$

or $$\mathbf{r} \,.\, (\mathbf{b} \times \mathbf{c}) = [\mathbf{abc}]$$

6.11. Vector equation of the plane containing a given line and perpendicular to a given plane

Let $\mathbf{r} = \mathbf{a} + t\mathbf{b}$ be the line that the required plane contains and let $\mathbf{r} \,.\, \mathbf{c} = q$ be the plane to which the required plane is perpendicular. Now **c** is perpendicular to $\mathbf{r} \,.\, \mathbf{c} = q$ which means the required plane is parallel to **c**. Thus the required plane passes through **a** and is parallel to **b** and **c**. Hence the equation of the plane from **art 6.9**

$$(\mathbf{r} - \mathbf{a}) \,.\, (\mathbf{b} \times \mathbf{c}) = 0 \text{ or } \mathbf{r} \,.\, (\mathbf{b} \times \mathbf{c}) = [\mathbf{abc}]$$

6.12. Vector equation of the plane containing the line r = a + tb and a given point

The given line is $\mathbf{r} = \mathbf{a} + t\mathbf{b}$, which passes through the point **a** and is parallel to **b**. Let **a** and **c** denote the points A and C and let any point P on the plane be **r**. Since the plane is parallel to $\vec{AC}$ and **b**.

$\therefore$ $\vec{AC} \times \mathbf{b}$ is normal to the plane. Hence equation to the required plane from **art 6.9** is $\vec{AP} \cdot (\vec{AC} \times \mathbf{b}) = 0$

or $\quad (\mathbf{r} - \mathbf{a}) \cdot \{(\mathbf{c} - \mathbf{a}) \times \mathbf{b}\} = 0$

or $\quad (\mathbf{r} - \mathbf{a}) \cdot (\mathbf{b} \times \mathbf{c} + \mathbf{a} \times \mathbf{b}) = 0$

or $\quad \mathbf{r} \cdot (\mathbf{b} \times \mathbf{c} + \mathbf{a} \times \mathbf{b}) = \mathbf{a} \cdot (\mathbf{b} \times \mathbf{c}) + \mathbf{a} \cdot (\mathbf{a} \times \mathbf{b})$

or $\quad \mathbf{r} \cdot (\mathbf{b} \times \mathbf{c} + \mathbf{a} \times \mathbf{b}) = [\mathbf{abc}] \qquad [\because \mathbf{aab}] = 0$

6.13. Vector equation of the plane passing through two given points a and b and parallel to a given straight line (or parallel to vector c)

Let the given straight line be parallel to vector **c**. It is given that the required plane is passing through the points **a** and **b** denoted by A and B. Therefore the plane passes through **a** and is parallel to $\vec{AB}$ and **c**. Let any point P on the plane be **r**. Hence the equation to the required plane from **art 6.9** is

$$\vec{AP} \cdot (AB \times \mathbf{c}) = 0$$

or $\quad (\mathbf{r} - \mathbf{a}) \cdot \{(\mathbf{b} - \mathbf{a}) \times \mathbf{c}\} = 0$

or $\quad (\mathbf{r} - \mathbf{a}) \cdot \{(\mathbf{b} \times \mathbf{c}) - (\mathbf{a} \times \mathbf{c})\} = 0$

or $\quad (\mathbf{r} - \mathbf{a}) \cdot \{(\mathbf{b} \times \mathbf{c}) - (\mathbf{c} \times \mathbf{a})\} = 0$

or $\quad \mathbf{r} \cdot \{(\mathbf{b} \times \mathbf{c}) + (\mathbf{c} \times \mathbf{a})\} = \mathbf{a} \cdot \{(\mathbf{b} \times \mathbf{c}) + (\mathbf{c} \times \mathbf{a})\}$

or $\quad \mathbf{r} \cdot \{(\mathbf{b} \times \mathbf{c}) + (\mathbf{c} \times \mathbf{a})\} = [\mathbf{abc}] \qquad [\because \mathbf{a}.(\mathbf{c} \times \mathbf{a}) = 0]$

6.14. Angle between two planes

Let the two planes be $\mathbf{r}.\mathbf{n}_1 = q$, and $\mathbf{r}.\mathbf{n}_2 = q_2$ where $\mathbf{n}_1$ and $\mathbf{n}_2$ are vectors perpendicular to the two planes. Let θ be the angle between the planes i.e., angle between the normals.

Then $\quad \mathbf{n}_1.\mathbf{n}_2 = |\mathbf{n}_1||\mathbf{n}_2| \cos\theta$

or $$\cos\theta = \frac{\mathbf{n}_1.\mathbf{n}_2}{|\mathbf{n}_1||\mathbf{n}_2|} \text{ or } \theta = \cos^{-1} \frac{\mathbf{n}_1.\mathbf{n}_2}{|\mathbf{n}_1||\mathbf{n}_2|}$$

6.15. Angle between a line and a plane

The angle between a line and a plane is compliment of the angle between the line and the normal to the plane.

Let **r** . **n** = q be the given plane and the given line be parallel to vector **b**. Let θ be the angle between the line and the plane. The if ϕ be the angle between the normal to the plane i.e., **n** and the given line (parallel to **b**) then we have

$$\mathbf{n}\,.\,\mathbf{b} = |\mathbf{n}|\,|\mathbf{b}|\cos\phi \qquad \therefore\ \phi = \cos^{-1}\left\{\frac{\mathbf{n.b}}{|\mathbf{n}|\,|\mathbf{b}|}\right\}$$

and

$$\theta = \frac{\pi}{2} - \phi = \frac{\pi}{2} - \cos^{-1}\left\{\frac{\mathbf{n.b}}{|\mathbf{n}|\,|\mathbf{b}|}\right\}$$

$$= \sin^{-1}\left\{\frac{\mathbf{n.b}}{|\mathbf{n}|\,|\mathbf{b}|}\right\}$$

6.16. Intercepts made by the plane on the rectangular coordinate axes

Let the equation of the given plane be

$$\mathbf{r}\,.\,\mathbf{n} = q \qquad ...(1)$$

Let a, b, c be the intercepts made by this plane on the co-ordinate axes and let **i, j, k** be the unit vectors along the axes of x, y, z respectively. Then the points a**i**, b**j** and c**k** all lie on the plane (1)

$$\therefore\ a\mathbf{i}\,.\,\mathbf{n} = q;\ b\mathbf{j}\,.\,\mathbf{n} = q \text{ and } c\mathbf{k}\,.\,\mathbf{n} = q$$

or

$$a = \frac{q}{\mathbf{i}\,.\,\mathbf{n}};\ b = \frac{q}{\mathbf{j}\,.\,\mathbf{n}};\ c = \frac{q}{\mathbf{k}\,.\,\mathbf{n}}$$

which are the required intercepts

6.17. The two sides of a plane

The points **a** *and* b *lie on the same or opposite sides of a plane* **r.n** = q *according as* **a.n** − q *and* **b.n** − q *one of the same or opposite signs*

Let A and B be the two points whose p.v's are a and b respectively. Let AB cuts the plane at P in AP : PB = l : m ratio

$$\text{Then p.v. of p} = \frac{l\mathbf{b} + m\mathbf{a}}{l + m}$$

But p lies on **r** . **n** = q

$$\therefore\ \left\{\frac{l\mathbf{b} + m\mathbf{a}}{l + m}\right\}.\,\mathbf{n} = q$$

or $\quad l(\mathbf{b}\,.\,\mathbf{n}) + m(\mathbf{a}\,.\,\mathbf{n}) = q(l + m)$

or $\quad m\{\mathbf{a.n} - q\} = -\,l\{\mathbf{b.n} - q\}$

or $$\frac{l}{m} = -\frac{\mathbf{a.n} - q}{\mathbf{b.n} - q} \qquad ...(1)$$

If the ratio $\frac{l}{m}$ is positive then $\mathbf{a.n} - q$ and $\mathbf{b.n} - q$ are of opposite signs and P divides AB internally and hence A and B lie on opposite sides of the plane. Again if $\frac{l}{m}$ is negative then $\mathbf{a.n} - q$ and $\mathbf{b.n} - q$ are of the same signs and P divides AB externally. Hence the points A and B lie on the same sides of the plane.

6.18. Perpendicular distance of a point from a plane

Let O be the origin of reference and let A be the given point whose p.v is **a.**

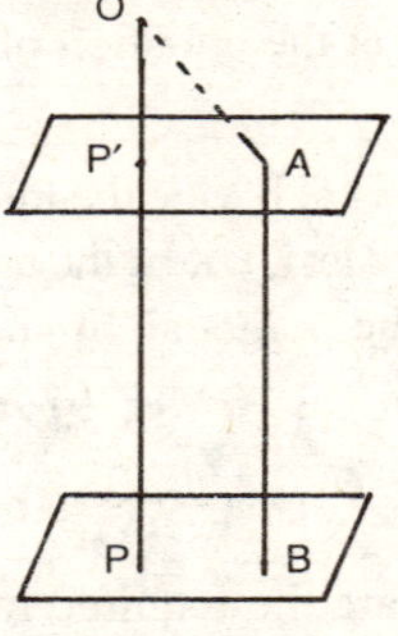

Fig. 6.10

Let the vector equation of the plane be $\mathbf{r.n} = q$ ...(1)

$\therefore$ Perpendicular distance from O on to it

$$= ON = p = \frac{q}{|\mathbf{n}|} \qquad ...(2)$$

Now we have to find the perpendicular distance from A on to the plane (1) i.e., AB. Consider a plane through A and parallel to plane (1). Then the vector equation of this plane (i.e. plane through A) is

$$(\mathbf{r} - \mathbf{a}).\mathbf{n} = 0 \quad \text{or} \quad \mathbf{r.n} = \mathbf{a.n} \qquad ...(3)$$

The length of perpendicular from the origin O to (3) is

$$= OP' = \frac{\mathbf{a.n}}{|\mathbf{n}|} \qquad ...(4)$$

$$\text{Now } AB = P'P = OP - OP' = \frac{q}{|\mathbf{n}|} - \frac{\mathbf{a.n}}{|\mathbf{n}|}$$

$$= \frac{q - \mathbf{a.n}}{|\mathbf{n}|}$$

Also $\overrightarrow{AB} = (AB)\,\hat{\mathbf{n}} = \frac{q - \mathbf{a.n}}{|\mathbf{n}|}\,\frac{\mathbf{n}}{|\mathbf{n}|}$

$$= \frac{(q - \mathbf{a.n})}{n^2}\,\mathbf{n}$$

Note. *The value of* AB *is positive for all such points which lie on the same side of the plane as the origin but it will be negative for the points on opposite sides*

6.19. Distance of a point from a plane measured in a given direction

Here we are to find the distance of the point P whose p.v w.r.t origin O is **a**, from the plane **r.n** = q in a given direction of unit vector **b**. Let PQ be the line through P, parallel to the unit vector **b** meeting the plane is Q such that PQ = k

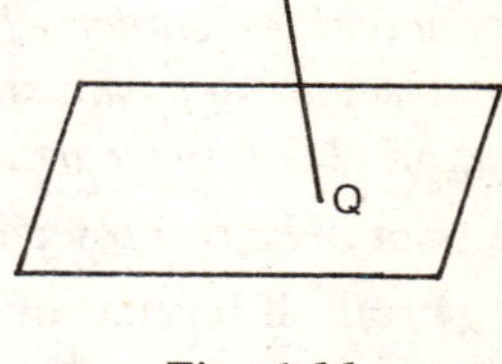

Fig. 6.11

Then $\overrightarrow{PQ} = k\mathbf{b}$.

Now the p.v. of Q is

$$\overrightarrow{OQ} = \overrightarrow{OP} + \overrightarrow{PQ} = \mathbf{a} + k\mathbf{b}.$$

But Q lies on the plane **r.n** = q

$$\therefore \quad (\mathbf{a} + k\mathbf{b}) \,.\, \mathbf{n} = q$$

or $$\mathbf{a.n} + k\,\mathbf{b.n} = q$$

or $$k = \frac{q - \mathbf{a.n}}{\mathbf{b.n}}$$ which is the required distance $\overline{PQ}$

6.20. Bisecting planes

To find the vector equations of the planes bisecting the angles between two given planes $\mathbf{r.n_1} = q_1$ and $\mathbf{r.n_2} = q_2$

Let **r** be the p.v of any point on the plane bisecting the angle between the given planes, then the perpendicular distance of this point from both planes should be equal.

$$\therefore \quad \frac{q_1 - \mathbf{r.n_1}}{|\mathbf{n_1}|} = \pm \frac{q_2 - \mathbf{r.n_2}}{|\mathbf{n_2}|} \quad ...(1)$$

or $$\frac{q_1}{|\mathbf{n}_1|} - \mathbf{r}.\hat{\mathbf{n}}_1 = \pm \frac{q_2}{|\mathbf{n}_2|} - \mathbf{r}.\hat{\mathbf{n}}_2 \qquad \left[\because \frac{\mathbf{n}_1}{|\mathbf{n}_1|} = \hat{\mathbf{n}}_1\right]$$

or $$\mathbf{r}.(\hat{\mathbf{n}}_1 + \hat{\mathbf{h}}_2) = \frac{q_1}{|\mathbf{n}_1|} + \frac{q_2}{|\mathbf{n}_2|} \qquad ...(2)$$

and $$\mathbf{r}.(\hat{\mathbf{n}}_1 - \hat{\mathbf{n}}_2) = \frac{q_1}{|\mathbf{n}_1|} - \frac{q_2}{|\mathbf{n}_2|} \qquad ...(3)$$

which are the vector equations of the planes bisecting the angles between the two given planes.

Note: *Since we consider the perpendicular distance as positive when measured in the direction from the origin to the plane, for the points on the plane bisecting the angle in which originlies, the perpendicular distance will have the same sign, but for points on the other bisector opposite signs. Due to this we have taken ± signs in equation (1) of above article*

Cor. *The two bisector planes are always perpendicular.*

Proof. If θ is the angle between the two bisector planes (2) and (3) i.e., angle between their normals

$$\cos\theta = \frac{(\hat{\mathbf{n}}_1 + \hat{\mathbf{n}}_2) . (\hat{\mathbf{n}}_1 - \hat{\mathbf{n}}_2)}{|\hat{\mathbf{n}}_1 + \hat{\mathbf{n}}_2||\hat{\mathbf{n}}_1 - \hat{\mathbf{n}}_2|} \qquad ...(4)$$

Since $(\hat{\mathbf{n}}_1 + \hat{\mathbf{n}}_2) . (\hat{\mathbf{n}}_1 - \hat{\mathbf{n}}_2) = \hat{\mathbf{n}}_1^2 - \hat{\mathbf{n}}_2^2 = 1 - 1 = 0$ we

Here $\hat{\mathbf{n}}_1, \hat{\mathbf{n}}_2$ are unit vectors

$\therefore\ (\hat{\mathbf{n}}_1 + \hat{\mathbf{n}}_2) . (\hat{\mathbf{n}}_1 - \hat{\mathbf{n}}_2) = \hat{\mathbf{n}}_1^2 - \hat{\mathbf{n}}_2^2 = 1 - 1 = 0$

Hence from (4), we get

$$\cos\theta = o \quad \text{or} \quad \theta = 90°$$

∴ The two bisector planes are always perpendicular

6.21. Vector equation of a plane through the intersection of two given planes

Let the two given planes be

$$\mathbf{r} . \mathbf{n}_1 = q_1 \qquad ...(1)$$

and $$\mathbf{r} . \mathbf{n}_2 = q_2 \qquad ...(2)$$

Let $(\mathbf{r.n_1} - q_1) - \lambda(\mathbf{r.n_2} - q_2) = 0$ ...(3)

or $\mathbf{r}.(\mathbf{n_1} - \lambda\mathbf{n_2}) = q_1 - \lambda q_2$ ...(4)

where λ is a scalar.

Now equation (4) is satisfied by all those points which satisfy the given equations (1) and (2) for any value of λ. Moreover it is of the form $r.n = q$. Hence equation (4) i.e., (3) represents vector equations of a plane through the intersection of two given planes (1) and (2).

λ is to be determined by an additional condition of the problem.

For example the plane passes through a

Then from (3) $\lambda = \dfrac{\mathbf{a.n_1} - q_1}{\mathbf{a.n_2} - q_2}$

6.22. Vector equation of line of intersection of two planes

Let $\mathbf{r.n_1} = q_1$ ...(1) and $\mathbf{r.n_2} = q_2$ ...(2) be the two given planes. The line of intersection of both the planes (1) and (2), is perpendicular to both the normals $\mathbf{n_1}$ and $\mathbf{n_2}$, is parallel to $\mathbf{n_1} \times \mathbf{n_2}$ Let OP be the perpendicular drawn from the origin O to the line of intersection of (1) and (2). Then $\overrightarrow{OP}$ is coplanar with the normals $\mathbf{n_1}$ and $\mathbf{n_2}$.

$\therefore$ $\overrightarrow{OP} = x\mathbf{n_1} + y\mathbf{n_2}$ ...(3)

where x and y are scalars. Since P lies on both the planes $\overrightarrow{OP}$ must satisfy (1) and (2).

$\therefore$ $(x\mathbf{n_1} + y\mathbf{n_2}).\mathbf{n_1} = q_1$ and $(x\mathbf{n_1} + y\mathbf{n_2}).\mathbf{n_2} = q_2$

or $x\mathbf{n_1^2} + y\mathbf{n_1.n_2} = q_1$ and $x\mathbf{n_1.n_2} + y\mathbf{n_2^2} = q_2$

Solving the equations, we get

$$x = \frac{q_1\mathbf{n_1^2} - q_2\,\mathbf{n_1.n_2}}{\mathbf{n_1^2n_2^2} - (\mathbf{n_1.n_2})^2},\quad y = \frac{q_2\mathbf{n_1^2} - q_1\,\mathbf{n_1.n_2}}{\mathbf{n_1^2n_2^2} - (\mathbf{n_1.n_2})^2} \qquad ...(4)$$

$\therefore$ The line of intersection of the planes is a line through the point parallel to the vector $\mathbf{n_1} \times \mathbf{n_2}$. Hence the required equation is

$$\mathbf{r} = x\mathbf{n_1} + y\mathbf{n_2} + t(\mathbf{n_1} \times \mathbf{n_2})$$

where t is a parameter and x, y are given by (4).

6.23. Condition for two straight lines to be coplanar

To find the condition of intersection of two straight lines.

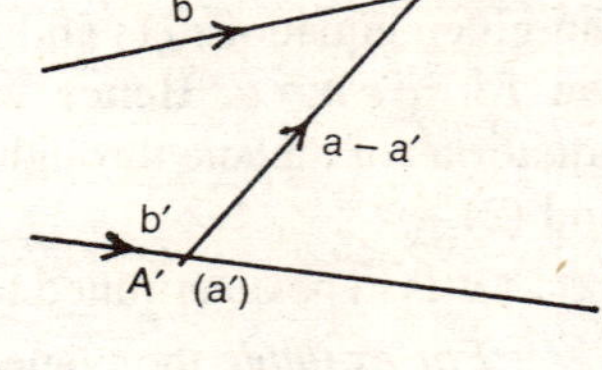

Fig. 6.12

Let the two lines be

$$\mathbf{r} = \mathbf{a} + t\mathbf{b} \quad ...(1)$$

and $$\mathbf{r} = \mathbf{a}' + s\mathbf{b}' \quad ...(2)$$

which pass through **a**, **a′** and parallel to **b**, **b′** respectively. If the lines intersect their common plane will be parallel to the vectors **b**, **b′** and must contain $\overrightarrow{A'A}$ i.e., $\mathbf{a} - \mathbf{a}'$.

$\Rightarrow$ $\quad \mathbf{a} - \mathbf{a}', \mathbf{b}, \mathbf{b}'$ are coplanar

i.e., $$(\mathbf{a} - \mathbf{a}') \cdot (\mathbf{b} \times \mathbf{b}') = 0$$

or $$\mathbf{a} \cdot (\mathbf{b} \times \mathbf{b}') = \mathbf{a}' \cdot (\mathbf{b} \times \mathbf{b}') \quad \text{or} \quad [\mathbf{a}\mathbf{b}\mathbf{b}'] = [\mathbf{a}'\mathbf{b}\mathbf{b}']$$

SOLVED PROBLEMS

Example 10. Find the equation of the plane through the point $\mathbf{i} + 2\mathbf{j} - \mathbf{k}$ and perpendicular to the vector $2\mathbf{i} - \mathbf{j} - 3\mathbf{k}$. Also find the perpendicular distance from the origin.

Solution. The equation of the plane through the point **a** and perpendicular to **n** is

$$(\mathbf{r} - \mathbf{a}).\mathbf{n} = 0 \quad \text{or} \quad \mathbf{r}.\mathbf{n} = \mathbf{a}.\mathbf{n} \quad ...(1)$$

Here $\quad \mathbf{a} = \mathbf{i} + 2\mathbf{j} - \mathbf{k}, \mathbf{n} = 2\mathbf{i} - \mathbf{j} - 3\mathbf{k},$

$$\mathbf{a}.\mathbf{n} = (\mathbf{i} + 2\mathbf{j} - \mathbf{k}) \cdot (2\mathbf{i} - \mathbf{j} - 3\mathbf{k}) = 2 - 2 + 3 = 3$$

and $$|\mathbf{n}| = \sqrt{(2)^2 + (-1)^2 + (-3)^2} = \sqrt{4 + 1 + 9} = \sqrt{14}$$

Hence the equation of the plane is

$$\mathbf{r}.(2\mathbf{i} - \mathbf{j} - 3\mathbf{k}) = 3 \quad \text{[from (1)]}$$

and the perpendicular distance from orign

$$= p = \frac{\mathbf{a} \cdot \mathbf{n}}{|\mathbf{n}|} = \frac{3}{\sqrt{14}}$$

Note: If $\mathbf{r} + x\mathbf{i} + y\mathbf{j} + z\mathbf{k}$ then the equation of the plane in cartesian form is $2x - y - 3z = 3$.

Example 11. Find the equation of the plane perpendicular to the line $\frac{x+1}{2} = \frac{y-3}{-3} = \frac{z}{1}$ and passing through (-1, 2, 6).

Solution: Given line is $\frac{x+1}{2} = \frac{y-3}{-3} = \frac{z}{1}$. the direction cosines are proportional to 2, - 3, 1.

⇒ The given line is parallel to the vector $2\mathbf{i} - 3\mathbf{j} + \mathbf{k}$.

∴ The required plane passes through the point (- 1, 2, 6) i.e., $\mathbf{a} = -\mathbf{i} + 2\mathbf{j} + 6\mathbf{k}$ and is perpendicular to the vector $\mathbf{n} = 2\mathbf{i} - 3\mathbf{j} + \mathbf{k}$

∴ Its equation is $(\mathbf{r} - \mathbf{a}).\mathbf{n} = 0$

or $\quad [\mathbf{r} - (-\mathbf{i} + 2\mathbf{j} + 6\mathbf{k})] . [2\mathbf{i} - 3\mathbf{j} + \mathbf{k}] = 0$

or $\quad \mathbf{r} . (2\mathbf{i} - 3\mathbf{j} + \mathbf{k}) = -2.$

Example 12. The position vectors of two points P and Q are $2\mathbf{i} + 3\mathbf{j} + 6\mathbf{k}$ and $\mathbf{i} + 3\mathbf{j} + 2\mathbf{k}$. Find the equation of the plane through Q and perpendicular to PQ.

Solution: $\overrightarrow{PQ}$ = p.v. of Q - p.v of P

$= (\mathbf{i} + 3\mathbf{j} + 2\mathbf{k}) - (2\mathbf{i} + 3\mathbf{j} + 6\mathbf{k}) = -\mathbf{i} = 4\mathbf{k}$

∴ $\mathbf{a} = \mathbf{i} + 3\mathbf{j} + 2\mathbf{k}$; ∵ $= -\mathbf{i} - 4\mathbf{k}$ and $\mathbf{a}.\mathbf{n} = -1 - 8 = -9$

Hence the required plane is

$$(\mathbf{r} - \mathbf{a}) . \mathbf{n} = 0 \quad \text{or} \quad \mathbf{r}.\mathbf{n} = \mathbf{a}.\mathbf{n}$$

or $\quad \mathbf{r}.(-\mathbf{i} - 4\mathbf{k}) = -9 \quad \text{or} \quad \mathbf{r}.(\mathbf{i} + 4\mathbf{k}) = 9.$

Example 13. Show that the points $A(\mathbf{i} + \mathbf{j} + \mathbf{k})$ and $B(\mathbf{i} + \mathbf{j} + 4\mathbf{k})$ are on opposite sides of the plane $\mathbf{r} . (3\mathbf{i} + 4\mathbf{j} + 5\mathbf{k}) = 19$. Find the perpendicular distance of the former from the plane and the vector through it perpendicular to the plane.

Solutio.: Given plane is $\mathbf{r}.(3\mathbf{i} + 4\mathbf{j} + 5\mathbf{k}) = 10$ ...(1)

$$\mathbf{n} = 3\mathbf{i} + 4\mathbf{j} + 5\mathbf{k} \therefore |\mathbf{n}| = \sqrt{9 + 16 + 25} = 5\sqrt{2}; \; q = 19$$

Now perpendicular distance of a point **a** form the plane $\mathbf{r} . \mathbf{n} = q$ is $= \frac{q - \mathbf{a}.\mathbf{n}}{|\mathbf{n}|}$

∴ Perpendicular distance from A i.e., $\mathbf{i} + \mathbf{j} + \mathbf{k}$ on to the plane (1) is

$$= \frac{19 - (\mathbf{i} + \mathbf{j} + \mathbf{k}) . (3\mathbf{i} + 4\mathbf{j} + 5\mathbf{k})}{5\sqrt{2}} = \frac{19 - (3 + 4 + 5)}{5\sqrt{2}} = \frac{7}{5\sqrt{2}}$$

also perpendicular distance from B i.e., $\mathbf{i} + \mathbf{j} + 4\mathbf{k}$ on to the plane (1) is

$$= \frac{19 - (\mathbf{i} + \mathbf{j} + 4\mathbf{k}) . (3\mathbf{i} + 4\mathbf{j} + 5\mathbf{k})}{5\sqrt{2}} = \frac{19 - (3 + 4 + 20)}{5\sqrt{2}} = \frac{-8}{5\sqrt{2}}$$

Again since q − **a** . **n** = 7 and q − **a.n** = − 8 i.e., they are of opposite signs therefore the two points are on opposite sides of the plane.

Now a vector through A(**i** + **j** + **k**), perpendicular to the plane is $\frac{q - \mathbf{a.n}}{|\mathbf{n}|^2}$ n = $\frac{7}{(5\sqrt{2})^2}$ (3**i** + 4**j** + 5**k**).

Example 14. Find the locus of a point which moves so that the difference of its distances from two given points is constant

Solution. w.r.t some origin O let the p.v's of the given points A and B be **a** and **b** and that of the variable point P be **r**. Then we are given that

$$\overline{AP}^2 - \overline{BP}^2 = c \text{ (some constant)}$$

$$\overrightarrow{AP}^2 - \overrightarrow{BP}^2 = c \text{ (square of a vector = square of its module)}$$

i.e., $(\mathbf{r} - \mathbf{a})^2 - (\mathbf{r} - \mathbf{b})^2 = \mathbf{c}$

or $(r^2 - 2\mathbf{r.a} + a^2) - (r^2 - 2\mathbf{r.b} = b^2) = c$

or $2\mathbf{r.(a - b)} = a^2 + b^2 - c$

or $\mathbf{r.(a - b)} = q$ where $q = \frac{1}{2}(a^2 + b^2 - c) =$ constant

This is of the form **r.n** = q and hence represents a plane whose normal is **a** − **b** i.e., $\overrightarrow{BA}$.

Example 15. Find the vector equation of the plane through the point **i** − 2**j** + 3**k** and parallel to the plane **r**.(2**i** + **j** + 2k) = 6. Also find the perpendicular distance of this plane from 3**i** − **j** + 4**k** and show that this point is opposite side from the origin.

Solution. Given plane is **r**.(2**i** + **j** + 2k) = 6 ...(1)

Normal vector **n** = 2**i** + **j** + 2k; | **n** | = $\sqrt{4 + 1 + 4}$ = 3

Let **a** = **i** − 2**j** + 3k. Now the vector equation of the plane through a and parallel to (1) is

$$(\mathbf{r} - \mathbf{a}) . \mathbf{n} = 0 \text{ i.e., } [\mathbf{r} - (\mathbf{i} + 2\mathbf{j} + 3\mathbf{k})] . (2\mathbf{i} + \mathbf{j} + 2\mathbf{k}) = 0$$

or **r**.(2**i** + **j** + 2**k**) = 6 ...(2)

Perpendicular distance of (2) from 3**i** − **j** + 4**k**

$$p_1 = \frac{6 - (3\mathbf{i} - \mathbf{j} + 4\mathbf{k}) \cdot (2\mathbf{i} + \mathbf{j} + 2\mathbf{k})}{3} = -\frac{7}{3}$$

Perpendicular distance of (2) from origin

$$p_2 = \frac{6 - 0}{3} = 2$$

$\therefore$ p_1 and p_2 are of opposite signs the point is on the opposite side from the origin.

Example 16. Find the locus of the point which is equidistant from the three planes $\mathbf{r}.\mathbf{n}_1 = q$, $\mathbf{r}.\mathbf{n}_2 = q_2$; $\mathbf{r}.\mathbf{n}_3 = q_3$.

Solution: Let $\mathbf{r}'$ be the p.v of the point P which is equidistant from the three given planes

Then $$\frac{q_1 - \mathbf{r}'.\mathbf{n}_1}{|\mathbf{n}_1|} = \frac{q_2 - \mathbf{r}'.\mathbf{n}_2}{|\mathbf{n}_2|} = \frac{q_3 - \mathbf{r}'.\mathbf{n}_3}{|\mathbf{n}_3|}$$

Hence the locus is

$$\frac{q_1 - \mathbf{r}.\mathbf{n}_1}{|\mathbf{n}_1|} = \frac{q_2 - \mathbf{r}.\mathbf{n}_2}{|\mathbf{n}_2|} = \frac{q_3 - \mathbf{r}.\mathbf{n}_3}{|\mathbf{n}_3|}.$$

Example 17. Prove that the sum of the reciprocals of the squares of the intercepts on rectangular axes made by a fixed plane is same for all systems of rectangular axes with a given origin.

Solution: Let the equation of the fixed plane be $\mathbf{r}.\mathbf{n} = q$ ...(1)
Let **a, b, c** be the intercepts made by (1) on any system of rectangular axes with a given origin.

Then the points of **ai, bj** and **ck** will satisfy the equation of (1)

i.e., $\mathbf{ai}.\mathbf{n} = q$; $\mathbf{bj}.\mathbf{n} = q$ and $\mathbf{ck}.\mathbf{n} = q$

$$\Rightarrow \quad a = \frac{q}{\mathbf{i}.\mathbf{n}};\ b = \frac{q}{\mathbf{j}.\mathbf{n}} = q \text{ and } c = \frac{q}{\mathbf{k}.\mathbf{n}}$$

$$\Rightarrow \quad \frac{1}{a^2} + \frac{1}{b^2} + \frac{1}{c^2} = \frac{1}{q^2}\{(\mathbf{i}.\mathbf{n})^2 + (\mathbf{j}.\mathbf{n})^2 + (\mathbf{k}.\mathbf{n})^2\} \quad \text{...(2)}$$

We know that $\mathbf{n} = (\mathbf{i}.\mathbf{n})\mathbf{i} + (\mathbf{j}.\mathbf{n})\mathbf{j} + (\mathbf{k}.\mathbf{n})\mathbf{k}$

i.e., $$n^2 = (\mathbf{i}.\mathbf{n})^2 + (\mathbf{j}.\mathbf{n})^2 + (\mathbf{k}.\mathbf{n})^2 \quad [\because\ \mathbf{i}^2 - \mathbf{j}^2 = \mathbf{k}^2 = 1]$$

$$\therefore \quad \frac{1}{a^2} + \frac{1}{b^2} + \frac{1}{c^2} = \frac{n^2}{q^2} = \text{constant}$$

($\therefore$ the plane is fixed and thus n and q are both constants)

Example 18. Find the vector equation of the plane through the point (1, -2, 1) and passing through the intersection of the planes $\mathbf{r}.(2\mathbf{i}+\mathbf{j}-\mathbf{k})=0$ and $\mathbf{r}.(3\mathbf{i}-\mathbf{j}-3\mathbf{k})=0$

Solution. Equation of any plane through the line of intersection of the given planes

$\mathbf{r}.(2\mathbf{i}+\mathbf{j}-\mathbf{k})=0$ and $\mathbf{r}.(3\mathbf{i}-\mathbf{j}-3\mathbf{k})=0$ in given by

$$\mathbf{r}.(2\mathbf{i}+\mathbf{j}-\mathbf{k})+\lambda[\mathbf{r}.(3\mathbf{i}-\mathbf{j}-3\mathbf{k})]=0$$

or $$\mathbf{r}.\{(2\mathbf{i}+\mathbf{j}-\mathbf{k})+\lambda(3\mathbf{i}-\mathbf{j}-3\mathbf{k})\}=0$$

or $$\mathbf{r}.\{(2+3\lambda)\mathbf{i}+(1-\lambda)\mathbf{j}-(1+3\lambda)\mathbf{k}\}=0 \qquad ...(1)$$

If it passes through the point (1, - 2, 1) i.e., $\mathbf{i}-2\mathbf{j}+\mathbf{k}$ then

$$(\mathbf{i}-2\mathbf{j}+\mathbf{k}).\{(2+3\lambda)\mathbf{i}+(1-\lambda)\mathbf{j}-(1+3\lambda)\mathbf{k}\}=0$$

or $$(2+3\lambda)-2(1-\lambda)-1.(1+3\lambda)=0 \text{ or } \lambda=\frac{1}{2}$$

Substituting in (1), we get after simplification

$$\mathbf{r}.\frac{1}{2}(7\mathbf{i}+\mathbf{j}-5\mathbf{k})=0 \text{ or } \mathbf{r}.(7\mathbf{i}+\mathbf{j}-5\mathbf{k})=0$$

which is the required equation.

Example 19. Find the vector equation of a plane through $(2\mathbf{i}+\mathbf{j}-2\mathbf{k})$ perpendicular to the line of intersection of the planes $\mathbf{r}.(3\mathbf{i}-5\mathbf{j}+7\mathbf{k})=13$ and $\mathbf{r}.(-\mathbf{i}+3\mathbf{j}+10\mathbf{k})+26=0$.

Solution. The line of intersection of the given planes

$$\mathbf{r}.(3\mathbf{i}-5\mathbf{j}+7\mathbf{k})=13 \qquad ...(1)$$

and $$\mathbf{r}.(-\mathbf{i}+3\mathbf{j}+10\mathbf{k})=-26 \qquad ...(2)$$

is common to both the planes (1) and (2).

Hence it is perpendicular to the vectors

$$\mathbf{n}_1=3\mathbf{i}-5\mathbf{j}+7\mathbf{k} \text{ and } \mathbf{n}_2=-\mathbf{i}+3\mathbf{j}+10\mathbf{k}$$

$\Rightarrow$ parallel to $\mathbf{n}_1\times\mathbf{n}_2$

$$\therefore \quad \mathbf{n}_1\times\mathbf{n}_2=\begin{vmatrix}\mathbf{i} & \mathbf{j} & \mathbf{k}\\ 3 & -5 & 7\\ -1 & 3 & 10\end{vmatrix}=-71\mathbf{i}-37\mathbf{j}+4\mathbf{k}$$

Thus we are to find the equation through $\mathbf{a}=2\mathbf{i}+\mathbf{j}-2\mathbf{k}$ and normal to $\mathbf{n}_1\times\mathbf{n}_2=-71\mathbf{i}-37\mathbf{j}+4\mathbf{k}$

$\therefore$ Required equation of the plane is

$$(\mathbf{r} - \mathbf{a}) \cdot (\mathbf{n_1} \times \mathbf{n_2}) = 0$$

$$[\mathbf{r} - (2\mathbf{i} + \mathbf{j} - 2\mathbf{k})] \cdot [-17\mathbf{i} - 37\mathbf{j} + 4\mathbf{k}] = 0$$

or $$\mathbf{r} \cdot [-71\mathbf{i} - 37\mathbf{j} + 4\mathbf{k}) + 187 = 0.$$

Example 20. Find the vector equation of a plane through the point (– 2, 3, 1) and perpendicular to each other planes $3x - 4y + 7z - 13 = 0$ and $6x + y - 13z + 19 = 0$.

Solution. The given planes are

$$3x - 4y + 7z - 13 = 0 \text{ or } \mathbf{r}.(3\mathbf{i} - 4\mathbf{j} + 7\mathbf{k}) - 13 = 0 \quad ...(1)$$

and $$6x + y - 13z + 19 = 0 \text{ or } \mathbf{r}.(6\mathbf{i} + \mathbf{j} - 13\mathbf{k}) + 19 = 0 \quad ...(2)$$

where $\mathbf{r} = x\mathbf{i} + y\mathbf{j} + z\mathbf{k}$

The required plane is perpendicular to both (1) and (2) i.e., parallel to each of the vectors $\mathbf{n_1} = 3\mathbf{i} - 4\mathbf{j} + 7\mathbf{k}$ and $\mathbf{n_2} = 6\mathbf{i} + \mathbf{j} - 13\mathbf{k}$

$\Rightarrow$ $\mathbf{n_1} \times \mathbf{n_2}$ is perpendicular to each $\mathbf{n_1}$ and $\mathbf{n_2}$ i.e. normal to the required plane

$$\therefore \quad \mathbf{n_1} \times \mathbf{n_2} = \begin{vmatrix} \mathbf{i} & \mathbf{j} & \mathbf{k} \\ 3 & -4 & 7 \\ 6 & 1 & -13 \end{vmatrix} = 45\mathbf{i} + 8\mathbf{j} + 27\mathbf{k}$$

It is passing through (– 2, 3, 1) i.e., $\mathbf{a} = -2\mathbf{i} + 3\mathbf{j} - \mathbf{k}$

$\therefore$ Required equation of the plane is

$$(\mathbf{r} - \mathbf{a}) \cdot (\mathbf{n_1} \times \mathbf{n_2}) = 0$$

$$[\mathbf{r} - (-2\mathbf{i} + 3\mathbf{j} + \mathbf{k})] \cdot (45\mathbf{i} + 81\mathbf{j} + 27\mathbf{k}) = 0$$

or $$\mathbf{r}.(45\mathbf{i} + 81\mathbf{j} + 27\mathbf{k}) = 160$$

or $$45x + 81y + 27z - 160 = 0.$$

Example 21. Find the equation of the plane which contain the line $\mathbf{r} = (\mathbf{i} + 2\mathbf{j} - 3\mathbf{k}) + t(2\mathbf{i} - 3\mathbf{j} + \mathbf{k})$ and is perpendicular to the plane $\mathbf{r} \cdot (3\mathbf{i} + 2\mathbf{j} + \mathbf{k}) = 6$. Find the position vector of the point where this plane meets the line $\mathbf{r} = t(\mathbf{i} + \mathbf{j} + 2\mathbf{k})$.

Solution: The required plane contains

$$\mathbf{r} = (\mathbf{i} + 2\mathbf{j} - 3\mathbf{k}) + t(2\mathbf{i} - 3\mathbf{j} + \mathbf{k}) \quad ...(1)$$

and is perpendicular to the plane

$$\mathbf{r}.(3\mathbf{i} + 2\mathbf{j} + \mathbf{k}) = 6 \quad ...(2)$$

$\therefore$ The required plane passes through the point $\mathbf{a} = \mathbf{i} + 2\mathbf{j} - 3\mathbf{k}$ and is parallel to the vectors $\mathbf{n_1} = 2\mathbf{i} - 3\mathbf{j} + \mathbf{k}$ and $\mathbf{n_2} = 2\mathbf{i} + 2\mathbf{j} + \mathbf{k}$

$\therefore$ normal to the required plane $= \mathbf{n_1} \times \mathbf{n_2}$

$$\mathbf{n_1} \times \mathbf{n_2} = \begin{vmatrix} \mathbf{i} & \mathbf{j} & \mathbf{k} \\ 2 & -3 & 1 \\ 3 & 2 & 1 \end{vmatrix} = -5\mathbf{i} + \mathbf{j} + 13\mathbf{k}$$

$\therefore$ Required plane is

$$(\mathbf{r} - \mathbf{a}) \cdot (\mathbf{n_1} \times \mathbf{n_2}) = 0$$

i.e., $$[\mathbf{r} - (\mathbf{i} + 2\mathbf{j} - 3\mathbf{k})] \cdot (-5\mathbf{i} + \mathbf{j} + 13\mathbf{k}) = 0$$

or $$\mathbf{r} \cdot (-5\mathbf{i} + \mathbf{j} + 13\mathbf{k}) + 42 = 0 \qquad ...(3)$$

Let $\mathbf{b}$ be the p.v. of the point of intersection of the line $\mathbf{r} = t(\mathbf{i} + \mathbf{j} + 2\mathbf{k})$ and the plane (3)

$$\therefore \quad t(\mathbf{i} + \mathbf{j} + 2\mathbf{k}) \cdot (-5\mathbf{i} + \mathbf{j} + 13\mathbf{k}) + 42 = 0$$

or $$t(22) + 42 = 0 \quad \text{or} \quad t = -\frac{21}{11}$$

Hence $\mathbf{b} = -\dfrac{21}{11}(\mathbf{i} + \mathbf{j} + 2\mathbf{k})$.

Example 22. Prove that the line of intersection of the planes $\mathbf{r} \cdot (\mathbf{i} + 2\mathbf{j} + 3\mathbf{k}) = 0$ and $\mathbf{r} \cdot (3\mathbf{i} + 2\mathbf{j} + \mathbf{k}) = 0$ is $\mathbf{r} = t(\mathbf{i} - 2\mathbf{j} + \mathbf{k})$. Show that this line is equally inclined to $\mathbf{i}$ and $\mathbf{k}$ and makes an angle $\dfrac{1}{2}\sec^{-1}3$ with $\mathbf{j}$.

Solution. Given planes are

$$\mathbf{r} \cdot (\mathbf{i} + 2\mathbf{j} + 3\mathbf{k}) = 0 \qquad ...(1)$$

and $$\mathbf{r} \cdot (3\mathbf{i} + 2\mathbf{j} + \mathbf{k}) = 0 \qquad ....(2)$$

Let $\mathbf{n_1} = \mathbf{i} + 2\mathbf{j} + 3\mathbf{k}$ and $\mathbf{n_2} = 3\mathbf{i} + 2\mathbf{j} + \mathbf{k}$

Then the line of intersection of the plane (1) and (2) is parallel to $\mathbf{n_1} \times \mathbf{n_2}$.

$$\therefore \quad \mathbf{n_1} \times \mathbf{n_2} = \begin{vmatrix} \mathbf{i} & \mathbf{j} & \mathbf{k} \\ 1 & 2 & 3 \\ 3 & 2 & 1 \end{vmatrix} = -4\mathbf{i} + 8\mathbf{j} - 4\mathbf{k}$$

Again both the planes (1) and (2) passes through the origin.

$\therefore$ The required line of intersection also passes through the origin.

Hence its equation is

$$\mathbf{r} = 0 + s(-4\mathbf{i} + 8\mathbf{j} - 4\mathbf{k})$$

or $\mathbf{r} = t(\mathbf{i} - 2\mathbf{j} + \mathbf{k})$ [where $-4s = t$] ...(3)

Let $\theta_1, \theta_2, \theta_3$ be the angles which the line makes with **i, j** and **k** respectively

$\therefore$ $\mathbf{i}.(\mathbf{i} - 2\mathbf{j} + \mathbf{k}) = 1.\sqrt{1+4+1}\cos\theta_1$

or $1 = \sqrt{6}\cos\theta$, or $\theta_1 = \cos^{-1}\frac{1}{\sqrt{6}}$

and $\mathbf{j}.(\mathbf{i} - 2\mathbf{j} + \mathbf{k}) = 1.\sqrt{1+4+1}\cos\theta_2$

or $-2 = \sqrt{6}\cos\theta_2$ or $\cos\theta_2 = \frac{-2}{\sqrt{6}}$

also $\mathbf{k}.(\mathbf{i} - 2\mathbf{j} + \mathbf{k}) = 1.\sqrt{1+4+1}\cos\theta_3$

or $1 = \sqrt{6}\cos\theta_3$ or $\cos\theta_3 = \frac{1}{\sqrt{6}}$

$\Rightarrow$ The line of intersection is equally inclined to **i** and **k**.

Now $\cos\theta_2 = -\frac{2}{\sqrt{6}}$ or $\cos^2\theta_2 = \frac{4}{6} = \frac{2}{3}$

or $2\cos^2\theta_2 = \frac{4}{3}$ or $2\cos^2\theta_2 - 1 = \frac{1}{3}$

or $\cos 2\theta_2 = \frac{1}{3}$ or $\theta_2 = \frac{1}{2}\sec^{-1}\frac{1}{3}$

$\therefore$ The line makes an angle of $\frac{1}{2}\sec^{-1}\frac{1}{3}$ with **j**.

Example 23. Find the equation of the line of intersection of the planes $\mathbf{r}.(4\mathbf{i} - 2\mathbf{j} + \mathbf{k}) = 3$ and $\mathbf{r}.(\mathbf{i} + \mathbf{j} - 3\mathbf{k}) = 1$.

Solution. The given planes are

$$\mathbf{r}.(4\mathbf{i} - 2\mathbf{j} + \mathbf{k}) = 3 \quad ...(1)$$

and $$\mathbf{r}.(\mathbf{i} + \mathbf{j} - 3\mathbf{k}) = 1 \quad ...(2)$$

let $\mathbf{n}_1 = 4\mathbf{i} - 2\mathbf{j} + \mathbf{k}$ and $\mathbf{n}_2 = \mathbf{i} + \mathbf{j} - 3\mathbf{k}$ be the normals

$\therefore$ The line of intersection of the given planes being perpendicular to each of the normal $\mathbf{n}_1$ and $\mathbf{n}_2$, is parallel to $\mathbf{n}_1 \times \mathbf{n}_2$.

$$\therefore\ \mathbf{n_1} \times \mathbf{n_2} = \begin{vmatrix} \mathbf{i} & \mathbf{j} & \mathbf{k} \\ 4 & -2 & 1 \\ 1 & 1 & -3 \end{vmatrix} = 5\mathbf{i} + 13\mathbf{j} + 6\mathbf{k}$$

Let **a** be the position vector of the foot of the perpendicular from origin to the line of intersection. Then **a** will be coplanar with $\mathbf{n_1}$ and $\mathbf{n_2}$ and hence can be expressed as

$$\mathbf{a} = x\mathbf{n_1} + y\mathbf{n_2}$$

i.e., $$\mathbf{a} = x\,(4\mathbf{i} - 2\mathbf{j} + \mathbf{k}) + y\,(\mathbf{i} + \mathbf{j} - 3\mathbf{k}) \qquad ...(3)$$

where x and y are scalars such that a lies on both the given planes (1) and (2)

$$\therefore\quad \mathbf{a}\,.\,(4\mathbf{i} - 2\mathbf{j} + \mathbf{k}) = 3 \text{ and } \mathbf{a}.(\mathbf{i} + \mathbf{j} - 3\mathbf{k}) = 1$$

i.e., $$[x\,(4\mathbf{i} - 2\mathbf{j} + \mathbf{k}) + y(\mathbf{i} + \mathbf{j} - 3\mathbf{k})]\,.\,(4\mathbf{i} - 2\mathbf{j} + \mathbf{k}) = 3$$

and $$[x\,(4\mathbf{i} - 2\mathbf{j} + \mathbf{k}) + y(\mathbf{i} + \mathbf{j} - 3\mathbf{k})]\,.\,(\mathbf{i} + \mathbf{j} - 3\mathbf{k}) = 1$$

or $$21x - y = 3 \text{ and } -x + 11y = 1$$

On solving $$x = \frac{17}{115} \text{ and } y = \frac{12}{115}$$

$\therefore$ From (3), we get

$$\mathbf{a} = \frac{17}{115}(4\mathbf{i} - 2\mathbf{j} + \mathbf{k}) + \frac{12}{115}(\mathbf{i} + \mathbf{j} - 3\mathbf{k})$$

$$= \frac{1}{115}(80\mathbf{i} - 22\mathbf{j} - 19\mathbf{k})$$

Now the line of intersection passes through the point **a** and parallel to $\mathbf{n_1} \times \mathbf{n_2}$ and its equation is $\mathbf{r} = \mathbf{a} + t\,(\mathbf{n_1} \times \mathbf{n_2})$ where t is some scalar.

$$\therefore\ \mathbf{r} = \frac{1}{115}(80\mathbf{i} - 22\mathbf{j} - 19\mathbf{k}) + t(5\mathbf{i} + 13\mathbf{j} + 6\mathbf{k}).$$

Example 24. Show that the planes

$$\mathbf{r}.(\mathbf{i} + \mathbf{j} - \mathbf{k}) = 0;\ \ \mathbf{r}.(2\mathbf{i} + \mathbf{j} + \mathbf{k}) + 3 = 0 \text{ and } \mathbf{r}.(\mathbf{i} + 2\mathbf{k}) + 3 = 0$$

have a common line of intersection.

Solution. Any plane through the line of intersection of the planes

$$\mathbf{r}.(\mathbf{i} + \mathbf{j} - \mathbf{k}) = 0 \text{ and } \mathbf{r}.(2\mathbf{i} + \mathbf{j} + \mathbf{k}) + 3 = 0 \text{ is given by}$$

$$\mathbf{r}.(\mathbf{i} + \mathbf{j} - \mathbf{k}) + \lambda\,\{\mathbf{r}.(2\mathbf{i} + \mathbf{j} + \mathbf{k}) + 3\} = 0$$

or $\quad \mathbf{r}.\{(1+2\lambda)\mathbf{i}+(1+\lambda)\mathbf{j}+(-1+\lambda)\mathbf{k}\},+3\lambda=0$ ...(1)

The third plane is $\quad \mathbf{r}.(\mathbf{i}+2\mathbf{k})+3=0$ ...(2)

Now the given planes will have a common line of intersection if for some value of λ (1) and (2) are identical

$\because$ Coefficient of $\mathbf{j} = 0$ in (2), we put $\lambda = -1$ in (1) which gives

$$\mathbf{r}.(-\mathbf{i}-2\mathbf{k})-3=0 \quad \text{or} \quad \mathbf{r}.(\mathbf{i}+2\mathbf{k})+3=0$$

i.e., equation (2). Hence the given three planes have a common line of intersection.

Example 25. Find the equation of the plane passing through three points $-\mathbf{i}+2\mathbf{j}+3\mathbf{k}$, $4\mathbf{i}-3\mathbf{j}-7\mathbf{k}$ and $-6\mathbf{i}+2\mathbf{j}-5\mathbf{k}$. Also express it in cartesian form.

Solution. Let $\vec{OA} = -\mathbf{i}+2\mathbf{j}+3\mathbf{k}$, $\vec{OB} = 4\mathbf{i}-3\mathbf{j}-7\mathbf{k}$ and $\vec{OC} = -6\mathbf{i}+2\mathbf{j}-5\mathbf{k}$ be the p.v.'s of points A, B, C w.r.t. some origin O. Let $\vec{OP} = r$ be the p.v. of a point P on the plane through the points A, B and C.

Then the vectors $\vec{AP}, \vec{AB}, \vec{AC}$ are coplanar

$$\vec{AP} = \vec{OP} - \vec{OA} = \mathbf{r} - (-\mathbf{i}+2\mathbf{j}+3\mathbf{k})$$

$$\vec{AB} = \vec{OB} - \vec{OA} = 5\mathbf{i}-5\mathbf{j}-10\mathbf{k}$$

$$\vec{AC} = \vec{OC} - \vec{OA} = -5\mathbf{i}+\mathbf{j}-8\mathbf{k}$$

$$\vec{AB} \times \vec{AC} = \begin{vmatrix} \mathbf{i} & \mathbf{i} & \mathbf{k} \\ 5 & -5 & -10 \\ -5 & 1 & -8 \end{vmatrix}$$

$$= 50\mathbf{i}+90\mathbf{j}-20\mathbf{k}$$

$\because \quad \vec{AP}, \vec{AB}, \vec{AC}$ are coplanar, we have

$$\vec{AP}\,.\,(\vec{AB} \times \vec{AC}) = 0$$

i.e., $\quad \{r-(-\mathbf{i}+2\mathbf{j}+3\mathbf{k})\}\,.\,\{50\mathbf{i}+90\mathbf{j}-20\mathbf{k}\} = 0$

or $\quad \{r-(-\mathbf{i}+2\mathbf{j}+3\mathbf{k})\}\,.\,\{5\mathbf{i}+9\mathbf{j}-2\mathbf{k}\} = 0$

or $\quad r\,.\,(5\mathbf{i}+9\mathbf{j}-2\mathbf{k}) = 7$

which is the required plane.

If $r = x\mathbf{i}+y\mathbf{j}+z\mathbf{k}$ then in cartesian form the plane equation is $5x+9y-2z-7=0$.

Example 26. Prove that the lines $\mathbf{r} = \mathbf{a} + t(\mathbf{b} \times \mathbf{c})$ and $\mathbf{r} = \mathbf{b} + s(\mathbf{c} \times \mathbf{a})$ will intersect if $\mathbf{a.c} = \mathbf{b.c}$.

Solution: The lines $\mathbf{r} = \mathbf{a} + t(\mathbf{b} \times \mathbf{c})$ ...(1)

and $\mathbf{r} = \mathbf{b} + s(\mathbf{c} \times \mathbf{a})$ ...(2)

will intersect if

$$[\mathbf{a} - \mathbf{b}, \mathbf{b} \times \mathbf{c}, \mathbf{c} \times \mathbf{a}] = 0$$

i.e., $(\mathbf{a} - \mathbf{b}) . \{(\mathbf{b} \times \mathbf{c}) \times (\mathbf{c} \times \mathbf{a})\} = 0$

or $\mathbf{a} . \{(\mathbf{b} \times \mathbf{c}) \times (\mathbf{c} \times \mathbf{a})\} = \mathbf{b} . \{(\mathbf{b} \times \mathbf{c}) \times (\mathbf{c} \times \mathbf{a})\}$

or $\mathbf{a}.[\{(\mathbf{b} \times \mathbf{c}).\mathbf{a}\} \mathbf{c} - \{(\mathbf{b} \times \mathbf{c}) . \mathbf{b}\}\mathbf{a}] = \mathbf{b}.[\{(\mathbf{b} \times \mathbf{c}).\mathbf{a}\}\mathbf{c} - \{(\mathbf{b} \times \mathbf{c}).\mathbf{b}\}\mathbf{a}]$

or $\mathbf{a}.\{[\mathbf{bca}]\mathbf{c} - [\mathbf{bcb}]\mathbf{a}\} = \mathbf{b}.\{[\mathbf{bca}]\mathbf{c} - [\mathbf{bcb}]\mathbf{a}\}$

or $(\mathbf{a} . \mathbf{c}) [\mathbf{bca}] = (\mathbf{b} . \mathbf{c}) [\mathbf{bca}]$ $(\because [\mathbf{bcb}] = 0)$

or $\mathbf{a}.\mathbf{c} = \mathbf{b}.\mathbf{c}$ ($\because$ $[\mathbf{bca}] \neq 0$ if **a, b, c** are non-coplanar)

Example 27. Find the equation of the plane containing the two parallel lines $\mathbf{r} = \mathbf{a} + s\mathbf{b}$ and $\mathbf{r} = \mathbf{a}' + t\mathbf{b}$

Solution. Here the plane passes through **a, a′** and parallel to **b**. Let r be the p.v of any variable point on the plane. Then $\mathbf{r} - \mathbf{a}$, $\mathbf{a}' - \mathbf{a}$ and **b** are coplanar.

$$\therefore \quad (\mathbf{r} - \mathbf{a}) . \{(\mathbf{a}' - \mathbf{a}) \times \mathbf{b}\} = 0$$

$$\mathbf{r}.\{(\mathbf{a}' - \mathbf{a}) \times \mathbf{b}\} = \mathbf{a}.\{(\mathbf{a}' - \mathbf{a}) \times \mathbf{b}\}$$

$$= \mathbf{a}.(\mathbf{a}' \times \mathbf{b} - \mathbf{a} \times \mathbf{b})$$

$$= [\mathbf{a}\,\mathbf{a}'\,\mathbf{b}] - [\mathbf{aab}]$$

$$= [\mathbf{a}\,\mathbf{a}'\,\mathbf{b}] \qquad (\because [\mathbf{aab}] = 0)$$

Hence equation of the required plane is

$$\mathbf{r}.\{(\mathbf{a}' - \mathbf{a}) \times \mathbf{b}\} = [\mathbf{aa}'\mathbf{b}].$$

Example 28. Find the plane containing the two straight lines $\mathbf{r} = \mathbf{a} + t\mathbf{a}'$ and $\mathbf{r} = \mathbf{a}' + s\mathbf{a}$.

Solution: The required plane contains the lines

$$\mathbf{r} = \mathbf{a} + t\mathbf{a}' \text{ and } \mathbf{r} = \mathbf{a}' + s\mathbf{a}$$

⇒ It passes through **a** and parallel to **a** and **a′**. If **r** be the p.v of a variable point P on the plane then $\mathbf{r} - \mathbf{a}$, **a** and **a′** should be coplanar.

$$\therefore \quad (\mathbf{r} - \mathbf{a}).(\mathbf{a} \times \mathbf{a}') = 0$$

or $\mathbf{r}.(\mathbf{a} \times \mathbf{a}') = \mathbf{a}.(\mathbf{a} \times \mathbf{a}')$

or $[\mathbf{raa}'] = 0$ ($\because$ $\mathbf{a}.(\mathbf{a} \times \mathbf{a}') = 0$)

which is the required plane.

Example 29. Find the vector equation of straight line through the point c intersecting both the lines $\mathbf{r} = \mathbf{a} + s\mathbf{b}$ and $\mathbf{r} = \mathbf{a}' + t\mathbf{b}'$.

Solution Let the required equation of the straight line through **c** and parallel to **d** be

$$\mathbf{r} = \mathbf{c} + k\mathbf{d} \qquad ...(1)$$

where k is some scalar.

Now the line (1) intersects each of the lines

$$\mathbf{r} = \mathbf{a} + s\mathbf{b} \text{ and } \mathbf{r} = \mathbf{a}' + t\mathbf{b}'$$

$$\mathbf{d}.\{\mathbf{b} \times (\mathbf{c} - \mathbf{a})\} = 0 \qquad ...(2)$$

and $$\mathbf{d}.\{\mathbf{b}' \times (\mathbf{c} - \mathbf{a}')\} \qquad ...(3)$$

It is clear that **d** is perpendicular to both $\mathbf{b} \times (\mathbf{c} - \mathbf{a})$ and $\mathbf{b}' \times (\mathbf{c} - \mathbf{a}')$ therefore parallel to $[\mathbf{b} \times (\mathbf{c} - \mathbf{a})] \times [\mathbf{b}' \times (\mathbf{c} - \mathbf{a}')]$. Hence from (1) the required equation of the plane is

$$\mathbf{r} = \mathbf{c} + k\{\mathbf{b} \times (\mathbf{c} - \mathbf{a}) \times \mathbf{b}' \times (\mathbf{c} - \mathbf{a}')\}.$$

Example 30. A variable plane is at a constant distance p from the origin and meets the axes in A, B and C. Through A, B and C planes are drawn parallel to the co-ordinate planes. Prove that the locus of their point of intersection is given by

$$x^{-2} + y^{-2} + z^{-2} = p^{-2}$$

Solution: Let $\mathbf{r}.\mathbf{n} = 1$ ...(1)

be the plane equation where $p = \dfrac{1}{|\mathbf{n}|}$

Let $\mathbf{n} = n_1\mathbf{i} + n_2\mathbf{j} + n_3\mathbf{k}$ $\quad \therefore \quad |\mathbf{n}|^2 = n^2 = n_1^2 + n_2^2 + n_3^2$

also $\mathbf{i}.\mathbf{n} = n_1$; $\mathbf{j}.\mathbf{n} = n_2$ and $\mathbf{k}.\mathbf{n} = n_3$

Intercept on x-axis $= \dfrac{q}{\mathbf{i}.\mathbf{n}} = \dfrac{q}{n_1}$

$\therefore \qquad A = \dfrac{q}{n_1}\mathbf{i}$

Similarly $B = \dfrac{q}{n_2}\mathbf{j}$ and $C = \dfrac{1}{n_2}\mathbf{k}$

Now any plane through A i.e., $\dfrac{q}{n_1}\mathbf{i}$ and parallel to **j-k** plane whose normal will be along **i** is given by

$$\left(\mathbf{r} - \frac{q}{n_1}\mathbf{i}\right).\mathbf{i} = 0 \text{ or } \mathbf{r}.\mathbf{i} = \frac{q}{n_1} \quad ...(2)$$

Similarly planes through B and C parallel to **k.i** and **i.j** planes are given by

$$\mathbf{r}.\mathbf{j} = \frac{q}{n_2} \quad ...(3)$$

and

$$\mathbf{r}.\mathbf{k} = \frac{q}{n_3} \quad ...(4)$$

If $\mathbf{r} = x\mathbf{i} + y\mathbf{j} + z\mathbf{k}$ then from (2), (3) and (4) we get

$$x = \frac{q}{n_1}, y = \frac{q}{n_2} \text{ and } z = \frac{q}{n_3}$$

$$\therefore \quad \frac{1}{x^2} + \frac{1}{y^2} + \frac{1}{z^2} = \frac{n_1^2 + n_2^2 + n_3^2}{q^2}$$

$$= \frac{n^2}{q^2} = \frac{1}{p^2}$$

or

$$x^{-2} + y^{-2} + z^{-2} = p^{-2}$$

Hence the result.

Example 31. OA, OB and OC are three mutually perpendicular lines. P is the length of the perpendicular from O to the plane ABC. Show that $p^{-2} = a^{-2} + b^{-2} + c^{-2}$ and the area of triangle ABC is $\frac{1}{2}\sqrt{b^2c^2 + c^2a^2 + a^2b^2}$, a, b, c being the lengths of OA, OB and OC.

Solution: Equation of the plane through **a, b, c** is

$$\mathbf{r}.\mathbf{n} = [\mathbf{abc}] \quad ...(1)$$

where

$$\mathbf{n} = \mathbf{b} \times \mathbf{c} + \mathbf{c} \times \mathbf{a} + \mathbf{a} \times \mathbf{b} \quad ...(2)$$

and

$$|\mathbf{n}| = 2\,\Delta\,ABC \text{ [Art } \mathbf{6.8}] \quad ...(3)$$

Now perpendicular distance from origin

$$p = \frac{[\mathbf{abc}]}{|\mathbf{n}|} \quad \therefore \quad p^{-2} = \frac{|\mathbf{n}|^2}{[\mathbf{abc}]^2} \quad ...(4)$$

Here $\mathbf{a} = a\mathbf{i}$, $\mathbf{b} = b\mathbf{j}$ and $\mathbf{c} = c\mathbf{k}$

$$\therefore \quad [\mathbf{abc}] = abc\,[\mathbf{ijk}] = abc \quad ...(5)$$

And $\mathbf{n} = bc\,(\mathbf{j} \times \mathbf{k}) + ca(\mathbf{k} \times \mathbf{i}) + ab\,(\mathbf{i} \times \mathbf{j})$ [from (2)]

$$= bc\mathbf{i} + ca\mathbf{j} + ab\mathbf{k}$$

$$\therefore \quad |\mathbf{n}| = \sqrt{b^2c^2 + c^2a^2 + a^2b^2} = 2\,\Delta ABC \qquad \text{[from (3)]}$$

$$\therefore \quad \Delta ABC = \frac{1}{2}\sqrt{b^2c^2 + c^2a^2 + a^2b^2}$$

$$\text{Also} \quad p^{-2} = \frac{b^2c^2 + c^2a^2 + a^2b^2}{a^2b^2c^2} \qquad \text{[from (4) and (5)]}$$

$$\text{or} \qquad p^{-2} = a^{-2} + b^{-2} + c^{-2}.$$

Example 32. Find the equation of the plane which contains the line $\mathbf{r} = t\mathbf{a}$ and is perpendicular to the plane containing $\mathbf{r} = t'\mathbf{b}$ and $\mathbf{r} = s\mathbf{c}$

Solution. The required plane contains the line $\mathbf{r} = t\mathbf{a}$

$\therefore$ It passes through the origin and parallel to **a** Also it is perpendicular to both the vectors **b** and **c**

$\Rightarrow$ parallel to the vector $\mathbf{b} \times \mathbf{c}$

$\therefore$ The required plane passes through the origin and parallel to the vectors **a** and $\mathbf{b} \times \mathbf{c}$.

If **r** be the p.v of any variable point, then **r**, **a** and $\mathbf{b} \times \mathbf{c}$ are coplanar.

Hence the required equation is

$$\mathbf{r} \,.\, [\mathbf{a} \times (\mathbf{b} \times \mathbf{c})] = 0.$$

Example 33. Find the straight line through the point c which is parallel to the plane $\mathbf{r} \,.\, \mathbf{a} = 0$ and intersects the line $\mathbf{r} = \mathbf{a}' + t\mathbf{b}$.

Solution. Let the required line be parallel to the vector **d** It is given that the required line is parallel to $\mathbf{r}.\mathbf{a} = 0$ $\therefore$ it is perpendicular to **a**

$$\Rightarrow \qquad \mathbf{d}.\mathbf{a} = 0 \qquad \text{...(1)}$$

Since the required line is passing through **c** and parallel to **d** its equation is

$$\mathbf{r} = \mathbf{c} + s\mathbf{d} \qquad \text{...(2)}$$

Now (2) intersects the line $\mathbf{r} = \mathbf{a}' + t\mathbf{b}$ ...(3)

$$\therefore \qquad \mathbf{d}.\{\mathbf{b} \times (\mathbf{a}' - \mathbf{c})\} = 0 \qquad \text{...(4)}$$

From (1) and (4) **d** is perpendicular to both **a** and $\mathbf{b} \times (\mathbf{a}' - \mathbf{c})$ $\Rightarrow$ parallel to $\mathbf{a} \times \{\mathbf{b} \times (\mathbf{a}' - \mathbf{c})\}$

Let $\mathbf{d} = k\,[\mathbf{a} \times \{\mathbf{b} \times (\mathbf{a}' - \mathbf{c})\}]$ where k is some scalar.

Hence from (2) the required line is

$$\mathbf{r} = \mathbf{c} + m\,[\mathbf{a} \times \{\mathbf{b} \times (\mathbf{a}' - \mathbf{c}\}]$$

where m = sk some scalar.

6.24. Shortest distance

To find the shortest distance between two skew (nonintersecting, non parallel) lines

Let the given skew lines be

$$\mathbf{r} = \mathbf{a} + t\mathbf{b} \quad \text{...(1)}$$

and $$\mathbf{r} = \mathbf{a}' + s\mathbf{b}' \quad \text{...(2)}$$

Now the shortest distance in the length of the line which is perpendicular to both the skew lines- Let PP′ be the shortest distance (S.D) between the lines

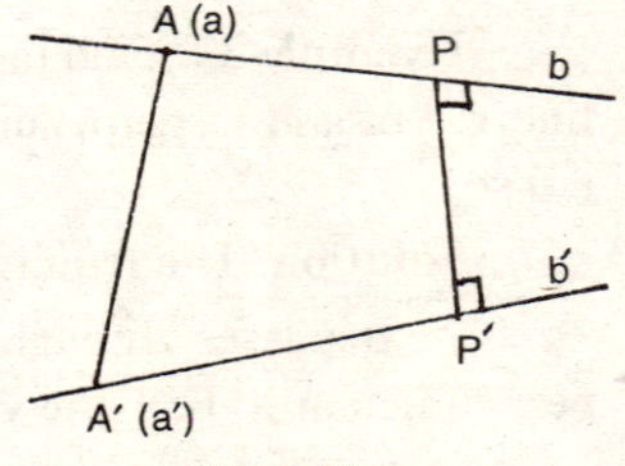

Fig. 613

∵ PP′ is perpendicular to both **b** and **b′**, it is parallel to **b** × **b′**. If A and A′ are the points a and **a′** then $\overrightarrow{A'A} = \mathbf{a} = \mathbf{a}'$ and the shortest distance **a**

$$PP' = \text{Projection of } \overrightarrow{A'A} \text{ on } PP'$$

$$= \text{Projection of } (\mathbf{a} - \mathbf{a}') \text{ on } (\mathbf{b} \times \mathbf{b}')$$

$$= \frac{(\mathbf{a} - \mathbf{a}') \cdot (\mathbf{b} \times \mathbf{b}')}{|\mathbf{b} \times \mathbf{b}'|}$$

$$= \frac{[\mathbf{a} - \mathbf{a}', \mathbf{b}, \mathbf{b}']}{|\mathbf{b} \times \mathbf{b}'|}$$

Working Method: To find S.D between two skew lines parallel to **b** and **b′** respectively, obtain the projection of the line joining any two points, one on each line, on the vector **b** × **b′** which is perpendicular to both the lines

Note: If the straight lines intersect then the shortest distance = 0.

$$\Rightarrow [\mathbf{a} - \mathbf{a}', \mathbf{b}, \mathbf{b}'] = 0 \text{ or } [\mathbf{a}\mathbf{b}\mathbf{b}'] = [\mathbf{a}'\,\mathbf{b}\,\mathbf{b}']$$

6.25. Equations of shortest distance (or common perpendicular)

The shortest distance (S.D) is the line of intersection of the planes containing AP, PP′ and A′P′, PP. Let **r** be the p.v of any point on the S.D i.e, on PP′. Then first plane contains **r** − **a**, **b**, **b** × **b**′, hence its equation is

$$[\mathbf{r}-\mathbf{a}, \mathbf{b}, \mathbf{b}\times\mathbf{b}'] = 0 \qquad ...(1)$$

the second plane contains **r** − **a**′, **b**′, **b** × **b**′ and hence its equation is

$$[\mathbf{r}-\mathbf{a}', \mathbf{b}', \mathbf{b}\times\mathbf{b}'] = 0 \qquad ...(2)$$

The equations (1) and (2) together represents the equation to S.D

SOLVED PROBLEMS

Examples 34. Find the shortest distance between the straight lines **r** = **a** + t**b** and **r** = s**c** and determine the equation of the line which cuts both the lines at right angles

Solution. The given two lines are

$$\mathbf{r} = \mathbf{a} + t\mathbf{b} \qquad ...(1)$$

and

$$\mathbf{r} = s\mathbf{c} \qquad ...(2)$$

The S.D between (1) and (2) is perpendicular to both **b** and **c** therefore parallel to **b** × **c**. Any point on (1) is **a** and on (2) is origin

∴ S.D = Projection of the joining **a** and origin on **b** and **c**.

$$= \frac{(\mathbf{a}-0)\,.\,(\mathbf{b}\times\mathbf{c})}{|\mathbf{b}\times\mathbf{c}|} = \frac{[\mathbf{abc}]}{|\mathbf{b}\times\mathbf{c}|}$$

Now the shortest distance is the line of intersection of the plane containing (1) and S.D and the plane containing (2) and S.D.

The equation of the plane containing (1) and S.D is

$$(\mathbf{r}-\mathbf{a})\,.\,[\mathbf{b}\times(\mathbf{b}\times\mathbf{c}) = 0 \qquad ...(3)$$

and the equation of the plane containing (2) and S.D is

$$(\mathbf{r}-\mathbf{0})\,.\,[\mathbf{c}\times(\mathbf{b}\times\mathbf{c})] = 0 \qquad ...(4)$$

Now the required equation of the line which cuts both (1) and (2) at right angles i.e., equation to S.D is the line of intersection of (3) and (4)

Example 35. Find the shortest distance between the straight lines through the points A (1, 3, 4) and B (− 2, 1, 6) in the directions (2, 1, 2) and (− 1, 3, 4). Also find the feet F_1 and F_2 of the common perpendicular.

or

Find the S.D between the straight lines

$$\frac{x-1}{2}=\frac{y-3}{1}=\frac{z-4}{2} \text{ and } \frac{x+2}{-1}=\frac{y-1}{3}=\frac{z-6}{4}.$$

Also find the points of intersection of S.D with the given lines.

Solution. The vector equation of the straight line through A(1, 3, 4) in the direction of (2, 1, 2) is $\mathbf{r} = \mathbf{a_1} + t\mathbf{b_1}$, where $\mathbf{a_1} = \mathbf{i} + 3\mathbf{j} + 4\mathbf{k}$ and $\mathbf{b_1} = 2\mathbf{i} + \mathbf{j} + 2\mathbf{k}$.

or $$\mathbf{r} = (\mathbf{i} + 3\mathbf{j} + 4\mathbf{k}) + t(2\mathbf{i} + \mathbf{j} + 2\mathbf{k}) \quad ...(1)$$

Again the vector equation of the straight line through B (- 2, 1, 6) in the direction of (- 1, 3, 4) is

$$\mathbf{r} = \mathbf{a_2} + s\mathbf{b_2} \text{ where } \mathbf{a_2} = -2\mathbf{i} + \mathbf{j} + 6\mathbf{k} \text{ and } \mathbf{b_2} = -\mathbf{i} + 3\mathbf{j} + 4\mathbf{k}$$

$$\mathbf{r} = (-2\mathbf{i} + \mathbf{j} + 6\mathbf{k}) + (-\mathbf{i} + 3\mathbf{j} + 4\mathbf{k}) \quad ...(2)$$

$$\text{S.D} = F_1F_2 = \text{Projection of } F_1F_2 \text{ on } \mathbf{b_1} \times \mathbf{b_2}$$

$$= \frac{(\mathbf{a_1} - \mathbf{a_2}) \cdot (\mathbf{b_1} \times \mathbf{b_2})}{|\mathbf{b_1} \times \mathbf{b_2}|} \quad ...(3)$$

$$\mathbf{b_1} \times \mathbf{b_2} = \begin{vmatrix} \mathbf{i} & \mathbf{j} & \mathbf{k} \\ 2 & 1 & 2 \\ -1 & 3 & 4 \end{vmatrix} = -2\mathbf{i} - 10\mathbf{j} + 7\mathbf{k}$$

$$|\mathbf{b_1} \times \mathbf{b_2}| = \sqrt{4 + 100 + 49} = \sqrt{153}$$

$$\mathbf{a_1} - \mathbf{a_2} = (\mathbf{i} + 3\mathbf{j} + 4\mathbf{k}) - (-2\mathbf{i} + \mathbf{j} + 6\mathbf{k}) = 3\mathbf{i} + 2\mathbf{j} - 2\mathbf{k}$$

∴ From (1)

$$\text{S.D.} = \frac{(3\mathbf{i} + 2\mathbf{j} - 2\mathbf{k}) \cdot (-2\mathbf{i} - 10\mathbf{j} = 7\mathbf{k})}{\sqrt{153}} = \frac{-40}{\sqrt{153}}$$

or $$\text{S.D.} = \frac{40}{\sqrt{153}} \quad (\because \text{ distance is positive})$$

Equation of the plane containing the first line and S.D i.e., plane AF_1F_2 is

$$[\mathbf{r} - \mathbf{a_1}, \mathbf{b_1}, \mathbf{b_1} \times \mathbf{b_2}] = 0$$

$$\begin{vmatrix} x-1 & y-3 & z-4 \\ 2 & 1 & 2 \\ -2 & -10 & 7 \end{vmatrix} = 0$$

or $$27x - 28y - 18z + 129 = 0$$

i.e., $$\mathbf{r} \cdot (27\mathbf{i} - 28\mathbf{j} - 18\mathbf{k}) + 129 = 0 \qquad ...(4)$$

Equations of the plane containing the second line and S.D. i.e., plane BF_2F_1

$$[\mathbf{r} - \mathbf{a}_2, \mathbf{b}_2, \mathbf{b}_1 \times \mathbf{b}_2] = 0$$

$$\begin{vmatrix} x+2 & y-1 & z-6 \\ -1 & 3 & 4 \\ -2 & -10 & 7 \end{vmatrix} = 0$$

or $$61x - y + 16z - 27 = 0$$

i.e., $$\mathbf{r} \cdot (61\mathbf{i} - \mathbf{j} + 16\mathbf{k}) - 27 = 0 \qquad ...(5)$$

To find the point F_1. The plane BF_2F_1 i.e., (5) meets the line (1) in F_1

$$[(\mathbf{i} + 3\mathbf{j} + 4\mathbf{k}) + t(2\mathbf{i} + \mathbf{j} + 2\mathbf{k})] \cdot (61\mathbf{i} - \mathbf{j} + 16\mathbf{k}) - 27 = 0$$

$$122 + 153t - 27 = 0$$

or $$t = -\frac{95}{153}$$

Putting $t = -\frac{95}{153}$ in (1) we get

$$F_1 = \frac{1}{153}(-37\mathbf{i} + 364\mathbf{j} + 422\mathbf{k})$$

or $$F_1 = \left(\frac{-37}{153}, \frac{364}{153}, \frac{422}{153}\right)$$

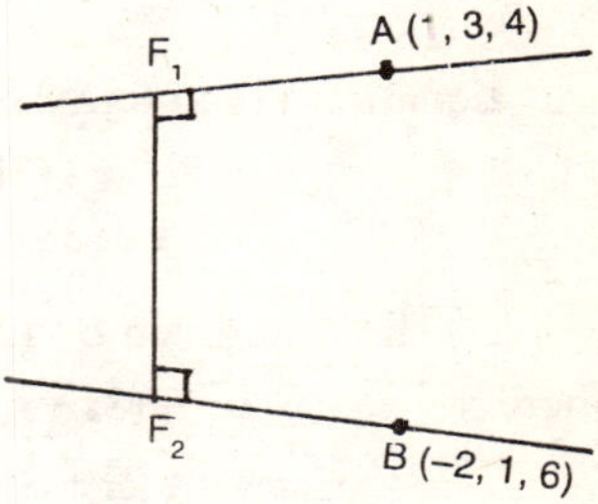

Fig. 6.14

To find the point F_2 The plane AF_1F_2 i.e., (4) meets the line (2) in F_2

$$[(-2\mathbf{i} + \mathbf{j} + 6\mathbf{k}) + (-\mathbf{i} + 3\mathbf{j} + 4\mathbf{k})] \cdot (27\mathbf{i} - 28\mathbf{j} - 18\mathbf{k}) + 129 = 0$$

$$-190 + s(-183) + 129 = 0 \text{ or } s = -\frac{1}{3}$$

Putting $s = -\frac{1}{3}$ in (2), we get

$$F_2 = \frac{1}{3}(-5\mathbf{i} + 14\mathbf{k}) \quad \text{or} \quad F_2 = \left(-\frac{5}{3}, 0, \frac{14}{3}\right)$$

Example 36. Find the shortest distance between the lines determined by the equations

$$3x - 4y - z + 5 = 0 = 3x - 6y - 2z + 13$$

and
$$3x + 4y - 3z + 2 = 0 = 3x - 2y + 6z + 17$$

Solution. Given
$$\left.\begin{array}{r} 3x - 4y - z + 5 = 0 \\ 3x - 6y - 2z + 13 = 0 \end{array}\right\} \quad ...(1)$$

Put $z = 0 \Rightarrow 3x - 4y + 5 = 0$ and $3x - 6y + 13 = 0$

On solving $x = \frac{11}{3}$, $y = 4$

$\therefore$ A point on this line is $A\left(\frac{11}{3}, 4, 0\right)$

In terms of unit vectors let it be

$$\mathbf{a}_1 = \frac{11}{3}\mathbf{i} + 4\mathbf{j}$$

Equation (1) in terms of vectors

$$\left.\begin{array}{r} \mathbf{r} \cdot (3\mathbf{i} - 4\mathbf{j} - \mathbf{k}) + 5 = 0 \\ \mathbf{r} \cdot (3\mathbf{i} - 6\mathbf{j} - 2\mathbf{k}) + 13 = 0 \end{array}\right\} \quad ...(2)$$

$\therefore$ The line given by (2) is parallel to $\mathbf{b}_1$ (say)

where
$$\mathbf{b}_1 = (3\mathbf{i} - 4\mathbf{j} - \mathbf{k}) \times (3\mathbf{i} - 6\mathbf{j} - 2\mathbf{k})$$
$$= 2\mathbf{i} + 3\mathbf{j} - 6\mathbf{k}$$

Similarly a point on other given line

$$\left.\begin{array}{r} 3x + 4y - 3z + 2 = 0 \\ 3x - 2y + 6z + 17 = 0 \end{array}\right\} \quad ...(3)$$

is $B\left(-4, \frac{5}{2}, 0\right)$ or in terms of unit vectors

$$\mathbf{a}_2 = -4\mathbf{i} + \frac{5}{2}\mathbf{j}$$

and
$$\mathbf{b}_2 = (3\mathbf{i} + 4\mathbf{j} - 3\mathbf{k}) \times (3\mathbf{i} - 2\mathbf{j} + 6\mathbf{k})$$
$$= 9(2\mathbf{i} - 3\mathbf{j} - 2\mathbf{k})$$

$$\text{S.D.} = \frac{(\mathbf{a}_1 - \mathbf{a}_2) \cdot (\mathbf{b}_1 \times \mathbf{b}_2)}{|\mathbf{b}_1 \times \mathbf{b}_2|} \quad ...(4)$$

$$\mathbf{a}_1 - \mathbf{a}_2 = \left(\frac{11}{3}\mathbf{i} + 4\mathbf{j}\right) - \left(-4\mathbf{i} + \frac{5}{2}\mathbf{j}\right)$$

$$= \frac{23}{3}\mathbf{i} + \frac{3}{2}\mathbf{j}$$

$$\mathbf{b_1} \times \mathbf{b_2} = \begin{vmatrix} \mathbf{i} & \mathbf{j} & \mathbf{k} \\ 2 & 3 & -6 \\ 18 & -27 & -18 \end{vmatrix}$$

$$= -36(6\mathbf{i} + 2\mathbf{j} + 3\mathbf{k})$$

$$|\mathbf{b_1} \times \mathbf{b_2}| = 36\sqrt{36 + 4 + 9} = 252$$

Substituting (4) the S.D between (1) and (3) is

$$= \frac{\left(\frac{23}{3}\mathbf{i} + \frac{3}{2}\mathbf{j}\right) \cdot -36(6\mathbf{i} + 2\mathbf{j} + 3\mathbf{k})}{252}$$

$$= -\frac{1}{7}[46 + 3] = -7$$

$\therefore$ S.D = 7 units.

Example 37. Find the shortest distance between a diagonal of a rectangular parallelopiped whose sides are a, b, c and the edges not meeting it are $\frac{bc}{\sqrt{b^2 + c^2}}, \frac{ca}{\sqrt{c^2 + a^2}}, \frac{ab}{\sqrt{a^2 + b^2}}$.

Solution. Let OA = a, OB = b and OC = c be the three coterminous edges of the rectangular parallelopiped w.r.t O as origin let **i**, **j**, and **k** be the unit vectors along OA, OB and OC.

$\therefore$ $\vec{OA} = a\mathbf{i}$, $\vec{OB} = b\mathbf{j}$ and $\vec{OC} = c\mathbf{k}$

Also $\vec{OF} = a\mathbf{i} + b\mathbf{j} + c\mathbf{k}$, the edges which do not meet the diagonal OF are AH, AD, BD and their parallels BE, CE, CH

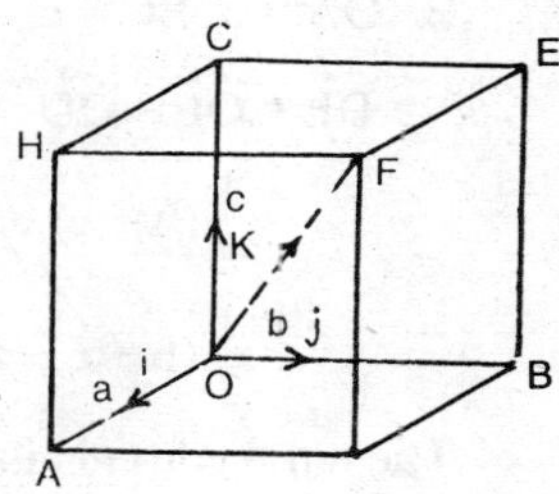

Fig. 6.15

To find the S.D between OF and BD

Equation of the diagonal OF is

$$\mathbf{r} = 0 + t(a\mathbf{i} + b\mathbf{j} + c\mathbf{k}) \quad ...(1)$$

$\because$ $\vec{BD} = \vec{OA} = a\mathbf{i}$ and $\vec{OB} = b\mathbf{j}$

$\therefore$ Equation to edge BD is

$$\mathbf{r} = b\mathbf{j} + s\mathbf{i} \qquad ...(2)$$

$\therefore$ The S.D between (1) and (2) is parallel to

$$\mathbf{n} = \mathbf{i} \times (a\mathbf{i} + b\mathbf{j} + c\mathbf{k}) = (b\mathbf{k} - c\mathbf{j})$$

Point on BD is $b\mathbf{j}$ and on OF is $\mathbf{0}$

$$\text{S.D.} = \frac{(b\mathbf{j} - \mathbf{0}) \cdot (b\mathbf{k} - c\mathbf{j})}{\sqrt{b^2 + c^2}} = \frac{-bc}{\sqrt{b^2 + c^2}}$$

$$= \frac{bc}{\sqrt{b^2 + c^2}} \qquad \text{(neglecting - ve sign)}$$

Similarly we can show that the S.D between OF and AD is $\dfrac{ca}{\sqrt{c^2 + a^2}}$ and that between OF and AH is $\dfrac{ab}{\sqrt{a^2 + b^2}}$.

Example 38. Prove that the S.D between pairs of opposite edges of an isosceles tetrahedron lies along the join of their mid points that their S.D's are perpendicular.

Solution: Let OABC be the tetrahedron. w.r.t O as origin the p.v's of A, B and C be **a, b** and **c** respectively. Let D, E be the mid points of edges OA and BC.

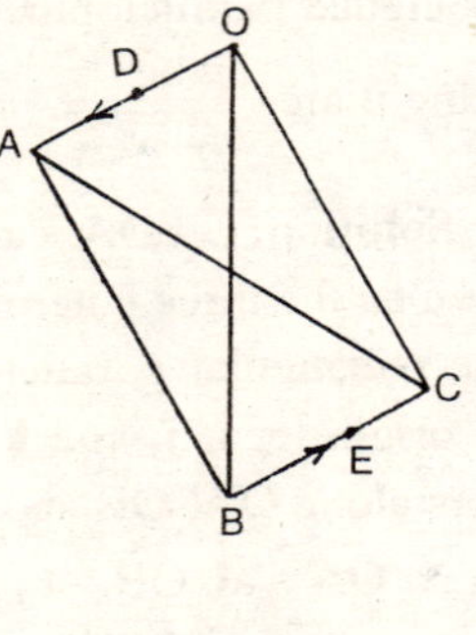

Fig. 6.16

$$\therefore \ \overrightarrow{OA} = \mathbf{a}, \ \overrightarrow{BC} = \mathbf{c} - \mathbf{b}$$

$$\overrightarrow{DE} = \overrightarrow{OE} - \overrightarrow{OD}$$

$$= \frac{1}{2}(\mathbf{b} + \mathbf{c}) - \frac{1}{2}\mathbf{a}$$

$$= \frac{1}{2}(\mathbf{b} + \mathbf{c} - \mathbf{a}) \qquad ...(1)$$

The tetrahedron is isosceles

$$\therefore \quad OA = BC, OB = CA \text{ and } OC = AB$$

i.e. $\quad \overrightarrow{OA}^2 = \overrightarrow{BC}^2; \ \overrightarrow{OB}^2 = \overline{CA}^2$ and $\overline{OC}^2 = \overline{AB}^2$

or $\quad \mathbf{a}^2 = (\mathbf{c} - \mathbf{b})^2, \ \mathbf{b}^2 = (\mathbf{a} - \mathbf{c})^2$ and $\mathbf{c}^2 = (\mathbf{b} - \mathbf{a})^2$

$$\therefore \quad 2\mathbf{b}.\mathbf{c} = \mathbf{b}^2 + \mathbf{c}^2 - \mathbf{a}^2 \qquad ...(2)$$

$$2\mathbf{a}.\mathbf{c} = \mathbf{c}^2 + \mathbf{a}^2 - \mathbf{b}^2 \qquad ...(3)$$

and $$2\mathbf{a}.\mathbf{b} = a^2 + b^2 - c^2 \qquad ...(4)$$

If DE is the S.D between OA and BC then DE must be perpendicular to OA and BC

$$\therefore \quad \frac{1}{2}(\mathbf{c} + \mathbf{b} - \mathbf{a}).\mathbf{a} = 0$$

and $$\frac{1}{2}(\mathbf{c} + \mathbf{b} - \mathbf{a}).(\mathbf{c} - \mathbf{b}) = 0$$

or $$\mathbf{c}.\mathbf{a} + \mathbf{c}.\mathbf{b} = a^2 \text{ and } c^2 - b^2 = \mathbf{a}.\mathbf{c} - \mathbf{a}.\mathbf{b}$$

which can also be deduced by adding and subtracting (3) and (4)

Hence DE is the S.D between OA and BC.

Similarly we can prove for other pairs of opposite edges of the tetrahedron.

The three S.D's along $\mathbf{b} + \mathbf{c} - \mathbf{a}$, $\mathbf{c} + \mathbf{a} - \mathbf{b}$ and $\mathbf{a} + \mathbf{b} - \mathbf{c}$ and these will be perpendicular if

$$(\mathbf{b} + \mathbf{c} - \mathbf{a}).(\mathbf{c} + \mathbf{a} - \mathbf{b}) = 0$$

or $$c^2 - (a - b)^2 = 0$$

or if $2\mathbf{a}.\mathbf{b} = a^2 + b^2 - c^2$ which is true by (4).

Similarly we can prove the other pairs to be perpendicular by (2) and (3)

Example 39. Prove that the shortest distance between opposite edges of a regular tetrahedron is equal to half the diagonal of the square described on an edge.

Solution: Let the unit vectors along OA, OB and OC be **i**, **j** and **k** respectively.

$$\therefore \quad \vec{OD} = \mathbf{i} + \mathbf{j} + \mathbf{k}$$

BACD is a regular tetrahedron and we are to find the shortest distance between pair of opposite edges say BC and AD.

$$\vec{BC} = \vec{OC} - \vec{OB} = \mathbf{k} - \mathbf{j}$$

$$\vec{AD} = \vec{OD} - \vec{OA}$$

$$= (\mathbf{i} + \mathbf{j} + \mathbf{k}) - \mathbf{i} = \mathbf{j} + \mathbf{k}$$

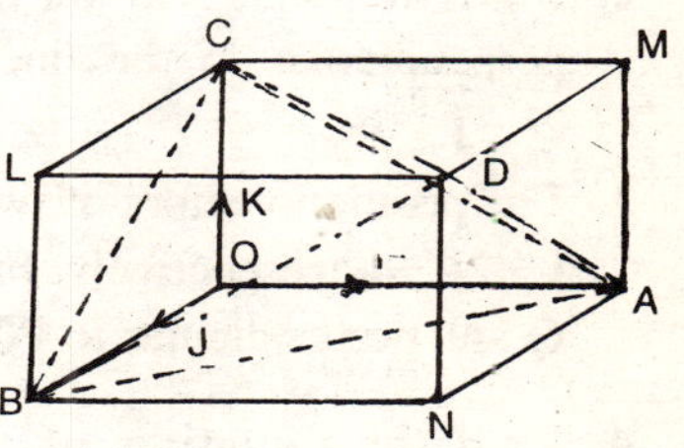

Fig. 6.17

$$\therefore \quad \vec{BC} \times \vec{AD} = \begin{vmatrix} \mathbf{i} & \mathbf{j} & \mathbf{k} \\ 0 & -1 & 1 \\ 0 & 1 & 1 \end{vmatrix} = -2\mathbf{i}$$

$$|\vec{BC} \times \vec{AD}| = 2$$

Now B is any point on BC and A point on AD

$$\therefore \quad \vec{AB} = \vec{OB} - \vec{OA} = \mathbf{j} - \mathbf{i}$$

$\therefore$ Requires S.D. = Projection of AB on $\vec{BC} \times \vec{AD}$

$$= \frac{\vec{AB}.(\vec{BC} \times \vec{AD})}{|\vec{BC} \times \vec{AD}|}$$

$$= \frac{(\mathbf{j} - \mathbf{i}).(-2\mathbf{i})}{2} = 1$$

Now $\quad AB = |\mathbf{j} - \mathbf{i}| = \sqrt{2}$

Diagonal of the square with one side AB

$$= \sqrt{2 + 2} = 2$$

$$\therefore \quad S.D = 1 = \frac{1}{2} . 2 = \frac{1}{2} \quad \text{(Diagonal of the square)}$$

EXERCISE

1. Find the equation of the plane through $2\mathbf{i} + 3\mathbf{j} - \mathbf{k}$ and perpendicular to the vector $3\mathbf{i} - 4\mathbf{j} + 7\mathbf{k}$. Also find the perpendicular distance from the origin to this plane.
2. Find the equation of the plane through $2\mathbf{i} - 3\mathbf{j} + \mathbf{k}$ and perpendicular to the line joining the points $3\mathbf{i} + 4\mathbf{j} - \mathbf{k}$ and $2\mathbf{i} - \mathbf{j} + 5\mathbf{k}$.
3. The position vectors of two points P and Q are $3\mathbf{i} + \mathbf{j} + 2\mathbf{k}$ and $\mathbf{i} - 2\mathbf{j} - 4\mathbf{k}$ respectively. Find the equation of the plane through Q and perpendicular to PQ.
4. Find the equation of the plane perpendicular to $\frac{x-2}{1} = \frac{y+3}{-1} = \frac{z-6}{2}$ and passing through the point (2, 3, 1).
5. Show that the points $\mathbf{i} - \mathbf{j} + 3\mathbf{k}$ and $3(\mathbf{i} + \mathbf{j} + \mathbf{k})$ are equidistant

from the plane **r**.(5**i** + 2**j** − 7**k**) + 9 = 0 and on opposite sides of it.

6. Find the distances of (3, 4, 7) and (2, − 3, − 5) from the plane x + 2y + 2z = 9. Show that they lie on same side of the plane.
7. Show that the locus of a point the sum of the squares of whose distances from the planes **r**.(**i** + **j** + **k**) = 0, **r**.(**i** − **k**) = 0 and **r**.(**i** − 2**j** + **k**) = 0 is 9 is | **r** | = 3.
8. Find the vector equation of the plane through the point (2**i** − **j** − 4**k**) and parallel to the plane **r**.(4**i** − 12**j** − 3**k**) − 7 = 0. Also find the perpendicular distance from the point (3**i** − 5**j** − 4**k**) to this plane and show that the point is on the opposite side from the origin.
9. Find the equations of the planes parallel to the plane **r**.(**i** − 2**j** + 2**k**) = 3, whose perpendicular distance from **i** + 2**j** + 3**k** is 1.
10. Find the intercepts made on the co-ordinate axes by the plane **r**.(2**i** + **j** + 2**k**) = 6. Find also the direction cosines of the normal to this plane.
11. (i) A variable plane passes through a fixed point (a, b, c) and meets the co-ordinate axes in A, B, C. Show that the locus of the point common to the planes through A, B, C parallel to the co-ordinate planes is $\frac{a}{x} + \frac{b}{y} + \frac{c}{z} = 1$.

    (ii) A plane is at a constant distance p from the origin and meets the axes in A, B and C. Show that the locus of the centroid of the tetrahedron OABC is $x^{-2} + y^{-2} + z^{-2} = 16p^{-2}$
12. Find the equation of the plane passing through the origin and the line of intersection of the planes **r.a** = p and **r.b** = q
13. Find the vector equation of the plane through the points 2**i** + 2**j** + **k**, **i** − 2**j** + 3**k** and parallel to the line joining the points 2**i** + **j** − 3**k** and **j** + 5**j** − 8**k**
14. Find the vector equation to the plane through the point 2**i** + **j** − **k** and passing through the line of intersection of the planes **r**.(**i** + 3**j** − **k**) = 0 and **r**.(**j** + 2**k**) = 0
15. Find the vector equation to the plane through origin and passing

through the line of intersection of the planes $x + 2y + 3z = 4$ and $4x + 3y + 2z + 1 = 0$.

16. Determine the plane through the point $\mathbf{i} + 2\mathbf{j} - \mathbf{k}$ which is perpendicular to the line of intersection of the planes $\mathbf{r}.(\mathbf{i} + \mathbf{j} - 2\mathbf{k}) = 5$ and $\mathbf{r}.(3\mathbf{i} - \mathbf{j} + 4\mathbf{k}) = 12$.

17. Find the vector equation of the plane through (1, 2, −1) which is perpendicular to the line of intersection of the planes $3x - y + z = 1$ and $x + 4y - 2z = 2$.

18. Find the vector equation of the plane through the point $5\mathbf{i} + 2\mathbf{j} - 3\mathbf{k}$ and perpendicular to each of the planes $\mathbf{r}.(2\mathbf{i} - \mathbf{j} + 2\mathbf{k}) = 0$ and $\mathbf{r}.(\mathbf{i} + 3\mathbf{j} - 5\mathbf{k}) + 3 = 0$.

19. Find the vector equation of the plane passing through the points

   (i) $\mathbf{j} + \mathbf{k}$, $\mathbf{i} + \mathbf{j} + 2\mathbf{k}$, $-\mathbf{i} + 2\mathbf{j} + 2\mathbf{k}$

   (ii) $-2\mathbf{i} + 6\mathbf{j} - 6\mathbf{k}$, $-3\mathbf{i} + 10\mathbf{j} - 9\mathbf{k}$ and $5\mathbf{i} - 6\mathbf{k}$

20. A plane passes through (0, 1, 3), (1, 0, − 4), (1, 1, − 1). Find the length of perpendicular to it from (2, 4, − 3).

21. Find the equation of the plane which contains the line $\mathbf{r} = 2\mathbf{i} + t(\mathbf{j} - \mathbf{k})$ and the perpendicular to the plane $\mathbf{r}.(\mathbf{i} + \mathbf{k}) = 3$. Find the position vector of the point where this plane meets the line $\mathbf{r} = t(2\mathbf{i} + 3\mathbf{j} + \mathbf{k})$

22. Find the equation of the plane bisecting the join of $2\mathbf{i} + 4\mathbf{j} + 6\mathbf{k}$ and $4\mathbf{i} + 6\mathbf{j} + 8\mathbf{k}$ at right angles.

23. Find the equation of the plane that bisects the line joining the points (1, 2, 3) and (3, 4, 5) at right angles.

24. Find the angle between the planes

$$\mathbf{r},(2\mathbf{i} - \mathbf{j} + 2\mathbf{k}) = 6 \text{ and } \mathbf{r}.(\mathbf{i} + 6\mathbf{j} + 2\mathbf{k}) = 9$$

25. Find the angle between the line

$$\mathbf{r}.(\mathbf{i} + \mathbf{j}) - 3 = 0 = \mathbf{r}.(\mathbf{i} + \mathbf{j} + \mathbf{k}) \text{ and the plane } \mathbf{r}.(\mathbf{j} - \mathbf{k}) + 2 = 0$$

26. Find the distance between the planes

$$\mathbf{r}.(2\mathbf{i} + \mathbf{j} + 2\mathbf{k}) + 3 = 0 \text{ and } \mathbf{r}.(4\mathbf{i} + 2\mathbf{j} + 4\mathbf{k}) - 9 = 0$$

27. Find the vector equation of the line of intersection of the planes

$$\mathbf{r}.(3\mathbf{i} - \mathbf{j} + \mathbf{k}) = 1 \text{ and } \mathbf{r}.(\mathbf{i} + 4\mathbf{j} - 2\mathbf{k}) = 2$$

28. Find the equation of the line through the point a parallel to the plane $\mathbf{r}.\mathbf{n} = q$ and perpendicular to the line $\mathbf{r} = \mathbf{b} + t\mathbf{c}$.

29. Show that the planes $\mathbf{r}.(2\mathbf{i} + 5\mathbf{j} + 3\mathbf{k}) = 0$ and $\mathbf{r}.(\mathbf{i} - \mathbf{j} + 4\mathbf{k}) = 2$ and $\mathbf{r}.(7\mathbf{j} - 5\mathbf{k}) + 4 = 0$ have a common line of intersection.

30. Show that the equation of the plane containing the line $\mathbf{r} = \mathbf{a} + t\mathbf{b}$ and parallel to another line $\mathbf{r} = \mathbf{c} + s\mathbf{d}$ is $(\mathbf{r} - \mathbf{a}).(\mathbf{b} \times \mathbf{d}) = 0$.

31. Show that the straight lines $\mathbf{r} \times \mathbf{a} = \mathbf{b} \times \mathbf{a}$ and $\mathbf{r} \times \mathbf{b} = \mathbf{a} \times \mathbf{b}$ intersect and the point of intersection is $\mathbf{a} + \mathbf{b}$.

32. Find the shortest distance between the straight lines through the points A (6, 2, 2) and B (- 4, 0, - 1) in the directions (1, - 2, 2) and (3, - 2, - 2). Also find the feet P and Q of the common perpendicular.

33. Find the magnitude and the equations of the shortest distance between the lines

$$\frac{x-6}{3} = \frac{y-7}{-1} = \frac{z-4}{1} \text{ and } \frac{x}{-3} = \frac{y+9}{z} = \frac{z-2}{4}.$$

34. Find the length and equation of the shortest distance between the two lines

$$\frac{x-1}{2} = \frac{y-3}{4} = \frac{z+2}{1} \text{ and } 3x - y - 2z + 4 = 0 = 2x + y + z + 1.$$

35. Find the length and equations of shortest distance between

$$3x - 9y + 5z = 0 = x + y - z \text{ and}$$
$$6x + 8y + 3z - 13 = 0 = x + 2y + z - 3.$$

36. The common perpendicular to two straight lines whose equations are $(\mathbf{r} - \mathbf{a}_1) \times \mathbf{b}_1 = \mathbf{0}$ and $(\mathbf{r} - \mathbf{a}_2) \times \mathbf{b}_2 = \mathbf{0}$ meets them at P and Q. A and B are the points with position vectors $\mathbf{a}_1$ and $\mathbf{a}_2$. By considering the projection of AB on PQ or otherwise prove that

$$\overrightarrow{PQ} = \frac{(\mathbf{a}_1 - \mathbf{a}_2)\mathbf{n}}{\mathbf{n}^2} \cdot \mathbf{n} \text{ where } \mathbf{n} = \mathbf{b}_1 \times \mathbf{b}_2.$$

37. Show that the shortest distance between any two opposite edges of the tetrahedron formed by the planes

$$y + z = 0, z + x = 0, x + y = 0 \text{ and } x + y + z = a \text{ is } \frac{2a}{\sqrt{6}}.$$

ANSWERS

(1) $\mathbf{r}.(-3\mathbf{i}+4\mathbf{j}-7\mathbf{k})=13;\ \frac{13}{\sqrt{74}}$

(2) $\mathbf{r}.(\mathbf{i}+5\mathbf{j}+6\mathbf{k})+19=0$

(3) $\mathbf{r}.(2\mathbf{i}+3\mathbf{j}+6\mathbf{k})+28=0$

(4) $\mathbf{r}.(\mathbf{i}-\mathbf{j}+2\mathbf{k})=1$

(6) 4, 1

(8) $\mathbf{r}.(4\mathbf{i}-12\mathbf{j}-3\mathbf{k})=32;\ 4$

(9) $\mathbf{r}.(\mathbf{i}-2\mathbf{j}+2\mathbf{k})=0;\ \mathbf{r}.(\mathbf{i}-2\mathbf{j}+2\mathbf{k})=6$

(10) 3, 6, 3; $\frac{2}{3}, \frac{1}{3}, \frac{2}{3}$

(12) $\mathbf{r}.(q\mathbf{a}-p\mathbf{b})=0$

(13) $\mathbf{r}.(12\mathbf{i}-11\mathbf{j}-16\mathbf{k})+14=0$

(14) $\mathbf{r}.(\mathbf{i}+9\mathbf{j}+11\mathbf{k})=0$

(15) $\mathbf{r}.(17\mathbf{i}+14\mathbf{j}+10\mathbf{k})=0$

(16) $\mathbf{r}.(\mathbf{i}-5\mathbf{j}-2\mathbf{k})+7=0$

(17) $\mathbf{r}.(2\mathbf{i}-7\mathbf{j}-13\mathbf{k})=1$

(18) $\mathbf{r}.(\mathbf{i}-12\mathbf{j}-7\mathbf{k})=2$

(19) (i) $(\mathbf{i}-2\mathbf{j}-\mathbf{k})-3=0$

(ii) $\mathbf{r}.(2\mathbf{i}-\mathbf{j}-2\mathbf{k})-2=0$

(20) $\frac{7}{\sqrt{26}}$

(21) $-(2\mathbf{i}+3\mathbf{j}+\mathbf{k})$

(22) $\mathbf{r}.(\mathbf{i}+\mathbf{j}+\mathbf{k})=15$

(23) $\mathbf{r}.(\mathbf{i}+\mathbf{j}+\mathbf{k})=9$

(24) $\cos^{-1}\left(\frac{4}{21}\right)$

(25) $\frac{\pi}{6}$

(26) $\frac{5}{2}$

(27) $\mathbf{r} = \frac{1}{222}(106\mathbf{i} + 73\mathbf{j} - 23\mathbf{k}) + t(-\mathbf{i} + 7\mathbf{j} + 13\mathbf{k})$

(28) $\mathbf{r} = \mathbf{a} + a(\mathbf{n} \times \mathbf{c})$

(32) S.D = 9 , P = (5, 4, 0), Q = (- 1, - 2, - 3),
APQ = $\mathbf{r}.(2\mathbf{i} - \mathbf{j} - 2\mathbf{k}) = 5$ and BQP = $\mathbf{r}.(2\mathbf{i} - 7\mathbf{j} + 10\mathbf{k}) + 18 = 0$

(33) $3\sqrt{30}$, $\mathbf{r}.(4\mathbf{i} - 5\mathbf{j} + 17\mathbf{k}) + 19 = 0 = \mathbf{r}.(22\mathbf{i} - 5\mathbf{j} + 19\mathbf{k}) - 83$

(34) $\frac{8}{\sqrt{14}}$; $\mathbf{r}.(\mathbf{i} - \mathbf{j} - 2\mathbf{k}) + 6 = 0 = \mathbf{r}.(19\mathbf{i} + 17\mathbf{j} + 20\mathbf{k}) + 2$

(35) $\frac{11}{3\sqrt{38}}$; $\mathbf{r}.(10\mathbf{i} - 29\mathbf{j} + 16\mathbf{k}) = 0 = \mathbf{r}.(13\mathbf{i} + 82\mathbf{j} + 55\mathbf{k}) - 109$

7

Volume of a Tetrahedron
(Vector Method)

7.1. Volume of a tetrahedron

Let the position vectors of the extremities of three coterminous edges OA, OB and OC of the tetrahedron OABC be **a, b, c,** respectively i.e., $\vec{OA} = \mathbf{a}$, $\vec{OB} = \mathbf{b}$ and $\vec{OC} = \mathbf{c}$ w.r.t origin O.

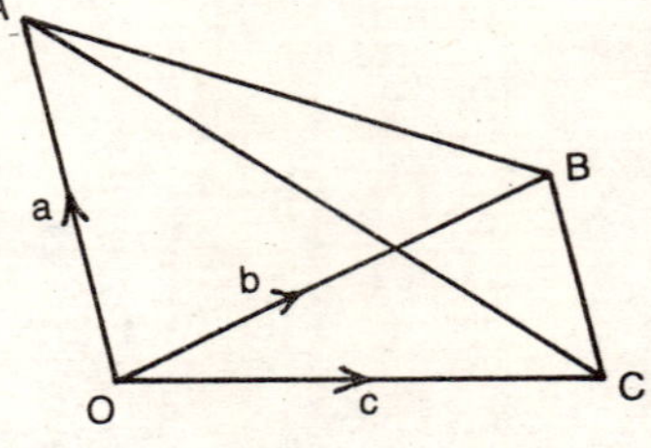

Fig. 7.1

Then the volume V of the tetrahedron in given by

$$V = \frac{1}{3} \text{ (area of } \Delta OBC)$$

(perpendicular distance from A on the plane OBC) ...(1)

If **n** be the unit vector perpendicular to the plane of the triangle OBC such that **b, c,** and **n** are in a right handed system then

$$\mathbf{n} = \frac{\mathbf{b} \times \mathbf{c}}{|\mathbf{b} \times \mathbf{c}|}$$

since **b, c,** and **b** × **c** are in right handed orientation.

∴ Length of perpendicular from A on the plane OBC

= length of the projection of OA on the perpendicular to the plane OBC in the direction of n

$$= \vec{OA} \cdot \mathbf{n} = \mathbf{a} \cdot \frac{\mathbf{b} \times \mathbf{c}}{|\mathbf{b} \times \mathbf{c}|} = \frac{\mathbf{a} \cdot (\mathbf{b} \times \mathbf{c})}{|\mathbf{b} \times \mathbf{c}|} = \frac{[\mathbf{abc}]}{|\mathbf{b} \times \mathbf{c}|}$$

Substituting in (1), we get

$$V = \frac{1}{3}(\text{area of OBC})\frac{[\mathbf{abc}]}{|\mathbf{b}\times\mathbf{c}|}$$

$$= \frac{1}{3}\frac{1}{2}|\vec{OB}\times\vec{OC}|\frac{[\mathbf{abc}]}{|\mathbf{b}\times\mathbf{c}|}$$

$$= \frac{1}{6}|\mathbf{b}\times\mathbf{c}|\frac{[\mathbf{abc}]}{|\mathbf{b}\times\mathbf{c}|} = \frac{1}{6}[\mathbf{abc}]$$

$\therefore$ **Volume of the tetrahedron = $\frac{1}{6}$ (Volume of the parallelopiped)**

Corollary 1. *Volume of the tetrahedron in terms of position vectors of the four vertices, neither of which is at origin*

Let $\vec{OA} = \mathbf{a}$, $\vec{OB} = \mathbf{b}$, $\vec{OC} = \mathbf{c}$ and $\vec{OD} = \mathbf{d}$ be the P.v's of the vertices A, B, C, and D of a tetrahedron w.r.t O.

Now volume of tetrahedron

$$= \frac{1}{6}[\vec{AB}, \vec{AC}, \vec{AD}]$$

$$= \frac{1}{6}[\mathbf{b}-\mathbf{a}, \mathbf{c}-\mathbf{a}, \mathbf{d}-\mathbf{a}]$$

$$= \frac{1}{6}(\mathbf{b}-\mathbf{a}).[(\mathbf{c}-\mathbf{a})\times(\mathbf{d}-\mathbf{a})]$$

$$= \frac{1}{6}(\mathbf{b}-\mathbf{a}).(\mathbf{c}\times\mathbf{d}-\mathbf{c}\times\mathbf{a}-\mathbf{a}\times\mathbf{d}+\mathbf{a}\times\mathbf{a})$$

$$= \frac{1}{6}(\mathbf{b}-\mathbf{a}).(\mathbf{c}\times\mathbf{d}-\mathbf{c}\times\mathbf{a}-\mathbf{a}\times\mathbf{d}) \quad [\because \mathbf{a}\times\mathbf{a}=0]$$

$$= \frac{1}{6}\{\mathbf{b}.(\mathbf{c}\times\mathbf{d}) - \mathbf{b}.(\mathbf{c}\times\mathbf{a}) - \mathbf{b}.(\mathbf{a}\times\mathbf{d}) - \mathbf{a}.(\mathbf{c}\times\mathbf{d}) + \mathbf{a}.(\mathbf{c}\times\mathbf{a}) + \mathbf{a}.(\mathbf{a}\times\mathbf{d})\}$$

$$= \frac{1}{6}\{[\mathbf{bcd}] - [\mathbf{bca}] - [\mathbf{bad}] - [\mathbf{acd}] + [\mathbf{aca}] + [\mathbf{aad}]\}$$

$$= \frac{1}{6}\{[\mathbf{bcd}] - [\mathbf{bca}] - [\mathbf{bad}] - [\mathbf{acd}]\}$$

$$[\because [\mathbf{aca}] = [\mathbf{aad}] = 0]$$

Corollary 2. *The four points* a, b, c, d *will be coplanar iff the volume of the tetrahedron formed by them is zero.*

i.e., $\frac{1}{6}[\mathbf{b}-\mathbf{a}, \mathbf{c}-\mathbf{a}, \mathbf{d}-\mathbf{a}] = 0$

or $[\mathbf{bcd}] - [\mathbf{bca}] - [\mathbf{bad}] - [\mathbf{acd}] = 0$

Corollary 3. *Volume of tetrahedron in terms of the co-ordinates of the vertices.* Let A (a_1, a_2, a_3), B (b_1, b_2, b_3), C(c_1, c_2, c_3), and D (d_1, d_2, d_3) be the vertices of a tetrahedron and let O be the point of reference.

$$\vec{OA} = a_1\,\mathbf{i} + a_2\,\mathbf{j} + a_3\,\mathbf{k}\ ;\ \vec{OB} = b_i\,\mathbf{i} + b_2\,\mathbf{j} + b_3\,\mathbf{k}$$

$$\vec{OC} = c_i\,\mathbf{i} + c_2\mathbf{j} + c_3\,\mathbf{k}\ ;\ \vec{OD} = di_1\,\mathbf{i} + d_2\,\mathbf{j} + d_3\,\mathbf{k}$$

$$\therefore\quad \vec{AB} = \vec{OB} - \vec{OA} = (b_1 - a_1)\,\mathbf{i} + (b_2 - a_2)\,\mathbf{j} + (b_3 - a_3)\,\mathbf{k}$$

$$\vec{AC} = \vec{OC} - \vec{OA} = (c_1 - a_1)\,\mathbf{i} + (c_2 - a_2)\,\mathbf{j} + (c_3 - a_3)\,\mathbf{k}$$

$$\vec{AD} = \vec{OD} - \vec{OA} = (d_1 - a_1)\,\mathbf{i} + (d_2 - a_2)\,\mathbf{j} + (d_3 - a_3)\,\mathbf{k}$$

$\therefore$ Volume of the tetrahedron ABCD

$$= \frac{1}{6}\vec{AB}\,.\,[\vec{AC} \times \vec{AD}]$$

$$= \frac{1}{6}\begin{vmatrix} b_1 - a_1 & b_2 - a_2 & b_3 - a_3 \\ c_1 - a_1 & c_2 - a_2 & c_3 - a_3 \\ d_1 - a_1 & d_2 - a_2 & d_3 - a_3 \end{vmatrix}$$

$$= \frac{1}{6}\begin{vmatrix} b_1 - a_1 & b_2 - a_2 & b_3 - a_3 & 0 \\ c_1 - a_1 & c_2 - a_2 & c_3 - a_3 & 0 \\ d_1 - a_1 & d_2 - a_2 & d_3 - a_3 & 0 \\ a_1 & a_2 & a_3 & 1 \end{vmatrix}$$

(Note that the value of the determinant has not changed by writing like this)

$$= \frac{1}{6}\begin{vmatrix} b_1 & b_2 & b_3 & 1 \\ c_1 & c_2 & c_3 & 1 \\ d_1 & d_2 & d_3 & 1 \\ a_1 & a_2 & a_3 & 1 \end{vmatrix}$$

or $$V = \frac{1}{6}\begin{vmatrix} a_1 & a_2 & a_3 & 1 \\ b_1 & b_2 & b_3 & 1 \\ c_1 & c_2 & c_3 & 1 \\ d_1 & d_2 & d_3 & 1 \end{vmatrix}$$

SOLVED PROBLEMS

Example 1. Find the volume of the tetrahedron whose vertices are A(1, 3, 7), B(2, 5, 1), C(1, 1, 5) and D(–2, 3, 2).

Solution: Let $\vec{OA} = \mathbf{i} + 3\mathbf{j} + 7\mathbf{k}$, $\vec{OB} = 2\mathbf{i} + 5\mathbf{j} + \mathbf{k}$

$$\vec{OC} = \mathbf{i} + \mathbf{j} + 5\mathbf{k} \text{ and } \vec{OD} = -2\mathbf{i} + 3\mathbf{j} + 2\mathbf{k}$$

be the p.v's of the vertices A, B, C and D w.r.t origins O. Then

$$\vec{AB} = \vec{OB} - \vec{OA} = (2\mathbf{i} + 5\mathbf{j} + \mathbf{k}) - (\mathbf{i} + 3\mathbf{j} + 7\mathbf{k}) = \mathbf{i} + 2\mathbf{j} - 6\mathbf{k}$$

$$\vec{AC} = \vec{OC} - \vec{OA} = (\mathbf{i} + \mathbf{j} + 5\mathbf{k}) - (\mathbf{i} + 3\mathbf{j} + 7\mathbf{k}) = -2\mathbf{j} - 2\mathbf{k}$$

and $$\vec{AD} = \vec{OD} - \vec{OA} = (-2\mathbf{i} - 3\mathbf{j} + 2\mathbf{k}) - (\mathbf{i} + 3\mathbf{j} + 7\mathbf{k}) = -3\mathbf{i} - 5\mathbf{k}$$

Now the required volume of tetrahedron is

$$= \frac{1}{6}\vec{AB} . \{\vec{AC} \times \vec{AD}\}$$

$$= \frac{1}{6}\begin{vmatrix} 1 & 2 & -6 \\ 0 & -2 & -2 \\ -3 & 0 & -5 \end{vmatrix}$$

$$= \frac{1}{6}\{1(10 - 0) - 2(0 - 6) - 6(0 - 6)\}$$

$$= \frac{1}{6}\{10 + 12 + 36\} = \frac{29}{3} \text{ cubic units}$$

Example 2. Show that the volume of a pyramid of which the vertex is a given point (x, y, z) and the base a triangle formed by joining the points (a, 0, 0) (0, b, 0) and (0, 0, c) in rectangular co-ordinates is

$$\frac{1}{6}abc\left\{\frac{x}{a} + \frac{y}{b} + \frac{z}{c} - 1\right\}.$$

Solution. A triangular pyramid is a tetrahedron w.r.t. O as origin Let the p.v's of the given four points be $\vec{OA} = a\mathbf{i}$, $\vec{OB} = b\mathbf{j}$, $\vec{OC} = c\mathbf{k}$ and $\vec{OD} = x\mathbf{i} + y\mathbf{j} + z\mathbf{k}$.

$$\text{Volume} = \frac{1}{6}\,\vec{AB}\,.\,\{(\vec{AC} \times \vec{AD}\}$$

$$= \frac{1}{6}(b\mathbf{j} - a\mathbf{i})\,.\,\{(c\mathbf{k} - a\mathbf{i}) \times (x - a)\,\mathbf{i} + y\mathbf{j} + z\mathbf{k}\}$$

$$= \frac{1}{6}\begin{vmatrix} -a & b & 0 \\ -a & 0 & c \\ x-a & y & z \end{vmatrix}$$

$$= \frac{1}{6}[-a\,(0 - yc) - b\,\{-az - c\,(x - a)\}]$$

$$= \frac{1}{6}\{acy + abz + bcx - abc\{$$

$$= \frac{abc}{6}\left\{\frac{x}{a} + \frac{y}{b} + \frac{z}{c} - 1\right\}$$

Example 3. Prove that the volume of a tetrahedron bounded by the four planes **r**.(m**j** + n**k**) = 0, **r**.(n**k** + l**i** = 0, **r**.(l**i** + m**j**) = 0 and **r**.(l**i** + m**j** + n**k**) = 0 is $\frac{2\,p^3}{3\,lmn}$.

Solution. Given planes are

$$\mathbf{r}\,.\,(m\mathbf{j} + n\mathbf{k}) = 0 \qquad ...(1)$$

$$\mathbf{r}\,.\,(n\mathbf{k} + l\mathbf{i}) = 0 \qquad ...(2)$$

$$\mathbf{r}\,.\,(l\mathbf{i} + m\mathbf{j}) = 0 \qquad ...(3)$$

$$\mathbf{r}\,.\,(l\mathbf{i} + m\mathbf{j} + n\mathbf{k}) = 0 \qquad ...(4)$$

It is obvious that the planes (1), (2), (3) intersect at origin O.

Let the point of intersection of (1), (2) and (4) be x**i** + y**j** + z**k**

$$\therefore \quad (x\mathbf{i} + y\mathbf{j} + z\mathbf{k})\,.\,(m\mathbf{j} + n\mathbf{k}) = 0 \text{ or } ym + zn = 0 \qquad ...(5)$$

$$(x\mathbf{i} + y\mathbf{j} + z\mathbf{k})\,.\,(n\mathbf{k} + l\mathbf{i}) = 0 \text{ or } lx + nz = 0 \qquad ...(6)$$

$$(x\mathbf{i} + y\mathbf{i} + z\mathbf{k})\,.\,(l\mathbf{i} + m\mathbf{j} + n\mathbf{k}) = 0 \text{ or } lx + my + nz = 0 \qquad ...(7)$$

Solving (5), (6) and (7) we get

$$x = \frac{p}{l},\ y = \frac{p}{m} \text{ and } z = \frac{-p}{n}$$

∴ Position vector of intersecting point

$$= \frac{p}{l}\,\mathbf{i} + \frac{p}{m}\,\mathbf{j} - \frac{p}{n}\,\mathbf{k} = \vec{OB} \text{ (say)}$$

Similarly the position vectors of intersections of (1), (3), (4) and (2), (3), (4) are

$$\frac{p}{l}\mathbf{i} - \frac{p}{m}\mathbf{j} + \frac{p}{n}\mathbf{k} = \vec{OC} \text{ (say)}$$

and
$$-\frac{p}{l}\mathbf{i} + \frac{p}{m}\mathbf{j} + \frac{p}{n}\mathbf{k} = \vec{OD} \text{ (say)}$$

$$\therefore \text{ Volume} = \frac{1}{6}[\vec{OB} \,.\, \{\vec{OC} \times \vec{OD}\}]$$

$$= \frac{1}{6}\begin{vmatrix} \frac{p}{l} & \frac{p}{m} & -\frac{p}{n} \\ \frac{p}{l} & -\frac{p}{m} & \frac{p}{n} \\ -\frac{p}{l} & \frac{p}{m} & \frac{p}{n} \end{vmatrix}$$

$$= \frac{p^3}{6\,lmn}\begin{vmatrix} 1 & 1 & -1 \\ 1 & -1 & 1 \\ -1 & 1 & 1 \end{vmatrix}$$

$$= \frac{p^3}{6\,lmn}\{1(-1-1) - 1(1+1) - 1(1-1)\} = \frac{p^3}{6\,lmn}(-4)$$

or $\text{volume} = \frac{2}{3}\frac{p^3}{lmn}$ (by neglecting negative sign)

Example 4. Prove the following formula for the volume V or a tetrahedron in terms of lengths of three concurrent edges and their mutual inclinations

$$V^2 = \frac{a^2 b^2 c^2}{36}\begin{vmatrix} 1 & \cos\phi & \cos\Psi \\ \cos\phi & 1 & \cos\theta \\ \cos\Psi & \cos\theta & 1 \end{vmatrix}$$

Solution. Let OABC be the tetrahedron such that

$O = (0, 0, 0)$, $A = (x_1, y_1, z_1)$, $B = (x_2, y_2, z_2)$, $C = (x_3, y_3, z_3)$

Then $\vec{OA} = \mathbf{a} = x_1\mathbf{i} + y_1\mathbf{j} + z_1\mathbf{k}$, $\vec{OB} = \mathbf{b} = x_2\mathbf{i} + y_2\mathbf{j} + z_2\mathbf{k}$

and $\vec{OC} = \mathbf{c} = x_3\,\mathbf{i} + y_3\,\mathbf{j} + z_3\,\mathbf{k}$

$$\therefore \quad OA^2 = a^2 = \mathbf{a} \,.\, \mathbf{a} = x_1^2 + y_1^2 + z_1^2 = \Sigma\, x_1^2$$

$$OB^2 = b^2 = \mathbf{b} \,.\, \mathbf{b} = \Sigma\, x_2^2 \text{ and } OC^2 = c^2 = \mathbf{c} \,.\, \mathbf{c} = \Sigma\, x_3^2$$

Again $\mathbf{a} \cdot \mathbf{b} = x_1x_2 + y_1y_2 + z_1z_2 = \Sigma\, x_1x_2 = ab \cos \phi$

$\mathbf{b} \cdot \mathbf{c} = \Sigma\, x_2x_3 = bc \cos \theta$

and $\mathbf{c} \cdot \mathbf{a} = \Sigma\, x_3x_1 = ca \cos \Psi$ where ϕ, θ and Ψ are angles between **a, b ; b, c and c, a**

Now $$V = \frac{1}{6} \overrightarrow{OA} \cdot \{\overrightarrow{OB} \times \overrightarrow{OC}\}$$

$$= \frac{1}{6} \begin{vmatrix} x_1 & y_1 & z_1 \\ x_2 & y_2 & z_2 \\ x_3 & y_3 & z_3 \end{vmatrix}$$

$$V^2 = \frac{1}{36} \begin{vmatrix} x_1 & y_1 & z_1 \\ x_2 & y_2 & z_2 \\ x_3 & y_3 & z_3 \end{vmatrix} \begin{vmatrix} x_1 & y_1 & z_1 \\ x_2 & y_2 & z_2 \\ x_3 & y_3 & z_3 \end{vmatrix}$$

$$= \frac{1}{36} \begin{vmatrix} \Sigma\, x_1^2 & \Sigma\, x_1x_2 & \Sigma\, x_1x_3 \\ \Sigma\, x_1x_2 & \Sigma\, x_2^2 & \Sigma\, x_2x_3 \\ \Sigma\, x_1x_3 & \Sigma\, x_2x_3 & \Sigma\, x_3^2 \end{vmatrix}$$

$$= \frac{1}{36} \begin{vmatrix} a^2 & ab\cos\phi & ac\cos\Psi \\ ab\cos\phi & b^2 & bc\cos\theta \\ ac\cos\Psi & bc\cos\theta & c^2 \end{vmatrix}$$

$$= \frac{abc}{36} \begin{vmatrix} a & b\cos\phi & c\cos\Psi \\ a\cos\phi & b & c\cos\theta \\ a\cos\Psi & b\cos\theta & c \end{vmatrix}$$

$$= \frac{a^2\, b^2\, c^2}{36} \begin{vmatrix} 1 & \cos\phi & \cos\Psi \\ \cos\phi & 1 & \cos\theta \\ \cos\Psi & \cos\theta & 1 \end{vmatrix}$$

Example 5. In the tetrahedron (OABC) prove that the volume V is given by the formula $\frac{1}{6}$ (AB) (OC) p sin θ. where p is the shortest distance between AB and OC.

Solution: Referring to origin O let **a, b, c** be the p.v's of A, B, and C respectively.

$\overrightarrow{OA} = \mathbf{a}, \overrightarrow{OB} = \mathbf{b}, \overrightarrow{OC} = \mathbf{c}$

$\therefore \quad \overrightarrow{AB} = \mathbf{b} - \mathbf{a}.$

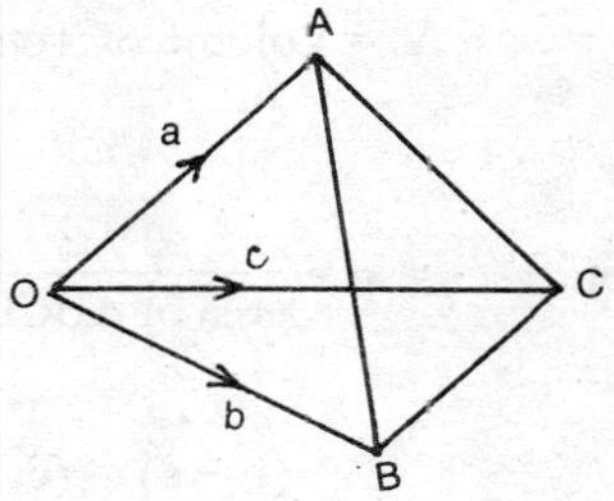

Fig. 7.2

Shortest distance between AB and OC is parallel to (b − a) × c

Also **a** and **c** are points on OA and OC.

∴ S.D is the projection of $\overrightarrow{CA}$ on $(\mathbf{b} - \mathbf{a}) \times \mathbf{c}$

∴ If P be the S.D. then

$$p = \frac{\overrightarrow{CA} \cdot \{(\mathbf{b} - \mathbf{a}) \times \mathbf{c}\}}{|(\mathbf{b} - \mathbf{a}) \times \mathbf{c}|}$$

$$= \frac{(\mathbf{a} - \mathbf{c}) \cdot \{(\mathbf{b} - \mathbf{a}) \times \mathbf{c}\}}{|(\mathbf{b} - \mathbf{a}) \times \mathbf{c}|}$$

$$= \frac{(\mathbf{a} - \mathbf{c}) \cdot (\mathbf{b} \times \mathbf{c} - \mathbf{a} \times \mathbf{c})}{(\mathbf{b} - \mathbf{a})\, \mathbf{c} \sin \theta}$$

$$= \frac{\mathbf{a} \cdot (\mathbf{b} \times \mathbf{c}) - \mathbf{a} \cdot (\mathbf{a} \times \mathbf{c}) - \mathbf{c} \cdot (\mathbf{b} \times \mathbf{c}) + \mathbf{c} \cdot (\mathbf{a} \times \mathbf{c})}{(AB)(OC) \sin \theta}$$

$$= \frac{[6V]}{(AB)(OC) \sin \theta} \qquad \because [\mathbf{aac}] = [\mathbf{cbc}] = [\mathbf{cac}] = 0]$$

or $$V = \frac{p}{6} (OB)(OC) \sin \theta . \qquad \because V = \frac{1}{6} [\mathbf{abc}]$$

Example 6. Show that the pependicular distance of a point whose position vector is a from the plane through three points with position vectors b, c, d is

$$\frac{[\mathbf{bcd}] + [\mathbf{cad}] + [\mathbf{abd}] - [\mathbf{abc}]}{|\mathbf{b} \times \mathbf{c} + \mathbf{c} \times \mathbf{d} + \mathbf{d} \times \mathbf{b}|}$$

Solution. Let ΔBCD be the tetrahedron. Let the position vectors of vertices A, B, C, D be $\overrightarrow{OA} = \mathbf{a}$, $\overrightarrow{OB} = \mathbf{b}$, $\overrightarrow{OC} = \mathbf{c}$ and $\overrightarrow{OD} = \mathbf{d}$ when O is point of reference. Let p be the perpendicular from A on the plane BCD.

$\therefore$ V = Volume of Tetrahedron = $\frac{1}{3}$ p area of ΔBCD

or
$$p = \frac{3V}{\text{area of } \Delta BCD} = \frac{3 \cdot \frac{1}{6}(\overrightarrow{AB} \cdot \{\overrightarrow{AC} \times \overrightarrow{AD}\})}{\frac{1}{2} \,|\, \overrightarrow{BC} \times \overrightarrow{CD} \,|}$$

$$= \frac{(\mathbf{b} - \mathbf{a}) \cdot \{(\mathbf{c} - \mathbf{a}) \times (\mathbf{d} - \mathbf{a})\}}{|(\mathbf{c} - \mathbf{b}) \times (\mathbf{d} - \mathbf{c})|}$$

$$= \frac{(\mathbf{b} - \mathbf{a}) \cdot \{\mathbf{c} \times \mathbf{d} - \mathbf{c} \times \mathbf{a} - \mathbf{a} \times \mathbf{d}\}}{|\mathbf{c} \times \mathbf{d} - \mathbf{b} \times \mathbf{d} + \mathbf{b} \times \mathbf{c}|}$$

$$= \frac{[\mathbf{bcd}] - [\mathbf{bca}] - [\mathbf{bad}] - [\mathbf{acd}]}{|\mathbf{c} \times \mathbf{d} + \mathbf{b} \times \mathbf{c} + \mathbf{d} \times \mathbf{b}|}$$

$$= \frac{[\mathbf{bcd}] - [\mathbf{abc}] - [\mathbf{abd}] + [\mathbf{cad}]}{|\mathbf{c} \times \mathbf{d} + \mathbf{b} \times \mathbf{c} + \mathbf{d} \times \mathbf{b}|}.$$

EXERCISE

1. **a, b, c** are p.v of A, B, C w.r.t. O. Show that the volume of the tetrahedron OABC is $\frac{1}{6}(\mathbf{a} \times \mathbf{b}).\mathbf{c}$.

2. Show that the volume of the tetrahedron whose vertices are O (0, 0, 0), A(a, 0, 0), B(0, a, 0) and C(0, 0, a) is $\frac{a^3}{6}$ cubic units.

3. Prove that four points with position vectors **a, b, c** and **d** are coplanar if and only if

$$[\mathbf{abc}] = [\mathbf{bcd}] + [\mathbf{cad}] + [\mathbf{abd}].$$

4. Find the volume of the tetrahedron whose vertices are (1, 1, 3), (- 2, 1, 4), (2, 3, - 6) and (3, 2, - 1). [**Ans.** 6 cubic units]

5. Find the volume of the tetrahedron whose vertices are (2, 3, - 1), (1, - 2, 3), (3, 4, - 2), (1, - 6, 6). [**Ans.** O]

6. Prove that each of the four faces of a tetrahedron subtends the same volume at the centroid.

7. G_1, G_2, G_3 are the controids of the triangular faces OBC, OCA, OAB of a tetrahedron OABC. Prove that the volume of the tetrahedron OABC in to the volume of the parallelopiped constructed with OG_1, OG_2 and OG_3 as coterminous edges as 9 : 4.

8. O is the centroid of the tetrahedron OABC; O′A′B′C′ is another tetrahedron such that OD′, AA′, BB′ and CC′ are all bisected at G. Show that G is also the centroid of the tetrahedron O′A′B′C′.

8

Vector Equation of a Sphere

8.1. Vector equation of a sphere

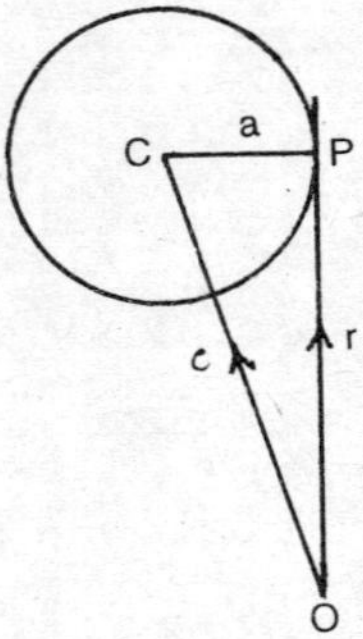

W.r.t O as origin let **c** and **r** be the position vectors of the points C and P.

Then $\vec{OC} = \mathbf{c}$ and $\vec{OP} = \mathbf{r}$

Now $\vec{CP} = \vec{OP} - \vec{OC} = \mathbf{r} - \mathbf{c}$

Let $\overline{CP} = |\vec{CP}| = a$

$\therefore |\mathbf{r} - \mathbf{c}| = a$

$$(\mathbf{r} - \mathbf{c}) \cdot (\mathbf{r} - \mathbf{c}) = |\mathbf{r} - \mathbf{c}|^2 = a^2$$

or $$\mathbf{r}.\mathbf{r} - \mathbf{r}.\mathbf{c} - \mathbf{c}.\mathbf{r} + \mathbf{c}.\mathbf{c} = a^2$$

or $$\mathbf{r}^2 - 2\mathbf{r}.\mathbf{c} + c^2 - a^2 = 0; \quad |\mathbf{c}| = c$$

Fig. 8.1

or $$\mathbf{r}^2 - 2\mathbf{r}.\mathbf{c} + k = 0 \qquad \ldots(1)$$

where $k = c^2 - a^2$ = constant

The equation (1) is satisfied by the position vector of every point on the surface of the sphere and by no other point, hence (1) represents the *Vector equation of the sphere.* Here the centre is the point **c** and radius = **a** = $\sqrt{c^2 - k}$.

The L.H.S of (1) may be regarded as a function of vector **r**. If we denote it by F(**r**) then

$$F(\mathbf{r}) = \mathbf{r}^2 - 2\mathbf{r}.\mathbf{c} + k = 0 \qquad \ldots(2)$$

Imp. Note. *The equation (1) may also be regarded as the vector equation of a circle if the point* **r** *is restricted to move in a plane.*

Particular cases

(i) *If the sphere (1) passes through the origin*

In this case $|\mathbf{c}| = c = a$ and hence $k = a^2 - a^2 = 0$

$\therefore$ Vector equation of the sphere with centre C and passing through the origin is $\mathbf{r}^2 - 2\mathbf{r}.\mathbf{c} = 0$

(ii) *If the centre of the sphere (1) is origin.*

In this case $\mathbf{c} = \mathbf{0}$ $\quad \therefore \ |\mathbf{c}| = c = 0$

and $\quad k = c^2 - a^2 = 0 - a^2 = -a^2$

Hence the vector equation of the sphere whose centre is origin and radius a is $\mathbf{r}^2 = a^2$.

8.2. Theorem.

The diameter of a sphere subtends a right angle at any point on the surface.

Let the origin be the centre C of the sphere and ACB be the diameter. Let P be any point on the surface of the sphere. Join AP, BP and CP.

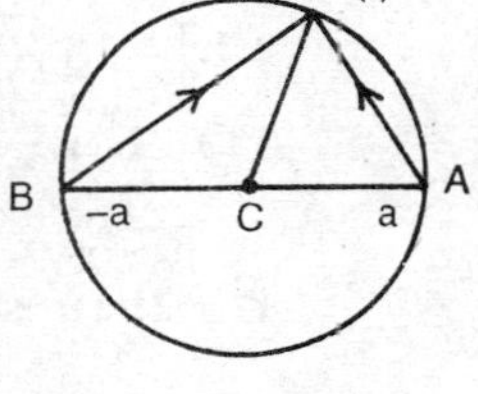

Fig. 8.2

Let $\vec{CP} = \mathbf{r}$, $\vec{CA} = \mathbf{a}$ then $\vec{CB} = -\mathbf{a}$

$\therefore \quad \vec{AP} = \vec{CP} - \vec{CA} = \mathbf{r} - \mathbf{a}$

and $\quad \vec{BP} = \vec{CP} - \vec{CB} = \mathbf{r} + \mathbf{a}$

$\because \quad CP = CA \quad \therefore \ |\vec{CP}| = |\vec{CA}|$

i.e., $\quad |\mathbf{r}| = |\mathbf{a}|$ or $\mathbf{r}^2 = \mathbf{a}^2$ or $\mathbf{r}^2 - \mathbf{a}^2 = 0$

or $\quad (\mathbf{r} - \mathbf{a}) . (\mathbf{r} + \mathbf{a}) = 0 \Rightarrow \vec{AP} . \vec{BP} = 0$

$\therefore$ AP and BP are at right angles to each other.

8.3. Vector equation of the sphere on the join of two given points as diameter

Let **a** and **b** be the p.v's of the extre mities A and B of any diameter. Let **r** be the P.V of any point P on the surface of the sphere

Then $\quad \vec{AP} = \mathbf{r} - \mathbf{a}$ and $\vec{BP} = \mathbf{r} - \mathbf{b}$

$\because$ $\vec{AP}$ and $\vec{BP}$ are perpendicular to each other **[Ref art 8.2]**

$$\vec{AP} . \vec{BP} = 0 \text{ i.e., } (\mathbf{r} - \mathbf{a}) . (\mathbf{r} - \mathbf{b}) = 0$$

which is the required equation. If $\mathbf{p} = (x, y, z)$, $\mathbf{a} = (x_1, y_1, z_1)$ and $\mathbf{b} = (x_2, y_2, z_2)$ then the castesian form of (1) is $(x - x_1)(x - x_2) + (y - y_1)(y - y_2) + (z - z_1)(z - z_2) = 0$.

8.4. Points of intersection of a line and space

Let the vector equation of the sphere with centre C, whose position vector is **c** be

$F(\mathbf{r}) = \mathbf{r}^2 - 2\mathbf{r}.\mathbf{c} + k = 0$ ...(1)

Let the equation of the line PAB be

$\mathbf{r} = \mathbf{a} + t\mathbf{b}$ ...(2)

which passes through the point P whose position vector is **a** and is parallel to unit vector **b**.

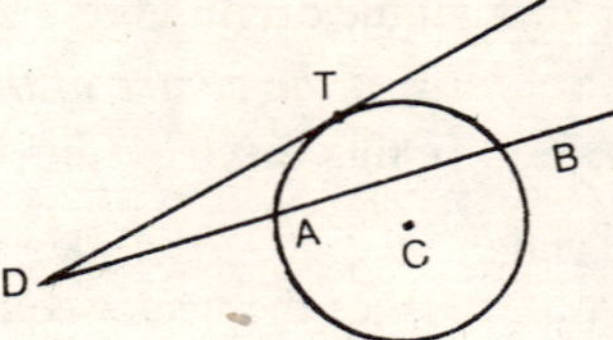

Fig. 8.3

For the points of intersection of (1) and (2) on eliminating **r** between (1) and (2), we obtain

$$(\mathbf{a} + t\mathbf{b})^2 - (2\perp\mathbf{a} + t\mathbf{b}).\mathbf{c} + k = 0$$

or $$t^2\mathbf{b}^2 = 2\mathbf{b} . (\mathbf{a} - \mathbf{c})t + (\mathbf{a}^2 - 2\mathbf{a} . \mathbf{c} + k) = 0$$

or $$t^2 + 2\mathbf{b}.(\mathbf{a} - \mathbf{c})t + (\mathbf{a}^2 - 2\mathbf{a}.\mathbf{c} + k) = 0$$

[$\because$ $b^2 = 1$, b being unit vector]

or $$t^2 + 2\mathbf{b}.(\mathbf{a} - \mathbf{c})\, t + F(\mathbf{a}) = 0 \quad ...(3)$$

[where $F(\mathbf{a}) = \mathbf{a}^2 - 2\mathbf{a}.\mathbf{c} + \mathbf{k}$]

Equation (3) is a quadratic in it. let its roots be t_1 and t_2 which are real if

$$[\mathbf{b} . (\mathbf{a} - \mathbf{c})]^2 > F(\mathbf{a})$$

Let (2) meet (1) in two points A and B

Then $PA = t_1$ and $PB = t_2$

$\therefore$ $PA.PB = t_1t_2$ = Product of the roots = $F(\mathbf{a})$

$$= \mathbf{a}^2 - 2\mathbf{a}.\mathbf{c} + k$$

which is independent of **b** thus is constant for all lines drawn through P in any direction to cut the surface of (1).

8.5. Length of the tangent

If the point A and B coincide at T, then the line PAB becomes tangent PT at T.

i.e., $PA = PB = PT$

$\therefore$ $PA.PB = (PT)^2 = F(\mathbf{a})$

which gives the square of length of a tangent from any point to the sphere.

Note. Square of the length of tangent from the origin $= F(\mathbf{0}) = k$. If the origin is within the sphere, k is negative and thus tangents from the origin are imaginary

8.6. Tangent plane at a given point

Let the equation of the sphere be

$$F(\mathbf{r}) = \mathbf{r}^2 - 2\mathbf{r}.\mathbf{c} + k = 0 \qquad ...(1)$$

Let **a** be the p.v of a given point on this sphere at which the tangent plane is required

$$\therefore \qquad F(\mathbf{a}) = \mathbf{a}^2 - 2\mathbf{a}.\mathbf{c} + k = 0 \qquad ...(2)$$

and let

$$\mathbf{r} = \mathbf{a} + t\mathbf{b} \qquad ...(3)$$

be any line through **a** where **b** is any variable unit vector. Eliminating **r** for the intersection of (1) and (3) we have

$$(\mathbf{a} + t\mathbf{b})^2 - 2(\mathbf{a} + t\mathbf{b}) . \mathbf{c} + k = 0$$

or $\quad t^2 + 2t\,\mathbf{b}. (\mathbf{a} - \mathbf{c}) + F(\mathbf{a}) = 0 \qquad [\because \mathbf{b}^2 = 1$, **b** being unit vector]

or $\quad t^2 + 2t\,\mathbf{b}.(\mathbf{a} - \mathbf{c}) = 0 \quad$ [using (2)] $\qquad ...(4)$

Now one root of the equation is zero.

$\therefore$ If the line (3) is to touch the sphere (1) at P(**a**) then the other root also must be zero [Ref. fig 8.3 PAB will be a tangent at P provided P lies on the sphere. PA = PB = 0]

$$\therefore \qquad \mathbf{b} . (\mathbf{a} - \mathbf{c}) = 0 \qquad ...(5)$$

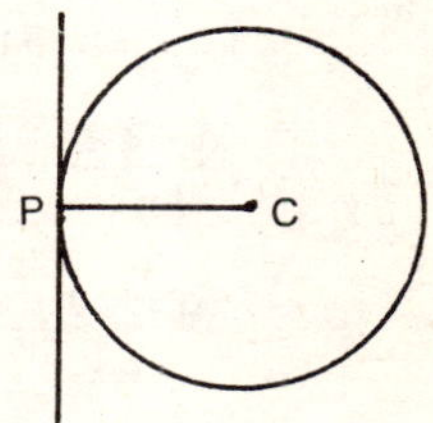

Fig. 8.4

Now C(**c**) is the centre of the sphere and P(**a**) be any point then CP is the radius of the sphere. Also $\overrightarrow{CP} = \mathbf{a} - \mathbf{c}$. Thus (5) shows that tangent line art P is perpendicular to CP. Eliminating **b** from (3) and (5) we get the locus of the tangent line at P(**a**) as

$$(\mathbf{r} - \mathbf{a}) . (\mathbf{a} - \mathbf{c}) = 0$$

which is the equations of a plane so called *tangent plane* at **a** and normal to the radius.

Adding the zero quantity $F(\mathbf{a}) = \mathbf{a}^2 - 2\mathbf{a}.\mathbf{c} + k$ from (2) to the left hand side of (6) then we get

$$(\mathbf{r} - \mathbf{a}) \cdot (\mathbf{a} - \mathbf{c}) + \mathbf{a}^2 - 2\mathbf{a}.\mathbf{c} + k = 0$$

or $$\mathbf{r}.\mathbf{a} - \mathbf{c}.(\mathbf{r} + \mathbf{a}) + k = 0 \qquad ...(7)$$

which is the standard form of the equation of the tangent plane at **a**.

Note. To write the equation of the tangent plane at the point **a** replace $\mathbf{r}^2$ by **r.a** and $2\mathbf{r}$ by **r + a.**

8.7. Condition for any plane to be a tangent plane

Let the equation of the sphere be

$$\mathbf{r}^2 - 2\mathbf{r}.\mathbf{c} + k = 0 \qquad ...(1)$$

with the centre C(**c**) and the plane be

$$\mathbf{r}.\mathbf{n} = q \qquad ...(2)$$

∵ Tangent plane is perpendicular to the radius CP of the sphere through the point of contact P we have

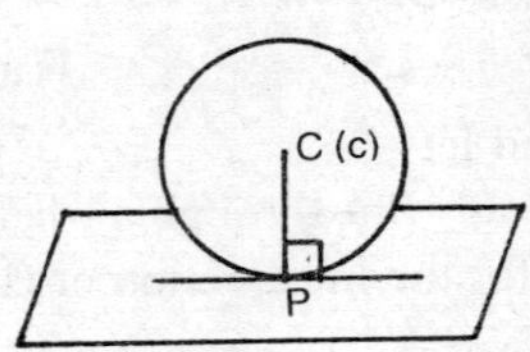

Fig. 8.5

$$\left\{\begin{array}{c}\text{Perpendicular distance from} \\ \text{centre C of the sphere} \\ \text{on to the plane}\end{array}\right\} = \text{(radius of the sphere)}$$

i.e., $$\frac{q - \mathbf{c}.\mathbf{n}}{|\mathbf{n}|} = a$$

or $$\frac{(q - \mathbf{c}.\mathbf{n})^2}{\mathbf{n}^2} = \mathbf{c}^2 - k \qquad \because \; a^2 = \mathbf{c}^2 - k; \; |\mathbf{n}^2| = \mathbf{n}^2$$

8.8. Orthogonality condition

To find the condition that the two given spheres may intersect orthogonally. Let the two spheres be

$$\mathbf{r}^2 - 2\mathbf{r}.\mathbf{c}_1 + k_1 = 0 \text{ where}$$

$$k_1 = c_1^2 - a_1^2 \qquad ...(1)$$

and $$\mathbf{r}^2 - 2\mathbf{r}.\mathbf{c}_2 + k_2 = 0$$

where $$k_2 = c_2^2 - a_2^2 \qquad ...(2)$$

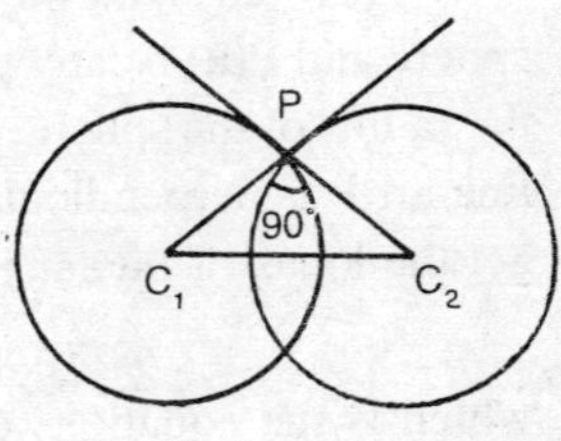

Fig. 8.6

If the two spheres (1) and (2) intersect each other orthogonally (i.e., cut each other at right angles) then the tangent plane to one of them at their common point of intersection will pass through the centre of the other.

i.e., (distance between centres)^2 = sum of squares of radii

$$\overline{C_1C_2}^2 = \overline{C_1P}^2 + \overline{C_2P}^2$$

$$(c_1 - c_2)^2 = a_1^2 + a_2^2$$

where a_1 and a_2 are radii of spheres (1) and (2)

or $$(c_1 - c_2)^2 = (c_1^2 + k_1) + (c_2^2 - k_2)$$

or $$c_1^2 + c_2^2 - 2\mathbf{c_1}.\mathbf{c_2} = c_1^2 + c_2^2 - k_1 - k_2$$

or $$2\mathbf{c_1}.\mathbf{c_2} = k_1 + k_2.$$

8.9 Polar plane

The polar plane of a given point w.r.t a sphere is the locus of points the target planes at which pass through the given point.

Let the equation of the sphere be

$$\mathbf{r}^2 - 2\mathbf{r.c} + k = 0 \qquad ...(1)$$

and the position vector of the given point be **d**.

The tangent plane to (1) at any point **a** is given by

$$\mathbf{r.a} - (\mathbf{r} + \mathbf{a}).\mathbf{c} + k = 0 \qquad ...(2)$$

If this tangent plane passes through **d**, then

$$\mathbf{d} \,.\, \mathbf{a} - (+ \mathbf{a}) \,.\, \mathbf{c} + k = 0 \qquad ...(3)$$

From (3) the locus of the point **a** in

$$\mathbf{d.r} - (\mathbf{d} + \mathbf{r}).\mathbf{c} + k = 0$$

or $$\mathbf{r.d} - (\mathbf{r} + \mathbf{d}).\mathbf{c} + k = 0$$

or $$\mathbf{r.(d - c)} = \mathbf{c.d} + k \qquad ...(4)$$

which is the equation of the polar plane of **d** w.r.t (1)

Note. The equation **r.(d − c) = c.d** + k shows that the plane is perpendicular to the line joining the centre **c** with the given point.

8.10. In verse points with respect to a given sphere

Let the sphere be

$$\mathbf{r}^2 - 2\mathbf{r.c} + k = 0 \qquad ...(1)$$

with centre C(**c**) and D(**d**) be the given point. Let the polar of D w.r.t the sphere meet the line CD at T. Then CT is perpendicular to the polar plane of D from the centre C [$\because$ CD is normal to the polar plane of D]

$$\therefore \quad CT = \frac{q - \mathbf{c}.\mathbf{n}}{|\mathbf{n}|} = \frac{(\mathbf{c}.\mathbf{d} - k) - \mathbf{c}.(\mathbf{d} - \mathbf{c})}{|\mathbf{d} - \mathbf{c}|}$$

$$= \frac{\mathbf{c}.\mathbf{d} - k - \mathbf{c}.\mathbf{d} + \mathbf{c}^2}{CD} \qquad \because \quad \vec{CD} = \mathbf{d} - \mathbf{c}$$

$$= \frac{c^2 - k}{CD} = \frac{a^2}{CD} \qquad \because \quad c^2 - a^2 = k$$

$\therefore$ CD.CT = a^2 i.e., the two points T and D are called the *inverse points* w.r.t the sphere.

Property 1. *If the polar plane of the point* P *w.r.t a sphere passes through Q, their polar plane of* Q *passes through* P.

Let the equation of the sphere be

$$\mathbf{r}^2 - 2\mathbf{r}.\mathbf{c} + k = 0 \qquad ...(1)$$

and let the p.v's of P and Q be **a** and **b**. Then the polar plane of P w.r.t (1) is

$$\mathbf{r}.\mathbf{a} - (\mathbf{r} + \mathbf{a}).\mathbf{c} + k = 0 \qquad ...(2)$$

If Q lies on (2) then we have

$$\mathbf{b}.\mathbf{a} - (\mathbf{b} + \mathbf{a}).\mathbf{c} + k = 0$$

The symmetry of this result shows that P(**a**) lies on the polar of Q(**b**).

Property 2. *The distances of two points P and Q from the centre of a given sphere are proportional to the distances of each from the polar plane of the other.*

Let $$\mathbf{r}^2 - 2\mathbf{r}.\mathbf{c} + k = 0 \qquad ...(1)$$

be the equation of the sphere and the p.v's of P and Q be **a** and **b**. Then the polar planes of P(**a**) and Q(**b**) w.r.t (1) are

$$\mathbf{r}.\mathbf{a} - (\mathbf{r} + \mathbf{a}).\mathbf{c} + k = 0 \qquad ...(2)$$

$$\mathbf{r}.\mathbf{b} - (\mathbf{r} + \mathbf{b}).\mathbf{c} + k = 0 \qquad ...(3)$$

The distance of Q (**b**) from the plane (2) is equal to

$$\frac{q - \mathbf{b.n}}{|\mathbf{n}|} = \frac{(\mathbf{c.a} - k) - \mathbf{b.(a - c)}}{|\mathbf{a - c}|} + \frac{\mathbf{c.a + b.d - b.a} - k}{|\mathbf{a - c}|} \quad ...(4)$$

and the distance of P(a) from the plane (3) is equal to

$$\frac{q - \mathbf{a.n}}{|\mathbf{n}|} = \frac{(\mathbf{c.b} - k) - \mathbf{a.(b - c)}}{|\mathbf{b - c}|} = \frac{\mathbf{c.b + a.c - a.b} - k}{|\mathbf{b - c}|} \quad ...(5)$$

Now dividing (4) by (5) we get

$$\frac{\text{distance of Q from polar plane of P}}{\text{distance of P from polar plane of Q}} = \frac{|\mathbf{b - c}|}{|\mathbf{a - c}|} = \frac{CB}{CA}$$

where C is the centre of the sphere.

8.11. Diamentral plane

The locus of points which bisect a system of parallel chords of a given sphere is called diametral plane.

Let the equation of the given sphere be

$$\mathbf{r}^2 - 2\mathbf{r.c} + k = 0 \quad ...(1)$$

and let
$$\mathbf{r} = \mathbf{a} + t\mathbf{b} \quad ...(2)$$

be one of the chords of the sphere (1), parallel to a given unit vector **b**. Eliminating **r** between (1) and (2), we get

$$(\mathbf{a} + t\mathbf{b})^2 - 2(\mathbf{a} + t\mathbf{b}).\mathbf{c} + k = 0$$

or
$$t^2 - 2\mathbf{b.(a - c)}t + (\mathbf{a}^2 - 2\mathbf{c.a} + k) = 0 \quad [\because \mathbf{b}^2 = 1] \quad ...(3)$$

If **a** is the midpoint of the chord (2) then the diametral plane is the locus of **a** and the two values of t given by the equation (3) must be equal and opposite in sign, and the condition for the same is $\mathbf{b.(a - c)} = 0$

$\therefore$ The locus of **a** i.e., the equation of the diametral plane is $\mathbf{b.(r - c)} = 0$ or $\mathbf{r.(b - c)} = 0$ which represents a plane passing through the centre **c** and perpendicular to **b** i.e., perpendicular to the chords.

8.12 Radical plane

The radical plane of any two given spheres is the locus of a point which moves so that the squares of the lengths of the tangents drawn from it to the given spheres are equal.

Let
$$\mathbf{r}^2 - 2\mathbf{r.c}_1 + k_1 = 0 \quad ...(1)$$

and
$$\mathbf{r}^2 - 2\mathbf{r.c}_2 + k_2 = 0 \quad ...(2)$$

be the two spheres and let **d** be any point on the radical plane. Then the squares of the tangents from it to (1) and (2) must be equal

$$\therefore \qquad \mathbf{d}^2 - \mathbf{d} \cdot \mathbf{c}_1 + k_1 = \mathbf{d}^2 - 2\mathbf{d} \cdot \mathbf{c}_2 + k_2$$

or

$$2\mathbf{d}.(\mathbf{c}_1 - \mathbf{c}_2) = k_1 - k_2.$$

Hence the locus of **d** i.e., the radical plane is

$$2\mathbf{r}.(\mathbf{c}_1 - \mathbf{c}_2) = k_1 - k_2 \qquad ...(3)$$

Note. Equation (3) shows that the radical plane is perpendicular to $\mathbf{c}_1 - \mathbf{c}_2$ i.e., the line of centres.

8.13. Co–axial system of spheres

Any system of spheres in which every pair has the same radical plane is called coaxial system of spheres.

8.14. Equation of a system of coaxial spheres

To show that the equation $\mathbf{r}^2 - 2\lambda\mathbf{r}.\mathbf{c} + k = 0$, *where* λ *is a parameter and k is a constant represents a system of coaxial spheres.*

Let

$$\mathbf{r}^2 - 2\lambda_1\mathbf{r}.\mathbf{c} + k = 0 \qquad ...(1)$$

and

$$\mathbf{r}^2 - 2\lambda_2\,\mathbf{r}.\mathbf{c} + k = 0 \qquad ...(2)$$

be two members of the system of spheres corresponding to values λ_1 and λ_2 of parameter λ Now the radical plane of (1) and (2) is

$$2\,(\lambda_1 - \lambda_2)\,\mathbf{r} \cdot \mathbf{c} = 0 \quad \text{or} \quad \mathbf{r} \cdot \mathbf{c} = 0 \qquad (\because\ \lambda_1 \neq \lambda_2)$$

which is independent of λ

$\therefore$ Whatever be the value of λ the radical plane of every two members will have the same radical plane $\mathbf{r}.\mathbf{c} = 0$

Thus $\mathbf{r}^2 - 2\lambda\,\mathbf{r}.\mathbf{c} + k = 0$ represents a coaxial system of spheres

8.15. Limiting points.

The members of coaxial system of zero radius are called limiting points.

If $\mathbf{r}^2 - 2\lambda\,\mathbf{r}.\mathbf{c} + k = 0$ is the coaxial system. Radius of the sphere of the system $= \sqrt{\lambda^2 c^2 - k}$

$$\text{Now } \sqrt{\lambda^2 c^2 - k} = 0 \Rightarrow \lambda = \pm\sqrt{\frac{k}{c}}$$

ence the limiting points are $\pm\sqrt{\dfrac{k}{c}}\,\mathbf{c}$

8.15. Equation of the sphere passing through the intersection of two spheres

Let the two spheres be

$$F_1(\mathbf{r}) = \mathbf{r}^2 - 2\mathbf{r}.\mathbf{c}_1 + k_1 = 0 \qquad ...(1)$$

$$F_2(\mathbf{r}) = \mathbf{r}^2 - 2\mathbf{r}.\mathbf{c}_2 + k_2 = 0 \qquad ...(2)$$

Then the equation $F_1(\mathbf{r}) + \lambda F_2(\mathbf{r})$ gives

$$\mathbf{r}^2 - 2\mathbf{r}.\mathbf{c}_1 + k_1 + \lambda[\mathbf{r}^2 - 2\mathbf{r}.\mathbf{c}_2 + k_2] = 0$$

or
$$\mathbf{r}^2(1+\lambda) - 2\mathbf{r}.(\mathbf{c}_1 + \lambda\mathbf{c}_2) + (k_1 + \lambda k_2) = 0$$

or
$$\mathbf{r}^2 - 2\mathbf{r}.\frac{\mathbf{c}_1 + \lambda\mathbf{c}_2}{1+\lambda} + \frac{k_1 + \lambda k_2}{1+\lambda} = 0 \qquad ...(3)$$

Equation (3) represents a sphere whose centre is the point $\frac{\mathbf{c}_1 + \lambda\mathbf{c}_2}{1+\lambda}$. Moreover all the points which satisfy (1) and (2) also satisfy (3). Hence (3) is the equation of the sphere passing through the intersection of the spheres (1) and (2).

Note 1. *Since two spheres intersect along a circle equation (3) also represents a sphere passing through the circle of intersection of (1) and (2).*

Note 2. *Equation (3) also represents a system of spheres where λ is a parameter and for any two of above system the radical plane is same. Hence (3) represents a coaxial system.*

SOLVED PROBLEMS

Example 1. Prove that the locus of a point, the sum of the squares of whose distances from n given points is constant is a sphere.

Solution. Let $P_1, P_2, ..., P_n$ be n given points with p.v.'s $\mathbf{a}_1, \mathbf{a}_2, ..., \mathbf{a}_n$ and let $\mathbf{r}$ be the p.v of a generic point P whose locus is to be found. From the problem

$$P_1P^2 + P_2P^2 + \ldots + P_nP^2 = \lambda \qquad \text{(constant)}$$

or
$$\overrightarrow{P_1P}^2 + \overrightarrow{P_2P}^2 + \ldots + \overrightarrow{P_nP}^2 = \lambda$$

or $\quad (\mathbf{r}-\mathbf{a}_1)^2+(\mathbf{r}-\mathbf{a}_2)^2+\ldots+(\mathbf{r}-\mathbf{a}_n)^2=\lambda$

or $\quad n\mathbf{r}^2-2\mathbf{r}.(\mathbf{a}_1+\mathbf{a}_2+\ldots+\mathbf{a}_n)+(\mathrm{a}_1^2+\mathrm{a}_2^2+\ldots+\mathrm{a}_n^2)=\lambda$

or $\quad \mathbf{r}^2-2\mathbf{r}\,.\,\dfrac{\Sigma\,\mathbf{a}_1}{n}+\dfrac{\Sigma\,\mathrm{a}_1^2-\lambda}{n}=0$

which is the required locus.

Hence centre of the sphere is centroid of the n points.

Example 2. A straight line is drawn from a point O to meet the sphere in P. In OP a point Q is taken so that OP : OQ is a fixed ratio. Prove that the locus of Q is a sphere.

Solution. Let $\mathbf{r}_1$ be the p.v of Q referred to O as origin. $\therefore$ OQ = $\mathbf{r}_1$

Let $\dfrac{OP}{OQ}=n$ (say) $\therefore\ \overrightarrow{OP}=n\,\overrightarrow{OQ}$ i.e., $\overrightarrow{OP}=n\mathbf{r}_1$

Let the equation of the sphere be

$$\mathbf{r}^2-2\mathbf{r}.\mathbf{c}+k=0 \qquad \ldots(1)$$

Since P lies on (1) we have

$$n^2\mathbf{r}_1^2-2n\mathbf{r}_1.\mathbf{c}+k=0 \quad \text{or} \quad \mathbf{r}_1^2-\frac{2}{n}\mathbf{r}_1.\mathbf{c}+\frac{k}{n^2}=0$$

$\therefore$ Locus of Q (i.e., $\mathbf{r}_1$) is

$$\mathbf{r}^2-\frac{2}{n}\mathbf{r}.\mathbf{c}+\frac{k}{n^2}=0.$$

Example 3. Obtain the equation of the sphere described on the join of the points A (2, – 3, 4) and B (– 5, 6, – 7) as a diameter.

Solution. Let O be the origin and **r** be position vector of any point P on the required sphere. Then the position vectors of A and B are

$$\mathbf{a}=\overrightarrow{OA}=2\mathbf{i}-3\mathbf{j}+4\mathbf{k};\ \mathbf{b}=\overrightarrow{OB}=-5\mathbf{i}+6\mathbf{j}-7\mathbf{k}$$

$$\overrightarrow{OP}=\mathbf{r}=x\mathbf{i}+y\mathbf{j}+z\mathbf{k}$$

Equation of the sphere joining AB as diameter is

$$(\mathbf{r}-\mathbf{a})\,.\,(\mathbf{r}-\mathbf{b})=0 \qquad \ldots(1)$$

$$(\mathbf{r}-\mathbf{a})=(x\mathbf{i}+y\mathbf{j}+z\mathbf{k})-(2\mathbf{i}-3\mathbf{j}+4\mathbf{k})$$
$$=(x-2)\mathbf{i}+(y+3)\mathbf{j}+(z-4)\mathbf{k}$$

$$(\mathbf{r} - \mathbf{b}) = (x\mathbf{i} + y\mathbf{j} + z\mathbf{k}) - (-5\mathbf{i} + 6\mathbf{j} - 7\mathbf{k})$$
$$= (x + 5)\mathbf{i} + (y - 6)\mathbf{j} + (z + 7)\mathbf{k}$$

Substituting in (1) we get

$$(x - 2)(x + 5) + (y + 3)(y - 6) + (z - 4)(z + 7) = 0$$

or $$x^2 + y^2 + z^2 + 3x - 3y + 3z - 56 = 0$$

or $$r^2 - 2\mathbf{r}, \left(-\frac{3}{2}\mathbf{i} + \frac{3}{2}\mathbf{j} - \frac{3}{2}\mathbf{k} \right) - 56 = 0$$

which is the required sphere. Here the p.v of the centre is

$$-\frac{3}{2}\mathbf{i} + \frac{3}{2}\mathbf{j} - \frac{3}{2}\mathbf{k}.$$

Example 4. Find the coordinates of the centre of the sphere inscribed in the tetrahedron bounded by the planes $\mathbf{r}.\mathbf{i} = 0$; $\mathbf{r}.\mathbf{j} = 0$, $\mathbf{r}.\mathbf{k} = 0$ and $\mathbf{r}.(\mathbf{i} + \mathbf{j} + \mathbf{k}) = a$. Also write the equation of the sphere.

Solution. Let the centre of the sphere be $x\mathbf{i} + y\mathbf{j} + z\mathbf{k}$ since the given planes are the tangent planes therefore lengths of the perpendicular from the centre of the sphere to the given tangent planes are equal (= radius)

Using the formula $\frac{q - \mathbf{c}.\mathbf{n}}{|\mathbf{n}|}$ length of the perpendicular from the point c on to the plane $\mathbf{r}.\mathbf{n} = q$ we get

$$\frac{0 - (x\mathbf{i} + y\mathbf{j} + z\mathbf{k}).\mathbf{i}}{|\mathbf{i}|} = \frac{0 - (x\mathbf{i} + y\mathbf{j} + z\mathbf{k}).\mathbf{j}}{|\mathbf{j}|}$$
$$= \frac{0 - (x\mathbf{i} + y\mathbf{j} + z\mathbf{k}).\mathbf{k}}{|\mathbf{k}|} = \frac{a - (x\mathbf{i} + y\mathbf{j} + z\mathbf{k}).(\mathbf{i} + \mathbf{j} + \mathbf{k})}{|\mathbf{i} + \mathbf{j} + \mathbf{k})}$$

by taking only positive values (since it is the length) we get

$$\frac{x}{1} + \frac{y}{1} + \frac{z}{1} = \frac{a - (x + y + z)}{\sqrt{3}} = k \text{ (say)}$$

$\therefore$ $$x = y = z = k \text{ and } a - ((x + y + z) = \sqrt{3}k$$

or $$a - (k + k + k) = \sqrt{3}k$$

or $$k = \frac{a}{3 + \sqrt{3}} = \frac{a(3 - \sqrt{3})}{6}$$

$\therefore\ x = y = z = \dfrac{a(3-\sqrt{3})}{6}$ = radius

Hence the p.v of centre = $\dfrac{a(3-\sqrt{3})}{6}$ $(\mathbf{i}+\mathbf{j}+\mathbf{k})$ and the equation of the sphere is $(\mathbf{i}+\mathbf{j}+\mathbf{k})$ and the equations of the sphere is $(\mathbf{r}-\mathbf{c})^2 = a^2$

i.e.,
$$\left[\mathbf{r} = \frac{a}{6}(3-\sqrt{3})(\mathbf{i}+\mathbf{j}+\mathbf{k})\right]^2 = \left[\frac{a}{6}(3-\sqrt{3})\right]^2.$$

Example 5. Find the equations of the sphere circumscribing the tetrahedron $\mathbf{r}.\mathbf{i} = 0$, $\mathbf{r}.\mathbf{j} = 0$, $\mathbf{r}.\mathbf{k} = 0$ and $\mathbf{r}.(\mathbf{i}+\mathbf{j}+\mathbf{k}) = a$.

Solution. Tetrahedron is bounded by the planes

$$\mathbf{r}.\mathbf{i} = 0 \qquad ...(1)$$
$$\mathbf{r}.\mathbf{j} = 0 \qquad ...(2)$$
$$\mathbf{r}.\mathbf{k} = 0 \qquad ...(3)$$

and
$$\mathbf{r}.(\mathbf{i}+\mathbf{j}+\mathbf{k}) = a \qquad ...(4)$$

The planes (1), (2) and (3) intersect at the origin O (0, 0, 0)

Let the point of intersection of planes (1), (2) and (4) be A (x, y, z) whose p.v is $\overrightarrow{OA} = x\mathbf{i} + y\mathbf{j} + z\mathbf{k}$. As it satisfies them, we have

$$(x\mathbf{i}+y\mathbf{j}+z\mathbf{k}).\mathbf{i} = 0 \Rightarrow x = 0$$
$$(x\mathbf{i}+y\mathbf{j}+z\mathbf{k}).\mathbf{j} = 0 \Rightarrow y = 0$$
$$(x\mathbf{i}+y\mathbf{j}+z\mathbf{k}).(\mathbf{i}+\mathbf{j}+\mathbf{k}) = a \Rightarrow x + y + z = a$$

Solving these we get x = 0, y = 0, z= a

$\therefore$ A = (0, 0, a) i.e., $\overrightarrow{OA} = a\mathbf{k}$

Similarly the p.v's of intersection of (2), (3) and (4) is $\overrightarrow{OB} = a\mathbf{i}$ and (1), (3) and (4) is $\overrightarrow{OC} = a\mathbf{j}$. Let these be p.v's of the four vertices O, A, B and C.

Let $a_1\mathbf{i} + a_2\mathbf{j} + a_2\mathbf{k}$ be the p.v of the centre D of the sphere circumscribing the tetrahedron and r be the radius of this sphere

$\therefore\ \overrightarrow{OD} = a_1\mathbf{i} + a_2\mathbf{j} + a_2\mathbf{k}$

Now r = distance of each vertex from centre D

$$\therefore \quad r^2 = \overline{OD}^2 = \overline{AD}^2 = \overline{BD}^2 = \overline{CD}^2$$

or
$$\mathbf{r}^2 = \overrightarrow{OD}^2 = \overrightarrow{AD}^2 = \overrightarrow{BD}^2 = \overrightarrow{CD}^2$$

$$\vec{OD}^2 = (a_1\mathbf{i} + a_2\mathbf{j} + a_3\mathbf{k})^2 = a_1^2 + a_2^2 + a_3^2$$

$$\vec{AD}^2 = [(a_1\mathbf{i} + a_2\mathbf{j} + a_3\mathbf{k}) - a\mathbf{i}]^2 = (a_1 - a)^2 - a_2^2 + a_3^2$$

Similarly $\vec{BD}^2 = a_1^2 + (a_2 - a)^2 + a_3^2$ and

$$\vec{CD}^2 = a_1^2 + a_2^2 + (a_3 - a)^2$$

From $\vec{OD}^2 = \vec{AD}^2$ we get

$$a_1^2 + a_2^2 + a_3^2 = (a_1 - a)^2 + a_2 + a_3^2$$

or $$2a_1a = a^2 \quad \text{or} \quad a_1 = \frac{a}{2} \qquad (\because\ a \neq 0)$$

Similarly $$a_2 = a_3 = \frac{a}{2}$$

$\therefore$ P.V of centre of the sphere $= \frac{a}{2}(\mathbf{i} + \mathbf{j} + \mathbf{k})$

$\because$ One vertex is the origin, the sphere passes through the origin. Hence the required equation of the sphere is

$$\mathbf{r}^2 - 2\mathbf{r} \cdot \frac{a}{2}(\mathbf{i} + \mathbf{j} + \mathbf{k}) = 0$$

or $$\mathbf{r}^2 - \mathbf{r} \cdot a(\mathbf{i} + \mathbf{j} + \mathbf{k}) = 0$$

or $$\mathbf{r} \cdot [\mathbf{r} - a(\mathbf{i} + \mathbf{j} + \mathbf{k})] = 0.$$

Example 6. If any targent plane to the sphere with origin as centre and radius d makes intercepts a, b, c on the axes prove by vector that $\frac{1}{a^2} + \frac{1}{b^2} + \frac{1}{c^2} = \frac{1}{d^2}$.

Solution. The centre of the sphere is origin and radius d, hence the sphere is

$$\mathbf{r}^2 = d^2 \qquad ...(1)$$

Any plane $\mathbf{r.n} = q$ will be tangent to (1) iff perpendicular distance from centre = radius d

i.e. $$\frac{q}{|\mathbf{n}|} = d \quad \text{or} \quad q = |\mathbf{n}|d \qquad ...(2)$$

intercepts made by this plane on the axes are

$$\frac{q}{\mathbf{i.n}} = a; \ \frac{q}{\mathbf{j.n}} = b, \ \frac{q}{\mathbf{k.n}} = c$$

$$\therefore \quad \frac{1}{a^2} + \frac{1}{b^2} + \frac{1}{c^2} = \frac{1}{q^2}\{(\mathbf{i.n})^2 + (\mathbf{j.n})^2 + (\mathbf{k.n})^2\} \qquad ...(2)$$

$$\text{RHS} = \frac{1}{q^2}\{(1.n.\cos\alpha)^2 + (1.n.\cos\beta)^2 + (1.n.\cos\gamma)^2\}$$

$$= \frac{n^2}{n^2d^2}\{\cos^2\alpha + \cos^2\beta + \cos^2\gamma\} \qquad \text{[using (2)]}$$

$$= \frac{1}{d^2}(1) \qquad [\because \cos^2\alpha + \cos^2\beta + \cos^2\gamma = 1 \text{ where } \alpha, \beta, \gamma$$

are the angles which normal makes with the axes]

Hence from (3)

$$\frac{1}{a^2} + \frac{1}{b^2} + \frac{1}{c^2} = \frac{1}{d^2}.$$

Example 7. From any point on the surface of a sphere, straight lines are drawn to extremities of any surface of a concentric sphere. Prove that the sum of the squares on these lines is constant

Solution: Let **c** be the line p.v of the centre C of the concentric spheres.

$$\mathbf{r}^2 - 2\mathbf{r.c} + k_1 = 0 \qquad ...(1)$$

and

$$\mathbf{r}^2 - 2\mathbf{r.c} + k_2 = 0 \qquad ...(2)$$

where $k_1 = c^2 - a_1^2$ and

$k_2 = c^2 - a_2^2$,

a_1, $a_2$being the radii.

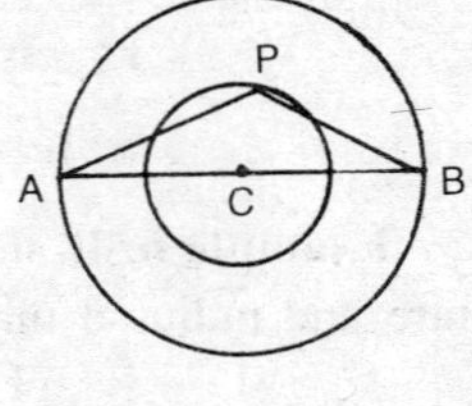

Fig. 8.7

Let P be any point on the inner sphere whose p.v is $\mathbf{r}_1$

$$\therefore \quad \mathbf{r}_1^2 - 2\mathbf{r}_1 . \mathbf{c} + k = 0 \qquad ...(3)$$

Let A, B be the extremities of the diameter of the outer sphere with p.v's **a** and **b**

∴ Equation to outer sphere is

$$(\mathbf{r} - \mathbf{a}) . (\mathbf{r} - \mathbf{b}) = 0$$

or

$$\mathbf{r}^2 - \mathbf{r}.(\mathbf{a} + \mathbf{b}) + \mathbf{a.b} = 0 \qquad ...(4)$$

which must be same as (2). Comparing (2) and (4), we get

$$\mathbf{a} + 2\mathbf{b} = 2\mathbf{c} \text{ and } \mathbf{a}.\mathbf{c} = k_2 \quad ...(5)$$

Now $$\overline{PA}^2 + \overline{PB}^2 = \overrightarrow{PA}^2 + \overrightarrow{PB}^2 \quad ...(6)$$

$$\text{RHS} = (\mathbf{a} - \mathbf{r_1})^2 + (\mathbf{b} - \mathbf{r_1})^2$$

$$= \mathbf{a^2} - 2\mathbf{r_1}.(\mathbf{a} + \mathbf{b}) + 2\mathbf{r_1^2} + \mathbf{b^2}$$

$$= (\mathbf{a} + \mathbf{b})^2 - 2\mathbf{a}.\mathbf{b} - 2\mathbf{r_1}.(2\mathbf{c}) + 2\mathbf{r_1^2} \qquad \text{[Using (5)]}$$

$$= (2\mathbf{c})^2 - 2k_2 - 2\mathbf{r_1}.2\mathbf{c} + 2\mathbf{r_1^2}$$

$$= 4c^2 - 2k_2 + 2(\mathbf{r_1^2} - 2\mathbf{r_1}.\mathbf{c}) \qquad [\because \mathbf{c^2} = c^2]$$

$$= 4c^2 - 2k_2 + 2(-k_1) \qquad \text{[Using (3)]}$$

$$= 2[(c^2 - k_1) + (c^2 - k_2)] = 2(a_1^2 + a_2^2)$$

= 2 [Sum of the squares of the radii of the spheres]

= Constant

Hence from (6) $\overline{PA}^2 + \overline{PB}^2$ = Constant

Hence proved.

Example 8. A plane passes through a fixed point (a, b, c) and cuts the axes in A, B, C. Prove by vector methods the locus of centre of sphere OABC is

$$\frac{a}{x} + \frac{b}{y} + \frac{c}{z} = 2.$$

Solution. Let O be the origin . Let the p.v of the fixed point (a, b, c) is $a\mathbf{i} + b\mathbf{j} + c\mathbf{k}$ and the normal be $\mathbf{n} = n_1\mathbf{i} + n_2\mathbf{j} + n_3\mathbf{k}$. Then equation to the plane passing through this fixed point is

$$\mathbf{r}.\mathbf{n} = (a\mathbf{i} + b\mathbf{j} + c\mathbf{k}).\mathbf{n}$$

or $$\mathbf{r}.(n_1\mathbf{i} + n_2\mathbf{j} + n_3\mathbf{k}) = an_1 + bn_2 + cn_3 = q \text{ (say)} \quad ...(1)$$

Let x_1 be the intercept on x-axis. Hence $x_1\mathbf{i}$ lies on (1)

$$\therefore \quad x_1\mathbf{i}.(n_1\mathbf{i} + n_2\mathbf{j} + n_3\mathbf{k}) = q \text{ or } x_1 = \frac{q}{n_1}$$

$$\therefore \quad OA = \frac{q}{n_1} \text{ and } \overrightarrow{OA} = \frac{q}{n_1}\mathbf{i}$$

Similarly $\overrightarrow{OB} = \frac{q}{n_2}\mathbf{j}$ and $\overrightarrow{OC} = \frac{q}{n_3}\mathbf{k}$

Let P (x, y, z) be the centre i.e., $\overrightarrow{OP} = x\mathbf{i} + y\mathbf{j} + z\mathbf{k}$

$\therefore \quad \overline{OP} = \overrightarrow{AP} = \overrightarrow{BP} = \overrightarrow{CP}$ = radius of sphere OABC

i.e., $|\mathbf{xi} + \mathbf{yj} + \mathbf{zk}| = \left|\left(x - \frac{q}{n_1}\right)\mathbf{i} + y\mathbf{j} + z\mathbf{k}\right|$

$$= \left|x\mathbf{i} + \left(y - \frac{q}{n_2}\right)\mathbf{j} + z\mathbf{k}\right| = \left|x\mathbf{i} + y\mathbf{j} + \left(z - \frac{q}{n_3}\right)\mathbf{k}\right|$$

From first two

$$x^2 + y^2 + z^2 = \left(x - \frac{q}{n_1}\right)^2 + y^2 + z^2$$

or $$\frac{q^2}{n_1^2} = 2x\frac{q}{n_1} \quad \text{or} \quad x = \frac{q}{2n_1}$$

Similarly $y = \dfrac{q}{2n_2}$ and $z = \dfrac{q}{2n_3}$

$$\therefore \quad \frac{a}{x} + \frac{b}{y} + \frac{c}{z} = \frac{2}{q}(an_1 + bn_2 + cn_3)$$

$$= \frac{2}{q}(q) = 2.$$

Example 9. Find the equation of the tangent plane to the sphere $\mathbf{r}^2 + 2\mathbf{r}.(\mathbf{i} - 3\mathbf{j}) + 1 = 0$ at $(1, 2, -2)$.

Solution. Let O be the origin and

$$\mathbf{r} = x\mathbf{i} + y\mathbf{j} + z\mathbf{k} \quad \mathbf{a} = \mathbf{i} + 2\mathbf{j} - 2\mathbf{k}$$

Given sphere is $r^2 + 2\mathbf{r}.(\mathbf{i} - 3\mathbf{j}) + 1 = 0$

Targent at **a** is

$$\mathbf{r}.\mathbf{a} + (\mathbf{r} + \mathbf{a}).(\mathbf{i} - 3\mathbf{j}) + 1 = 0$$

or $$\mathbf{r}.\mathbf{a} + \mathbf{r}.(\mathbf{i} - 3\mathbf{j}) + \mathbf{a}.(\mathbf{i} - 3\mathbf{j}) + 1 = 0$$

or $$\mathbf{r}.(\mathbf{a} + \mathbf{i} - 3\mathbf{j}) + \mathbf{a}.(\mathbf{i} - 3\mathbf{j}) + 1 = 0$$

or $$\mathbf{r}.(\mathbf{i} + 2\mathbf{j} - 2\mathbf{k} + \mathbf{i} - 3\mathbf{j}) + (\mathbf{i} + 2\mathbf{j} - 2\mathbf{k}).(\mathbf{i} - 3\mathbf{j}) + 1 = 0$$

or $$\mathbf{r}.(2\mathbf{i} - \mathbf{j} - 2\mathbf{k}) + (1 - 6) + 1 = 0$$

or $$\mathbf{r}.(2\mathbf{i} - \mathbf{j} - 2\mathbf{k}) - 4 = 0$$

which is required tangent plane in cartesian form as

$$2x - y - 2z - 4 = 0.$$

Example 10. Show that the two spheres

$\mathbf{r}^2 + 2\mathbf{r}.(3\mathbf{j} + \mathbf{k}) + 8 = 0$ and $\mathbf{r}^2 + 2\mathbf{r}.(3\mathbf{i} + 4\mathbf{j} + 2\mathbf{k}) + 20 = 0$

are orthogonal.

Solution. We know that if two spheres $\mathbf{r}^2 - 2\mathbf{r}.\mathbf{c_1} + k_1 = 0$ and $\mathbf{r}^2 - 2\mathbf{r}.\mathbf{c_2} + k_2 = 0$ intersects orthogonally then $2\mathbf{c_1}.\mathbf{c_2} = k_1 + k_2$.

The given spheres are $\mathbf{r}^2 + 2\mathbf{r}.(3\mathbf{j} + \mathbf{k}) + 8 = 0$

and $\qquad \mathbf{r}^2 + 2\mathbf{r}.(3\mathbf{i} + 4\mathbf{j} + 2\mathbf{k}) + 20 = 0$

$$\mathbf{c_1} = -3\mathbf{j} - \mathbf{k};\ \mathbf{c_2} = -3\mathbf{i} - 4\mathbf{j} - 2\mathbf{k},\ k_1 = 8;\ k_2 = 20$$

Now $\qquad 2\mathbf{c_1}.\mathbf{c_2} = 2(0 + 12 + 2) = 28$

$$= 8 + 20 = k_1 + k_2$$

Hence they are orthogonal.

Example 11. The sphere which cuts $F_1(\mathbf{r}) = 0$ and $F(\mathbf{r}) = 0$ orthogonally also cuts $F_1(\mathbf{r}) - \lambda\, F_2(\mathbf{r}) = 0$ orthogonally.

Solution. Let $\qquad F_1(\mathbf{r}) = \mathbf{r}^2 - 2\mathbf{r}.\mathbf{c_1} + k_1 = 0 \qquad$ ...(1)

$$F_2(\mathbf{r}) = \mathbf{r}^2 - 2\mathbf{r}.\mathbf{c_2} + k_2 = 0 \qquad \text{...(2)}$$

Then $F_1(\mathbf{r}) - \lambda F_2(\mathbf{r}) = 0$ gives

$$\mathbf{r}^2 - 2\mathbf{r}\,.\,\frac{\mathbf{c}_1 - \mathbf{c}_2}{1 - \lambda} + \frac{k_2 - \lambda k_2}{1 - \lambda} = 0 \qquad \text{...(3)}$$

Let the sphere $\mathbf{r}^2 - 2\mathbf{r}\,.\,\mathbf{c} + k = 0$ cut the spheres (1) and (2) orthogonally. T hen

$$2\mathbf{c}.\mathbf{c_1} = k + k_1 \qquad \text{...(4)}$$

and $\qquad 2\mathbf{c}.\mathbf{c_2} = k + k_2 \qquad$ ...(5)

Now (5) λ – (4) gives

$$2\mathbf{c}\,.\,(\mathbf{c_1} - \lambda\mathbf{c_2}) = k(1 - \lambda) + (k_1 - \lambda k_2)$$

or

$$2\mathbf{c}\,.\left[\frac{\mathbf{c_1} - \lambda\mathbf{c_2}}{1 - \lambda}\right] = k + \left[\frac{k_1 - \lambda k_2}{1 - \lambda}\right]$$

which implies that the sphere $\mathbf{r}^2 - 2\mathbf{r}.\mathbf{c} + k = 0$ cuts the sphere (3) orthogonally.

Example 12. The locus of a point which moves so that its distances from two fixed points are in a constant ratio n : 1 is a sphere. Prove that also that all such spheres, for different values of n have a common radical plane.

Solution. Consider the mid point of the line joining the two fixed points A and B (say) as origin and let the p.v's of A and B be **a** and − **a**. Let **r** be the p.v of a point P. Then

$$\vec{AP} = \mathbf{r} - \mathbf{a} \quad \text{and} \quad \vec{BP} = \mathbf{r} + \mathbf{a}$$

We are given that $\frac{AP}{BP} = \frac{n}{1}$ i.e. $\overline{AP}^2 = n^2\,\overline{BP}^2$

$$\therefore \quad (\mathbf{r} - \mathbf{a})^2 = n^2 (\mathbf{r} + \mathbf{a})^2$$

or $$\mathbf{r}^2 - 2\mathbf{r}.\mathbf{a} + \mathbf{a}^2 = n^2 (\mathbf{r}^2 + 2\mathbf{r}.\mathbf{a} + \mathbf{a}^2)$$

or $$\mathbf{r}^2(1 - n^2) - 2\mathbf{r}.\mathbf{a}(1 + n^2) + \mathbf{a}^2(1 - n^2) = 0$$

or $$\mathbf{r}^2 - 2\mathbf{r}.\mathbf{a}\left(\frac{1+n^2}{1-n^2}\right) + a^2 = 0 \qquad [\because \mathbf{a}^2 = a^2] \qquad ...(1)$$

which evidently represents a sphere.

By giving two values n_1 and n_2 to n we get two spheres as

$$\mathbf{r}^2 - 2\mathbf{r}.\mathbf{a}\left(\frac{1+n^2}{1-n_1^2}\right) + a^2 = 0 \qquad ...(2)$$

and $$\mathbf{r}^2 - 2\mathbf{r}.\mathbf{a}\left(\frac{1+n_2^2}{1-n_2^2}\right) + a^2 = 0 \qquad ...(3)$$

The radical plane of (2) and (3) is [by (3) - (2)]

$$2\mathbf{r}.\mathbf{a}\left[\frac{1+n_2^2}{1-n_2^2} - \frac{1+n_1^2}{1-n_1^2}\right] = 0$$

or **r.a** = 0 which is independent of n and is same for every pair of spheres obtained by giving different values of n in (1).

Example 13. Prove that any straight line drawn from a point O to intersect a sphere is cut harmonically by the surface and the polar plane of O.

Solution. Let the equation to the sphere be

$$\mathbf{r}^2 - 2\mathbf{r}.\mathbf{c} + k = 0 \qquad ...(1)$$

and the point O be taken as origin. Any line through O parallel to the unit vector b is r = tb where t stands for the distance of any point on it from O. If this line meets the sphere at P and Q then they are given by

$$t^2b^2 - 2t\,\mathbf{b.c} + k = 0$$

or $$t^2 - 2t\,\mathbf{b.c} + k = 0 \qquad [\because\ \mathbf{b}^2 = 1 \text{ unit vector}]$$

Let t_1, t_2 be the roots

$$\therefore \quad t_1 + t_2 = 2\mathbf{b.c} \quad \text{and} \quad t_1t_2 = k$$

Also $t_1 = OP$ and $t_2 = OQ$

$$\therefore \quad \frac{1}{OP} + \frac{1}{OQ} = \frac{1}{t_1} + \frac{1}{t_2} = \frac{t_1 + t_2}{t_1t_2} = \frac{2\mathbf{b.c}}{k} \qquad ...(2)$$

Now polar plane of the origin O w.r.t (1) is

$$\mathbf{r.0} - (\mathbf{r} + \mathbf{0})\,.\,\mathbf{c} + k = 0 \text{ or } \mathbf{r.c} = k \qquad ...(3)$$

Also if the line $\mathbf{r} = t\mathbf{b}$ meets the plane (3) in R then $t\mathbf{b.c} = k$ where $t = OR$.

$$\therefore \quad \mathbf{b.c} = \frac{k}{OR}$$

Substituting in (2) we get $\dfrac{1}{OP} + \dfrac{1}{OQ} = \dfrac{2}{OR}$

which shows that $\dfrac{1}{OP}, \dfrac{1}{OQ}$ and $\dfrac{1}{OR}$ are in A.P $\Rightarrow$ OP, OQ and OR are in H.P.

Example 14. Find the centre and radius of the circumsphere of the tetrahedron (O ABC) where the p.v's of A, B and C are a, b and c referred to O as origin.

Solution. Let D be the centre of the circumsphere whose p.v is **r** and radius be d units. Then

$$OD = AD = BD = CD = d$$

or $$d^2 = r^2 = (r - a)^2 = (r - b)^2 = (r - c)^2 \qquad ...(1)$$

or $$2\mathbf{r.a} = a^2, \quad 2\mathbf{r.b} = b^2 \quad \text{and} \quad 2\mathbf{r.c} = c^2 \qquad ...(2)$$

Also **r** can be expressed in terms of non-coplanar vectors **a, b, c** as

$$\mathbf{r} = l\mathbf{a} + m\mathbf{b} + n\mathbf{c} \qquad ...(3)$$

Multiplying (3), scalarly by **a, b** and **c** and using (2) we get

$$\mathbf{r.a} = \frac{1}{2}\mathbf{a}^2 = l\mathbf{a.a} + m\mathbf{a.b} + n\mathbf{a.c} \qquad ...(4)$$

$$\mathbf{r.b} = \frac{1}{2}\mathbf{b}^2 = \mathbf{lb.a + mb.b + nb.c} \qquad ...(5)$$

$$\mathbf{r.c} = \frac{1}{\mathbf{c}^2} = \mathbf{lc.a + mc.b + nc.c} \qquad ...(6)$$

Eliminating l, m, and n between (3), (4), (5) and (6) we get

$$\begin{vmatrix} \mathbf{a} & \mathbf{b} & \mathbf{c} & \mathbf{r} \\ \mathbf{a.a} & \mathbf{a.b} & \mathbf{a.c} & \frac{1}{2}\mathbf{a}^2 \\ \mathbf{b.a} & \mathbf{b.b} & \mathbf{b.c} & \frac{1}{1}\mathbf{b}^2 \\ \mathbf{c.a} & \mathbf{c.b} & \mathbf{c.c} & \frac{1}{2}\mathbf{c}^2 \end{vmatrix} = 0$$

which gives **r** the p.v of centre of the sphere

Again multiplying (3) scalarly by **r** we get

$$\mathbf{r.r} = \mathbf{lr.a + mr.b + nr.c}$$

or
$$d^2 = \mathrm{l}\frac{1}{2}\mathbf{a}^2 + \mathrm{m}\frac{1}{2}\mathbf{b}^2 + \mathrm{n}\frac{1}{2}\mathbf{c}^2 \quad \text{[from (1) and (2)]} \qquad ...(7)$$

Eliminating l, m, and n between (7), (4), (5) and (6) we get

$$\begin{vmatrix} \frac{1}{2}\mathbf{a}^2 & \frac{1}{2}\mathbf{b}^2 & \frac{1}{2}\mathbf{c}^2 & \mathbf{d}^2 \\ \mathbf{a.a} & \mathbf{a.b} & \mathbf{a.c} & \frac{1}{2}\mathbf{a}^2 \\ \mathbf{b.a} & \mathbf{b.b} & \mathbf{b.c} & \frac{1}{2}\mathbf{b}^2 \\ \mathbf{c.a} & \mathbf{c.b} & \mathbf{c.c} & \frac{1}{2}\mathbf{c}^2 \end{vmatrix} = 0$$

which gives the radius d of the sphere

EXERCISE

1. Find the equation of the sphere whose centre is (2, 5, 3) and radius is 3.

2. Find the p.v of the centre and radius of the sphere $\mathbf{r}^2 = \mathbf{r}.(-6\mathbf{i} + 8\mathbf{j} - 10\mathbf{k}) + 1 = 0$.

3. A point moves so that the sum of the squares of its distances from the six faces of a cube is constant. Show that its locus is a sphere.
4. A is a point (1, 3, 4) and B the point (1, - 2, - 1). A point P moves so that 3PA = 2PB. Prove by vector methods that the locus of P is the sphere.
5. If O is the centre of a sphere of radius unity and A, B are two points in a line with O such that OA.OB = 1. If P be any variable point on the sphere show that PA : PB = constant.
6. Show that any diameter of a sphere subtends a right angle at a point on the sphere using vector method.
7. Using vector method find the equation of the sphere joining (1, 0, 1) and (5, 4, 5) as diameter
8. A plane through a fixed point (1, 1, 1) cuts the axes in A, B, and C. Show that the locus of the centre of the sphere OABC where O is origin is

$$\frac{1}{x} + \frac{1}{y} + \frac{1}{z} = 2.$$

9. A sphere of constant radius 2R passes through the origin and meets the axes in A, B and C. Prove that the locus of the centroid of the tetrahedron OABC is the sphere $\mathbf{r}^2 = R^2$ where $\mathbf{r} = x\mathbf{i} + y\mathbf{j} + z\mathbf{k}$.
10. If the mid point of six edges of a tetrahedron lie on a sphere of radius r then prove that centre of the sphere is centroid of the tetrahedron.
11. Show that the radius of the circular section of the sphere $|\mathbf{r}| = 5$ by the plane $\mathbf{r}.(\mathbf{i} + \mathbf{j} + \mathbf{k}) = 3\sqrt{3}$ is 4 units.
12. Find the equations of the sphere through the circle $|\mathbf{r}| = 3$; $\mathbf{r}.(2\mathbf{i} + 3\mathbf{j} + 4\mathbf{k}) = 5$ at the point (1, 2, 3).
13. Find the position vector of the centre and radius of the circle given by $\mathbf{r}.(\mathbf{i} + 2\mathbf{j} + 3\mathbf{k}) = 15$ and $\mathbf{r}^2 = 2\mathbf{r}.(\mathbf{j} + 2\mathbf{k}) - 11 = 0$.
14. Prove that the sum of squares of intercepts made by a given sphere on any three mutually perpendicular lines through a fixed point is constant.

15. Find the equation of the tangent plane to the sphere $3r^2 - \mathbf{r}.(2\mathbf{i} + 3\mathbf{j} + 4\mathbf{k}) - 22 = 0$ at (1, 2, 3).

16. Show that the plane $\mathbf{r}.(2\mathbf{i} + \mathbf{j} - \mathbf{k}) = 12$ touches the sphere $|\mathbf{r}| = 2\sqrt{6}$ and the position vector of point of contact.

17. Obtain the condition that the sphere

$$a\mathbf{r}^2 + 2\mathbf{r}.(l\mathbf{i} + m\mathbf{j} + n\mathbf{k}) + p = 0 \text{ and } b\mathbf{r}^2 = \mathbf{k}^2$$

may cut orthogonally.

18. Show that the locus of the straight lines which intersects the sphere $\mathbf{r}^2 - 2\mathbf{r}.\mathbf{c} + k = 0$ and are bisected at a given point $\mathbf{d}$ is $(\mathbf{r} - \mathbf{d}) . (\mathbf{d} - \mathbf{c}) = 0$.

19. If the polar plane of the point H passes through G then show that the polar plane of G passes through H.

20. If the line joining the centre C of the sphere to any point P meets the polar plane of P in Q prove that CP.CQ = a^2 where a is the radius of the sphere

21. Prove that the every sphere that passes through the limiting points of a co-axial system cuts every sphere of the system orthogonally

ANSWERS

(1) $|\mathbf{r} - (2\mathbf{i} + 5\mathbf{j} + 3\mathbf{k})| = 3$

(2) $3\mathbf{i} - 4\mathbf{j} + 5\mathbf{k}, 7$

(7) $\mathbf{r}^2 - 2\mathbf{r}.(3\mathbf{i} + 2\mathbf{j} + 3\mathbf{k}) + 10 = 0$

(12) $3r^2 - \mathbf{r}.(2\mathbf{i} + 3\mathbf{j} + 4\mathbf{k}) - 22 = 0$

(13) $\mathbf{i} + 3\mathbf{j} + 4\mathbf{k}, r = \sqrt{7}$

(15) $\mathbf{r}.(4\mathbf{i} + 9\mathbf{j} + 14\mathbf{k}) = 64$

(16) $4\mathbf{i} + 2\mathbf{j} - 2\mathbf{k}$

(17) $ak^2 = bp$.

9

Application to Mechanics (Statics and Dynamics)

9.1. Force and its representation.

Force is defined as a cause which tends to change or changes the state of rest or of uniform motion of a body. The action of force is characterised by (i) *magnitude* (ii) *Direction* and (iii) *its point of application* . Hence force is a *localised vector*.

A force is represented by a directed line segment with an arrow head as shown in figure.

Let $\vec{OA}$ be the force vector. Here O is the point of action, and the arrow denotes the direction of force with magnitude $|\vec{OA}|$. We can also denote bold face-type capital letters to denote forces like **F**, **P**, ... etc.

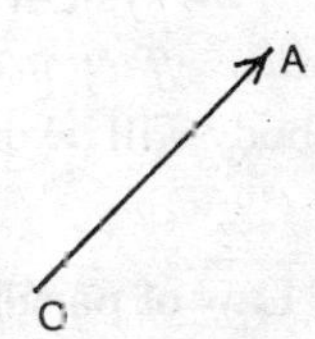

Fig. 9.1

Note. (1) In the previous chapters, we have studies about free vectors only. i.e., origin is immaterial. Here the origin is important because it is the point of action.

(2) The line along which the force acts is defined as the line of action of the force.

9.2. Concurrent Forces

Forces whose line of action are concurrent i.e., when all forces act on a single point are called *concurrent forces*.

9.3. Resultant force

If a number of forces $\mathbf{F_1}, \mathbf{F_2}, \mathbf{F_3}, \ldots, \mathbf{F_n}$ act on a body, if there exists a single force say **R** which has the same effect on the body as all the above forces $\mathbf{F_1}, \mathbf{F_2}, \mathbf{F_3}, \ldots, \mathbf{F_n}$ taken together, then R is called the resultant of the forces $\mathbf{F_1}, \mathbf{F_2}, \mathbf{F_3}, \ldots, \mathbf{F_n}$. This single force R can be determined by *Vector law of addition.*

$$\therefore \quad \mathbf{R} = \mathbf{F_1} + \mathbf{F_2} + \mathbf{F_3} + \ldots + \mathbf{F_n} = \Sigma\mathbf{F}$$

Here $\mathbf{F_1}, \mathbf{F_2}, \mathbf{F_3}, \ldots, \mathbf{F_n}$ are also called as components of **R** and the addition is not the mere sum of the magnitudes of $\mathbf{F_1}, \mathbf{F_2}, \mathbf{F_3}, \ldots, \mathbf{F_n}$.

9.4. Equilibrium

If a number of forces acting on a body keep the body at rest, then they are said to be in equilibrium. i.e., if $\mathbf{R} = 0$ then $\mathbf{F_1}, \mathbf{F_2}, \mathbf{F_3}, \ldots, \mathbf{F_n}$ are said to be in equilibrium.

9.5. Equivalence of two system of forces

Two system of forces are said to be equivalent *if they produce the same effect on a rigid body when applied separately* i.e., the state of a body will be unaltered if one such system is replaced by the other one.

9.6. Law of parallelogram of forces

If two forces, acting at a point be represented in magnitude and direction by two adjacent sides of a parallelogram drawn through their point of application, the resultant will be represented by the diagonal of the parallelogram through that point.

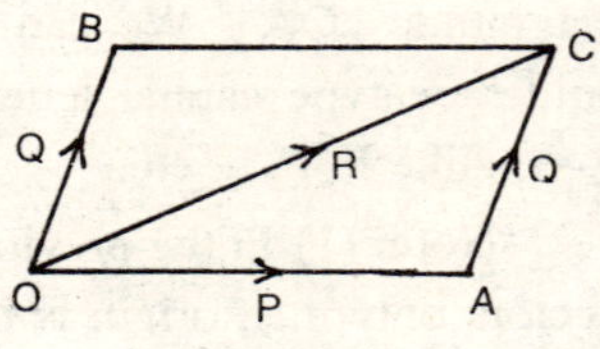

Fig. 9.2

Let P and Q be two forces acting at O be represented in magnitude and direction by the sides OA and OB of the parallelogram OACB. Then the resultant R of the forces P and Q will be represented by the diagonal OC. Using vector addition

$$\overrightarrow{OA} + \overrightarrow{OB} = \overrightarrow{OA} + \overrightarrow{AB} = \overrightarrow{OC}$$

$\therefore \qquad \mathbf{P} + \mathbf{Q} = \mathbf{R}$

Note. *The proof of above theorem being experimental is not given here.*

Magnitude and direction of the resultant of the forces P and Q acting at a given angle α:

We know that

$$\vec{OA} = \mathbf{P},\ \vec{OB} = \mathbf{Q} \text{ and } \vec{OC} = \mathbf{R} = \mathbf{P} + \mathbf{Q}$$

$$\therefore \qquad \mathbf{R}.\mathbf{R} = (\mathbf{P} + \mathbf{Q}).(\mathbf{P} + \mathbf{Q})$$

$$\mathbf{R}^2 = \mathbf{P}^2 + 2\mathbf{P}.\mathbf{Q} + \mathbf{Q}^2$$

$$R^2 = P^2 + Q^2 + 2PQ \cos \alpha$$

$$[\because \mathbf{R}^2 = R^2, \mathbf{P}^2 = P^2, \mathbf{Q}^2 = Q^2]$$

which gives the magnitude of **R** of forces **P** and **Q**.

Let **R** makes angle θ with **P** and CL be the perpendicular from C on to OA produced

Then $\tan \theta = \dfrac{CL}{AL} = \dfrac{AC \sin \alpha}{OA + AL}$

$$= \frac{Q \sin \alpha}{P + Q \cos \alpha} \qquad \text{[Ref. fig 9.2]}$$

or $\theta = \tan^{-1}\left\{\dfrac{Q \sin \alpha}{P + Q \cos \alpha}\right\}$

which gives the direction of the resultant **R**.

9.7. Triangle law of forces

If three forces acting at a point be represented in magnitude and direction by the sides of a triangle, taken in order, they are in equilibrium.

Let the forces be **P, Q** and **R** acting at a point O be represented in magnitude and direction by the sides AB, BC and CA respectively of triangle ABC as shown in the following figure 9.3

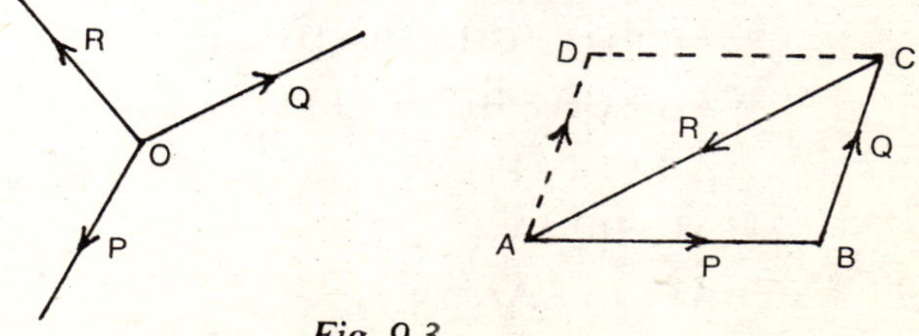

Fig. 9.3

Complete the parallelogram ABCD. Now AD = BC and also AD is parallel to BC $\therefore\ \vec{AD} = \vec{BC} = \mathbf{Q}$

Now we have by the parallelogram law of forces

$$\vec{AB} + \vec{AD} = \vec{AC} \Rightarrow \vec{AB} + \vec{BC} = \vec{AC} \qquad ...(1)$$

or $$\vec{AB} + \vec{BC} = -\vec{CA} \text{ or } \vec{AB} + \vec{BC} + \vec{CA} = 0$$

i.e., $\mathbf{P} + \mathbf{Q} + \mathbf{R} = 0 \Rightarrow$ the forces are in equilibrium.

Remark. From (1) $\vec{AB} + \vec{BC} = \vec{AC}$ so called *triangular law of addition.* Keeping it in mind the triangular law of forces can also be stated as

If two forces acting at a point be represented in magnitude and direction by the two adjacent sides of a triangle then their resultant is given in magnitude by the third side of the triangle taken in order

9.8. Polygon law of forces.

If any number of forces acting at a point be represented in magnitude and direction by the sides of a closed polygon, taken in order, they are in equilibrium.

Let $\mathbf{F}_1, \mathbf{F}_2, \mathbf{F}_3, \mathbf{F}_4, \mathbf{F}_5, \mathbf{F}_6, \mathbf{F}_7$ and $\mathbf{F}_8$ be the forces acting at O be represented in magnitude and direction by the sides AB, BC, CD, DE, EG, GH, HI and IJ respectively of a closed polygon ABCDEGHIJ.

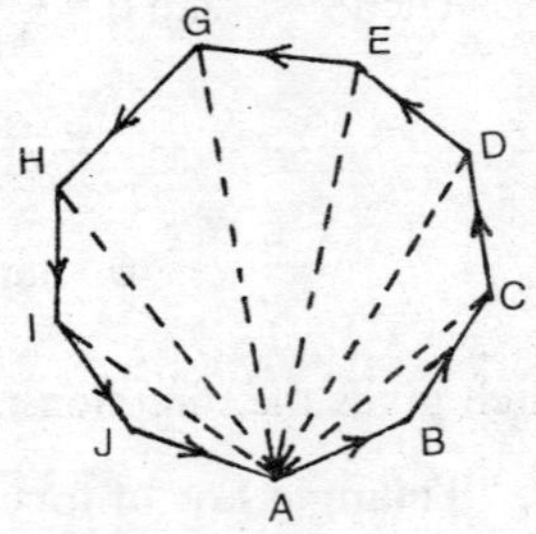

Fig. 9.4

Join AC, AD, AE, AG, AH and AI as shown in the figure.

Now consider

$$\vec{AB} + \vec{BC} + \vec{CD} + \vec{DE} + \vec{EG} + \vec{GH} + \vec{HI} + \vec{IJ} + \vec{JA}$$
$$= \vec{AC} + \vec{CD} + \vec{DE} + \vec{EG} + \vec{GH} + \vec{HI} + \vec{IJ} + \vec{JA}$$
$$= \vec{AD} + \vec{DE} + \vec{EG} + \vec{GH} + \vec{HI} + \vec{IJ} + \vec{JA}$$
$$= \vec{AE} + \vec{EG} + \vec{GH} + \vec{HI} + \vec{IJ} + \vec{JA}$$
$$= \vec{AH} + \vec{GH} + \vec{HI} + \vec{IJ} + \vec{JA}$$
$$= \vec{AH} + \vec{HI} + \vec{IJ} + \vec{JA}$$
$$= \vec{AI} + \vec{IJ} + \vec{JA}$$
$$= \vec{AJ} + \vec{JA} = \vec{AA} = 0$$

i.e. $\mathbf{F_1 + F_2 + F_3 + F_4 + F_5 + F_6 + F_7 + F_8 = 0}$

$\Rightarrow$ The given forces are in equilibrium

Note. In above the reduction at every stage is due to the triangle law of forces as explained is **art 9.6.**

Remark. From above

$$\overrightarrow{AB} + \overrightarrow{BC} + \overrightarrow{CD} + \overrightarrow{DE} + \overrightarrow{EG} + \overrightarrow{GH} + \overrightarrow{HI} + \overrightarrow{IJ} = -\overrightarrow{JA} + \overrightarrow{AJ}$$

which is usual polygon law of vector addition.

Theorem 1. *The resultant of the forces* $\lambda.\overrightarrow{OA}$ *and* $\mu.\overrightarrow{OB}$ *acting at a point* O *is given by* $(\lambda + \mu)\overrightarrow{OC}$ *where* C *is a point in* AB *such that* $\lambda AC = \mu\, CB$.

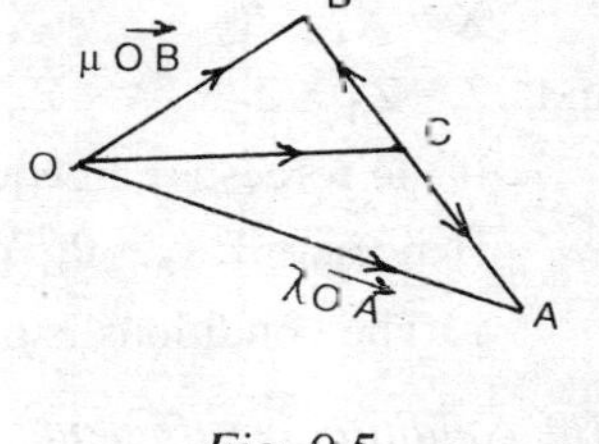

Fig. 9.5

Proof. We know that

$\overrightarrow{OA} = \overrightarrow{OC} + \overrightarrow{CA}$ and

$\overrightarrow{OB} = \overrightarrow{OC} + \overrightarrow{CB}$

$$\therefore \quad \lambda\overrightarrow{OA} + \mu\overrightarrow{OB} = \lambda(\overrightarrow{OC} + \overrightarrow{CA}) + \mu(\overrightarrow{OC} + \overrightarrow{CB})$$

$$= (\lambda + \mu)\,\overrightarrow{OC} + \lambda\overrightarrow{CA} + \mu\overrightarrow{CB}$$

$$= (\lambda + \mu)\,\overrightarrow{OC} + \lambda\overrightarrow{CA} - \mu\overrightarrow{BC} \qquad \text{...(1)}$$

But given that $\lambda AC = \mu CB \Rightarrow \lambda\overrightarrow{AC} = \mu\overrightarrow{CB}$

Using this in (1) we get

$$\lambda\overrightarrow{OA} + \mu\overrightarrow{OB} = (\lambda + \mu)\,\overrightarrow{OC}$$

which proves the theorem. It is also called λ-μ theorem.

Note. If $\lambda = \mu$ then $\overrightarrow{OA} + \overrightarrow{OB} = 2\overrightarrow{OC}$ i.e., C is the midpoint of AB $\Rightarrow$ the resultant of the forces $\overrightarrow{OA}$ and $\overrightarrow{OB}$ acting at a point O is given by $2\overrightarrow{OC}$, where C is the mid point of AB.

Theorem 2. *The necessary and sufficient conditions that a system of forces acting at apoint be in equilibrium is that the algebraic sums of the components of the forces in any three non–coplanar directions (not necessarily mutually perpendicular directions) vanish separately.*

Proof. *The condition is necessary*

Let there be a number of forces $\mathbf{F_1}, \mathbf{F_2}, ..., \mathbf{F_n}$ acting at O which are in equilibrium. Let **a**, **b**, **c** be unit vectors along any three non-coplanar directions

Let X_1, Y_1, Z_1; $X_2, Y_2, Z_2, \ldots, X_n, Y_n, Z_n$ be the components of $\mathbf{F_1}, \mathbf{F_2}, \ldots, \mathbf{F_n}$ along these directions. Let **R** be their resultant whose components are X, Y, Z.

Then $\quad \mathbf{R} = \mathbf{F_1} + \mathbf{F_2} + \ldots + \mathbf{F_n}$

i.e., $X\mathbf{a} + Y\mathbf{b} + Z\mathbf{c} = (X_1\mathbf{a} + Y_1\mathbf{b} + Z_1\mathbf{c}) + (X_2\mathbf{a} + Y_2\mathbf{b} + Z_2\mathbf{c}) + \ldots + (X_n\mathbf{a} + Y_n\mathbf{b} + Z_n\mathbf{c})$

Equating the co-efficients of like vectors on both sides, we get

$X = X_1 + X_2 + \ldots + X_n = \Sigma X_n$; $Y = Y_1 + Y_2 + \ldots + Y_n = \Sigma Y_n$

and $Z = Z_1 + Z_2 + \ldots + Z_n = \Sigma Z_n$.

If the forces are in equilibrium then the resultant $\mathbf{R} = \mathbf{0}$

Hence $x = \Sigma X_n = 0$; $Y = \Sigma Y_n = 0$; $Z = \Sigma Z_n = 0$

$\therefore$ The conditions is necessary

The condition is sufficient

Again $\Sigma X_n = 0$, $\Sigma Y_n = 0$, $\Sigma Z_n = 0$

then $\quad X = 0$, $Y = 0$ and $Z = 0 \quad \therefore R = 0$

$\therefore$ The forces are in equilibrium.

Hence the condition is also sufficient.

9.10. Lamis theorem.

If three concurrent forces are in equilibrium, then each force is proportional to the sine of the angle between the other two.

Proof: Let $\mathbf{F_1}, \mathbf{F_2}, \mathbf{F_3}$ be three forces acting at a point O be in equilibrium. Let $\mathbf{e_1}, \mathbf{e_2}, \mathbf{e_3}$ be unit vectors in the direction of the forces whose magnitudes are F_1, F_2, F_3 respectively so that $\mathbf{F_1} = F_1\mathbf{e_1}$; $\mathbf{F_2} = F_2\mathbf{e_2}$ and $\mathbf{F_3} = F_3\mathbf{e_3}$. Since the forces are in equilibrium their vector sum is zero

$$\mathbf{F_1} + \mathbf{F_2} + \mathbf{F_3} = \mathbf{0}$$

i.e.,
$$F_1\mathbf{e_1} + F_2\mathbf{e_2} + F_3\mathbf{e_3} = \mathbf{0} \qquad \ldots(1)$$

Multiplying (1) vectorically by $\mathbf{e_1}$, we get

$$F_1(\mathbf{e_1} \times \mathbf{e_1}) + F_2(\mathbf{e_2} \times \mathbf{e_1}) + F_3(\mathbf{e_3} \times \mathbf{e_1}) = \mathbf{0}$$

or $$F_2(\mathbf{e}_1 \times \mathbf{e}_2) = F_3 (\mathbf{e}_3 \times \mathbf{e}_1)$$

or $$\frac{F_2}{\mathbf{e}_3 \times \mathbf{e}_1} = \frac{F_3}{\mathbf{e}_1 \times \mathbf{e}_2} \qquad ...(2)$$

Again multiplying (1) vectorially by $\mathbf{e}_2$, we get

$$F_1(\mathbf{e}_1 \times \mathbf{e}_2) + F_2(\mathbf{e}_2 \times \mathbf{e}_2) + F_3(\mathbf{e}_3 \times \mathbf{e}_2) = \mathbf{0}$$

or $$F_1(\mathbf{e}_1 \times \mathbf{e}_2) = F_3(\mathbf{e}_2 \times \mathbf{e}_3)$$

or $$\frac{F_1}{\mathbf{e}_2 \times \mathbf{e}_3} = \frac{F_3}{\mathbf{e}_1 \times \mathbf{e}_2} \qquad ...(3)$$

From (2) and (3), we get

$$\frac{F_1}{\mathbf{e}_2 \times \mathbf{e}_3} = \frac{F_2}{\mathbf{e}_3 \times \mathbf{e}_1} = \frac{F_3}{\mathbf{e}_1 \times \mathbf{e}_2} \qquad ...(4)$$

Let θ_1, θ_2 and θ_3 be the angles between F_1 and F_2 ; F_3 and F_1; F_1 and F_2 i.e.,$\mathbf{e}_2$ and $\mathbf{e}_3$; $\mathbf{e}_3$ and $\mathbf{e}_1$; $\mathbf{e}_1$ and $\mathbf{e}_2$ respectively.

We know that $|\mathbf{e}_1| = |\mathbf{e}_2| = |\mathbf{e}_3| = 1$

$\therefore$ $\mathbf{e}_2 \times \mathbf{e}_3 = \sin\theta_1$; $\mathbf{e}_3 \times \mathbf{e}_1 = \sin\theta_2$; $\mathbf{e}_1 \times \mathbf{e}_2 = \sin\theta_3$

Substituting in (4), we get

$$\frac{F_1}{\sin\theta_1} = \frac{F_2}{\sin\theta_2} = \frac{F_3}{\sin\theta_3}$$

i.e. each of the forces F_1, F_2, F_3 is proportional to the sine of the angles between the other two.

9.11. Four forces in equilibrium.

If F_1, F_2, F_3 *and* F_4 *are the magnitudes of four forces in equilibrium and* $\mathbf{e}_1$, $\mathbf{e}_2$, $\mathbf{e}_3$ *and* $\mathbf{e}_4$ *are the unit vectors in the directions of these forces respectively then each force is proportional to the scalar triple product of unit vectors in the direction of the other three.*

i.e., $$\frac{F_1}{[\mathbf{e}_2\mathbf{e}_3\mathbf{e}_4]} = \frac{F_2}{[\mathbf{e}_3\mathbf{e}_1\mathbf{e}_4]} + \frac{F_3}{[\mathbf{e}_1\mathbf{e}_2\mathbf{e}_4]} = \frac{-F_4}{[\mathbf{e}_1\mathbf{e}_2\mathbf{e}_3]}$$

Proof: The forces $F_1\mathbf{e_1}$, $F_2\mathbf{e_2}$, $F_3\mathbf{e_3}$ and $F_4\mathbf{e_4}$ are in equilibrium

$$\therefore \quad F_1\mathbf{e}_1 + F_2\mathbf{e}_2 + F_3\mathbf{e}_3 + F_4\mathbf{e}_4 = 0 \qquad ...(1)$$

Multiplying (1) scalarly by $\mathbf{e_3} \times \mathbf{e_4}$, we get

$$F_1 [\mathbf{e}_1\mathbf{e}_3\mathbf{e}_4] + F_2 [\mathbf{e}_2\mathbf{e}_3\mathbf{e}_4] + F_3[\mathbf{e}_3\mathbf{e}_3\mathbf{e}_4] + F_4[\mathbf{e}_4\mathbf{e}_3\mathbf{e}_4] = 0$$

or $$F_1[\mathbf{e}_1\mathbf{e}_3\mathbf{e}_4] + F_2[\mathbf{e}_2\mathbf{e}_3\mathbf{e}_4] = 0 \qquad [\because [\mathbf{e}_3\mathbf{e}_3\mathbf{e}_4] = [\mathbf{e}_4\mathbf{e}_3\mathbf{e}_4] = 0]$$

or $$F_1[\mathbf{e}_3\mathbf{e}_1\mathbf{e}_4] = F_2[\mathbf{e}_2\mathbf{e}_3\mathbf{e}_4]$$

or $$\frac{F_1}{[\mathbf{e}_2\mathbf{e}_3\mathbf{e}_4]} = \frac{F_2}{[\mathbf{e}_3\mathbf{e}_1\mathbf{e}_4]} \qquad ...(2)$$

Similarly multiplying (1) scalarly by $(\mathbf{e_1} \times \mathbf{e_3})$ and $(\mathbf{e_1} \times \mathbf{e_2})$, we get

$$\frac{F_2}{[\mathbf{e}_3\mathbf{e}_1\mathbf{e}_4]} = \frac{-F_4}{[\mathbf{e}_1\mathbf{e}_2\mathbf{e}_3]} \qquad ...(3)$$

and $$\frac{F_3}{[\mathbf{e}_1\mathbf{e}_2\mathbf{e}_4]} = \frac{-F_4}{[\mathbf{e}_1\mathbf{e}_2\mathbf{e}_3]} \qquad ...(4)$$

Hence from (2), (3) and (4) we can write

$$\frac{F_1}{[\mathbf{e}_2\mathbf{e}_3\mathbf{e}_4]} = \frac{F_2}{[\mathbf{e}_3\mathbf{e}_1\mathbf{e}_4]} = \frac{F_3}{[\mathbf{e}_1\mathbf{e}_2\mathbf{e}_4]} = \frac{-F_4}{[\mathbf{e}_1\mathbf{e}_2\mathbf{e}_3]}$$

9.12. Vector moment of a force about a point

The vector moment or torque **M** of a force **F** about any given point O is the measure of its turning effect about O and in magnitude equal to the product of force and the perpendicular distance of its line of action from that point.

Let **r** be the position vector relative to O of any point P on the line of action of the force F. Then the moment M of the force **F** about O is

$$\mathbf{M} = \mathbf{r} \times \mathbf{F}$$

$$\therefore \ |\mathbf{M}| = |\mathbf{r} \times \mathbf{F}| = \mathbf{d}.\overline{OP} \sin \theta$$

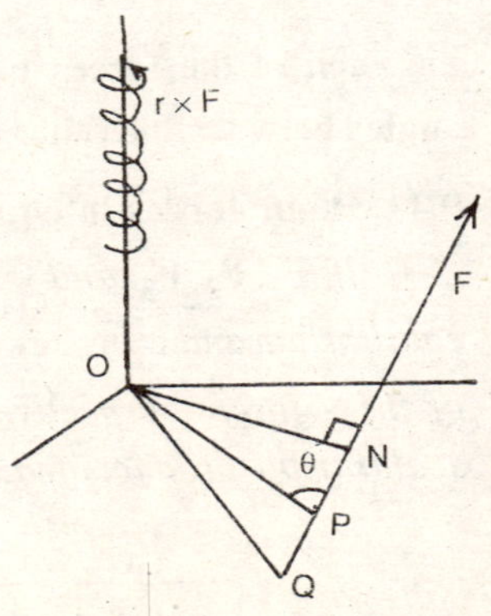

Fig. 9.6

$= F\, r \sin\theta;\ r = |\mathbf{r}|;\ F = |\mathbf{F}|$

where θ is the angle between OP and the force **F**.

But $r \sin\theta = \overline{OP} \sin\theta = ON$, the length of perpendicular from O on the line of action of force **F**

$\therefore \qquad |\mathbf{M}| = ON.F$

The direction of vector moment is to be regarded as positive relative to a rotation from **r** to **F**. [Thus in the figure the direction of the vector moment is along the normal drawn through O on the plane of paper and pointing towards the reader]

The moment of a force **F** about a point O is said to be positive or negative according as the rotation of the force about the point is anticlockwise or clockwise.

Theorem 1. *The moment vector about O is independent of the choice of the point P on the line of action of the force.*

Proof. Let Q be any other point on the line of action of the force F. Then we have

$$\vec{OQ} \times \mathbf{F} = (\vec{OP} + \vec{PQ}) \times \mathbf{F} \qquad [\because \vec{OQ} = \vec{OP} + \vec{PQ}]$$

$$= \vec{OP} \times \mathbf{F} + \vec{PQ} \times \mathbf{F} \qquad \text{...(1)}$$

Now P and Q are points on the line of action of the force F. Therefore $\vec{PQ}$ and **F** are parallel

i.e., $\vec{PQ} \times \mathbf{F} = 0$ Substituting in (1), we get

$\vec{OQ} \times \mathbf{F} = \vec{OP} \times \mathbf{F}$. Hence the proof.

Theorem 2. The algebraic sum of the moments of a system of forces about any point is equal to the moment of their resultant about the same point.

Proof. Let $\vec{OP} = \mathbf{r}$. Then the algebraic sum of moments of a system of forces say $\mathbf{F}_1, \mathbf{F}_2, \ldots, \mathbf{F}_n$ about O is

$$= \mathbf{r} \times \mathbf{F}_1 + \mathbf{r} \times \mathbf{F}_2 + \ldots + \mathbf{r} \times \mathbf{F}_n$$

$$= \mathbf{r} \times (\mathbf{F}_1 + \mathbf{F}_2 + \ldots + \mathbf{F}_n)$$

$$= \mathbf{r} \times \mathbf{F}$$

(where **F** is the resultant of the forces $\mathbf{F}_1, \mathbf{F}_2, \ldots, \mathbf{F}_n$)

= vector moment about O of the resultant **F**.

Hence the proof.

9.13. Moment about a line

The moment **M** of a force **F** acting at a point P about a line L is given by $\mathbf{q} = (\mathbf{r} \times \mathbf{F}) \cdot \hat{\mathbf{e}}$, which is a scalar. Here $\hat{\mathbf{e}}$ is a unit vector in the direction of the line and $\overrightarrow{OP} = \mathbf{r}$, O being any point on the line.

⇒ Moment of a force **F** about any straight line in $\mathbf{M}.\hat{\mathbf{e}}$ represents the resolved part of the vector moment along the given line.

9.14. Work done by a force

A force acting on a particle is said to do work if the particle is displaced in a direction which is not perpendicular to the force. Work done by a force is a scalar quantity and is defined as *the product of the force and the displacement of its point of application in its direction.*

Let there is a force **F** acting at O in the direction of OA and let it displaces the point of application from O to B. If W be the work done then

$$W = \mathbf{F}.\overrightarrow{\mathbf{OB}} = F\, OB \cos\theta,$$

where θ is the angle between the direction of the force and the direction of the displacement.

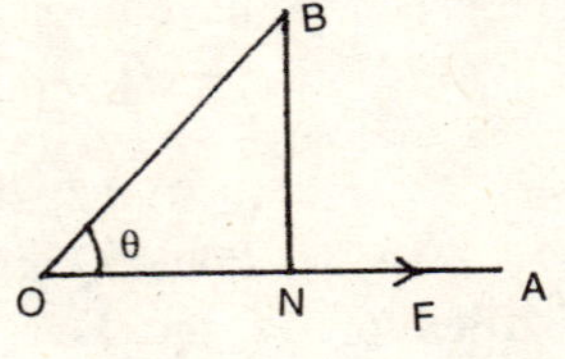

Fig. 9.7

Again $F \cos\theta$ is the component of the force **F** in the direction of displacement

$$\therefore \quad W = (F \cos\theta)\, OB$$

= (component of the force in the direction of displacement) (distance moved OB)

Note: If **F** is perpendicular to $\overline{OB}$ then $W = 0$.

Theorem. The total work done by a number of forces acting on a particle, is equal to the work done by the resultant of all these forces.

Proof: Let $\mathbf{F}_1, \mathbf{F}_2, \ldots, \mathbf{F}_n$ be the n forces acting on a particle and let **d** be the displacement vector then the total work done by these forces

$$= \mathbf{F_1.d} + \mathbf{F_2.d} + \dots + \mathbf{F_n.d}$$

$$= (\mathbf{F_1} + \mathbf{F_2} + \dots + \mathbf{F_n}) . \mathbf{d}$$

$$= \mathbf{R.d} \text{ where } \mathbf{R} \text{ is resultant of } \mathbf{F_1}, \mathbf{F_2}, \dots, \mathbf{F_n}.$$

Note. Let $\mathbf{d_1}, \mathbf{d_2}, \dots, \mathbf{d_n}$ be n displacements caused by the action of a force $\mathbf{F}$ on a particle then the work done is

$$W = \mathbf{F}.(\mathbf{d_1} + \mathbf{d_2} + \dots + \mathbf{d_n})$$

SOLVED PROBLEMS

Example 1. If the position vectors of n points represent n concurrent forces, show that the forces will be in equilibrium if the centroid of the points coincides with the origin.

Solution. Let the p.v's of n points be **a, b, c,** ... w.r.t an origin O and let G be their centroid. Then by definition of centroid

$$\overrightarrow{OG} = \frac{\mathbf{a} + \mathbf{b} + \mathbf{c} + \dots}{n} \quad \dots(1)$$

But the centroid G coincides with origin O.

$\therefore \quad \overrightarrow{OG} = \overrightarrow{OO} = \mathbf{0}$. Hence from (1)

$$\frac{\mathbf{a} + \mathbf{b} + \mathbf{c} + \dots}{n} = \mathbf{0} \Rightarrow \mathbf{a} + \mathbf{b} + \mathbf{c} + \dots = \mathbf{0}$$

i.e., the n concurrent forces are in equilibrium

Example 2. Two forces at the corner A, of a quadrilateral ABCD are represented by $\overrightarrow{AB}$ and $\overrightarrow{AD}$ and two at C represents by $\overrightarrow{CB}$ and $\overrightarrow{CD}$. Show that their resultant is represented by $4\overrightarrow{PQ}$ where P, Q are the middle points of AC, BD respectively.

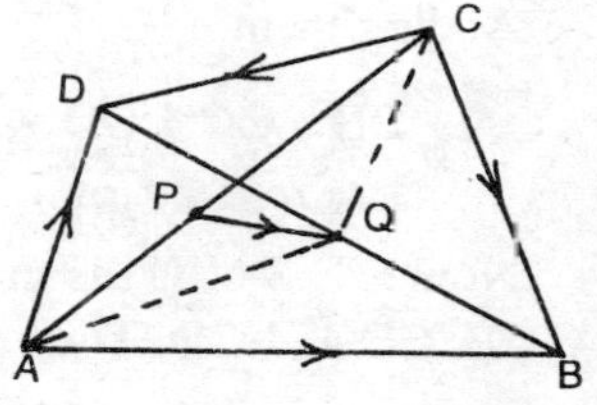

Fig. 9.8

Solution. Since Q is the mid point of BD

$\therefore \quad 2\overrightarrow{AQ} = \overrightarrow{AB} + \overrightarrow{AD}$

Also $2\overrightarrow{CQ} = \overrightarrow{CB} + \overrightarrow{CD}$

Now the forces $2\overrightarrow{AQ}$ and $3\overrightarrow{CQ}$ are concurrent forces

$\therefore$ The resultant is

$$2\vec{AQ} + 2\vec{CQ} = -2(\vec{QA} + \vec{QC})$$

$$= -2(2\vec{QP}) = -2(-2\vec{PQ}) = 4\vec{PQ}.$$

[$\because$ P is the mid point of AC]

Example 3. Three forces P, 2P, 3P act along the sides AB, BC, CA of an equilateral triangle ABC. Find their resultant.

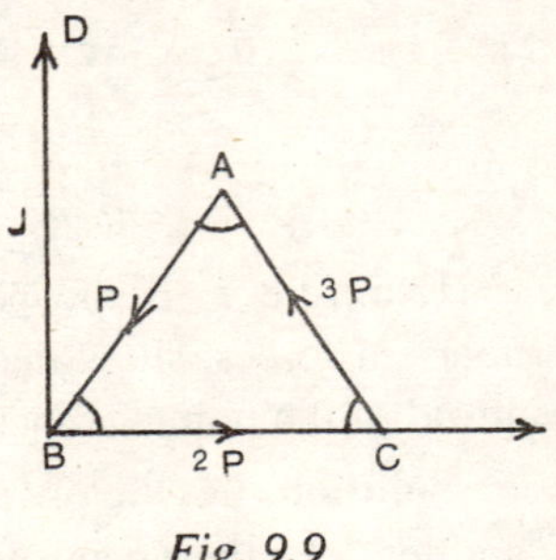

Fig. 9.9

Solution. Consider the origin as B. Let **i**, **j** be the unit vectors along BC and BD. Let R be the resultant

$$\therefore \quad \mathbf{R} = 2P\mathbf{i} + (-3P\cos 60°\mathbf{i} + 3P\sin 60°\mathbf{j}) + (-P\cos 60°\mathbf{i} - P\sin 60°\mathbf{j})$$

$$= \sqrt{3}\,P\mathbf{j}$$

and magnitude = $\sqrt{3}$P whose direction is perpendicular to BC.

Example 4. ABCD is a quadrilateral. Find by the vector method the position of a point O inside the quadrilateral such that the forces represented by $\vec{OA}$, $\vec{OB}$, $\vec{OC}$ and $\vec{OD}$ may be in equilibrium.

Solution. ABCD is a quadrilateral and let O be a point inside of it. Let E, F be the mid points of the sides AB and CD respectively.

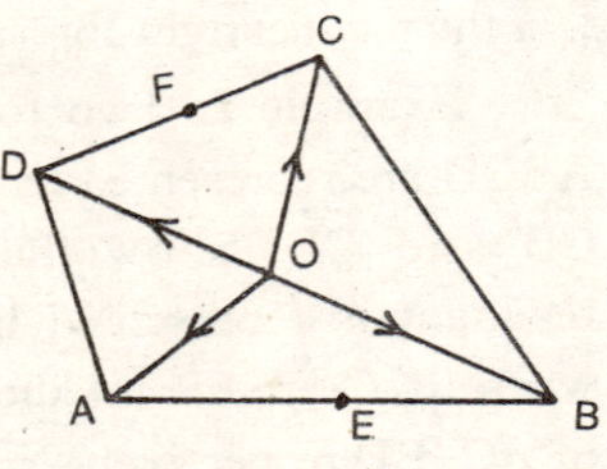

Fig. 9.10

$$\vec{OA} + \vec{OB} = 2\vec{OE} \text{ and}$$

$$\vec{OC} + \vec{OD} = 2\vec{OF}$$

Adding them

$$\therefore \vec{OA} + \vec{OB} + \vec{OC} + \vec{OD} = 2\vec{OE} + 2\vec{OF}$$

$$= 2(\vec{OE} + \vec{OF}) \quad \text{...(1)}$$

Now L.H.S of (1) is the resultant of the forces represented by $\vec{OA}$, $\vec{OB}$, $\vec{OC}$ and $\vec{OD}$. They are in equilibrium.

$$\therefore \quad \vec{OA} + \vec{OB} + \vec{OC} + \vec{OD} = \mathbf{0}$$

So from (1) $\vec{OE} + \vec{OF} = \mathbf{0} \Rightarrow \vec{OE} = \vec{FO}$

Hence O is the midpoint of the line joining the middle points of the opposite sides of the quadrilateral.

Example 5. If the resultant of two forces is equal in magnitude to one of the components and perpendicular to it in direction. find the other component.

Solution. Let **R** be the resultant of the two forces **P** and **Q** which are included at an angle θ to each other

Let **i** and **j** be the unit vectors along the directions of the forces P and R.

Now R = P and the direction of **R** is perpendicular to the direction of **P** as shown in the figure (given)

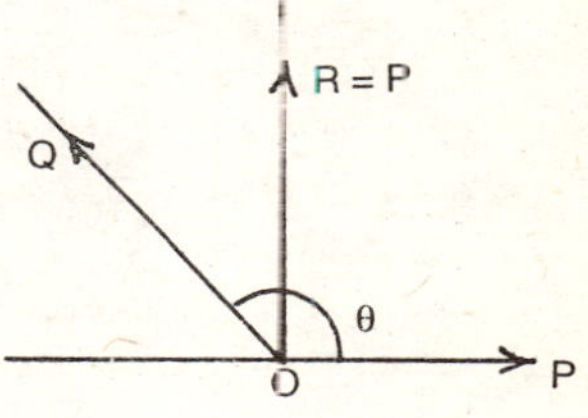

Fig. 9.11

$$\therefore \qquad \mathbf{P} = P\mathbf{i},\ \mathbf{R} = R\mathbf{j} = P\mathbf{j} \quad (\because\ R = P) \text{ and}$$

$$\mathbf{Q} = Q\cos\theta\,\mathbf{i} + Q\sin\theta\,\mathbf{j}.$$

But $\qquad \mathbf{R} = \mathbf{P} + \mathbf{Q}$

i.e., $\qquad P\mathbf{j} = P\mathbf{i} + Q\cos\theta\,\mathbf{i} + Q\sin\theta\,\mathbf{j}$

Equating the co-efficients of i and j on both sides we get O = P + Q cos θ and P = Q sin θ

Equating the co-efficients of i and j on both sides we get

$$0 = P + Q\cos\theta \text{ and } P = Q\sin\theta$$

$$\therefore \qquad Q\cos\theta = -P \text{ and } Q\sin\theta = P$$

$$\Rightarrow \qquad \tan\theta = -1 \quad \text{or} \quad \theta = \frac{3\pi}{4}$$

Hence $\qquad Q^2 = 2P^2$ i.e., $\quad Q = \sqrt{2}P.$

Example 6. If forces $\mathbf{F_1}, \mathbf{F_2}, ..., \mathbf{F_n}$ acting in a plane at O are in equilibrium and any transversal cuts their lines of action in points $L_1, L_2, ..., L_n$ and length OL is positive when in the same direction as F then $\Sigma \dfrac{F}{OL} = 0$.

Solution. Let the forces $\mathbf{F_1}, \mathbf{F_2}, ..., \mathbf{F_n}$ cut the transversal AB in $L_1, L_2, ..., L_n$ making angles $\theta_1, \theta_2, ..., \theta_n$ respectively as shown in the figure. Let **i, j** be the unit vectors along and perpendicular to AB so that

$$\mathbf{F_r} = F_r \cos\theta_r\, \mathbf{i} + F_r \sin\theta_r\, \mathbf{j},\ r = 1, 2, \ldots, n$$

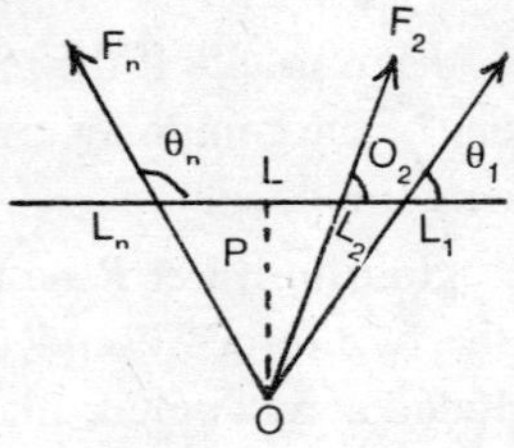

Since the forces are in equilibrium

$$\therefore \quad \sum_{r=1}^{n} \mathbf{F_r} = 0$$

i.e., $$\sum_{r=1}^{n} (F_r \cos\theta_r\, \mathbf{i} + F_r \sin\theta_r \mathbf{j}) = 0$$

Fig. 9.12

$$\Rightarrow \quad \sum_{r=1}^{n} F_r \cos\theta_r = 0 \text{ and } \sum_{r=1}^{n} F_r \sin\theta_r = 0$$

If P is the length of perpendicular from O and AB then

$\sin\theta_r = \dfrac{p}{OL_r}$ and hence from the second equation

$$\sum_{r=1}^{n} F_r \frac{P}{OL_r} = 0 \text{ or } \Sigma \frac{F}{OL} = 0.$$

By cancelling p (≠ 0) and dropping the suffixes. Hence the result.

Example 7. ABC is a triangle and D, E, F are the mid points of the sides BC, CA and AB. Forces represented by AD, $\frac{2}{3}$ BE and $\frac{1}{3}$ CF act on a particle at point where AD and BE meet. Show that the resultant is represented both in magnitude and direction by $\frac{1}{2}$ AC, and the line of action of the resultant divides BC in the ration 2 : 1.

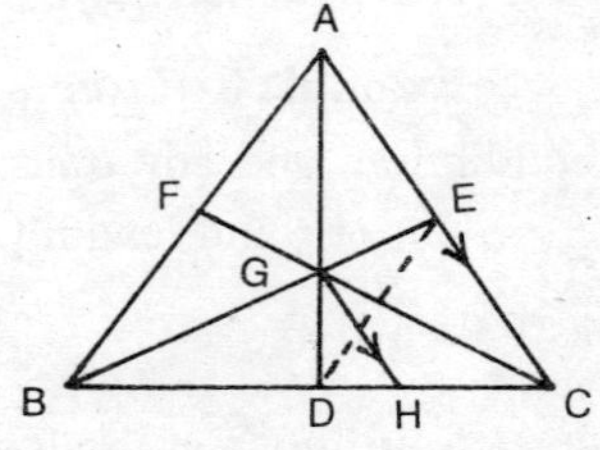

Fig. 9.13

Solution. Given that D, E, F are midpoints of sides BC, CA and AB of a triangle ABC. Now the medians AD and BE meet at the centroid G in 2 : 1 ratio. Forces represented by AD, $\frac{2}{3}$ BE and $\frac{1}{3}$ CF act on a particle at a point where AD and BE meet. i.e, at G.

$$\therefore \quad \frac{2}{3}\text{BE} = \text{BG and } \frac{1}{3}\text{CF} = \text{GF}$$

i.e., $\frac{2}{3}\overrightarrow{BE} = \overrightarrow{BG}$ and $\frac{1}{3}\overrightarrow{CF} = \overrightarrow{GF}$. Now the resultant is

$$\therefore \quad \overrightarrow{AD} + \frac{2}{3}\overrightarrow{BE} = \frac{1}{3}\overrightarrow{CF} = \overrightarrow{AD} + \overrightarrow{BG} + \overrightarrow{GF}$$

$$= \overrightarrow{AD} + \overrightarrow{BF} \qquad (\Delta\text{lc law})$$

$$= \overrightarrow{AD} + \frac{1}{2}\overrightarrow{BA}$$

$$= \overrightarrow{AD} + \overrightarrow{DE} \qquad \left(\text{DE} = \frac{1}{2}\text{BA}\right)$$

$\because$ Forces act at G $\Rightarrow$ Resultant must act at G and must be parallel to AC. If GH is parallel to AC, we have

$$\text{BH : HC} = \text{BG : GE} = 1 : 2$$

$\Rightarrow$ H divides BC in 2 : 1 ratio.

Example 8. Forces act at the vertices of a tetrahedron along the lines joining them to the centroids of opposite faces and proportional to the lengths of the same. Show that the system is in equilibrium.

Solution. Let OABC be the tetrahedron, the vertex O being the origin. Let **a**, **b**, **c** be the p.v's of A, B, C w.r.t O. Let G_1, G_2, G_3 and G_4 be the centroids of faces ABC, BCO, COA and OAB.

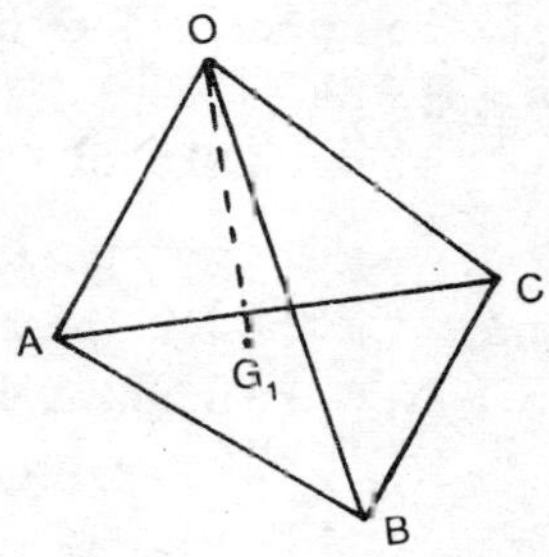

Fig. 9.14

$$\therefore \quad \overrightarrow{OG_1} = \frac{\mathbf{a}+\mathbf{b}+\mathbf{c}}{3}, \; \overrightarrow{OG_2} = \frac{\mathbf{b}+\mathbf{c}}{3},$$

$$\overrightarrow{OG_3} = \frac{\mathbf{a}+\mathbf{c}}{3} \text{ and } \overrightarrow{OG_4} = \frac{\mathbf{a}+\mathbf{b}}{3}.$$

The forces are

$\lambda\,\overrightarrow{OG_1}$, $\lambda\,\overrightarrow{AG_2}$, $\lambda\,\overrightarrow{BG_3}$ and $\lambda\,\overrightarrow{CG_4}$ [as per given statement]

$\therefore$ Resultant = $\mathbf{R} = \lambda\overrightarrow{OG_1} + \lambda\overrightarrow{AG_2} + \lambda\overrightarrow{BG_3} + \lambda\overrightarrow{CG_4}$

$$\therefore \quad \mathbf{R} = \lambda\left\{\overrightarrow{OG_1} + \overrightarrow{AG_2} + \overrightarrow{BG_3} + \overrightarrow{CG_4}\right\}$$

$$= \lambda\left\{\overrightarrow{OG_1} + (\overrightarrow{OG_2} - \overrightarrow{OA}) + (\overrightarrow{OG_3} - \overrightarrow{OB}) + (\overrightarrow{OG_4} - \overrightarrow{OC})\right\}$$

$$= \lambda\left\{\frac{\mathbf{a}+\mathbf{b}+\mathbf{c}}{3} + \frac{\mathbf{b}+\mathbf{c}}{3} - \mathbf{a} + \frac{\mathbf{a}+\mathbf{c}}{3} - \mathbf{b} + \frac{\mathbf{a}+\mathbf{b}}{3} - \mathbf{c}\right\}$$

$$= \lambda(\mathbf{0}) = \mathbf{0}$$

$\because$ Resultant is zero, the given forces are in equilibrium.

Example 9. ABCD is a parallelogram and P any point. Then show that the forces PA and PC are equivalent to the forces represented by $\overrightarrow{PB}$ and $\overrightarrow{PD}$.

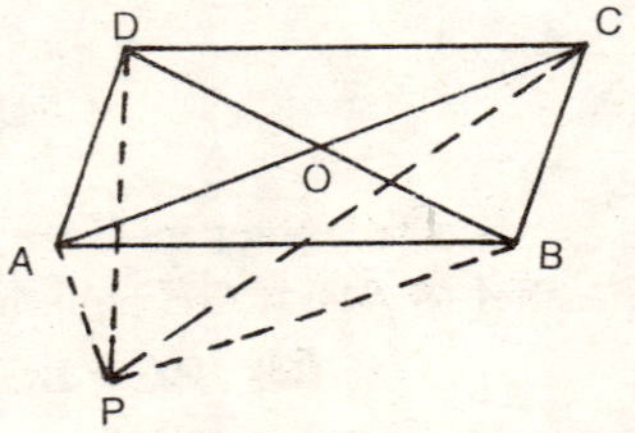

Solution. ABD is a parallelogram. Let the diagonals AC and BD intersect at O. Since the diagonals bisects each other in a parallelogram.

$$\therefore \quad \overrightarrow{PA} + \overrightarrow{PC} = 2\overrightarrow{PO}$$

also $$\overrightarrow{PB} + \overrightarrow{PD} = 2\overrightarrow{PO}$$

Fig. 9.15

$\Rightarrow$ the two systems are equivalent.

Example 10. Find the moment about the point $\mathbf{i} + 3\mathbf{j} + \mathbf{k}$ of a force represented by $3\mathbf{j} + 2\mathbf{k}$ acting through the point $5\mathbf{i} + 4\mathbf{j} - 3\mathbf{k}$.

Solution: Let O be the point $\mathbf{i} + 3\mathbf{j} + \mathbf{k}$ and P the point $5\mathbf{i} + 4\mathbf{j} - 3\mathbf{k}$

Then $$\overrightarrow{OP} = (5\mathbf{i} + 4\mathbf{j} - 3\mathbf{k}) - (\mathbf{i} + 3\mathbf{j} + \mathbf{k})$$

$$= 4\mathbf{i} + \mathbf{j} - 4\mathbf{k}$$

and $$\mathbf{F} = 3\mathbf{j} + 2\mathbf{k}$$

Now the moment about O of the force $\mathbf{F}$ acting through P is

$$= \overrightarrow{OP} \times \mathbf{F}$$

$$= (4\mathbf{i} + \mathbf{j} - 4\mathbf{k}) \times (3\mathbf{i} + 2\mathbf{k})$$

$$= \begin{vmatrix} \mathbf{i} & \mathbf{j} & \mathbf{k} \\ 4 & 1 & -4 \\ 0 & 3 & 2 \end{vmatrix}$$

$= (2 + 12)\mathbf{i} - (8 - 0)\mathbf{j} + (12 - 0)\mathbf{k}$

$= 14\mathbf{i} - 8\mathbf{j} + 12\mathbf{k}.$

Example 11. Find the torque about the point $\mathbf{i} + 2\mathbf{j} + 3\mathbf{k}$ of a force represented by $\mathbf{i} + \mathbf{j} + \mathbf{k}$ acting through the point $-2\mathbf{i} + 3\mathbf{j} + \mathbf{k}$.

Solution. Let O be the point $\mathbf{i} + 2\mathbf{j} + 3\mathbf{k}$ and P the point $-2\mathbf{i} + 3\mathbf{j} + \mathbf{k}$.

Then $\overrightarrow{OP} = (-2\mathbf{i} + 3\mathbf{j} + \mathbf{k}) - (\mathbf{i} + 2\mathbf{j} + 3\mathbf{k})$

$= -3\mathbf{i} + \mathbf{j} - 2\mathbf{k}$

and $\mathbf{F} = \mathbf{i} + \mathbf{j} + \mathbf{k}$

Now the torque (vector moment) about O of the force **F** acting through P.

$= \overrightarrow{OP} \times \mathbf{F}$

$= (-3\mathbf{i} + \mathbf{j} - 2\mathbf{k}) \times (\mathbf{i} + \mathbf{j} + \mathbf{k})$

$$= \begin{vmatrix} \mathbf{i} & \mathbf{j} & \mathbf{k} \\ -3 & 1 & -2 \\ 1 & 1 & 1 \end{vmatrix}$$

$= (1 + 2)\mathbf{i} - (-3 + 2)\mathbf{j} + (-3 - 1)\mathbf{k}$

$= 3\mathbf{i} + \mathbf{j} - 4\mathbf{k}.$

Example 12. Find the vector moment of the three forces $\mathbf{i} + 2\mathbf{j} - 3\mathbf{k}$, $2\mathbf{i} + 3\mathbf{j} + 4\mathbf{k}$ and $-\mathbf{i} - \mathbf{j} + \mathbf{k}$ acting on a particle at point P(0, 1, 2) about the point A(1, – 2, 0).

Solution: If **R** is the resultant of the forces then

$\mathbf{R} = (\mathbf{i} + 2\mathbf{j} - 3\mathbf{k}) + (2\mathbf{i} + 3\mathbf{j} + 4\mathbf{k}) + (-\mathbf{i} - \mathbf{j} + \mathbf{k})$

$= 2\mathbf{i} + 4\mathbf{j} + 2\mathbf{k}.$

Now $\overrightarrow{AP}$ = p.v of p – p.v of A

$= (\mathbf{j} + 2\mathbf{k}) - (\mathbf{i} - 2\mathbf{j}) = -\mathbf{i} + 3\mathbf{j} + 2\mathbf{k}$

Required moment of forces about A(1, -2, 0)

= moment of R at P about A

$= \overrightarrow{AP} \times \mathbf{R}$

$= (-\mathbf{i} + 3\mathbf{j} + 2\mathbf{k}) \times (2\mathbf{i} + 4\mathbf{j} + 2\mathbf{k})$

$$= \begin{vmatrix} \mathbf{i} & \mathbf{j} & \mathbf{k} \\ -1 & 3 & 2 \\ 2 & 4 & 2 \end{vmatrix} = -2\mathbf{i} + 6\mathbf{j} - 10\mathbf{k}$$

Example 13. Find the work done in moving an object along a vector $\overrightarrow{AB} = 3\mathbf{i} + 2\mathbf{j} - 5\mathbf{k}$ if the force applied is $\mathbf{F} = 2\mathbf{i} - \mathbf{j} - \mathbf{k}$.

Solution: Work done $= \mathbf{F}.\overrightarrow{AB}$

$$= (3\mathbf{i} + 2\mathbf{j} - 5\mathbf{k}) . (2\mathbf{i} - \mathbf{j} - \mathbf{k})$$

$$= 6 - 2 + 5 = 9 \text{ units.}$$

Example 14. Find the work done in moving an object along a straight line from (– 3, 2, – 4) to (2, – 1, 5) in a force field given by $\mathbf{F} = 3\mathbf{i} + 2\mathbf{j} + \mathbf{k}$.

Solution Let P = (3, 2, – 4) and Q = (2, – 1, 5)

Then $\overrightarrow{PQ}$ = p.v of Q – p.v of P

$$= (2\mathbf{i} - \mathbf{j} + 5\mathbf{k}) - (-3\mathbf{i} + 2\mathbf{j} - 4\mathbf{k})$$

$$= 5\mathbf{i} - 3\mathbf{j} + 9\mathbf{k}$$

∴ Work done $= \mathbf{F}.\overrightarrow{PQ}$

$$= (3\mathbf{i} + 2\mathbf{j} + \mathbf{k}) . (5\mathbf{i} - 3\mathbf{j} + 9\mathbf{k})$$

$$= 15 - 6 + 9 = 18 \text{ units.}$$

Example 15. Concurrent forces

$$\mathbf{F}_1 = \mathbf{i} - \mathbf{j} + \mathbf{k}, \ \mathbf{F}_2 = -\mathbf{i} + 2\mathbf{j} - \mathbf{k} \text{ and } \mathbf{F}_3 = \mathbf{j} - \mathbf{k}$$

act on a particle at a point A. Determine the work done when the particle is displaced from the position A to B where the position vectors of A and B are $4\mathbf{i} - 3\mathbf{j} - 2\mathbf{k}$ and $6\mathbf{i} + \mathbf{j} - 3\mathbf{k}$ respectively.

Solution: If **R** be the resultant of the forces $\mathbf{F}_1$, $\mathbf{F}_2$ and $\mathbf{F}_3$ then

$$\mathbf{R} = \mathbf{F}_1 + \mathbf{F}_2 + \mathbf{F}_3$$

$$= (\mathbf{i} - \mathbf{j} + \mathbf{k}) + (-\mathbf{i} + 2\mathbf{j} - \mathbf{k}) + (\mathbf{j} - \mathbf{k}) = 2\mathbf{j} - \mathbf{k}$$

Also $\overrightarrow{AB}$ = p.v of B – p.v of A

$$= (6\mathbf{i} + \mathbf{j} - 3\mathbf{k}) - (4\mathbf{i} - 3\mathbf{j} - 2\mathbf{k})$$

$$= 2\mathbf{i} + 4\mathbf{j} - \mathbf{k}$$

∴ Required work done $= \mathbf{R}.\overrightarrow{AB}$

$$= (2\mathbf{j} - \mathbf{k}) . (2\mathbf{i} + 4\mathbf{j} - \mathbf{k})$$

$$= 8 + 1 = 9 \text{ units.}$$

Example 16. Forces acting on a particle have magnitudes 5, 3, 1 lbs. wt. and act in the directions of the vectors $6\mathbf{i} + 2\mathbf{j} + 3\mathbf{k}$, $3\mathbf{i} - 2\mathbf{j} + 6\mathbf{k}$ and $2\mathbf{i} - 3\mathbf{j} - 6\mathbf{k}$ respectively. These remain constant while the particle is displaced from the point A (2**i** − **j** + 3**k**) to B (5**i** − **j** + **k**). Find the total work done by the forces.

Solution. Unit vectors in the direction of the given forces are

$$\frac{6\mathbf{i}+2\mathbf{j}+3\mathbf{k}}{7},\ \frac{3\mathbf{i}-2\mathbf{j}+6\mathbf{k}}{7} \text{ and } \frac{2\mathbf{i}-3\mathbf{j}-6\mathbf{k}}{7}$$

They have magnitudes 5, 3, 1. Hence they are

$$\frac{5}{7}(6\mathbf{i}+2\mathbf{j}+3\mathbf{k}),\ \frac{3}{7}(3\mathbf{i}-2\mathbf{j}+6\mathbf{k}) \text{ and } \frac{1}{7}(2\mathbf{i}-3\mathbf{j}-6\mathbf{k})$$

Let **R** be the resultant

$$\therefore \qquad \mathbf{R} = \frac{5}{7}(6\mathbf{i}-2\mathbf{j}+3\mathbf{k}) + \frac{3}{7}(3\mathbf{i}-2\mathbf{j}+6\mathbf{k}) + \frac{1}{7}(2\mathbf{i}-3\mathbf{j}-6\mathbf{k})$$

$$= \frac{1}{7}(41\mathbf{i}+\mathbf{j}+27\mathbf{k})$$

Now $\overrightarrow{AB}$ = p.v of B − p.v of A

$$= (5\mathbf{i}-\mathbf{j}+\mathbf{k}) - (2\mathbf{i}-\mathbf{j}+3\mathbf{k}) = 3\mathbf{i}+4\mathbf{k}$$

∴ Work done = $\mathbf{R}.\overrightarrow{AB}$

$$= \frac{1}{7}(41\mathbf{i}+\mathbf{j}+27\mathbf{k}).(3\mathbf{i}+4\mathbf{k})$$

$$= \frac{1}{7}[123 + 108] = 33 \text{ units.}$$

EXERCISE

1. Five forces act at one vertex A of a regular hexagon in the directions of the other vertices and proportional to the distances of these vertices from A. Find this resultant
2. A, B, C are fixed points and P a variable point such that the resultant of forces at P represented by $\overrightarrow{PA}$ and $\overrightarrow{PB}$ always passes through C. Find the locus of P
3. Forces P, Q act at O and have a resultant R. If any transversal cuts their lines of action at A, B, C respectively show that $\frac{P}{OA} + \frac{Q}{OC} = \frac{R}{RC}$.
4. Forces 3P, 7P, 5P act along the sides AB, BC, CA of an equilateral triangle ABC. Find the magnitude, direction and line

of action of the resultant.

5. D, E, F are the mid points of the sides of the triangle ABC. Show that, for any point O, the system of concurrent forces represented by $\vec{OA}, \vec{OB}, \vec{OC}$ is equivalent to the system represented by $\vec{OD}, \vec{OE}$ and $\vec{OF}$.

6. A particle at the corner of a cube is acted by the forces 1, 2, 3 kg wt respectively along the diagonals of the cube which meet the particle, Find their resultant.

7. Find the horizontal force and the force included at 60° to the vertical whose resultant is a vertical force P lb. wt.

8. ABCD is a quadrilateral and E, F are the mid points of the sides AB and CD. Show that the resultant of forces PA, PB, PC, PD is 4PG where G is the middle point of EF where ever the point may be. When will these forces be in equilibrium.

9. A particle is acted on by a number of centres of forces, some of which attract and some repel the force in each case varying as the distance and the intensities for different centres being different. Show that the resultant passes through a fixed point for all positions of the particle.

10. ABC ... is a polygon of n sides and forces act at a point parallel and proportional to AB, 2BC, 3CD, ... etc. Show that their resultant is parallel and proportional to (n-1) OA, where O is the centroid of points B, C, D ... excluding A.

11. Find the moment about the point $\mathbf{i} + 2\mathbf{j} - \mathbf{k}$ of a force represented by $\mathbf{i} + 2\mathbf{j} + \mathbf{k}$ acting the through the point $2\mathbf{i} + 3\mathbf{j} + \mathbf{k}$.

12. Find the torque about the point $\mathbf{i} + 2\mathbf{j} - \mathbf{k}$ of a force represented by $3\mathbf{i} + \mathbf{k}$ acting through the point $2\mathbf{i} - 3\mathbf{j} + 3\mathbf{k}$.

13. Find the moment about the point $\mathbf{i} + 2\mathbf{j} + 3\mathbf{k}$ of a force represented by $\mathbf{j} + 2\mathbf{k}$ acting at the point $\mathbf{i} + \mathbf{j} + \mathbf{k}$.

14. Find the vector moment of the three forces $3\mathbf{j} - 6\mathbf{k}$, $\mathbf{i} + 2\mathbf{j} + 3\mathbf{k}$ and $3\mathbf{i} + 2\mathbf{j} - 4\mathbf{k}$ acting on a particle at (1, - 1, 2) about the point (2, - 1, 3).

15. A force of magnitude 5 acts along a vector directed from the point (3, - 1, 6) to the point (8, 4, - 1). Find the moment of the

force about the point (1, 2, – 3).

16. Find the work done in moving an object along a straight line form (3, 2, – 1) to (2, – 1, 4) in a force field given by $\mathbf{F} = 4\mathbf{i} + 3\mathbf{j} + 3\mathbf{k}$.

17. A particle acted on by the force $3\mathbf{i} + 2\mathbf{j} - \mathbf{k}$ is displaced from $2\mathbf{i} - \mathbf{j} + \mathbf{k}$ to $3\mathbf{i} - 2\mathbf{j} + 4\mathbf{k}$. Find the work done by the force.

18. A particle is displaced from the point whose position vector is $5\mathbf{i} - 5\mathbf{j} - 7\mathbf{k}$ to the point whose position vector is $6\mathbf{i} + 2\mathbf{j} - 2\mathbf{k}$ under the action of constant forces $10\mathbf{i} - \mathbf{j} + 11\mathbf{k}$, $4\mathbf{i} + 5\mathbf{j} + 6\mathbf{k}$ and $-2\mathbf{i} + \mathbf{j} - 9\mathbf{k}$. Find the total work done

19. A particle is displaced from (1, 2, 3) to (5, 4, 6) and again from (5, 4, 6) to (2, – 1, 4) under the action of constant forces $2\mathbf{i} - 5\mathbf{j} + 6\mathbf{k}$, $2\mathbf{i} + 7\mathbf{j} + \mathbf{k}$ find the total work done

20. Show that twice the vector area of a closed plane polygon is equal to the sum of the torques about any point of forces represented by sides of the polygon taken in order.

ANSWERS

(1) $3\overrightarrow{AD}$ (2) straight line joining C and D

(4) $\frac{\sqrt{37}\,P}{2}$, $BO = \frac{5}{2}\,BC$ or $CO = \frac{3}{2}\,BC$

(6) 5 kg wt (7) $P = \frac{S}{2}$, $Q = \frac{\sqrt{3}\,S}{2}$

(8) P coincides with Q (11) $-3\mathbf{i} + \mathbf{j} + \mathbf{k}$

(12) $-3\mathbf{i} + 11\mathbf{j} + 9\mathbf{k}$ (13) 0

(14) $-7\mathbf{i} + 11\mathbf{j} + 7\mathbf{k}$ (15) $\frac{5}{3\sqrt{11}}(24\mathbf{i} - 59\mathbf{j} - 25\mathbf{k})$

(16) 2 units (17) – 2 units

(18) 87 units (19) 5 units.

9.15. Displacement

The displacement of a moving particle is its change of position. If a particle moves from A to B in an interval of time then the distance AB and the direction AB together defines displacement. Hence it is a vector quantity and we write it as $\overrightarrow{AB}$. Ab denotes the magnitude. Two successive displacements of a point are compounded according to as vector law of addition. If the particle is displaced from A to B and then from B to C then the resultant displacement is $\overrightarrow{AC}$.

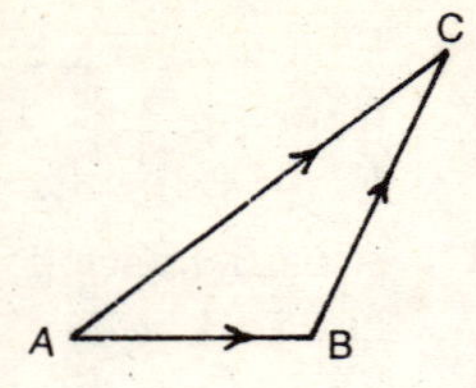

$$\therefore \qquad \overrightarrow{AC} = \overrightarrow{AB} + \overrightarrow{BC}$$

Fig. 9.16

Relative position. If the two particles A and B are in motion then relative position of B w.r.t A is a vector quantity and is denoted by $\overrightarrow{AB}$.

9.17. Relative displacement

If during any interval of time P moves to P′ and Q moves to Q′, then the relative displacement of P to Q during that interval is given by the change in their relative position during that interval and is determined by the vector $\overrightarrow{P'Q'} - \overrightarrow{PQ}$.

9.18. Relative velocity

The velocity of a moving particle is defined as the rate of displacement. It is a vector quantity. The relative velocity of a particle Q w.r.t P is defined as the rate of change of position of Q relative to P.

To show that the relative velocity of Q *w.r.t* P *is the vector difference of the velocities of Q and P referred to any fixed point O.*

Let U and V be the uniform velocity vectors of P and Q relative to the fixed point O. Let these particles move in unit time from P to P′ and Q to Q′.

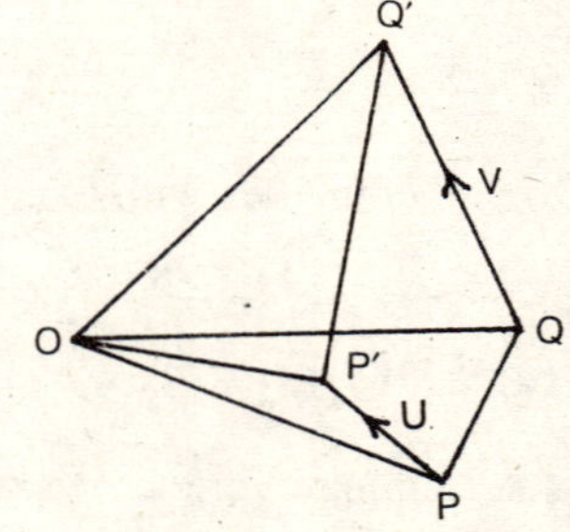

Fig. 9.17

Then $\overrightarrow{PP'} = \mathbf{u}$ and $\overrightarrow{QQ'} = \mathbf{v}$

Since u and v are uniform, the relative velocity of Q w.r.t P

$$= \overrightarrow{P'Q'} - \overrightarrow{PQ}$$

$$= (\overrightarrow{OQ'} - \overrightarrow{OP'}) - (\overrightarrow{OQ} - \overrightarrow{OP})$$

$$= \vec{OQ'} - \vec{OP'} - \vec{OQ} + \vec{OP}$$

$$= (\vec{OQ'} - \vec{OQ}) - (\vec{OP'} - \vec{OP})$$

$$= \vec{OQ'} - \vec{PP'} = \mathbf{v} - \mathbf{u}$$

9.19. Angular velocity of a rigid body about a fixed axis

Let a rigid body be rotating with an angular velocity ω about a fixed axis ON. Then the angular velocity ω is a vector having magnitude ω and direction parallel to the axis of rotation and in the positive sense relative to the rotation. Let $\mathbf{r}$ be the p.v of any fixed point P of the body w.r.t O. Draw PN perpendicular to ON and let $\angle PON = \theta$

$\therefore \quad OP = |\mathbf{r}| = r$

Fig. 9.18

and $\quad PN = OP \sin\theta = r \sin\theta.$

Let the point P moves with velocity $\mathbf{v}$ along the target to the circular path with N as the centre and PN as radius. Then the velocity vector $\mathbf{v}$ of P is perpendicular to the plane OPN with magnitude

$$= \omega \times PN$$

$$= \omega r \sin\theta = |\omega \times \mathbf{r}|$$

$$\therefore \quad \mathbf{v} = \omega \times \mathbf{r}$$

SOLVED PROBLEMS

Example 17. A man is travelling east at a rate of 3 km per hour finds that the wind seems to blow directly from the north. On doubling his speed he finds that it appears to come from north-east. Find the velocity of the wind.

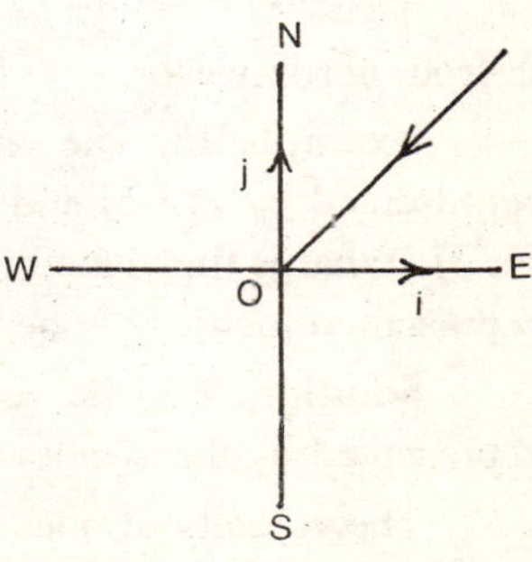

Fig. 9.19

Solution. Let $\mathbf{i}$ and $\mathbf{j}$ be the unit vectors towards east and north from O respectively.

Let the velocity of the wind be $\mathbf{V_1} = x\mathbf{i} + y\mathbf{j}$

Case 1. Let $\mathbf{v_2}$ be the velocity of man who is travelling towards cast at 3 km per hour

$\therefore \quad V_2 = 8\mathbf{i}$

$\therefore$ Velocity of wind relative to the man

$\therefore \quad \mathbf{V_1} = \mathbf{V_1} - \mathbf{V_2} = (x - 3)\mathbf{i} + y\mathbf{j}$

But it is given that wind appears to blow directly from the north i.e., it is parallel to $- j$

$\therefore \quad \mathbf{V_1} = - a\mathbf{j}$ and

$\therefore \quad (x - 3)\mathbf{i} + y\mathbf{j} = - a\mathbf{j} \Rightarrow x - 3 = 0 \quad \therefore x = 3$ ...(1)

Case 2. Let $\mathbf{V_3}$ be the double velocity of the man i.e., $\mathbf{V_3} = 2\mathbf{V_2} = 6\mathbf{i}$. Now velocity of the wind relative to the man

$$\mathbf{V_2} = \mathbf{V_1} - \mathbf{V_3} = (x - 6)\mathbf{i} + y\mathbf{j}$$

But it is given that now the wind appears to blow directly from the north east i.e., parallel to $- (\mathbf{i} + \mathbf{j})$

$\therefore \quad \mathbf{V_2} = - b\,(\mathbf{i} + \mathbf{j})$

Thus $\quad (x - y)\,\mathbf{i} + y\mathbf{j} = - b(\mathbf{i} + \mathbf{j})$

$\therefore \quad x - 6 = - b\,;\; y = - b \Rightarrow x - 6 = y$

$\therefore \quad y = 3 - 6 = - 3 \quad \therefore \quad y = - 3$ ...(2)

Thus velocity of the wind

$\mathbf{V_1} = x\mathbf{i} + y\mathbf{j} = 3\mathbf{i} - 3\mathbf{j}$ [Using (1) and (2)]

and its magnitude is

$= \sqrt{3^3 + (-3)^2} = 3\sqrt{2}$ km per hour and the direction of the wind is from north-west.

Example 18. The velocity of a boat relative to the water is represented by $2\mathbf{i} + 5\mathbf{j}$ and that of the water relative to the earth by $\mathbf{i} + 3\mathbf{j}$. What is the velocity of the boat relative to the earth if **i** and **j** represent velocities of one mile an hour east and north respectively.

Solution. Let the velocities of the boat, water and earth be represented by the vectors **u, v, w** respectively.

The velocity of boat relative to water is

$$\mathbf{u} - \mathbf{v} = 2\mathbf{i} + 5\mathbf{j} \quad \text{...(1)}$$

The velocity of water relative to earth is

$$\mathbf{v} - \mathbf{w} = \mathbf{i} - 3\mathbf{j} \qquad ...(2)$$

Adding (1) and (2) we get

$$\mathbf{u} - \mathbf{w} = 3\mathbf{i} + 2\mathbf{j} \qquad ...(3)$$

which represents velocity of boat relative to earth.

Magnitude of the velocity

$$= \sqrt{9+4} = \sqrt{13} \text{ miles/hour}$$

The direction of this velocity makes an angle

$$= \tan^{-1}\left(\frac{\text{Coeft of } \mathbf{j}}{\text{Coeff of } \mathbf{i}}\right) = \tan^{-1}\left(\frac{2}{3}\right) \text{ north of east.}$$

Example 19. Two particles instantaneously at A and B respectively 25 mts apart, are moving with uniform velocities, the former towards B at 8 mts/sec and the later perpendicular to AB at 5 mts/sec. Find the relative velocity, their shortest distance apart, and the instant when they are nearest.

Solution. Let AB and BC be two straight lines perpendicular to each other. Let **i** and **j** be unit vectors representing velocities of 1 mt/sec along AB and BC respectively. Then the velocity of A and B are 8**i** and 5**j**.

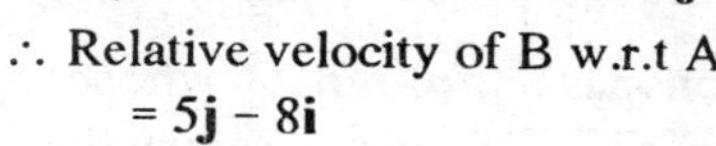

∴ Relative velocity of B w.r.t A

$$= 5\mathbf{j} - 8\mathbf{i}$$

∴ Magnitude

Fig. 9.20

$$= \sqrt{(-8)^2 + (5)^2} = \sqrt{89} \text{ mts/sec.}$$

Let A and B be the initial position of the particles and let P and Q be their position t seconds after the start.

Given AB = 25

$$\therefore \quad \overrightarrow{PB} = (25 - 8t)\,\mathbf{i} \text{ and } \overrightarrow{BQ} = 5t\mathbf{j}$$

$$\overrightarrow{PQ} = \overrightarrow{PB} + \overrightarrow{BQ} = (25 = 8t)\mathbf{i} + 5t\mathbf{j}$$

$$\text{Magnitude } PQ = \sqrt{(25 - 8t)^2 + (5t)^2}$$

$$= \sqrt{89t^2 - 400t + 625}$$

$$= \left\{ \left(\sqrt{89}\,t - \frac{200}{\sqrt{89}} \right)^2 + \frac{125}{\sqrt{89}} \right\}^{1/2}$$

showing that PQ is least when

$$\sqrt{89}\,t - \frac{200}{\sqrt{89}} = 0 \text{ i.e., } t = \frac{200}{89} \text{ seconds}$$

and the required shortest distance

$$= \left\{ \frac{125}{\sqrt{89}} \right\}^{1/2} = \sqrt{13.249974} = 3.64 \text{ mts.}$$

Example 20. A body is spinning with angular velocity of 3 radians per second about an axis parallel to $\mathbf{i} - 2\mathbf{j} + 2\mathbf{k}$ passing through the point $2\mathbf{i} - \mathbf{j} - \mathbf{k}$. Find the velocity and speed of the particle (of body) at the point $2\mathbf{i} + 3\mathbf{j} - 4\mathbf{k}$

Solution: Unit vector along ON

$$= \frac{\mathbf{i} - 2\mathbf{j} + 2\mathbf{k}}{\sqrt{1+4+4}} = \frac{1}{3}(\mathbf{i} - 2\mathbf{j} + 2\mathbf{k})$$

$\therefore$ angular velocity ω

$$= 3.\frac{1}{3}(\mathbf{i} - 2\mathbf{j} + 2\mathbf{k})$$

$$= \mathbf{i} - 2\mathbf{j} + 2\mathbf{k}$$

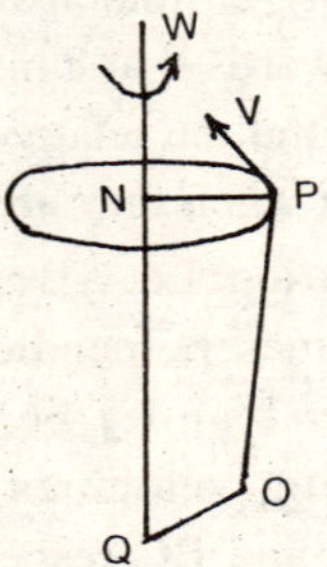

Fig. 9.21

Let the axis of rotation pass through Q whose p.v is $2\mathbf{i} - \mathbf{j} - \mathbf{k}$. Let P be the position of the given particle of the rigid body whose position vector is $2\mathbf{i} + 3\mathbf{j} - 4\mathbf{k}$.

Then velocity v of P is given by

$$\mathbf{v} = \omega \times \overrightarrow{QP}$$

$$= (\mathbf{i} - 2\mathbf{j} + 2\mathbf{k}) \times \{(2\mathbf{i} + 3\mathbf{j} - 4\mathbf{k}) - (2\mathbf{i} - \mathbf{j} - \mathbf{k})\}$$

$$= (\mathbf{i} - 2\mathbf{j} + 2\mathbf{k}) \times (4\mathbf{j} - 3\mathbf{k})$$

$$= \begin{vmatrix} \mathbf{i} & \mathbf{j} & \mathbf{k} \\ 1 & -2 & 2 \\ 0 & 4 & -3 \end{vmatrix}$$

$$= (6 - 8)\mathbf{i} - (-3 - 0)\mathbf{j} + (4 - 0)\mathbf{k}$$

$$= -2\mathbf{i} + 3\mathbf{j} + 4\mathbf{k}$$

Also required speed

$$= |\mathbf{v}| = \sqrt{4 + 9 + 16} = \sqrt{29}.$$

EXERCISE

1. A man travelling east at 8 miles an hour finds that the wind seems to blow directly from the north. On doubling his speed he finds that it appears to come from north-east. Show that the actual velocity of the wind is $8\sqrt{2}$ miles per hour and is from north-west
2. Two particles, instantaneously at P and Q respectively, 15 feet apart, are moving with uniform velocities, the former towards Q at 5 ft/sec and the later perpendicular to PQ at $3\frac{3}{4}$ ft/sec. Find their relative velocity, then short distance apart and the instant when they are nearest.
3. The velocity of a boat relative to the water is represented by $3\mathbf{i} + 4\mathbf{j}$ and that of the water relative to the earth by $\mathbf{i} - 3\mathbf{j}$. What is the velocity of the boat relative to the earth if **i** and **j** represent velocities of one mile an hour east and north respectively.
4. A rigid body is rotating with an angular velocity of 4 radians per second about an axis parallel to $3\mathbf{i} - \mathbf{k}$ passing through the point $\mathbf{i} + 3\mathbf{j} - \mathbf{k}$. Find the velocity and speed of the particle at the point $4\mathbf{i} - 2\mathbf{j} + \mathbf{k}$.

ANSWERS

(2) $\frac{15}{4}\mathbf{j} - 5\mathbf{i}$, $t = \frac{48}{25}$, S.D. = 9

(3) $4\mathbf{i} + \mathbf{j}$, $\tan^{-1}\left(\frac{1}{4}\right)$ north of east

(4) $\frac{4}{\sqrt{10}}(\mathbf{i} - 3\mathbf{j} - 9\mathbf{k})$; $\frac{4\sqrt{91}}{\sqrt{10}}$.

10

Vector Differentiation

10.1. Vector function of a scalar variable.

We know that in parametric form the equation of ellipse is x = a cos t, y = b sin t, z = 0, and any vector **r** can be written as **r** = x **i** + y **j** + z**k**, where **i, j, k** are three mutually perpendicular non-coplanar unit vectors.

$\therefore$ **r** = a cos t**i** + b sin t**j** + **0 k**

Now the vector **r** can be completely determined, as soon as the scalar variable t is known. Hence we say **r** is a *vector function of scalar variable t* and we write it as **r** = **F** (t)

$\therefore$ **r** = **F** (t) = a cos t**i** + b sin t**j**

Definition: Let **R** be the set of all real numbers and let **S** $\subseteq$ **R**. If to each t ε S, we associate by some rule a unique vector **F** (t), then this rule defines a *vector function of the scalar variable t.* Here **F** (t) is a *vector quantity* and **F** is a *vector function.*

If for every value of t there corresponds one and only one value **r** then we say that **r** is a *single valued vector function of t.*

Let the components of **F** (t) be f_1 (t), f_2 (t), f_3 (t) along the three axes. Then **F** (t) can be expressed as

$$\mathbf{F}(r) = f_1(t)\,\mathbf{i} + f_2(t)\,\mathbf{j} + f_3(t)\,\mathbf{k}.$$

Here f_1 (t), f_2 (t), f_3 (t) are known as scalar functions of scalar variable t.

Definition : Let **R** be the set of all real numbers and let **S** $\subseteq$ **R**. If to each t ε **S**, we associate by some rule a unique real number ϕ (t) then this rule defines a *scalar function of scalar variable t.*

Here ϕ (t) is a *scalar quantity* and ϕ is a *scalar function.*

For example $\phi(t) = (4\cos^2 t + 2e^t t^3)$ is a scalar function.

10.2. Limit of a vector function

A vector function **F** (t) is said to tend a limit **L**, when t tends to t*, if for any given positive number ε, however small, there corresponds a positive number δ such that

$$|\mathbf{F}(t) - \mathbf{L}| < \varepsilon, \text{ when ever } 0 < |t - t^*| < \delta.$$

We write it as $\lim\limits_{t \to t^*} \mathbf{F}(t) = \mathbf{L}$

10.3. Continuity of a vector function

A vector function F (t) is said continuous at t = t*, iff

(i) **F** (t*) exists

(ii) For any given positive number ε, however small there corresponds a positive number δ such that

$$|\mathbf{F}(t) - \mathbf{F}(t^*)| < \varepsilon, \text{ where } |t - t^*| < \delta.$$

Further a vector function **F** (t) is said to be continuous for all values of t for which **F**(t) is defined.

10.4. Theorems about limits and continuity of vector function: *(without proof)*

1. Let $\mathbf{F}(t) = f_1(t)\,\mathbf{i} + f_2(t)\,\mathbf{j} + f_3(t)\,\mathbf{k}$ and

$\mathbf{L} = l_1\,\mathbf{i} + l_2\,\mathbf{j} + l_3\,\mathbf{k}$. Then the necessary and sufficient condition that

$$\lim_{t \to t^*} \mathbf{F}(t) = \mathbf{L}$$

are $$\lim_{t \to t^*} f_1(t) = l_1 \,;\; \lim_{t \to t^*} f_2(t) = l_2 \,;\; \lim_{t \to t^*} f_3(t) = l_3.$$

2. Let **F** (t) and **G** (t) be vector functions of a scalar variable t, and let ϕ (t) be a scalar function of a scalar variable t.

Then (i) $\lim\limits_{t \to t^*} [\mathbf{F}(t) \pm \mathbf{G}(t)] = \lim\limits_{t \to t^*} \mathbf{F}(t) \pm \lim\limits_{t \to t^*} \mathbf{G}(t)$

(ii) $\lim\limits_{t \to t^*} \mathbf{F}(t) \cdot \mathbf{G}(t) = \lim\limits_{t \to t^*} \mathbf{F}(t) \cdot \lim\limits_{t \to t^*} \mathbf{G}(t)$

(iii) $\lim_{t \to t^*} \mathbf{F}(t) \times \mathbf{G}(t) = \lim_{t \to t^*} \mathbf{F}(t) \times \lim_{t \to t^*} \mathbf{G}(t)$

(iv) $\lim_{t \to t^*} \phi(t)\, \mathbf{F}(t) = \left[\lim_{t \to t^*} \phi(t)\right]\left[\lim_{t \to t^*} \mathbf{F}(t)\right]$

(v) $\lim_{t \to t^*} |\mathbf{F}(t)| = \left|\lim_{t \to t^*} \mathbf{F}(t)\right|$

3. The necessary and sufficient condition for a vector function **F** (t) to be continuous at t = t* is

$$\lim_{t \to t^*} \mathbf{F}(t) = \mathbf{F}(t^*)$$

4. If $\mathbf{F}(t) = f_1(t)\,\mathbf{i} + f_2(t)\,\mathbf{j} + f_3(t)\,\mathbf{k}$, then **F** (t) is continuous if $f_1(t)$, $f_2(t)$ and $f_3(t)$ are all continuous.

10.5. Derivative of a vector function

Let $$\mathbf{r} = \mathbf{F}(t) \qquad \text{...(1)}$$

be a single valued and continuous vector function of a scalar variable t with O as origin. Let it be represented by $\overrightarrow{OA}$ for a certain value of t. Let δ be a small increment in t and δ**r**, the corresponding increment in **r**. In the figure $\overrightarrow{OB}$ represents **r** + δ**r**

$$\therefore \quad \mathbf{r} + \delta\mathbf{r} = \mathbf{F}(t + \delta t) \qquad \text{...(2)}$$

From the figure

$$\overrightarrow{OB} - \overrightarrow{OA} = \overrightarrow{AB} = \delta\mathbf{r}$$

Now (2) - (1) gives

$$\delta\mathbf{r} = \mathbf{F}(t + \delta t) - \mathbf{F}(t)$$

Dividing both sides by δt, we get

$$\frac{\delta\mathbf{r}}{\delta t} = \frac{\mathbf{F}(t + \delta t) - \mathbf{F}(t)}{\delta t}$$

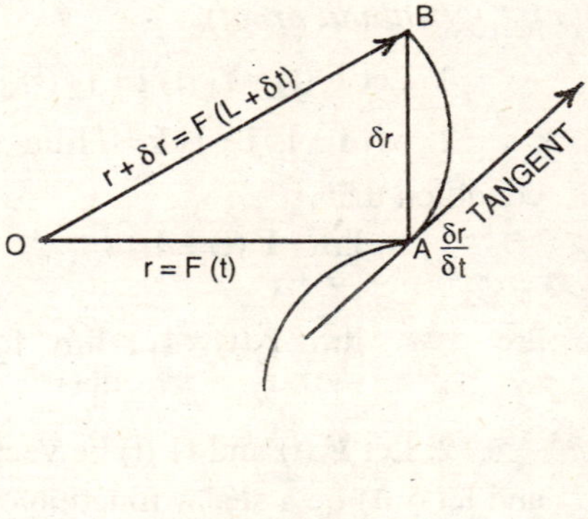

Fig 10.1

Taking limit δt → 0 on both sides, we get

$$\lim_{\delta t \to 0} \frac{\delta\mathbf{r}}{\delta t} = \lim_{\delta t \to 0} \frac{\mathbf{F}(t + \delta t) - \mathbf{F}(t)}{\delta t}$$

If the limit of R.H.S exists, then $\lim\limits_{\delta t \to 0} \frac{\delta \mathbf{r}}{\delta t}$ is called the derivative of the vector function w.r.t the scalar variable t, and is denoted $\frac{d\mathbf{r}}{dt}$.

$$\therefore \quad \frac{d\mathbf{r}}{dt} = \lim_{\delta t \to 0} \frac{\mathbf{F}(t + \delta t) - \mathbf{F}(t)}{\delta t}$$

Since $\frac{d\mathbf{r}}{dt}$ is itself a vector depending on t, we can consider its derivative w.r.t the scalar variable t. If this derivative exists it is denoted by $\frac{d^2\mathbf{r}}{dt^2}$. Similarly the higher order derivatives are described.

Note: If $\mathbf{r} = \mathbf{F}(t) = f_1(t)\,\mathbf{i} + f_2(t)\,\mathbf{j} + f_3(t)\,\mathbf{k}$

then
$$\frac{d\mathbf{r}}{dt} = \frac{d\mathbf{F}}{dt} = \frac{df_1}{dt}\mathbf{i} + \frac{df_2}{dt}\mathbf{j} + \frac{df_3}{dt}\mathbf{k}$$

10.6. Curves in space:

Let **O** be the origin of reference.

Let the position vector of a point A (x, y, z) be given by $\mathbf{r} = \mathbf{F}(t)$.

$$\therefore \quad \mathbf{r}(t) = x(t)\,\mathbf{i} + y(t)\,\mathbf{j} + z(t)\,\mathbf{k} = \mathbf{F}(t).$$

As the scalar parameter t takes different values, the point A traces a *space curve* having parametric equations.

$$x = x(t), \; y = y(t), \; z = z(t)$$

Then $\frac{\delta \mathbf{r}}{\delta t} = \frac{\mathbf{F}(t + \delta t) - \mathbf{F}(t)}{\delta t}$ is a vector in the direction of $\delta \mathbf{r}$ (See Fig. 10.1)

If $\lim\limits_{\delta t \to 0} \frac{\delta \mathbf{r}}{\delta t} = \frac{d\mathbf{r}}{dt}$ exists, the limit will be a vector in the direction of the tangent to the space curve at A(x, y, z) and is given by

$$\frac{d\mathbf{r}}{dt} = \frac{dx}{dt}\mathbf{i} + \frac{dy}{dt}\mathbf{j} + \frac{dz}{dt}\mathbf{k}$$

10.7. Velocity and acceleration

If the scalar variable t denotes the time, and **r** the position vector of a moving particle A (x, y, z), then $\delta \mathbf{r}$ is the displacement of the particle in time δt. But the time rate change of displacement is the velocity of a moving particle.

$$\therefore \text{ Velocity} = \lim_{\delta t \to 0} \frac{\delta \mathbf{r}}{\delta t} = \frac{\delta \mathbf{r}}{dt} = \mathbf{v}$$

Again since the time rate change of velocity **v** is the acceleration of a moving particle.

$$\text{Acceleration} = \lim_{\delta t \to 0} \frac{\delta \mathbf{v}}{\delta t} = \frac{d\mathbf{v}}{\delta t} = \mathbf{a}$$

or $$\mathbf{a} = \frac{d^2\mathbf{r}}{dt^2}$$

10.8. Differentiable formulae

(I) If **F, G** are differentiable vector functions of a scalar variable t and ϕ is a differentiable scalar function of the same scalar variable t, then

(i) $$\frac{d}{dt}(\mathbf{F} \pm \mathbf{G}) = \frac{d\mathbf{F}}{dt} \pm \frac{d\mathbf{G}}{dt}$$

(ii) $$\frac{d}{dt}(\mathbf{F} \cdot \mathbf{G}) = \mathbf{F} \cdot \frac{d\mathbf{G}}{dt} + \frac{d\mathbf{F}}{dt} \cdot \mathbf{G}$$

(iii) $$\frac{d}{t}(\mathbf{F} \times \mathbf{G}) = \mathbf{F} \times \frac{d\mathbf{G}}{dt} + \frac{d\mathbf{F}}{dt} \times \mathbf{G}$$

(iv) $$\frac{d}{dt}(\phi \mathbf{F}) = \phi \frac{d\mathbf{F}}{dt} + \frac{d\phi}{dt} \mathbf{F}$$

Proof. (i) Let $\mathbf{F}(t) = \mathbf{F}(t) \pm \mathbf{G}(t)$ ...(1)

$$\therefore \quad \frac{\mathbf{f}(t+\delta t) - \mathbf{f}(t)}{\delta t} = \frac{[\mathbf{F}(t+\delta t) \pm \mathbf{G}(t+\delta t)] - [\mathbf{F}(t) \pm \mathbf{G}(t)]}{\delta t}$$

or $$\frac{\mathbf{f}(t+\delta t) - \mathbf{f}(t)}{\delta t} = \left[\frac{\mathbf{F}(t+\delta t) - \mathbf{F}(t)}{\delta t}\right] \pm \left[\frac{\mathbf{G}(t+\delta t) - \mathbf{G}(t)}{\delta t}\right]$$

Taking limits on both sides as $\delta t \to 0$, we get

$$\lim_{\delta t \to 0} \frac{\mathbf{f}(t+\delta t) - \mathbf{f}(t)}{\delta t} = \left[\lim_{\delta t \to 0} \frac{\mathbf{F}(t+\delta t) - \mathbf{F}(t)}{\delta t}\right]$$

$$\pm \left[\lim_{\delta t \to 0} \frac{\mathbf{G}(t + \delta t) - \mathbf{G}(t)}{\delta t} \right]$$

or $$\frac{df}{dt} = \frac{d\mathbf{F}}{dt} \pm \frac{d\mathbf{G}}{dt}$$

i.e., The derivative of the sum (difference) of two vectors is equal to the sum (difference) of their derivatives.

(ii) Let $\mathbf{f} = \mathbf{F} . \mathbf{G}$

$$\therefore \quad \frac{\mathbf{f}(t + \delta t) - \mathbf{f}(t)}{\delta t} = \frac{\mathbf{F}(t + \delta t).\mathbf{G}(t + \delta t) - \mathbf{F}(t).\mathbf{G}(t)}{\delta t}$$

$$= \frac{[\mathbf{F}(t + \delta t).[\mathbf{G}(t + \delta t) - \mathbf{G}(t)]}{\delta t}$$

$$+ \frac{[\mathbf{F}(t + \delta t) - \mathbf{F}(t)].\mathbf{G}(t)}{\delta t}$$

Taking limit as $\delta t \to 0$ on both sides, we get

$$\lim_{\delta t \to 0} \frac{\mathbf{f}(t + \delta t) - \mathbf{f}(t)}{\delta t} = \left[\lim_{\delta t \to 0} \mathbf{F}(t + \delta t) \right] \cdot \left[\lim_{\delta t \to 0} \frac{\mathbf{G}(t + \delta t) - \mathbf{G}(t)}{\delta t} \right]$$

$$+ \left[\lim_{\delta t \to 0} \frac{\mathbf{F}(t + \delta t) - \mathbf{F}(t)}{\delta t} \right] \cdot \mathbf{G}(t)$$

or $$\frac{df}{dt} = \mathbf{F}(t) \cdot \frac{d\mathbf{G}(t)}{dt} + \frac{d\mathbf{F}(t)}{dt} \cdot \mathbf{G}(t)$$

or $$\frac{d}{dt}(\mathbf{F}.\mathbf{G}) = \mathbf{F} \cdot \frac{d\mathbf{G}}{dt} + \frac{d\mathbf{F}}{dt} \cdot \mathbf{G}$$

Note: In above **f** is not a vector function but it is a scalar function of t as **F.G** is a scalar quantity.

(iii) Let $\mathbf{f} = \mathbf{F} \times \mathbf{G}$. Here **f** is a vector quantity as $\mathbf{F} \times \mathbf{G}$ is a vector

$$\therefore \quad \frac{\mathbf{f}(t + \delta t) - \mathbf{f}(t)}{\delta t} = \frac{\mathbf{F}(t + \delta t) \times \mathbf{G}(t + \delta t) - \mathbf{F}(t) \times \mathbf{G}(t)}{\delta t}$$

$$= \mathbf{F}(t + \delta t) \times \left[\frac{\mathbf{G}(t + \delta t) - \mathbf{G}(1)}{\delta t} \right] + \left[\frac{\mathbf{F}(t + \delta t) - \mathbf{F}(t)}{\delta t} \right] \times \mathbf{G}(t)$$

Taking limits as $\delta t \to 0$ on both sides, we get

$$\lim_{\delta t \to 0} \frac{\mathbf{f}(t + \delta t) - \mathbf{f}\, t)}{\delta t}$$

$$= \left[\lim_{\delta t \to 0} \mathbf{F}(t + \delta t)\right] \times \left[\lim_{\delta t \to 0} \frac{\mathbf{G}(t + \delta t) - \mathbf{G}(t)}{\delta t}\right]$$

$$+ \left[\lim_{\delta t \to 0} \frac{\mathbf{F}(t + \delta t) - \mathbf{F}(t)}{\delta t}\right] \times \mathbf{G}(t)$$

or $$\frac{d\mathbf{f}}{dt} = \mathbf{F}(t) \times \frac{d\,\mathbf{G}(t)}{dt} + \frac{d\,\mathbf{F}(t)}{dt} \times \mathbf{G}(t)$$

or $$\frac{d}{dt}(\mathbf{F} \times \mathbf{G}) = \mathbf{F} \times \frac{d\mathbf{G}}{dt} + \frac{d\mathbf{F}}{dt} \times \mathbf{G}.$$

(iv) Let $\mathbf{f} = \phi\, \mathbf{F}$

$$\therefore \quad \frac{\mathbf{f}(t + \delta t) - \mathbf{f}(t)}{\delta t} = \frac{\phi(t + \delta t)\, \mathbf{F}(t + \delta t) - \phi(t)\, \mathbf{F}(t)}{\delta t}$$

$$= \frac{\phi(t + \delta t)\, [\mathbf{F}(t + \delta t) - \mathbf{F}(t)]}{\delta t}$$

$$+ \frac{[\phi(t + \delta t) - \phi(t)]\, \mathbf{F}(t)}{\delta t}$$

Taking limit as $\delta t \to 0$ on both sides, we get

$$\lim_{\delta t \to 0} \frac{\mathbf{f}(t + \delta t) - \mathbf{f}(t)}{\delta t}$$

$$= \left[\lim_{\delta t \to 0} \phi(t + \delta t)\right] \left[\lim_{\delta t \to 0} \frac{\mathbf{F}(t + \delta t) - \mathbf{F}(t)}{\delta t}\right]$$

$$+ \left[\lim_{\delta t \to 0} \frac{\phi(t + \delta t) - \phi(t)}{\delta t}\right] \mathbf{F}(t)$$

or $$\frac{d\mathbf{f}(t)}{dt} = \phi(t) \frac{d\mathbf{F}(t)}{dt} + \frac{d\phi(t)}{dt} \mathbf{F}(t)$$

or $$\frac{d}{dt}(\phi\, \mathbf{F}) = \phi \frac{d\mathbf{F}}{dt} + \frac{d\phi}{dt} \mathbf{F}$$

II. If **F, G,** and **H** are differentiable vector functions of a scalar variable t then

(i) $$\frac{d}{dt}[\mathbf{F}\ \mathbf{G}\ \mathbf{H}] = \left[\frac{d\mathbf{F}}{dt}\mathbf{G}\mathbf{H}\right] + \left[\mathbf{F}\frac{d\mathbf{G}}{dt}\mathbf{H}\right] + \left[\mathbf{F}\mathbf{G}\frac{d\mathbf{H}}{dt}\right]$$

(ii) $$\frac{d}{dt}[\mathbf{F} \times (\mathbf{G} \times \mathbf{H})] = \frac{d\mathbf{F}}{dt} \times (\mathbf{G} \times \mathbf{H}) + \mathbf{F} \times \left(\frac{d\mathbf{G}}{dt} \times \mathbf{H}\right) + \mathbf{F} \times \left(\mathbf{G} \times \frac{d\mathbf{H}}{dt}\right)$$

Proof: (i) $[\mathbf{F}\ \mathbf{G}\ \mathbf{H}] = \mathbf{F}.(\mathbf{G} \times \mathbf{H})$ (= Scalar triple product)

$$\therefore \quad \frac{d}{dt}[\mathbf{F}\ \mathbf{G}\ \mathbf{H}] = \frac{d}{dt}[\mathbf{F}.(\mathbf{G} \times \mathbf{H})]$$

$$= \mathbf{F}.\frac{d}{dt}(\mathbf{G} \times \mathbf{H}) + \frac{d\mathbf{F}}{dt}.(\mathbf{G} \times \mathbf{H})$$

$$= \mathbf{F}.\left(\mathbf{G} \times \frac{d\mathbf{H}}{dt} + \frac{d\mathbf{G}}{dt} \times \mathbf{H}\right) + \frac{d\mathbf{F}}{dt}.(\mathbf{G} \times \mathbf{H})$$

$$= \mathbf{F}.\left(G \times \frac{d\mathbf{H}}{dt}\right) + \mathbf{F}.\left(\frac{d\mathbf{G}}{dt} \times \mathbf{H}\right) + \frac{d\mathbf{F}}{dt}.(\mathbf{G} \times \mathbf{H})$$

$$= \left[\mathbf{F}\mathbf{G}\frac{d\mathbf{H}}{dt}\right] + \left[\mathbf{F}\frac{d\mathbf{G}}{dt}\mathbf{H}\right] + \left[\frac{d\mathbf{F}}{dt}\mathbf{G}\ \mathbf{H}\right]$$

or $$\frac{d}{dt}[\mathbf{F}\ \mathbf{G}\ \mathbf{H}] = \left[\frac{d\mathbf{F}}{dt}\mathbf{G}\mathbf{H}\right] + \left[\mathbf{F}\frac{d\mathbf{G}}{dt}\mathbf{H}\right] + \left[\mathbf{F}\mathbf{G}\frac{d\mathbf{H}}{dt}\right]$$

(ii) $$\frac{d}{dt}[\mathbf{F} \times (\mathbf{G} \times \mathbf{H})] = \mathbf{F} \times \frac{d}{dt}(\mathbf{G} \times \mathbf{H}) + \frac{d\mathbf{G}}{dt} \times (\mathbf{G} \times \mathbf{H})$$

$$= \mathbf{F} \times \left(\mathbf{G} \times \frac{d\mathbf{H}}{dt} + \frac{d\mathbf{G}}{dt} \times \mathbf{H}\right) + \frac{d\mathbf{F}}{dt} \times (\mathbf{G} \times \mathbf{H})$$

$$= \mathbf{F} \times \left(\mathbf{G} \times \frac{d\mathbf{H}}{dt}\right) + \mathbf{F} \times \left(\frac{d\mathbf{G}}{dt} \times \mathbf{H}\right) + \frac{d\mathbf{F}}{dt} \times (\mathbf{G} \times \mathbf{H})$$

or $$\frac{d}{dt}[\mathbf{F} \times (\mathbf{G} \times \mathbf{H})] = \frac{d\mathbf{F}}{dt} \times (\mathbf{G} \times \mathbf{H}) + \mathbf{F} \times \left(\frac{d\mathbf{G}}{dt} \times \mathbf{H}\right) + \mathbf{F} \times \left(\mathbf{G} \times \frac{d\mathbf{H}}{dt}\right)$$

10.9. Derivative of function of a function

Let **F** be a vector function of scalar variable s and s be a continuous function of another scalar variable t *i.e.,* s = ϕ (t). Then **F** is a derivable function of t.

$$\therefore \quad \frac{d\mathbf{F}}{dt} = \frac{d\mathbf{F}}{ds} \cdot \frac{d\phi}{dt}$$

Proof: Let δt be a small increment in t which produces corresponding increments $\delta\mathbf{F}$ and δs in **F** and s respectively.

$\delta\mathbf{F}$ and δs tends to zero as $\delta t \to 0$

$$\therefore \quad \lim_{\delta t \to 0} \frac{\delta\mathbf{F}}{\delta t} = \left(\lim_{\delta t \to 0} \frac{\delta\mathbf{F}}{\delta s}\right)\left(\lim_{\delta t \to 0} \frac{\delta s}{\delta t}\right)$$

or

$$\frac{d\mathbf{F}}{dt} = \frac{d\mathbf{F}}{ds} \cdot \frac{ds}{dt}$$

$$= \frac{d\mathbf{F}}{ds} \cdot \frac{d\phi}{dt}. \quad [\because \; s = \phi(t)]$$

10.10. Derivative of a constant vector

A vector is said to be constant only if both its magnitude and direction do not change i.e. fixed.

Let δt be a small increment in t, let F(t) be a constant vector function of t

$$\therefore \quad \mathbf{F}(t + \delta t) = \mathbf{F}(t) \quad \text{or} \quad \mathbf{F}(t + \delta t) - \mathbf{F}(t) = \mathbf{0}$$

or

$$\lim_{\delta t \to 0} \frac{\mathbf{F}(t + \delta t) - \mathbf{F}(t)}{\delta t} = \mathbf{0} \quad \therefore \quad \frac{d\mathbf{F}}{dt} = \mathbf{0}$$

i.e. the derivative of a constant vector function is a null vector.

10.11. Few theorems

Theorem 1: The necessary and sufficient condition for F (t) to be constant is $\frac{d\mathbf{F}}{dt} = \mathbf{0}$

Proof: *The condition is necessary.*

Given **F** (t) is constant vector function of the scalar variable t

Since **F** (t) is a constant vector function

$$\therefore \quad \mathbf{F}(t + \delta t) = \mathbf{F}(t)$$

$$\Rightarrow \quad \mathbf{F}(t + \delta t) - \mathbf{F}(t) = \mathbf{0}$$

$$\therefore \quad \frac{d\mathbf{F}(t)}{dt} = \lim_{\delta t \to 0} \frac{\mathbf{F}(t + \delta t) - \mathbf{F}(t)}{\delta t} = \mathbf{0},$$

which is the necessary condition.

The condition is sufficient.

Let $$\frac{d\mathbf{F}(t)}{dt} = \mathbf{0}$$

Then to prove that **F** (t) is a constant vector function, let $\mathbf{F}(t)\, f_1(t)\,\mathbf{i} + f_2(t)\,\mathbf{k}$ where $f_1(t)$, $f(t)$, $f_3(t)$ are components

$$\therefore \quad \frac{d\mathbf{F}(t)}{dt} = \frac{df_1(t)}{dt}\mathbf{i} + \frac{df_2(t)}{dt}\mathbf{j} + \frac{df_2(t)}{dt}\mathbf{k}$$

$$\Rightarrow \quad \mathbf{0} = \frac{df_1(t)}{dt}\mathbf{i} + \frac{df_2(t)}{dt}\mathbf{j} + \frac{df_3(t)}{dt}\mathbf{k}$$

$$\Leftrightarrow \quad \frac{df_1(t)}{dt} = 0; \quad \frac{df_2(t)}{dt} = 0; \quad \frac{df_3(t)}{dt} = 0$$

Hence $f_1(t), f_2(t), f_3(t)$ are constant scalars i.e., they are independent of t

$\therefore$ **F** (t) is a constant vector function, which is the sufficient condition.

Theorem 2: The necessary and sufficient condition for the vector function **F** (t) to have constant magnitude is

$$\mathbf{F} \cdot \frac{d\mathbf{F}}{dt} = 0$$

Proof: *The condition is necessary.*

Let **F** (t) be a vector function of scalar variable t of constant magnitude

$$\therefore \quad |\mathbf{F}| = \text{constant and } \mathbf{F}.\mathbf{F} = |\mathbf{F}|^2 = \text{constant}$$

$$\therefore \quad \frac{d}{dt}(\mathbf{F}.\mathbf{F}) = \mathbf{0} \Rightarrow \mathbf{F}.\frac{d\mathbf{F}}{dt} + \frac{d\mathbf{F}}{dt}.\mathbf{F} = \mathbf{0}$$

$$\Rightarrow \quad 2\,\mathbf{F}.\frac{d\mathbf{F}}{dt} = \mathbf{0} \text{ or } \mathbf{F}.\frac{d\mathbf{F}}{dt} = \mathbf{0},$$

which is the required necessary condition.

The condition is sufficient

Let $\mathbf{F}.\frac{d\mathbf{F}}{dt} = \mathbf{0} \quad \therefore \quad \mathbf{F}.\frac{d\mathbf{F}}{dt} + \frac{d\mathbf{F}}{dt}.\mathbf{F} = \mathbf{0}$

or $\frac{d}{dt}(\mathbf{F.F}) = \mathbf{0}$ or $\mathbf{F.F}$ = constant

or $|\mathbf{F}|^2$ = constant or $|\mathbf{F}|$ = constant,

which is the required condition.

Theorem 3. The necessary and sufficient condition for the vector function **F** (t) to have an constant direction is $F \times \frac{d\mathbf{F}}{dt} = \mathbf{0}$

Proof: Let **F** (t) be a vector function of the scalar variable t. Let ϕ (t) denotes the magnitude of **F** (t) and let **f** (t) be a unit vector function in the direction of **F** (t) then $\mathbf{F} = \phi\,\mathbf{f}$.

$$\therefore \quad \frac{d\mathbf{F}}{dt} = \phi\frac{d\mathbf{f}}{dt} + \frac{d\phi}{dt}\mathbf{f}$$

Thus $\mathbf{F} \times \frac{d\mathbf{F}}{dt} = (\phi\mathbf{f}) \times \left(\phi\frac{d\mathbf{f}}{dt} + \frac{d\phi}{dt}\mathbf{f}\right)$

$$= \phi^2\,\mathbf{f} \times \frac{d\mathbf{f}}{dt} + \phi\frac{d\phi}{dt} \; (\therefore \mathbf{f} \times \mathbf{f} = 0) \quad ...(1)$$

The condition is necessary: Suppose **F** (t) has a constant direction. Then **F** is a constant vector because **F** has a constant direction and magnitude.

$$\therefore \quad \frac{d\mathbf{f}}{dt} = 0$$

From (1) $\mathbf{F} \times \frac{d\mathbf{F}}{dt} = \phi^2\,\mathbf{f} \times \frac{d\mathbf{f}}{dt} = \mathbf{0}$

which is the required necessary condition.

The condition is sufficient

$$\because \quad F \times \frac{d\mathbf{F}}{dt} = \mathbf{0} \quad \phi^2\mathbf{f} \times \frac{d\mathbf{f}}{dt} = \mathbf{0} \text{ From (1)}$$

$$\Rightarrow \quad \mathbf{f} \times \frac{d\mathbf{f}}{dt} = \mathbf{0} \quad ...(2)$$

[$\because \phi$ is not constantly zero]

$\because$ **f** is of constant magnitude $\mathbf{f} \cdot \frac{d\mathbf{f}}{dt} = \mathbf{0}$ ...(3)

From (2) and (3), we get $\frac{d\mathbf{f}}{dt} = \mathbf{0}$

$\Rightarrow$ **f** is a constant vector function both in magnitude and direction.

But **F** = ϕ [from 1]

$\therefore$ Direction of **F** is also constant, which is the required sufficient condition.

SOLVED PROBLEMS

Example 1. Verify the formulae

(i) $\frac{d}{dt}(\mathbf{F}.\mathbf{G}) = \mathbf{F}.\frac{d\mathbf{G}}{dt} + \frac{d\mathbf{F}}{dt}.\mathbf{G}$

(ii) $\frac{d}{dt}(\mathbf{F} \times \mathbf{G}) = \mathbf{F} \times \frac{d\mathbf{G}}{dt} + \frac{d\mathbf{F}}{dt} \times \mathbf{G}$,

for $\mathbf{F} = 5\,t^2\mathbf{i} + t\mathbf{j} - t^3\mathbf{k}$

and $\mathbf{G} = \sin t\,\mathbf{i} - \cos t\,\mathbf{j}$

Solution : (i) $\mathbf{F}.\mathbf{G} = (5t^2\,\mathbf{i} + t\,\mathbf{j} - t^3\mathbf{k}).(\sin t\,\mathbf{i} - \cos t\,\mathbf{j})$

$= (5t^2 \sin t) + (-\,t \cos t) + 0$

$[\because \mathbf{i}.\mathbf{i} = \mathbf{j}.\mathbf{j} = \mathbf{k}.\mathbf{k} = 1]$

$= 5\,t^2 \sin t - t \cos t$

$\therefore \quad \frac{d}{dt}(\mathbf{F}.\mathbf{G}) = \frac{d}{dt}(5\,t^2 \sin t - t \cos t)$

$= 5(t^2 \cos t + 2t \sin t) - (-\,t \sin t + \cos t)$

$= (5\,t^2 - 1) \cos t + 11t \sin t$ ...(1)

Now $\frac{d\mathbf{F}}{dt} = \frac{d}{dt}(5\,t^2\mathbf{i} + t\,\mathbf{j} - t^3\,\mathbf{k}) = 10\,t\,\mathbf{i} + \mathbf{j} - 3\,t^2\,\mathbf{k}$

and $\frac{d\mathbf{G}}{dt} = \frac{d}{dt}(\sin t\,\mathbf{i} - \cos t\,\mathbf{j}) = \cos t\,\mathbf{i} + \sin t\,\mathbf{j}$

$$\therefore \quad \mathbf{F}.\frac{d\mathbf{G}}{dt} + \frac{d\mathbf{F}}{dt}.\mathbf{G} = (5t^2\,\mathbf{i} + t\mathbf{j} - t^3\mathbf{k}).(\cos t\,\mathbf{i} + \sin t\,\mathbf{j})$$

$$+ (10\,t\,\mathbf{i} + \mathbf{j} - 3\,t^2\,\mathbf{k}).(\sin t\,\mathbf{i} - \cos t\,\mathbf{j})$$

$$= (5\,t^2 \cos t + t \sin t + 0) + (10\,t \sin t - \cos t)$$

$$= (5\,t^2 - 1) \cos t + 11\,t \sin t \qquad ...(2)$$

$\therefore$ From (1) and (2)

$$\frac{d}{dt}(\mathbf{F}.\mathbf{G}) = \mathbf{F}.\frac{d\mathbf{G}}{dt} + \frac{d\mathbf{F}}{dt}.\mathbf{G}$$

Hence the verification.

$$\text{(ii) } \mathbf{F} \times \mathbf{G} = \begin{vmatrix} \mathbf{i} & \mathbf{j} & \mathbf{k} \\ 5\,t^2 & t & -t^3 \\ \sin t & -\cos t & 0 \end{vmatrix}$$

$$= (-t^3 \cos t)\,\mathbf{i} - (t^3 \sin t)\mathbf{j} + (-5\,t^2 \cos t - t \sin t)\,\mathbf{k}$$

$$\therefore \quad \frac{d}{dt}(\mathbf{F} \times \mathbf{G})$$

$$= -(3\,t^2 \cos t - t^3 \sin t)\,\mathbf{i}$$

$$- (3\,t_2 \sin t + t_3 \cos t)\,\mathbf{j}$$

$$- [5\,(2\,t \cos t - t^2 \sin t) + (t \cos t + \sin t)]\,\mathbf{k}$$

$$= (t^3 \sin t - 3\,t^2 \cos t)\,\mathbf{i} - (3\,t^2 \sin t + t^3 \cos t)\,\mathbf{j}$$

$$+ (5\,t^2 \sin t - 11\,t \cos t - \sin t)\,\mathbf{k} \qquad ...(1)$$

$$\text{Now } \mathbf{F} \times \frac{d\mathbf{G}}{dt} + \frac{d\mathbf{F}}{dt} \times \mathbf{G} = (5\,t^2\,\mathbf{i} + t\,\mathbf{j} - t^3\,\mathbf{k}) \times (\cos t\,\mathbf{i} + \sin t\,\mathbf{j})$$

$$+ (10\,t\,\mathbf{i} + \mathbf{j} \;\; 3t^2\,\mathbf{k}) \times (\sin t\,\mathbf{i} - \cos t\,\mathbf{j})$$

$$= \begin{vmatrix} \mathbf{i} & \mathbf{j} & \mathbf{k} \\ 5\,t^2 & t & -t^3 \\ \cos t & \sin t & 0 \end{vmatrix} + \begin{vmatrix} \mathbf{i} & \mathbf{j} & \mathbf{k} \\ 10\,t & 1 & -3\,t^2 \\ \sin t & -\cos t & 0 \end{vmatrix}$$

$$= [(t^3 \sin t)\,\mathbf{i} - (t^3 \cos t)\,\mathbf{j} + (5\,t^2 \sin t - t \cos t)\,\mathbf{k}]$$

$$+ [(-3\,t^2 \cos t)\mathbf{i} + (-3\,t^2 \sin t)\,\mathbf{j} - (10\,t \cos t + \sin t)\,\mathbf{k}]$$

$$= (t^3 \sin t - 3\,t^2 \cos t)\,\mathbf{i} - (t^3 \cos t + 3\,t^2 \cos t)\,\mathbf{j}$$

$$+ (5\,t^2 \sin t - 11\,t \cos t - \sin t)\,\mathbf{k} \qquad ...(2)$$

$\therefore$ From (1) and (2)

$$\frac{d}{dt}(\mathbf{F} \times \mathbf{G}) = \mathbf{F} \times \frac{d\mathbf{G}}{dt} + \frac{d\mathbf{F}}{dt} \times \mathbf{G}$$

Hence the verification.

Example 2. If $\mathbf{r} = a \cos t\, \mathbf{i} + a \sin t\, \mathbf{j} + at \tan \alpha\, \mathbf{k}$

Show that (i) $\left| \frac{d\mathbf{r}}{dt} \times \frac{d^2\mathbf{r}}{dt^2} \right| = a^2 \sec \alpha$

(ii) $\left(\frac{d\mathbf{r}}{dt} \times \frac{d^2\mathbf{r}}{dt^2} \right) \cdot \frac{d^3\mathbf{r}}{dt^3} = a^3 \tan \alpha$

Solution: Given $\mathbf{r} = a \cos t\, \mathbf{i} + a \sin t\, \mathbf{j} + at \tan \alpha\, \mathbf{k}$

$$\therefore \quad \frac{d\mathbf{r}}{dt} = \frac{d}{dt}(a \cos t\, \mathbf{i} + a \sin t\, \mathbf{j} + at \tan \alpha\, \mathbf{k})$$

$$= (-a \sin t)\, \mathbf{i} + (a \cos t)\, \mathbf{j} + (a \tan \alpha)\, \mathbf{k}$$

$$\frac{d^2\mathbf{r}}{dt^2} = \frac{d}{dt}(-a \sin t)\, \mathbf{i} + \frac{d}{dt}(a \cos t)\, \mathbf{j} + \frac{d}{dt}(a \tan \alpha)\, \mathbf{k}$$

$$= -(a \cos t\, \mathbf{i} + a \sin t\, \mathbf{j})$$

and
$$\frac{d^3\mathbf{r}}{dt^3} = -\left[\frac{d}{dt}(a \cos t)\, \mathbf{i} + \frac{d}{dt}(a \sin t)\, \mathbf{j} \right]$$

$$= a \sin t\, \mathbf{i} - a \cos t\, \mathbf{j}$$

(i)
$$\frac{d\mathbf{r}}{dt} \times \frac{d^2\mathbf{r}}{dt^2} = \begin{vmatrix} \mathbf{i} & \mathbf{j} & \mathbf{k} \\ -a \sin t & a \cos t & a \tan \alpha \\ -a \cos t & -a \sin t & 0 \end{vmatrix}$$

$$= a^2 \sin t \tan \alpha\, \mathbf{i} - a^2 \cos t \tan \alpha\, \mathbf{j} + a^2\, \mathbf{k}$$

$$= a^2 (\sin t \tan \alpha\, \mathbf{i} - \cos t \tan \alpha\, \mathbf{j} + \mathbf{k})$$

$$\left| \frac{d\mathbf{r}}{dt} \times \frac{d^2\mathbf{r}}{dt^2} \right| = a^2 \sqrt{(\sin t \tan \alpha)^2 + (-\cos t \tan \alpha)^3 + (1)^2}$$

$$= a^2 \sqrt{\tan^2 \alpha\, (\sin^2 t + \cos^2 t) + 1}$$

$$= a^2 \sqrt{\tan^2 \alpha + 1} = a^2 \sec \alpha$$

$$\therefore \quad \left| \frac{d\mathbf{r}}{dt} \times \frac{d^2\mathbf{r}}{dt^2} \right| = a^2 \sec \alpha$$

(ii) $\left(\frac{d\mathbf{r}}{dt} \times \frac{d^2\mathbf{r}}{dt^2}\right) \cdot \frac{d^3\mathbf{r}}{dt^3}$

$= a^2 (\sin t \tan \alpha\, \mathbf{i} - \cos t \tan \alpha\, \mathbf{j} + \mathbf{k}).(a \sin t\, \mathbf{i} - a \cos t\, \mathbf{j})$

$= a^3 (\sin^2 t \tan \alpha + \cos^2 t \tan \alpha)$

$= a^3 \tan \alpha (\sin^2 t + \cos^2 t)$

$= a^3 \tan \alpha$

Hence the results.

Example 3. Show that if $\mathbf{r} = \mathbf{a} \cos \omega t + \mathbf{b} \sin \omega t$, where **a, b,** ω are constants, then

$$\frac{d^2\mathbf{r}}{dt^2} = -\omega^2 \mathbf{r} \text{ and } \mathbf{r} \times \frac{d\mathbf{r}}{dt} = -\omega\, \mathbf{a} \times \mathbf{b}$$

Solution: Here $\mathbf{r} = \mathbf{a} \cos \omega t + \mathbf{b} \sin \omega t$ ...(1)

$\therefore \quad \frac{d\mathbf{r}}{dt} = \mathbf{a} (-\sin \omega t)\, \omega + \mathbf{b} (\cos \omega t)\, \omega$

$= \omega (-\mathbf{a} \sin \omega t + \mathbf{b} \cos \omega t)$ ...(2)

Now $\frac{d^2\mathbf{r}}{dt^2} = \omega (-\mathbf{a} (\cos \omega t)\, \omega + \mathbf{b} (-\sin \omega t)\, \omega)$

$= -\omega^2 (\mathbf{a} \cos \omega t + \mathbf{b} \sin \omega t)$

$= -\omega^2 \mathbf{r}$ [from (1)]

$\therefore \quad \frac{d^2\mathbf{r}}{dt^2} = -\omega^2 \mathbf{r}$ Hence the first result.

Consider

$\mathbf{r} \times \frac{d\mathbf{r}}{dt} = (\mathbf{a} \cos \omega t + \mathbf{b} \sin \omega t) \times \omega (-\mathbf{a} \sin \omega t + \mathbf{b} \cos \omega t)$

$= \omega [(\mathbf{a} \times \mathbf{b}) \cos^2 \omega t + (-\mathbf{b} \times \mathbf{a}) \sin^2 \omega t]$

$\because \mathbf{a} \times \mathbf{a} = \mathbf{0};\ \mathbf{b} \times \mathbf{b} = \mathbf{0}$

$= \omega [(\mathbf{a} \times \mathbf{b}) \cos^2 \omega t + (\mathbf{a} \times \mathbf{b}) \sin^2 \omega t]$

$(\because \mathbf{a} \times \mathbf{b} = -\mathbf{b} \times \mathbf{a})$

$= \omega (\cos^2 \omega t + \sin^2 \omega t) (\mathbf{a} \times \mathbf{b})$

$= \omega\, \mathbf{a} \times \mathbf{b}$ $(\because \sin^2 \omega t + \cos^2 \omega t = 1)$

$$\therefore \quad \mathbf{r} \times \frac{d\mathbf{r}}{dt} = \omega\, \mathbf{a} \times \mathbf{b}$$

Hence the second result.

Example 4: If $\frac{d\mathbf{U}}{dt} = \mathbf{W} \times \mathbf{U}$ and $\frac{d\mathbf{V}}{dt} = \mathbf{W} \times \mathbf{V}$,

prove that $$\frac{d}{dt}(\mathbf{U} \times \mathbf{V}) = \mathbf{W} \times (\mathbf{U} \times \mathbf{V})$$

Solution: $\frac{d}{dt}(\mathbf{U} \times \mathbf{V}) = \mathbf{U} \times \frac{d\mathbf{V}}{dt} + \frac{d\mathbf{U}}{dt} \times \mathbf{V}$

$$\begin{aligned}
\textbf{R.H.S} &= \mathbf{U} \times (\mathbf{W} \times \mathbf{V}) + (\mathbf{W} \times \mathbf{U}) \times \mathbf{V} \\
&= [(\mathbf{U}.\mathbf{V})\, \mathbf{W} - (\mathbf{U}.\mathbf{W})\, \mathbf{V}] + [(\mathbf{V}.\mathbf{W})\mathbf{U} - (\mathbf{V}.\mathbf{U})\, \mathbf{W}] \\
&= (\mathbf{V}.\mathbf{W})\, \mathbf{U} - (\mathbf{U}.\mathbf{W})\, \mathbf{V} \qquad (\because \mathbf{U}.\mathbf{V} = \mathbf{V}.\mathbf{U}) \\
&= (\mathbf{W}.\mathbf{V})\, \mathbf{U} - (\mathbf{W}.\mathbf{U})\mathbf{V} = \mathbf{W} \times (\mathbf{U} \times \mathbf{V})
\end{aligned}$$

$$\therefore \quad \frac{d}{dt}(\mathbf{U} \times \mathbf{V}) = \mathbf{W} \times (\mathbf{U} \times \mathbf{V})$$

Hence the result.

Example 5: Find the derivative of $\mathbf{F} = \frac{\mathbf{r}}{r}$ with respect to **t**, where $\mathbf{r} = \mathbf{r}\,(t)$ and $|\,\mathbf{r}\,| = r$, then show that

$$\mathbf{F} \times \frac{d\mathbf{F}}{dt} = \frac{\mathbf{r}}{r^2} \times \frac{d\mathbf{r}}{dt}$$

Solution: We have $\mathbf{F} = \frac{\mathbf{r}}{r}$

$$\therefore \quad \frac{d\mathbf{F}}{dt} = \frac{d}{dt}\left(\frac{\mathbf{r}}{r}\right) = \frac{1}{r}\frac{d\mathbf{r}}{dt} - \frac{\mathbf{r}}{r^2}\frac{dr}{dt}$$

$$\begin{aligned}
\therefore \quad \mathbf{F} \times \frac{d\mathbf{F}}{dt} &= \frac{\mathbf{r}}{r} \times \left(\frac{1}{r}\frac{d\mathbf{r}}{dt} - \frac{\mathbf{r}}{r^2}\frac{dr}{dt}\right) \\
&= \frac{\mathbf{r}}{r^2} \times \frac{d\mathbf{r}}{dt} - \frac{\mathbf{r} \times \mathbf{r}}{r^3}\frac{dr}{dt} \\
&= \frac{\mathbf{r}}{r^2} \times \frac{d\mathbf{r}}{dt} \qquad [\because \mathbf{r} \times \mathbf{r} = 0]
\end{aligned}$$

$$\therefore \quad \mathbf{F} \times \frac{d\mathbf{F}}{dt} = \frac{\mathbf{r}}{r^2} \times \frac{d\mathbf{r}}{dt}$$

Example 6: Show that $\mathbf{r} = \mathbf{a}e^{mt} + \mathbf{b}e^{nt}$ is the solution of the differential equation

$$\frac{d^2\mathbf{r}}{dt^2} - (m + n)\frac{d\mathbf{r}}{dt} + m\,n\,\mathbf{r} = \mathbf{0}$$

where **a** and **b** are constant vectors.

Solution: We have $\mathbf{r} = \mathbf{a}e^{mt} + \mathbf{b}e^{nt}$

$$\therefore \qquad \frac{d\mathbf{r}}{dt} = \mathbf{a}\,me^{mt} + \mathbf{b}\,ne^{nt}$$

$$\frac{d^2\mathbf{r}}{dt^2} = \mathbf{a}\,m^2\,e^{mt} + \mathbf{b}\,n^2\,e^{nt}$$

$$\therefore \qquad \frac{d^2\mathbf{r}}{dt^2} - (m + n)\frac{d\mathbf{r}}{dt} + m\,n\,\mathbf{r}$$

$$= (\mathbf{a}\,m^2\,e^{mt} + \mathbf{b}\,n^2\,e^{nt}) - (m + n)(\mathbf{a}\,m\,e^{mt} + \mathbf{b}\,n\,e^{nt}) + m\,n(\mathbf{a}\,e^{mt} + \mathbf{b}\,e^{nt})$$

$$= \mathbf{a}\,[m^2 - (m + n)\,m + mn]\,e^{mt} + \mathbf{b}\,[n^2 - (m + n)\,n + mn]e^{nt}$$

$$= \mathbf{0}$$

Hence $\mathbf{r} = \mathbf{a}\,e^{mt} + \mathbf{b}\,e^{nt}$ is the solution of the differential equation

$$\frac{d^2\mathbf{r}}{dt^2} - (m + n)\frac{d\mathbf{r}}{dt} + m\,n\,\mathbf{r} = \mathbf{0}.$$

Example 7: If **F** has a constant magnitude, and if $\frac{d\mathbf{F}}{dt}$ is not a zero vector then prove that $\mathbf{F}.\frac{d\mathbf{F}}{dt} = 0$

Solution: Suppose **F** has a constant magnitude

$$\therefore \qquad |\mathbf{F}| = f = \text{constnat}$$

$$\mathbf{F.F} = |\mathbf{F}|\,|\mathbf{F}|\cos 0° = |\mathbf{F}|^2 = f^2$$

$$\therefore \qquad \frac{d}{dt}(\mathbf{F.F}) = \frac{d}{dt}(f^2)$$

$$= 0 \quad [\because \; f^2 \text{ is constant}]$$

or $$\mathbf{F}.\frac{d\mathbf{F}}{dt} + \frac{d\mathbf{F}}{dt}.\mathbf{F} = 0$$

or $$\mathbf{F}.\frac{d\mathbf{F}}{dt} = 0 \qquad \text{iff} \qquad \left|\frac{d\mathbf{F}}{dt}\right| \neq 0$$

which means that **F** and $\frac{d\mathbf{F}}{dt}$ are perpendicular to each other.

Example 7: If **F** is a derivable vector function of the scalar variable t, then prove that

$$\frac{d}{dt}\left(\mathbf{F} \times \frac{d\mathbf{F}}{dt}\right) = \mathbf{F} \times \frac{d^2\mathbf{F}}{dt^2}$$

Solution: $$\frac{d}{dt}\left(\mathbf{F} \times \frac{d\mathbf{F}}{dt}\right) = \mathbf{F} \times \frac{d}{dt}\left(\frac{d\mathbf{F}}{dt}\right) + \frac{d\mathbf{F}}{dt} \times \frac{d\mathbf{F}}{dt}$$

$$= \mathbf{F} \times \frac{d^2\mathbf{F}}{dt^2} + \frac{d\mathbf{F}}{dt} \times \frac{d\mathbf{F}}{dt}$$

$$= \mathbf{F} \times \frac{d^2\mathbf{F}}{dt^2} + \mathbf{0}$$

Hence $$\frac{d}{dt}\left(\mathbf{F} \times \frac{d\mathbf{F}}{dt}\right) = \mathbf{F} \times \frac{d^2\mathbf{F}}{dt^2}.$$

Example 9: Evaluate (i) $\frac{d}{dt}\left(\mathbf{r}.\frac{d\mathbf{r}}{dt} \times \frac{d^2\mathbf{r}}{dt^2}\right)$

(ii) $\frac{d^2}{dt^2}\left[\mathbf{r}\ \frac{d\mathbf{r}}{dt}\ \frac{d^2\mathbf{r}}{dt^2}\right]$

Solution: We know that

$$\frac{d}{dt}[\mathbf{F}\ \mathbf{G}\ \mathbf{H}] = \left[\frac{d\mathbf{F}}{dt}\mathbf{G}\ \mathbf{H}\right] + \left[\mathbf{F}\ \frac{d\mathbf{G}}{dt}\ \mathbf{H}\right] + \left[\mathbf{F}\ \mathbf{G}\ \frac{d\mathbf{H}}{dt}\right]$$

and we also know that $[\mathbf{F}\ \mathbf{G}\ \mathbf{H}] = \mathbf{F}.(\mathbf{G} \times \mathbf{H})$

$$\therefore \quad \frac{d}{dt}\left(\mathbf{r}.\frac{d\mathbf{r}}{dt} \times \frac{d^2\mathbf{r}}{dt^2}\right) = \frac{d}{dt}\left[\mathbf{R}\ \frac{d\mathbf{r}}{dt}\ \frac{d^2\mathbf{r}}{dt^2}\right]$$

$$= \left[\frac{d\mathbf{r}}{dt}\frac{d\mathbf{r}}{dt}\frac{d^2\mathbf{r}}{dt^2}\right] + \left[\mathbf{r}\frac{d^2\mathbf{r}}{dt^2}\frac{d^2\mathbf{r}}{dt^2}\right] + \left[\mathbf{r}\frac{d\mathbf{r}}{dt}\frac{d^3\mathbf{r}}{dt^3}\right]$$

$$= \mathbf{0} + \mathbf{0} + \left[\mathbf{r}\frac{d\mathbf{r}}{dt}\frac{d^3\mathbf{r}}{dt^3}\right]$$

$$\therefore \quad \frac{d}{dt}\left(\mathbf{r}\cdot\frac{d\mathbf{r}}{dt}\times\frac{d^2\mathbf{r}}{dt^2}\right) = \left[\mathbf{r}\frac{d\mathbf{r}}{dt}\frac{d^3\mathbf{r}}{dt^3}\right] \qquad ...(1)$$

(ii) Differentiating (1) w.r.t once again

$$\frac{d^2}{dt^2}\left[\mathbf{r}\frac{d\mathbf{r}}{dt}\frac{d^2\mathbf{r}}{dt^2}\right] = \frac{d}{dt}\left[\mathbf{r}\frac{d\mathbf{r}}{dt}\frac{d^3\mathbf{r}}{dt^3}\right]$$

$$= \left[\frac{d\mathbf{r}}{dt}\frac{d\mathbf{r}}{dt}\frac{d^3\mathbf{r}}{dt^3}\right] + \left[\mathbf{r}\frac{d^2\mathbf{r}}{dt^2}\frac{d^3\mathbf{r}}{dt^3}\right] + \left[\mathbf{r}\frac{d\mathbf{r}}{dt}\frac{d^4\mathbf{r}}{dt^4}\right]$$

$$= 0 + \left[\mathbf{r}\frac{d^2\mathbf{r}}{dt^2}\frac{d^3\mathbf{r}}{dt^3}\right] + \left[\mathbf{r}\frac{d\mathbf{r}}{dt}\frac{d^4\mathbf{r}}{dt^4}\right]$$

$$\therefore \quad \frac{d^2}{dt^2}\left[\mathbf{r}\frac{d\mathbf{r}}{dt}\frac{d^2\mathbf{r}}{dt^2}\right] = \left[\mathbf{r}\frac{d^2\mathbf{r}}{dt^2}\frac{d^3\mathbf{r}}{dt^2}\right] + \left[\mathbf{r}\frac{d\mathbf{r}}{dt}\frac{d^4\mathbf{r}}{dt^4}\right]$$

Example 10. Evaluate $\dfrac{d^2}{dt^2}\left[\mathbf{r}\times\left(\dfrac{d\mathbf{r}}{dt}\times\dfrac{d^2\mathbf{r}}{dt^2}\right)\right]$

Solution: Let $\mathbf{F} = \mathbf{r}\times\left(\dfrac{d\mathbf{r}}{dt}\times\dfrac{d^2\mathbf{r}}{dt^2}\right)$

$$\therefore \quad \frac{d\mathbf{F}}{dt} = \frac{d}{dt}\left[\mathbf{r}\times\left(\frac{d\mathbf{r}}{dt}\times\frac{d^2\mathbf{r}}{dt^2}\right)\right]$$

$$= \frac{d\mathbf{r}}{dt}\times\left(\frac{d\mathbf{r}}{dt}\times\frac{d^2\mathbf{r}}{dt^2}\right) + \mathbf{r}\times\left(\frac{d^2\mathbf{r}}{dt^2}\times\frac{d^2\mathbf{r}}{dt^2}\right) + \mathbf{r}\times\left(\frac{d\mathbf{r}}{dt}\times\frac{d^3\mathbf{r}}{dt^3}\right)$$

$$= \frac{d\mathbf{r}}{dt}\times\left(\frac{d\mathbf{r}}{dt}\times\frac{d^2\mathbf{r}}{dt^2}\right) + \mathbf{r}\times\left(\frac{d\mathbf{r}}{dt}\times\frac{d^3\mathbf{r}}{dt^3}\right)$$

[The second term = 0, $\therefore$ $\mathbf{A}\times\mathbf{A} = \mathbf{0}$]

Now $\dfrac{d^2\mathbf{F}}{dt^2} = \dfrac{d}{dt}\left(\dfrac{d\mathbf{F}}{dt}\right)$

$$= \frac{d}{dt}\left[\frac{d\mathbf{r}}{dt} \times \left(\frac{d\mathbf{r}}{dt} \times \frac{d^2\mathbf{r}}{dt^2}\right)\right] + \frac{d}{dt}\left[\mathbf{r} \times \left(\frac{d\mathbf{r}}{dt} \times \frac{d^3\mathbf{r}}{dt^3}\right)\right]$$

$$= \left[\frac{d^2\mathbf{r}}{dt^2} \times \left(\frac{d\mathbf{r}}{dt} \times \frac{d^2\mathbf{r}}{dt^3}\right) + \frac{d\mathbf{r}}{dt} \times \left(\frac{d^2\mathbf{r}}{dt^2} \times \frac{d^2\mathbf{r}}{dt^2}\right)\right.$$

$$\left. + \frac{d\mathbf{r}}{dt} \times \left(\frac{d\mathbf{r}}{dt} \times \frac{d^3\mathbf{r}}{dt^3}\right)\right] + \left[\frac{d\mathbf{r}}{dt} \times \left(\frac{d\mathbf{r}}{dt} \times \frac{d^3\mathbf{r}}{dt^3}\right)\right.$$

$$\left. + \mathbf{r} \times \left(\frac{d^2\mathbf{r}}{dt^2} \times \frac{d^3\mathbf{r}}{dt^3}\right) + \mathbf{r} \times \left(\frac{d\mathbf{r}}{dt} \times \frac{d^4\mathbf{r}}{dt^4}\right)\right]$$

$$= \frac{d^2\mathbf{r}}{dt^2} \times \left(\frac{d\mathbf{r}}{dt} \times \frac{d^2\mathbf{r}}{dt^2}\right) + \frac{d\mathbf{r}}{dt} \times \left(\frac{d\mathbf{r}}{dt} \times \frac{d^3\mathbf{r}}{dt^3}\right)$$

$$+ \frac{d\mathbf{r}}{dt} \times \left(\frac{d\mathbf{r}}{dt} \times \frac{d^3\mathbf{r}}{dt^3}\right) + \mathbf{r} \times \left(\frac{d^2\mathbf{r}}{dt^2} \times \frac{d^3\mathbf{r}}{dt^3}\right)$$

$$+ \mathbf{r} \times \left(\frac{d\mathbf{r}}{dt} \times \frac{d^4\mathbf{r}}{dt^4}\right).$$

Example 11: If **r** is a vector function of a scalar t and **a** is a constant vector, differentiate the following w.r.t 't'.

(i) $\mathbf{r}^2 + \dfrac{1}{\mathbf{r}^2}$ (ii) $\dfrac{\mathbf{r} \times \mathbf{a}}{\mathbf{r}.\mathbf{a}}$ (iii) $\dfrac{\mathbf{r} + \mathbf{a}}{\mathbf{r}^2 + \mathbf{a}^2}$

Solution: (i) $\dfrac{d}{dt}\left(\mathbf{r}^2 + \dfrac{1}{\mathbf{r}^2}\right) = \dfrac{d}{dt}(\mathbf{r}^2) + \dfrac{d}{dt}\left(\dfrac{1}{\mathbf{r}^2}\right)$

$$= \frac{d}{dt}(r^2) + \frac{d}{dt}\left(\frac{1}{r^2}\right) \qquad [\because \mathbf{r}^2 = r^2]$$

$$= 2\,r\frac{dr}{dt} - \frac{2}{r^3}\frac{dr}{dt}$$

(ii) $\dfrac{d}{dt}\left(\dfrac{\mathbf{r} \times \mathbf{a}}{\mathbf{r}.\mathbf{a}}\right) = \dfrac{d}{dt}(\mathbf{r} \times \mathbf{a})\dfrac{1}{\mathbf{r}.\mathbf{a}} + (\mathbf{r} \times \mathbf{a})\dfrac{d}{dt}\left(\dfrac{1}{\mathbf{r}.\mathbf{a}}\right)$

$$= \left(\frac{d\mathbf{r}}{dt} \times \mathbf{a} + \mathbf{r} \times \frac{d\mathbf{a}}{dt} \right) \cdot \frac{1}{\mathbf{r.a}}$$

$$+ (\mathbf{r} \times \mathbf{a}) \left(\frac{-1}{(\mathbf{r.a})^2} \right) \left(\mathbf{r}.\frac{d\mathbf{a}}{dt} + \frac{d\mathbf{r}}{dt}.\mathbf{a} \right)$$

$$= \frac{\left(\frac{d\mathbf{r}}{dt} \times \mathbf{a} \right)}{(\mathbf{r} . \mathbf{a})} - \frac{(\mathbf{r} \times \mathbf{a}) \left(\frac{d\mathbf{r}}{dt} . \mathbf{a} \right)}{(\mathbf{r} . \mathbf{a})^2}$$

(iii) $$\frac{d}{dt} \left(\frac{\mathbf{r} + \mathbf{a}}{\mathbf{r}^2 + \mathbf{a}^2} \right) = \frac{(\mathbf{r}^2 + \mathbf{a}^2) \frac{d}{dt} (\mathbf{r} + \mathbf{a}) - (\mathbf{r} + \mathbf{a}) \frac{d}{dt} (\mathbf{r}^3 + \mathbf{a}^3)}{(\mathbf{r}^2 + \mathbf{a}^2)^2}$$

$$= \frac{(\mathbf{r}^2 + \mathbf{a}^2) \left(\frac{d\mathbf{r}}{dt} + \frac{d\mathbf{a}}{dt} \right) - (\mathbf{r} + \mathbf{a}) \left(\frac{d\mathbf{r}^2}{dt} + \frac{d\mathbf{a}^2}{dt} \right)}{(\mathbf{r}^2 + \mathbf{a}^2)^2}$$

$$= \frac{(\mathbf{r}^2 + \mathbf{a}^2) \frac{d\mathbf{r}}{dt} - (\mathbf{r} + \mathbf{a}) \left(2\, \mathbf{r} . \frac{d\mathbf{r}}{dt} \right)}{(\mathbf{r}^2 + \mathbf{a}^2)^2}$$

$$= \frac{\frac{d\mathbf{r}}{dt}}{(\mathbf{r}^2 + \mathbf{a}^2)} - \frac{2\, \mathbf{r} \frac{d\mathbf{r}}{dt} (\mathbf{r} + \mathbf{a})}{(\mathbf{r}^2 + \mathbf{a}^2)^2}.$$

Example 12: A particle moves along curve $x = e^{-t}$, $y = 2 \cos 3t$, $z = 2 \sin 3t$ where t is the time variable. Determine its velocity and acceleration vectors and also the magnitudes of velocity and acceleration at $t = 0$.

Solution: Let **r** be the position vector of the particle at tiem t.

Then $\mathbf{r} = x\mathbf{i} + y\mathbf{j} + 2\mathbf{k}$

$$= e^{-t}\,\mathbf{i} + 2 \cos 3t\,\mathbf{j} + 2 \sin 3t\mathbf{k}$$

If **v** and **a** be the velocity and acceleration respectively of the particle at time t.

Then $\mathbf{v} = \frac{d\mathbf{r}}{dt} = -e^{-t}\,\mathbf{i} - 6 \sin 3t\mathbf{j} + 6 \cos 3t\mathbf{k}$

and $\mathbf{a} = \frac{dV}{dt} = \frac{d^2\mathbf{r}}{dt^2}$

$= e^{-t}\,\mathbf{i} - 18 \cos 3\,t\mathbf{j} - 18 \sin 3\,t\mathbf{k}$

Magnitude of velocity:

$$|\mathbf{v}| = \sqrt{(-e^{-t})^2 + (-6 \sin 3\,t)^2 + (6 \cos 3\,t)^2}$$

$$= \sqrt{e^{-2t} + 36(\sin^2 3\,t + \cos^2 3\,t)}$$

at $t = 0$

$$|\mathbf{v}| = \sqrt{1 + 36(1)} = \sqrt{37}$$

Magnitude of acceleration:

$$|\mathbf{a}| = \sqrt{(e^{-t})^2 + (-18 \cos 3\,t)^2 + (-18 \sin 3t)^2}$$

at $t = 0$

$$|\mathbf{a}| = \sqrt{1 + 324(1)} = \sqrt{325.}$$

Example 13: The position vector of a particle at time t is $\mathbf{r} = \cos(t-1)\,\mathbf{i} + \sin h\,(t-1)\,\mathbf{j} + \propto t^3\,\mathbf{k}$. Find the condition imposed on $\propto$ by requiring that at time t = 1 the acceleration is normal to the position vector.

Solution: The position vector of a particle at time t is

$$\mathbf{r} = \cos(t-1)\,\mathbf{i} + \sin h\,(t-1)\,\mathbf{j} + \propto t^3\,\mathbf{k}$$

$$\text{Velocity} = v = \frac{d\mathbf{r}}{dt}$$

$$= -\sin(t-1)\,\mathbf{i} + \cos h\,(t-1)\,\mathbf{j} + 3 \propto t^2\,\mathbf{k}$$

$$\text{Acceleration} = \mathbf{a} = \frac{d^2\,\mathbf{r}}{dt^2}$$

$$= -\cos(t-1)\,\mathbf{i} + \sin h\,(t-1)\,\mathbf{j} + 6 \propto t\,\mathbf{k}$$

Now $\mathbf{a.r} = \cos^2(t-1) + \sin h^2(t-1) + 6 \propto^2 t^4 = 0$

$$[\mathbf{a}\,.\,\mathbf{r}]_{t=1} = \cos^2(0) + \sin h^2(0) + 6 \propto^2 = 0$$

or $-1 + 0 + 6 \propto^2 = 0$

or $\propto = \pm \frac{1}{\sqrt{6}}$

Example 14: A particle moves along the curve $\mathbf{x} = t^3 + 1$, $y = t^2$, $z = 2t + 5$, where t is the time. Find the component of its velocity and acceleration at time t = 1 in the direction of $\mathbf{i} + \mathbf{j} + 3\mathbf{k}$.

Solution: Here $\mathbf{r} = (t^3 + 1)\,\mathbf{i} + t^2\,\mathbf{j} + (2t + 5)\,\mathbf{k}$

(i) Velocity $= \mathbf{v} = \dfrac{d\mathbf{r}}{dt} = 3t^2\,\mathbf{i} + 2t\,\mathbf{j} + 2\,\mathbf{k}$

$= 3\,\mathbf{i} + 2\,\mathbf{j} + 2\,\mathbf{k}$ when t = 1

Unit vector in the direction of $\mathbf{i} + \mathbf{j} + 3\,\mathbf{k}$

$$= \frac{\mathbf{i} + \mathbf{j} + 3\,\mathbf{k}}{\sqrt{1^2 + 1^2 + 3^2}} = \frac{1}{\sqrt{11}}(\mathbf{i} + \mathbf{j} + 3\,\mathbf{k})$$

∴ Required component of the velocity

$$= (3\mathbf{i} + 2\mathbf{j} + 2\mathbf{k}) \cdot \frac{1}{\sqrt{11}}(\mathbf{i} + \mathbf{j} + 3\mathbf{k})$$

$$= \frac{3 + 2 + 6}{\sqrt{11}} = \sqrt{11}$$

(ii) Acceleration

$$\frac{d^2\mathbf{r}}{dt^2} = \mathbf{a} = \frac{d\mathbf{v}}{dt} = 6t\,\mathbf{i} + 2\mathbf{j}$$

$= 6\mathbf{i} + 2\mathbf{j}$ at t = 1

∴ Required component of acceleration

$$= (6\mathbf{i} + 2\mathbf{j}) \cdot \frac{\mathbf{i} + \mathbf{j} + 3\,\mathbf{k}}{\sqrt{11}}$$

$$= \frac{6 + 2}{\sqrt{11}} = \frac{8}{\sqrt{11}}$$

Example 15: A particle moves so that its position vector is given by $\mathbf{r} = \cos\omega t\,\mathbf{i} + \sin\omega t\,\mathbf{j}$ Show that the velocity **v** of the particle is perpendicular to **t** and $\mathbf{r} \times \mathbf{v}$ is a constant vector.

Solution: Here $\mathbf{r} = \cos\omega t\,\mathbf{i} + \sin\omega t\,\mathbf{j}$ where ω is a constant vector

(i) Velocity $= \dfrac{d\mathbf{r}}{dt} = \mathbf{v} = (-\sin\omega t)\,\omega\mathbf{i} + (\cos\omega t)\,\omega\mathbf{j}$

$$\mathbf{v}.\mathbf{r} = (-\,\omega\sin\omega t\,\mathbf{i} + \omega\cos\omega t\,\mathbf{j}) \cdot (\cos\omega t\,\mathbf{i} + \sin\omega t\,\mathbf{j})$$

$$= -\,\omega\sin\omega t\cos\omega t + \omega\cos\omega t\sin\omega t = 0$$

∴ **v** is perpendicular to **r**

$$\mathbf{r} \times \mathbf{v} = \begin{vmatrix} \mathbf{i} & \mathbf{j} & \mathbf{k} \\ \omega \cos t & \omega \sin t & 0 \\ -\omega \sin \omega t & \omega \cos \omega t & 0 \end{vmatrix}$$

$$= \omega^2 (\cos^2 \omega t + \sin^2 \omega t)\mathbf{k}$$

$$= \omega^2 \mathbf{k} \text{ which is a constant vector.}$$

Example 16: A particle (position vector **r**) is moving in a circle with constant angular velocity ω. Show by vector methods that the acceleration is equal to $-\omega^2 \mathbf{r}$.

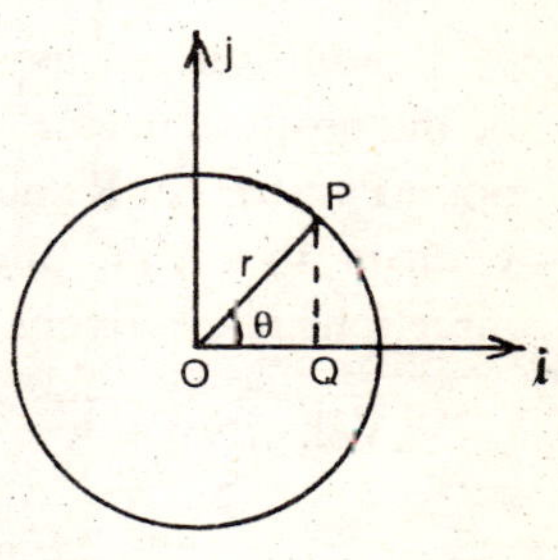

Fig. 10.2

Solution: Let **i** and **j** be unit vectors along two perpendicular radii of the circle centred at **O**. Let **P** be any point on the circle such that **OP** makes an angle θ with **i**. Then the position vector of **P** is given by

$$\mathbf{r} = \vec{OP} = \vec{OQ} + \vec{QP} = (\rho \cos \theta)\, \mathbf{i} + (\rho \sin \theta)\, \mathbf{j} \qquad \text{...(1)}$$

where ρ is the radius of the circle (constant)

$$\therefore \quad \frac{d\mathbf{r}}{dt} = -\rho \sin \theta \frac{d\theta}{dt} \mathbf{i} + \rho \cos \theta \frac{d\theta}{dt} \mathbf{j}$$

$$= (-\rho \sin \theta\, \mathbf{i} + \rho \cos \theta\, \mathbf{j})\, \omega, \text{ where}$$

$$\omega = \frac{d\theta}{dt} \text{ is the constant angular velocity}$$

$$\text{Hence } \mathbf{a} = \frac{d^2\mathbf{r}}{dt^2} = \left(-\rho \cos \theta \frac{d\theta}{dt} \mathbf{i} - \rho \sin \theta \frac{d\theta}{dt} \mathbf{j}\right) \omega$$

$$= -(\rho \cos \theta\, \mathbf{i} + \rho \sin \theta\, \mathbf{j}) \left(\frac{d\theta}{dt}\right) \omega$$

$$= (\rho \cos \theta\, \mathbf{i} + \rho \sin \theta\, \mathbf{j})\, (\text{bold}\omega)\, (\text{bold}\omega)$$

$$= -\mathbf{r}\, \omega^2$$

Hence acceleration = $\mathbf{r}\, \omega^2$.

Example 17. Find the radial and transverse accelerations of a particle moving in a plane curve.

or

Show that the radial and transverse velocities of a particle moving in a plane are respectively $\frac{dr}{dt}$ and $r\frac{d\theta}{dt}$ and acceleration in the same directions are respectively

$$\frac{d^2r}{dt^2} - r\left(\frac{d\theta}{dt}\right)^2 \text{ and } \frac{1}{r}\frac{d}{dt}\left(r^2\frac{d\theta}{dt}\right)$$

Solution: At any time t, let **r** be the position vector of the moving point **P** (r, θ). Let **R** and **T** be the unit vectors in radial and transverse directions respectively. Now **r** = r**R**

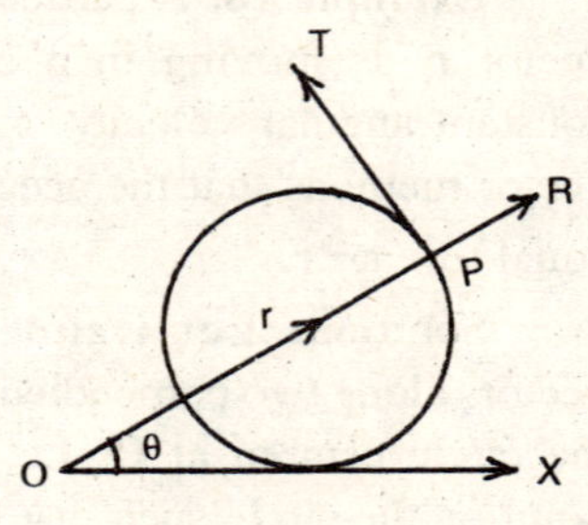

Fig. 10.3

$$\text{Velocity } \mathbf{v} = \frac{d\mathbf{r}}{dt}$$

$$= \frac{dr}{dt}\mathbf{R} + r\frac{d\mathbf{R}}{dt}$$

$$= \frac{dr}{dt}\mathbf{R} + r\frac{d\mathbf{R}}{d\theta}\frac{d\theta}{dt}$$

$$= \frac{dr}{dt}\mathbf{R} + r\frac{d\theta}{dt}\mathbf{R} \qquad \left(\because \frac{d\mathbf{R}}{d\theta} = \mathbf{T}\right)$$

⇒ The components of velocity in the radial and transverse directions are $\frac{dr}{dt}$ and $r\frac{d\theta}{dt}$

$$\text{Acceleration } \mathbf{a} = \frac{d\mathbf{v}}{dt}$$

$$= \left(\frac{d^2r}{dt^2}\mathbf{R} + \frac{dr}{dt}\frac{d\mathbf{R}}{dt}\right)$$

$$+ \left(\frac{dt}{dt}\frac{d\theta}{dt}\mathbf{T} + r\frac{d^2\theta}{dt^2}\mathbf{T} + r\frac{d\theta}{dt}\frac{d\mathbf{T}}{dt}\right)$$

$$= \frac{d^2r}{dt^2}\mathbf{R} + \frac{dr}{dt}\frac{d\mathbf{R}}{d\theta}\frac{d\theta}{dt} + \left(\frac{dr}{dt}\frac{d\theta}{dt} + r\frac{d^2\theta}{dt^2}\right)\mathbf{T}$$

$$+ r\frac{d\theta}{dt}\frac{d\mathbf{T}}{d\theta}\frac{d\theta}{dt}$$

$$= \frac{d^2r}{dt^2}\mathbf{R} + \frac{dr}{dt}\frac{d\theta}{dt}\mathbf{T} + \left(\frac{dr}{dt}\frac{d\theta}{dt} + r\frac{d^2\theta}{dt^2}\right)\mathbf{T} - r\left(\frac{d\theta}{dt}\right)^2\mathbf{R}$$

$$\left(\because \frac{d\mathbf{R}}{d\theta} = \mathbf{T} \text{ and } \frac{d\mathbf{T}}{d\theta} = -\mathbf{R}\right)$$

$$= \left[\frac{d^2r}{dt^2} - r\left(\frac{d\theta}{dt}\right)^2\right]\mathbf{R} + \left(2\frac{dr}{dt}\frac{d\theta}{dt} + r\frac{d^2\theta}{dt^2}\right)\mathbf{T}$$

$$= \left[\frac{d^2r}{dt^2} - r\left(\frac{d\theta}{dt}\right)^2\right]\mathbf{R} + \frac{1}{r}\frac{d}{dt}\left(r^2\frac{d\theta}{dt}\right)\mathbf{T}$$

Hence the radial and transverse components of the acceleration are

$$\left[\frac{d^2r}{dt^2} - r\left(\frac{d\theta}{dt}\right)^2\right] \text{ and } \frac{1}{r}\frac{d}{dt}\left(r^2\frac{d\theta}{dt}\right)$$

Example 18. If two particles are moving on two different curves find the relative velocity and acceleration.

Solution: Let the two particles $\mathbf{P}_1$ and $\mathbf{P}_2$ moving along the curves $\mathbf{C}_1$ and $\mathbf{C}_2$ have position vectors $\mathbf{r}_1$ and $\mathbf{r}_2$ at time t so that

$$\mathbf{r} = \overrightarrow{P_1P_2} = \overrightarrow{OP_2} - \overrightarrow{OP_1} = \mathbf{r}_2 - \mathbf{r}_1$$

Differentiating w.r.t t, we get

$$\frac{d\mathbf{r}}{dt} = \frac{d\mathbf{r}_2}{dt} - \frac{d\mathbf{r}_1}{dt} \quad ...(1)$$

Here $\frac{d\mathbf{r}}{dt}$ defines the relative velocity (vector) of P_2 w.r.t P_1

$\therefore$ Velocity of P_2 relative to P_1 = velocity of P_2 − velocity of P_1

Again differentiating (1) w.r.t t, we have

$$\frac{d^2\mathbf{r}}{dt^2} = \frac{d^2\mathbf{r}_2}{dt^2} - \frac{d^2\mathbf{r}_1}{dt^2}$$

i.e. Acceleration of P_2 relative to P_1

= acceleration of P_2 − acceleration of P_1.

10.12. Partial derivatives of vectors

So far we have studied the differentiation of vector functions of a scaler variable t only. But a vector may be function of more than one scaler variable.

Suppose $$\mathbf{r} = \mathbf{F}(x, y, z) \qquad \text{...(1)}$$

be a function of three independent variables x, y and z.

Then the partial derivative of **r** w.r.t **x** is defined as

$$\frac{\partial \mathbf{r}}{\partial x} = \lim_{\delta x \to 0} \frac{\mathbf{F}(x + \delta x, y, z) - \mathbf{F}(x, y, z)}{\delta x}$$

$\therefore \frac{\partial \mathbf{r}}{\partial x}$ is nothing but the ordinary derivative of **r** w.r.t x, provided the other variables **y** and **z** are constants.

Similarly the partial derivatives

$\frac{\partial \mathbf{r}}{\partial y}$ and $\frac{\partial \mathbf{r}}{\partial z}$ can be defined.

i.e. $$\frac{\partial \mathbf{r}}{\partial y} = \lim_{\delta y \to 0} \frac{\mathbf{F}(x, y + \delta y, z) - \mathbf{F}(x, y, z)}{\delta y}$$

and $$\frac{\partial \mathbf{r}}{\partial z} = \lim_{\delta z \to 0} \frac{\mathbf{F}(x, y, z + \delta z) - \mathbf{F}(x, y, z)}{\delta z}$$

The partial derivatives $\frac{\partial \mathbf{r}}{\partial x}, \frac{\partial \mathbf{r}}{\partial y}, \frac{\partial \mathbf{r}}{\partial z}$ may be differentiated again in order to get higher order partial derivatives

$$\frac{\partial^2 \mathbf{r}}{\partial x^2} = \frac{\partial}{\partial x}\left(\frac{\partial \mathbf{r}}{\partial x}\right), \quad \frac{\partial^2 \mathbf{r}}{\partial x \partial y} = \frac{\partial}{\partial x}\left(\frac{\partial \mathbf{r}}{\partial y}\right)$$

$$\frac{\partial^2 \mathbf{r}}{\partial y^2} = \frac{\partial}{\partial y}\left(\frac{\partial \mathbf{r}}{\partial y}\right), \quad \frac{\partial^2 \mathbf{r}}{\partial \mathbf{x} \partial y} = \frac{\partial}{\partial y}\left(\frac{\partial \mathbf{r}}{\partial x}\right)$$

$$\frac{\partial^2 \mathbf{r}}{\partial z^2} = \frac{\partial}{\partial z}\left(\frac{\partial \mathbf{r}}{\partial z}\right)$$

If **r** has continuous partial derivatives of second order at least then

$$\frac{\partial^2 \mathbf{r}}{\partial \mathbf{x}\, \partial y} = \frac{\partial^2 \mathbf{r}}{\partial y\, \partial x}$$

10.13. Total Differential

Let $$\mathbf{r} = \mathbf{F}(x, y, z) \qquad ...(1)$$
be a vector function of independent variables x, y, and z.

Let δx, δy and δz be small increments in x, y and z.

Let δ**r** be the corresponding increase in **r**

Then $$\mathbf{r} + \delta\mathbf{r} = \mathbf{F}(x + \delta x, y + \delta y, z + \delta z) \qquad ...(2)$$

$$\therefore \quad \delta\mathbf{r} = (x + \delta x, y + \delta y, z + \delta z) - \mathbf{F}(x, y, z)$$

$$= \mathbf{F}(x + \delta x, y + \delta y, z + \delta z) - \mathbf{F}(x, y + \delta y, z + \delta z)$$
$$+ \mathbf{F}(x, y + \delta y, z + \delta z) - \mathbf{F}(x, y, z + \delta z)$$
$$+ \mathbf{F}(x, y, z + \delta z) - \mathbf{F}(x, y, z)$$

$$= \frac{\mathbf{F}(x + \delta x, y + \delta y, z + \delta z) - \mathbf{F}(x, y + \delta y, z + \delta z)}{\delta x}\,\delta y$$
$$+ \frac{\mathbf{F}(x, y + \delta y, z + \delta z) - \mathbf{F}(x, y, z + \delta z)}{\delta y}\,\delta y$$
$$+ \frac{\mathbf{F}(x, y, z + \delta z) - \mathbf{F}(x, y, z)}{\delta z}\,\delta z \qquad ...(3)$$

By taking limits when δx → 0, δy → 0, δz → 0 then δ**r** → 0

Then the coefficients of δx, δy, δz in (3) tend to

$$\frac{\partial \mathbf{r}}{\partial x}, \frac{\partial \mathbf{r}}{\partial x} \text{ and } \frac{\partial \mathbf{r}}{\partial x}.$$

dr, dx, dy and dz denote the small quantities whose quotients are equal to the limiting values of the quotients of δr, δx and δz.

∴ The expression (3) in limit becomes

$$d\mathbf{r} = \frac{\partial \mathbf{r}}{\partial x}dx + \frac{\partial \mathbf{r}}{\partial y}dy + \frac{\partial \mathbf{r}}{\partial z}dz.$$

Here dr is called the *total differential* of **r**.

Note: If **r** = **F** (x, y, z)

and $$x = x(s, t),\ y = y(s, t),\ z = z(s, t)$$

Then $$\frac{\partial \mathbf{r}}{\partial s} = \frac{\partial \mathbf{r}}{\partial x}\frac{\partial x}{\partial s} + \frac{\partial \mathbf{r}}{\partial y}\frac{\partial y}{\partial s} + \frac{\partial \mathbf{r}}{\partial z}\frac{\partial z}{\partial t}$$

and $$\frac{\partial \mathbf{r}}{\partial t} = \frac{\partial \mathbf{r}}{\partial x}\frac{\partial x}{\partial t} + \frac{\partial \mathbf{r}}{\partial y}\frac{\partial y}{\partial t} + \frac{\partial \mathbf{r}}{\partial z}\cdot\frac{\partial z}{\partial t}$$

10.14. Rules of partial Differentiation

Let **F, G**, ϕ be the functions of more than one scalar variable

Then

(i) $\frac{\partial}{\partial t}(\mathbf{F} \pm \mathbf{G}) = \frac{\partial \mathbf{F}}{\delta t} \pm \frac{\partial \mathbf{G}}{\delta t}$

(ii) $\frac{\partial}{\partial t}(\mathbf{F}.\mathbf{G}) = \frac{\partial \mathbf{F}}{\partial t} . \mathbf{G} + \mathbf{F}.\frac{d\mathbf{G}}{\partial t}$

(iii) $\frac{\partial}{\partial t}(\mathbf{F} \times \mathbf{G}) = \frac{\partial \mathbf{F}}{\partial t} \times \mathbf{G} + \mathbf{F} \times \frac{\partial \mathbf{G}}{\partial t}$

(iv) $\frac{\partial}{\partial t}(\phi\, \mathbf{F}) = \frac{\partial \phi}{\partial f}\mathbf{F} + \phi \frac{\partial \mathbf{F}}{\partial t}$

(v) If $\mathbf{F} = f_1\, \mathbf{i} + f_2\, \mathbf{j} + f_3\, \mathbf{k}$ where f_1, f_2, f_3 are scalar functions of more than one variable then

$$\frac{\partial \mathbf{F}}{\delta t} = \frac{\partial f_1}{\partial t}\mathbf{i} + \frac{\partial f_2}{\partial t}\mathbf{j} + \frac{\partial f_3}{\partial t}\mathbf{k}$$

Example 19. If $\mathbf{F} = \cos xy\, \mathbf{i} + (3\,xy - 2x^2)\,\mathbf{j} - (3x + 2y)\,\mathbf{k}$, then find $\frac{\partial^2 \mathbf{F}}{\partial x^2}, \frac{\partial^2 \mathbf{F}}{\partial y^2}$ and show that $\frac{\partial^2 \mathbf{F}}{\partial x\, \partial y} = \frac{\partial^2 \mathbf{F}}{\partial y\, \partial x}$

Solution: Given $\mathbf{F} \cos xy\, \mathbf{i} + (3\,xy - 2\,x^2)\,\mathbf{j} - (3x + 2y)\,\mathbf{k}$

$$\therefore \quad \frac{\partial \mathbf{F}}{\partial t} = -\,y \sin xy\, \mathbf{i} + (3y - 4\,\mathbf{x})\,\mathbf{j} - 3\mathbf{k}$$

$$\frac{\partial^2 \mathbf{F}}{\partial x^2} = \frac{\partial}{\partial x}\left(\frac{\partial \mathbf{F}}{\partial x}\right) = -\,y^2 \cos xy\, \mathbf{i} - 4\,\mathbf{j}$$

$$\frac{\partial \mathbf{F}}{\partial y} = -x \sin xy\, \mathbf{i} + 3x\, \mathbf{j} - 2\, \mathbf{k}$$

$$\frac{\partial^2 \mathbf{F}}{\partial y^2} = \frac{\partial}{\partial y}\left(\frac{\partial \mathbf{F}}{\partial y}\right) = -\,x^2 \cos xy\, \mathbf{i} + 3\mathbf{j}$$

$$\frac{\partial^2 \mathbf{F}}{\partial x\, \partial y} = \frac{\partial}{\partial x}\left(\frac{\partial \mathbf{F}}{\partial y}\right) = -\,[\sin yx + yx \cos xy\,]\,\mathbf{i} + 3\,\mathbf{j}$$

$$\frac{\partial^2 \mathbf{F}}{\partial x\, \partial y} = \frac{\partial}{\partial\, y}\left(\frac{\partial\, \mathbf{F}}{\partial x}\right) = -\,[yx \cos xy + \sin xy]\,\mathbf{i} + 3\,\mathbf{j}$$

Hence $\frac{\partial^2 \mathbf{F}}{\partial x\, \partial y} = \frac{\partial^2 \mathbf{F}}{\partial y\, \partial x}$

Example 20. If $\mathbf{F} = 2x^2\,\mathbf{i} - 3yz\,\mathbf{j} + xz^2\,\mathbf{k}$ and $\phi = 2z - x^3y$,

then find $\mathbf{F}.\left[\mathbf{i}\frac{\partial \phi}{\partial x} + \mathbf{j}\frac{\partial \phi}{\partial y} + \mathbf{k}\frac{\partial \phi}{\partial z}\right]$

and $\mathbf{F} \times \left[\mathbf{i}\frac{\partial \phi}{\partial x} + \mathbf{j}\frac{\partial \phi}{\partial y}\mathbf{k} + \frac{\partial \phi}{\partial z}\right]$ at (1, –1, 1).

Solution: Given $\mathbf{F} = 2x^2\,\mathbf{i} - 3yz\,\mathbf{j} + xz^2\,\mathbf{k}$ and $\phi = 2z - x^3y$

$$\frac{\partial \phi}{\partial x} = -3x^2y = 3 \text{ at } (-1, 1, 1)$$

$$\frac{\partial \phi}{\partial y} = -x^3 = -1 \text{ at } (-1, 1, 1)$$

$$\frac{\partial \phi}{\partial z} = 2 = 2 \text{ at } (-1, 1, 1)$$

$$\mathbf{F} = 2\,\mathbf{i} + 3\mathbf{j} + \mathbf{k} \text{ at } (-1, 1, 1)$$

$$\therefore \quad \mathbf{F}.\left(\mathbf{i}\frac{\partial \phi}{\partial x} + \mathbf{j}\frac{\partial \phi}{\partial y} + \mathbf{k}\frac{2\phi}{\partial z}\right)$$

$$= (2\,\mathbf{i} + 3\,\mathbf{j} + \mathbf{k}) . (3\,\mathbf{i} - \mathbf{j} + 2\,\mathbf{j}) = 6 - 3 + 2 = 5$$

$$\mathbf{F} \times \left(\mathbf{i}\frac{\partial \phi}{\partial x} + \mathbf{j}\frac{\partial \phi}{\partial y} + \mathbf{k}\frac{\partial \phi}{\partial z}\right)$$

$$= (2\,\mathbf{i} + 3\mathbf{j} + \mathbf{k}) \times (3\mathbf{i} - \mathbf{j} + 2\mathbf{k})$$

$$= \begin{vmatrix} \mathbf{i} & \mathbf{j} & \mathbf{k} \\ 2 & 3 & 1 \\ 3 & -1 & 2 \end{vmatrix} = 7\,\mathbf{i} - \mathbf{j} - 11\,\mathbf{k}.$$

Example 21: If $\mathbf{r} = x\,\mathbf{i} + y\,\mathbf{j} + z\,\mathbf{k}$ and a is constant vector prove that

(i) $\frac{\partial}{\partial x}(\mathbf{a}.\mathbf{r})\,\mathbf{i} + \frac{\partial}{\partial y}(\mathbf{a}.\mathbf{r})\,\mathbf{j} + \frac{\partial}{\partial z}(\mathbf{a}.\mathbf{r})\,\mathbf{k} = \mathbf{a}$

(ii) $\frac{\partial}{\partial x}(\mathbf{a} \times \mathbf{r}).\mathbf{i} + \frac{\partial}{\partial y}(\mathbf{a} \times \mathbf{r}).\mathbf{j} + \frac{\partial}{\partial z}(\mathbf{a} \times \mathbf{r})\,.\,\mathbf{k} = \mathbf{0}$

(iii) $\frac{\partial}{\partial x}(\mathbf{a} \times \mathbf{r}) \times \mathbf{i} + \frac{\partial}{\partial y}(\mathbf{a} \times \mathbf{r}) \times \mathbf{j} + \frac{\partial}{\partial z}(\mathbf{a} \times \mathbf{r}) \times \mathbf{k} = -2\mathrm{a}$

Solution: Given $\mathbf{r} = x\,\mathbf{i} + y\,\mathbf{j} + z\,\mathbf{k}$ and $\mathbf{a}$ is constant vector

Let $\mathbf{a} = a_1\,\mathbf{i} + a_2\,\mathbf{j} + a_3\,\mathbf{k}$

(i) $(\mathbf{a}.\mathbf{r}) = (a_1 x + a_2 y + a_3 z)$

Then $\frac{\partial}{\partial x}(\mathbf{a}.\mathbf{r})\,\mathbf{i} + \frac{\partial}{\partial y}(\mathbf{a}.\mathbf{r})\,\mathbf{j} + \frac{\partial}{\partial z}(\mathbf{a}.\mathbf{r})\,\mathbf{k}$

$$= a_1\,\mathbf{i} + a_2\,\mathbf{j} + a_3\,\mathbf{k} = \mathbf{a}$$

(ii) $$\mathbf{a} \times \mathbf{r} = \begin{vmatrix} \mathbf{i} & \mathbf{j} & \mathbf{k} \\ a_1 & a_2 & a_3 \\ x & y & z \end{vmatrix}$$

$$= (a_2 z - y a_3)\,\mathbf{i} + (x a_3 - a_1 z)\,\mathbf{j} + (a_1 y - x a_2)\,\mathbf{k}$$

then $$\frac{\partial}{\partial x}(\mathbf{a} \times \mathbf{r}) \cdot \mathbf{i} + \frac{\partial}{\partial y}(\mathbf{a} \times \mathbf{r}) \cdot \mathbf{j} + \frac{\partial}{\partial z}(\mathbf{a} \times \mathbf{r}) \cdot \mathbf{k}$$

$$= \frac{\partial}{\partial x}(a_2 z - y a_3) + \frac{\partial}{\partial y}(x a_3 - a_1 z) + \frac{\partial}{\partial z}(a_1 y - x a_2)$$

$$= 0 + 0 + 0 = 0$$

(iii) $$(\mathbf{a} \times \mathbf{r}) \times \mathbf{i} = \begin{vmatrix} \mathbf{i} & \mathbf{j} & \mathbf{k} \\ a_2 z - y a_3 & x a_3 - a_1 z & a_1 y - x a_2 \\ 1 & 0 & 0 \end{vmatrix}$$

$$= (a_1 y - x a_2)\,\mathbf{j} - (x a_3 - a_1 z)\,\mathbf{k}$$

$$\frac{\partial}{\partial x}[(\mathbf{a} \times \mathbf{r}) \times \mathbf{i}] = (-a_2\,\mathbf{j} - a_3\,\mathbf{k}) = -(a_2\,\mathbf{j} + a_3\,\mathbf{k})$$

Similarly $\frac{\partial}{\partial y}[(\mathbf{a} \times \mathbf{r}) \times \mathbf{j}] = -(a_1\,\mathbf{i} + a_3\,\mathbf{k})$

and $$\frac{\partial}{\partial z}[(\mathbf{a} \times \mathbf{r}) \times \mathbf{k}] = -(a_1\,\mathbf{i} + a_2\,\mathbf{j})$$

$$\therefore \quad \frac{\partial}{\partial x}[(\mathbf{a} \times \mathbf{r}) \times \mathbf{i}] + \frac{\partial}{\partial y}[(\mathbf{a} \times \mathbf{r}) \times \mathbf{j}] + \frac{\partial}{\partial x}[(\mathbf{a} \times \mathbf{r}) \times \mathbf{k}] = -(a_2\,\mathbf{j} + a_3\,\mathbf{k}) - (a_1\,\mathbf{i} + a_3\,\mathbf{k}) - (a_1\,\mathbf{i} + a_2\,\mathbf{j})$$

$$= -2(a_1\mathbf{i} + a_2\mathbf{j} + a_3\mathbf{k})$$

$$= -2\mathbf{a}$$

EXERCISE

1. IF $\mathbf{F}(t) = t^2\,\mathbf{i} - t\,\mathbf{j} + (2t + 1)\,\mathbf{k}$
   and $\mathbf{G}(t) = (2t - 3)\,\mathbf{i} + \mathbf{j} - t\,\mathbf{i}$ then verify
   (i) $\frac{d}{dt}(\mathbf{F}.\mathbf{G}) = \mathbf{F}.\frac{d\mathbf{G}}{dt} + \frac{d\mathbf{F}}{dt}.\mathbf{G}$
   (ii) $\frac{d}{dt}(\mathbf{F} \times \mathbf{G}) = \mathbf{F} \times \frac{d\mathbf{G}}{dt} + \frac{d\mathbf{F}}{dt} \times \mathbf{G}$
2. If $\mathbf{U} = \sin y\,\mathbf{i} + \cos t\,\mathbf{j} + t\,\mathbf{k}$
   $\mathbf{V} = \cos t\,\mathbf{i} - \sin t\,\mathbf{j} - 3\,\mathbf{k}$
   and $\mathbf{W} = 2\,\mathbf{i} + 3\,\mathbf{j} - \mathbf{k}$, Find $\frac{d}{dt}[\mathbf{U} \times (\mathbf{V} \times \mathbf{W})]$ at $t = 0$.
3. If $\mathbf{r} = (t + 1)\,\mathbf{i} + (t^2 + t + 1)\,\mathbf{j} + (t^3 + t^2 + t + 1)\mathbf{k}$,
   find $\frac{d\mathbf{r}}{dt}$; $\frac{d^2\mathbf{r}}{dt^2}$ and $\left|\frac{d\mathbf{r}}{dt}\right|$, $\left|\frac{d^2\mathbf{r}}{dt^2}\right|$ at $t = 1$.
4. If $\mathbf{r} = \mathbf{a}\,e^{nt} + \mathbf{b}\,e^{-nt}$ where **a** and **b** are constant vectors show that $\frac{d^2\mathbf{r}}{dt^2} - n^2\,\mathbf{r} = 0$.
5. If $\mathbf{F}(t) = 3\,t^2\,\mathbf{i} - (t + 4)\,\mathbf{j} + (t^2 - 2t)\,\mathbf{k}$
   $\mathbf{G}(t) = \sin t\,\mathbf{i} + 3\,e^{-t}\,\mathbf{j} - 3\cos t\,\mathbf{k}$ find $\frac{d^2}{dt^2}(\mathbf{F} \times \mathbf{G})$ at $t = 0$
6. If **r** is a vector function of a scalar variable t and **a** and **b** are constant vectors then find
   (i) $\frac{d}{dt}(\mathbf{r}.\mathbf{a})$ (ii) $\frac{d}{dt}(\mathbf{r} \times \mathbf{b})$
   (iii) $\frac{d}{dt}\left(\mathbf{r}.\frac{d\mathbf{r}}{dt}\right)$ (iv) $\frac{d}{dt}\left(\mathbf{r} \times \frac{d\mathbf{r}}{dt}\right)$
   (v) $r^2\,\mathbf{r} + (\mathbf{a}.\mathbf{r})\,\mathbf{b}$ (vi) $\frac{\mathbf{r}}{r^2} + \frac{\mathbf{r}\,\mathbf{b}}{\mathbf{a}.\mathbf{r}}$
7. If **F** is a derivable vector function of the scalar variable t and if $|\mathbf{F}| = f$, then show that

(i) $\frac{d}{dt}(\mathbf{F}^2) = 2\mathbf{F}.\frac{d\,\mathbf{F}}{dt}$

(ii) $\mathbf{F}.\frac{d\mathbf{F}}{dt} = f\frac{df}{dt}$

8. A particle moves along the curve

$$\mathbf{R} = (t^3 - 4\,t)\mathbf{i} + (t^2 + 4t)\,\mathbf{j} + (8\,t^2 - 3\,t^3)\,\mathbf{k},$$

where t denotes the time. find the magnitudes of acceleration at t = 2.

9. Show that if **a, b, c** are constant vectors then $\mathbf{r} = \mathbf{a}\,t^2 + \mathbf{b}\,t + \mathbf{c}$ is the path of a point moving with constant acceleration.

10. A particle moves on the curve $x = 2\,t^2$, $y = t^2 - 4\,t$, $z = 3\,t - 5$ where t is the time. find the components of velocity and acceleration at t = 1 in the direction of $\mathbf{i} - 3\,\mathbf{j} + 2\,\mathbf{k}$.

11. The velocity of a boat relative to water is represented by $3\,\mathbf{i} + 4\,\mathbf{j}$ and that of water relative to the earth is $\mathbf{i} - 3\,\mathbf{j}$, What is the velocity of the boat relative to the earth if **i** and **j** represent 1 km/hour east and north respectively.

12. If $\mathbf{F} = x^2\,yz\,\mathbf{i} - 2\,xz^3\,\mathbf{j} + xz^2\,\mathbf{k}$ and

$\mathbf{G} = 2\,z\,\mathbf{i} + y\,\mathbf{j} - x^2\,\mathbf{k}$, show that

$$\frac{\partial^2}{\partial x\,\partial y}(\mathbf{F} \times \mathbf{G}) \text{ at } (1, 0, 2) \text{ is } -4\,\mathbf{i} - 8\mathbf{j}$$

13. If $\mathbf{F} = yz\,\mathbf{i} + xz\,\mathbf{j} + xy\,\mathbf{k}$, prove that

$$\mathbf{i} \times \frac{\partial \mathbf{F}}{\partial x} + \mathbf{j} \times \frac{\partial \mathbf{F}}{\partial y} + \mathbf{k} \times \frac{\partial \mathbf{F}}{\partial z} = 0$$

14. If $\mathbf{F} = e^{xy}\,\mathbf{i} + (2\,x - y)\,\mathbf{j} + y \sin x\,\mathbf{k}$ evaluate $\frac{\partial^2 \mathbf{F}}{\partial x^2}$ and $\frac{\partial^2 \mathbf{F}}{\partial y^2}$

and verify $\frac{\partial^2 \mathbf{F}}{\partial x\,\partial y} = \frac{\partial^2 \mathbf{F}}{\partial y\,\partial x}$

15. Verify $\frac{\partial^2 \mathbf{F}}{\partial y\,\partial x} = \frac{\partial^2 \mathbf{F}}{\partial x\,\partial y}$ for the vector function

$\mathbf{F} = x^2\,(2y - x^2)\,\mathbf{i} + (e^{xy} - y \sin x)\,\mathbf{j} + x^2 \cos y\,\mathbf{k}$.

16. If $\phi = 2\,xz^4 - x^2\,y$ then show that the value of

$$\left| \frac{\partial \phi}{\partial x}\mathbf{i} + \frac{\partial \phi}{\partial y}\mathbf{j} + \frac{\partial \phi}{\partial z}\mathbf{k} \right| \text{ at } (2, -2, -1) \text{ is } 2\sqrt{93}.$$

ANSWERS

(2) $7\,\mathbf{i} + 6\mathbf{j} - 6\,\mathbf{k}$

(3) $\mathbf{i} + 3\,\mathbf{j} + 6\,\mathbf{k} \quad 2\,\mathbf{j} + 8\,\mathbf{k}, \quad \sqrt{46}, \sqrt{68}$

(6) (i) $\frac{d\mathbf{r}}{dt}\,.\,\mathbf{a}$ (ii) $\frac{d\mathbf{r}}{dt} \times \mathbf{b}$ (iii) $\left(\frac{d\mathbf{r}}{dt}\right)^2 + \mathbf{r}.\frac{d\mathbf{r}}{dt}$

(iv) $\mathbf{r} \times \frac{d^2\,\mathbf{r}}{dt^2}$ (v) $2\,\mathbf{r}\frac{d\mathbf{r}}{dt}\mathbf{r} + \mathbf{r}^2\frac{d\mathbf{r}}{dt} + \left(\mathbf{a}.\frac{d\mathbf{r}}{dt}\right)\mathbf{b}$

(8) $12, 2\sqrt{137}$ **(10)** $\frac{8\sqrt{14}}{7}, \frac{-\sqrt{14}}{7}$

(11) $\sqrt{17}$ miles per hour in the direction of $\tan^{-1}\left(\frac{1}{4}\right)$ north of east.

11

Differential Vector Geometry

11.1. Tangent Vector of a curve

Let $\mathbf{r} = \mathbf{F}\ (t)$ be the vector function of a curve in space. Let $\mathbf{r}$ and $\mathbf{r} + \delta\mathbf{r}$ be the position vectors of two neighbouring points **P** and **Q** on this curve.

Thus we have $\overrightarrow{OP} = \mathbf{r} = F\ (t)$

$\overrightarrow{OQ} = \mathbf{r} + \delta\mathbf{r} = F\ (t + \delta t)$

$\overrightarrow{PQ} = \overrightarrow{OQ} - \overrightarrow{OP} = (\mathbf{r} + \delta\mathbf{r}) - \mathbf{r} = \delta\mathbf{r}$

Thus $\dfrac{\delta\mathbf{r}}{\delta t}$ is a vector parallel to the chord PQ

As Q → P *i.e.,* as δt → 0, chord PQ →tangent at P to the curve.

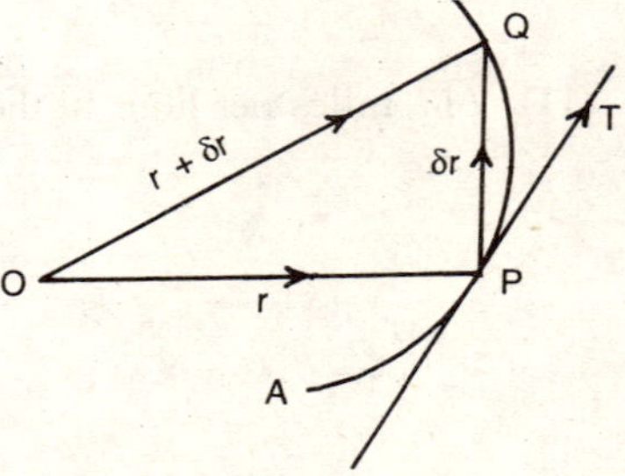

Fig. 11.1

$$\therefore \quad \lim_{\delta t \to 0} \frac{\delta\mathbf{r}}{\delta t} = \frac{d\mathbf{r}}{dt}$$

$\therefore\ \dfrac{d\mathbf{r}}{dt}$ is a vector parallel to the tangent at P to the curve $\mathbf{r} = F\ (t)$.

11.2. Unit tangent vector

Let **A** be any fixed point on the curve. Suppose if we take parameter s (arc length measured along the curve from any fixed point, here say **A**) as scalar variable in place of t, then the position vector **P** with respect to **O** is function of s.

If P and Q are two points corresponding to the values s and s + δs *ie* $\widehat{AP} = s$, $\widehat{AQ} = s + \delta s$

Then $\qquad \overrightarrow{OP} = \mathbf{r}\,(s)$ and $\overrightarrow{OQ}\ \mathbf{r} = (s + \delta s).$

$\Rightarrow \qquad \overrightarrow{OQ} - \overrightarrow{OP} = \overrightarrow{PQ}$

$\Rightarrow \qquad \mathbf{r}\,(s + \delta s) - \mathbf{r}\,(s) = \overrightarrow{PQ}$

$$\Rightarrow \qquad \frac{\mathbf{r}\,(s + \delta s) - \mathbf{r}\,(s)}{\delta s} = \frac{\overrightarrow{PQ}}{\delta s}$$

If the limit as $Q \to P$ the vector $\overrightarrow{PQ}$ becomes a tangent vector at P to the curve.

Let T be the unit tangent vector along the direction of tangent to the curve at P.

$$\therefore \qquad \lim_{\delta s \to 0} \frac{\mathbf{r}\,(s + \delta s) - \mathbf{r}\,(s)}{\delta s} = \lim_{Q \to P} \frac{\overrightarrow{PQ}}{\delta s}$$

or
$$\frac{d\mathbf{r}}{ds} = \lim_{Q \to P} \frac{\mathbf{T}\,(\text{chord PQ})}{\text{arc} = \text{PQ}}$$

$$= \left(\lim_{Q \to P} \mathbf{T}\right)\left(\lim_{Q \to P} \frac{\text{chord PQ}}{\text{arc PQ}}\right) = \mathbf{T}$$

$$\therefore \qquad \frac{d\mathbf{r}}{ds} = \mathbf{T} = \mathbf{r}'$$

Hence the unit tangent vector to the space curve **r**(s) is $\mathbf{T} = \dfrac{d\mathbf{r}}{ds} = \mathbf{r}'$ and is along the tangent at **P** to the curve in direction of increasing s.

If x, y, z are cartesian co-ordinates of **P**, we have

$$\mathbf{r} = x\,\mathbf{i} + y\,\mathbf{j} + z\,\mathbf{k} \text{ so that}$$

$$\mathbf{T} = \frac{d\mathbf{r}}{ds} = \frac{dx}{ds}\mathbf{i} + \frac{dy}{ds}\mathbf{j} + \frac{dz}{ds}\mathbf{k}$$

$$= \left(\frac{dx}{dt}\right)\left(\frac{dt}{ds}\right)\mathbf{i} + \left(\frac{dy}{dt}\right)\left(\frac{dt}{ds}\right)\mathbf{j} + \left(\frac{dz}{dt}\right)\left(\frac{dt}{ds}\right)\mathbf{k}$$

$$|\mathbf{T}| = \left|\frac{d\mathbf{r}}{dt}\right| = \sqrt{\left\{\left(\frac{dx}{dt}\right)^2 + \left(\frac{dy}{dt}\right)^2 + \left(\frac{dz}{dt}\right)^2\right\}\left(\frac{dt}{ds}\right)^2}$$

As $\quad |\mathbf{T}| = 1$

$$\Rightarrow \quad \left(\frac{ds}{dt}\right)^2 = \left(\frac{dx}{dt}\right)^2 + \left(\frac{dy}{dt}\right)^2 + \left(\frac{dx}{dt}\right)^2 = \left|\frac{dr}{dt}\right|^2$$

where t is any parameter.

$$\therefore \quad \frac{dz}{dt} = \left|\frac{dr}{dt}\right|$$

Note: We represent in future differentiation w.r.t. arc length s by using primes and differentiation w.r.t any other scalar variable t by dots.

i.e. $\quad \mathbf{r}', \mathbf{r}'', \mathbf{r}''', \ldots\ldots.$ for $\frac{dr}{ds}, \frac{d^2\mathbf{r}}{ds^2}, \frac{d^3\mathbf{r}}{dz^3}, \ldots\ldots$

$\quad \dot{\mathbf{r}}, \ddot{\mathbf{r}}, \dddot{\mathbf{r}} \quad \ldots\ldots.$ for $\frac{dr}{dt}, \frac{d^2\mathbf{r}}{dt^2}, \frac{d^3\mathbf{r}}{dt^3}, \ldots\ldots$

11.3. Principal normal

$\because \quad \frac{dr}{ds}$ is the unit tangent vector, we have

$$\frac{dr}{ds} \cdot \frac{dr}{ds} = 1$$

Differentiating $\frac{dr}{ds} \cdot \frac{d^2\mathbf{r}}{ds^2} = 0 \quad \Rightarrow \quad \mathbf{T} \cdot \frac{d^2\mathbf{r}}{ds^2} = 0$

$\therefore \quad \frac{d^2\mathbf{r}}{ds^2}$ is orthogonal to T and therefore normal to the curve at P.

The direction determined by $\frac{d^2\mathbf{r}}{ds^2}$ is called the *positive direction* of the *principal normal.* Let N be the unit vector along the principal normal and let k be a positive scalar such that

$$\frac{d^2\mathbf{r}}{ds^2} = kN \ (K > 0) \qquad i.e. \quad \frac{d}{ds}\left(\frac{dr}{ds}\right) = kN$$

or

$$\frac{dT}{ds} = kN \qquad \ldots(I)$$

The scalar k s equal to $\left|\frac{dT}{ds}\right|$, called the *curvature* of the curve at the point P and its reciprocal. $\rho = \frac{1}{k}$ is called the *radius of curvature*, at P. The vector $\mathbf{T}'$ or $\mathbf{r}''$ is some times called the *curvature vector* at P.

11.4. Binormal

Let **B** be the unit vector perpendicular to both the tangent vector **T** and the principal normal vector **N**, such that

$$\mathbf{T} \times \mathbf{N} = \mathbf{B}, \mathbf{N} \times \mathbf{B} = \mathbf{T}, \mathbf{B} \times \mathbf{T} = \mathbf{N}$$

i.e. they form a right handed system of unit vectors.

Here **B** is called the unit *binormal vector* and these vectors **T**, **N** and **B** are called the trihedral or triad at a point.

$\because$ **B** is a unit vector $\mathbf{B}.\mathbf{B} = 1 \Rightarrow \mathbf{B} . \frac{d\mathbf{B}}{ds} = 0$ $\quad (\because \mathbf{B} = \mathbf{B}(s))$

i.e., **B** is perpendicular to $\frac{d\mathbf{B}}{ds}$ ...(1)

As **T**, **N** and **B** form right handed system of unit vectors

$$\mathbf{B}.\mathbf{T} = 0$$

Differentiating w.r.t s

$$\mathbf{B} . \frac{d\mathbf{T}}{ds} + \frac{d\mathbf{B}}{ds} . \mathbf{T} = 0$$

$$\Rightarrow \quad \mathbf{B} . (k\mathbf{N}) + \frac{d\mathbf{B}}{ds} . \mathbf{T} = 0$$

$$\Rightarrow \quad \frac{d\mathbf{B}}{ds} . \mathbf{T} = 0 \qquad (\because \mathbf{B}.\mathbf{N} = 0)$$

$\Rightarrow \quad \frac{d\mathbf{B}}{ds}$ is perpendicular to **T** ...(2)

From (1) and (2), $\frac{d\mathbf{B}}{ds}$ is perpendicular to both **B** and **T**

$\therefore$ Parallel to **N**. Hence there exists a scalar τ such that

$$\frac{d\mathbf{B}}{ds} = -\tau \mathbf{N} \qquad \text{...(II)}$$

The negative sign in this equation is only a matter of convention. Moreo ver $|\tau| = \left|\frac{d\mathbf{B}}{ds}\right| \because |\mathbf{N}| = 1$.

τ is called *torsion* and $\frac{1}{\tau} = \sigma$ is known as *radius of torsion.*

Now consider $\mathbf{N} = \mathbf{B} \times \mathbf{T}$

$$\frac{d\mathbf{N}}{ds} = \mathbf{B} \times \frac{d\mathbf{T}}{ds} + \frac{d\mathbf{B}}{ds} \times \mathbf{T}$$

$$= \mathbf{B} \times k\mathbf{N} - \tau\mathbf{N} \times \mathbf{T}$$

$$= k(\mathbf{B} \times \mathbf{N}) - \tau(\mathbf{N} \times \mathbf{T})$$

$$= -k\mathbf{T} + \tau\mathbf{B}$$

$$\therefore \quad \frac{d\mathbf{N}}{ds} = \tau\mathbf{B} - k\mathbf{T} \qquad \text{...(III)}$$

11.5. Frenet—Serret formulae

The set of relations involving derivatives of the fundamental vectors **T, N** and **B** *i.e*, (I), (II) and (III) are

$$\frac{d\mathbf{T}}{ds} = k\mathbf{N}$$

$$\frac{d\mathbf{B}}{ds} = -\tau\mathbf{N}$$

$$\frac{d\mathbf{N}}{ds} = \tau\mathbf{B} - k\mathbf{T}$$

These three equations are called Frenet-Serret formulae. These can be represented in the form of matrix equation as follows

$$\begin{bmatrix} \frac{d\mathbf{T}}{ds} \\ \frac{d\mathbf{N}}{ds} \\ \frac{d\mathbf{B}}{ds} \end{bmatrix} = \begin{bmatrix} 0 & k & 0 \\ -k & 0 & \tau \\ 0 & -\tau & 0 \end{bmatrix} \begin{bmatrix} \mathbf{T} \\ \mathbf{N} \\ \mathbf{B} \end{bmatrix}$$

Corpllory: If $\quad \omega = \tau\mathbf{T} + k\mathbf{B}$

then
$$\omega \times \mathbf{T} = (\tau\mathbf{T} + k\mathbf{B}) \times \mathbf{T}$$

$$= \tau(\mathbf{T} \times \mathbf{T}) + k(\mathbf{B} \times \mathbf{T})$$

$$= \tau(0) + k\mathbf{N}$$

$$= 0 + \frac{d\mathbf{T}}{ds}$$

$$\therefore \qquad \frac{d\mathbf{T}}{ds} = \omega \times \mathbf{T}$$

Similarly $\dfrac{d\mathbf{B}}{ds} = \omega \times \mathbf{B}$

$$\frac{d\mathbf{N}}{ds} = \omega \times \mathbf{N}$$

Here ω is known as *Darboux vector*.

11.6. Equations of tangent line to a curve at a given point

Let the equation of curve C be $\mathbf{r} = \mathbf{F}(t)$

Let **R** be the position vector of any current point Q on the tangent to the curve C at the point P on it with position vector **r**.

$$\therefore \quad \overrightarrow{OP} = \mathbf{r}, \overrightarrow{OQ} = \mathbf{R}$$

As $\overrightarrow{PQ}$ is parallel to the tangent vector.

$\overrightarrow{PQ} = \lambda\mathbf{T}$ where λ is a scalar parameter such that

$$|\overrightarrow{PQ}| = |\lambda| \text{ and } \mathbf{T} = \frac{d\mathbf{r}}{ds}$$

is the unit tangent vector to the curve at **P**.

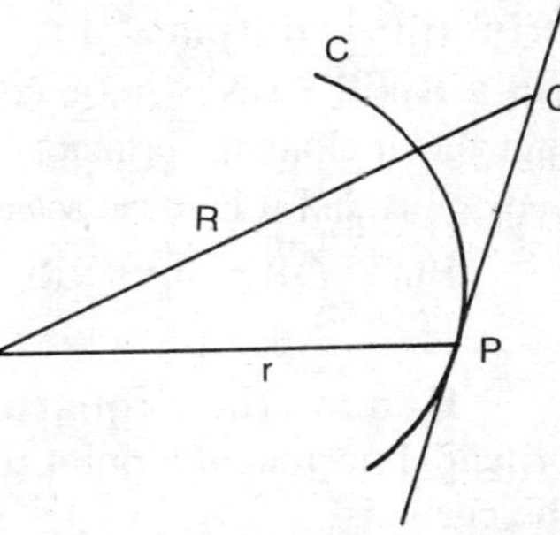

Fig. 11.2

Now we have $\overrightarrow{OQ} = \overrightarrow{OP} + \overrightarrow{PQ}$

i.e $\qquad \mathbf{R} = \mathbf{r} + \lambda\mathbf{T} \qquad$ ...(1)

or $\qquad \mathbf{R} = \mathbf{r} + \lambda\mathbf{r}' \qquad$ ...(2)

which is the equation of the tangent line.

$\because \mathbf{r} = \dfrac{d\mathbf{r}}{dt}$ is parallel to the vector **T**, the equation (1) or (2) may be written as

$$\mathbf{R} = \mathbf{r} + \lambda\mathbf{r} \qquad \text{...(3)}$$

where λ is a scalar parameter.

Cartesian form: Let P = **P** (x, y, z) referred to fixed rectangular axes and Q = **Q** (X, Y, Z).

Then equation (1) or (2) is equivalent to

$$X\mathbf{i} + Y\mathbf{j} + Z\mathbf{k} = x\mathbf{i} + y\mathbf{j} + z\mathbf{k} + \lambda\,(x'\,\mathbf{i} + y'\,\mathbf{j} + z'\,\mathbf{k})$$

$$\Leftrightarrow \qquad \frac{X-x}{x'} = \frac{Y-y}{y'} = \frac{Z-z}{x'} = \lambda$$

∴ The direction cosines of the tangent are x', y', z' and the cartesian form of (3) is

$$\frac{X-x}{\dot{x}} = \frac{Y-y}{\dot{y}} = \frac{Z-z}{\dot{z}} = \lambda$$

where the direction cosines of the tangent are proportional to $\dot{x}$, $\dot{y}$ $\dot{z}$.

11.7. Equations of Principal normal and Binormal

Let **r** be the position vector of a point **P** on the curve **C**.

∴ $\overrightarrow{OP} = \mathbf{r}$ Let R be the position vector of the current point R on the principle normal $\overrightarrow{PR}$ then $\overrightarrow{OR} = \mathbf{R}$, $\overrightarrow{PR} = \mu\mathbf{N}$ where **N** is the unit vector along the principal normal vector $\overrightarrow{PR}$ and μ is some scalar.

But $\overrightarrow{OR} = \overrightarrow{OP} + \overrightarrow{PR}$

∴ $\mathbf{R} = \mathbf{r} + \mu\,\mathbf{N}$

Hence the equation of principal normal at a point **p** (**r**) on the curve is

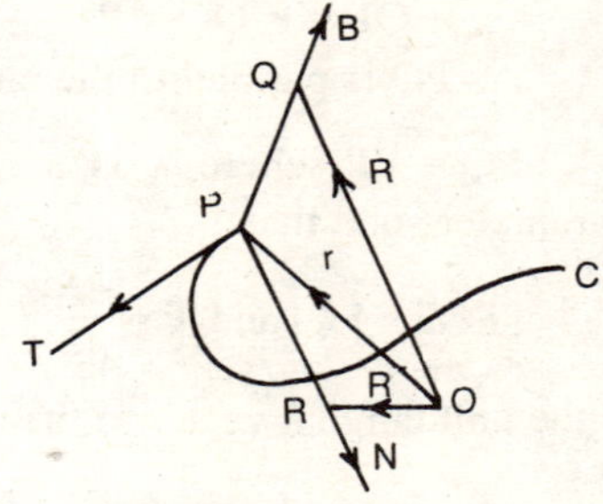

Fig. 11.3

$$\mathbf{R} = \mathbf{r} + \mu\mathbf{N}$$

Similarly if R denotes the position vector of a current point Q on the binormal PQ then

$$\overrightarrow{OR} = \overrightarrow{OP} + \overrightarrow{PQ} \quad \therefore \quad \mathbf{R} = \mathbf{r} + \nu\mathbf{B}$$

where **B** is the unit vector along the binormal vector $\overrightarrow{PQ}$ and ν is some scalar.

Hence the equation of binormal at a point **Q** (**r**) on the curve is

$$\mathbf{R} = \mathbf{r} + \nu\mathbf{B}$$

11.8. Fundamental planes

In the study of differential geometry the three unit vectors **T**, **N** and **B** play an important part.

∴ The triad **T, N, B** form a right handed arthogonal system of axes.

$$\mathbf{T.N} = \mathbf{N.B} = \mathbf{B.T} = 0$$
$$\mathbf{T} = \mathbf{N} \times \mathbf{B}\ ;\ \mathbf{B} = \mathbf{T} \times \mathbf{N}\ ;\ \mathbf{N} = \mathbf{B} \times \mathbf{T}$$

Now at each point of the curve there are three mutually perpendicular planes.

They are

(i) The *osculating plane* containing **T** and **N**

Its equation is

$(\mathbf{R}-\mathbf{r}) \cdot \mathbf{B} = 0$

(ii) The *normal plane* containing **N** and **B**

Its equation is

$(\mathbf{R}-\mathbf{r}) \cdot \mathbf{T} = 0$

(iii) The *rectifying plane* containing **B** and **T**

Its equation is

$(\mathbf{R}-\mathbf{r}) \cdot \mathbf{N} = 0$

where the point P(**r**) is on the space curve, **R** lies in these planes. These planes are also called *fundamental planes.*

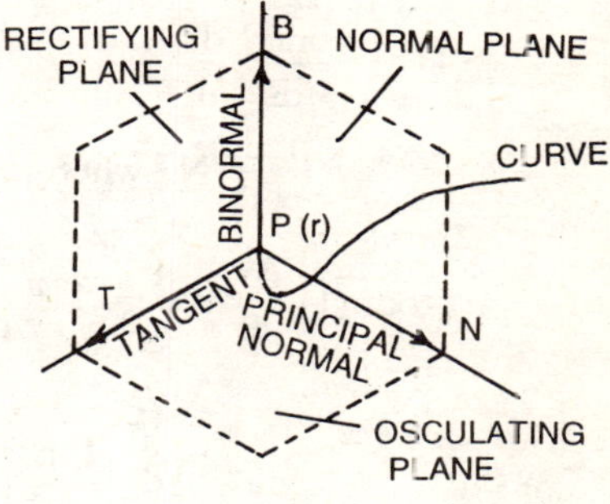

Fig. 11.4

SOLVED PROBLEMS

Example 1. Find the tangential and normal accelerations of a point moving in a plane curve.

Solution : Let **r** be the position vector of a point **P** at any time t.

Its velocity

$$\mathbf{V} = \frac{d\mathbf{r}}{dt} = \frac{d\mathbf{r}}{ds} \cdot \frac{ds}{dt}$$

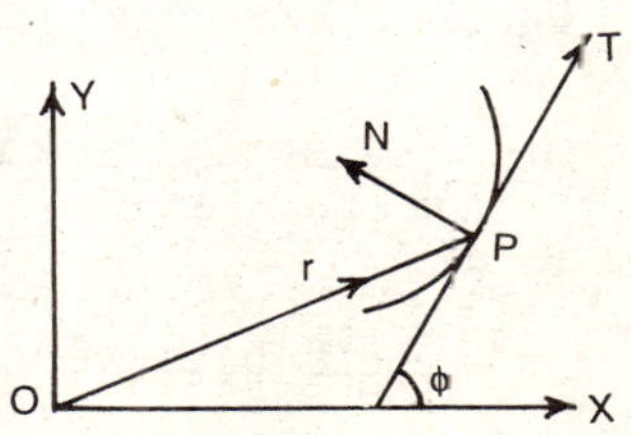

Fig. 11.5

$\because \quad \frac{d\mathbf{r}}{ds} = \mathbf{T}$ is a unit vector along the tangent **P**.

$$\therefore \quad \mathbf{V} = \frac{ds}{dt}\mathbf{T}$$

$$\frac{d\mathbf{V}}{dt} = \frac{d^2s}{dt^2}\mathbf{T} + \frac{ds}{dt}\frac{d\mathbf{T}}{dt} \qquad \text{...(1)}$$

Now $\frac{d\mathbf{T}}{dt} = \frac{d\mathbf{T}}{d\phi}\frac{d\phi}{dt} = \frac{d\phi}{dt}\mathbf{N}$

where **N** is a unit vector along the normal **P**.

$$= \frac{d\phi}{ds} \cdot \frac{ds}{dt}\mathbf{N}$$

$= \frac{1}{\rho}\frac{ds}{dt}\mathbf{N}$, where ρ is the radius of curvature at **P**.

$$\therefore \text{ From (1) } \frac{d\mathbf{V}}{dt} = \frac{d^2s}{dt^2}\mathbf{T} + \frac{1}{\rho}\left(\frac{ds}{dt}\right)^2\mathbf{N}$$

$$= \frac{d\mathbf{V}}{dt}\mathbf{T} + \frac{\mathbf{V}^2}{\rho}\mathbf{N} \text{ where } \mathbf{V} = \frac{ds}{dt}$$

Hence the tangential and normal accelerations (components along **T** and **N**) are

$$\frac{d\mathbf{V}}{dt} \text{ or } \frac{d^2s}{dt^2} \text{ and } \frac{\mathbf{V}^2}{\rho}$$

Example 2: Find the angle between the tangents to the curve $\mathbf{r} = t^2\mathbf{i} + 2t\mathbf{j} - t^3\mathbf{K}$ at the points $\mathbf{t} = \pm 1$.

Solution: Given $\mathbf{r} = t^2\mathbf{i} + 2t\mathbf{j} - t^3\mathbf{k}$ is the curve.

Then $\frac{d\mathbf{r}}{dt} = 2t\mathbf{i} + 2\mathbf{j} - 3t^2\mathbf{k}$ is a vector along the tangent at any point 't'.

Let $\mathbf{T_i}$ and $\mathbf{T_2}$ be the vector along the tangents to the curve at t = + 1 and t = − 1.

$\therefore \mathbf{T_1} = 2\mathbf{i} + 2\mathbf{j} - 3\mathbf{k}$ and $\mathbf{T_2} = -2\mathbf{i} + 2\mathbf{j} - 3\mathbf{k}$

If θ is the angle between $\mathbf{T_1}$ and $\mathbf{T_2}$ then

$$\cos\theta = \frac{\mathbf{T_1} \cdot \mathbf{T_2}}{|\mathbf{T_1}||\mathbf{T_2}|}$$

$$= \frac{(2)(-2) + (2)(2) + (-3)(-3)}{\sqrt{4+4+9}\sqrt{4+4+9}} = \frac{9}{17}$$

$$\therefore \qquad \theta = \cos^{-1}\frac{9}{17}$$

Example 3. Find the unit tangent vector at any point on the curve $x = t^2 + 2$, $y = 4t - 5$, $z = 2t^2 - 6t$ where t is any variable. Also determine the unit tangent vector at $t = 2$.

Solution: If $\mathbf{r}$ is the position vector of any point **P** (x, y, z) on the given curve then

$$\mathbf{r} = x\mathbf{i} + y\mathbf{j} + z\mathbf{k}$$

$$= (t^2 + 2)\,\mathbf{i} + (4t - 5)\,\mathbf{j} + (2t^2 - 6t)\mathbf{k}$$

Then the vector $\frac{d\mathbf{r}}{dt}$ is along the tangent at the point **P** (x, y, z) to the given curve

$$\therefore \qquad \frac{d\mathbf{r}}{dt} = 2t\mathbf{i} + 4\mathbf{j} + (4t - 6)\,\mathbf{k}$$

$$\left|\frac{d\mathbf{r}}{dt}\right| = \sqrt{(2t)^2 + (4)^2 + (4t-6)^2}$$

$$= 2\sqrt{5t^2 - 12t + 13}$$

$\therefore$ The unit tangent vector

$$\mathbf{T} = \frac{\frac{d\mathbf{r}}{dt}}{\left|\frac{d\mathbf{r}}{dt}\right|} = \frac{t\mathbf{i} + 2\mathbf{j} + (2t-3)\,\mathbf{k}}{\sqrt{5t^2 - 12t + 13}}$$

At $t = 2$ the unit tangent vector is

$$= \frac{2\mathbf{i} + 2\mathbf{j} + \mathbf{k}}{\sqrt{9}} = \frac{1}{3}(2\mathbf{i} + 2\mathbf{j} + \mathbf{k})$$

Example 4. For the space curve x = 3 cos t, y = 3 sin t, z = 4t, find (i) the unit tangent **T**. (ii) the principal normal **N**, curvature **k** and radius of curvature ρ, (iii) the binormal **B**, torsion τ and radius of torsion.

Solution: Here the Space curve is circular helix.

∴ The position vector for any point on the curve is

$$\mathbf{r} = (3\cos t)\,\mathbf{i} + (3\sin t)\,\mathbf{j} + (4t)\,\mathbf{k}$$

$$\therefore \quad \frac{d\mathbf{r}}{dt} = (-3\sin t)\,\mathbf{i} + (3\cos t)\,\mathbf{j} + 4\mathbf{k}$$

$$\text{(i)} \quad \frac{d\mathbf{r}}{ds} = \mathbf{T} \Rightarrow \left|\frac{d\mathbf{r}}{ds}\right| = |\mathbf{T}| = 1$$

$$\Rightarrow \quad \left|\frac{\left(\frac{d\mathbf{r}}{dt}\right)}{\left(\frac{ds}{dt}\right)}\right| = 1$$

$$\Rightarrow \quad \frac{ds}{dt} = \left|\frac{d\mathbf{r}}{dt}\right| = \sqrt{9\sin^2 t + 9\cos^2 t + 16} = 5$$

$$\Rightarrow \quad \mathbf{T} = \frac{d\mathbf{r}}{dt} = \frac{\left(\frac{d\mathbf{r}}{dt}\right)}{\left(\frac{ds}{dt}\right)}$$

$$= \frac{1}{5}[-3\sin t\,\mathbf{i} + 3\cos t\,\mathbf{j} + 4\mathbf{k}]$$

$$\text{(ii)} \quad \frac{d\mathbf{T}}{dt} = \frac{1}{5}[-3\cos t\,\mathbf{i} - 3\sin t\,\mathbf{j}]$$

$$\frac{d\mathbf{T}}{ds} = \frac{\left(\frac{d\mathbf{T}}{dt}\right)}{\left(\frac{ds}{dt}\right)} = \frac{1}{25}[-3\cos t\,\mathbf{i} - 3\sin t\,\mathbf{j}]$$

$$\because \quad \frac{d\mathbf{T}}{ds} = k\mathbf{N}, \left|\frac{d\mathbf{T}}{ds}\right| = |k\mathbf{N}| = k \qquad [\because |\mathbf{N} = 1]$$

$$\therefore \quad k = \left|\frac{d\mathbf{T}}{ds}\right| = \frac{1}{25}\sqrt{(-3\cos t)^2 + (|\,3\sin t\,|)^2}$$

$$\therefore \text{ Curvature } k = \frac{3}{25}$$

$$\text{Radius of curvature} = \rho = \frac{1}{k} = \frac{25}{3}$$

Again $\quad \dfrac{d\mathbf{T}}{ds} = k\mathbf{N} \Leftrightarrow \mathbf{N} = \dfrac{1}{k}\dfrac{d\mathbf{T}}{ds}$

$$\mathbf{N} = \frac{25}{3}\left[\frac{1}{25}(-3\cos t\,\mathbf{i} - 3\sin t\,\mathbf{j})\right]$$

$$= -(\cos t\,\mathbf{i} + \sin t\,\mathbf{j})$$

(iii) $\mathbf{B} = \mathbf{T} \times \mathbf{N} = \begin{vmatrix} \mathbf{i} & \mathbf{j} & \mathbf{k} \\ -\frac{3}{5}\sin t & \frac{3}{5}\cos t & 4 \\ -\cos t & -\sin t & 0 \end{vmatrix}$

$$= \frac{1}{5}(\mathbf{r}\sin t\,\mathbf{i} - \mathbf{r}\cos t\,\mathbf{j} + 3\mathbf{k})$$

$$\frac{d\mathbf{B}}{ds} = \frac{1}{5}(4\cos t\,\mathbf{i} + 4\sin t\,\mathbf{j})$$

$$\frac{d\mathbf{B}}{ds} = \frac{\left(\dfrac{d\mathbf{B}}{dt}\right)}{\left(\dfrac{ds}{dt}\right)} = +\frac{4}{25}(\cos t\mathbf{i} + \sin t\mathbf{j})$$

Now $\quad \dfrac{d\mathbf{B}}{ds} = -\tau\,\mathbf{N}$

$$\therefore \quad \left|\frac{d\mathbf{B}}{ds}\right| = \tau = \frac{4}{25}\sqrt{\cos^2 t + \sin^2 t} = \frac{4}{25}$$

$$\therefore \text{ Torsion } = \tau = \frac{4}{25}$$

$$\text{Radius of Torsion } \sigma = \frac{1}{\tau} = \frac{24}{5}$$

Example 5. Given the curve

$$x = \frac{\sin t}{e^t}, y = \frac{\cos t}{e^t}, z = \frac{1}{e^t}$$

find at any point 't' of this curve

(i) **T** (ii) equation of the tangent (iii) k (iv) **n** (v) equation to the principal normal (vi) **B** (vii) equation to the binormal (viii) τ.

Solution: The position vector for any point on the curve is

$$\mathbf{r} = e^{-t} \sin t\, \mathbf{i} + e^{-t} \cos t\, \mathbf{j} + e^{-t} \mathbf{k} \qquad \text{...(1)}$$

$$\therefore \quad \frac{d\mathbf{r}}{dt} = e^{-t}(\cos t - \sin t)\, \mathbf{i} - e^{-t}(\cos t + \sin t)\, \mathbf{j} - e^{-t} \mathbf{k} \qquad \text{...(2)}$$

Now $\frac{ds}{dt} = \left| \frac{d\mathbf{r}}{dt} \right|$

$= \sqrt{[e^{-t}(\cos t - \sin t)]^2 + [-e^{-t}(\cos t + \sin t)]^2 + [-e^{-t}]^2}$

$= e^{-t} \sqrt{2 + 1} = \sqrt{3}\, e^{-t}$

Hence

$$\text{(i)} \quad \mathbf{T} = \frac{d\mathbf{r}}{ds} = \frac{\left(\frac{d\mathbf{r}}{dt}\right)}{\left(\frac{ds}{dt}\right)}$$

$$= \frac{1}{\sqrt{3}} [(\cos t - \sin t)\, \mathbf{i} - (\sin t + \cos t)\, \mathbf{j} - \mathbf{k}] \qquad \text{...(3)}$$

(ii) The equation of the tangent line to the curve at the point t is

$$\mathbf{R} = \mathbf{r} + \lambda \mathbf{T}$$

or

$$\mathbf{R} = (e^{-t} \sin t\, \mathbf{i} + e^{-t} \cos t\, \mathbf{j} + e^{-t} \mathbf{k}) + \lambda . \frac{1}{\sqrt{3}} [(\cos t - \sin t)\, \mathbf{i} - (\sin t + \cos t)\, \mathbf{j} - \mathbf{k}]$$

If **R** = (X, Y, Z), the equations of the tangent line in cartesian form are

$$\frac{X - e^{-t} \sin t}{\cos t - \sin t} = \frac{Y - e^{-t} \cos t}{(-\sin t + \cos t)} = \frac{Z - e^{-t}}{-1}$$

(iii) Differentiating (3), we get

$$\frac{d\mathbf{T}}{dt} = \frac{1}{\sqrt{3}} [-(\sin t + \cos t)\, \mathbf{i} + (\sin t - \cos t)\, \mathbf{j}]$$

$$\frac{d\mathbf{T}}{ds} = \frac{\left(\frac{d\mathbf{T}}{dt}\right)}{\left(\frac{ds}{dt}\right)} = \frac{e^t}{3}\left[-(\sin t + \cos t)\,\mathbf{i} + (\sin t - \cos t)\,\mathbf{j}\right]$$

But $\frac{d\mathbf{T}}{ds} = k\mathbf{N}$ $k = \left|\frac{d\mathbf{T}}{ds}\right|$ as $|\mathbf{N}| = 1$

$$= \frac{e^t}{3}\sqrt{(\sin t + \cos t)^2 + (\sin t - \cos t)^2}$$

$\therefore$ $k = \text{curvature} = \frac{\sqrt{2}\,e^t}{3}$

(iv) $\mathbf{N} = \frac{1}{k}\frac{d\mathbf{T}}{ds}$

$$= +\frac{1}{\sqrt{2}}\left[-(\sin t + \cos t)\,\mathbf{i} + (\sin t - \cos t)\,\mathbf{j}\right] \qquad ...(4)$$

(v) The equation of the principal normal **N** at the point 't' is $\mathbf{R} = \mathbf{r} + \mu\,\mathbf{N}$

or
$$\mathbf{R} = (e^{-t}\sin t\,\mathbf{i} + e^{-t}\cos t\,\mathbf{j} + e^{-t}\mathbf{k}) + \mu\frac{1}{\sqrt{2}}\left[-(\sin t + \cos t)\,\mathbf{i} + (\sin t - \cos t)\,\mathbf{j}\right]$$

In cartesian form the equations of the principal normal are

$$\frac{X - e^{-t}\sin t}{\sin t + \cos t} = \frac{Y - e^{-t}\cos t}{\cos t - \sin t} = \frac{Z - e^{-t}}{0}$$

(vi) Now $\mathbf{B} = \mathbf{T} \times \mathbf{N}$

$$= \begin{vmatrix} \mathbf{i} & \mathbf{j} & \mathbf{k} \\ \frac{\cos t - \sin t}{\sqrt{3}} & \frac{-(\sin t + \cos t)}{\sqrt{3}} & -\frac{1}{\sqrt{3}} \\ -\frac{(\sin t + \cos t)}{\sqrt{2}} & -\frac{(\cos t - \sin t)}{\sqrt{2}} & 0 \end{vmatrix}$$

$$= -\frac{1}{\sqrt{6}}\left[(\cos t - \sin t)\,\mathbf{i} - (\cos t + \sin t)\,\mathbf{j} + 2\mathbf{k}\right] \qquad ....(5)$$

(vii) The equation to the binormal at the point 't' is

$$\mathbf{R} = \mathbf{r} + \nu\,\mathbf{B}$$

or $$\mathbf{R} = (e^{-t}\sin t + \mathbf{i} + e^{-t}\cos t\,\mathbf{j} + e^{-t}\mathbf{k})$$
$$-\frac{v}{\sqrt{6}}[(\cos t - \sin t)\,\mathbf{i} - (\cos t + \sin t)\,\mathbf{j} + 2\mathbf{k}]$$

In cartesian form the equations of the binormal are

$$\frac{X - e^{-t}\sin t}{\cos t - \sin t} = \frac{Y - e^{-t}\cos t}{-(\sin t + \cos t)} = \frac{Z - e^{-t}}{2}$$

(viii) Differentiating (5) w.r.t t, we get

$$\frac{d\mathbf{D}}{dt} = \frac{1}{\sqrt{6}}[(\sin t + \cos t)\,\mathbf{i} + (\cos t - \sin t)\,\mathbf{j}]$$

$$\frac{d\mathbf{B}}{ds} = \frac{\left(\frac{d\mathbf{B}}{dt}\right)}{\left(\frac{ds}{dt}\right)} = \frac{e^{t}}{3\sqrt{2}}[(\sin t + \cos t)\,\mathbf{i} + (\cos t - \sin t)\,\mathbf{j}]$$

But $\frac{d\mathbf{B}}{ds} = -\tau\mathbf{N} \Rightarrow \tau = \left|\frac{d\mathbf{B}}{ds}\right|$

$$= \frac{e^{t}}{3\sqrt{2}}\sqrt{(\sin t + \cos t)^2 + (\cos t - \sin t)^2}$$

$$= \frac{1}{3}e^{t}$$ which is the torsion.

Example 6: For the curve $\mathbf{r} = \mathbf{F}(t)$, prove that

$$k = \frac{|\dot{\mathbf{r}} \times \ddot{\mathbf{r}}|}{|\dot{\mathbf{r}}|^3}$$

and $$\tau = \frac{[\dot{\mathbf{r}}\ \ddot{\mathbf{r}}\ \dddot{\mathbf{r}}]}{|\dot{\mathbf{r}} \times \ddot{\mathbf{r}}|^2}$$ where dots stand for 't' derivatives

Solution: Let $\mathbf{r} = \mathbf{F}(t) = \mathbf{F}[s(t)]$, ...(1)

where s is the arc length.

$$\therefore \quad \frac{d\mathbf{r}}{dt} = \mathbf{r} = \frac{d\mathbf{F}[s(t)]}{ds}\frac{ds}{dt}$$

$$= \frac{ds}{dt}\mathbf{T} \qquad ...(2)$$

and because $$|\mathbf{T}| = 1, |\dot{\mathbf{r}}| = \left|\frac{ds}{dt}\right| \qquad ...(3)$$

Now $$\frac{d\mathbf{T}}{dt} = \frac{d\mathbf{T}}{ds}\frac{ds}{dt} = k\mathbf{N}\frac{ds}{dt} \qquad ...(4)$$

$$\left(\because \frac{d\mathbf{T}}{dt} = k\mathbf{N}\right)$$

$$\therefore \quad \frac{d^2\mathbf{r}}{dt^2} = \ddot{\mathbf{r}} = \frac{d}{dt}\left(\mathbf{T}\frac{ds}{dt}\right)$$

$$= \frac{d^2s}{dt^2}\mathbf{T} + \frac{ds}{dt}\frac{d\mathbf{T}}{dt}$$

$$\ddot{\mathbf{r}} = \frac{d^2s}{dt^2}\mathbf{T} + \left(\frac{ds}{dt}\right)^2 k\mathbf{N} \qquad ...(5)$$

$$\therefore \quad \frac{d\mathbf{r}}{dt} \times \frac{d^2\mathbf{r}}{dt^2}$$

$$= \left(\frac{ds}{dt}\mathbf{T}\right) \times \left[\frac{d^2s}{dt^2}\mathbf{T} + \left(\frac{ds}{dt}\right)^2 k\mathbf{N}\right]$$

$$= \left(\frac{ds}{dt}\right)^3 k\mathbf{T} \times \mathbf{N}$$

ie $$\dot{\mathbf{r}} \times \ddot{\mathbf{r}} = \left(\frac{ds}{dt}\right)^3 k\mathbf{B} \qquad ...(6)$$

$$\therefore \quad |\mathbf{B}| = 1$$

$$|\dot{\mathbf{r}}(t) \times \ddot{\mathbf{r}}(t)| = \left|\frac{ds}{dt}\right|^3 k$$

$$= |\dot{\mathbf{r}}(t)|^3 k$$

Hence k = curvature $= \dfrac{|\dot{\mathbf{r}} \times \ddot{\mathbf{r}}|}{|\dot{\mathbf{r}}|^3}$

Now differentiating (5) w.r.t. t

$$\dddot{\mathbf{r}}(t) = \frac{d}{dt}[\ddot{\mathbf{r}}(t)] = \frac{d}{dt}\left[\frac{d^2s}{dt^2}\mathbf{T} + \left(\frac{ds}{dt}\right)^2 k\mathbf{N}\right]$$

$$= \frac{d^3s}{dt^3}\mathbf{T} + \frac{d^2s}{dt^2}\frac{d\mathbf{T}}{dt} + 2\frac{ds}{dt}\frac{d^2s}{dt^2}k\mathbf{N}$$

$$+ \left(\frac{ds}{dt}\right)^2 \frac{dk}{dt}\mathbf{N} + \left(\frac{ds}{dt}\right)^2 k\frac{d\mathbf{N}}{dt} \qquad \text{...(7)}$$

We know that from (4)

$$\frac{d\mathbf{T}}{dt} = k\mathbf{N}\frac{ds}{dt}$$

and from Freent-Serret formula

$$\frac{d\mathbf{N}}{dt} = \frac{d\mathbf{N}}{ds}\frac{ds}{dt} = \frac{ds}{dt}(\tau\mathbf{B} - k\mathbf{T})$$

Substituting these results in (7) and rearranging the terms, we obtain

$$\dddot{\mathbf{r}}(t) = \left[\frac{d^3s}{dt^3} - \left(\frac{ds}{dt}\right)^3 k^2\right]\mathbf{T}$$

$$+ \left[3\frac{ds}{dt}\frac{d^2s}{dt^2}k + \left(\frac{ds}{dt}\right)^2\left(\frac{dk}{dt}\right)\right]\mathbf{N}$$

$$+ \left(\frac{ds}{dt}\right)^3 k\,\tau\,\mathbf{B} \qquad \text{...(8)}$$

Now from (6) and (8) we have

$$[\dot{\mathbf{r}}\,\ddot{\mathbf{r}}\,\dddot{\mathbf{r}}] = \dddot{\mathbf{r}} \times (\dot{\mathbf{r}} \times \ddot{\mathbf{r}})$$

$$= \left(\frac{ds}{dt}\right)^6 k^2 \tau$$

$$\therefore \qquad \tau = \frac{[\dot{\mathbf{r}}\,\ddot{\mathbf{r}}\,\dddot{\mathbf{r}}]}{\left(\frac{ds}{dt}\right)^6 k^2}$$

Substituting the value of k obtained in above analysis, we get

$$\tau = \frac{[\dot{\mathbf{r}}\ \ddot{\mathbf{r}}\ \dddot{\mathbf{r}}]}{|\dot{\mathbf{r}} \times \ddot{\mathbf{r}}|^2}$$

Example 7. For the space curve $x = t - \frac{t^3}{3}, y = t^2, x = t + \frac{t^3}{3}$, whow that $\mathbf{k} = \tau = \frac{1}{(1+t^2)^2}$.

Solution: Let **r** be the position vector of any point P (x, y, z) on the given curve, then

$$\mathbf{r} = x\mathbf{i} + y\mathbf{j} + z\mathbf{k}$$

i.e $$\mathbf{r} = \left(t - \frac{t^3}{3}\right)\mathbf{i} + t^2\mathbf{j} + \left(t + \frac{t^3}{3}\right)\mathbf{k}$$

$$\therefore \quad \frac{d\mathbf{r}}{dt} = \dot{\mathbf{r}} = (1 - t^2)\,\mathbf{i} + 2\,t\mathbf{j} + (1 + t^2)\,\mathbf{k}$$

$$\frac{d^2\mathbf{r}}{dt^2} = \ddot{\mathbf{r}} = -2\,t\,\mathbf{i} + 2\,t\mathbf{j} + 2\,t\,\mathbf{k}$$

$$\frac{d^3\mathbf{r}}{dt^3} = \dddot{\mathbf{r}} = -2\mathbf{i} + 2\mathbf{k}$$

$$\therefore \quad \dot{\mathbf{r}} \times \ddot{\mathbf{r}} = \begin{vmatrix} \mathbf{i} & \mathbf{j} & \mathbf{k} \\ 1 - t^2 & 2t & 1 + t^2 \\ -2t & 2 & 2t \end{vmatrix}$$

$$= [4\,t^2 - 2\,(1 + t^2)]\,\mathbf{i} - [2t\,(1 - t^2) + 2t\,(1 + t^2)]\,\mathbf{j} + [2\,(1 - t^2) + 4\,t^2]\,\mathbf{k}$$

$$= 2\,(t^2 - 1)\,\mathbf{i} - (4t)\,\mathbf{j} + 2\,(1 + t^2)\,\mathbf{k}$$

$$|\dot{\mathbf{r}} \times \ddot{\mathbf{r}}| = \sqrt{\{2\,(t^2 - 1)\}^2 + (-4t)^2 + \{2\,(t^2 + 1)\}^2}$$

$$= 2\sqrt{2\,(t^4 + 1) + 4t^2}$$

$$= 2\sqrt{2}\,\sqrt{(t^2 + 1)^2} = 2\,\sqrt{2}\,(t^2 + 1)$$

$$|\dot{\mathbf{r}}| = \sqrt{(1 - t^2)^2 + 4t^2 + (1 + t^2)^2}$$

$$= \sqrt{2}\,(1 + t^2)$$

$$k = \frac{|\dot{\mathbf{r}} \times \ddot{\mathbf{r}}|}{|\dot{\mathbf{r}}|^3} = \frac{2\sqrt{2}\,(t^2 + 1)}{2\sqrt{2}\,(1 + t^2)^3} = \frac{1}{(1 + t^2)^2}$$

$$\text{Now } [\dot{\mathbf{r}}\ \ddot{\mathbf{r}}\ \dddot{\mathbf{r}}] = \begin{vmatrix} 1 - t^2 & 2t & 1 + t^2 \\ -2t & 2 & 2t \\ -2 & 0 & 2 \end{vmatrix}$$

$$= 4\,(1 - t^2) - (0)\,2t + 4\,(1 + t^2)$$
$$= 8$$

$$\therefore\ \tau = \frac{[\dot{\mathbf{r}}\ \ddot{\mathbf{r}}\ \dddot{\mathbf{r}}]}{|\dot{\mathbf{r}} \times \ddot{\mathbf{r}}|^2} = \frac{8}{[2\sqrt{2}\,(1 + t^2)]^2} = \frac{1}{(1 + t^2)^2}$$

Hence $k = \tau \dfrac{1}{(1 + t^2)^2}$.

Example 8. For the curve $x = e^t$, $y = -e^{-t}$, $z = \sqrt{2}\,t$ show that

$$k = \tau = \frac{\sqrt{2}}{(e^t + e^{-t})^2}$$

Solution: Let **r** be the position vector of any point **P** (x, y, z) on the curve, then

$$\mathbf{r} = x\mathbf{i} + y\mathbf{j} + z\mathbf{k}$$

i.e $\qquad \mathbf{r} = e^t\,\mathbf{i} - e^{-t}\,\mathbf{j} = \sqrt{2}\,t\,\mathbf{k}$

$$\dot{\mathbf{r}} = e^t\,\mathbf{i} + e^{-t}\,\mathbf{j} + \sqrt{2}\,\mathbf{k}$$

$$\ddot{\mathbf{r}} = e^t\,\mathbf{i} - e^{-t}\,\mathbf{j}$$

$$\dddot{\mathbf{r}} = e^t\,\mathbf{i} + e^{-t}\,\mathbf{j}$$

$$\therefore \quad \dot{\mathbf{r}} \times \ddot{\mathbf{r}} = \begin{vmatrix} \mathbf{i} & \mathbf{j} & \mathbf{k} \\ e^t & e^{-t} & \sqrt{2} \\ e^t & -e^{-t} & 0 \end{vmatrix}$$

$$= (0 + \sqrt{2}\,e^{-t})\,\mathbf{i} - (0 - \sqrt{2}\,e^t)\,\mathbf{j} + (-e^{-t}\,e^t - e^t \,.\, e^{-t})\,\mathbf{k}$$

$$= \sqrt{2}\,e^{-t}\,\mathbf{i} + \sqrt{2}\,e^t\,\mathbf{j} - 2\,\mathbf{k}$$

$$= \sqrt{2}\,[e^{-t}\,\mathbf{i} + e^t\,\mathbf{j} - \sqrt{2}\,\mathbf{k}]$$

$$|\dot{\mathbf{r}} \times \ddot{\mathbf{r}}| = \sqrt{2}\sqrt{(e^{-t})^2 + (e^t)^2 + (-\sqrt{2})^2}$$

$$= \sqrt{2}\sqrt{e^{-2t} + e^{2t} + 2}$$

$$= \sqrt{2}\sqrt{(e^{-t} + e^t)^2}$$

$$= \sqrt{2}\,(e^{-t} + e^t)$$

$$|\dot{\mathbf{r}}| = \sqrt{(e^{-t})^2 + (e^t)^2 + (-\sqrt{2})^2}$$

$$= (e^{-t} + e^t)$$

$$\therefore \qquad \mathbf{k} = \frac{|\dot{\mathbf{r}} \times \ddot{\mathbf{r}}|}{|\dot{\mathbf{r}}|^3} = \frac{\sqrt{2}\,(e^{-t} + e^t)}{(e^{-t} + e^t)^3}$$

$$= \frac{\sqrt{2}}{(e^t + e^{-t})^2} \qquad \text{...(1)}$$

$$\text{Now } [\,\dot{\mathbf{r}}\,\ddot{\mathbf{r}}\,\dddot{\mathbf{r}}\,] = \begin{vmatrix} e^t & -e^{-t} & \sqrt{2} \\ e^t & -e^{-t} & 0 \\ e^t & e^{-t} & 0 \end{vmatrix}$$

$$= \sqrt{2}\,(e^t e^{-t} + e^t e^{-t}) = 2\sqrt{2}$$

[expanding along last column]

$$\therefore \qquad \tau = \frac{[\,\dot{\mathbf{r}}\,\ddot{\mathbf{r}}\,\dddot{\mathbf{r}}\,]}{|\dot{\mathbf{r}} \times \ddot{\mathbf{r}}|^2} = \frac{2\sqrt{2}}{[\sqrt{2}\,(e^t + e^{-t})]^2}$$

$$= \frac{\sqrt{2}}{(e^t + e^{-t})^2} \qquad \text{...(2)}$$

Hence from (1) and (2)

$$\mathbf{k} = \tau = \frac{\sqrt{2}}{(e^t + e^{-t})^2}.$$

EXERCISE

1. A particle moves along the curve

$$\mathbf{r} = t\,(t^2 - 4)\,\mathbf{i} + t\,(t + 4)\,\mathbf{j} + t^2\,(8 - 3t)\,\mathbf{k}$$

where t denotes time. Find the magnitudes of acceleration along the tangent and normal at t = 2.

2. Find the equations for the (i) tangent (ii) principal normal and (iii) binormal to the curve x = 3 cos t ; y = 3 sin t, z = 4t at the point t = π.

3. Find the unit tangent vector, principal normal and binormal for the space curve

$$x = t - \frac{t^3}{3}, y = t^2, z = t + \frac{t^3}{3}.$$

4. Show that the torsion of the curve

$$x = \frac{2t+1}{t-1}, y = \frac{t^2}{t-1}, z = t + 2 \text{ is zero.}$$

5. Find k and τ for the space curve

$$x + t - \sin t, y = 1 - \cos t, z = 4 \sin \frac{t}{2}.$$

6. Show that the radius of curvature for the plane curve y = f (x), z = 0, is

$$k = \frac{y''}{\{1 + (y')^2\}^{3/2}}.$$

ANSWERS

1. 16, $2\sqrt{73}$

2. Tangent : $\mathbf{r} = -2\mathbf{i} - \frac{3}{5} t\,\mathbf{j} + \left(\frac{4}{5} t + 4\pi\right)\mathbf{k}$

   Principal normal : $\mathbf{r} = -3\mathbf{i} + 4\pi\mathbf{j} + t\mathbf{k}$

   Binormal : $\mathbf{r} = -3\mathbf{i} + \left(4\pi + \frac{4}{5} \text{trigh}\right) \mathbf{j} + \frac{3}{5}\mathbf{k}$

3. $\mathbf{T} = \frac{1}{\sqrt{2}\,(1+t^2)} \{(1 - t^2)\,\mathbf{i} + 2t\mathbf{j} + (1 + t^2)\,\mathbf{k}\}$

   $\mathbf{N} = \frac{1}{(1+t)^2} \{-2t\,\mathbf{i} + (1 - t^2)\,\mathbf{j}\}$

   $\mathbf{B} = \frac{1}{\sqrt{2}\,(1+t^2)} \{(t^2 - 1)\,\mathbf{i} - 2\,t\,\mathbf{j} + (t^2 + 1)\,\mathbf{k}\}$

5. $k = \dfrac{\sqrt{6 - 2\cos t}}{8}$,

$$\tau = \frac{\left[(3 + \cos t)\left(\cos\frac{t}{2}\right) + 2\sin t \sin\left(\frac{t}{2}\right)\right]}{4\,(3\cos t - 1)}.$$

12

Differential Operators (Gradient, Divergence and Curl)

12.1. Point function

A variable quantity whose value at any point in a region of space depends upon the position of the point is called a *point function.* There are two types of point functions.

(i) Scalar point function *(ii) Vector point function*

(i) Scalar point function: If corresponding to each point **P** of a region **R**, there corresponds a scalar denoted by Ø(**P**), then Ø is said to be a scalar point function for the region **R**. If the co-ordinates of **P** are (x, y, z) then Ø (**P**) = Ø (x, y, z).

e.g. Let Ø(**P**) be the temperature at any point of a body occupying a certain region **R** then Ø is a *scalar point function* defined over **R**.

(ii) Vector point function: If corresponding to each point **P** of a region **R**, there corresponds a vector denoted by **F** (**P**), then **F** is said to be a *vector point function* for the region **R**.

If the co-ordinates of **P** are (x, y, z) then

$$\mathbf{F}(\mathbf{P}) = \mathbf{F}(x, y, z) = \mathbf{F_1}(x, y, z)\,\mathbf{i} + \mathbf{F_2}(x, y, z)\,\mathbf{j} + \mathbf{F_3}(x, y, z)\,\mathbf{k}$$

e.g. Let **F**(**P**) denote the velocity of a particle at a point **P** in a region **R**, then **F** is a vector point function defined in the region **R**.

12.2. Scalar field

If to each point **P**(x, y, z) of a region **R**, there is associated a scalar point function Ø (x, y, z), then the set of points of the region

R, together with the functional values Ø (x, y, z) at these points is said to be a *scalar field* over **R**.

e.g. (i) Temperature distribution in a medium, the gravitational potential of a system of masses etc.

(ii) Ø (x, y, z) = $4x^2 y - z^2 x$ defines a scalar field.

12.3. Vector field

If to each point **P** (x, y, z) of a region **R**, there is associated a vector point function **F**(x, y, z), then the set of points of the region **R** together with the functional values of **F** (x, y, z) at these points is said to constitute a *vector field* over **R**

e.g. (i) The distribution of electric or magnetic field intensity, velocity of a moving fluid at any instant etc.

(ii) $\mathbf{F}(x, y, z) = 4x\,\mathbf{i} + (x^2 + y^2)\,\mathbf{j} + 2z\,\mathbf{k}$ defines a vector field

Note: (i) A scalar field which is independent of time is called a steady state or stationary scalar field

(ii) A vector field which is independent of time is called a steady state or stationary vector field.

12.4. The vector differential operator Del:

It is denoted by ∇ and is defined by

$$\nabla = \mathbf{i}\frac{\partial}{\partial x} + \mathbf{j}\frac{\partial}{\partial y} + \mathbf{k}\frac{\partial}{\partial z} = \frac{\partial}{\partial x}\mathbf{i} + \frac{\partial}{\partial y}\mathbf{j} + \frac{\partial}{\partial z}\mathbf{k}$$

This vector operator possesses properties analogous to those of ordinary vectors. The operator ∇ is also known as *nabla.*

12.5. Gradient of a scalar field:

Let ϕ (x, y, z) be defined and differentiable at each point (x, y, z) in a certain region of space (*i.e.,* ϕ defines a differentiable scalar field).

Then the gradient of ϕ is denoted by $\nabla \phi$ or Grad ϕ and defined by

$$\nabla \phi \left(\frac{\partial}{\partial x}\mathbf{i} + \frac{\partial}{\partial y}\mathbf{j} + \frac{\partial}{\partial z}\right)\phi = \frac{\partial \phi}{\partial x}\mathbf{i} + \frac{\partial \phi}{\partial y}\mathbf{j} + \frac{\partial \phi}{\partial z}\mathbf{k}$$

Note: The gradient of a scalar field is a vector field.

12.6. Geometrical interpretation of Gradient

If a surface ϕ (x, y, z) = C is drawn through any point **P(r)** such that at each point on it, the function has the same value as at P, then such a surface is called *level surface* of the function ϕ through **P**.

Consider the level surface through **P (r)** at which the function has value ϕ and another level surface through a neighboring point Q (**r** + δ **r**) where the value is $\phi + \delta\phi$.

$\therefore \quad \overline{PQ} = \delta \mathbf{r}$

Now

$$\nabla\phi.\delta\,\mathbf{r} = \left(\mathbf{i}\frac{\delta\phi}{\delta x} + \mathbf{j}\frac{\partial\phi}{\partial y} + \mathbf{k}\frac{\partial\phi}{\partial z}\delta z\right).(\mathbf{i}\,\delta x + \mathbf{j}\,\delta y + \mathbf{k}\,\delta z)$$

$$= \frac{\partial\phi}{\partial x}\delta x + \frac{\partial\phi}{\partial y}\delta y + \frac{\partial\phi}{\partial z}\delta z$$

$$= \delta\phi \qquad ...(1)$$

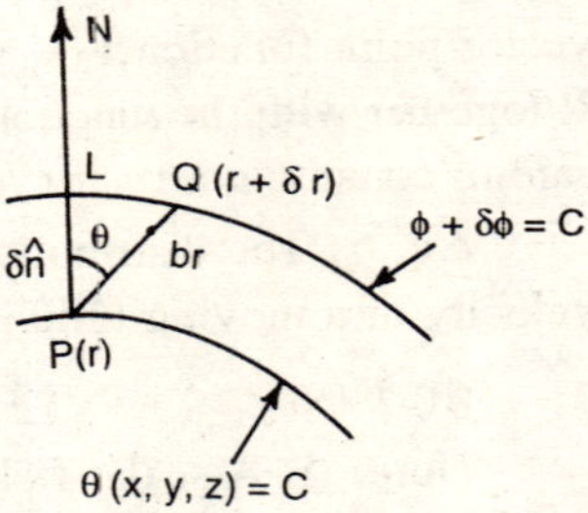

Fig. 12.1

If Q lies on the same level surface as at P then $\delta\phi = 0$.

ie $\qquad \nabla\phi.\delta\mathbf{r} = 0$

This means that $\nabla\phi$ is perpendicular to every $\delta\mathbf{r}$ lying on this surface.

Hence $\nabla\phi$ *is normal to the surface* $\phi(x, y, z) = C$

$\therefore \qquad \nabla\phi = |\nabla\phi|\,\mathbf{N}$

where **N** is a unit vector normal to this surface. If **PL** = **δn** be the perpendicular distance between the two level surfaces through P and Q, then the rate of change of ϕ normal to the surface through P.

$$= \frac{\partial\phi}{\partial n} = \underset{\delta n \to 0}{\text{Lt}}\frac{\delta\phi}{\delta n} = \underset{\delta n \to 0}{\text{Lt}}\frac{\nabla\phi.\delta\mathbf{r}}{\delta n} \qquad \text{(by (1)}$$

$$= \underset{\delta n \to 0}{\text{Lt}}\frac{|\nabla\phi|\,\mathbf{N}.\delta\mathbf{r}}{\delta n} = |\nabla\phi|$$

$[\therefore\ \mathbf{N}.\delta\mathbf{r} = |\mathbf{N}|\,|\delta\mathbf{r}|\cos\theta = |\delta\mathbf{r}|\cos\theta = \delta n]$

$$\therefore \quad |\nabla\phi| = \frac{\partial\phi}{\partial n}$$

Hence the gradient of a scalar field ϕ *is a vector normal to the surface* ϕ *(x, y, z) = C and has* ***a*** *magnitude equal to the rate of* *change of ϕ along this normal.*

12.7. Directional derivative

Let **PQ** = $\delta \mathbf{r}$

Then $\underset{\delta r \to 0}{\mathrm{Lt}} \dfrac{\delta\phi}{\delta r} = \dfrac{\partial\phi}{\partial r}$ is called the directional derivative of ϕ at P in the direction PQ.

Let N′ be a unit vector in the direction PQ.

$$\because \qquad \delta \mathbf{r} = \frac{\delta n}{\cos\theta} = \frac{\delta n}{N.N'}$$

$$\therefore \qquad \frac{\delta\phi}{\partial r} = \underset{\partial r \to 0}{\mathrm{Lt}} \left(N.N' \frac{\partial\phi}{\partial n} \right) = N.N' \frac{\partial\phi}{\partial n}$$

$$= N'.N \,|\, \nabla \phi \,|$$

$$= N'.\nabla \phi$$

Thus the directional derivative $\dfrac{\partial\phi}{\partial r}$ *is the resolved part of* $\nabla\phi$ in the direction **N′**

$$\therefore \qquad \frac{\partial\phi}{\partial r} = N'.\nabla\phi = |\nabla\phi| \cos\theta \le |\nabla\phi|$$

$\therefore$ $\nabla\phi$ *gives the maximum rate of change of* ϕ *and the magnitude of this maximum is* $|\nabla\phi|$.

* *Conclusions:* From above we conclude that

(i) If ϕ (x, y, z) be a scalar field over a region R, then through any point of R there passes one and only one level surface.

(ii) $\nabla\phi$ is normal to the surface ϕ (x, y, z) = k, where k is constant.

(iii) The directional derivative of a scalar field ϕ at a point **P** (x, y, z) in the direction of a unit vector e is $\nabla\phi$. e

i.e. (Grad ϕ). (unit vector)

(iv) The maximum rate of change of ϕ is $\nabla\phi$ and the magnitude of this maximum is $|\nabla\phi|$.

or The directional derivative is maximum along the normal to the surface. Its maximum value is $|\nabla\phi|$.

12.8. Gradient of a constant

The necessary and sufficient condition for a scalar point function ϕ to be constant is that $\nabla\phi = 0$.

Proof: *Necessary condition*

Let ϕ (x, y, z) be constant.

Then $\frac{\partial\phi}{\partial x} = \frac{\partial\phi}{\partial y} = \frac{\partial\phi}{\partial z} = 0$

By definition $\nabla\phi = \mathbf{i}\frac{\partial\phi}{\partial x} + \mathbf{j}\frac{\partial\phi}{\partial y} + \mathbf{k}\frac{\partial\phi}{\partial z}$

$$= \mathbf{i}(0) + \mathbf{j}(0) + \mathbf{k}(0)$$

$= 0$ which is the necessary condition.

Sufficient condition: Let $\nabla\phi = 0$

$$\mathbf{i}\frac{\partial\phi}{\partial x} + \mathbf{j}\frac{\partial\phi}{\partial y} + \mathbf{k}\frac{\partial\phi}{\partial z} = 0$$

This is possible iff $\frac{\partial\phi}{\partial x} = 0;\ \frac{\partial\phi}{\partial y} = 0;\ \frac{\partial\phi}{\partial z} = 0$

$\Rightarrow \phi$ is independent of x, y and z

$\therefore \phi$ is constant which is sufficient.

12.9. Properties of Gradient

If ϕ and ψ are two scalar point functions then

(i) $\nabla(\phi \pm \psi) = \nabla\phi \pm \nabla\psi$

(ii) $\nabla(C_1\phi \pm C_2\psi) = C_1\nabla\phi \pm C_2\nabla\psi$, where C_1 and C_2 are constants.

(iii) $\nabla(\phi\psi) = \phi\nabla\psi + \psi\nabla\phi$

(iv) $\nabla\left(\frac{\phi}{\Psi}\right) = \frac{\Psi\nabla\phi - \phi\nabla\Psi}{\Psi^2},\ \Psi \neq 0$

Proof: We have

(i) $\nabla(\phi \pm \psi) = \left(\mathbf{i}\frac{\partial}{\partial x} + \mathbf{j}\frac{\partial}{\partial y} + \mathbf{k}\frac{\partial}{\partial z}\right)(\phi \pm \psi)$

$$= \mathbf{i}\frac{\partial}{\partial x}(\phi \pm \psi) + \mathbf{j}\frac{\partial}{\partial y}(\phi \pm \psi) + \mathbf{k}\frac{\partial}{\partial z}(\phi \pm \psi)$$

$$= \mathbf{i}\frac{\partial\phi}{\partial x} \pm \mathbf{i}\frac{\partial\phi}{\partial y} + \mathbf{j}\frac{\partial\psi}{\partial y} \pm \mathbf{j}\frac{\partial\psi}{\partial z} + \mathbf{k} - \frac{\partial\phi}{\partial z} \pm \mathbf{k}\frac{\partial\psi}{\partial z}$$

$$= \left(\mathbf{i}\frac{\partial\phi}{\partial x} + \mathbf{j}\frac{\partial\phi}{\partial y} + \mathbf{k}\frac{\partial\phi}{\partial z}\right) \pm \left(\mathbf{i}\frac{\partial\psi}{\partial x} + \mathbf{j}\frac{\partial\psi}{\partial y} + \mathbf{k}\frac{\partial\psi}{\partial z}\right)$$

$$= \left(\mathbf{i}\frac{\partial}{\partial x} + \mathbf{j}\frac{\partial}{\partial y} + \mathbf{k}\frac{\partial}{\partial z}\right)\phi \pm \left(\mathbf{i}\frac{\partial}{\partial x} + \mathbf{j}\frac{\partial}{\partial y} + \mathbf{k}\frac{\partial}{\partial z}\right)\psi$$

$$= \nabla\phi \pm \nabla\psi$$

$$\therefore\ \nabla(\phi \pm \psi) = \nabla\phi \pm \nabla\psi.$$

i.e. Grad $(\phi \pm \psi)$ = Grad ϕ ± Grad ψ.

(ii) It can be easily proved as (i)

(iii) We have

$$\nabla(\phi\,\psi) = \left(\mathbf{i}\frac{\partial}{\partial x} + \mathbf{j}\frac{\partial}{\partial y} + \mathbf{k}\frac{\partial}{\partial z}\right)(\phi\psi)$$

$$= \mathbf{i}\frac{\partial}{\partial x}(\phi\,\psi) + \mathbf{j}\frac{\partial}{\partial y}(\phi\psi) + \mathbf{k}\frac{\partial}{\partial z}(\phi\psi)$$

$$= \mathbf{i}\left(\phi\frac{\partial\psi}{\partial y} + \psi\frac{\partial\phi}{\partial x}\right) + \mathbf{j}\left(\phi\frac{\partial\psi}{\partial y} + \psi\frac{\partial\phi}{\partial y}\right) + \mathbf{k}\left(\phi\frac{\partial\psi}{\partial z} + \psi\frac{\partial\phi}{\partial z}\right)$$

$$= \phi\left(\mathbf{i}\frac{\partial\psi}{\partial x} + \mathbf{j}\frac{\partial\psi}{\partial y} + \mathbf{k}\frac{\partial\psi}{\partial z}\right) + \psi\left(\mathbf{i}\frac{\partial\phi}{\partial x} + \mathbf{j}\frac{\partial\phi}{\partial y} + \mathbf{k}\frac{\partial\phi}{\partial z}\right)$$

$$= \phi\,\nabla\psi + \psi\,\nabla\phi.$$

$$\therefore\ \nabla(\phi\,\psi) = \phi\,\nabla\Psi + \psi\,\nabla\phi$$

i.e. Grad $(\phi\psi) = \phi$ Grad $\psi + \psi$ Grad ϕ

(iv) We have

$$\nabla\left(\frac{\phi}{\Psi}\right)\left(\mathbf{i}\frac{\partial}{\partial x} + \mathbf{j}\frac{\partial}{\partial y} + \mathbf{k}\frac{\partial}{\partial z}\right)\left(\frac{\phi}{\Psi}\right)$$

$$= \mathbf{i}\frac{\partial}{\partial x}\left(\frac{\phi}{\Psi}\right) + \mathbf{j}\frac{\partial}{\partial y}\left(\frac{\phi}{\Psi}\right) + \mathbf{k}\frac{\partial}{\partial z}\left(\frac{\phi}{\Psi}\right)$$

$$= \frac{1}{\Psi^2}\left[\mathbf{i}\left(\Psi\frac{\partial\phi}{\partial x} - \phi\frac{\partial\Psi}{\partial x}\right) + \mathbf{j}\left(\Psi\frac{\partial\phi}{\partial y} - \phi\frac{\partial\Psi}{\partial y}\right)\right.$$

$$+ \mathbf{k}\left(\Psi \frac{\partial \phi}{\partial z} - \phi \frac{\partial \Psi}{\partial z}\right)\Bigg]$$

$$= \frac{1}{\Psi^2}\left[\Psi\left(\mathbf{i}\frac{\partial \phi}{\partial x} + \mathbf{j}\frac{\partial \phi}{\partial y} + \mathbf{k}\frac{\partial \phi}{\partial z}\right)\right.$$

$$\left. - \phi\left(\mathbf{i}\frac{\partial \Psi}{\partial x} + \mathbf{j}\frac{\partial \Psi}{\partial y} + \mathbf{k}\frac{\partial \Psi}{\partial z}\right)\right]$$

$$= \frac{1}{\Psi^2}\{\Psi\, \nabla \phi - \phi\, \nabla \Psi\}$$

$$\therefore \quad \nabla\left(\frac{\phi}{\Psi}\right) = \frac{1}{\Psi^2}\{\Psi\, \nabla \phi - \phi\, \nabla \Psi\}$$

i.e. $\quad \text{Grad}\left(\frac{\phi}{\Psi}\right) = \frac{1}{\Psi^2}\{\Psi \text{ Grad } \phi - \phi \text{ Grad } \Psi\}$

12.10. Tangent plane and Normal to a level surface:

Let ϕ (x, y, z) = C be the equation of a level surface.

Let $\mathbf{r} = x\mathbf{i} + y\mathbf{j} + z\mathbf{k}$ be the position vector of any point A (x, y, z) on this surface.

Then $\quad \nabla \phi = \frac{\partial \phi}{\partial x}\mathbf{i} + \frac{\partial \phi}{\partial y}\mathbf{j} + \frac{\partial \phi}{\partial z}\mathbf{k} \quad$ ...(1)

is along the normal to the surface at A.

i.e. $\nabla\phi$ is perpendicular to the tangent plane at A.

Tangent plan at A:

Let $\mathbf{R} = X\mathbf{i} + Y\mathbf{j} + Z\mathbf{k}$ be the position vector of any current point B (X, Y, Z) on the tangent plane at A to the surface.

The vector $\vec{\mathbf{AB}} = (\mathbf{R} - \mathbf{r})$ which lies in the tangent plane at A

$\therefore \quad (\mathbf{R} - \mathbf{R} - \mathbf{r}) \cdot \nabla\phi = 0 \quad$ ...(2)

But $\quad \mathbf{R} - \mathbf{r} = (X - x)\,\mathbf{i} + (Y - y)\,\mathbf{j} + (Z - z)\,\mathbf{k} \quad$ ...(3)

Using (1) and (3), equation (2) becomes

$$(X - x)\frac{\partial \phi}{\partial x} + (Y - y)\frac{\partial \phi}{\partial y} + (Z - z)\frac{\partial \phi}{\partial z} = \mathbf{0} \quad \text{...(4)}$$

which is the equation of the tangent plane at A.

Normal at A

Let $\mathbf{R} = X\mathbf{i} + Y\mathbf{j} + Z\mathbf{k}$ be the position vector of any current point B (X, Y, Z) on the normal at A to the surface.

$\therefore$ $\vec{AB} = (\mathbf{R} - \mathbf{r})$ lies along the normal at A to the surface.

$\therefore$ $\vec{AB}$ is parallel to $\nabla \phi$

i.e
$$(\mathbf{R} - \mathbf{r}) \times \nabla \phi = 0 \quad ...(5)$$
which is the vector equation of normal at **A** to the given surface.

or $\mathbf{R} - \mathbf{r} = a\nabla\phi$ where a is some scalar.

or
$$(X - x)\mathbf{i} + (Y - y)\mathbf{j} + (Z - z)\mathbf{k} = a\left(\frac{\partial\phi}{\partial x}\mathbf{i} + \frac{\partial\phi}{\partial y}\mathbf{j} + \frac{\partial\phi}{\partial z}\mathbf{k}\right)$$
$$\Leftrightarrow \quad X - x = a\frac{\partial\phi}{\partial x}; \; Y - y = a\frac{\partial\phi}{\partial y}; \; Z - z = a\frac{\partial\phi}{\partial z}$$

or
$$\frac{X - x}{\dfrac{\partial\phi}{\partial x}} = \frac{Y - y}{\dfrac{\partial\phi}{\partial z}} = \frac{Z - z}{\dfrac{\partial\phi}{\partial z}}$$
are the equations of the normal at A.

SOLVED PROBLEMS

Example 1. Find $\nabla\phi$ where

(i) $\phi(x, y, z) = x^3 + y^3 + 3xyz$

(ii) $\phi(x, y, z) = x^2 y + y^2 x + z^2$ at the point (1, 1, 1)

(iii) $\phi(x, y, z) = \log(x^2 + y^2 + z^2)$

Solution: (i)
$$\nabla\phi = \left(\frac{\partial}{\partial x}\mathbf{i} + \frac{\partial}{\partial y}\mathbf{j} + \frac{\partial}{\partial z}\mathbf{k}\right)(x^3 + y^3 + 3xyz)$$
$$= \frac{\partial}{\partial x}(x^3 + y^3 + 3xyz)\,\mathbf{i} + \frac{\partial}{\partial y}(x^3 + y^3 + 3xyz)\,\mathbf{j} + \frac{\partial}{\partial z}(x^3 + y^3 + 3xyz)\,\mathbf{k}$$
$$= (3x^2 + 3yz)\,\mathbf{i} + (3y^2 + 3xz)\,\mathbf{j} + (3xy)\,\mathbf{k}$$
$$= 3(x^2 + yz)\,\mathbf{i} + 3(y^2 + xz)\,\mathbf{j} + 3xy\,\mathbf{k}$$

$$\text{(ii) } \nabla \phi = \left(\frac{\partial}{\partial x}\mathbf{i} + \frac{\partial}{\partial y}\mathbf{j} + \frac{\partial}{\partial z}\mathbf{k}\right)(x^2y + y^2x + z^2)$$

$$= (2\,xy + y^2)\,\mathbf{i} + (x^2 + 2\,yx)\,\mathbf{j} + (2\,z)\,\mathbf{k}$$

$$\left[\because \quad \frac{\partial\phi}{\partial x} = 2\,xt + y^2;\ \frac{\partial\phi}{\partial y} = x^2 + 2\,yx \text{ and } \frac{\partial\phi}{\partial z} = 2z\right]$$

$$\nabla \phi \text{ at } (1, 1, 1) = (2 + 1)\,\mathbf{i} + (1 + 2)\,\mathbf{j} + 2\,\mathbf{k}$$

$$= 3\,\mathbf{i} + 3\,\mathbf{j} + 2\,\mathbf{k}$$

$$\text{(iii) } \nabla \phi = \left(\frac{\partial}{\partial x}\mathbf{i} + \frac{\partial}{\partial y}\mathbf{j} + \frac{\partial}{\partial z}\mathbf{k}\right)\log(x^2 + y^2 + z^2)$$

$$= \left(\frac{2x}{x^2 + y^2 + z^2}\right)\mathbf{i} + \left(\frac{2\,y}{x^2 + y^2 + z^2}\right)\mathbf{j} + \left(\frac{2\,z}{x^2 + y^2 + z^2}\right)\mathbf{k}$$

$$= \left(\frac{2}{x^2 + y^2 + z^2}\right)(x\mathbf{i} + y\mathbf{j} + z\mathbf{k})$$

Example 2. If $r = |\mathbf{r}|$ where $\mathbf{r} = x\mathbf{i} + y\mathbf{j} + z\mathbf{k}$, prove that

(i) $\nabla r = \dfrac{\mathbf{r}}{r}$ (ii) $\nabla\left(\dfrac{1}{r}\right) = -\dfrac{\mathbf{r}}{r^3}$

(iii) $\nabla(r^n) = n\,r^{n-2}\,\mathbf{r}$ (iv) $\Delta \log|\mathbf{r}| = \dfrac{\mathbf{r}}{r^2}$.

$$\text{(i)} \quad \nabla r = \frac{\partial r}{\partial x}\mathbf{i} + \frac{\partial r}{\partial y}\mathbf{j} + \frac{\partial r}{\partial z}\mathbf{k} \qquad \text{...(1)}$$

$$r = |\mathbf{r}| = \sqrt{x^2 + y^2 + z^2}$$

$$\therefore \quad r^2 = x^2 + y^2 + z^2 \qquad \text{...(2)}$$

Differentiating (2) partially w.r.t x, we get

$$2\,r\frac{\partial r}{\partial x} = 2\,x \text{ or } \frac{\partial r}{\partial x} = \frac{x}{r} \qquad \text{...(3)}$$

Similarly differentiating (2), partially w.r.t y and z, we get

$$\frac{\partial r}{\partial y} = \frac{y}{r} \qquad \text{...(4)}$$

and

$$\frac{\partial r}{\partial z} = \frac{z}{r}$$

Substitution in (1), we get

$$\nabla r = \frac{x}{r}\mathbf{i} + \frac{y}{r}\mathbf{j} + \frac{z}{r}\mathbf{k}$$

$$= \frac{1}{r}(x\mathbf{i} + y\mathbf{j} + z\mathbf{k}) = \frac{\mathbf{r}}{r}$$

(ii) $$\nabla\left(\frac{1}{r}\right) = \frac{\partial}{\partial x}\left(\frac{1}{r}\right)\mathbf{i} + \frac{\partial}{\partial y}\left(\frac{1}{r}\right)\mathbf{j} + \frac{\partial}{\partial z}\left(\frac{1}{r}\right)\mathbf{k}$$

$$= \left(-\frac{1}{r^2}\right)\frac{\partial r}{\partial x}\mathbf{i} + \left(-\frac{1}{r^2}\right)\frac{\partial r}{\partial y}\mathbf{j} + \left(-\frac{1}{r^2}\right)\frac{\partial r}{\partial z}\mathbf{k}$$

$$= \left(-\frac{1}{r^2}\right)\left(\frac{x}{r}\mathbf{i} + \frac{y}{r}\mathbf{j} + \frac{z}{r}\mathbf{k}\right)$$

$$= \left(-\frac{1}{r^3}\right)\left(\frac{\mathbf{r}}{r}\right) = \frac{-\mathbf{r}}{r^3}$$

(iii) $$\nabla(r^n) = \frac{\partial}{\partial x}(r^n)\mathbf{i} + \frac{\partial}{\partial y}(r^n)\mathbf{j} + \frac{\partial}{\partial z}(r^n)\mathbf{k}$$

$$= nr^{n-1}\frac{\partial r}{\partial x}\mathbf{i} + nr^{n-1}\frac{\partial r}{\partial y}\mathbf{j} + nr^{n-1}\frac{\partial r}{\partial z}\mathbf{k}$$

$$= nr^{n-1}\left(\frac{x}{r}\mathbf{i} + \frac{y}{r}\mathbf{j} + \frac{z}{r}\mathbf{k}\right)$$

$$= nr^{n-1}\left(\frac{\mathbf{r}}{r}\right) = nr^{n-2}\,\mathbf{r}$$

(iv) $$\nabla \log|\mathbf{r}| = \frac{\partial}{\partial x}\log|\mathbf{r}|\,\mathbf{i} + \frac{\partial}{\partial y}\log|\mathbf{r}|\,\mathbf{j} + \frac{\partial}{\partial z}\log|\mathbf{r}|\,\mathbf{k}$$

$$= \frac{\partial}{\partial x}\log r\,\mathbf{i} + \frac{\partial}{\partial y}\log r\,\mathbf{j} + \frac{\partial}{\partial z}\log r\,\mathbf{k} \qquad (\because\ |\mathbf{r}| = r)$$

$$= \frac{1}{r}\frac{\partial r}{\partial x}\mathbf{i} + \frac{1}{r}\frac{\partial r}{\partial y}\mathbf{j} + \frac{1}{r}\frac{\partial r}{\partial z}\mathbf{k}$$

$$= \frac{1}{r}\left(\frac{x}{r}\mathbf{i} + \frac{y}{r}\mathbf{j} + \frac{z}{r}\mathbf{k}\right)$$

$$= \frac{1}{r}\frac{\mathbf{r}}{r} = \frac{\mathbf{r}}{r^2}$$

Note: (i) $\mathbf{F}.\nabla\left(\frac{1}{r}\right) = -\frac{\mathbf{F.r}}{r^3}$

(ii) $\partial f(r) \times \mathbf{r} = f'(r)\frac{\mathbf{r}}{r} \times \mathbf{r} = \frac{f'(r)}{r}(\mathbf{r} \times \mathbf{r}) = 0$

Example 3: If $f = x + y + z,\ g = x^2 + y^2 + z^2$ and $h = xy + yz + zx$, prove that $[\nabla f, \nabla g, \nabla h] = 0$.

Solution: Given $f = x + y + z$

$$\therefore \quad \nabla f = \frac{\partial f}{\partial x}\mathbf{i} + \frac{\partial f}{\partial y}\mathbf{j} + \frac{\partial f}{\partial z}\mathbf{k} = \mathbf{i} + \mathbf{j} + \mathbf{k}$$

Given $g = x^2 + y^2 + z^2$

$$\therefore \quad \nabla g = \frac{\partial g}{\partial x}\mathbf{i} + \frac{\partial g}{\partial y}\mathbf{j} + \frac{\partial g}{\partial z} = 2x\,\mathbf{i} + 2y\,\mathbf{j} + 2z\,\mathbf{k}$$

Given $h = xy + yz + zx$

$$\therefore \quad \Delta h = \frac{\partial h}{\partial x}\mathbf{i} + \frac{\partial h}{\partial y}\mathbf{j} + \frac{\partial h}{\partial z}\mathbf{k}$$

$$= (y + z)\,\mathbf{i} + (x + z)\,\mathbf{j} + (y + x)\,\mathbf{k}$$

$$[\Delta f, \Delta g, \Delta h] = \Delta f.(\Delta g \times \Delta h)$$

$$= \begin{vmatrix} 1 & 1 & 1 \\ 2x & 2y & 2z \\ y+z & x+z & x+y \end{vmatrix}$$

$$= \begin{vmatrix} 1 & 0 & 0 \\ 2x & 2(y-x) & 2(z-x) \\ y+z & -(y+x) & -(z-x) \end{vmatrix}$$

$$= (-2)\begin{vmatrix} y-x & z-x \\ y-x & z-x \end{vmatrix}$$

$$\therefore \quad [\nabla f, \nabla g, \nabla h] = 0.$$

4.11 Operator A. ∇ where A is constant vector.

Let $\mathbf{A} = \mathbf{A_1}\,\mathbf{i} + A_2\,\mathbf{j} + A_3\,\mathbf{k}$

$$\therefore \quad \mathbf{A}.\nabla = (\mathbf{A_1\,i} + \mathbf{A_2\,j} + \mathbf{A_3\,k}).\left(\frac{\partial}{\partial x}\mathbf{i} + \frac{\partial}{\partial y}\mathbf{j} + \frac{\partial}{\partial z}\mathbf{k}\right)$$

$$= \mathbf{A_1}\frac{\partial}{\partial x} + \mathbf{A_2}\frac{\partial}{\partial y} + \mathbf{A_3}\frac{\partial}{\partial z}$$

Example 4. If **A** is a constant vector and $\mathbf{r} = x\,\mathbf{i} + y\,\mathbf{j} + z\,\mathbf{k}$, prove that

$$(\mathbf{A}.\nabla)\,\mathbf{r} = \mathbf{A}$$

Solution: We know that

$$\mathbf{A}.\nabla = A_1 \frac{\partial}{\partial x} + A_2 \frac{\partial}{\partial y} + A_3 \frac{\partial}{\partial z}$$

where $\mathbf{A} = A_1\,\mathbf{i} + A_2\,\mathbf{j} + A_3\,\mathbf{k}$ (constant vector) and $\mathbf{r} = x\,\mathbf{i} + y\,\mathbf{j} + z\,\mathbf{k}$

$$(\mathbf{A}.\nabla)\,\mathbf{r} = A_1 \frac{\nabla \mathbf{r}}{\partial x} + A_2 \frac{\partial \mathbf{r}}{\partial y} + A_3 \frac{\partial \mathbf{r}}{\partial z}$$

$$= A_1\,\mathbf{i} + A_2\,\mathbf{j} + A_3\,\mathbf{k} = \mathbf{A}$$

$$\left(\because\ \frac{\partial \mathbf{r}}{\partial x} = \mathbf{i},\ \frac{\partial \mathbf{r}}{\partial y} = \mathbf{j};\ \frac{\partial \mathbf{r}}{\partial z} = \mathbf{k}\right)$$

$\therefore$ $(\mathbf{A}.\nabla)\,\mathbf{r} = \mathbf{A}$

Note: If **F** is any vector then $(\mathbf{F}.\nabla)\,\mathbf{r} = \mathbf{F}$.

Example 5. If **A** and **B** be the vectors joining the fixed points $\mathbf{P}(x_1, y_1, z_1)$ and $\mathbf{Q}(x_2, y_2, z_2)$ respectively to a variable point $\mathbf{R}(x, y, z)$, prove that $\nabla(\mathbf{A.B}) = \mathbf{A} + \mathbf{B}$.

Solution: $\mathbf{A} = \overrightarrow{PR} = (x - x_1)\,\mathbf{i} + (y - y_1)\,\mathbf{j} + (z - z_1)\,\mathbf{k}$

$$\mathbf{B} = \overrightarrow{QR} = (x - x_2)\,\mathbf{i} + (y - y_2)\,\mathbf{j} + (z - z_2)\,\mathbf{k}$$

$$\therefore\ \mathbf{A.B} = (x - x_1)(x - x_2) + (y - y_1)(y - y_2) + (z - z_1)(z - z_2)$$

$$= [x^2\, x(x_1 + x_2) + x_1 x_2] + [y^2 - y(y_1 + y_2) + y_1 y_2]$$
$$+ [z^2 - z(z_1 + z_2) + z_1 z_2]$$

$$\nabla(A.B) = \frac{\partial}{\partial x}(\mathbf{A.B})\,\mathbf{i} + \frac{\partial}{\partial y}(\mathbf{A.B})\,\mathbf{j} + \frac{\partial}{\partial z}(A.B)\,\mathbf{k}$$

$$= [2x - (x_1 + x_2)]\,\mathbf{i} + [2y - (y_1 + y_2)]\mathbf{j} + [2z - (z_1 + z_2)]\mathbf{k}$$

$$= [(x - x_1) + (x - x_2)]\,\mathbf{i} + (y - y_1) + (y - y_2)]\,\mathbf{j}$$
$$+ [2z - (z_1 + z_2)]\,\mathbf{k}$$

$$= (x - x_1)\,\mathbf{i} + (y - y_1)\,\mathbf{j} + (z - z_1)\,\mathbf{k}] + [(x - x_2)\,\mathbf{i}$$
$$+ (y - y_2)\,\mathbf{j} + (z - z_2)\,\mathbf{k}]$$

$$= \mathbf{A} + \mathbf{B}$$

$$\therefore\ \nabla(\mathbf{A.B}) = \mathbf{A} + \mathbf{B}$$

Example 6. Find a unit vector normal to the surface

$$x^3 + y^3 + z^3 + 3\,xyz = 4 \text{ at the point } (1, -2, -1).$$

Solution. The equation of the level surface is

$$\phi\,(x, y, z) = x^3 + y^3 + z^3 + 3\,xyz = 4$$

The vector $\nabla\phi$ is along the normal to the surface at the point (x, y, z)

$$\therefore \quad \nabla\,\phi = \frac{\partial\phi}{\partial x}\,\mathbf{i} + \frac{\partial\phi}{\partial y}\,\mathbf{j} + \frac{\partial\phi}{\partial z}\,\mathbf{k}$$

$$= (3\,x^2 + 3\,yz)\,\mathbf{i} + (3\,y^2 + 3\,xz)\,\mathbf{j} + (3\,z^2 + 3\,xy)\,\mathbf{k}$$

At (1, - 2, - 1)

$$\nabla\phi = 9\,\mathbf{i} + 9\,\mathbf{j} - 3\mathbf{k} = 3\,(3i + 3j - k)$$

Unit normal at (1, —2, —1) vector $= \dfrac{\nabla\phi}{|\nabla\phi|}$ at (1, — 2, —1)

$$= \frac{3\,(3\,\mathbf{i} + 3\,\mathbf{j} - \mathbf{k})}{|3\,(3\,\mathbf{i} + 3\,\mathbf{j} - \mathbf{k})|} = \frac{1}{\sqrt{19}}(3 + 3\mathbf{j} - \mathbf{k}).$$

Example 7. Find the directional derivative of

$$\phi\,(x, y, z) = x^2\,yz + 4\,xz^2 \text{ at the point } (1, -2, -1)$$

(i) in the direction of $2\,\mathbf{i} - \mathbf{j} - 2\,\mathbf{k}$.

In what direction will the directional derivative be maximum. Find the magnitude of this maximum.

(ii) In the direction of normal to the surface

$$f(x, y, z) = x \log z - y^2 \text{ at } (-1, 2, -1)$$

(iii) in the direction of $\overrightarrow{PQ}$ where P = (1, 2, −1) and

$$Q = (-1, 2, 3)$$

Solution: We have $\phi\,(x, y, z) = x^2\,yz + 4\,xz^2$

$$\nabla\,\phi = \frac{\partial\phi}{\partial x}\,\mathbf{i} + \frac{\partial\phi}{\partial y}\,\mathbf{j} + \frac{\partial\phi}{\partial z}\,\mathbf{k}$$

$$= (2\,xyz + 4z^2)\,\mathbf{i} + x^2 z\,\mathbf{j} + (x^2 y + 8\,xz)\,\mathbf{k}$$

At (1, — 2, — 1)

$$\nabla\phi = 8\,\mathbf{i} - \mathbf{j} - 10\mathbf{k} \qquad ...(1)$$

(i) Let e be the unit vector in the given direction, then

$$e = \frac{2\,\mathbf{i} - \mathbf{j} - 2\,\mathbf{k}}{\sqrt{4 + 1 + 4}} = \frac{1}{2}(2\,\mathbf{i} - \mathbf{j} - 2\,\mathbf{k})$$

$\therefore$ The required directional derivative is

$$= \nabla\phi.\mathbf{e} \text{ at } (1, -2, -1)$$

$$= (8\,\mathbf{i} - \mathbf{j} - 10\,\mathbf{k}).\frac{1}{2}(2\,\mathbf{i} - \mathbf{j} - 2\,\mathbf{k})$$

$$= \frac{1}{3}(16 + 1 + 20) = \frac{37}{3}$$

The directional derivative is maximum in the direction of $\nabla\phi$ and the magnitude of this maximum is

$$|\nabla\phi| = \sqrt{8^2 + (-1)^2 + (-10)^2} = \sqrt{165}$$

(ii) Normal to the surface $f(x, y, z) = x \log z - y^2$

is $\nabla f = \frac{\partial f}{\partial x}\mathbf{i} - \frac{\partial f}{\partial y}\mathbf{j} + \frac{\partial f}{\partial z}\mathbf{k}$

$$= \log z\,\mathbf{i} - 2\,y\,\mathbf{j} + \frac{x}{z}\mathbf{k}$$

At $(-1, 2, 1)$ $\nabla f = (\log 1)\,\mathbf{i} - 4\,\mathbf{j} - \mathbf{k} = -4\,\mathbf{j} - \mathbf{k}$

$\therefore$ The unit vector in the given direction is

$$\mathbf{e} = \frac{\nabla f}{|\nabla f|} = \frac{-4\,\mathbf{j} - \mathbf{k}}{\sqrt{16 + 1}} = \frac{-4\,\mathbf{j} - \mathbf{k}}{\sqrt{17}}$$

$\therefore$ The required directional derivative is

$$= \nabla\phi.\mathbf{e} \quad \text{at } (1, -2, -1)$$

$$= 8\,\mathbf{i} - \mathbf{j} - 10\,\mathbf{k}).\frac{-4\,\mathbf{j} - \mathbf{k}}{\sqrt{17}}$$

$$= \frac{1}{\sqrt{17}}(4 + 10) = \frac{14}{\sqrt{17}}$$

(iii) $\overrightarrow{PQ} = \overrightarrow{OQ} - \overrightarrow{OP}$ (O is the origin)

$$= (-\,\mathbf{i} + 2\,\mathbf{j} + 3\mathbf{k}) - (\mathbf{i} + 2\mathbf{j} - \mathbf{k})$$

$$= -2\,\mathbf{i} + 4\,\mathbf{k} = 2(-\,\mathbf{i} + 2\,\mathbf{k})$$

$\therefore$ The unit vector in the given direction is

$$\mathbf{e} = \frac{\overrightarrow{PQ}}{|\overrightarrow{PQ}|} = \frac{2(-\mathbf{i} + 2\,\mathbf{k})}{2\sqrt{1 + 4}} = \frac{-\,\mathbf{i} + 2\mathbf{k}}{\sqrt{5}}$$

$\therefore$ The required directional derivative is

$$= \nabla\phi.\mathbf{e} \text{ at } (1, -2, -1)$$

$$= (8\,\mathbf{i} - \mathbf{j} - 10\,\mathbf{k}).\frac{-\,\mathbf{i} + 2\,\mathbf{k}}{\sqrt{5}}$$

$$= \frac{-8-20}{\sqrt{5}} = -\frac{28}{\sqrt{5}}.$$

Example 8. (i) Find the directional derivative of the function $\phi(x, y, z)\, x^3 + y^3 - 3\,xyz$ along the tangent to the curve $x=t$, $y = t^2$, $z = t^3$ at (1, 1, 1).

(ii) What is the greatest rate of increase of $\phi(x, y, z)$k at (1, 1, 1).

Solution: $\nabla\phi = (3x^2 - 3yz)\,\mathbf{i} + (3y^2 - 3xz)\,\mathbf{j} - 3xy\,\mathbf{k}$

$= (0)\,\mathbf{i} + (0)\,\mathbf{j} - 3\,\mathbf{k}$ at (1, 1, 1)

$\therefore \quad \nabla\phi = -3\,\mathbf{k}$ at (1, 1, 1)

Let **r** be the position vector of any point on the given curve than

$$\mathbf{r} = x\,\mathbf{i} + y\,\mathbf{j} + z\,\mathbf{k}$$

$$= t\,\mathbf{i} + t^2\,\mathbf{j} + t^3\,\mathbf{k}$$

$$\frac{d\mathbf{r}}{dt} = \mathbf{i} + 2\,\mathbf{j} + 3\,\mathbf{k})$$

$$= (\mathbf{i} + 2\mathbf{j} + 3\mathbf{k} \text{ at } t = 1.$$

$\frac{d\mathbf{r}}{dt}$ is the vector along the tangent to the curve.

$$\therefore \text{ Unit tangent vector} = \mathbf{T} = \frac{\left(\frac{d\mathbf{r}}{dt}\right)}{\left|\frac{d\mathbf{r}}{dt}\right|}$$

$$= \frac{\mathbf{i} + 2\mathbf{j} + 3\mathbf{k}}{\sqrt{1+4+9}}$$

$$= \frac{1}{\sqrt{14}}(\mathbf{i} + 2\mathbf{j} + 3\,\mathbf{k})$$

$\therefore$ The directional derivative along the tangent

$$= \nabla\phi.\mathbf{T}$$

$$= (-3\mathbf{k}).\frac{1}{\sqrt{14}}(\mathbf{i} + 2\mathbf{j} + 3\mathbf{k})$$

$$= -\frac{9}{\sqrt{14}}$$

(ii) We have $\nabla\phi = -3\mathbf{k}$ at (1, 1, 1)

The greatest rate of increase of ϕ (x, y, z) at (1, 1, 1)

= The maximum value of directional derivative of ϕ at the point (1, 1, 1)

= $|\nabla\phi|$ at (1, 1, 1)

= $|-3\mathbf{k}| = 3$

Example 9. (i) Find the angle between the surfaces $xy^2z = 3x + z^2$ and, $3x^2 - y^2 + 2z = 1$ at (1, — 2, 1).

(ii) Find the angle between the tangent planes to the surfaces $x \log z = y^2 - 1$, $x^2y = 2 - z$ at the point (1, 1, 1)

Solution: Angle between two surfaces at a point is the angle between the normals to the surfaces at that point.

Let $\phi_1 = xy^2z - 3x - z^2$ and $\phi_2 = 3x^2 - y^2 + 2z - 1$

Then $\nabla\phi_1 = (y^2z - 3)\,\mathbf{i} + (2xyz)\,\mathbf{j} + (xy^2 - 2z)\,\mathbf{k}$

$= \mathbf{i} - 4\,\mathbf{j} + 2\,\mathbf{k}$ at (1, - 2, 1)

and $\nabla\phi_2 = 6x\,\mathbf{i} - 2y\,\mathbf{j} + 2\mathbf{k}$

$= 6\,\mathbf{i} \div 4\,\mathbf{j} + 2\,\mathbf{k}$ at (1, - 2, 1)

Now the vectors $\nabla\phi_1$ and $\nabla\phi_2$ are along normals to two surfaces at (1, - 2, 1). If θ be the angle between these surfaces, then

$$\cos\theta = \frac{\nabla\phi_1 . \nabla\phi_2}{|\nabla\phi_2||\nabla\phi_2|}$$

$$= \frac{(\mathbf{i} - 4\,\mathbf{j} + 2\,\mathbf{k}).(6\,\mathbf{i} + 4\,\mathbf{j} + 2\,\mathbf{k})}{\{\sqrt{1 + 16 + 4}\}\{\sqrt{36 + 16 + 4}\}}$$

$$= \frac{6 - 16 + 4}{\sqrt{21 \times 56}}$$

$$= \frac{-6}{14\sqrt{6}} = \frac{3}{7\sqrt{6}}$$

$$\therefore \qquad \theta = \cos^{-1}\left(-\frac{3}{7\sqrt{6}}\right)$$

(ii) Angle between the tangent planes to the surfaces $Ø_1$ and $Ø_2$ at A means, the angle between the normals to the surfaces $Ø_1$ and $Ø_2$ at **A**.

$\because$ If Ø is the surface then $\nabla Ø$ at **A** is normal to the tangent plane at **A**.

Let $\varnothing_1 = x \log z - y^2 + 1$ and $\varnothing_2 = x^2 y - 2 + z$

Then $\nabla\varnothing_1 = \log z\,\mathbf{i} - 2y\,\mathbf{j} + \dfrac{x}{z}\mathbf{k}$

$= 2\,\mathbf{j} - \mathbf{k}$ at (1, 1, 1)

and $\nabla\varnothing_2 = 2\,xy\,\mathbf{i} + x^2\,\mathbf{j} + \mathbf{k}$

$= 2\,\mathbf{i} + \mathbf{j} + \mathbf{k}$ at (1, 1, 1)

Let θ be the required angle

$$\therefore \quad \cos\theta = \frac{\nabla\varnothing_1.\nabla\varnothing_2}{|\nabla\varnothing_1||\nabla\varnothing_2|}$$

$$= \frac{(-2j + k).(2\,\mathbf{i} + \mathbf{j} + \mathbf{k})}{\{\sqrt{4+1}\}\,\{\sqrt{4+1+1}\}}$$

$$= \frac{0-2+1}{\sqrt{5\times 6}} = \frac{1}{\sqrt{30}}$$

or $$\theta = \cos^{-1}\left(-\frac{1}{\sqrt{30}}\right).$$

Example 10. (i) Find the equations of the tangent plane and normal to a surface $x + \log z = y^2$ at the point (1, 1, 1).

(ii) Given the curve $x^2 + y^2 + z^2 = 1, x + y + z = 1$ (intersection of two surfaces). Find the equation of the tangent line at the point (1, 0, 0).

Solution: (i) Here $\varnothing = x + \log z - y^2$

$$\nabla\varnothing = \mathbf{i} - 2\,y\mathbf{j} + \frac{1}{z}\mathbf{k}$$

$$= \mathbf{i} - 2\,\mathbf{j} + \mathbf{k} \text{ at } (1, 1, 1)$$

$\therefore$ i − 2 j + k is a vector along the normal to the surface at point (1, 1, 1)

r = The position vector of the point (1, 1, 1)

$\therefore \quad \mathbf{r} = \mathbf{i} + \mathbf{j} + \mathbf{k}$

If **R** = **X**i + **Y** j + **Z** k is the position vector of any current point (X, Y, Z) on the tangent plane at (1, 1, 1) then (**R** − **r**) is perpendicular to ∇Ø

$$\mathbf{R} - \mathbf{r}) = (X-1)\mathbf{i} + (Y-1)\mathbf{j} + (Z-1)\mathbf{k}$$

$\therefore$ The equation of the tangent plane to the surface ϕ at (1, 1, 1) is

$$(\mathbf{R} - \mathbf{r}).\nabla\varnothing = 0$$

i.e. $[(X - 1)\,\mathbf{i} + (Y - 1)\mathbf{j} + (Z - 1)\,\mathbf{k}].[\mathbf{i} - 2\,\mathbf{j} + \mathbf{k}] = 0$

ie. $(X - 1) - 2(Y - 1) + (Z - 1) = 0$

or $X - 2\,Y + Z = 0$

The equation of normal to the surface at (1, 1, 1) is

$$\frac{X-1}{\frac{\partial\varnothing}{\partial x}} = \frac{Y-1}{\frac{\partial\varnothing}{\partial y}} = \frac{Z-1}{\frac{\partial\varnothing}{\partial z}}$$

ie
$$\frac{X-1}{1} = \frac{Y-1}{-2} = \frac{Z-1}{1}$$

(ii) Let $\mathbf{R} = X\,\mathbf{i} + Y\,\mathbf{j} + Z\,\mathbf{k}$, be the position vector of any point **P** on the tangent at Q (1, 0, 0) whose position vector is say $\mathbf{r} = \mathbf{i}$

Thus $\overrightarrow{QP} = \mathbf{R} - \mathbf{r} = (X - 1)\mathbf{i} + Y\mathbf{j} + Z\mathbf{k}$, which is a vector along the tangent line.

Let $\varnothing_1 = x^2 + y^2 + z^2 - 1$ and $\varnothing_2 = x + y + z - 1$

Now a normal to the surface $\varnothing_1$ is

$$\mathbf{n_1} = \nabla\varnothing_2 = x\,\mathbf{i} + 2\,y\,\mathbf{j} + 2\,z\,\mathbf{k}$$
$$= 2\,\mathbf{i} \quad \text{at } (1, 0, 0)$$

and a normal to the surface $\varnothing_2$ is

$$\mathbf{n_2} = \nabla\varnothing_2 = \mathbf{i} + \mathbf{j} + \mathbf{k} \quad \text{at } (1, 0, 0)$$

Now we find that $\overrightarrow{QP}$ is perpendicular to both $\mathbf{n_1}$ and $\mathbf{n_2}$.

Hence $\overrightarrow{QP}$ will be parallel to $\mathbf{n_1} \times \mathbf{n_2}$.

$$\therefore \quad \overrightarrow{QP} \times (\mathbf{n_1} \times \mathbf{n_2}) = 0$$

Now $\mathbf{n_1} \times \mathbf{n_2} = \nabla\varnothing_1 \times \nabla\varnothing_2$

$$= (2\,\mathbf{i}) \times (\mathbf{i} + \mathbf{j} + \mathbf{k})$$
$$= -2\,\mathbf{j} + 2\,\mathbf{k}$$

$$\therefore \quad \overrightarrow{QP} \times (\mathbf{n_1} \times \mathbf{n_2}) = \begin{vmatrix} \mathbf{i} & \mathbf{j} & \mathbf{k} \\ X-1 & Y & Z \\ 0 & -2 & 2 \end{vmatrix} = 0$$

$$(2\,Y + 2\,Z)\,\mathbf{i} - 2(X - 1)\,\mathbf{j} - 2\,(X - 1)\,\mathbf{k} = 0$$

$$\Rightarrow \quad 2\,Y + 2\,Z = 0;\; 2\,(X - 1) = 0$$

$\therefore \quad X = 1, \; Y + Z = 0$

or $\dfrac{X-1}{0} = \dfrac{Y}{1} = \dfrac{Z}{-1}$ is the required equation of the tangent line to the curve of intersection of the given surfaces.

Example 11. (i) If f and **G** are point functions, prove that the components of the later normal and tangential to the surface f=0 are

$$\frac{(\mathbf{G}.\nabla f)\,\nabla f}{(\nabla f)^2} \text{ and } \frac{\nabla f \times (\mathbf{G} \times \nabla f)}{(\nabla f)^2}$$

(ii) Find the values of the constants a, b, c so that the directional derivative of $\text{Ø} = axy^2 + byz + cz^2x^3$ at (1, 2, –1) has a maximum magnitude 64 in a direction parallel to z - axiz.

Solution: (i) Let the vector **r** be inclined at an angle θ to the vector **A**.

$\vec{OQ} = (\mathbf{r} \cos\theta)\,\mathbf{e}$

$= (\mathbf{r.e})\,\mathbf{e}$ where $\mathbf{e} = \mathbf{A} \,|\, \mathbf{A} \,|$

$\vec{OR} = \vec{QP} = \vec{OP} - \vec{OQ}.$

$= \mathbf{r} - (\mathbf{r.e})\mathbf{e}.$

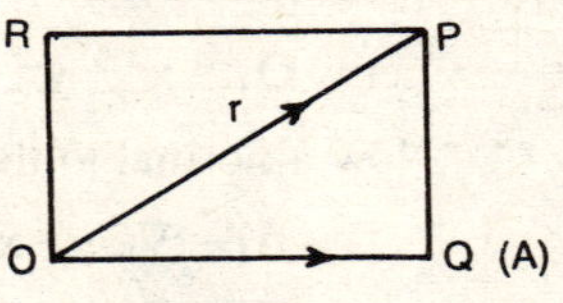

Fig. 11.2

∴ Component of **r** parallel to

$\mathbf{A} = (\mathbf{r.e})\,\mathbf{e}$ ...(1)

and component of **r** perpendicular to

$\mathbf{A} = \mathbf{r} - (\mathbf{r.e})\,\mathbf{e}$ ...(2)

The unit normal vector to the surface f = 0 is

$$= \frac{\nabla f}{|\nabla f|}$$

∴ The magnitude of the component of **G** along the normal

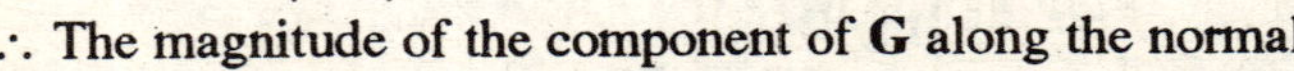

$$= \mathbf{G}.\frac{\nabla f}{|\nabla f|}$$

∴ The component of **G** along the normal

$$= \left(\mathbf{G}.\frac{\nabla f}{|\nabla f|}\right)\frac{\nabla f}{|\nabla f|}. \qquad \text{[equation (1)]}$$

$$= \frac{(\mathbf{G}.\nabla f)\,\nabla f}{(\nabla f)^2} \qquad (\because \; |\nabla f|^2 = (\nabla f)^2)$$

The tangential component of **G** is [equation (2)]

$$\mathbf{G} - \frac{(\mathbf{G}.\nabla f)\,\nabla f}{(\nabla f^2)} = \frac{(\nabla f.\nabla f)\,\mathbf{G} - (\mathbf{G}.\nabla f)\,\nabla f}{(\nabla f)^2}$$

$$= \frac{\nabla f \times (\mathbf{G} \times \nabla f)}{(\nabla f)^2}$$

(iii) The directional derivative is maximum along the normal *i.e.* along ∇Ø

$$\nabla Ø = (ay^2 + 3\,cz^2\,x^2)\,\mathbf{i} + (2\,axy + bz)\,\mathbf{j} + (by + 2\,cz\,x^3)\,\mathbf{k}$$
$$= (4a + 3c)\,\mathbf{i} + (4a - b)\,\mathbf{j} + (2b - 2c)\,\mathbf{k} \text{ at } (1, 2, -1)$$

But directional derivative is maximum along z - axis given. Hence coefficients of **i** and **j** should be zero.

$\therefore$ $4a + 3c = 0$ ...(1)

$4a - b = 0$ ...(2)

$\therefore$ $\nabla\phi = (2b - 2c)\,\mathbf{k}$

Maximum value of directional derivative is = | ∇Ø |

$\therefore$ $64 = |\,2\,(b - c)\,|$ or $32 = (b - c)$ ...(3)

From (1) and (2) B + 3c = 0 ...(4)

From (4) and (3) b = 24, c = – 8 so that a = 6.

EXERCISE

1. Find ∇ϕ and unit normal vector to ϕ. where

(i) $Ø(x,y,z) = x^2y + y^2z + z^2x$ at (1, 2, 1)

(ii) $\{Ø(x, y, z) = \log\sqrt{x^2 + y^2 + z^2} + xyz$ at (– 1, 1, 1)

(iii) $Ø(x, y, z) = x^3 + y^3 + z^3 - 3\,xyz$ at (– 1, 1, —1)

(iv) $Ø(x,y,z) = (x^2 + y^2 + z^2)\exp\{x^2 + y^2 + z^2\}$ at (1, 1, 1).

2. If $\mathbf{F} = \left(y\frac{\partial f}{\partial z} - z\frac{\partial f}{\partial y}\right)\mathbf{i} + \left(z\frac{\partial f}{\partial x} - x\frac{\partial f}{\partial z}\right)\mathbf{j} + \left(x\frac{\partial f}{\partial y} - y\frac{\partial f}{\partial x}\right)\mathbf{k}$

prove that

(i) $\mathbf{F} - (\mathbf{r} \times \nabla f) = 0$ (ii) $\mathbf{F}.\mathbf{r} = 0$

(iii) $\mathbf{F}.\nabla f = 0$

3. Prove that

(i) $\nabla f(r) = f'(r)\,\nabla r$ (ii) $\nabla Ø.dr = d\,Ø$

(iii) $\mathbf{F}.\nabla\left(\frac{1}{r}\right) = \frac{\mathbf{F.r}}{r^3}$ (iv) $\nabla(\mathbf{r.a}) = \mathbf{a}$

(v) ∇ **[rab]** = **a** × **b** where **a, b** are constant vectors.

4. Find the directional derivative of

$$f(x, y, z) = x^3 + y^3 + z^3 - 3\,xyz \text{ at } (-1, 1, -1)$$

in the direction of the vector **i** + 2 **j** + **k**.

5. Find the directional derivative of

$$f(x, y, z) = \log\sqrt{x^2 + y^2 + z^2} + xyz$$

in the direction of normal to the surface

$$f(x, y, z) = xy + yz + zx, \text{ at } (1, 1, 1)$$

6. Find the direction derivative of the function

$$f(x, y, z) = 4\,xy^2 + 2\,x^2yz$$

at the point A (1, 2, 3) in the direction of the line $\overline{AB}$, where **B** is the point (5, 0, 4).

7. For problems (4), (5) and (6), in what direction the directional derivative will be maximum and what is its magnitude?

8. Find the directional derivative of the function

$$f(x, y, z) = x^3 + y^3 + z^3 + x^2y + y^2z + z^2x + xyz$$

along the tangent to the curve $x = t$, $y = t^2$, $z = t^3$ at (1, 1, 1). What is the greatest rate of increase of f (x, y, z) at (1, 1, 1).

9. Find the angle between the normals to the surfaces $x^2 + y^2 + z^2 = 4$ and $z = x^2 + y^2 + 3$ at the point (2, –1, 2).

10. Find the angle between the normals to the surface $x^2 + xy + y^2z + 3\,xyz = 4$ at (1, 2, 1) and (– 1, 1, – 1).

11. Find the equations of the tangent plane and normal to a surface

$$2\,xz^2 + 7 + 3\,xy + 4\,x \text{ at } (1, -1, 2).$$

12. Find the constants a and b so that the surfaces $ax^2 - byz = (a + 2)\,x$ will be orthogonal to the surfaces $4(x^2y - 1) + z^3 = 0$ at the point (1, – 1, 2).

13. Find the values of the constants a, b, c so that the directional derivative of $f(x, y, z) = ax^2 + by^2 + cz^2$ at (1, 1, 2) has a maximum magnitude 4 in the direction parallel to y-axis.

ANSWERS

(1) (i) $\nabla\phi = 5\,\mathbf{i} + 5\,\mathbf{j} + 6\,\mathbf{k}$

Unit normal vector $= \dfrac{\nabla\phi}{|\nabla\phi|} = \dfrac{1}{\sqrt{86}}\{5\,\mathbf{i} + 5\,\mathbf{j} + 6\,\mathbf{k}\}$

(ii) $\nabla\phi = \dfrac{4}{3}(\mathbf{i} + \mathbf{j} + \mathbf{k})$

Unit normal vector $= \dfrac{\nabla\phi}{|\nabla\phi|} = \dfrac{1}{\sqrt{3}}(\mathbf{i} + \mathbf{j} + \mathbf{k})$

(iii) $\nabla\phi = 6(\mathbf{i} + \mathbf{k})$

Unit normal vector $= \dfrac{\nabla\phi}{|\nabla\phi|} = \dfrac{1}{\sqrt{2}}(\mathbf{i} + \mathbf{k})$

(iv) $\nabla\phi = 8\,e^3(\mathbf{i} + \mathbf{j} + \mathbf{k})$

Unit normal vector $= \dfrac{\nabla\phi}{|\nabla\phi| = \dfrac{1}{\sqrt{3}}(\mathbf{i} + \mathbf{j} + \mathbf{k})}$

(4) $2\sqrt{6}$ **(5)** $\dfrac{4}{\sqrt{3}}$ **(6)** $\dfrac{69}{\sqrt{21}}$

(7) The directional derivative of f is maximum in the direction of ∇f and magnitude $= |\nabla f|$ and the magnitudes are

(i) $6\sqrt{2}$, (ii) $\dfrac{4}{\sqrt{3}}$ (iii) $10\sqrt{21}$

(8) $\dfrac{42}{\sqrt{14}}$, $7\sqrt{3}$ **(9)** $\cos^{-1}\dfrac{16}{6\sqrt{21}}$

(10) $\cos^{-1}\left(-\sqrt{\dfrac{15}{22}}\right)$

(11) $7(X-1) - 3(Y+1) + 8(Z-2) = 0$

$\dfrac{X-1}{7} = \dfrac{Y+1}{-3} = \dfrac{Z-2}{8}$

(12) $a = \dfrac{k}{2}$, $b = 1$ **(13)** $a = 0$, $b = 2$, $c = 0$.

12.12. Divergence of a vector

Let **F** be any continuously differentiable vector point function. Then

$$\mathbf{i} \cdot \frac{\partial \mathbf{F}}{\partial x} + \mathbf{j} \cdot \frac{\partial \mathbf{F}}{\partial y} + \mathbf{k} \cdot \frac{\partial \mathbf{F}}{\partial z} \text{ is called}$$

divergence of **F** and is written as div **F**

$$\therefore \quad \text{div } \mathbf{F} = \mathbf{i} \cdot \frac{\partial \mathbf{F}}{\partial x} + \mathbf{j} \cdot \frac{\partial \mathbf{F}}{\partial y} + \mathbf{k} \cdot \frac{\partial \mathbf{F}}{\partial z}$$

$$= \left(\mathbf{i} \cdot \frac{\partial}{\partial x} + \mathbf{j} \cdot \frac{\partial}{\partial y} + \mathbf{k} \cdot \frac{\partial}{\partial z} \right) \mathbf{F}$$

$$= \left(\mathbf{i} \frac{\partial}{\partial x} + \mathbf{j} \frac{\partial}{\partial y} + \mathbf{k} \frac{\partial}{\partial z} \right) . \mathbf{F}$$

$$= \nabla . \mathbf{F}$$

$$\therefore \text{div } \mathbf{F} = \nabla . \mathbf{F}$$

Obviously $\nabla.\mathbf{F}$ is a scalar quantity. Thus the divergence of a vector point function is a scalar point function.

If $\mathbf{F} = F_1 \mathbf{i} + F_2 \mathbf{j} + F_3 \mathbf{k}$, then

$$\nabla.\mathbf{F} = \left(\mathbf{i} \frac{\partial}{\partial x} + \mathbf{j} \frac{\partial}{\partial y} + \mathbf{k} \frac{\partial}{\partial z} \right) . (F_1 \mathbf{i} + F_2 \mathbf{j} + F_3 \mathbf{k})$$

$$= \frac{\partial F_1}{\partial x} + \frac{\partial F_2}{\partial y} + \frac{\partial F_3}{\partial z}$$

12.13. Physical interpretation of divergence:

Consider the motion of fluid having velocity

$$\mathbf{V} = V_x \mathbf{i} + V_y \mathbf{j} + V_z \mathbf{k} \text{ at a point } \mathbf{A}\ (x, y, z)$$

Consider a small parallelopiped with edges δx, δy, δz parallel to the axes in the mass of fluid, with one of its corners at A.

Consider the flow parallel to y - axis.

i.e. across the faces **AHFG** and **BCDE,**

Velocity of fluid perpendicular to face **AHFG** = V_y,

Flow per unit time throught face AHFG

$$= V_y \, \delta z \, \delta x \ (\text{area} = \delta z \, \delta x)$$

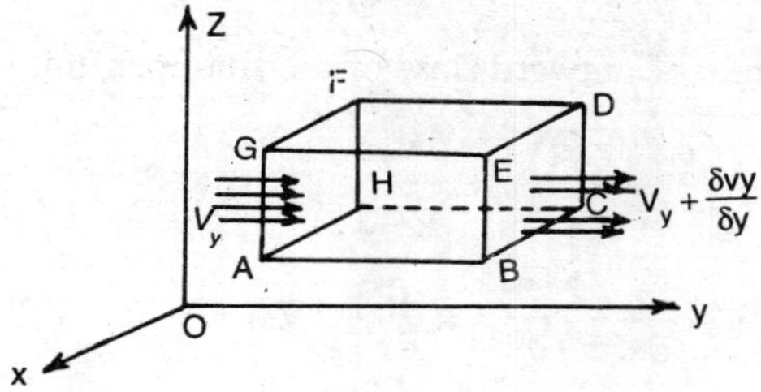

Fig. 12.3

Similarly, velocity of fluid at the face.

$$BCDE = V_y + \frac{\partial V_y}{\partial y} \cdot \partial y$$

Flow per unit time across face BCDE

$$= \left(V_y + \frac{\partial V_y}{\partial y} \right) \delta z \, \delta x.$$

Hence the total rate of flow outward from the element of volume along y-direction

$$= \left[V_y + \frac{\partial V_y}{\partial y} \delta y \right] \delta z \, \delta x - V_y \, \delta z \, \delta x$$

$$= \frac{\partial V_y}{\partial y} \delta x \, \delta y \, \delta z \qquad ...(1)$$

Similarly the total rate of flow outward from the element of volume along x—direction

$$= \frac{\partial V_x}{\partial x} \delta x \, \delta y \, \delta z \qquad ...(2)$$

and the total rate of flow outward from the element of volume along z—direction

$$= \frac{\partial V_z}{\partial z} \partial x \, \partial y \, \partial z \qquad ...(3)$$

∴ The total rate of outward flow of the fluid

$$= \left(\frac{\partial V_z}{\partial x} + \frac{\partial V_y}{\partial y} + \frac{\partial V_z}{\partial z} \right) \delta x \, \delta y \, \delta z \qquad ...(4)$$

so that the total rate of outward flow of the fluid per unit volume

$$= \left(\frac{\partial V_x}{\partial x} + \frac{\partial V_y}{\partial y} + \frac{\partial V_z}{\partial z} \right)$$

$$= \left(\mathbf{i} \frac{\partial}{\partial x} + \mathbf{j} \frac{\partial}{\partial y} + \mathbf{k} \frac{\partial}{\partial z} \right) . (V_x \mathbf{i} + V_y \mathbf{j} + V_z \mathbf{k})$$

$$= \nabla . \mathbf{V}$$

Note 1: In general, the divergence of a vector point function representing any physical quantity gives at each point the rate per unit volume at which the physical quantity is issuing from that point.

Note 2: If the fluid is incompressible, there can be no gain or loss in volume element.

Hence $\nabla.\mathbf{V} = 0$, which is known in Hydrodynamics as the *equation of continuity* in incompressible fluids.

Definition: If the flux entering any element of space is the same as that leaving it *i.e.* $\nabla.\mathbf{F} = 0$ everywhere, then **F** is called a *solenoidal vector* function.

12.14. Curl of a vector

Let **F** be any continuously differentiable vector point function. Then the vector function defined by

$$\mathbf{i} \times \frac{\partial \mathbf{F}}{\partial x} + \mathbf{j} \times \frac{\partial \mathbf{F}}{\partial y} + \mathbf{k} \times \frac{\partial \mathbf{F}}{\partial z}$$

is called curl of **F** and is denoted by $\nabla \times \mathbf{F}$

$$\therefore \text{ Curl } \mathbf{F} = \mathbf{i} \times \frac{\partial \mathbf{F}}{\partial x} + \mathbf{j} \times \frac{\partial \mathbf{F}}{\partial y} + \mathbf{k} \times \frac{\partial \mathbf{F}}{\partial z}$$

$$= \left(\mathbf{i} \times \frac{\partial}{\partial x} + \mathbf{j} \times \frac{\partial}{\partial y} + \mathbf{k} \times \frac{\partial}{\partial z} \right) \mathbf{F}$$

$$= \left(\mathbf{i} \frac{\partial}{\partial x} + \mathbf{j} \frac{\partial}{\partial y} + \mathbf{k} \frac{\partial}{\partial z} \right) \times \mathbf{F}$$

$$= \nabla \times \mathbf{F}$$

Curl F in terms of components of F:

Let $\mathbf{F} = \mathbf{F_1\, i} + \mathbf{F_2 j} + \mathbf{F_3 k}$

where F_1, F_2, F_3 are scalar functions of x, y and z.

$$\therefore \text{ Curl F} = \nabla \times F = \Sigma \mathbf{i} \times \frac{\partial \mathbf{F}}{\partial x}$$

$$= \Sigma \mathbf{i} \times \frac{\partial}{\partial x}(F_1\mathbf{i} + F_2\mathbf{j} + F_3\,\mathbf{k})$$

$$= \Sigma\left(\frac{\partial F_2}{\partial x}\mathbf{k} - \frac{\partial F_3}{\partial x}\right)$$

$$\therefore\ \nabla \times \mathbf{F} = \left(\frac{\partial F_2}{\partial x}\mathbf{k} - \frac{\partial F_3}{\partial x}\mathbf{j}\right) + \left(\frac{\partial F_2}{\partial y}\mathbf{i} - \frac{\partial F_1}{\partial y}\mathbf{k}\right) + \left(\frac{\partial F_1}{\partial z}\mathbf{j} - \frac{\partial F_2}{\partial z}\mathbf{i}\right)$$

$$= \left(\frac{\partial F_3}{\partial y} - \frac{\partial F_2}{\partial z}\right)\mathbf{i} + \left(\frac{\partial F_1}{\partial z} - \frac{\partial F_3}{\partial x}\right)\mathbf{j} + \left(\frac{\partial F_2}{\partial x} - \frac{\partial F_1}{\partial y}\right)\mathbf{k}$$

which can also be written in the form

$$\nabla \times \mathbf{F} = \begin{vmatrix} \mathbf{i} & \mathbf{j} & \mathbf{k} \\ \frac{\partial}{\partial x} & \frac{\partial}{\partial y} & \frac{\partial}{\partial z} \\ F_1 & F_2 & F_3 \end{vmatrix}$$

Note: If **F** is constant then curl F = 0

12.15. Physical interpretation of curl:

Consider the motion of a rigid body rotating about a fixed axis through O

If ω be its angular velocity, then the velocity **V** of any particle **P(r)** of the body is given by $\mathbf{V} = \omega \times \mathbf{r}$

If $\quad \omega = \omega_1\,\mathbf{i} + \omega_2\,\mathbf{j} + \omega_3\,\mathbf{k}$

and $\quad \mathbf{r} = x\,\mathbf{i} + y\,\mathbf{i} + z\,\mathbf{k}$

then

$$\omega \times \mathbf{r} = \mathbf{V} = \begin{vmatrix} \mathbf{i} & \mathbf{j} & \mathbf{k} \\ \omega_1 & \omega_2 & \omega_3 \\ x & y & z \end{vmatrix}$$

$$= (\omega_2 z - \omega_3 y)\,\mathbf{i} + (\omega_3 x - \omega_1 z)\,\mathbf{j} + (\omega_1 y - \omega_2 x)\,\mathbf{k}$$

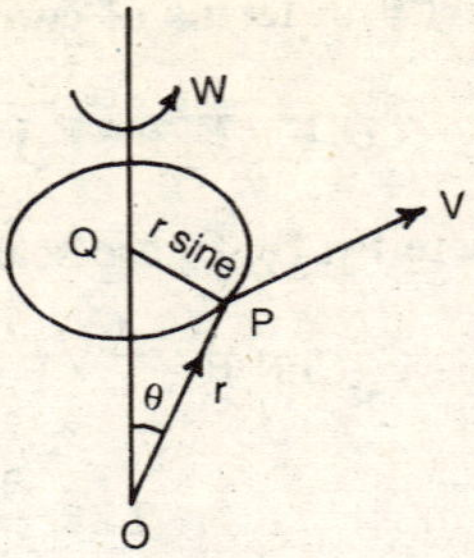

Fig. 12.4

$$\nabla \times \mathbf{V} = \begin{vmatrix} \mathbf{i} & \mathbf{j} & \mathbf{k} \\ \dfrac{\partial}{\partial x} & \dfrac{\partial}{\partial y} & \dfrac{\partial}{\partial z_2} \\ \omega_2 z - \omega_3 y & \omega_3 x - \omega_1 z & \omega_1 y - \omega z\, x \end{vmatrix}$$

$$= (\omega_1 + \omega_1)\,\mathbf{i} + (\omega_2 + \omega_2)\,\mathbf{j} + (\omega_2 + \omega_3)\,\mathbf{k}$$

$$= 2\,(\omega_1\,\mathbf{i} + \omega_2\,\mathbf{j} + \omega_3\,\mathbf{k})$$

$$= 2\,\omega$$

$$\therefore \quad \omega = \frac{1}{2}(\nabla \times \mathbf{V})$$

Thus the angular velocity of rotation ω at any point is equal to half the curl of the velocity vector **V** which justifies the word *curl of vector.*

In general, the curl of any vector point function gives the measure of the angular velocity at any point of the vector field.

Definition: A vector **F** is said to be irrotational if $\nabla \times \mathbf{F} = 0$

12.16. The Laplacian Operator ∇^2:

The Laplacian operator is defined as

$$\nabla^2 = \frac{\partial^2}{\partial x^2} + \frac{\partial^2}{\partial y^2} + \frac{\partial^2}{\partial z^2}.$$

If Ø is a scalar point function, then

$$\nabla^2 \varnothing = \frac{\partial^2 \varnothing}{\partial x^2} + \frac{\partial^2 \varnothing}{\partial y^2} + \frac{\partial^2 \varnothing}{\partial z^2} \quad \text{(a scalar quantity)}$$

If $\nabla^2 \varnothing = 0$, the equation is known as Laplace's equation.

A function which satisfies Laplace's equation is called a *harmonic function.*

12.17. Vector Identities:

(I) If **F** and **G** are vector point functions then

(i) $\nabla.(\mathbf{F} \pm \mathbf{G}) = \nabla.\mathbf{F} \pm \nabla.\mathbf{G}$

i.e. div $(\mathbf{F} \pm \mathbf{G})$ = div $\mathbf{F}$ ± div $\mathbf{G}$

(ii) $\nabla \times (\mathbf{F} \pm \mathbf{G}) = \nabla \times \mathbf{F} \pm \nabla \times \mathbf{G}$

i.e. curl $(\mathbf{F} \pm \mathbf{G})$ = curl $\mathbf{F}$ ± curl $\mathbf{G}$

Proof: (i) $\nabla.(\mathbf{F} \pm \mathbf{G})$

$$= \left(\mathbf{i}\frac{\partial}{\partial x} = \mathbf{j}\frac{\partial}{\partial y} + \mathbf{k}\frac{\partial}{\partial z}\right).(\mathbf{F} \pm \mathbf{G})$$

$$= \mathbf{i}.\frac{\partial}{\partial x}(\mathbf{F} \pm \mathbf{G}) + \mathbf{j}.\frac{\partial}{\partial y}(\mathbf{F} \pm \mathbf{G}) + \mathbf{k}\frac{\partial}{\partial z}(\mathbf{F} \pm \mathbf{G})$$

$$= \mathbf{i}.\left(\frac{\partial \mathbf{F}}{\partial x} \pm \frac{\partial \mathbf{G}}{\partial x}\right) + \mathbf{j}.\left(\frac{\partial \mathbf{F}}{\partial y} \pm \frac{\partial \mathbf{G}}{\partial y}\right) + \mathbf{k}.\left(\frac{\partial \mathbf{F}}{\partial z} \pm \frac{\partial \mathbf{G}}{\partial z}\right)$$

$$= \left(\mathbf{i}.\frac{\partial \mathbf{F}}{\partial x} + \mathbf{j}.\frac{\partial \mathbf{F}}{\partial y} + \mathbf{k}.\frac{\partial \mathbf{F}}{\partial z}\right) \pm \left(\mathbf{i}.\frac{\partial \mathbf{G}}{\partial x} + \mathbf{j}.\frac{\partial \mathbf{G}}{\partial y} + \mathbf{k}.\frac{\partial \mathbf{G}}{\partial z}\right)$$

$$= \nabla.\mathbf{F} \pm \nabla.\mathbf{G}$$

Hence the proof.

(ii) $\nabla \times (\mathbf{F} \pm \mathbf{G})$

$$= \left(\mathbf{i}\frac{\partial}{\partial x} + \mathbf{j}\frac{\partial}{\partial y} + \mathbf{k}\frac{\partial}{\partial z}\right) \times (\mathbf{F} \pm \mathbf{G})$$

$$= \mathbf{i} \times \frac{\partial}{\partial x}(\mathbf{F} \pm \mathbf{G}) + \mathbf{j} \times \frac{\partial}{\partial y}(\mathbf{F} \pm \mathbf{G}) + \mathbf{k} \times \frac{\partial}{\partial z}(\mathbf{F} \pm \mathbf{G})$$

$$= \mathbf{i} \times \left(\frac{\partial \mathbf{F}}{\partial x} \pm \frac{\partial \mathbf{G}}{\partial x}\right) + \mathbf{j} \times \left(\frac{\partial \mathbf{F}}{\partial y} \pm \frac{\partial \mathbf{G}}{\partial y}\right) + \mathbf{k} \times \left(\frac{\partial \mathbf{F}}{\partial z} \pm \frac{\partial \mathbf{G}}{\partial z}\right)$$

$$= \left(\mathbf{i} \times \frac{\partial \mathbf{F}}{\partial x} + \mathbf{j} \times \frac{\partial \mathbf{F}}{\partial y} + \mathbf{k} \times \frac{\partial \mathbf{F}}{\partial z}\right) \pm \left(\mathbf{i} \times \frac{\partial \mathbf{G}}{\partial x} + \mathbf{j} \times \frac{\partial \mathbf{G}}{\partial y} + \mathbf{k} \times \frac{\partial \mathbf{G}}{\partial z}\right)$$

$$= \nabla \times \mathbf{F} \pm \nabla \times \mathbf{G}$$

Hence the proof.

(II) If **F** is a differentiable vector function and Ø is a differentiable scalar function, then

(i) $\nabla.(Ø\mathbf{F}) = \nabla Ø.\mathbf{F} + Ø\,(\nabla.\mathbf{F})$

i.e. $\text{div}\,(Ø\mathbf{F}) = (\nabla Ø)\,.\,\mathbf{F} + Ø\,(\text{div}\,\mathbf{F})$

(ii) $\nabla \times (Ø\mathbf{F}) = \nabla Ø \times \mathbf{F} + Ø\,(\nabla \times \mathbf{F})$

i.e. $\text{curl}\,(Ø\mathbf{F}) = (\nabla\,Ø) \times \mathbf{F} + Ø\,(\text{curl}\,\mathbf{F})$

Proof: (i) Let $\mathbf{F} = \mathbf{F_1 i} + \mathbf{F_2 j} + \mathbf{F_3 k}$

Then $Ø\mathbf{F} = Ø\mathbf{F_1 i} + Ø\mathbf{F_2 j} + Ø\mathbf{F_3 k}$

$$\therefore\ \nabla.(Ø\mathbf{F}) = \frac{\partial}{\partial x}(ØF_1) + \frac{\partial}{\partial y}(Ø\mathbf{F_2}) + \frac{\partial}{\partial z}(Ø\mathbf{F_3})$$

$$= \left(\frac{\partial Ø}{\partial x}\mathbf{F_1} + Ø\frac{\partial F_1}{\partial x}\right)\left(\frac{\partial Ø}{\partial y}F_2 + Ø\frac{\partial F_2}{\partial y}\right) + \left(\frac{\partial Ø}{\partial z}\mathbf{F_3} + Ø\frac{\partial F_3}{\partial z}\right)$$

$$= \left(\frac{\partial Ø}{\partial x}\mathbf{F_1} + \frac{\partial Ø}{\partial y}\mathbf{F_2} + \frac{\partial Ø}{\partial z}\mathbf{F_3}\right) + \phi\left(\frac{\partial F_1}{\partial x} + \frac{\partial F_2}{\partial y} + \frac{\partial F_3}{\partial z}\right)$$

$$= \left(\frac{\partial Ø}{\partial x}\mathbf{i} + \frac{\partial Ø}{\partial y}\mathbf{j} + \frac{\partial Ø}{\partial z}\mathbf{k}\right).(\mathbf{F_1 i} + \mathbf{F_2 j} + \mathbf{F_3\,k}) + Ø\left(\frac{\partial}{\partial x}\mathbf{i} + \frac{\partial}{\partial y}\mathbf{j} + \frac{\partial}{\partial z}\mathbf{k}\right).(\mathbf{F_1 i} + \mathbf{F_2 j} + \mathbf{F_3 k})$$

$$= \nabla Ø.\mathbf{F} + Ø\,(\nabla.\mathbf{F})$$

Hence the proof.

(ii) $Ø\mathbf{F} = Ø\mathbf{F_1\,i} + ØF_2\mathbf{j} + ØF_3\mathbf{k}$

$$\nabla \times (Ø\mathbf{F}) = \begin{vmatrix} \mathbf{i} & \mathbf{j} & \mathbf{k} \\ \frac{\partial}{\partial x} & \frac{\partial}{\partial y} & \frac{\partial}{\partial z} \\ ØF_1 & ØF_2 & ØF_3 \end{vmatrix}$$

$$= \left[\frac{\partial}{\partial y}(ØF_3) - \frac{\partial}{\partial z}(ØF_2)\right]\mathbf{i}$$

$$+\left[\frac{\partial}{\partial z}(ØF_1)-\frac{\partial}{\partial x}(ØF_3)\right]\mathbf{j}+\left[\frac{\partial}{\partial x}(ØF_2)-\frac{\partial}{\partial y}(ØF_1)\right]\mathbf{k}$$

$$=\left[Ø\frac{\partial F_3}{\partial y}+\frac{\partial Ø}{\partial y}F_3-Ø\frac{\partial F_2}{\partial z}-\frac{\partial Ø}{\partial z}F_2\right]\mathbf{i}$$

$$+\left[Ø\frac{\partial F_1}{\partial z}+\frac{\partial Ø}{\partial z}F_1-Ø\frac{\partial F_3}{\partial x}-\frac{\partial Ø}{\partial x}\right]\mathbf{j}$$

$$+\left[Ø\frac{\partial F_2}{\partial x}+\frac{\partial Ø}{\partial x}F_2-Ø\frac{\partial F_1}{\partial y}-\frac{\partial Ø}{\partial y}F_1\right]\mathbf{k}$$

$$=Ø\left[\left(\frac{\partial F_3}{\partial y}-\frac{\partial F_2}{\partial z}\right)\mathbf{i}+\left(\frac{\partial F_1}{\partial z}-\frac{\partial F_3}{\partial x}\right)\mathbf{j}\right.$$

$$\left.+\left(\frac{\partial F_2}{\partial x}-\frac{\xi F_1}{\partial y}\right)\mathbf{k}\right]+\left[\left(\frac{\partial Ø}{\partial y}F_3-\frac{\partial Ø}{\partial z}F_2\right)\mathbf{i}\right.$$

$$\left.+\left(\frac{\partial Ø}{\partial z}F_1-\frac{\partial Ø}{\partial x}F_3\right)\mathbf{j}+\left(\frac{\partial Ø}{\partial x}F_2-\frac{\partial Ø}{\partial y}F_1\right)\mathbf{k}\right]$$

$$=Ø(\nabla\times\mathbf{F})+\begin{vmatrix}\mathbf{i} & \mathbf{j} & \mathbf{k}\\ \frac{\partial Ø}{\partial x} & \frac{\partial Ø}{\partial y} & \frac{\partial \phi}{\partial z}\\ F_1 & F_2 & F_3\end{vmatrix}$$

$$=Ø(\nabla\times\mathbf{F})+(\nabla Ø)\times\mathbf{F}$$

Hence the proof.

(III) If **F** and **G** are vector point functions

Then (i) $\nabla.(\mathbf{F}\times\mathbf{G})=\mathbf{G}.(\nabla\times\mathbf{F})-\mathbf{F}.(\nabla\times\mathbf{G})$

i.e. div $(\mathbf{F}\times\mathbf{G})=\mathbf{G}.(\text{curl }\mathbf{F})-\mathbf{F}.(\text{curl }\mathbf{G})$

(ii) $\nabla\times(\mathbf{F}\times\mathbf{G})=\mathbf{F}(\nabla.\mathbf{G})-\mathbf{G}(\nabla.\mathbf{F})+(\mathbf{G}.\nabla)\mathbf{F}-(\mathbf{F}.\nabla)\mathbf{G}$

i.e. Curl $(\mathbf{F}\times\mathbf{G})=\mathbf{F}\text{ div }\mathbf{G}-\mathbf{G}\text{ div }\mathbf{F}+(\mathbf{G}.\nabla)\mathbf{F}=(\mathbf{F}.\nabla)\mathbf{G}$

(iii) $\nabla(\mathbf{F}.\mathbf{G})=(\mathbf{G}.\nabla)\mathbf{F}+(\mathbf{F}.\nabla)\mathbf{G}+\mathbf{G}\times(\nabla\times\mathbf{F})+\mathbf{F}\times(\nabla\times\mathbf{G})$

i.e. $\nabla(\mathbf{F}.\mathbf{G})=(\mathbf{G}.\nabla)\mathbf{F}+(\mathbf{F}.\nabla)\mathbf{G}+\mathbf{G}\times(\text{curl }\mathbf{F})+\mathbf{F}\times(\text{curl }\mathbf{G})$.

Proof: (i) We have

$$\nabla.(\mathbf{F}\times\mathbf{G})=\Sigma\left[\mathbf{i}.\frac{\partial}{\partial x}(\mathbf{F}\times\mathbf{G})\right]$$

$$= \Sigma\left[\mathbf{i}\,.\left(\frac{\partial \mathbf{F}}{\partial x}\times \mathbf{G} + \mathbf{F}\times\frac{\partial \mathbf{G}}{\partial x}\right)\right]$$

$$= \Sigma\left[\mathbf{i}\,.\left(\frac{\partial \mathbf{F}}{\partial x}\times \mathbf{G}\right)\right] + \Sigma\left[\mathbf{i}\,.\left(\mathbf{F}\times\frac{\partial \mathbf{G}}{\partial x}\right)\right]$$

$$= \Sigma\left[\left(\mathbf{i}\,.\frac{\partial \mathbf{F}}{\partial x}\right)\times \mathbf{G}\right] - \Sigma\left[\mathbf{i}\,.\left(\frac{\partial \mathbf{G}}{\partial x}\times\mathbf{F}\right)\right]$$

$[\because \mathbf{A}.(\mathbf{B}\times\mathbf{C}) = (\mathbf{A}\times\mathbf{B})\,.\,\mathbf{C}$ and $\mathbf{A}\,.\,(\mathbf{B}\times\mathbf{C}) = -\,\mathbf{A}.(\mathbf{C}\times\mathbf{B})]$

$$= \left[\Sigma\left(\mathbf{i}\times\frac{\partial \mathbf{F}}{\partial x}\right)\right].\mathbf{G} - \left[\Sigma\left(\mathbf{i}\times\frac{\partial \mathbf{G}}{\partial x}\right)\right].\mathbf{F}$$

$$= \left[\Sigma\left(\mathbf{i}\times\frac{\partial \mathbf{F}}{\partial x}\right)\right].\mathbf{G} - \left[\Sigma\left(\mathbf{i}\times\frac{\partial \mathbf{G}}{\partial x}\right)\right].\mathbf{F}$$

$$= (\nabla\times\mathbf{F}).\mathbf{G} - (\nabla\times\mathbf{G}).\mathbf{F}$$

or $$= \mathbf{G}.\,(\nabla\times\mathbf{F}) - \mathbf{G}.\,(\nabla\times\mathbf{G})$$

Hence the proof.

(ii) We have

$$\nabla\times(\mathbf{F}\times\mathbf{G}) = \Sigma\left[\mathbf{i}\times\frac{\partial}{\partial x}(\mathbf{F}\times\mathbf{G})\right]$$

$$= \Sigma\left[\mathbf{i}\times\left(\mathbf{F}\times\frac{\partial \mathbf{G}}{\partial x} + \frac{\partial \mathbf{F}}{\partial x}\times\mathbf{G}\right)\right]$$

$$= \Sigma\left[\mathbf{i}\times\left(\mathbf{F}\times\frac{\partial \mathbf{G}}{\partial x}\right)\right] + \Sigma\left[\mathbf{i}\times\left(\frac{\partial \mathbf{F}}{\partial x}\times\mathbf{G}\right)\right]$$

$$= \Sigma\left[\left(\mathbf{i}.\frac{\partial \mathbf{G}}{\partial x}\,\mathbf{F} - (\mathbf{i}.\mathbf{F})\,\frac{\partial \mathbf{G}}{\partial x}\right)\right] + \Sigma\left[(\mathbf{i}.\mathbf{G})\,\frac{\partial \mathbf{F}}{\partial x} - \left(\mathbf{i}.\frac{\partial \mathbf{F}}{\partial x}\right)\mathbf{G}\right]$$

$$= \Sigma\left[\left(\mathbf{i}.\,\frac{\partial \mathbf{G}}{\partial x}\right)\mathbf{F}\right] - \Sigma\left[(\mathbf{F}.\mathbf{i})\,\frac{\partial \mathbf{G}}{\partial x}\right]$$

$$+ \Sigma\left[(\mathbf{i}.\mathbf{G})\,\frac{\partial \mathbf{F}}{\partial x} - \left(\mathbf{i}.\frac{\partial \mathbf{F}}{\partial x}\right)\mathbf{G}\right]$$

$$= \left[\Sigma \left(\mathbf{i}.\frac{\partial \mathbf{G}}{\partial x} \right) \right] \mathbf{F} - \left[\mathbf{F}.\Sigma \mathbf{i} \frac{\partial}{\partial x} \right] \mathbf{G}$$

$$+ \left[\mathbf{G}.\Sigma \mathbf{i} \frac{\partial}{\partial x} \right] \mathbf{F} - \left[\Sigma \left((\mathbf{i}.\frac{\partial \mathbf{F}}{\partial x} \right) \right] \mathbf{G}$$

$$= (\nabla.\mathbf{G})\mathbf{F} - (\mathbf{F}.\nabla)\,\mathbf{G} + (\mathbf{G}.\nabla)\,\mathbf{F} - (\nabla.\mathbf{F})\,\mathbf{G}$$

Hence the proof.

(iii) We have

$$\nabla\,(\mathbf{F}.\mathbf{G})\,\Sigma\,\mathbf{i}\,\frac{\partial}{\partial x}\,(\mathbf{F}.\mathbf{G})$$

$$= \Sigma \mathbf{i} \left(\mathbf{F}.\frac{\partial \mathbf{G}}{\partial x} + \frac{\partial \mathbf{F}}{\partial x}.\mathbf{G} \right)$$

$$= \Sigma \left[\left(\mathbf{F}.\frac{\partial \mathbf{G}}{\partial x} \right) \mathbf{i} \right] + \Sigma \left[\left(\mathbf{G}.\frac{\partial \mathbf{F}}{\partial x} \right) \mathbf{i} \right] \quad \text{...(1)}$$

We know that

$$\mathbf{A} \times (\mathbf{B} \times \mathbf{C}) = (\mathbf{A}.\mathbf{C})\,\mathbf{B} - (\mathbf{A}.\mathbf{B})\,\mathbf{C}$$

$$\therefore \quad (\mathbf{A}.\mathbf{B})\,\mathbf{C} = (\mathbf{A}.\mathbf{C})\,\mathbf{B} - \mathbf{A} \times (\mathbf{B} \times \mathbf{C})$$

$$\therefore \quad \left(\mathbf{F}.\frac{\partial \mathbf{G}}{\partial x} \right) \mathbf{i} = (\mathbf{F}.\mathbf{i})\,\frac{\partial \mathbf{G}}{\partial x} - \mathbf{F} \times \left(\frac{\partial \mathbf{G}}{\partial x} \times \mathbf{i} \right)$$

$$= (\mathbf{F}.\mathbf{i})\,\frac{\partial \mathbf{G}}{\partial x} + \mathbf{F} \times \left(\mathbf{i} \times \frac{\partial \mathbf{G}}{\partial x} \right)$$

Thus $\Sigma \left[\left(\mathbf{F}.\frac{\partial \mathbf{G}}{\partial x} \right) \mathbf{i} \right]$

$$= \Sigma \left[(\mathbf{F}.\mathbf{i})\,\frac{\partial \mathbf{G}}{\partial x} \right] + \Sigma \left[\mathbf{F} \times \left(\mathbf{i} \times \frac{\partial \mathbf{G}}{\partial x} \right) \right]$$

$$= \left[\mathbf{F}\,\Sigma\,\mathbf{i}\,\frac{\partial}{\partial x} \right] \mathbf{G} + \mathbf{F} \times \Sigma \left(\mathbf{i} \times \frac{\partial \mathbf{G}}{\partial x} \right)$$

$$= (\mathbf{F}.\nabla)\,\mathbf{G} + \mathbf{F} \times (\nabla \times \mathbf{G}) \quad \text{...(2)}$$

Similarly

$$\Sigma \left[\left(\mathbf{G}.\frac{\partial \mathbf{F}}{\partial x} \right) \mathbf{i} \right] = (\mathbf{G}.\nabla)\,\mathbf{F} + \mathbf{G} \times (\nabla \times \mathbf{F}) \quad \text{...(3)}$$

Substituting (2) and (3) in (1), we get

$\nabla(\mathbf{F}.\mathbf{G}) = (\mathbf{F}.\nabla)\,\mathbf{G} + \mathbf{F}\times(\nabla\times\mathbf{G}) + (\mathbf{G}.\nabla)\mathbf{F} + \mathbf{G}\times(\nabla\times\mathbf{F})$.

Hence the proof.

Note: If **G** is replaced by **F**, then

$\nabla(\mathbf{F}.\mathbf{F}) = 2\,(\mathbf{F}.\nabla)\mathbf{F} + 2\,\mathbf{F}\times(\nabla\times\mathbf{F})$

or $\frac{1}{2}\nabla(f^2) = (\mathbf{F}.\nabla)\mathbf{F} + \mathbf{F}\times(\nabla\times\mathbf{F})$ where $\mathbf{F}.\mathbf{F} = f^2$

(IV) If ϕ is any scalar point function and **F** is any vector point function then

(i) $\nabla.(\nabla\phi) = \nabla^2\phi$

div (grad ϕ) = Laplacian ϕ

(ii) $\nabla\times(\nabla\,\phi) = 0$

curl of grad $\phi = 0$

(iii) $\nabla.(\nabla\times\mathbf{F}) = 0$

div curl **F** = 0

(iv) $\nabla\times(\nabla\times\mathbf{F}) = \nabla(\nabla.\mathbf{F}) - \nabla^2\mathbf{F}$

curl **F** = grad (div **F**) - Laplacian **F**.

Proof: (i) $\nabla.\,(\nabla\phi)$

$$= \left(\frac{\partial}{\partial x}\mathbf{i} + \frac{\partial}{\partial y}\mathbf{j} + \frac{\partial}{\partial z}\mathbf{k}\right).\left(\frac{\partial\phi}{\partial x}\mathbf{i} + \frac{\partial\phi}{\partial y}\mathbf{j} + \frac{\partial\phi}{\partial z}\mathbf{k}\right)$$

$$= \frac{\partial}{\partial x}\left(\frac{\partial\phi}{\partial x}\right) + \frac{\partial}{\partial y}\left(\frac{\partial\phi}{\partial y}\right) + \frac{\partial}{\partial z}\left(\frac{\partial\phi}{\partial z}\right)$$

$$= \frac{\partial^2\phi}{\partial x^2} + \frac{\partial^2\phi}{\partial y^2} + \frac{\partial^2\phi}{\partial z^2}$$

$$= \left(\frac{\partial^2}{\partial x^2} + \frac{\partial^2}{\partial y^2} + \frac{\partial^2}{\partial z^2}\right)\phi$$

$$= \nabla^2\phi$$

Hence the proof.

(ii) $\nabla\times(\nabla\phi)$

$$= \left(\frac{\partial}{\partial x}\mathbf{i} + \frac{\partial}{\partial y}\mathbf{j} + \frac{\partial}{\partial z}\mathbf{k} \right) \times \left(\frac{\partial \phi}{\partial x}\mathbf{i} + \frac{\partial \phi}{\partial y}\mathbf{j} + \frac{\partial \phi}{\partial z}\mathbf{k} \right)$$

$$= \begin{vmatrix} \mathbf{i} & \mathbf{j} & \mathbf{k} \\ \frac{\partial}{\partial x} & \frac{\partial}{\partial y} & \frac{\partial}{\partial z} \\ \frac{\partial \phi}{\partial x} & \frac{\partial \phi}{\partial y} & \frac{\partial \phi}{\partial z} \end{vmatrix}$$

$$= \left(\frac{\partial^2 \phi}{\partial y\, \partial z} - \frac{\partial^2 \phi}{\partial z\, \partial y} \right)\mathbf{i} + \left(\frac{\partial^2 \phi}{\partial z \partial x} - \frac{\partial^2 \phi}{\partial x\, \partial z} \right)\mathbf{j}$$

$$+ \left(\frac{\partial^2 \phi}{\partial x\, \partial y} - \frac{\partial^2 \phi}{\partial y\, \partial x} \right)\mathbf{k}$$

$$= (0)\,\mathbf{i} + (0)\,\mathbf{j} + (0)\,\mathbf{k}$$

Hence the proof.

Note: Since curl (gradient $\phi = 0$, if curl F = 0, then the vector F can be expressed as gradient of a scalar function, *ie.*if $\nabla \times \mathbf{F} = 0$ then $F = \nabla \phi$.

(iii) Let $\mathbf{F} = \mathbf{F_1 i} + \mathbf{F_2 j} + \mathbf{F_3 k}$

$$\text{Then } \nabla \times \mathbf{F} = \begin{vmatrix} \mathbf{i} & \mathbf{j} & \mathbf{k} \\ \frac{\partial}{\partial x} & \frac{\partial}{\partial y} & \frac{\partial}{\partial z} \\ F_1 & F_2 & F_3 \end{vmatrix}$$

$$= \left(\frac{\partial F_3}{\partial y} - \frac{\partial F_2}{\partial z} \right)\mathbf{i} + \left(\frac{\partial F_1}{\partial z} - \frac{\partial F_3}{\partial x} \right)\mathbf{j}$$

$$+ \left(\frac{\partial F_2}{\partial x} - \frac{\partial F_1}{\partial y} \right)\mathbf{k}$$

$$= \Sigma \left(\frac{\partial F_3}{\partial y} - \frac{\partial F_2}{\partial z} \right)\mathbf{i}$$

$$\text{Now } \nabla.(\nabla \times \mathbf{F}) = \left(\Sigma \mathbf{i} \frac{\partial}{\partial x} \right) . \left[\Sigma \left(\frac{\partial F_3}{\partial y} - \frac{\partial F_2}{\partial z} \right)\mathbf{i} \right]$$

$$= \Sigma \frac{\partial}{\partial x}\left(\frac{\partial F_3}{\partial y} - \frac{\partial F_2}{\partial z}\right)$$

$$= \frac{\partial}{\partial x}\left(\frac{\partial F_3}{\partial y} - \frac{\partial F_2}{\partial z}\right) + \frac{\partial}{\partial y}\left(\frac{\partial F_1}{\partial z} - \frac{\partial F_3}{\partial x}\right) + \frac{\partial}{\partial z}\left(\frac{\partial F_2}{\partial x} - \frac{\partial F_1}{\partial y}\right)$$

$$= \frac{\partial^2 F_3}{\partial x\,\partial y} - \frac{\partial^2 F_3}{\partial x\,\partial z} + \frac{\partial^2 F_1}{\partial y\,\partial z} - \frac{\partial^2 F_3}{\partial y\,\partial x} + \frac{\partial^2 F_2}{\partial z\,\partial x} - \frac{\partial^2 F_1}{\partial z\,\partial y}$$

= 0 (Assuming that **F** has continuous partial derivatives)

Hence the proof

(iv) Let $\mathbf{F} = F_1\mathbf{i} + F_2\mathbf{j} + F_3\mathbf{k}$

Then $\nabla \times \mathbf{F} = \left(\frac{\partial F_3}{\partial y} - \frac{\partial F_2}{\partial z}\right)\mathbf{i} + \left(\frac{\partial F_1}{\partial z} - \frac{\partial F_3}{\partial x}\right)\mathbf{j}$

$$+ \left(\frac{\partial F_2}{\partial x} - \frac{\partial F_1}{\partial y}\right)\mathbf{k}$$

$\therefore \quad \nabla \times (\nabla \times \mathbf{F})$

$$= \begin{vmatrix} \mathbf{i} & \mathbf{j} & \mathbf{k} \\ \frac{\partial}{\partial x} & \frac{\partial}{\partial y} & \frac{\partial}{\partial z} \\ \left(\frac{\partial F_3}{\partial y} - \frac{\partial F_2}{\partial z}\right) & \left(\frac{\partial F_1}{\partial z} - \frac{\partial F_3}{\partial x}\right) & \left(\frac{\partial F_3}{\partial x} - \frac{\partial F_1}{\partial y}\right) \end{vmatrix}$$

$$= \Sigma\left[\left\{\frac{\partial}{\partial y}\left(\frac{\partial F_2}{\partial x} - \frac{\partial F_1}{\partial y}\right) - \frac{\partial}{\partial z}\left(\frac{\partial F_1}{\partial z} - \frac{\partial F_3}{\partial x}\right)\right\}\mathbf{i}\right]$$

$$= \Sigma\left[\left\{\left(\frac{\partial^2 F_2}{\partial y\,\partial x} + \frac{\partial^2 F_3}{\partial z\,\partial x}\right) - \left(\frac{\partial^2 F_1}{\partial y^2} + \frac{\partial^2 F_1}{\partial z^2}\right)\right\}\mathbf{i}\right]$$

$$= \Sigma\left[\left\{\frac{\partial}{\partial x}\left(\frac{\partial F_2}{\partial y} + \frac{\partial F_3}{\partial z}\right) - \left(\frac{\partial^2 F_1}{\partial y^2} + \frac{\partial^2 F_1}{\partial z^2}\right)\right\}\mathbf{i}\right]$$

$$= \Sigma\left[\left\{\frac{\partial}{\partial x}\left(\frac{\partial F_1}{\partial x} + \frac{\partial F_2}{\partial y} + \frac{\partial F_3}{\partial z}\right)\right.\right.$$

$$-\left(\frac{\partial^2 F_1}{\partial x^2}+\frac{\partial^2 F_1}{\partial y^2}+\frac{\partial^2 F_1}{\partial z^2}\right)\bigg\}\,\mathbf{i}\Bigg]$$

$$=\Sigma\left[\left\{\frac{\partial}{\partial x}(\nabla.\mathbf{F})-\nabla^2\,\mathbf{F_1}\right\}\mathbf{i}\right]$$

$$=\Sigma\left[\left\{\frac{\partial}{\partial x}(\nabla.\mathbf{F})\right\}\mathbf{i}\right]-\nabla^2\,\Sigma\,F_1\,\mathbf{i}$$

$$=\left(\Sigma\frac{\partial}{\partial x}\mathbf{i}\right)(\nabla.\mathbf{F})-\nabla^2(F_1\mathbf{i}+F_2\mathbf{j}+F_3\mathbf{k})$$

$$=\nabla(\nabla.\mathbf{F})-\nabla^2\mathbf{F}$$

Hence the proof.

12.18. Meaning of (F.∇) G

$$\nabla=\mathbf{i}\frac{\partial}{\partial x}+\mathbf{j}\frac{\partial}{\partial y}+\mathbf{k}\frac{\partial}{\partial z}=\text{a vector operator}$$

Let $\mathbf{F}=\mathbf{F_1 i}+\mathbf{F_2 j}+\mathbf{F_3 k}$

Then $(\mathbf{F}.\nabla)=\mathbf{F_1}\frac{\partial}{\partial x}+F_2\frac{\partial}{\partial y}+F_3\frac{\partial}{\partial z}$

$$\therefore\ (\mathbf{F}.\nabla)\,\mathbf{G}=\left(F_1\frac{\partial}{\partial x}+F_2\frac{\partial}{\partial y}+F_3\frac{\partial}{\partial z}\right)\mathbf{G}$$

$$=F_1\frac{\partial\mathbf{G}}{\partial x}+F_2\frac{\partial\mathbf{G}}{\partial y}+F_3\frac{\partial\mathbf{G}}{\partial z}$$

∴ **(F.∇) G** means **F** associated with ∇ operates on **G**. It can also be written as **F.∇G.**

Similarly we shall now prove that)**F**.∇) φ = **F.∇** φ

As before

$$\mathbf{F}.\nabla=F_1\frac{\partial}{\partial x}+F_2\frac{\partial}{\partial y}+F_3\frac{\partial}{\partial z}$$

Hence $(\mathbf{F}.\nabla)\phi=F_1\frac{\partial\phi}{\partial x}+F_2\frac{\partial\phi}{\partial y}+F_3\frac{\partial\phi}{\partial z}$

$$=(F_1\mathbf{i}+F_2\mathbf{j}+F_3\mathbf{k}).\left(\frac{\partial\phi}{\partial x}\mathbf{i}+\frac{\partial\phi}{\partial y}\mathbf{j}+\frac{\partial\phi}{\partial z}\mathbf{k}\right)$$

$= \mathbf{F}.\nabla\phi$

Hence $(\mathbf{F}.\nabla)\phi = \mathbf{F}.\nabla\phi$

Example 12. Show that

(i) $\nabla.\mathbf{r} = 3$ (ii) $\nabla \times \mathbf{r} = 0$

(iii) $\nabla.(\mathbf{C} \times \mathbf{r}) = 0$ (iv) $\nabla \times (\mathbf{C} \times \mathbf{r}) = 2\,\mathbf{C}$ and

(v) $\nabla \times [(\mathbf{C}.\mathbf{r})\mathbf{r}] = 0$

where **r** is the position vector of any point (x, y, z) and **C** is a constant vector.

Solution: Given $\mathbf{r} = x\mathbf{i} + y\mathbf{j} + z\mathbf{k}$

and $$\mathbf{C} = C_1\mathbf{i} + C_2\mathbf{j} + C_3\mathbf{k}$$

(i) $$\nabla.\mathbf{r} = \left(\mathbf{i}\frac{\partial}{\partial x} + \mathbf{j}\frac{\partial}{\partial y} + \mathbf{k}\frac{\partial}{\partial z}\right).(x\mathbf{i} + y\mathbf{j} + z\mathbf{k})$$

$$= \frac{\partial}{\partial x}(x) + \frac{\partial}{\partial y}(y) + \frac{\partial}{\partial z}(z)$$

$$= 1 + 1 + 1 = 3$$

(ii) $$\nabla \times \mathbf{r} = \begin{vmatrix} \mathbf{i} & \mathbf{j} & \mathbf{k} \\ \frac{\partial}{\partial x} & \frac{\partial}{\partial y} & \frac{\partial}{\partial z} \\ x & y & z \end{vmatrix}$$

$$= \Sigma\left[\frac{\partial}{\partial y}(z) - \frac{\partial}{\partial z}(y)\right]\mathbf{i} = \Sigma\,(0)\,\mathbf{i}$$

$$= (0)\,\mathbf{i} + (0)\,\mathbf{j} + (0)\,\mathbf{k} = \mathbf{0}$$

(iii) $\nabla.(\mathbf{C} \times \mathbf{r})$

$$\mathbf{C} \times \mathbf{r} = \begin{vmatrix} \mathbf{i} & \mathbf{j} & \mathbf{k} \\ C_1 & C_2 & C_3 \\ x & y & z \end{vmatrix}$$

$$= (C_2z - yC_2)\,\mathbf{i} + (C_3x - C_1z)\,\mathbf{j} + (C_1y - C_2x)\,\mathbf{k}$$

$$\therefore \nabla.(\mathbf{C} \times \mathbf{r}) = \frac{\partial}{\partial x}(C_2z - yC_3) + \frac{\partial}{\partial y}(C_3x - C_1z) + \frac{\partial}{\partial z}(C_1y - C_2x)$$

$= 0 + 0 + 0 = 0$

(iv) $\nabla \times (\mathbf{C} \times \mathbf{r}) = \begin{vmatrix} \mathbf{i} & \mathbf{j} & \mathbf{k} \\ \frac{\partial}{\partial x} & \frac{\partial}{\partial y} & \frac{\partial}{\partial z} \\ C_2z - yC_3 & C_3x - C_1z & C_1y = C_2x \end{vmatrix}$

$= \Sigma \left[\frac{\partial}{\partial y}(C_1y - C_2x) - \frac{\partial}{\partial z}(C_3x - C_1z) \right] \mathbf{i}$

$= \Sigma (C_1 + C_1)\, \mathbf{i} = 2\, \Sigma C_1\, \mathbf{i}$

$= 2\,(C_1\, \mathbf{i} + C_2\mathbf{j} + C_3\, \mathbf{k}) = 2\, \mathbf{C}$

(v) $\nabla \times [(\mathbf{C}.\mathbf{r})\, \mathbf{r}]$

$= (\mathbf{C}.\mathbf{r})\, (\nabla \times \mathbf{r}) = 0 \qquad (\because \nabla \times \mathbf{r} = 0)$

Example 13. Evaluate

(i) $\nabla.(3x^2\, \mathbf{i} + 5xy^2\, \mathbf{j} + xyz^3\, \mathbf{k})$

(ii) $\nabla \times [xyz\, \mathbf{i} + 3x^2y\, \mathbf{j} + (xz^2 - y^2z)\, \mathbf{k}]$

at the point (1, 2, 3).

Solution: (i) $\nabla.(3x^2\, \mathbf{i} + 5xy^2\, \mathbf{j} + xyz^3\, \mathbf{k})$

$= \frac{\partial}{\partial x}(3x^2) + \frac{\partial}{\partial y}(5xy^2) + \frac{\partial}{\partial z}(xyz^3)$

$= 6x + 10xy + 3xyz^2$

$= 6\,(1) + 10\,(1)\,(2) + 3\,(1)\,(2)\,(3)^2$ at (1, 2, 3)

$= 6 + 20 + 54 = 80$

(ii) $\nabla \times [xyz\, \mathbf{i} + 3x^2y\, \mathbf{j} + (xz^2 - y^2z)\, \mathbf{k}]$

$= \begin{vmatrix} \mathbf{i} & \mathbf{j} & \mathbf{k} \\ \frac{\partial}{\partial x} & \frac{\partial}{\partial y} & \frac{\partial}{\partial z} \\ xyz & 3x^2y & (xz^2 - y^2z) \end{vmatrix}$

$= \left[\frac{\partial}{\partial y}(xz^2 - y^2z) = \frac{\partial}{\partial z}(3x^2y) \right] \mathbf{i}$

$+ \left[\frac{\partial}{\partial z}(xyz) - \frac{\partial}{\partial x}(xz^2 - y^2z) \right] \mathbf{j}$

$$+\left[\frac{\partial}{\partial x}(3x^2y)\,\frac{\partial}{\partial y}(xyz)\right]\mathbf{k}$$

$$= (-2yz(\ \mathbf{i} + (xy - z^2)\,\mathbf{j} + (6xy - xz)\,\mathbf{k}$$

$$= (-2)(2)(3)\,\mathbf{i} + (1)(2) - (3)^2\,\mathbf{j} + \{6(1)(2) - (1)(3)\}\,\mathbf{k}$$

at (1, 2, 3)

$$= -12\,\mathbf{i} - 7\,\mathbf{j} + 9\,\mathbf{k}.$$

Example 14. (i) If $\mathbf{F} = \nabla(x^3 + y^3 + z^3 - 3xyz)$ find

(a) $\nabla.\mathbf{F}$ and (b) $\nabla \times \mathbf{F}$.

(ii) If $\mathbf{F} = \frac{\mathbf{r}}{r}$, show that

(a) $\nabla.\mathbf{F} = \frac{2}{r}$ and (b) $\nabla \times \mathbf{F} = 0$.

(iii) If $\mathbf{F} = (x + 3y)\,\mathbf{i} + (y - 3z)\,\mathbf{j} + (x - 2z)\,\mathbf{k}$

show that $\nabla.\mathbf{F} = 0$.

(iv) If $\mathbf{F} = (x + y + 1)\,\mathbf{i} + \mathbf{j} - (x + y)\,\mathbf{k}$, show that $\mathbf{F}.(\nabla \times \mathbf{F}) = 0$

(v) Show that $\nabla \times (\mathbf{r} \sin r) = 0$

where $\mathbf{r} = x\mathbf{i} + y\mathbf{j} + z\mathbf{k}$ and $r = |\,\mathbf{r}\,|$

Solution: (i) $F = \nabla(x^3 + y^3 + z^3 - 3\,xyz)$

$$= \frac{\partial}{\partial x}(x^3 + y^3\ z^3 - 3\,xyz)\,\mathbf{i}$$

$$+ \frac{\partial}{\partial y}(x^3 + y^3 + z^3 - 3\,xyz)\,\mathbf{j}$$

$$+ \frac{\partial}{\partial z}(x^3 + y^3 + z^3 - 3\,xyz)\,\mathbf{k}$$

$$= 3(x^2 - yz)\,\mathbf{i} + 3(y^2 - xz)\,\mathbf{j} + 3(z^2 - xy)\,\mathbf{k}$$

(a) $$\nabla.\mathbf{F} = \frac{\partial}{\partial x}\left[3(x^2 - yz\right] + \frac{\partial}{\partial y}\left[3(y^2 - xz)\right]$$

$$+ \frac{\partial}{\partial z}\left[3(z^2 - xy)\right]$$

$$= 6x + 6y + 6z.$$

$$= 6(x + y + z).$$

(b) $\nabla \times \mathbf{F} = \begin{vmatrix} \mathbf{i} & \mathbf{j} & \mathbf{k} \\ \dfrac{\partial}{\partial x} & \dfrac{\partial}{\partial y} & \dfrac{\partial}{\partial z} \\ 3(x^2 - yz) & 3(y^2 - xz) & 3(z^2 - xy) \end{vmatrix}$

$$= \left[\frac{\partial}{\partial y} [3(z^2 - xy)] - \frac{\partial}{\partial z} [3(y^2 - xz)] \right] \mathbf{i}$$

$$+ \left[\frac{\partial}{\partial z} 93(x^2 - yz)] = \frac{\partial}{\partial x} [3(z^2 - xy)] \right] \mathbf{j}$$

$$+ \left[\frac{\partial}{\partial x} [3(y^2 - xz)] - \frac{\partial}{\partial y} [3(x^2 - yz)] \right] \mathbf{k}$$

$$= (-3x + 3x)\,\mathbf{i} + (-3y + 3y)\,\mathbf{j} + (-3z + 3z)\,\mathbf{k}$$

$$= (0)\,\mathbf{i} + (0)\,\mathbf{j} + (0)\,\mathbf{k} = 0$$

(ii) Given $\mathbf{F} = \dfrac{\mathbf{r}}{\mathbf{r}}$

(a) $$\nabla.\mathbf{F} = \nabla.\left(\frac{\mathbf{r}}{\mathbf{r}} \right)$$

$$= \nabla\left(\frac{1}{r} \right).\mathbf{r} + \frac{1}{r}(\nabla.\mathbf{r})$$

$$= \left(-\frac{\mathbf{r}}{r^3} \right).\mathbf{r} + \frac{1}{r}(3)$$

$$\left[\because \ \nabla\left(\frac{1}{r} \right) = \frac{-\mathbf{r}}{r^3} \text{ and } \nabla.\mathbf{r} = 3 \right]$$

$$= -\frac{1}{r^3}(\mathbf{r}.\mathbf{r}) + \frac{3}{r}$$

$$= -\frac{1}{r} + \frac{3}{r} = \frac{2}{r} \qquad (\because \mathbf{r}.\mathbf{r} = r^2)$$

(b) $$\nabla \times \mathbf{F} = \nabla \times \left(\frac{\mathbf{r}}{\mathbf{r}} \right)$$

$$= \nabla \times \left(\frac{x\mathbf{i} + y\mathbf{j} + z\mathbf{k}}{r} \right)$$

$$= \begin{vmatrix} \mathbf{i} & \mathbf{j} & \mathbf{k} \\ \frac{\partial}{\partial x} & \frac{\partial}{\partial y} & \frac{\partial}{\partial z} \\ \frac{x}{r} & \frac{y}{r} & \frac{z}{r} \end{vmatrix}$$

$$= \left\{\frac{\partial}{\partial y}\left(\frac{z}{r}\right) - \frac{\partial}{\partial z}\left(\frac{y}{r}\right)\right\}\mathbf{i} + \left\{\frac{\partial}{\partial z}\left(\frac{x}{r}\right) - \frac{\partial}{\partial x}\left(\frac{z}{r}\right)\right\}\mathbf{j}$$

$$+ \left\{\frac{\partial}{\partial x}\left(\frac{y}{r}\right) - \frac{\partial}{\partial y}\left(\frac{x}{r}\right)\right\}\mathbf{k}$$

$$= \left(\frac{-z}{r^2}\cdot\frac{\partial r}{\partial y} + \frac{y}{r^2}\cdot\frac{\partial r}{\partial z}\right)\mathbf{i} + \left(\frac{-x}{r^2}\cdot\frac{\partial r}{\partial z} + \frac{z}{r^2}\cdot\frac{\partial r}{\partial x}\right)\mathbf{j}$$

$$+ \left(\frac{-y}{r^2}\cdot\frac{\partial r}{\partial x} + \frac{x}{r^2}\cdot\frac{\partial r}{\partial y}\right)\mathbf{k}$$

$$= \left\{\frac{-z}{r^2}\left(\frac{y}{r}\right) + \frac{v}{r^2}\left(\frac{z}{r}\right)\right\}\mathbf{i} + \left\{\frac{-x}{r^2}\left(\frac{z}{r}\right) + \frac{z}{r^2}\left(\frac{x}{r}\right)\right\}\mathbf{j}$$

$$+ \left\{\frac{-y}{r^2}\left(\frac{x}{r}\right) + \frac{x}{r^2}\left(\frac{y}{r}\right)\right\}\mathbf{k}$$

$$= (0)\,\mathbf{i} + (0)\,\mathbf{j} + (0)\,\mathbf{k} = \mathbf{0}$$

$\therefore$ $\nabla \times \mathbf{F} = 0$ Hence here F in irrotational.

(iii) $\nabla.\mathbf{F} = \frac{\partial}{\partial x}(x + 3y) + \frac{\partial}{\partial y}(y - 3z) + \frac{\partial}{\partial z}(x - 2z)$

$$= 1 + 1 - 2 = 0$$

$\therefore$ $\nabla.\mathbf{F} = 0$ Hence here F is solenoidal.

(iv) $\nabla \times \mathbf{F} = \begin{vmatrix} \mathbf{i} & \mathbf{j} & \mathbf{k} \\ \frac{\partial}{\partial x} & \frac{\partial}{\partial y} & \frac{\partial}{\partial z} \\ x + y + 1 & 1 & -(x + y) \end{vmatrix}$

$$= \left\{\frac{\partial}{\partial y}(-x - y) - \frac{\partial}{\partial z}(1)\right\}\mathbf{i}$$

$$+\left\{\frac{\partial}{\partial z}(x+y+1)+\frac{\partial}{\partial x}(x+y)\right\}\mathbf{j}$$

$$+\left\{\frac{\partial}{\partial x}(1)-\frac{\partial}{\partial y}(x+y+1)\right\}\mathbf{k}$$

$$=(-1)\,\mathbf{i}+(1)\,\mathbf{j}-(1)\,\mathbf{k}$$

$$=-\mathbf{i}+\mathbf{j}-\mathbf{k}$$

$$\therefore\ \mathbf{F}.(\nabla\times\mathbf{F})=[(x+y+1)\,\mathbf{i}+\mathbf{j}-(x+y)\,\mathbf{k}]\,.\,[-\mathbf{i}+\mathbf{j}-\mathbf{k}]$$

$$=-(x+y+1)+1+(x+y)=0$$

$$\therefore\ \mathbf{F}.(\nabla\times\mathbf{F})=0$$

(v) $\nabla\times(\mathbf{r}\sin r)=\nabla\times[(\sin r)\,(x\mathbf{i}+y\mathbf{j}+z\mathbf{k})]$

$$=\begin{vmatrix}\mathbf{i} & \mathbf{j} & \mathbf{k}\\ \frac{\partial}{\partial x} & \frac{\partial}{\partial y} & \frac{\partial}{\partial z}\\ x\sin r & y\sin r & z\sin r\end{vmatrix}$$

$$=\left[\frac{\partial}{\partial y}(z\sin r)-\frac{\partial}{\partial z}(y\sin r)\right]\mathbf{i}$$

$$+\left[\frac{\partial}{\partial z}(x\sin r)-\frac{\partial}{\partial x}(z\sin r)\right]\mathbf{j}$$

$$+\left[\frac{\partial}{\partial x}(y\sin r)-\frac{\partial}{\partial y}(x\sin r)\right]\mathbf{k}$$

$$=\left(z\cos r\frac{\partial r}{\partial y}-y\cos r\frac{\partial r}{\partial z}\right)\mathbf{i}$$

$$+\left(x\cos r\frac{\partial r}{\partial z}-z\cos r\frac{\partial r}{\partial x}\right)\mathbf{j}$$

$$+\left(y\cos r\frac{\partial r}{\partial x}-x\cos r\frac{\partial r}{\partial y}\right)\mathbf{k}$$

$$=\left(\frac{\cos r}{r}\right)[(zy-yz)\,\mathbf{i}+(xz-zx)\,\mathbf{j}+(yx-xy)\,\mathbf{k}]$$

$= (0)\,\mathbf{i} + (0)\,\mathbf{j} + (0)\,\mathbf{k} = \mathbf{0}$

Example 15. (i) If $\mathbf{F} = (x + 3y)\,\mathbf{i} + (y - 2z)\,\mathbf{j} + (x + pz)\,\mathbf{k}$ is solenoidal, find p.

(ii) Show that the vector

$\mathbf{F} = (\sin y + z)\,\mathbf{i} + (x \cos y - z)\,\mathbf{j} + (x - y)\,\mathbf{k}$ is irrotational.

(iii) If $\mathbf{F} = z\,\mathbf{i} + x\,\mathbf{j} + y\,\mathbf{k}$, prove that $\nabla \times (\nabla \times \mathbf{F}) = \mathbf{0}$

Solution: (i) Given $\mathbf{F} = (x + 3y)\,\mathbf{i} + (y - 2z)\,\mathbf{j} + (x + pz)\,\mathbf{k}$ and $\mathbf{F}$ is solenoidal i.e. $\nabla.\mathbf{F} = 0$

$$\Leftrightarrow \frac{\partial}{\partial x}(x + 3y) + \frac{\partial}{\partial y}(y - 2z) + \frac{\partial}{\partial z}(x + pz) = 0$$

$$1 + 1 + p = 0$$

or $$p = -2$$

(ii) Given $F = (\sin y + z)\,\mathbf{i} + (x \cos y - z)\,\mathbf{j} + (x - y)\,\mathbf{k}$

$$\text{Now } \nabla \times \mathbf{F} = \begin{vmatrix} \mathbf{i} & \mathbf{j} & \mathbf{k} \\ \dfrac{\partial}{\partial x} & \dfrac{\partial}{\partial x} & \dfrac{\partial}{\partial x} \\ (\sin y + z) & x \cos y - z) & (x - y) \end{vmatrix}$$

$$= \left\{\frac{\partial}{\partial y}(x - y) - \frac{\partial}{\partial z}(x \cos y - z)\right\}\mathbf{i}$$

$$+ \left\{\frac{\partial}{\partial z}(\sin y + z)\frac{\partial}{\partial x}(x - y)\right\}\mathbf{j}$$

$$+ \left\{\frac{\partial}{\partial x}(x \cos y - z) - \frac{\partial}{\partial y}(\sin y + z)\right\}\mathbf{k}$$

$$= [-1 - (-1)]\,\mathbf{i} + [1 - 1]\,\mathbf{j} + [\cos y - \cos y]\,\mathbf{k}$$

$$= (0)\,\mathbf{i} + (0)\,\mathbf{j} + (0)\,\mathbf{k} = 0$$

$\because \quad \nabla \times \mathbf{F} = 0 \Rightarrow \mathbf{F}$ is irrotational.

(iii) Given $\mathbf{F} = z\,\mathbf{i} + x\,\mathbf{j} + y\,\mathbf{k}$

$$\nabla \times \mathbf{F} = \begin{vmatrix} \mathbf{i} & \mathbf{j} & \mathbf{k} \\ \dfrac{\partial}{\partial x} & \dfrac{\partial}{\partial y} & \dfrac{\partial}{\partial z} \\ z & x & y \end{vmatrix}$$

$$= \mathbf{i} + \mathbf{j} + \mathbf{k}$$

$$\nabla \times (\nabla \times \mathbf{F}) = \begin{vmatrix} \mathbf{i} & \mathbf{j} & \mathbf{k} \\ \frac{\partial}{\partial x} & \frac{\partial}{\partial y} & \frac{\partial}{\partial z} \\ 1 & 1 & 1 \end{vmatrix}$$

$$= (0)\,\mathbf{i} + (0)\,\mathbf{j} + (0)\,\mathbf{k}$$

$$= \mathbf{0}$$

$\therefore\ \nabla \times (\nabla \times \mathbf{F}) = \mathbf{0}.$

Example 16. Show that

(i) $\nabla . \nabla\,(r^n) = n\,(n + 1)\,r^{n-2}$ (ii) $\nabla \times (\nabla\,(r^n)) = 0$

(iii) $\nabla . \nabla\left(\frac{1}{r}\right) = 0$, where $\mathbf{r} = x\mathbf{i} + y\mathbf{j} + z\mathbf{k}$ and $r = |\,\mathbf{r}\,|$.

Solution: (i) $\nabla\,(r^n) = \left(\mathbf{i}\frac{\partial}{\partial x} + \mathbf{j}\frac{\partial}{\partial y} + \mathbf{k}\frac{\partial}{\partial z}\right) r^n$

$$= nr^{n-1}\left[\mathbf{i}\frac{\partial r}{\partial x} + \mathbf{j}\frac{\partial r}{\partial y} + \mathbf{k}\frac{\partial r}{\partial z}\right]$$

$$= nr^{n-1}\left[\mathbf{i}\frac{x}{r} + \mathbf{j}\frac{y}{r} + \mathbf{k}\frac{z}{r}\right]$$

$$\left[\because\ r^2 = x^2 + y^2 + z^2,\ \frac{\partial r}{\partial x} = \frac{x}{r},\ \frac{\partial r}{\partial y} = \frac{y}{r},\ \frac{\partial r}{\partial z} = \frac{z}{r}\right]$$

$$= nr^{n-2}\,(x\,\mathbf{i} + y\,\mathbf{j} + z\,\mathbf{k})$$

Now

$$\nabla . \nabla(r^n) = \frac{\partial}{\partial x}(nr^{n-2}\,x) + \frac{\partial}{\partial y}(nr^{n-2}\,y) + \frac{\partial}{\partial z}(nr^{n-2}\,z)$$

$$= n\left[r^{n-2} + x\,(n-2)r^{n-3}.\frac{\partial r}{\partial x}\right]$$

$$+ n\left[r^{n-2} + y\,(n-2)\,r^{n-3}\,\frac{\partial r}{\partial y}\right]$$

$$+ n\left[r^{n-2} + z\,(n-2)\,r^{n-3}\,\frac{\partial r}{\partial z}\right]$$

$$= 3n\,r^{n-2} + n\,(n-2)\,r^{n-3}\left[x\frac{\partial r}{\partial z} + y\frac{\partial r}{\partial y} + z\frac{\partial r}{\partial z}\right]$$

$$= 3n\,r^{n-2} + n\,(n-2)\,r^{n-3}\left[\frac{x^2+y^2+z^2}{r}\right]$$

$$= 3n\,r^{n-2} + n\,(n-2)\,r^{n-3}\left(\frac{r^2}{r}\right)$$

$$= 3n\,r^{n-2} + n\,(n-2)\,r^{n-2}$$

$$= n\,(n+1)\,r^{n-2}$$

(ii) $\nabla \times \nabla\,(r^n)$

$$= \nabla \times [n\,r^{n-2}\,(x\,\mathbf{i} + y\,\mathbf{j} + z\,\mathbf{k})]$$

$$= \begin{vmatrix} \mathbf{i} & \mathbf{j} & \mathbf{k} \\ \dfrac{\partial}{\partial x} & \dfrac{\partial}{\partial y} & \dfrac{\partial}{\partial z} \\ n\,r^{n-2}\,x & n\,r^{n-2}\,y & n\,r^{n-2}\,z \end{vmatrix}$$

$$= n\Sigma\left\{\frac{\partial}{\partial y}\left(r^{n-1}z\right) - \frac{\partial}{\partial z}\left(r^{n-2}y\right)\right\}\mathbf{i}$$

$$= n\Sigma\left\{z\,(n-2)\,r^{n-3}\frac{\partial r}{\partial y} - y\,(n-2)\,r^{n-3}\frac{\partial r}{\partial z}\right\}\mathbf{i}$$

$$= n\,(n-2)\,r^{n-3}\,\Sigma\left\{z\frac{y}{r} - y\frac{z}{r}\right\}\mathbf{i}$$

$$= n\,(n-2)\,r^{n-2}\,\{(0)\,\mathbf{i} + (0)\,\mathbf{j} + (0)\,\mathbf{k}\} = \mathbf{0}.$$

(iii) $\nabla.\nabla\left(\dfrac{1}{r}\right)$

$$\nabla\left(\frac{1}{r}\right) = \left(\mathbf{i}\frac{\partial}{\partial x} + \mathbf{j}\frac{\partial}{\partial y} + \mathbf{k}\frac{\partial}{\partial z}\right)\left(\frac{1}{r}\right)$$

$$= \mathbf{i}\frac{\partial}{\partial x}\left(\frac{1}{r}\right) + \mathbf{j}\frac{\partial}{\partial y}\left(\frac{1}{r}\right) + \mathbf{k}\frac{\partial}{\partial z}\left(\frac{1}{r}\right)$$

$$= \mathbf{i}\left(-\frac{1}{r^2}\right)\left(\frac{x}{r}\right) + \mathbf{j}\left(-\frac{1}{r^2}\right)\left(\frac{y}{r}\right) + \mathbf{k}\left(-\frac{1}{r^2}\right)\left(\frac{z}{r}\right)$$

$$= -\frac{1}{r^3}(\mathbf{xi} + \mathbf{yj} + \mathbf{zk})$$

$$\nabla.\nabla\left(\frac{1}{r}\right) = -\left[\frac{\partial}{\partial x}\left(\frac{x}{r^3}\right) + \frac{\partial}{\partial y}\left(\frac{y}{r^3}\right) + \frac{\partial}{\partial z}\left(\frac{z}{r^3}\right)\right]$$

$$= \left[-r^{-3} + 3xr^{-4}\frac{\partial r}{\partial x}\right] + \left[-r^{-3} + 3yr^{-4}\frac{\partial r}{\partial y}\right]$$

$$+ \left[-r^{-3} + 3z\,r^{-4}\frac{\partial r}{\partial z}\right]$$

$$= -3r^{-3} + 3r^{-4}\left(x\frac{\partial r}{\partial x} + y\frac{\partial r}{\partial x} + z\frac{\partial r}{\partial z}\right)$$

$$= -3r^{-3} + 3r^{-4}\left(\frac{x^2 + y^2 + z^2}{r}\right)$$

$$= -3r^{-3} + 3r^{-4}\left(\frac{r^2}{r}\right)$$

$$= -3r^{-3} + 3r^{-3} = 0.$$

Example 17. If C is a constant vector,

$$\mathbf{r} = x\mathbf{i} + y\mathbf{j} + z\mathbf{k} \text{ and } r = |\mathbf{r}|$$

then prove that

(i) $\nabla \times [(\mathbf{C} \times \mathbf{r})r^n] = (n + 2)\,r^n\mathbf{C} - nr^{n-2}(\mathbf{C.r})\,\mathbf{r}$

(ii) $\nabla \times \dfrac{(\mathbf{C} \times \mathbf{r})}{r^n} = \dfrac{2-n}{r^n}\mathbf{C} + \dfrac{n(\mathbf{C.r})\,\mathbf{r}}{r^{n+2}}$

Solution: Let $\mathbf{C} = C_1\mathbf{i} + C_2\mathbf{j} + C_3\mathbf{k}$, $\mathbf{r} = x\mathbf{i} + y\mathbf{i} + z\mathbf{k}$)

$$(\mathbf{C} \times \mathbf{r}) = (C_1\mathbf{i} + C_2\mathbf{j} + C_3\mathbf{k}) \times (x\mathbf{i} + y\mathbf{j} + z\mathbf{k})$$

$$= \begin{vmatrix} \mathbf{i} & \mathbf{j} & \mathbf{k} \\ C_1 & C_2 & C_3 \\ x & y & z \end{vmatrix}$$

$$= (C_2z - C_{3y})\,\mathbf{i} + (C_3x - C_1z)\,\mathbf{j} + (C_1y - C_2x)\,\mathbf{k}$$

$$\therefore \quad (\mathbf{C} \times \mathbf{r})\,r^n = (C_2z - C_3y)\,r^n\,\mathbf{i} + (C_3x - C_1z)\,r^n\,\mathbf{j}$$

$$+ (C_1y - C_2x)r^n\,\mathbf{k}$$

Now $\nabla \times (\mathbf{C} \times \mathbf{r})r^n$

$$= \begin{vmatrix} \mathbf{i} & \mathbf{j} & \mathbf{k} \\ \dfrac{\partial}{\partial x} & \dfrac{\partial}{\partial y} & \dfrac{\partial}{\partial z} \\ (C_2z - C_3y)r^n & (C_3x - C_1z)r^n & (C_1y - C_2x)r^n \end{vmatrix}$$

$$= \Sigma \left\{ \frac{\partial}{\partial y}(C_1y - C_2x)r^n - \frac{\partial}{\partial z}(C_3x - C_1z)r^n \right\} \mathbf{i}$$

$$= \Sigma \left\{ C_1 r^n + (C_1y - C_2x)\, n\, r^{n-1} \frac{\partial r}{\partial y} \right.$$

$$\left. + C_1 r^n - (C_2x - C_1z)\, nr^{n-1} \frac{\partial r}{\partial z} \right\} \mathbf{i}$$

$$= 2\, r^n\, \Sigma\, C_1\, \mathbf{i} + nr^{n-1}\, \Sigma \left\{ (C_1y - C_2x)\frac{y}{r} - (C_3x - C_1z)\frac{z}{r} \right\} \mathbf{i}$$

$$= 2r^n\, \mathbf{C} + nr^{n-2}\, \Sigma\, \{C_1\, (y + z^2) - x\, (C_1\, y + C_3\, z)\}\, \mathbf{i}$$

$$= 2\, r^n\, \mathbf{C} + nr^{n-2}\, \Sigma\, \{C_1\, (x^2 + y^2 + z^2)$$

$$- x\, (C_1x + C_2\, y + C_3z)\, \}\, \mathbf{i}$$

$$= 2r^n\, \mathbf{C} + nr^{n-2}\, \{\, (\Sigma C_1\, \mathbf{i})\, r^2 - (\Sigma x\, \mathbf{i})\, (\mathbf{C} \,.\, \mathbf{r})\, \}$$

$$= 2\, r^n\, \mathbf{C} + nr^n\, \mathbf{C} - nr^{n-2}\, (\mathbf{C.r})\, \mathbf{r}$$

$$= (n + 2)\, r^n\, \mathbf{C} - nr^{n-2}\, (\mathbf{C.r})\mathbf{r}.$$

(ii) $$\frac{(\mathbf{C} \times \mathbf{r})}{r^n} = \frac{1}{r^n}\Big[\, (C_2\, z - C_3\, y(\, \mathbf{i} + (C_3\, x - C_1\, z)\, \mathbf{j}$$

$$+ (C_1\, y - C_2\, x)\, \mathbf{k}\, \Big]$$

$$\nabla \times \frac{(\mathbf{C} \times \mathbf{r})}{r^n} = \begin{vmatrix} \mathbf{i} & \mathbf{j} & \mathbf{k} \\ \dfrac{\partial}{\partial x} & \dfrac{\partial}{\partial y} & \dfrac{\partial}{\partial z} \\ \dfrac{C_2z - C_3y}{r^n} & \dfrac{C_3x - C_1z}{r^n} & \dfrac{C_1y - C_2x}{r^n} \end{vmatrix}$$

$$= \Sigma \left\{ \frac{\partial}{\partial y}\left(\frac{C_1y - C_2x}{r^n} \right) - \frac{\partial}{\partial z}\left(\frac{C_3x - C_1z}{r^n} \right) \right\} \mathbf{i}$$

$$= \Sigma \left\{ \frac{\partial}{\partial y}(C_1 y - C_2 x)\, r^{-n} - \frac{\partial}{\partial z}(C_3 x - C_1 z)\, r^{-n} \right\} \mathbf{i}$$

$$= \Sigma \left[2\, C_1\, r^{-n} + n r^{-n-1} \left\{ (C_1 y - C_2 x) \frac{y}{r} + (C_3 x - C_1 z) \frac{z}{r} \right\} \right] \mathbf{i}$$

$$= 2 r^{-n} \Sigma\, C_1\, \mathbf{i} + n r^{-n-2} \Sigma \{ x (C_2 y + C_3 z) - C_1 (y^2 + z^2) \}\, \mathbf{i}$$

$$= 2\, r^{-n}\, \mathbf{C} + n\, r^{-n-2} \Sigma \{ x (\mathbf{C}.\mathbf{r}) - C_1\, r^2 \}\, \mathbf{i}$$

$$= 2\, r^{-n}\, \mathbf{C} + n\, r^{-n-2} (\mathbf{C}.\mathbf{r}) \Sigma\, x\, \mathbf{i} - n\, r^{-n} \Sigma\, C_1\, \mathbf{i}$$

$$= 2\, r^{-1}\, \mathbf{C} + n\, r^{-n-2} (\mathbf{C}.\mathbf{r})\, \mathbf{r} - n\, r^{-n}\, \mathbf{C}$$

$$= (2 - n)\, r^{-n}\, \mathbf{C} + n\, r^{-(n+2)} (\mathbf{C}.\mathbf{r})\, \mathbf{r}$$

$$= \frac{(2 - n)}{r^n} \mathbf{C} + \frac{n (\mathbf{C}.\mathbf{r})\, \mathbf{r}}{r^{n+2}}$$

Hence the proof.

Example 18. Show that

$$\nabla^2 f(r) = f''(r) + \frac{2}{r} f'(r)$$

Solution: $\nabla^2 f(r) = \frac{\partial^2}{\partial x^2} f(r) + \frac{\partial^2}{\partial y^2} f(r) + \frac{\partial^2}{\partial z^2} f(r)$

$$= \Sigma \frac{\partial^2}{\partial x^2} f(r)$$

$$= \Sigma \frac{\partial}{\partial x} \left[\frac{\partial}{\partial x} f(r) \right]$$

$$= \Sigma \frac{\partial}{\partial x} \left[f'(r) \frac{\partial r}{\partial x} \right]$$

$$= \Sigma \frac{\partial}{\partial x} \left[f'(r) \frac{x}{r} \right] \qquad \left(\because\ \frac{\partial r}{\partial x} = \frac{x}{r} \right)$$

$$= \Sigma \left[f''(r) \frac{x}{r} \cdot \frac{\partial r}{\partial x} + \frac{f'(r)}{r} - \frac{f'(r)\, x}{r^2} \cdot \frac{\partial r}{\partial x} \right]$$

$$= \Sigma\left[f''(r)\frac{x^2}{r^2} + \frac{f'(r)}{r} - \frac{f'(r)\,x^2}{r^3} \right]$$

$$= \frac{f''(r)}{r^2}[x^2 + y^2 + z^2] + \frac{3\,f'(r)}{r} - \frac{f'(r)}{r^3}(x^2 + y^2 + z^2)$$

$$= f'(r) + \frac{2}{r} f'(r) \qquad (\because x^2 + y^2 + z^2 = r^2)$$

Hence the result.

Example 19. Prove that

(i) $\nabla \times (\mathbf{F} \times \mathbf{r}) = 2\,\mathbf{F} - (\nabla.\mathbf{F})\,\mathbf{r} + (\mathbf{r}.\nabla)\,\mathbf{F}$

(ii) $\mathbf{B}.\nabla\left(A\,.\nabla\frac{1}{r}\right) = \frac{3\,(\mathbf{A.r})\,(\mathbf{B.r})}{r^5} - \frac{\mathbf{A.B}}{r^3}$

where **A** and **B** are constant vectors.

Solution: (i) We know that.

$$\nabla \times (\mathbf{A} \times \mathbf{B}) = (\mathbf{B}.\nabla)\,\mathbf{A} - \mathbf{B}.(\nabla.\mathbf{A}) + \mathbf{A}\,(\nabla.\mathbf{B}) - (\mathbf{A}.\nabla)\,\mathbf{B}$$

Putting A = F and B = r, this reduces to

$$\nabla \times (F \times r) = (\mathbf{r}.\nabla)\,\mathbf{F} - \mathbf{r}\,(\nabla.\mathbf{F}) + \mathbf{F}\,(\nabla.\mathbf{r}) - (\mathbf{F}.\nabla)\,\mathbf{r} \qquad ...(1)$$

Now $\qquad \nabla.\mathbf{r} = 3 \qquad ...(2)$

Again if $\mathbf{F} = F_1\mathbf{i} + F_2\,\mathbf{j} + F_3\mathbf{k}$ then

$$\mathbf{F}.\nabla = F_1\frac{\partial}{\partial x} + F_2\frac{\partial}{\partial y} + F_3\frac{\partial}{\partial z}$$

$$(\mathbf{F}.\nabla)\,\mathbf{r} = \left(F_1\frac{\partial}{\partial x} + F_2\frac{\partial}{\partial y} + F_3\frac{\partial}{\partial z}\right).(x\mathbf{i} + y\mathbf{j} + z\mathbf{k})$$

$$= F_1\mathbf{i} + F_2\mathbf{j} + F_3\mathbf{k} = \mathbf{F} \qquad ...(3)$$

From (1), (2) and (3) we get

$$\nabla \times (F \times \mathbf{r}) = (\mathbf{r}.\nabla)\,\mathbf{F} - \mathbf{r}\,(\nabla.\mathbf{F}) + 3\,\mathbf{F} - \mathbf{F}$$

$$= 2\,\mathbf{F}\,(\nabla.\mathbf{F})\,\mathbf{r} + (\mathbf{r}.\nabla)\,\mathbf{F}$$

Hence proved.

(ii) $\nabla\left(\frac{1}{r}\right) = -\frac{1}{r^2}\nabla\mathbf{r} = -\frac{1}{r^2}\cdot\frac{1}{r}\,\mathbf{r}$

$$= -\frac{\mathbf{r}}{r^3}$$

$$\therefore \quad \mathbf{A}.\nabla\left(\frac{1}{r}\right) = \mathbf{A}.\left(-\frac{1}{r^3}\mathbf{r}\right) = -\frac{\mathbf{A.r}}{r^3}$$

$$\therefore \quad \mathbf{B}.\nabla\left(\mathbf{A}.\nabla\frac{1}{r}\right) = \mathbf{B}.\nabla\left(-\frac{\mathbf{A.r}}{r^3}\right)$$

$$= \mathbf{B}.\Sigma\left\{\frac{\partial}{\partial x}\left(-\frac{\mathbf{A.r}}{r^3}\right)\right\}\mathbf{i}$$

$$= \mathbf{B}.\Sigma\left\{-\frac{1}{r^3}\frac{\partial}{\partial x}(\mathbf{A.r}) + (\mathbf{A.r})\frac{\partial}{\partial x}\left(-\frac{1}{r^3}\right)\right\}\mathbf{i}$$

$$= \mathbf{B}.\Sigma\left\{-\frac{1}{r^3}\left(\mathbf{A}.\frac{\partial \mathbf{r}}{\mathbf{x}}\right) + 3\frac{(\mathbf{A.r})}{r^4}\frac{\partial r}{\partial x}\right\}\mathbf{i} \qquad \left[\therefore \quad \frac{\partial \mathbf{r}}{\partial x} = 0\right]$$

$$= \mathbf{B}.\Sigma\left\{-\frac{\mathbf{A.i}}{r^3} + \frac{3\,x}{r^5}(\mathbf{A.r})\right\}\mathbf{i} \qquad \left[\therefore \quad \frac{\partial \mathbf{r}}{\partial x} = \mathbf{i}, \frac{\partial r}{\partial x} = \frac{x}{r}\right]$$

$$= \mathbf{B}.\Sigma\left\{-\frac{1}{r^3}(\mathbf{A.i})\,\mathbf{i} + \frac{3}{r^5}(\mathbf{A.r})\,x\mathbf{i}\right\}$$

$$= \mathbf{B}.\left(-\frac{1}{r^3}\right)\Sigma\,(\mathbf{A.i})\,\mathbf{i} + \mathbf{B}.\frac{3}{r^3}(\mathbf{A.r})\Sigma\,x\mathbf{i}$$

$$= -\frac{\mathbf{B.A}}{r^3} + \frac{3\,(\mathbf{A.r})\,(\mathbf{B.r})}{r^5}.$$

Hence the result. $\qquad [\because \ \Sigma\,(\mathbf{A.i})\,\mathbf{i} = \mathbf{A}, \ \Sigma x\,\mathbf{i} + \mathbf{r}]$

Example 20. For a solenoidal vector **F**,

Show that $\nabla \times (\nabla \times (\nabla \times (\nabla \times \mathbf{F}))) = \nabla^4\mathbf{F}$

or

Curl Curl Curl Curl $\mathbf{F} = \nabla^4\mathbf{F}$

Solution: Given **F** is a solenoidal vector

$$\therefore \qquad \nabla.\mathbf{F} = 0 \qquad \text{...(1)}$$

We also know that

$$\nabla \times (\nabla \times \mathbf{F}) = \nabla\,(\nabla.\mathbf{F}) - \nabla^2\mathbf{F}$$

$$= -\nabla^2\mathbf{F} \ \text{[using (1)]} \qquad \text{...(2)}$$

Now $\nabla \times (\nabla \times (\nabla \times (\nabla \times \mathbf{F}))) = \nabla \times \nabla \times (-\nabla^2\mathbf{F})$

$$= \nabla [\nabla.(-\nabla^2\mathbf{F})] - \nabla^2(-\nabla^2\mathbf{F})$$

$$= -\nabla [\nabla.\nabla^2\mathbf{F}] + \nabla^2(\nabla^2\mathbf{F}) \qquad ...(3)$$

But $\nabla^2\mathbf{F} = \frac{\partial^2\mathbf{F}}{\partial x^2} + \frac{\partial^2\mathbf{F}}{\partial y^2} + \frac{\partial^2\mathbf{F}}{\partial z^2}$

$$= \frac{\partial^2}{\partial x^2}(F_1\,\mathbf{i} + F_2\,\mathbf{j} + F_3\,\mathbf{k}) + \frac{\partial^2}{\partial y^2}(F_1\,\mathbf{i} + F_2\,\mathbf{j} + F_3\,\mathbf{k}) + \frac{\partial^2}{\partial z^2}(F_1\,\mathbf{i} + F_2\,\mathbf{j} + F_3\,\mathbf{k})$$

$$= \left(\frac{\partial^2 F_1}{\partial x^2} + \frac{\partial^2 F_1}{\partial y^2} + \frac{\partial^2 F_1}{\partial z^2}\right)\mathbf{i} + \left(\frac{\partial^2 F_2}{\partial x^2} + \frac{\partial^2 F_2}{\partial y^2} + \frac{\partial^2 F_2}{\partial z^2}\right)\mathbf{j} + \left(\frac{\partial^2 F_3}{\partial x^2} + \frac{\partial^2 F_3}{\partial y^2} + \frac{\partial^2 F_3}{\partial z^2}\right)\mathbf{k}$$

$\therefore$ $\nabla.\nabla^2\mathbf{F} = \frac{\partial}{\partial x}\left(\frac{\partial^2 F_1}{\partial x^2} + \frac{\partial^2 F_1}{\partial y^2} + \frac{\partial^2 F_1}{\partial z^2}\right) + \frac{\partial}{\partial y}\left(\frac{\partial^2 F_2}{\partial x^2} + \frac{\partial^2 F_2}{\partial y^2} + \frac{\partial^2 F_2}{\partial z^2}\right) + \frac{\partial}{\partial z}\left(\frac{\partial^2 F_3}{\partial x^2} + \frac{\partial^2 F_3}{\partial y^2} + \frac{\partial^2 F_3}{\partial z^2}\right)$

$$= \frac{\partial^2}{\partial x^2}\left(\frac{\partial F_1}{\partial x} + \frac{\partial F_2}{\partial y} + \frac{\partial F_3}{\partial z}\right) + \frac{\partial^2}{\partial y^2}\left(\frac{\partial F_1}{\partial x} + \frac{\partial F_2}{\partial y} + \frac{\partial F_3}{\partial z}\right) + \frac{\partial^2}{\partial z^2}\left(\frac{\partial F_1}{\partial x} + \frac{\partial F_2}{\partial y} + \frac{\partial F_3}{\partial z}\right)$$

$$= \frac{\partial^2}{\partial x^2}(\nabla.\mathbf{F}) + \frac{\partial^2}{\partial y^2}(\nabla.\mathbf{F}) + \frac{\partial^2}{\partial z^2}(\nabla.\mathbf{F})$$

$$= 0 + 0 + 0 = \qquad \text{[using (1)]}$$

i.e., $\nabla.\nabla^2\,\mathbf{F} = 0$ Substituting it in (3), we get

$$\nabla \times (\nabla \times (\nabla \times (\nabla \times \mathbf{F})) = -\nabla(0) + \nabla^2(\nabla^2\mathbf{F})$$
$$= \nabla^4 \mathbf{F}$$

Hence the result.

Example 21. In electromagnetic theory, we have

$$\nabla.\mathbf{D} = \rho, \quad \nabla.\mathbf{H} = 0$$

$$\nabla \times \mathbf{D} = \frac{1}{C}\frac{\partial \mathbf{H}}{\partial t}; \quad \nabla \times \mathbf{H} = \frac{1}{C}\left(\rho\mathbf{V} + \frac{\partial \mathbf{D}}{\partial t}\right),$$

then prove that

$$\nabla^2\mathbf{D} - \frac{1}{C^2}\frac{\partial^2 \mathbf{D}}{\partial t^2} = \nabla\rho + \frac{1}{C^2}\frac{\partial}{\partial t}(\rho\mathbf{V})$$

and
$$\nabla^2\mathbf{H} - \frac{1}{C^2}\frac{\partial^2 \mathbf{H}}{\partial t^2} = -\frac{1}{C}\nabla \times (\rho\mathbf{V})$$

Solution: Given : $\nabla.\mathbf{D} = \rho$ ...(1)

$$\nabla.\mathbf{H} = 0 \quad ...(2)$$

$$\nabla \times \mathbf{D} = -\frac{1}{C}\frac{\partial \mathbf{H}}{\partial t} \quad ...(3)$$

$$\nabla \times \mathbf{H} = \frac{1}{C}\left(\rho\mathbf{V} + \frac{\partial \mathbf{D}}{\partial t}\right) \quad ...(4)$$

(i) We have

$$\frac{1}{C^2}\frac{\partial^2 \mathbf{D}}{\partial t^2} + \frac{1}{C^2}\frac{\partial}{\partial t}(\rho\mathbf{V})$$

$$= \frac{1}{C}\frac{\partial}{\partial t}\left\{\frac{1}{C}\left(\frac{\partial \mathbf{D}}{\partial t} + \rho V\right)\right\}$$

$$= \frac{1}{C}\frac{\partial}{\partial t}(\nabla \times \mathbf{H}) \quad \text{[Using (4)]}$$

$$= \frac{1}{C}\nabla \times \frac{\partial \mathbf{H}}{\partial t}$$

$$= \frac{1}{C}\nabla \times (-\mathbf{C}(\nabla \times \mathbf{D})) \quad \text{[Using (3)]}$$

$$= -\nabla \times (\nabla \times \mathbf{D})$$

$$= -\nabla(\nabla.\mathbf{D}) + \nabla^2 \mathbf{D}$$

$$= -\nabla\rho + \nabla^2 \mathbf{D} \qquad \text{[Using (1)]}$$

Hence

$$\frac{1}{C^2}\frac{\partial^2 \mathbf{D}}{\partial t^2} + \frac{1}{C^2}\frac{\partial}{\partial t}(\rho \mathbf{V}) = -\nabla\rho + \nabla^2 \mathbf{D}$$

or
$$\nabla^2 \mathbf{D} - \frac{1}{C^2}\frac{\partial^2 \mathbf{D}}{\partial t^2} = \nabla\rho + \frac{1}{C^2}\frac{\partial}{\partial t}(\rho \mathbf{V})$$

Hence the result (i)

(ii)
$$\nabla^2 \mathbf{H} - \frac{1}{C^2}\frac{\partial^2 \mathbf{H}}{\partial t^2}$$

$$= \nabla^2 \mathbf{H} + \frac{1}{C}\frac{\partial}{\partial t}\left(-\frac{1}{C}\frac{\partial \mathbf{H}}{\partial t}\right)$$

$$= \nabla^2 \mathbf{H} + \frac{1}{C}\frac{\partial}{\partial t}(\nabla \times \mathbf{D})$$

$$= \nabla^2 \mathbf{H} + \frac{1}{C}\nabla \times \frac{\partial \mathbf{D}}{\partial t}$$

$$= \nabla^2 \mathbf{H} + \nabla \times \left(\frac{1}{c}\frac{\partial \mathbf{D}}{\partial t}\right)$$

$$= \nabla^2 \mathbf{H} + \nabla \times \left(\nabla \times \mathbf{H} - \frac{1}{C}\rho \mathbf{V}\right) \qquad \text{[Using (4)]}$$

$$= \nabla^2 \mathbf{H} + \nabla \times (\nabla \times \mathbf{H}) - \frac{1}{C}\nabla \times (\rho \mathbf{V})$$

$$= \nabla^2 \mathbf{H} + \nabla(\nabla.\mathbf{H}) - \nabla^2 \mathbf{H} - \frac{1}{C}\nabla \times (\rho \mathbf{V})$$

$$= \nabla(\nabla.\mathbf{H}) - \frac{1}{C}\nabla \times (\rho \mathbf{V})$$

$$= \nabla(0) - \frac{1}{C}\nabla \times (\rho \mathbf{V}) \qquad \text{[Using (1)]}$$

$$= -\frac{1}{C}\nabla \times (\rho \mathbf{V})$$

Hence the result (ii).

Example 22. (i) If f and ϕ are differentiable scalar point functions, show that

$$\nabla.(\nabla f \times \nabla\phi) = 0 \quad (\text{or } \nabla f \times \nabla\phi \text{ is solenoidal}).$$

(ii) If **F** and **G** are irrotational prove that **F** × **G** is solenoidal.

(iii) Prove that

$$\nabla.(\mathbf{A}\nabla\mathbf{B} - \mathbf{B}\nabla\mathbf{A}) = \mathbf{A}\nabla^2\mathbf{B} - \mathbf{B}\nabla^2\mathbf{A}$$

Solution: (i) We know that

$$\nabla.(\mathbf{F} \times \mathbf{G}) = \mathbf{G}.(\nabla \times \mathbf{F}) - \mathbf{F}.(\nabla \times \mathbf{G})$$

$$\therefore \quad \nabla.(\nabla f \times \nabla\phi) = \nabla \mathbf{f}.(\nabla \times \nabla f) - \nabla f.(\nabla \times \nabla\phi)$$

$$= \nabla\phi.(0) - \nabla f.(0)$$

$$= 0$$

$[\therefore$ curl (grad (scalar point function)) = 0]

Hence $\nabla.(\nabla f \times \nabla\phi) = 0$

(ii) Given **F** and **G** are irrotational

$$\Rightarrow \quad \nabla \times \mathbf{F} = 0$$

and

$$\nabla \times \mathbf{G} = 0$$

Now $\nabla.(\mathbf{F} \times \mathbf{G}) = \mathbf{G}.(\nabla \times \mathbf{F}) - \mathbf{F}.(\nabla \times \mathbf{G})$

$$= \mathbf{G}.(0) - \mathbf{F}.(0)$$

$$= 0$$

$\therefore$ **F** × **G** is solenoidal.

(iii) We have

$$\nabla.(\mathbf{A}\nabla\mathbf{B} - \mathbf{B}\nabla\mathbf{A}) = \nabla.(\mathbf{A}\nabla\mathbf{B}) - \nabla.(\mathbf{B}\nabla\mathbf{A})$$

$$= \mathbf{A}(\nabla.\nabla\mathbf{B}) - \nabla \mathbf{A}.\nabla\mathbf{B}$$

$$- \{\mathbf{B}(\nabla.\nabla\mathrm{A}) - \nabla\mathbf{B}.\nabla\mathrm{A}\}$$

$$= \mathbf{A}\,\nabla^2\,\mathbf{B} - \mathbf{B}\,\nabla^2\,\mathbf{A}$$

Hence the proof.

Example 23: (i) Show that $\nabla \times (\phi\,\mathbf{F}) = \phi\,(\nabla \times \mathbf{F}) + (\nabla\phi) \times \mathbf{F}$, where ϕ is a scalar point function.

(ii) Prove that $\nabla \times (\phi\nabla\phi) = 0$

(iii) If f and g are scalar point functions and $\mathbf{A} = f\nabla g$, prove that $\mathbf{A}.(\nabla \times \mathbf{A}) = \mathbf{0}$

Solution: (i) Let $\mathbf{F} = F_1\mathbf{i} + F_2\mathbf{j} + F_3\mathbf{k}$

$$\therefore \quad \phi\,\mathbf{F} = \phi\,F_1\mathbf{i} + \phi\,F_2\mathbf{j} + \phi\,F_3\mathbf{k}$$

$$\therefore\ \nabla \times (\phi\, \mathbf{F}) = \begin{vmatrix} \mathbf{i} & \mathbf{j} & \mathbf{k} \\ \dfrac{\partial}{\partial x} & \dfrac{\partial}{\partial y} & \dfrac{\partial}{\partial z} \\ \phi F_1 & \phi F_2 & \phi F_3 \end{vmatrix}$$

$$= \Sigma \left\{ \frac{\partial}{\partial y}(\phi F_3) - \frac{\partial}{\partial z}(\phi F_2) \right\} \mathbf{i}$$

$$= \Sigma \left\{ \phi \frac{\partial F_3}{\partial y} + F_3 \frac{\partial \phi}{\partial y} - \phi \frac{\partial F_2}{\partial z} - F_2 \frac{\partial \phi}{\partial z} \right\} \mathbf{i}$$

$$= \Sigma \left\{ \phi \left(\frac{\partial F_3}{\partial y} - \frac{\partial F_2}{\partial z} \right) + \left(F_3 \frac{\partial \phi}{\partial y} - F_2 \frac{\partial \phi}{\partial z} \right) \right\} \mathbf{i}$$

$$= \phi\, \Sigma \left\{ \left(\frac{\partial F_3}{\partial y} - \frac{\partial F_2}{\partial z} \right) \mathbf{i} \right\} + \Sigma \left\{ \left(F_3 \frac{\partial \phi}{\partial y} - F_2 \frac{\partial \phi}{\partial z} \right) \mathbf{i} \right\}.$$

$$= \phi \begin{vmatrix} \mathbf{i} & \mathbf{j} & \mathbf{k} \\ \dfrac{\partial}{\partial x} & \dfrac{\partial}{\partial y} & \dfrac{\partial}{\partial z} \\ F_1 & F_2 & F_3 \end{vmatrix} + \begin{vmatrix} \mathbf{i} & \mathbf{j} & \mathbf{k} \\ \dfrac{\partial \phi}{\partial x} & \dfrac{\partial \phi}{\partial y} & \dfrac{\partial \phi}{\partial z} \\ F_1 & F_2 & F_3 \end{vmatrix}$$

$$= \phi\,(\nabla \times \mathbf{F}) + (\nabla\phi) \times \mathbf{F}$$

$$\therefore\ \nabla \times (\phi \mathbf{F}) = \phi\,(\nabla \times \mathbf{F}) + (\nabla\phi) \times \mathbf{F}$$

Hence the proof

(ii) $\nabla \times (\phi\nabla\phi) = \phi\,(\nabla \times \nabla\,\phi) + (\nabla\phi \times \nabla\phi)$

$$= \phi\,(0) + 0 = 0$$

(iii) $\nabla \times \mathbf{A} = \nabla \times (f\nabla g)$

$$= f\,(\nabla \times \nabla g) + (\nabla f \times \nabla g)$$

$$= f\,(0) + (\nabla f \times \nabla g)$$

$$= \nabla f \times \nabla g$$

$$\mathbf{A}.(\nabla \times \mathbf{A}) = f\nabla g . \nabla f \times \nabla g$$

$$= [f\,\nabla g, \nabla f, \nabla g]$$

$$= f\,[\nabla g, \nabla f, \nabla g]$$

$$= 0 \qquad [\because\ [a\ b\ a] = 0]$$

Example 24: (i) If u**F** = ∇v, where u and v are scalar fields and **F** is a vector field, show that **F**.curl **F** = 0.

(ii) If **r** is the distance of a point (x, y, z) from the origin prove that

$$\text{Curl}\left(\mathbf{k} \times \text{grad}\frac{1}{r}\right) + \text{grad}\left(\mathbf{k}.\text{grad}\frac{1}{r}\right) = 0,$$

wher **k** is a unit vector in the direction of oz.

(iii) Prove that

$$\mathbf{V} \times (\nabla \times \mathbf{V}) = \frac{1}{2}\nabla V^2 - (\mathbf{V}.\nabla)\,\mathbf{V}$$

Solution: (i) Given u**F** = ∇v $\therefore$ $F = \frac{1}{u}\nabla v$

$$\text{Now} \quad \nabla \times \mathbf{F} = \nabla \times \left(\frac{1}{u}\nabla v\right)$$

$$= \nabla\frac{1}{u} \times \nabla v + \frac{1}{u}(\nabla \times (\nabla v))$$

$$= \nabla\frac{1}{u} \times \nabla v + \frac{1}{u}(0)$$

$$= \nabla\frac{1}{u} \times \nabla v$$

Hence

$$\mathbf{F}.(\nabla \times \mathbf{F}) = \left(\frac{1}{u}\nabla v\right).\left(\nabla\frac{1}{u} \times \nabla v\right)$$

$$= \left[\frac{1}{u}\nabla v,\ \nabla\frac{1}{u},\ \nabla v\right]$$

$$= \frac{1}{u}\left[\nabla v,\ \nabla\frac{1}{u},\ \nabla v\right] = 0$$

$\therefore$ $\mathbf{F}.(\nabla \times \mathbf{F}) = 0.$

(ii) Given $\mathbf{r} = x\mathbf{i} + y\mathbf{j} + z\mathbf{k}$

$$r = |\mathbf{r}| = \sqrt{x^2 + y^2 + z^2} \qquad \text{...(1)}$$

We know that grad $\frac{1}{r} = \frac{-\mathbf{r}}{r^3}$

$$= \frac{-(x\mathbf{i} + y\mathbf{j} + z\mathbf{k})]}{r^3}$$

$$k \times \text{grad}\,\frac{1}{r} = k \times \frac{[-(x\mathbf{i} + y\mathbf{j} + z\mathbf{k})]}{r^3}$$

$$= \frac{-x\,(\mathbf{k} \times \mathbf{i}) - y\,(\mathbf{k} \times \mathbf{j}) - z\,(\mathbf{k} \times \mathbf{k})}{r^3}$$

$$= \frac{-x\mathbf{j} + y\mathbf{i}}{r^3}$$

$$\text{Curl}\left(\mathbf{k} \times \text{grad}\,\frac{1}{r}\right) = \begin{vmatrix} \mathbf{i} & \mathbf{j} & \mathbf{k} \\ \dfrac{\partial}{\partial x} & \dfrac{\partial}{\partial y} & \dfrac{\partial}{\partial z} \\ \dfrac{-y}{r^3} & \dfrac{-x}{r^3} & 0 \end{vmatrix}$$

$$= \frac{\partial}{\partial z}\left(\frac{x}{r^3}\right)\mathbf{i} + \frac{\partial}{\partial z}\left(\frac{y}{r^3}\right)\mathbf{j} + \left[\frac{\partial}{\partial x}\left(-\frac{x}{r^3}\right) - \frac{\partial}{\partial y}\left(\frac{y}{r^3}\right)\right]\mathbf{k}$$

$$= x\left(-\frac{3}{r^4}\frac{\partial r}{\partial z}\right)\mathbf{i} + y\left(-\frac{3}{r^4}\cdot\frac{\partial r}{\partial z}\right)\mathbf{j}$$

$$+ \left[\left\{(-x)\left(-\frac{3}{r^4}\frac{\partial r}{\partial x}\right) + \frac{1}{r^3}(-1)\right\}\right.$$

$$\left. - \left\{y\left(-\frac{3}{r^4}\right)\frac{\partial r}{\partial y} + \frac{1}{r^3}\right\}\right]\mathbf{k}$$

$$= -\frac{3xz}{r^5}\mathbf{i} - \frac{3yz}{r^5}\mathbf{j} + \left(\frac{3x^2}{r^5} + \frac{3y^2}{r^5} - \frac{2}{r^3}\right)\mathbf{k}$$

$$= -\frac{3xz}{r^5}\mathbf{i} - \frac{3yz}{r^5}\mathbf{j} + \left[\frac{3}{r^5}(x^2 + y^2) - \frac{2}{r^3}\right]\mathbf{k}$$

$$= -\frac{3xz}{r^5}\mathbf{i} - \frac{3yz}{r^5}\mathbf{j} + \left[\frac{3}{r^5}(r^2 - z^2) - \frac{2}{r^3}\right]\mathbf{k}$$

$$= -\frac{3xz}{r^5}\mathbf{i} - \frac{3yz}{r^5}\mathbf{j} + \left(-\frac{3z^2}{r^5} + \frac{1}{r^3}\right)\mathbf{k} \quad ...(1)$$

Now $\mathbf{k}.\ \text{grad}\,\frac{1}{r} = \frac{\mathbf{k}.(x\mathbf{i} + y\mathbf{j} + z\mathbf{k})]}{r^3} = -\frac{z}{r^3}$

$$\text{Grad}\left(\mathbf{k}.\ \text{grad}\,\frac{1}{r}\right)$$

$$= \left(\frac{\partial}{\partial x}\mathbf{i} + \frac{\partial}{\partial y}\mathbf{j} + \frac{\partial}{\partial z}\mathbf{k}\right)\left(-\frac{z}{r^3}\right)$$

$$= -\frac{\partial}{\partial x}\left(\frac{z}{r^3}\right)\mathbf{i} - \frac{\partial}{\partial y}\left(\frac{z}{r^3}\right)\mathbf{j} - \frac{\partial}{\partial z}\left(\frac{z}{r^3}\right)\mathbf{k}$$

$$= \frac{3z}{r^4}\frac{\partial r}{\partial x}\mathbf{i} + \frac{3z}{r^4}\frac{\partial r}{\partial y}\mathbf{j} - \left\{\frac{-3z}{r^4}\frac{\partial r}{\partial z} + \frac{1}{r^3}\right\}\mathbf{k}$$

$$= \frac{3zx}{r^5}\mathbf{i} + \frac{3zy}{r^5}\mathbf{j} + \left(\frac{3z^2}{r^5} + \frac{1}{r^3}\right)\mathbf{k} \quad ...(2)$$

Adding (1) and (2), we get

$$\therefore \text{Curl}\left(\mathbf{k} \times \text{grad}\,\frac{1}{r}\right) + \text{grad}\left(\mathbf{k}.\text{grad}\,\frac{1}{r}\right) = 0.$$

(iii) Let $\mathbf{V} = v_1\mathbf{i} + v_2\mathbf{j} + v_3\mathbf{k}$

$$\text{Then } \nabla \times \mathbf{V} = \begin{vmatrix} \mathbf{i} & \mathbf{j} & \mathbf{k} \\ \frac{\partial}{\partial x} & \frac{\partial}{\partial y} & \frac{\partial}{\partial z} \\ v_1 & v_2 & v_3 \end{vmatrix}$$

$$= \left(\frac{\partial v_3}{\partial y} - \frac{\partial v_2}{\partial z}\right)\mathbf{i} + \left(\frac{\partial v_1}{\partial z} - \frac{\partial v_3}{\partial x}\right)\mathbf{j} + \left(\frac{\partial v_2}{\partial x} - \frac{\partial v_1}{\partial y}\right)\mathbf{k}$$

$$\mathbf{V} \times (\nabla \times \mathbf{V}) = \begin{vmatrix} \mathbf{i} & \mathbf{j} & \mathbf{k} \\ v_1 & v_2 & v_3 \\ \frac{\partial v_3}{\partial y} - \frac{\partial v_2}{\partial z} & \frac{\partial v_1}{\partial z} - \frac{\partial v_3}{\partial x} & \frac{\partial v_2}{\partial x} - \frac{\partial v_1}{\partial y} \end{vmatrix}$$

$$= \Sigma \left\{ v_2 \left(\frac{\partial v_2}{\partial x} - \frac{\partial v_1}{\partial y} \right) - v_3 \left(\frac{\partial v_1}{\partial z} - \frac{\partial v_3}{\partial x} \right) \right\} \mathbf{i}$$

$$= \Sigma \left\{ \left(v_2 \frac{\partial v_2}{\partial x} + v_3 \frac{\partial v_3}{\partial x} \right) - \left(v_2 \frac{\partial v_1}{\partial y} + v_3 \frac{\partial v_1}{\partial z} \right) \right\} \mathbf{i}$$

$$= \Sigma \left\{ \left(v_1 \frac{\partial v_1}{\partial x} + v_3 \frac{\partial v_2}{\partial x} + v_3 \frac{\partial v_3}{\partial x} \right) \right.$$

$$\left. - \left(v_1 \frac{\partial v_1}{\partial x} + v_2 \frac{\partial v_1}{\partial y} + v_3 \frac{\partial v_1}{\partial z} \right) \right\} \mathbf{i}$$

$$= \Sigma \left\{ \frac{1}{2} \frac{\partial}{\partial x} (v_1^2 + v_2^2 + v_3^2) - \left(v_1 \frac{\partial v_1}{\partial x} + v_2 \frac{\partial v_1}{\partial y} + v_3 \frac{\partial v_1}{\partial z} \right) \right\} \mathbf{i}$$

$$= \Sigma \left\{ \frac{1}{2} \frac{\partial}{\partial x} (\mathbf{V})^2 - \left(v_1 \frac{\partial v_1}{\partial x} + v_2 \frac{\partial v_1}{\partial y} + v_3 \frac{\partial v_1}{\partial z} \right) \right\} \mathbf{i}$$

$$= \frac{1}{2} \Sigma \left\{ \frac{\partial}{\partial x} (\mathbf{V})^2 \mathbf{i} \right\} - \Sigma \left\{ v_1 \frac{\partial v_1}{\partial x} + v_2 \frac{\partial v_1}{\partial y} + v_3 \frac{\partial v_1}{\partial z} \right\} \mathbf{i}$$

$$= \frac{1}{2} \left[\left(\frac{\partial}{\partial x} \mathbf{i} + \frac{\partial}{\partial y} \mathbf{j} + \frac{\partial}{\partial z} \mathbf{k} \right) (\mathbf{V})^2 \right]$$

$$- \left[v_1 \left(\frac{\partial v_1}{\partial x} \mathbf{i} + \frac{\partial v_2}{\partial y} \mathbf{j} + \frac{\partial v_3}{\partial z} \mathbf{k} \right) \right.$$

$$\left. + v_2 \left(\frac{\partial v_1}{\partial x} \mathbf{i} + \frac{\partial v_2}{\partial y} \mathbf{j} + \frac{\partial v_3}{\partial z} \mathbf{k} \right) + v_3 \left(\frac{\partial v_1}{\partial x} \mathbf{i} + \frac{\partial v_2}{\partial y} \mathbf{j} + \frac{\partial v_3}{\partial z} \mathbf{k} \right) \right]$$

$$= \frac{1}{2} \nabla (\mathbf{V})^2 - \left(v_1 \frac{\partial}{\partial x} (v_1 \mathbf{i} + v_2 \mathbf{j} + v_3 \mathbf{k}) \right.$$

$$\left. + v_2 \frac{\partial}{\partial y} (v_1 \mathbf{i} + v\, \mathbf{j} + v_3\, \mathbf{k}) + v_3 \frac{\partial}{\partial z} (v_1 \mathbf{i} + v_2 \mathbf{j} + v_3 \mathbf{k}) \right]$$

$$= \frac{1}{2} \nabla (\mathbf{V})^2 - \left[v_1 \frac{\partial}{\partial x} + v_2 \frac{\partial}{\partial y} + v_3 \frac{\partial}{\partial z} \right] \mathbf{V}$$

$$(\because \mathbf{V} = v_1 \mathbf{i} + v_1 \mathbf{j} + v_3 \mathbf{k})$$

$$= \frac{1}{2} \nabla (\mathbf{V})^2 - (\mathbf{V}.\nabla)\,\mathbf{V}$$

Example 25. (i) Prove that $\nabla \times [(\mathbf{r} \times \mathbf{A}) \times \mathbf{B}] = \mathbf{B} \times \mathbf{A}$, where **A** and **B** are constant vectors and **r** is a position vector.

(ii) Prove that $\nabla.[(\mathbf{r} \times \mathbf{A}) \times \mathbf{B}] = -2\,\mathbf{B}.\mathbf{A}$

(iii) If $\phi = \frac{1}{r} e^{\lambda r}$ prove that $(\nabla^2 - \lambda^2)\,\phi = 0$

(iv) If **r** (p) be the vector from the origin **O** to a point p in the xy-plane, then show that the plane scalar field u (p) = log **r** satisfies that equation $\nabla^2 u = 0$

Solution: (i) $\nabla \times [(\mathbf{r} \times \mathbf{A}) \times \mathbf{B}]$

$$= \nabla \times [-\{\mathbf{B} \times (\mathbf{r} \times \mathbf{A})\}]$$

$$= \nabla \times [-\{(\mathbf{B}.\mathbf{A})\,\mathbf{r} - (\mathbf{B}.\mathbf{r})\,\mathbf{A}\}]$$

$$= \nabla \times [-(\mathbf{B}.\mathbf{r}\,\mathbf{A} - (\mathbf{B}.\mathbf{A})\,\mathbf{r}]$$

$$= \nabla \times \{(\mathbf{B}.\mathbf{r})\,\mathbf{A}\} - \nabla \times \{(\mathbf{B}.\mathbf{A})\mathbf{r}\}$$

$$= [(\mathbf{r}.\mathbf{B})(\nabla \times \mathbf{A}) + \nabla(\mathbf{r}.\mathbf{B}) \times \mathbf{A}] - (\mathbf{B}.\mathbf{A})(\nabla \times \mathbf{r}) + \nabla)(\mathbf{B}.\mathbf{A}) \times \mathbf{r}]$$

$[\because \nabla \times (\phi\mathbf{F}) = \phi(\nabla \times \mathbf{F}) + \nabla\phi \times \mathbf{F}]$

$$= \nabla(\mathbf{r}.\mathbf{B}) \times \mathbf{A}$$

$[\because \nabla \times \mathbf{A} = 0;\ \nabla \times \mathbf{r} = 0,\ \nabla(\mathbf{B}.\mathbf{A}) = 0$ as **A** and **B** are constant vectors]

$$= \left[\Sigma \frac{\partial}{\partial x}(\mathbf{r}.\mathbf{B})\,\mathbf{i}\right] \times \mathbf{A}$$

$$= \left[\Sigma\left(\frac{\partial}{\partial x}.\mathbf{B}\right)\mathbf{i}\right] \times \mathbf{A}$$ [∴ **B** is constant vector]

$$= [\Sigma\,(\mathbf{i}.\mathbf{B})\,\mathbf{i}] \times \mathbf{A}$$

$$= (\Sigma\,\mathbf{B_1 i}) \times \mathbf{A}$$ [∴ $\mathbf{B} = B_1 i + B_2 j + B_3 k$]

$$= \mathbf{B} \times \mathbf{A}$$

Hence the proof.

(ii) $\nabla.[(\mathbf{r} \times \mathbf{A}) \times \mathbf{B}]$

$$= \nabla.[\{\mathbf{B} \times (\mathbf{r} \times \mathbf{A})\}]$$

$$= \nabla.[-\{(\mathbf{B}.\mathbf{A})\,\mathbf{r} - (\mathbf{B}.\mathbf{r})\,\mathrm{A}\}]$$

$$= \nabla.[(\mathbf{B}.\mathbf{r}\ \mathbf{A} - (\mathbf{B}.\mathbf{A})\,\mathbf{r}]$$

$$= \{\nabla(\mathbf{r}.)\,.\mathbf{A} + (\mathbf{r}.\mathbf{B})\,\nabla.\mathbf{A}\} - \{\nabla(\mathbf{B}.\mathbf{A}).\mathbf{r} - (\mathbf{B}.\mathbf{A})\,\nabla.\mathbf{r}\}$$

$$[\because\ \nabla.(\Phi\mathbf{F}) = \nabla\phi.\mathbf{F} + \phi\,(\nabla.\mathbf{F})]$$

$$= \nabla(\mathbf{r}.\mathbf{B}).\mathbf{A} - (\mathbf{B}.\mathbf{A})\,\nabla.\mathbf{r}$$

$$[\because\ \nabla.\mathbf{A} = 0,\ \nabla(\mathbf{B}.\mathbf{A}) = 0]$$

$$= \left\{\left(\frac{\partial}{\partial x}\mathbf{i} + \frac{\partial}{\partial y}\mathbf{j} + \frac{\partial}{\partial z}\mathbf{k}\right)(B_1x + B_1y + B_2z)\right\}.\mathbf{A} - (\mathbf{B}.\mathbf{A})\left\{\frac{\partial}{\partial x}(z) + \frac{\partial}{\partial y}(y) + \frac{\partial}{\partial z}(z)\right\}$$

$$= (B_1\mathbf{i} + B_2\mathbf{j} + B_3\mathbf{k}).\mathbf{A} - (\mathbf{B}.\mathbf{A})\,(3)$$

$$= \mathbf{A}.\mathbf{A}\ \ 3\,(\mathbf{B}.\mathbf{A}) = -2\,(\mathbf{B}.\mathbf{A}) = -2\,\mathbf{B}.\mathbf{A}$$

(iii) Given $\phi = \frac{1}{r}e^{\lambda r}$

where $r^2 = x^2 + y^2 + z^2.$

$$\Leftrightarrow \quad \frac{\partial r}{\partial x} = \frac{x}{r},\ \frac{\partial r}{\partial y} = \frac{y}{r} \text{ and } \frac{\partial r}{\partial z} = \frac{z}{r}$$

Now
$$\frac{\partial \phi}{\partial x} = \frac{\partial}{\partial x}\left[\frac{e^{\lambda r}}{r}\right] = \frac{d}{dr}\left[\frac{e^{\lambda r}}{r}\right]\frac{\partial r}{\partial x}$$

$$= \frac{r\lambda e^{\lambda r} - e^{\lambda r}}{r^2}\cdot\frac{x}{r}$$

$$= \frac{e^{\lambda r}(\lambda r - 1)x}{r^3}$$

$$\frac{\partial^2\phi}{\partial x^2} = \frac{\partial}{\partial x}\left(\frac{\partial\phi}{\partial x}\right) = \frac{\partial}{\partial x}\left[\frac{(r\lambda - 1)\,xe^{\lambda r}}{r^3}\right]$$

$$= \left(\frac{r\lambda - 1}{r^3}\right)e^{\lambda r} + \left(\frac{r\lambda - 1}{r^3}\right)\left(\lambda e^{\lambda r}\frac{\partial r}{\partial x}\right)x + \left\{\frac{r^3(\lambda) - (\lambda r - 1)(3r^2)}{r^6}\cdot\frac{\partial r}{\partial x}\right\}\ e^{\lambda r}\,x$$

$$= \frac{(r\lambda - 1)}{r^3} e^{\lambda r} + \frac{\lambda(r\lambda - 1)}{r^3} e^{\lambda r} \cdot \frac{x}{r} \cdot x + \frac{3 - 2r\lambda}{r^4} \cdot \frac{x}{r} \cdot e^{\lambda r} x$$

$$= \frac{(r\lambda - 1)e^{\lambda r}}{r^3} + \left\{ \frac{\lambda(r\lambda - 1)}{r^4} + \frac{3 - 2r\lambda}{r^5} \right\} x^2 e^{\lambda r}$$

$$\therefore \quad \frac{\partial^2 \phi}{\partial x^2} = \frac{(r\lambda - 1)e^{\lambda r}}{r^3} + \frac{r^2\lambda^2 - 3\lambda r + 3)\ e^{\lambda r} x^2}{r^5}$$

Similarly

$$\frac{\partial^2 \phi}{\partial y^2} = \frac{(r\lambda - 1)e^{\lambda r}}{r^3} + \frac{(r^2\lambda^2 - 3\lambda r + 3)\ e^{\lambda r} y^2}{r^5}$$

$$\frac{\partial^2 \phi}{\partial y^2} = \frac{(r\lambda - 1)e^{\lambda r}}{r^3} + \frac{(r^2\lambda^2 - 3\lambda r + 3)\ e^{\lambda r} z^2}{r^5}$$

Adding, we have

$$\frac{\partial^2 \phi}{\partial x^2} + \frac{\partial^2 \phi}{\partial y^2} + \frac{\partial^2 \phi}{\partial z^2}$$

$$= \frac{3(r\lambda - 1)e^{\lambda r}}{r^3} + \frac{(r^2\lambda^2 - e\lambda r + 3)e^{\lambda r}}{r^5} (x^2 + y^2 + z^2)$$

$$= \frac{3(r\lambda - 1)e^{\lambda r}}{r^3} + \frac{(r^2\lambda^2 - 3\lambda r + 3)e^{\lambda r}}{r^5} \cdot r^2$$

$$= \frac{\lambda^2 e^{\lambda r}}{r}$$

$$= \lambda^2 \phi$$

$\therefore\ \nabla^2\phi = \lambda^2\phi$ *or* $(\nabla^2 - \lambda^2)\phi = 0.$

Hence the result.

Example 26. If f = (**r** × **A**).(**r** × **B**) then show that

$$\nabla f = 2\,[\mathbf{B} \times (\mathbf{r} \times \mathbf{A}) + \mathbf{A} \times (\mathbf{r} \times \mathbf{B})],$$

where **A** and **B** are constant vectors.

Solution: We know that

$$\nabla(\mathbf{A}.\mathbf{B}) = (\mathbf{A}.\nabla)\mathbf{B} + (\mathbf{B}.\nabla)\mathbf{A} + \mathbf{A} \times (\nabla \times \mathbf{B}) + \mathbf{B} \times (\nabla \times \mathbf{A})$$

$$\nabla(\mathbf{F}.\mathbf{G}) = (\mathbf{F}.\nabla)\mathbf{G} + (\mathbf{G}.\nabla)\mathbf{F} + \mathbf{F} \times (\nabla \times \mathbf{G}) + \mathbf{G} \times (\nabla \times \mathbf{F})$$

Putting **F** = **r** × **A** and **G** = **r** × **B**, we get

$$\nabla f = \nabla\,[(\mathbf{r}\times\mathbf{A}).(\mathbf{r}\times\mathbf{B}]$$

$$= [(\mathbf{r}\times\mathbf{A}).\nabla]\,[\mathbf{r}\times\mathbf{B}] + [(\mathbf{r}\times\mathbf{B}).\nabla]\,[\mathbf{r}\times\mathbf{A}]$$
$$+ [\mathbf{r}\times\mathbf{A}]\times[\nabla\times(\mathbf{r}\times\mathbf{B})] + (\mathbf{r}\times\mathbf{B})\times[\nabla\times(\mathbf{r}\times\mathbf{A})] \quad ...(1)$$

Now $(\mathbf{r}\times\mathbf{A}).\nabla = \nabla.(\mathbf{r}\times\mathbf{A})$

$$= \mathbf{A}.(\nabla\times\mathbf{r}) - \mathbf{r}.(\nabla\times\mathbf{A})$$

$$[\because\ \nabla.(\mathbf{A}\times\mathbf{B}) = \mathbf{B}.(\nabla\times\mathbf{A}) - \mathbf{A}.(\nabla\times\mathbf{B})]$$

$$= \mathbf{A}.(0) - \mathbf{r}.(0)$$

$[\because\ \mathbf{A}$ is constant vector and $\nabla\times\mathbf{r} = 0]$

or $\quad (\mathbf{r}\times\mathbf{A}).\nabla = \mathbf{0} \quad ...(2)$

Similarly $\quad (\mathbf{r}\times\mathbf{B}).\nabla = \mathbf{0} \quad ...(3)$

Again $\nabla\times(\mathbf{r}\times\mathbf{A}) = \Sigma\mathbf{i}\times\dfrac{\partial}{\partial x}(\mathbf{r}\times\mathbf{A})$

$$= \Sigma\mathbf{i}\times\left(\frac{\partial\mathbf{r}}{\partial x}\times\mathbf{A}\right)$$

$$= \Sigma\,[\mathbf{i}\times(\mathbf{i}\times\mathbf{A})] \qquad \left[\because\ \frac{\partial\mathbf{r}}{\partial x} = \mathbf{i}\right]$$

$$= \Sigma\,[(\mathbf{i}.\mathbf{A})\,\mathbf{i} - (\mathbf{i}.\mathbf{i})\,\mathbf{A}]$$

$$= \Sigma\,[A_1\mathbf{i} - \mathbf{A}] \qquad [\because\ \mathbf{A} = A_1\mathbf{i} + A_2\mathbf{j} + A_3\mathbf{k}]$$

$$= \Sigma A_1\,\mathbf{i} - \Sigma\mathbf{A}$$

$\therefore \quad \nabla\times(\mathbf{r}\times\mathbf{A}) = \mathbf{A} - 3\mathbf{A} = -2\mathbf{A} \quad ...(4)$

Similarly

$$\nabla\times(\mathbf{r}\times\mathbf{B}) = -2\mathbf{B} \quad ...(5)$$

Substituting the values from (2), (3), (4) and (5) in (1), we get

$$\nabla f = \mathbf{0} + \mathbf{0} + (\mathbf{r}\times\mathbf{A})\times(-2\mathbf{B}) + (\mathbf{r}\times\mathbf{B})\times(-2\mathbf{A})$$

$$= -2\,[(\mathbf{r}\times\mathbf{A})\times\mathbf{B} + (\mathbf{r}\times\mathbf{B})\times\mathbf{A}]$$

$$= 2\,[\mathbf{B}\times(\mathbf{r}\times\mathbf{A}) + \mathbf{A}\times(\mathbf{r}\times\mathbf{B})]$$

Hence the result.

12.19. Invariance

Transformation equations between the co-ordinates of a point in two systems.

Let the two rectangular co-ordinate systems Oxyz and $Ox^1y^1z^1$ having the same origin be rotated with respect to each other.

Let P be any point in space whose co-ordinates are (x, y, z) or (x′, y′, z′) with respect to the systems of axes i.e. Ox, Oy, Oz and Ox′, Oy′, Oz′ respectively.

Let **r** and **r**′ be the position vectors of P in the two systems.

The **r** = **r**^{1}

i.e.

$x\mathbf{i} + y\mathbf{j} + z\mathbf{k} = x'\mathbf{i}' + y'\mathbf{j}' + z'\mathbf{k}'$ ...(1)

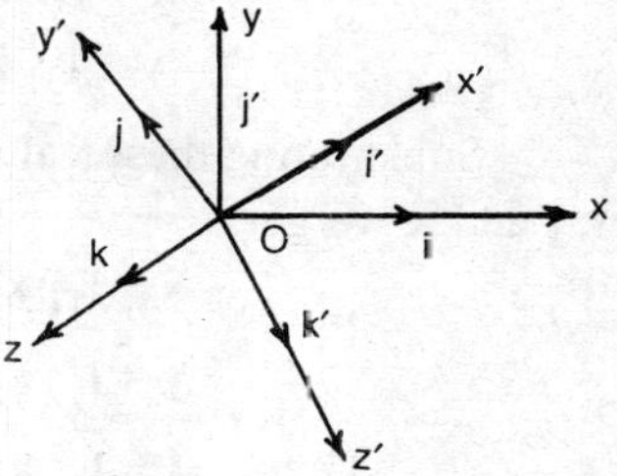

Fig 12.5

Also we know that

$$\mathbf{A} = (\mathbf{A}.\mathbf{i}')\,\mathbf{i}' + (\mathbf{A}.\mathbf{j}')\,\mathbf{j}' + (\mathbf{A}.\mathbf{k}')\,\mathbf{k}' \quad ...(2)$$

Then letting **A** = **i, j, k** in succession in (2), we get

$$\mathbf{i} = (\mathbf{i}.\mathbf{i}')\,\mathbf{i}' + (\mathbf{i}.\mathbf{j}')\,\mathbf{j}' + (\mathbf{i}.\mathbf{k}')\,\mathbf{k}'$$
$$= l_{11}\,\mathbf{i}' + l_{31}\,\mathbf{j}' + l_{21}\,\mathbf{k}' \quad ...(3.1)$$

$$\mathbf{j} = (\mathbf{j}.\mathbf{i}')\,\mathbf{i}' + (\mathbf{j}.\mathbf{j}')\,\mathbf{j}' + (\mathbf{j}\,\mathbf{k}')\,\mathbf{k}'$$
$$= l_{12}\,\mathbf{i}' + l_{22}\,\mathbf{j}' = l_{32}\,\mathbf{k}' \quad ...(3.2)$$

$$\mathbf{k} = (\mathbf{k}.\mathbf{i}')\,\mathbf{i}' + (\mathbf{k}\,\mathbf{j}')\,\mathbf{j}' + (\mathbf{k}\,\mathbf{k}')\,\mathbf{k}'$$
$$= l_{13}\,\mathbf{i}' + l_{23}\,\mathbf{j}' + l_{33}\,\mathbf{k}' \quad ...(3.3)$$

Substituting **i, j, k** in (1) and equating the coefficients of **i**′, **j**′ and **k**′, we get

$$x' = l_{11}\,x + l_{12}y + l_{13}z \quad ...(4.1)$$

$$y' = l_{21}\,x + l_{22}y + l_{23}z \quad ...(4.2)$$

$$z' = l_{31}\,x + l_{32}\,y + l_{33}z \quad ...(4.3)$$

which are the required transformation equations

Again as before we know that

$$\mathbf{A} = (\mathbf{A}.\mathbf{i})\,\mathbf{i} + (\mathbf{A}.\mathbf{j})\,\mathbf{j} + (\mathbf{A}\,\mathbf{k})\,\mathbf{k} \quad ...(5)$$

Letting **A** = **i**′, **j**′, **k**′ in succession in (5), we get

$$\mathbf{i}' = (\mathbf{i}'.\mathbf{i})\,\mathbf{i} - (\mathbf{i}'_2 \cdot \mathbf{j})\,\mathbf{j} + (\mathbf{i}'.\mathbf{k})\,\mathbf{k}$$
$$= l_{11}\,\mathbf{i} + l_{12}\,\mathbf{j} + l_{13}\,\mathbf{k} \quad ...(6.1)$$

$$\mathbf{j}' = (\mathbf{j}'.\,\mathbf{i})\,\mathbf{i} + (\mathbf{j}'.\,\mathbf{j})\,\mathbf{j} + (\mathbf{j}'.\mathbf{k})\,\mathbf{k}$$
$$= l_{21}\,\mathbf{i} + l_{22}\,\mathbf{j} + l_{23}\,\mathbf{k} \quad ...(6.2)$$

$$\mathbf{k}' = (\mathbf{k}'.\mathbf{i})\,\mathbf{i} + (\mathbf{k}'.\mathbf{j})\,\mathbf{j} + (\mathbf{k}'.\mathbf{k})\,\mathbf{k}$$
$$= l_{31}\,\mathbf{i} + l_{32}\,\mathbf{j} + l_{33}\,\mathbf{k} \qquad ...(6.3)$$

Substituting these values in (1) and equating the coefficients of **i, j** and **k** we get

$$x = l_{11}\,x' + l_{21}\,y' + l_{31}\,z'$$
$$y = l_{12}\,x' + l_{22}\,y' + l_{32}\,z'$$
$$z = l_{13}\,x' + l_{23}\,y' + l_{33}\,z'$$

which are the required transformations.

Example 27. If $\phi\,(x, y, z)$ is a scalar invariant with respect to a rotation of axes, prove that $\nabla\phi$ is a vector invariant under this transformation.

Solution: Since $\phi\,(x, y, z)$ is a scalar invariant with respect to rotation of axes, we have

$$\phi\,(x, y, z) = \phi'\,(x', y', z') \qquad \text{[By hypothesis]}$$

We have to prove that

$$\nabla\,\phi\,(x, y, z) = \nabla\,\phi'\,(x', y', z')$$

i.e.
$$\frac{\partial\phi}{\partial x}\,\mathbf{i} + \frac{\partial\phi}{\partial y}\,\mathbf{j} + \frac{\partial\phi}{\partial z}\,\mathbf{k} = \frac{\partial\phi^1}{\partial x^1}\,\mathbf{i}^1 + \frac{\partial\phi^1}{\partial y^1}\,\mathbf{j}^1 + \frac{\partial\phi^1}{\partial z^1}\,\mathbf{k}^1$$

We know by chain rule of differentiation that

$$\frac{\partial\phi}{\partial x} = \frac{\partial\phi^1}{\partial x^1}\cdot\frac{\partial x^1}{\partial x} + \frac{\partial\phi^1}{\partial y^1}\cdot\frac{\partial y^1}{\partial x} + \frac{\partial\phi^1}{\partial z^1}\,\frac{\partial z^1}{\partial x}$$
$$= \frac{\partial\phi^1}{\partial x^1}\,l_{11} + \frac{\partial\phi^1}{\partial y^1}\,l_{21} + \frac{\partial\phi^1}{\partial z^1}\,l_{31} \qquad ...(1)$$

Similarly

$$\frac{\partial\phi}{\partial y} = \frac{\partial\phi'}{\partial x'}\,l_{12} + \frac{\partial\phi'}{\partial y'}\,l_{22} + \frac{\partial\phi'}{\partial z'}\,l_{32} \qquad ...(2)$$

and
$$\frac{\partial\phi}{\partial z} = \frac{\partial\phi'}{\partial x'}\,l_{13} + \frac{\partial\phi'}{\partial y'}\,l_{23} + \frac{\partial\phi'}{\partial z'}\,l_{33} \qquad ...(3)$$

Multiplying (1), (2), (3) by **i, j, k** and adding using the results (6.1) to (6.3) of art [4.19], we get

$$\frac{\partial\phi}{\partial x}\,\mathbf{i} + \frac{\partial\phi}{\partial y}\,\mathbf{j} + \frac{\partial\phi}{\partial z}\,\mathbf{k} = \frac{\partial\phi'}{\partial x'}\,\mathbf{i}' - \frac{\partial\phi'}{\partial y'}\,\mathbf{j}' + \frac{\partial\phi'}{\partial z'}\,\mathbf{k}'$$

i.e. Grad ϕ = Grad ϕ'. Hence proved.

Example 28. If $\mathbf{F}(x, y, z)$ be a vector function in variant with respect to a rotation of axes, prove that $\nabla.\mathbf{F}$ is a scalar invariant under this transformation.

Solution: $\because$ $\mathbf{F}(x, y, z)$ is invariant with respect to rotation of axes, we have

$$\mathbf{F}(x, y, z) = \mathbf{F}'(x', y', z')$$

We have to prove that

$$\nabla.\mathbf{F}(x, y, z) = \nabla.\mathbf{F}'(x', y', z')$$

By chain rule of differentiation we get

$$\frac{\partial \mathbf{F}}{\partial x} = \frac{\partial \mathbf{F}'}{\partial x'} \cdot \frac{\partial x'}{\partial x} + \frac{\partial \mathbf{F}'}{\partial y'} \cdot \frac{\partial y'}{\partial x} + \frac{\partial \mathbf{F}'}{\partial z'} \cdot \frac{\partial z'}{\partial x}$$

$$= \frac{\partial \mathbf{F}'}{\partial x'} l_{11} + \frac{\partial \mathbf{F}'}{\partial y'} l_{21} + \frac{\partial \mathbf{F}'}{\partial z'} l_{31} \qquad ...(1)$$

Similarly

$$\frac{\partial \mathbf{F}}{\partial y} = l_{12} \frac{\partial \mathbf{F}'}{\partial x'} + l_{22} \frac{\partial \mathbf{F}'}{\partial y'} + l_{21} \frac{\partial \mathbf{F}'}{\partial z'} \qquad ...(2)$$

$$\frac{\partial \mathbf{F}}{\partial x} = l_{13} \frac{\partial \mathbf{F}'}{\partial x'} + l_{23} \frac{\partial \mathbf{F}'}{\partial y'} + l_{33} \frac{\partial \mathbf{F}'}{\partial z'} \qquad ...(3)$$

Multiplying (1), (2) and (3) by **i, j, k** respectively then adding and using the results (6.1) to (6.3) of art [4.19], we get

$$\frac{\partial \mathbf{F}}{\partial x} \mathbf{i} + \frac{\partial \mathbf{F}}{\partial y} \mathbf{j} + \frac{\partial \mathbf{F}}{\partial z} \cdot \mathbf{k} = \frac{\partial \mathbf{F}'}{\partial x'} \cdot \mathbf{i}' + \frac{\partial \mathbf{F}'}{\partial y'} . \mathbf{j}' + \frac{\partial \mathbf{F}'}{\partial z'} . \mathbf{k}'$$

ie $\nabla.\mathbf{F} = \nabla.\mathbf{F}'$ Hence proved.

EXERCISE

1. If $F = 3xyz^2 \mathbf{i} + 2xy^3 \mathbf{j} - x^2yz \mathbf{k}$ and $\phi = 3x^2 - yz$, find

(i) $\nabla.\mathbf{F}$ (ii) $\mathbf{F}.\nabla\phi$

(iii) $\nabla.(\phi\mathbf{F})$ (iv) $\nabla.(\nabla\phi)$ at the point (1, - 1, 1).

2. Prove that the vector

(i) $\mathbf{F} = 3y^4z^2 \mathbf{i} + 4x^3z^3 \mathbf{j} + 3x^2y^2 \mathbf{k}$ is solenoidal.

(ii) $\mathbf{F} = 2x(x + 4y^2z)\mathbf{i} + 3xy(x^2 - 1)\mathbf{j} - 2z(2y^2z + x^3)\mathbf{k}$ < is not solenoidal but $\mathbf{G} = xyz^2 \mathbf{F}$ is solenoidal.

3. Show that

(i) $\nabla.(r^n\mathbf{r}) = (n + 3)\, r^n$ (ii) $\nabla \times (r^n\mathbf{r}) = 0$

(iii) $.\nabla\left(\frac{\mathbf{r}}{r}\right) = \frac{2}{r}$ (iv) $\nabla\left[\nabla.\left(\frac{\mathbf{r}}{r}\right)\right] = -\frac{2}{r^3}\mathbf{r}$

(v) $\nabla.\left[\frac{f(r)\,\mathbf{r}}{r}\right] = \frac{1}{r^2}\frac{d}{dr}(r^2 f)$

where $\mathbf{r} = x\mathbf{i} + y\mathbf{j} + z\mathbf{k}$ and $r = |\mathbf{r}|$

4. (i) Find $\nabla \times \mathbf{F}$ and $\nabla \times (\nabla \times \mathbf{F})$ of the vector valued function

$$\mathbf{F} = (yx + yz)\,\mathbf{i} + (yz + xz)\,\mathbf{j} + (zx + yx)\,\mathbf{k}$$

(ii) If $\mathbf{F} = 2xz^2\,\mathbf{i} - yz\,\mathbf{j} + 3xz^3\,\mathbf{k}$ and $f = x^2yz$, find

(a) $\nabla \times \mathbf{F}$ (b) $\nabla \times (f\mathbf{F})$

(c) $\nabla \times (\nabla \times \mathbf{F})$ (d) $\nabla\,[\mathbf{F}.(\nabla \times \mathbf{F})]$

(e) $\nabla \times \nabla\,(f\mathbf{F})$ at (1, 1, 1).

(iii) If $\mathbf{F} = 3x\,\mathbf{i} + 4z\,\mathbf{j} = xy\,\mathbf{k}$,

$\mathbf{G} = yz^2\,\mathbf{i} - 3xz^2\,\mathbf{j} + 2xyz\,\mathbf{k}$ evaluate

(a) $(\nabla \times \mathbf{G}) \times \mathbf{F}$ (b) $\mathbf{F}.(\nabla \times \mathbf{G})$

5. Prove that

(i) $\nabla.\left(\frac{\mathbf{r}}{r^3}\right) = 0$ (ii) $\nabla \times \left(\frac{\mathbf{r}}{r^2}\right) = 0$

(iii) $\nabla.(\mathbf{F} \times \mathbf{r}) = 0$ If $\nabla \times \mathbf{F} = 0$

(iv) $\nabla.(\mathbf{C} \times \mathbf{r}) = \mathbf{r}.(\nabla \times C)$

6. (i) If $\phi_1 = x^2yz$, $\phi_2 = xy - 3z^2$, find

(a) $\nabla\,[\nabla\phi_1 . \nabla\phi_2]$ (b) $\nabla . [\nabla\phi_1 \times \nabla\phi_2]$

(c) $k\,\nabla \times [\nabla\phi \times \nabla\phi_2]$

(ii) If V_1 and V_2 be the vectors joining the fixed points (x_1, y_1, z_1) and (x_2, y_2, z_2) respectively to a variable point (x, y, z) prove that

(a) $\nabla . (\mathbf{V_1} \times \mathbf{V_2}) = 0$ (b) $\nabla \times (\mathbf{V_1} \times \mathbf{V_2}) = 2\,(\mathbf{V_1} - \mathbf{V_2})$

7. Prove that

(i) $\nabla\,(\mathbf{A}.\mathbf{B}) = (\mathbf{A}.\nabla)\,\mathbf{B} + (\mathbf{B}.\nabla)\,\mathbf{A} + \mathbf{A} \times (\nabla \times \mathbf{B}) + \mathbf{B} \times (\nabla \times \mathbf{A})$.

(ii) $\nabla\,(\mathbf{A}^2) = 2\,(\mathbf{A}.\nabla)\,\mathbf{A} + 2\mathbf{A} \times (\nabla \times \mathbf{A})$.

8. (i) Prove that if **A** is a constant vector and **F** is a vector point function then

$$\nabla(\mathbf{A}.\mathbf{F}) = (\mathbf{A}.\nabla)\,\mathbf{F} + \mathbf{A} \times (\nabla \times \mathbf{F})$$

(ii) Prove that $(\mathbf{A} \times \nabla) . \mathbf{B} = \mathbf{A}.(\nabla \times \mathbf{B})$

(iii) Prove that $\nabla . \left[\mathbf{r} \nabla \left(\frac{1}{r^3} \right) \right] = \frac{3}{r^4}$.

9. (i) Show that $\nabla \times [\mathbf{r} \times (\mathbf{A} \times \mathbf{r})] = 3\mathbf{r} \times \mathrm{A}$, where **A** being a constant vector.

(ii) Show that $\nabla.[\mathbf{A} \times (\mathbf{r} \times \mathbf{A})] = 2\mathbf{A}^2$

(ii) Show that $\nabla \times \left[\frac{\mathbf{A} \times \mathbf{r}}{r^3} \right] = -\frac{\mathbf{A}}{r^3} + \frac{3\mathbf{r}\,(\mathbf{A}.\mathbf{r})}{r^5}$

10. If $\nabla \times \mathbf{E} = -\frac{1}{C} \frac{\partial \mathbf{H}}{\partial t}$, $\nabla \times \partial \mathrm{H} = \frac{1}{C} \frac{\partial \mathbf{E}}{\partial t}$ show that **E** and **H** satisfy

$$\nabla^2 \mathbf{A} = \frac{1}{C^2} \frac{\partial^2 \mathbf{A}}{\partial t^2}.$$

11. (i) Prove that $\phi(x, y, z) = x^2 + y^2 + z^2$ is invariant under a rotation of axes.

(ii) If **F** (x, y, z) be a vector function invariant with respect to a rotation of axeis, then $\nabla \times \mathbf{F}$ is a vector invariant under this transformation.

ANSWERS

(1) (i) 4 (ii) −15 (iii) 1 (iv) 6

(4) (i) $\mathbf{i} + \mathbf{j} + \mathbf{k}$ (ii) (a) $\mathbf{i} + \mathbf{j}$ (b) $5\mathbf{i} - 3\mathbf{j} - 4\mathbf{k}$

(c) $5\mathbf{i} + 3\mathbf{k}$ (d) $-2\mathbf{i} + \mathbf{j} + 8\mathbf{k}$ (e) $\mathbf{0}$.

(iii) (a) $16\,z^3\,\mathbf{i} + 4\,xz\,(2\,xy - 3\,z)\,\mathbf{j} + 32\,xz^2\,\mathbf{k}$

(b) $4xz\,(6x + 4yz)$

(6) (i) (a) $z\,(2\,y^2 + 3\,x^2 - 12\,xy)\mathbf{i} + 2\,xz\,(2\,y - 3\,x)\,\mathbf{j} + x\,(2\,y^2 + x^2 - 6\,xy)\,\mathbf{k}$ (b) $\mathbf{0}$

(c) $xz\,(x - 24\,y)\,\mathbf{i} - 2\,xz\,(6\,x + y)\,\mathbf{j} + (2\,xy^2 + 12\,yz^2 + x^3)\,\mathbf{k}$

13

Vector Integration

13.1. Basic definitions

Integration is the inverse operation (or reverse process) of differentiation.

Let F (t) be a differential vector function of a scaler variable t such that

$$\frac{d}{dt}\mathbf{F}(t) = \mathbf{G}(t) \quad \text{then} \quad \int \mathbf{G}(t)\,dt = \mathbf{F}(t)$$

If C is any arbitrary constant vector

$$\int \mathbf{G}(t)\,dt = \mathbf{F}(t) + \mathbf{C}$$

$\left[\because \frac{d}{dt}(c) = 0\right]$ is called *indefinite integral* **F** (t)

Hence indefinite integral of **F** (t) is not unique.

If $\quad \frac{d}{dt}\mathbf{F}(t) = \mathbf{G}(t) \ \forall\ t \in [a, b]$

then $\int_a^b \mathbf{G}(t)\,dt = [\mathbf{F}(t)]_a^b = \mathbf{F}(b) - \mathbf{F}(a)$ which is called the *definite integral* of **F** (t).

This integral can also be defined as a limit of a sum in a manner analogous to that of ordinary integral calculus.

Note : If $\quad \mathbf{F}(t) = \mathbf{F}_1(t)\,\mathbf{i} + \mathbf{F}_2(t)\,\mathbf{j} + \mathbf{F}_3(t)\,\mathbf{k}$ then

$$\int \mathbf{F}(t)\,dt = \left[\int \mathbf{F}_1(t)\,dt\right]\mathbf{i} + \left[\int \mathbf{F}_2(t)\,dt\right]\mathbf{j} + \left[\int \mathbf{F}_3(t)\,dt\right]\mathbf{k}$$

and $\int_a^b \mathbf{F}(t)\,dt = \left[\int_a^b \mathbf{F}_1(t)\,dt\right]\mathbf{i} + \left[\int_a^b \mathbf{F}_2(t)\,dt\right]\mathbf{j} + \left[\int_a^b \mathbf{F}_3(t)\,dt\right]\mathbf{k}$

13.2. Standard results

From the important and standard results which we have obtained in differentiation, we can deduce the corresponding results in integration.

(i) We have

$$\frac{d}{dt}(\mathbf{F}\,.\,\mathrm{G}) = \frac{d\mathbf{F}}{dt}\,.\,\mathbf{G} + \mathrm{F}\,.\,\frac{d\mathrm{G}}{dt}$$

$$\therefore \qquad \int\left(\frac{d\mathrm{F}}{dt}\,.\,\mathbf{G} + \mathrm{F}\,.\,\frac{d\mathrm{G}}{dt}\right)dt = \mathbf{F}\,.\,\mathrm{G} + \mathrm{C}$$

(ii) We have $\frac{d}{dt}(\mathrm{F}^2) = 2\mathbf{F}\,.\,\frac{d\mathrm{F}}{dt}$

$$\therefore \qquad \int\left(2\mathbf{F}\,.\,\frac{d\mathbf{F}}{dt}\right)dt = \frac{d}{dt}(\mathbf{F}^2) + \mathrm{C}$$

(iii) We have

$$\frac{d}{dt}\left(\frac{d\mathrm{F}}{dt}\right)^2 = \frac{d\mathbf{F}}{dt}\,.\,\frac{d^2\mathbf{F}}{dt^2}$$

$$\therefore \qquad \int\left(2\,\frac{d\mathbf{F}}{dt}\,.\,\frac{d^2\mathbf{F}}{dt^2}\right)dt = \left(\frac{d\mathbf{F}}{dt}\right)^2 + \mathbf{C}$$

(iv) We have

$$\frac{d}{dt}(\mathbf{F} \times \mathbf{G}) = \frac{d\mathbf{F}}{dt} \times \mathbf{G} + \mathbf{F} \times \frac{d\mathbf{G}}{dt}$$

$$\therefore \qquad \int\left(\frac{d\mathbf{F}}{dt} \times \mathbf{G} + \mathbf{F} \times \frac{d\mathbf{G}}{dt}\right)dt = \mathbf{F} \times \mathbf{G} + \mathbf{C}$$

(v) If **A** is constant vector, we have

$$\frac{d}{dt}(\mathbf{A} \times \mathbf{F}) = \mathbf{A} \times \frac{d\mathbf{F}}{dt}$$

$$\therefore \qquad \int\left(\mathrm{A} \times \frac{d\mathbf{F}}{dt}\right)dt = \mathbf{A} \times \mathbf{F} + \mathbf{C}$$

(vi) We know that

$$\frac{d}{dt}\left(\mathbf{F} \times \frac{d\mathbf{F}}{dt}\right) = \mathbf{F} \times \frac{d^2\mathbf{F}}{dt^2}$$

$$\therefore \qquad \int\left(\mathbf{F} \times \frac{d^2\mathbf{F}}{dt^2}\right) dt = \mathbf{F} \times \frac{d\mathbf{F}}{dt} + \mathbf{C}$$

(vii) If $f = |\mathbf{F}|$, then

$$\frac{d}{dt}\left(\frac{\mathbf{F}}{f}\right) = \frac{1}{f}\frac{d\mathbf{F}}{dt} - \frac{1}{f^2}\frac{df}{dt}\mathbf{F}$$

$$\therefore \qquad \int\left(\frac{1}{f}\frac{d\mathbf{F}}{dt} - \frac{1}{f^2}\frac{df}{dt}\mathbf{F}\right) dt = \frac{\mathbf{F}}{f} + \mathbf{C}$$

(viii) If k is a scalar constant then.

$$\int k\mathbf{F}\, dt = k\int \mathbf{F}\, dt.$$

(ix) $\int (\mathbf{F} \pm \mathbf{G})\, dt = \int \mathbf{F}\, dt \pm \int \mathbf{G}\, dt$ where **F** and **G** are any two vector functions and C is any scalar constant.

Example 1. (i) If $\mathbf{F}(t) = t\mathbf{i} + (t^2 - 2t)\,\mathbf{j} + (3t^2 + 3t^3)\,\mathbf{k}$ find $\int_i^2 \mathbf{F}(t)\, dt.$

(ii) If $\mathbf{F}(t) = t\mathbf{i} - 3\mathbf{j} + 2t\mathbf{k}$, $\mathbf{G}(t) = \mathbf{i} - 2\mathbf{j} + 2\mathbf{k}$,

$\mathbf{H}(t) = 3\mathbf{i} + t\mathbf{j} - \mathbf{k}$ then find

(a) $\int_1^2 [\mathbf{FGH}]\, dt$ (b) $\int_0^2 [\mathbf{F} \times (\mathbf{G} \times \mathbf{H})]\, dt.$

Solution: (i) Given $\mathbf{F}(t) = t\mathbf{i} + (t^2 - 2t)\,\mathbf{j} + (3t^2 + 3t^3)\,\mathbf{k}$

$$\int_1^2 \mathbf{F}(t)\, dt = \left(\int_1^2 t\, dt\right)\mathbf{i} + \left(\int_1^2 (t^2 - 2t)\, dt\right)\mathbf{j} + \left(\int_1^2 (3t^2 + 3t^2)\, dt\right)\mathbf{k}$$

$$= \left[\frac{t^2}{2}\right]_1^2 \mathbf{i} + \left[\frac{t^3}{3} - t^2\right]_1^2 \mathbf{j} + \left[t^3 + \frac{3t^4}{4}\right]_1^2 \mathbf{k}$$

$$= \frac{3}{2}\mathbf{i} - \frac{2}{3}\mathbf{j} + \frac{73}{4}\mathbf{k}$$

(ii) Given $\mathbf{F}(t) = t\mathbf{i} - 3\mathbf{j} + 2t\mathbf{k}$, $\mathbf{G}(t) = \mathbf{i} - 2\mathbf{j} + 2\mathbf{k}$

and $\mathbf{H}(t) = 3\mathbf{i} + t\mathbf{j} - \mathbf{k}$

(a) $[\mathbf{FGH}] = \mathbf{F} \cdot (\mathbf{G} \times \mathbf{H})$

$$= \begin{vmatrix} t & -3 & 2t \\ 1 & -2 & 2 \\ 3 & t & -1 \end{vmatrix} = (14t - 21)$$

$$\int_1^2 [\mathbf{FGH}]\, dt = \int_1^2 (14t - 21)\, dt$$

$$= \left[7t^2 - 21t \right]_1^2 = 42 - 42 = 0$$

(b) $\mathbf{F} \times (\mathbf{G} \times \mathbf{H}) = \mathbf{G}(\mathbf{F} \cdot \mathbf{H}) - \mathbf{H}(\mathbf{F} \cdot \mathbf{G})$

$$\mathbf{F} \cdot \mathbf{H} = (t\mathbf{i} - 3\mathbf{j} + 2t\mathbf{k}) \cdot (3\mathbf{i} + t\mathbf{j} - \mathbf{k})$$

$$= 3t - 3t - 2t = -2t$$

$$\therefore \quad \mathbf{G}(\mathbf{F} \cdot \mathbf{H}) = -2t(\mathbf{i} - 2\mathbf{j} + 2\mathbf{k})$$

$$= -2t\mathbf{i} + 4t\mathbf{j} + 4t\mathbf{k}$$

$$\mathbf{F} \cdot \mathbf{G} = (t\mathbf{i} - 3\mathbf{j} + 2t\mathbf{k}) \cdot (\mathbf{i} - 2\mathbf{j} + 2\mathbf{k})$$

$$= t + 6 + 4t = 5t + 6$$

$$\mathbf{H}(\mathbf{F} \cdot \mathbf{G}) = (5t + 6)[3\mathbf{i} + t\mathbf{j} - \mathbf{k}]$$

$$= (15t + 18)\mathbf{i} + (5t^2 + 6t)\mathbf{j} - (5t + 6)\mathbf{k}$$

$$\therefore \quad \mathbf{F} \times (\mathbf{G} \times \mathbf{H}) = [-2t\mathbf{i} + 4t\mathbf{j} - 4t\mathbf{k}]$$

$$- [15t + 18)\mathbf{i} + (5t^2 + 6t)\mathbf{j} - (5t + 6)\mathbf{k}]$$

$$= -(17t + 18)\mathbf{i} + (5t^2 + 10t)\mathbf{j} + (t + 6)\mathbf{k}$$

Hence $\int_0^2 [\mathbf{F} \times (\mathbf{G} \times \mathbf{H})]\, dt$

$$= -\left[\int_0^2 (17t + 18)\, dt\right]\mathbf{i} - \left[\int_0^2 (5t^2 + 10t)\, dt\right]\mathbf{j} + \left[\int_0^2 (t - 6)\, dt\right]\mathbf{k}$$

$$= -\left[\frac{17t^2}{2} + 18t\right]_0^2 \mathbf{i} - \left[\frac{5t^3}{3} + \frac{10t^2}{2}\right]_0^2 \mathbf{j} + \left[\frac{t^2}{2} + 6t\right]_0^2 \mathbf{k}$$

$$= -70\mathbf{i} - \frac{100}{3}\mathbf{j} + 14\,\mathbf{k}$$

Example 2. If $\mathbf{F}(t) = 3t^2\mathbf{i} + t\mathbf{j} + 2\mathbf{k}$

$\mathbf{G}(t) = 6t^3\mathbf{i} + (t-1)\,\mathbf{j} + 3t\mathbf{k}$, then evaluate

(i) $\int_0^1 \left[\frac{d\mathbf{F}}{dt} \times \mathbf{G} + \mathbf{F} \times \frac{d\mathbf{G}}{dt}\right] dt$

(ii) $\int_0^1 \left[\frac{d\mathbf{F}}{dt}\mathbf{G} + \mathbf{F}\,.\,\frac{d\mathbf{G}}{dt}\right] dt$ (iii) $\int \left[\mathbf{F} \times \frac{d^2\mathbf{F}}{dt^2}\right] dt$

Solution: Given : $\mathbf{F}(t) = 3t^2\mathbf{i} + t\mathbf{j} + 2\mathbf{k}$

and $\mathbf{G}(t) = 6t^2\mathbf{i} + (t-1)\,\mathbf{j} + 3t\mathbf{k}$

(i) We Know that

$$\int_0^1 \left(\frac{d\mathbf{F}}{dt} \times \mathbf{G} + \mathbf{F} \times \frac{d\mathbf{G}}{dt}\right) dt = [\,\mathbf{F} \times \mathbf{G}]_0^1$$

$$\mathbf{F} \times \mathbf{G} = \begin{vmatrix} \mathbf{i} & \mathbf{j} & \mathbf{k} \\ 3t^2 & t & 2 \\ 6t^3 & (t-1) & 3t \end{vmatrix}$$

$$= 3t^2 - 2t + 2)\,\mathbf{i} + 3t^3\mathbf{j} + (3t^3 - 3t^2 - 6t^4)\,\mathbf{k}$$

$$[\mathbf{F} \times \mathbf{G}]_0^1 = [(3 - 2 + 2) - (2)]\,\mathbf{i} + [3 - 0]\,\mathbf{j} + (3 - 3 - 6) - 0]\mathbf{k}$$

$$= \mathbf{i} + 3\mathbf{j} - 6\mathbf{k}.$$

(ii) We know that

$$\int_0^1 \left[\frac{d\mathbf{F}}{dt}\,.\,\mathbf{G} + \mathbf{F}\,.\,\frac{d\mathbf{G}}{dt}\right] dt = [\,\mathbf{F}\,.\,\mathbf{G}]_0^1$$

$$\mathbf{F}\,.\,\mathbf{G} = (3t^2\mathbf{i} + t\mathbf{j} + 2\mathbf{k})\,.\,[6t^3\mathbf{i} + (t-1)\,\mathbf{j} + 3t\mathbf{k}]$$

$$= [18t^5 + (t^2 - t) + 6t]$$

$$= (18^t5 + t^2 + 5t)$$

$$[\,\mathbf{F}\,.\,\mathbf{G}]_0^1 = (18 + 1 + 5) = 24$$

(iii) We know that

$$\int \left(\mathbf{F} \times \frac{d^t\mathbf{F}}{dt^2}\right) dt = \mathbf{F} \times \frac{d\mathbf{F}}{dt} + \mathbf{C}$$

$$\frac{d\mathbf{F}}{dt} = 6t\,\mathbf{i} + \mathbf{j}$$

$$\mathbf{F} \times \frac{d\mathbf{F}}{dt} = \begin{vmatrix} \mathbf{i} & \mathbf{j} & \mathbf{k} \\ 3t^2 & t & 2 \\ 6t & t & 0 \end{vmatrix}$$

$$= -2\mathbf{i} + 12t\mathbf{j} - 3t^2\mathbf{k}$$

$$\therefore \int \left(\mathbf{F} \times \frac{d^2\mathbf{F}}{dt^2} \right) dt = -2\mathbf{i} + 12t\mathbf{j} - 3t^2\mathbf{k} + \mathbf{C}$$

Example 3. (i) If $\frac{d^2\mathbf{r}}{dt^2} = 6t\mathbf{i} - 24t^2\,\mathbf{j} + 4\sin t\,\mathbf{k}$

Find **r** given that $\mathbf{r} = 2\mathbf{i} + \mathbf{j}$ and $\frac{d\mathbf{r}}{dt} = -\mathbf{i} - 3\mathbf{k}$ at $t = 0$.

(ii) The acceleration of a particle at any time $t \geq 0$ is given by

$$\mathbf{a} = 12\cos 2t\,\mathbf{i} - 8\sin 2t\,\mathbf{j} + 16t\,\mathbf{k}$$

If the velocity **V** and displacement **r** are zero at t=0, find **V** and **r** at any time.

Solution : (i) Given $\frac{d^2\mathbf{r}}{dt^2} = 6t\,\mathbf{i} - 24t^2\,\mathbf{j} + 4\sin t\,\mathbf{k}$

Integrating on both sides w.r.t t, we get

$$\frac{d\mathbf{r}}{dt} = 3t^2\mathbf{i} - 8t^3\mathbf{j} + 4\cos t\mathbf{k} + \mathbf{C}_1 \qquad \text{...(1)}$$

But $\frac{d\mathbf{r}}{dt} = -\mathbf{i} - 3\mathbf{k}$ at $t = 0$. Hence From (1)

$$-\mathbf{i} - 3\mathbf{k} = 4\mathbf{k} + \mathbf{c}_1 \qquad \therefore \qquad \mathbf{C}_1 = -\mathbf{i} - 7\mathbf{k}$$

Substituting in (1), we get

$$\frac{d\mathbf{r}}{dt} = (3t^2 - 1)\,\mathbf{i} - 8t^3\,\mathbf{j} + (4\cos t - 7)\,\mathbf{k}$$

Now integrating it once again w.r.t t, we get

$$\mathbf{r} = (t^3 - t)\,\mathbf{i} - 2t^4\mathbf{j} - (4\sin t + 7)\,\mathbf{k} + \mathbf{C}_2 \qquad \text{...(2)}$$

But $\mathbf{r} = 2\mathbf{i} + \mathbf{j}$ at $t = 0$ Hence from (2)

$$2\mathbf{i} + \mathbf{j} = -7\mathbf{k} + \mathbf{C}_2 \quad \therefore \qquad \mathbf{C}_2 = 2\mathbf{i} + \mathbf{j} + 7\mathbf{k}$$

Substituting in (2), we get

$$\mathbf{r} = (t^3 - t + 2)\,\mathbf{i} - (2t^4 - 1)\,\mathbf{j} - (4 \sin t + 14)\,\mathbf{k}$$

(ii) Given $\mathbf{a} = \dfrac{d\mathbf{V}}{dt} = 12 \cos 2t\,\mathbf{i} - 8 \sin 2t\,\mathbf{j} + 16t\,\mathbf{k}$

Integrating on both sides, we get

$$\mathbf{V} = 6 \sin 2t\,\mathbf{i} + 4 \cos 2t\,\mathbf{j} + 8t^2\,\mathbf{k} + \mathbf{C_1} \qquad ...(1)$$

But $\quad \mathbf{V} = 0$ at $t = 0$

Hence from (1) $\therefore\ 0 = 4\mathbf{j} + \mathbf{C_1}$ or $\mathbf{C_1} = 4\mathbf{j}$,

Substituting in (1), we get

$$\therefore \quad \mathbf{V} = \frac{d\mathbf{r}}{dt} = 6 \sin 2t\,\mathbf{i} + 4 (\cos 2t - 1)\,\mathbf{j} + 8t^2\,\mathbf{k}$$

Integrating on both sides, once again, we get

$$\mathbf{r} = -3 \cos 2t\,\mathbf{i} + (2 \sin 2t - 4t)\,\mathbf{j} + \frac{8t^3}{3}\,\mathbf{k} + \mathbf{C_2} \quad ...(2)$$

But $\quad \mathbf{r} = 0$ at $t = 0$

Hence from (2),

$$\mathbf{0} = -3\mathbf{i} + \mathbf{C_2} \quad \text{or} \quad \mathbf{C_2} = 3\mathbf{i}$$

Substituting in (2), we get

$$\mathbf{r} = 3(1 - \cos 2t)\,\mathbf{i} + 2(\sin 2t - 2t)\,\mathbf{j} + \frac{8t^3}{3}\,\mathbf{k}$$

EXERCISE

1. (i) If $\mathbf{F}(t) = t(t-1)\,\mathbf{i} + 2t^3\mathbf{j} - 3\mathbf{k}$, find $\int_1^2 \mathbf{F}(t)\,dt$.

(ii) Evaluate $\int_0^1 (e^t\,\mathbf{i} + e^{-2t}\,\mathbf{j} + t\,\mathbf{k})\,dt$

(iii) If $\mathbf{r}(t) = 5t^2\,\mathbf{i} + t\,\mathbf{j} - t^3\,\mathbf{k}$, find $\int_1^2 \left(\mathbf{r} \times \dfrac{d^2\mathbf{r}}{dt^2} \right) dt$

(iv) Given $\mathbf{r}(t) = 2\mathbf{i} - \mathbf{j} + 2\mathbf{k}$, when $t = 2$

$= 4\mathbf{i} - 2\mathbf{j} + 3\mathbf{k}$ when $t = 3$,

show that $\quad \int_2^3 \left(\mathbf{r} \cdot \dfrac{d\mathbf{r}}{dt} \right) dt = 10.$

2. (i) Given $\frac{d\mathbf{F}}{dt} = 12 \cos 2t\, \mathbf{i} - 8 \sin 2t\, \mathbf{j} + 16t\, \mathbf{k}$, and $\mathbf{F}(0) = 0$, find F(t).

   (ii) The acceleration of a particle at time t is given by $\mathbf{F} = 8 \sin 2t\, \mathbf{i} - 18 \cos 3t\, \mathbf{j} + 6t\, \mathbf{k}$.

   If the velocity **v** and displacement **r** be initially zero find **v** and **r** at any time.

3. If $\mathbf{r} \times d\mathbf{r} = 0$, prove that $\frac{\mathbf{r}}{|\mathbf{r}|}$ is a constant vector, where $\mathbf{r} = x\mathbf{i} + y\mathbf{j} + z\mathbf{k}$.

4. Show that the necessary and sufficient condition that a vector F(t) be of

   (i) constant magnitude is $\mathbf{F} \cdot \frac{d\mathbf{F}}{dt} = 0$.

   (ii) fixed direction is $\mathbf{F} \times \frac{d\mathbf{F}}{dt} = 0$.

ANSWERS

1. (i) $-\frac{5}{6}\mathbf{i} + \frac{15}{2}\mathbf{j} - 3\mathbf{k}$ (ii) $(e-1)\mathbf{i} - \frac{1}{2e^2}(1-e^2)\,\mathbf{j} + \frac{1}{2}\mathbf{k}$

   (iii) $-14\mathbf{i} + 75\mathbf{j} - 15\mathbf{k}$

2. (i) $\mathbf{F}(t) = 6 \sin 2t\, \mathbf{i} + 4(\cos 2t - 1)\, \mathbf{j} + t^2\, \mathbf{k}$

   (ii) $\mathbf{V} = 4(1 - \cos t)\, \mathbf{i} - 6 \sin 3t\, \mathbf{j} + 3t^2\, \mathbf{k}$

   $\mathbf{r} = 2(2t - \sin 2t)\, \mathbf{i} + 2(\cos 3t - 1)\, \mathbf{j} + t^3\, \mathbf{k}$.

13.3. Definitions

(i) Oriented curve : Let C be a curve in space for which **P** be the initial point and **Q** be the terminal point.

Let us orient **C** by taking one of the two directions say **P** to **Q** along **C** as *possitive direction,* the opposite

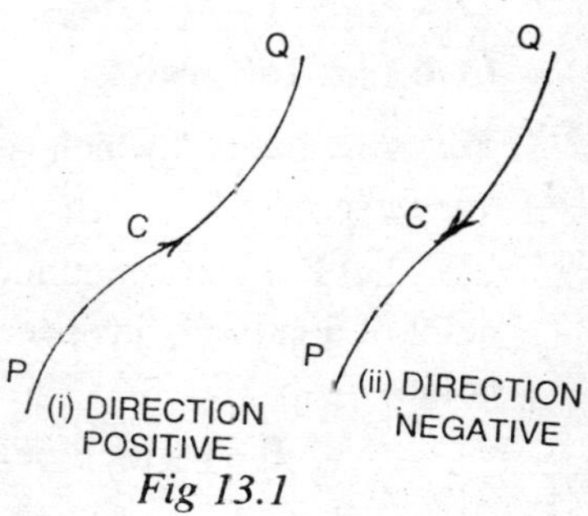

Fig 13.1

direction along **C** i.e. **Q** to **P** is then called *negative direction.*

(ii) Closed curve : If the initial point **P** and the terminal point **Q** of curve **C** under the chosen orientation coincide, the curve **C** is called a *closed curve.*

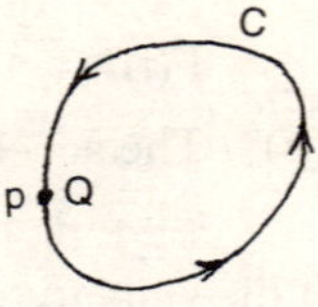

Fig 13.2

(iii) Smooth curve : A curve **r** = **F** (t) is called a *smooth curve* if **F** (t) is continuously differentiable.

i.e. The curve possesses a unique tangent at each of its points.

(iv) Piece wise smooth curve : A curve is said to be *piecewise smooth curve* if it consists of a finite number of smooth curves.

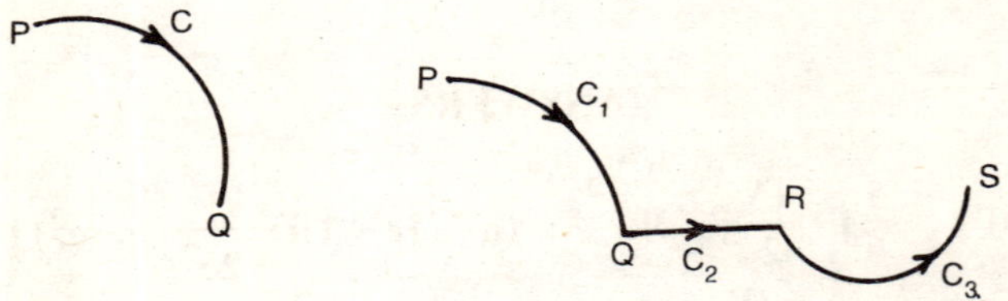

(i) Smooth curve (ii) Piecewise smooth curve

Fig 13.3

In above Figure 3.3 : (i) represents smooth curve and (ii) represents piece wise smooth curve consisting of smooth curves $\mathbf{C_1}$, $\mathbf{C_2}$, $\mathbf{C_3}$.

13.4. Line Integrals

Any integral which is to be evaluated along a curve is called a *line integral..*

Let **F** (**r**) be a continuous vector point function defined at every point of a curve **C** in space. Divide the curve **C** into n parts by the points.

$$\mathbf{P} = p_0, p_1, p_2 \,\ldots\ldots\ldots\ldots\, p_{i-1}, p_i, \,\ldots\ldots\ldots\, p_n = \mathbf{Q}$$

and let

$$\mathbf{r_0}, \mathbf{r_1}, \mathbf{r_2} \ldots \mathbf{r_{i-1}}, \mathbf{r_1}, \ldots \mathbf{r_n}$$

be the position vectors of these points. Let ξ_i be the position vector of any point on the arc $p_{i-1} p_i$.

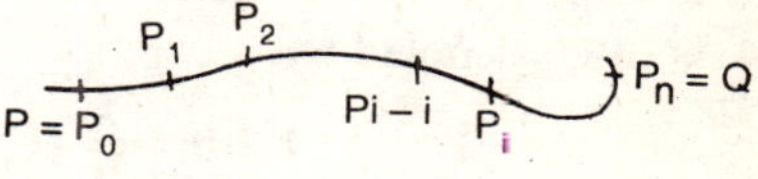

Fig 13.4

Then the limit of the sum

$$\sum_{i=1}^{n} \mathbf{F}(\xi_i) \,.\, \nabla \mathbf{r_i} \text{ where } \nabla \mathbf{r_i} = \mathbf{r_i} - \mathbf{r_{i-1}}$$

as $n \to \infty$ and every $|\nabla \mathbf{r_i}| \to 0$, if it exists, is called a *line integral* of **F** along **C** and is denoted by

$$\int_C \mathbf{F} \,.\, d\mathbf{r} \text{ or } \int_C \mathbf{F} \,.\, \frac{d\mathbf{r}}{dt}\, dt$$

The integral is also known as *tangential line integral* of **F** along **C**.

When the path of integration is a closed curve we write the integral $\int_C$ as $\oint_C$.

If $\mathbf{F(r)} = f_i(x, y, z)\,\mathbf{i} + f_2(x, y, z)\,\mathbf{j} + f_3(x, y, z)\,\mathbf{k}$

and $\mathbf{r} = x\mathbf{i} + y\mathbf{j} + z\mathbf{k}$ then $d\mathbf{r} = dx\mathbf{i} + dy\mathbf{j} + dz\mathbf{k}$

$$\therefore \quad \int_C \mathbf{F} \,.\, d\mathbf{r} = \int_C (f_i\, dx + f_2\, dy + f_3\, dz).$$

If the parametric equations of the curve **C** are $x = x(t)$, $y = y(t)$, $z = z(t)$ and $t = t_i$ at **P** and $t = t_2$ at Q, then

$$\int_C \mathbf{F} \,.\, d\mathbf{r} = \int_{t_1}^{t_2} \left(f_i \frac{dx}{dt} + f_2 \frac{dy}{dt} + f_3 \frac{dx}{dt} \right) dt$$

* The integral $\int_C \mathbf{F} \,.\, d\mathbf{r}$ is a scalar.

* The other type of line integrals are

$$\int_C \mathbf{F} \times d\mathbf{r} \text{ and } \int_C \phi\, d\mathbf{r} \text{ which are vectors}$$

13.5. Physical interpretations

(i) Circulation : If **C** is a simple closed curve then the tangent line integral of **F** around **C** is called the *Circulation* of **F** about **C**. It is often denoted by

$$\oint_C \mathbf{F} \cdot d\mathbf{r} = \oint_C (\mathbf{F_i} + dx + \mathbf{F_2}\, dy + \mathbf{F_3}\, dz)$$

When the circulation of **F** round every closed curve in a region **R*** vanishes then **F** is said to be *irrotational* in **R***.

(ii) Work done by a force: Let **F** represent the force acting on a particle moving along an are **PQ**. The work done during the small displacement **δr** is **F . δr**.

The total work done by **F** during displacement from **P** to **Q** is given by $\int_P^Q \mathbf{F} \cdot d\mathbf{r}$.

If the force **F** is conservative, then there exists a scalar function φ such that

$$\mathbf{F} = \nabla \phi = \frac{\partial \phi}{\partial x}\mathbf{i} + \frac{\partial \phi}{\partial y}\mathbf{j} + \frac{\partial \phi}{\partial z}\mathbf{k}$$

∴ The work done by **F** during displacement from **P** to **Q**

$$= \int_P^Q \mathbf{F} \cdot d\mathbf{r}$$

$$= \int_P^Q \left(\frac{\partial \phi}{\partial x}\mathbf{i} + \frac{\partial \phi}{\partial y}\mathbf{j} + \frac{\partial \phi}{\partial z}\mathbf{k} \right) \cdot (dx\mathbf{i} + dy\mathbf{j} + dz\mathbf{k})$$

$$= \int_P^Q \left(\frac{\partial \phi}{\partial x} dx + \frac{\partial \phi}{\partial y} dy + \frac{\partial \phi}{\partial z} dz \right) = \int_P^Q d\phi$$

$$= \phi(\mathbf{Q}) - \phi(\mathbf{P}).$$

Thus, in a conservative field, the work done depends on the value of φ at the end points **P** and **Q**, and not on the path joining **P** and **Q**.

Example 4. Evaluate $\int_C \mathbf{F} \cdot d\mathbf{r}$ where $\mathbf{F} = x^2y^2\mathbf{i} + y\mathbf{j}$ and the curve **C** in $y^2 = 4x$ in the xy-plane from (0, 0) to (4, 4).

Solution: Method 1 : In the xy - plane we have

$$\mathbf{r} = x\mathbf{i} + y\mathbf{j} \therefore d\mathbf{r} = dx\mathbf{i} + dy\mathbf{j}$$

Given $\mathbf{F} = x^2y^2\mathbf{i} + y\mathbf{j}$

$$\therefore \quad \mathbf{F} \,.\, d\mathbf{r} = (x^2y^2\,\mathbf{i} + y\mathbf{j}) \,.\, (dx\mathbf{i} + dy\mathbf{j})$$

$$= x^2y^2\,dx + y\,dy$$

$$\therefore \quad \int_C \mathbf{F}\,d\mathbf{r} = \int_C (x^2y^2dx + ydy)$$

Now along the curve $y^2 = 4x \;\therefore\; 2y\,dy = 4\,dx$

Case (i): When y dy=2dx

$$\text{Then } \int_C \mathbf{F} \,.\, d\mathbf{r} = \int_{x=0}^{4} [x^2\,(4x)\,dx + 2\,dx]$$

$$= \int_{x=0}^{4} (4x^3 +)\,dx$$

$$= (x^4 + 2x)_0^4 = (4)^4 + 2\,(4) = 264$$

Case (ii) : When $dx = \frac{1}{2}ydy$

Then $\int_C \mathbf{F} \,.\, d\mathbf{r}$

$$= \int_{y=0}^{4} \left[\left(\frac{v^2}{4} \right)^2 y^2 \left(\frac{1}{2}ydy \right) + ydy \right]$$

$$= \int_{y=0}^{4} \left[\frac{1}{32}y^7 + y \right] dy$$

$$= \left[\frac{y^8}{256} + \frac{y^2}{2} \right]_0^4$$

$$= 256 + 8 = 264$$

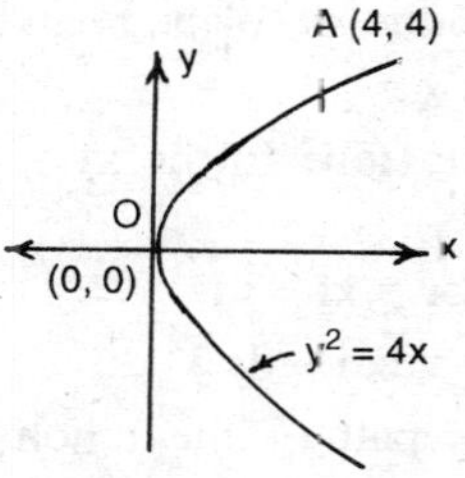

Fig. 13.5

Method 2: The curve **C** is a parabola $y^2 = 4x$ from (0, 0) to (4, 4).

Let $x = t^2$; then $y = 2t$

If **r** is the position vector of any point (x, y) on **C**, then

$$\mathbf{r}\,(t) = x\mathbf{i} + y\mathbf{j} = t^2\mathbf{i} + 2t\mathbf{j}$$

$$\therefore \quad \frac{d\mathbf{r}}{dt} = 2t\mathbf{i} + 2\mathbf{j}$$

Also in terms of t

$$\mathbf{F}\,(t) = (t^2)^2\,(2t)^2\,\mathbf{i} + 2t\mathbf{j} = 4\,t^6\,\mathbf{i}\;2t\,\mathbf{j}$$

At the point (0, 0) : t = 0 ; At the point (4, 4) : t = 2

$$\therefore \quad \int_C \mathbf{C}\,.\,d\mathbf{r} = \int_C \left[\mathbf{F}\,.\,\frac{d\mathbf{r}}{dt}\right] dt$$

$$= \int_{t=0}^{2} [4t^6\,\mathbf{i} + 2t\,\mathbf{j})\,.\,(2t\,\mathbf{i} + 2\,\mathbf{j})\,dt$$

$$= \int_{t=0}^{2} (8t^7 + 4t)\,dt = [t^8 + 2t^2]_0^2$$

$$= 2^8 + 2\,(2)^2 = 256 + 8 = 264.$$

Example 5. If $\mathbf{F} = x\,(x + y)\,\mathbf{i} + (x^2 + y^2)\,\mathbf{j}$, evaluate $\int_C \mathbf{F}\,.\,d\mathbf{r}$ where **C** is the square in the xy-plane bounded by the lines $x = \pm\,1, y = \pm\,1$.

Solution: In the xy - plane z = 0

$\therefore\;\; \mathbf{r} = x\mathbf{i} + y\mathbf{j}$

$d\mathbf{r} = dx\,\mathbf{i} + dy\,\mathbf{j}$

The path of integration **C** has been shown in the figure. it consists of the straight lines $\vec{AB}, \vec{BC}, \vec{CD}$ and $\vec{DA}$.

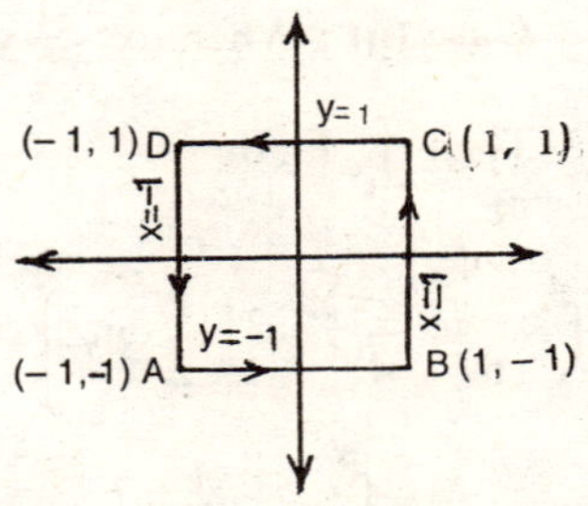

Fig. 13.6

We have

$$\int_C \mathbf{F}\,.\,d\mathbf{r} = \int_C [x\,(x + y)\,\mathbf{i} + (x^2 + y^2)\,\mathbf{j}]\,.\,[dx\mathbf{i} + dy\mathbf{j}]$$

$$= \int_C x\,(x + y)\,dx + (x^2 + y^2)\,dy$$

On $\vec{AB}$ $y = -\,1, dy = 0, -\,1 \le x \le 1$

On $\vec{BC}$ $x = 1, dx = 0, -\,1 \le y \le 1$

On $\vec{CD}$ $y = 1, dy = 0, 1 \le x \le -\,1$

On $\vec{DA}$ $x = -\,1, dx = 0, -\,1 \le y \le -.1$

$$\therefore \int_C \mathbf{F}\,.\,d\mathbf{r} = \int_{\overrightarrow{AB}} \mathbf{F}\,.\,d\mathbf{r} + \int_{\overrightarrow{BC}} \mathbf{F}\,.\,d\mathbf{r} + \int_{\overrightarrow{CD}} \mathbf{F}\,.\,d\mathbf{r} + \int_{\overrightarrow{DA}} \mathbf{F}\,.\,d\mathbf{r}$$

$$= \int_{-1}^{1} x\,(x-1)\,dx + \int_{-1}^{1} (1+y^2)\,dy$$

$$+ \int_{1}^{-1} x\,(x+1)\,dx + \int_{1}^{-1} (1+y^2)\,dy$$

$$= \frac{2}{3} + \frac{8}{3} - \frac{2}{3} - \frac{8}{3} = 0.$$

Example 6. A vector field is given by $\mathbf{F} = (\sin y)\mathbf{i} + x\,(1 + \cos y)\mathbf{j}$. Evaluate the line integral over the circular path by $x^2 + y^2 = a^2$, $z = 0$.

Solution: The parametric equations of the circular path are $x = a \cos t$, $y = a \sin t$, $z = 0$, where $0 \le t \le 2\pi$

$\because$ The particle moves in the xy-plane ($z = 0$)

$$\mathbf{r} = x\mathbf{i} + y\mathbf{j} \text{ so that } d\mathbf{r} = dx\,\mathbf{i} + dy\,\mathbf{j}$$

$$\therefore \oint_C \mathbf{F}\,.\,d\mathbf{r} = \oint_C [\sin y\mathbf{i} + x\,(1 + \cos y)\,\mathbf{j}]\,.\,[dx\mathbf{i} + dy\mathbf{j}]$$

$$= \oint_C [\sin y\, dx + x\,(1 + \cos y)\, dy]$$

$$= \oint_C [(\sin y\, dx + x \cos y\, dy) + x\, dy]$$

$$= \oint_C [d\,(x\sin y) + x\, dy]$$

$$= \oint_C d\,(x \sin y) + \oint_C x\, dy$$

$$= \int_0^{2\pi} d\,(x\sin y) + \int_0^{2\pi} x dy$$

$$= \int_0^{2\pi} d\,(a \cos t \sin (a \sin t)) + \int_0^{2\pi} a \cos t . a \cos t\, dt$$

$$= [\, a \cos t \sin (a \sin t)]_0^{2\pi} + a^2.2.2. \int_0^{\pi/2} \cos^2 t\, dt$$

$$= 0 + 4a^2 \left(\frac{1}{2} \cdot \frac{\pi}{2} \right) = \pi\, a^2$$

Example 7. If $\mathbf{F} = (3x^2 + 6y)\,\mathbf{i} - 14yz\mathbf{j} + 20xz^2\,\mathbf{k}$, evaluate the integral $\int \mathbf{F}\,.\,d\mathbf{r}$ along the straight line joining (0, 0, 0) and (1, 1, 1).

Solution: The equations of the straight line joining (0, 0, 0) and (1, 1, 1) are

$$\frac{x-0}{1-0} = \frac{y-0}{1-0} = \frac{z-0}{1-0} = t \text{ (say)}$$

Then along straight line **C** (say)

$$x = t, y = t, z = t$$

Also $\quad \mathbf{r} = x\mathbf{i} + y\mathbf{j} + z\mathbf{k} = t\mathbf{i} + t\mathbf{j} + t\mathbf{k}$

$\therefore \quad d\mathbf{r} = (\mathbf{i} + \mathbf{j} + \mathbf{k})\, dt$

Also along **C, F** = $(3t^2 + 6t)\,\mathbf{i} - 14t^2\mathbf{j} + 20t^3\mathbf{k}$.

At (0, 0, 0), t = 0 and at (1, 1, 1), t = 1

$$\therefore \quad \int_C \mathbf{F} \cdot d\mathbf{r} = \int_{t=0}^{1} [(3t^2 + 6t) - 14t^2 + 20t^3]\, dt = \frac{13}{3}$$

Example 8. (i) Evaluate $\int \mathbf{F} \cdot d\mathbf{r}$ where $\mathbf{F} = (x^2 + y^2)\,\mathbf{j} + xy\mathbf{i}$ and the curve **C** is the are of $y + 4 = x^2$ from (2, 0) to (4, 12).

(ii) Evaluate $\int (x\, dy - y\, dx)$ around the circle $x^2 + y^2 = 1$.

(iii) If $\mathbf{F} = (2x + y)\,\mathbf{i} + (3y - x)\,\mathbf{j}$, evaluate $\int_C \mathbf{F} \cdot d\mathbf{r}$ where **C** is the curve in the xy-plane consisting of the straight lones from (0, 0) to (2, 0) and then to (3, 2).

Solution: (i) We have

$$\int_C \mathbf{F} \cdot d\mathbf{r} = \int_C [xy\mathbf{i} + (x^2 + y^2)\,\mathbf{j}] \cdot [dx\mathbf{i} + dy\mathbf{j}]$$

$$= \int_C [xy\, dx + (x^2 + y^2)\, dy]$$

$$= \int_C xy\, dx + \int_C (x^2 + y^2)\, dy$$

Along $\quad$ **C**, $y = x^2 - 4$ and $x^2 = y + 4$

$$\int_C \mathbf{F} \cdot d\mathbf{r} = \int_{x=2}^{4} x\,(x^2 - 4)\, dx + \int_{y=0}^{12} (y + 4 + y^2)\, dy$$

$$= \left[\frac{x^4}{4} - 2x^2\right]_2^4 + \left[\frac{y^2}{2} + 4y + \frac{y^{13}}{3}\right]_0^{12}$$

$$= \left[\frac{(4)^4}{4} - 2(4)^2\right] - \left[\frac{(2)^4}{4} - 2(2)^2\right]$$

$$+ \left[\frac{(12)^2}{2} + 4(12) + \frac{(12)^3}{3}\right] = 732.$$

(ii) Let **C** denote the circle $x^2 + y^2 = 1$, $x = \cos t$, $y = \sin t$ are the parametric equations.

$$\oint_C (xdy - ydx) = \int_0^{2\pi} \left(x\frac{dy}{dt} - y\frac{dx}{dt} \right) dt$$

$$= \int_0^{2\pi} (\cos^2 t + \sin^2 t)\, dt = 2\pi$$

(iii) The path of integration **C** is from **O**(0, 0) to **P**(2, 0) and then **P**(2, 0) to **Q**(3, 2)

ie. It consists of straight lines **OP** and **PQ**.

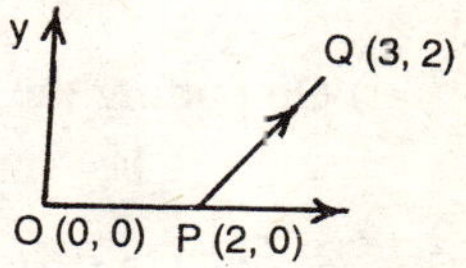

Fig. 13.7

$$\therefore \int_C \mathbf{F.dr} = \int_{\overline{OP}} \mathbf{F}\,.\,\mathbf{dr} + \int_{\overline{PQ}} \mathbf{F}\,.\,\mathbf{dr}$$

$$\mathbf{F} = (2x + y)\,\mathbf{i} + (3y - x)\,\mathbf{j}$$

$$\mathbf{r} = x\mathbf{i} + y\mathbf{j}$$

$$\mathbf{dr} = dx\mathbf{i} + dy\mathbf{j}$$

$$\int_C \mathbf{f}\,.\,\mathbf{dr} = \int_C [(2x + y)\,dx + (3y - x)\,dy]$$

Now along $\overline{OP}$ $y = 0$, $dy = 0$ and $0 \le x \le 2$

Equation of $\overline{PQ}$ is $(y - 0) = \left(\frac{2-0}{3-2}\right)(x - 2)$

ie. $y = 2x - 4.$

$\because$ Along $\overline{PQ}$ $y = 2x - 4$, $dy - 2\,dx$, $2 \le x \le 3$

$$\therefore \quad \int_C \mathbf{F}\,.\,\mathbf{dr} = \int_0^2 [(2x + 0)\,dx + 0]$$

$$+ \int_0^3 [(2x + 2x - 4)\,dx + (6x - 12 - x)\,2dx]$$

$$= \left[x^2 \right]_0^2 + \int_2^3 (14x - 28)\, dx$$

$$= 4 + 14 \int_2^3 (x - 2)\, dx$$

$$= 4 + 1 . \left[\frac{(x-2)^2}{2} \right]_2^3 = 4 + 7 = 11.$$

Example 9. If $\mathbf{f} = (2y + 3)\,\mathbf{i} + xz\,\mathbf{j} + (yz - x)\,\mathbf{k}$ evaluate $\int_C \mathbf{F} . \mathbf{dr}$ along the curve **C** given

(i) $x = 2t^2, y = t, z = t^3 = 0, t = 1.$

(ii) The straight lines from (0, 0, 0) to (0, 0, 1) then to (0, 1, 1) and then to (2, 1, 1).

Solution: Given $\mathbf{F} = (2y + 3)\,\mathbf{i} + xz\,\mathbf{j} + (yz - x)\,\mathbf{k}$

and $$\mathbf{r} = x\mathbf{i} + y\mathbf{j} + z\mathbf{k}$$

(i) On putting the values of x, y, z in terms of t

$$\mathbf{F} = (2t + 3)\,\mathbf{i} + 2t^5\,\mathbf{j} + (t^4 - 2t^2)\,\mathbf{k}$$

$$\mathbf{r} = 2t^2\,\mathbf{i} + t\mathbf{j} + t^3\,\mathbf{k}$$

$$\therefore \quad \mathbf{dr} = (4t\mathbf{i} + \mathbf{j} + 3t^2\mathbf{k})\, dt$$

$$\int_C \mathbf{F} . \mathbf{dr} = \int_0^1 [(2t + 3)(4t) + 2t^5 + (t^4 - 2t^2)(3t^2)]\, dt$$

$$= \int_0^1 [(12t + 8t^2 - 6t^4 + 2t^5 + 3t^6)]\, dt$$

$$= \left(12 + \frac{8}{3} - \frac{6}{5} + \frac{2}{6} + \frac{3}{7} \right) = \frac{288}{35}$$

(ii) Again

$$\int_{C_1} \mathbf{F} . \mathbf{dr} = \int_C [(2y + 3)\, dx + xz\, dy + (yz - x)\, dz] \qquad ...(1)$$

Along **C**, line joining (0, 0, 0) to (0, 0, 1) x = 0, dx = 0, dy = 0 and $0 \le z \le 1$.

$$\therefore \int_{C_1} \mathbf{F} . \mathbf{dr} = \int_0^1 (yz - x)\, dz = \int_0^1 (0)\, dz = 0 \quad (\because x = 0, y = 0)$$

Along C_2 the line joining (0, 0, 1) to (0, 1, 1), x = 0, z = 1, dx = 0, dz = 0, and $0 \leq x \leq 1$

$$\therefore \quad \int_{C_2} \mathbf{F} \cdot d\mathbf{r} = \int_0^1 xz\, dy = 0 \qquad (\because x = 0)$$

Along C_3 the line joining (0, 1, 1) to (2, 1, 1)

$$y = 1, z = 1, dy = 0, dz = 0 \text{ and } 0 \leq x \leq 2$$

$$\therefore \int_{C_3} \mathbf{F} \cdot d\mathbf{r} = \int_0^2 (2y + 3)\, dx = \int_0^2 5\, dx \quad (\because y=1)$$

$$= 5\,[\,x\,]_0^2 = 5\,(2) = 10.$$

$$\therefore \int_C \mathbf{F} \cdot d\mathbf{r} = \int_{C_1} \mathbf{F} \cdot d\mathbf{r} + \int_{C_2} \mathbf{F} \cdot d\mathbf{r} + \int_{C_3} \mathbf{F} \cdot d\mathbf{r}$$

$$= 0 + 0 + 10 = 10.$$

Example 10. Find the circulation of $\mathbf{F} = (2x - y + 2z)\,\mathbf{i} + (x + y - z)\,\mathbf{j} + (3x - 2y - 5z)\,\mathbf{k}$ along the circle $x^2 + y^2 = 4$, $z = 0$.

Solutions: Equation to the curve **C**

$$x^2 + y^2 = 4, z = 0 \;\therefore\; dz = 0$$

In parametric form x = cos 2t y = sin 2t z = 0.

The circulation of **F** along **C** is

$$\oint_C \mathbf{F} \cdot d\mathbf{r} = \oint_C (2x - y + 2z)\, dx + (x + y - z)\, dy + (3x - 2y - 5z)\, dz$$

$$= \oint_C (2x - y)\, dx + (x + y)\, dy$$

$$= \int_{t=0}^{2\pi} [(2 \cos 2t - \sin 2t)(-2 \sin 2t) + (\cos 2t + \sin 2t)(2 \cos 2t)]\, dt$$

$$= \int_0^{2\pi} (-\sin 4t + 2)\, dt = 4\pi$$

Example 11: Find the work done by the force $\mathbf{F} = x\mathbf{i} - z\mathbf{j} + 2y\mathbf{k}$ in the displacement along the closed path **C** consisting of the segments C_1, C_2 and C_3 where

On $C_1 : 0 \le x \le 1, y = x, z = 0$

On $C_2 : 0 \le z \le 1, x = 1, y = 1$

On $C_3 : 1 \ge x \ge 1, y = z = x$

Solution: Total work done $\oint_C \mathbf{F} \,.\, d\mathbf{r} = \mathbf{W}$

$$\mathbf{W} = \oint_C (x\mathbf{i} - z\mathbf{j} + 2y\mathbf{k}) \,.\, (dx\mathbf{i} + dy\mathbf{j} + dz\mathbf{k})$$

$$= \oint_C (xdx - zdy + 2ydz)$$

$$= \int_{C_1} (xdx - zdy + 2ydz) + \int_{C_2} (xdx - zdy + 2ydz) + \int_{C_3} (xdx - zdy + 2ydz)$$

$$= \mathbf{W}_1 + \mathbf{W}_2 + \mathbf{W}_3,$$

where $\mathbf{W}_1, \mathbf{W}_2, \mathbf{W}_3$ denote the work done in displacement along C_1, C_2 and C_3 respectively.

On $C_1 : 0 \le x \le 1, y = x, z = 0, dy = dx, dz = 0$

$$\therefore \quad \mathbf{W}_1 = \int_{C_1} xdx = \int_0^1 xdx = \frac{1}{2}$$

On $C_2 : 0 \le z \le 1, x = 1, y = 1, dx = 0, dy = 0$

$$\mathbf{W}_2 = \int_{C_2} 2dz = 2\int_0^1 dz = 2$$

On $C_3 : 1 \ge x \ge 0, y = z = x; dy = dz = dx$

$$\mathbf{W}_3 = \int_{C_3} xdx - xdx + 2xdx$$

$$= 2\int_1^0 xdx = -2$$

$$\therefore \mathbf{W} = \text{Total work done} = \mathbf{W}_1 + \mathbf{W}_2 + \mathbf{W}_3$$

$$= \frac{1}{2} + 2 - 2 = \frac{1}{2}.$$

Example 12. Show that the following vector function $\mathbf{F} = (x^2 - yz)\,\mathbf{i} + (y^2 - xz)\mathbf{j} + (z^2 - xy)\,\mathbf{k}$ is irrotational and find the corresponding scalar ϕ such that $\mathbf{F} = \nabla\,\phi$.

Solution:

$$\text{Curl } \mathbf{F} = \begin{vmatrix} \mathbf{i} & \mathbf{j} & \mathbf{k} \\ \dfrac{\partial}{\partial x} & \dfrac{\partial}{\partial y} & \dfrac{\partial}{\partial z} \\ x^2 - yz & y^2 - xz & z^2 - xy \end{vmatrix}$$

$$= (-x + x)\,\mathbf{i} - (-y + y)\,\mathbf{j} + (-z + z)\,\mathbf{k}$$

$$= \mathbf{0}$$

$\therefore$ $\nabla \times \mathbf{F} = 0$. Hence $\mathbf{F}$ is irrotational.

$\therefore$ We can write $\mathbf{F} = \nabla \phi$.

$$\mathbf{F} \cdot d\mathbf{r} = \nabla \phi \cdot d\mathbf{r} = d\phi$$

or

$$d\phi = \mathbf{F} \cdot d\mathbf{r}$$

$$= (x^2 - yz)\,dx + (y^2 - zx)\,dy + (z^2 - xy)\,dz$$

$$= \frac{1}{3}(3x^2\,dx + 3y^2\,dy + 3z^2\,dz)$$

$$- (yz\,dx + zx\,dy + xy\,dz)$$

$$= \frac{1}{3}\,d\,(x^3 + y^3 + z^3) - d\,(xyz)$$

$$\therefore \quad \phi = \frac{1}{3}(x^3 + y^3 + z^3) - xyz + C$$

Example 13. If $\mathbf{F} = 2y\mathbf{i} - z\mathbf{j} + x\mathbf{k}$, evaluate $\int_C \mathbf{F} \times d\mathbf{r}$ along the curve $x = \cos t$, $y = \sin t$, $z = 2\cos t$ from $t = 0$ to $t = \pi/2$.

Solution: $\mathbf{F} = 2y\mathbf{i} - z\mathbf{j} + x\mathbf{k}$

$$\mathbf{r} = x\mathbf{i} + y\mathbf{j} + z\mathbf{k} \;\therefore\; d\mathbf{r} = dx\mathbf{i} + dy\mathbf{j} + dz\mathbf{k}$$

$$\mathbf{F} \times d\mathbf{r} = \begin{vmatrix} \mathbf{i} & \mathbf{j} & \mathbf{k} \\ 2y & -z & x \\ dx & dy & dz \end{vmatrix}$$

$$= [-z\,dz - x\,dy]\mathbf{i} + [x\,dx - 2y\,dz]\mathbf{j} + [2y\,dy + z\,dx]\mathbf{k}$$

$$= [-2\cos t\,(-2\sin t)\,dt - \cos t\,(\cos t)\,dt]\,\mathbf{i}$$

$$+ [\cos t\,(-\sin t)\,dt - 2\sin t\,(-2\sin t)\,dt]\,\mathbf{j}$$

$$+ [2\sin t\,(\cos t)\,dt + 2\cos t\,(-\sin t)\,dt]\,\mathbf{k}$$

$$= [4\cos t\sin t - \cos^2 t]\,\mathbf{i} + [4\sin^2 t - \cos t\sin t]\,\mathbf{j}$$

$$\int_C \mathbf{F} \times d\mathbf{r} = \int_0^{\pi/2} [(4 \cos t \sin t - \cos^2 t)]\, \mathbf{i} + [(4 \sin^2 t - \cos t \sin t)]\, \mathbf{j}\, dt$$

$$= \left(4 \cdot \frac{1}{2} - \frac{1}{2} \cdot \frac{\pi}{2}\right)\mathbf{i} + \left(4 \cdot \frac{1}{2} \cdot \frac{\pi}{2} - \frac{1}{2}\right)\mathbf{j}$$

$$= \left(2 - \frac{\pi}{4}\right)\mathbf{i} + (\pi - \frac{1}{2})\,\mathbf{j}$$

Exmaple 14. If $\phi = x^2y\, z^3$ then evaluate $\int_C \phi\, d\mathbf{r}$ along the curve x = t, y = 2t, z = 3t from t = 0 to t = 1.

Solution: $\phi = x^2y\, z^3$

$$\mathbf{r} = x\mathbf{i} + y\mathbf{j} + z\mathbf{k} \quad \therefore \quad d\mathbf{r} = dx\mathbf{i} + dy\mathbf{j} + dz\mathbf{k}$$

$$\phi\, d\mathbf{r} = x^2\, yz^3\, (dx\mathbf{i} + dy\mathbf{j} + dz\mathbf{k})$$

$$= t^2 \,.\, 2t \,.\, (3t)^3\, [dt\, \mathbf{i} + 2dt\, \mathbf{j} + 3dt\, \mathbf{k}]$$

$$= 54t^6\, [\mathbf{i} + 2\mathbf{j} + 3\mathbf{k})\, dt]$$

$$\int_C \phi\, d\mathbf{r} = \int_0^1 [54t^6\, (\mathbf{i} + 2\mathbf{j} + 3\mathbf{k})\, dt]$$

$$= \left[54\, \frac{t^7}{7} (\mathbf{i} + 2\mathbf{j} + 3\mathbf{k})\right]_0^1 = \frac{54}{7} (\mathbf{i} + 2\mathbf{j} + 3\mathbf{k})$$

13.6. Definitions :

(i) Smooth surface : Suppose ϕ is a surface which has a unique normal at each of its points and the direction of this normal depends continuously on the points of ϕ. Then ϕ is called a *Smooth surface.*

e.g. The surface of sphere is a smooth surface.

(ii) Piecewise smooth surface : If a surface ϕ is not smooth but can be subdivided into a finite number of smooth surfaces, then it is called *Piecewise smooth surface.*

e.g. The surface of a cube is piecewise smooth.

(iii) Classification of regions :

Simply and multiply connected regions : A region **R** in which every closed curve can be contracted to a point without passing out of the region is called a *simply connected region.* Otherwise the region **R** is *multiply connected.*

e.g. (1) The region interior to a circle is a simply connected plane region.

(2) The region interior to a sphere is a simply-connected region in space.

(3) The region between two concentric circles lying in the same plane is a multiply connected plane region.

(4) The region between two infinitely long coaxial cylinders is a multiply-connected region in space.

13.7. Surface integral

Any integral which is to be evaluated over a surface is called a *surface integral.*

Let **F** (**r**) be a continuous vector point function-defined over the smooth surface $\mathbf{r} = \mathbf{f}(u, v)$

Let **S** be a region of the surface. Divide the region into n sub-regions of areas $\delta S_1, \delta S_2, \delta S_3 \ldots\ldots \delta S_i \ldots\ldots\ldots \delta S_n$. $\mathbf{P_i}$ be the point of S_i and N_i be the unit normal to δS_i at $\mathbf{P_i}$

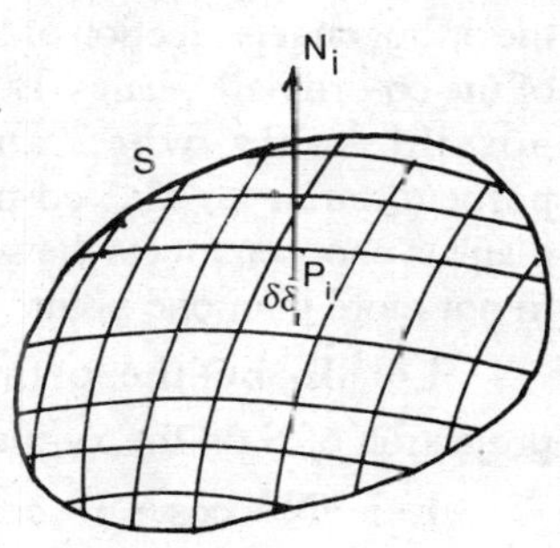

Fig. 13.8

Let $\delta \mathbf{A_i}$ be vector area of δS_i

then $$\delta \mathbf{A_i} = \mathbf{N_i}\, \delta S_i$$

Now the limit of the sum

$$\sum_{i=1}^{n} \mathbf{F}(\mathbf{r_i}) \cdot \mathbf{N_1}\, \delta S_i$$

or $$\sum_{i=1}^{n} \mathbf{F}(\mathbf{r_i}) \cdot \delta \mathbf{A_i}$$

Let $n \to \infty$ in such a way that each $\delta S_i \to 0$. if the limit exists is called *normal surface integral,* of **F** (**r**) over the region **S** of the **r** = f (u,v) and is denoted by

$$\int_C \mathbf{F}(\mathbf{r}) \cdot \mathbf{N}\, ds \ \text{ or } \int_C \mathbf{F}(\mathbf{r}) \cdot d\mathbf{A}$$

The other type of surface integrals are

$\int_C \mathbf{F} \times d\mathbf{A}$ or $\int_C \phi\, d\mathbf{A}$, where **F** is a continuous vector and ϕ a continuous scalar point function.

13.8 Flux

If **F** represents the velocity of a fluid at any point **P** on a closed surface **S**, then **F.N** is the normal component of **F** at **P** and $\int_S \mathbf{F} \cdot \mathbf{N} ds = \int_S \mathbf{F} \cdot d\mathbf{A}$ is a measure of volume emerging from **S** per unit time i.e., it measures the flux of **F** over **S**.

13.9. Cartesian form

In order to evaluate surface integrals, it is convenient to express them as double integrals taken over the orthogonal projection of **S** on one of the co-ordinate planes. But this is possible only when any line perpendicular to the co-ordinate plane is choosen meets the surface **S** in not more than one point.

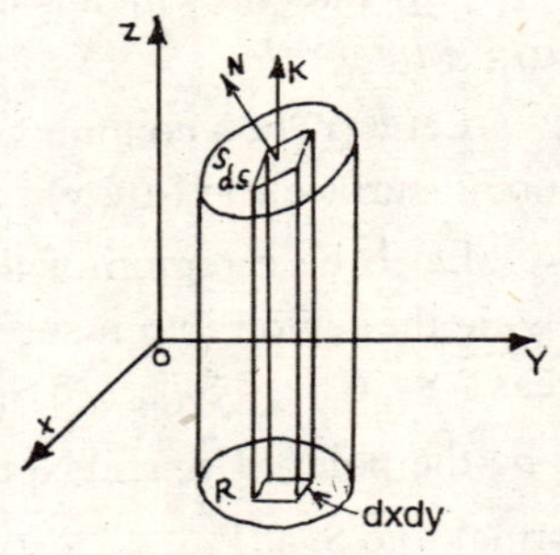

Fig 13.9

Let $\mathbf{R_i}$ be the orthogonal projection of **S** on the xy-plane.

Let $\mathbf{N} = \cos\alpha\, \mathbf{i} + \cos\beta\, \mathbf{j} + \cos\gamma\, \mathbf{k}$, where $\cos\alpha$, $\cos\beta$, $\cos\gamma$ are the direction cosines of **N**.

Now dx dy = Projection of ds on the xy plane

$$= ds \cos\gamma$$

$$\Rightarrow \qquad ds = \frac{dx\, dy}{\cos\gamma}$$

Also $|\mathbf{k.N}| = \cos\gamma$

$$\therefore \qquad ds = \frac{dx\, dy}{|\mathbf{k} \cdot \mathbf{N}|}$$

Hence $\displaystyle\int_S \mathbf{F} \cdot \mathbf{N}\, ds = \iint_{R^1} \mathbf{F} \cdot \mathbf{N} \frac{dx\, dy}{|\mathbf{k} \cdot \mathbf{N}|}$

Similarly $\int_S \mathbf{F}\,.\,\mathbf{N}\,ds = \iint_{R_2} \mathbf{F}\,.\,\mathbf{N}\,\dfrac{dx\,dy}{|\mathbf{N}\,.\,\mathbf{i}|}$

$$= \iint_{R^3} \mathbf{F}\,.\,\mathbf{N}\,\frac{dz\,dx}{|\mathbf{N}\,.\,\mathbf{j}|}$$

Example 15. Evaluate $\int_S \mathbf{F}\,.\,\mathbf{n}\,ds$, where

$\mathbf{F} = 18z\mathbf{i} - 12\mathbf{j} + 3y\mathbf{k}$ and **S** is the part of the plane $2x + 3y + 6z = 12$ located in the first octant.

Solution: Let $\phi = 2x + 3y + 6z - 12$

We know that $\nabla\phi$ is along the normal to the surface $\phi(x, y, z) = 0$

$$\therefore\ \nabla\phi = \nabla(2x + 3y + 6z - 12) = 2\mathbf{i} + 3\mathbf{j} + 6\mathbf{k}$$

N = Unit normal to the surface ϕ

$$= \frac{\nabla\phi}{|\nabla\phi|} = \frac{2\mathbf{i} + 3\mathbf{j} + 6\mathbf{k}}{\sqrt{4 + 9 + 36}} = \frac{1}{7}(2\mathbf{i} + 3\mathbf{j} + 6\mathbf{k}).$$

$$\mathbf{F}\,.\,\mathbf{N} = (18z\,\mathbf{i} - 12\,\mathbf{j} + 3y\,\mathbf{k})\,.\,\frac{1}{7}(2\mathbf{i} + 3\mathbf{j} + 6\mathbf{k})$$

$$= \frac{18}{7}(2z - 2 + y) = \frac{18}{7}[-2 + (2z + y)]$$

$$\mathbf{N}\,.\,\mathbf{k} = \frac{1}{7}(2\mathbf{i} + 3\mathbf{j} + 6\mathbf{k})\,.\,\mathbf{k} = \frac{6}{7}$$

Given surface S is $2x + 3y + 6z = 12$

or $$(y + 2z) = \frac{1}{3}(12 - 2x) = \frac{2}{3}(6 - x)$$

Now $\int_S \mathbf{F}\,.\,\mathbf{N}\,ds = \iint_R \mathbf{F}\,.\,\mathbf{N}\,\dfrac{dx\,dy}{|\mathbf{N}\,.\,\mathbf{k}|}$

$$= \iint_R \frac{18}{7}[-2 + (2z + y)]\,.\,\frac{dx\,dy}{\left(\frac{6}{7}\right)}$$

$$= 3\iint_R [-2 + (2z + y)]\,dx\,dy$$

$$= 3\iint_R \left[-2 + \frac{2(6-x)}{3}\right] dx\, dy$$

$$\left[\because\ 2z + y = \frac{2}{3}(6-x)\right]$$

$$= 2\iint_R (3-x)\, dx\, dy$$

$$= 2\int_0^6 (3-x)\left[\int_0^{\frac{2}{3}(6-x)} dy\right] dx$$

$$= 2\int_0^6 \frac{2}{3}(3-x)(6-x)\, dx$$

$$= \frac{4}{3}\int_0^6 (18 - 9x + x^2)\, dx$$

$$= \frac{4}{3}\left[18x - \frac{9x^2}{2} + \frac{x^3}{3}\right]_0^6 = 24$$

Note: We could also write

$$\int_S \mathbf{F} \cdot \mathbf{N}\, ds = \iint_R \mathbf{F} \cdot \mathbf{N} \frac{dy\, dz}{|\mathbf{N} \cdot \mathbf{i}|}$$

So that $0 \le z \le \frac{4-y}{2}$ and $0 \le y \le 4$.

Example 16. Evaluate $\int_S \mathbf{F} \cdot \mathbf{N}\, ds$, where $\mathbf{F} = z\mathbf{i} + x\mathbf{j} + 3y^2 z\mathbf{k}$ and **S** is the surface of the cylinder $x^2 + y^2 = 16$ included in the first octant between $z = 0$ and $z = 5$.

Solution: A vector normal to surface **S** given by

$$\nabla(x^2 + y^2) = 2x\mathbf{i} + 2y\mathbf{j}$$

N Unit normal to the surface

$$\mathbf{S} = \frac{2x\mathbf{i} + 2y\mathbf{j}}{\sqrt{4(x^2 + y^2)}}$$

$$= \frac{1}{4}(x\mathbf{i} + y\mathbf{j})$$

Let **R** be the projection of **S** on yz – plane

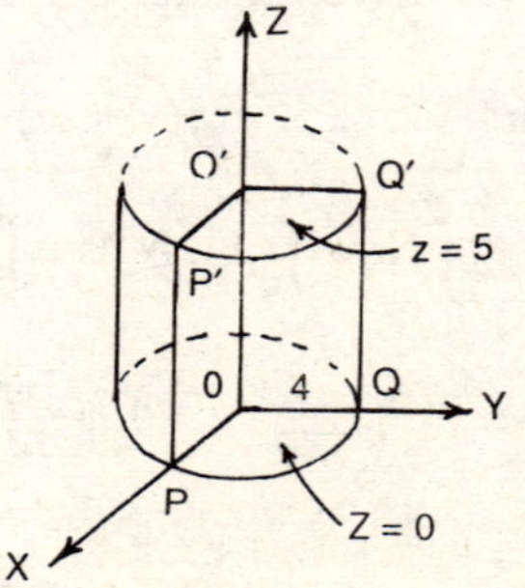

Fig 13.10

then $$\int_S \mathbf{F}\,.\,\mathbf{N}\,ds = \iint_R \mathbf{F}\,.\,\mathbf{N}\frac{dy\,dz}{|\mathbf{N}\,.\,\mathbf{i}|}$$

The Region **R** is 'O Q Q' enclosed by y = 0 to y = 4, and z = 0 to z = 5

Now $\mathbf{N}\,.\,\mathbf{i} = \frac{1}{4}(x\mathbf{i} + y\mathbf{j})\,.\,\mathbf{i} = \frac{1}{4}x.$

$$\mathbf{F}\,.\,\mathbf{N} = (z\mathbf{i} + x\mathbf{j} - 3y^2 z\mathbf{k})\,.\,\frac{1}{4}(x\mathbf{i} + y\mathbf{j}) = \frac{1}{4}x\,(y. + z)$$

Hence $\int_S \mathbf{F}\,.\,\mathbf{N}\,ds = \iint_R \mathbf{F}\,.\,\mathbf{N}\frac{dy\,dz}{|\mathbf{N}\,.\,\mathbf{i}|}$

$$= \int_0^5\int_0^4 \frac{1}{4}x\,(y + z)\,.\,\frac{dy\,dz}{\left(\frac{1}{4}x\right)} = \int_0^5\left[\frac{y^2}{2} + zy\right]_0^4 dz$$

$$= \int_0^5 (8 + 4z)\,dz = (8z + 2z^2)_0^5 = 40 + 50 = 90$$

Note: We could also write

$$\int_S \mathbf{F}\,.\,\mathbf{N}\,ds = \iint_R \mathbf{F}\,.\,\mathbf{N}\frac{dx\,dz}{|\mathbf{N}\,.\,\mathbf{j}|}$$

So that $0 \le x \le 4$

$0 \le z \le 5$

Example 17. Evaluate $\int_S \mathbf{F}\,.\,\mathbf{N}\,ds$ where

$$\mathbf{F} = (x + y^2)\,\mathbf{i} - 2x\mathbf{j} + 2yz\mathbf{k}$$

and **S** is the surface of the plane 2x + y +2z = 6 in the first octant.

Solution: A Vector normal to the surface **S** is given by

$$\nabla\,(2x + y + 2z) = 2\mathbf{i} + \mathbf{j} + 2\mathbf{k}$$

N Unit vector normal to surface $\mathbf{S} = \frac{2\mathbf{i} + \mathbf{j} + 2\mathbf{k}}{\sqrt{4 + 1 + 4}}$

$$= \frac{1}{3}(2\mathbf{i} + \mathbf{j} + 2\mathbf{k}).$$

$$\mathbf{N}\,.\,\mathbf{K} = \frac{2}{3}$$

$$\mathbf{F}\cdot\mathbf{n} = [(x+y^2)\,\mathbf{i} - 2x\mathbf{j} + 2y\,z\mathbf{k}]\cdot\left[\frac{1}{3}(2\mathbf{i}+\mathbf{j}+2\mathbf{k})\right]$$

$$= \frac{2}{3}(x+y^2) - \frac{2}{3}x + \frac{4}{3}yz$$

$$= \frac{2}{3}y^2 + \frac{4}{3}yz$$

$$= \frac{2}{3}y^2 + \frac{4}{3}y\left(\frac{6-2x-y}{2}\right)$$

$$\left[\because \text{On the plane } 2x+y+2z=6;\ z = \frac{6-2x-y}{2}\right]$$

$$= \frac{4}{3}y\,(3-x)$$

Hence $\int_S \mathbf{F}\cdot\mathbf{N}\,ds = \iint_R \mathbf{F}\cdot\mathbf{N}\,\frac{dx\,dy}{|\mathbf{N}\cdot\mathbf{k}|}$

$$= \iint_R \frac{4}{3}(3-x)\cdot\frac{3}{2}\,dx\,dy$$

$$= \int_0^3\int_0^{6-2x} 2y\,(3-x)\,dy\,dx$$

$$= \int_0^3 2\,(3-x)\left[\frac{y^2}{2}\right]_0^{6-2x} dx$$

$$= 4\int_0^3 (3-x)^3\,dx = 4\cdot\left[\frac{(3-x)^4}{4(-1)}\right]_0^3 = 81.$$

Example 18. Evaluate $\int_S \mathbf{F}\cdot\mathbf{N}\,ds$ where $\mathbf{F} = 4x\mathbf{i} - 2y^2\mathbf{j} + z^2\mathbf{k}$ and S is the surface bounding the region $x^2 + y^2 = 4$, $z = 0$, $z = 3$.

Solution: The surface S will consist of the base S_i in $z = 0$; the top S_3 in $z = 3$ and convex portion. For S_1; $z = 0$; $\mathbf{F} = 4x\mathbf{i} - 2y^2\mathbf{j}$.

A unit normal N is clearly **k** (outward drawn normal)

$$\mathbf{F}\cdot\mathbf{N} = (4x\mathbf{i} - 2y^2\mathbf{j})\cdot(-\mathbf{k}) = 0$$

$$\therefore \quad \int_{S_1} \mathbf{F}\cdot\mathbf{N}\,ds = 0$$

For $\mathbf{S_3}$, $z = 3$ $\mathbf{F} = 4x\mathbf{i} - 2y^2\mathbf{j} + 9\mathbf{k}$.

A unit normal $\mathbf{N}$ is $\mathbf{k}$

$\therefore$ $\mathbf{F.N} = (4x\mathbf{i} - 2y^2\mathbf{j} + 9\mathbf{k}).\mathbf{k} = 9$

$$\therefore \int_{S_3} \mathbf{F} \cdot \mathbf{N}\, ds = 9\int ds = 9\,(4\pi)$$

$$= 36\pi$$

[Because for $\mathbf{S_3}$ the surface area $2\pi r = 2\pi\,(2) = 4\pi$]

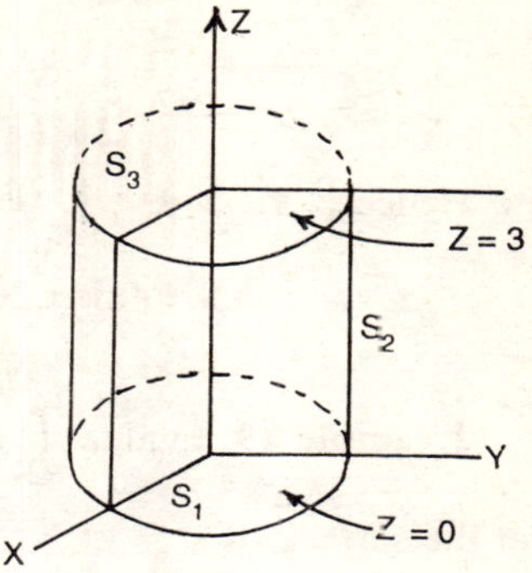

Fig 13.11

For $\mathbf{S_2}$ (convex portion)

$$\nabla \phi = 2x\mathbf{i} + 2y\mathbf{j}$$

$$\mathbf{N} = \frac{2x\mathbf{i} + 2y\,\mathbf{j}}{\sqrt{4\,(x^2 + y^2)}} = \frac{1}{2}(x\mathbf{i} + y\mathbf{j})$$

$$\mathbf{F} \cdot \mathbf{N} = (4x\mathbf{i} - 2y^2\mathbf{j} + z^2\mathbf{k}) \cdot \frac{1}{2}(x\mathbf{i} + y\mathbf{j})$$

$$= (2x^2 - y^3)$$

$$\text{Also} \int_S \mathbf{F} \cdot \mathbf{N}\, ds = \iint_R \mathbf{F} \cdot \mathbf{N} \frac{dy\,dz}{|\mathbf{N} \cdot \mathbf{i}|}$$

$$\mathbf{N} \cdot \mathbf{i} = \frac{1}{2}(x\mathbf{i} + y\mathbf{j}) \cdot \mathbf{i} = \frac{x}{2}$$

$$\therefore \int_{S_2} \mathbf{F} \cdot \mathbf{N}\, ds = \iint (2x^2 - y^3)\frac{2dy\,dz}{x}$$

Chose $x = 2\cos\theta$, $y = 2\sin\theta$

$$dy = 2\cos\theta\, d\theta$$

$$\therefore \int_{S_2} \mathbf{F} \cdot \mathbf{N}\, ds = \int_0^{2\pi}\int_0^3 \frac{8\cos^2\theta - 8\sin^2\theta}{2\cos\theta}(2\cos\theta\, c\,\theta)\, dz$$

$$= 16\int_0^{2\pi} (\cos^2\theta - \sin^3\theta)\, d\theta \cdot \int_0^3 dz$$

$$= 16\left\{2\int_0^{\pi} \cos^2\theta\, d\theta + 0\right\} \;(3)$$

$$= 96.2\int_0^{\pi/2} \cos^2\theta\, d\theta$$

$$= 192 \cdot \frac{1}{2} \cdot \frac{\pi}{2} = 48\pi.$$

Hence $\int_S \mathbf{F} \cdot \mathbf{N}\, ds = \int_{S_1} \mathbf{F} \cdot \mathbf{N}\, ds + \int_{S_2} \mathbf{F} \cdot \mathbf{N}\, ds + \int_{S_3} \mathbf{F} \cdot \mathbf{N}\, ds$

$$= 0 + 48\pi + 36\pi$$

$$= 84\pi.$$

Example 19. Evaluat $\int_S \mathbf{F} \cdot \mathbf{N}\, ds$, where $\mathbf{F} = yz\,\mathbf{i} + zx\,\mathbf{j} + xy\,\mathbf{k}$ over the surface of $x^2 + y^2 + z^2 = 1$, in the first octant.

Solution: A vector normal to the surface **S** is given by

$$\nabla (x^2 + y^2 + z^2) = 2\,(x\mathbf{i} + y\mathbf{j} + z\mathbf{k}).$$

N = Unit normal at (x, y, z) of **S**

$$= \frac{2\,(x\mathbf{i} + y\mathbf{j} + z\mathbf{k})}{2\sqrt{x^2 + y^2 + z^2}} = x\mathbf{i} + y\mathbf{j} + z\mathbf{k}$$

$$(\because x^2 + y^2 + z^2 = 1 \text{ on } \mathbf{S})$$

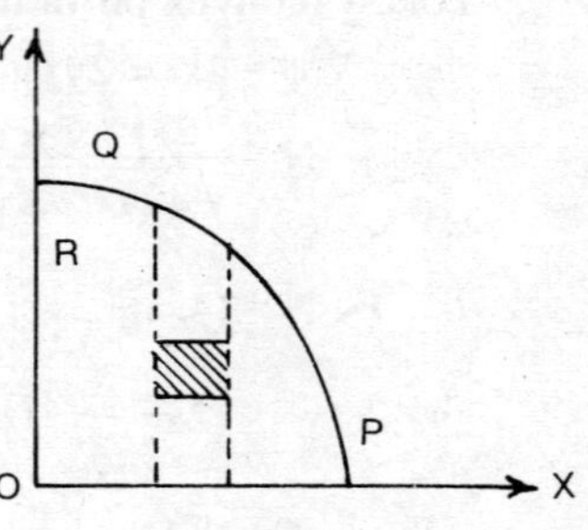

Fig 13.12

Now $\int_S \mathbf{F} \cdot \mathbf{N}\, ds$

$$= \iint_R \mathbf{F} \cdot \mathbf{N} \frac{dx\, dy}{|\mathbf{N} \cdot \mathbf{k}|},$$

where **R** is the projection of **S** on xy-plane.

$$\mathbf{F} \cdot \mathbf{N} = (yz\mathbf{i} + zx\mathbf{j} + xy\mathbf{k}) \cdot (x\mathbf{i} + y\mathbf{j} + z\mathbf{k}) = 3xyz$$

$$\mathbf{N} \cdot \mathbf{k} = (x\mathbf{i} + y\mathbf{j} + z\mathbf{k}) \cdot \mathbf{k} = z.$$

$$\therefore \quad \int_S \mathbf{F} \cdot \mathbf{N}\, ds = \iint_R 3xyz \frac{dx\, dy}{z}$$

$$= \int_0^1 \int_0^{\sqrt{1 - x^2}} xy\, dx\, dy$$

$$= 3\int_0^1 x \left(\frac{y^2}{2}\right)_0^{\sqrt{1 - x^2}} dx$$

$$= \frac{3}{2}\int_0^1 x\,(1 - x^2)\, dx$$

$$= \frac{3}{2}\left[\frac{x^2}{2} - \frac{x^4}{4}\right]_0^1 = \frac{3}{2}\left(\frac{1}{2} - \frac{1}{4}\right) = \frac{3}{8}.$$

Alternative method.

We have $\int_S \mathbf{F} \cdot \mathbf{N}\, ds = 3 \iint_R xy\, dx\, dy$

$$x = r\cos\theta,\ y = r\sin\theta$$

So that $dx\, dy = r\, dr\, d\theta$ and $0 \le r \le 1$; $0 \le \theta \le \pi/2$

$$\therefore \int_S \mathbf{F} \cdot \mathbf{N}\, ds = 3\int_0^1 \int_0^{\pi/2} (r\cos\theta)(r\sin\theta)\, r\, dr\, d\theta$$

$$= 2\left(\int_0^1 r^3\, dr\right)\left(\int_0^{\pi/2} \sin\theta\cos\theta\, d\theta\right)$$

$$= 3\left(\frac{r^4}{4}\right)_0^1 \left(\frac{1}{2}\right) = \frac{3}{8}$$

Example 20. Evaluat $\int_S \mathbf{F} \cdot \mathbf{N}\, ds$ where $\mathbf{F} = 4xz\,\mathbf{i} - y^2\,\mathbf{j} + yz\,\mathbf{k}$ where **S** is the surface of the cube bounded by $x = 0$, $x = 1$, $y = 0$, $y = 1$, $z = 0$, $z = 1$.

Solution: Since the cube has six faces, here we shall calculate $\int_S \mathbf{F} \cdot \mathbf{N}\, ds$ for all the six faces and then add the result.

1. **Face PQRS:**

A unit normal $\mathbf{N} = \mathbf{i}$ and $x = 1$

$\therefore\ \mathbf{F} = 4z\mathbf{i} - y^2\mathbf{j} + yz\mathbf{k}$ $(\because x = 1)$

$\mathbf{F}.\mathbf{N} = (4z\,\mathbf{i} - y^2\,\mathbf{j} + yz\,\mathbf{k}) \cdot \mathbf{i} = 4z$

$$ds = \frac{dy\, dz}{|\mathbf{N} \cdot \mathbf{i}|}\ \frac{dt\, dz}{|\mathbf{i} \cdot \mathbf{i}|} = dy\, dz$$

$$\therefore \int_{S_1} \mathbf{F} \cdot \mathbf{N}\, ds = \int_0^1 \int_0^1 4z\, dy\, dz$$

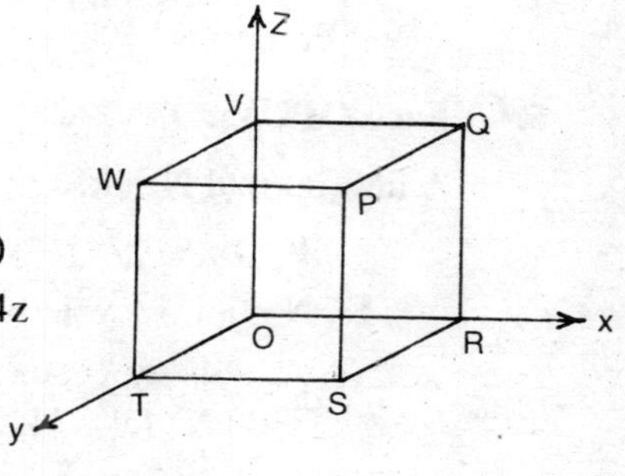

Fig 13.13

$$= \int_0^1 (2z^2)_0^1 \, dy = \int_0^1 2dy = 2$$

2. Face OTUV :

A unit normal $\mathbf{N} = -\mathbf{i}$, $x = 0$.

$\therefore \quad \mathbf{F} = -y^2\mathbf{j} + yz\mathbf{k} \quad (\because x = 0)$

$\mathbf{F} \cdot \mathbf{N} = (-y^2\mathbf{j} + yz\mathbf{k}) \cdot (-\mathbf{i}) = 0$

$\therefore \quad \int_{S_2} \mathbf{F} \cdot \mathbf{N} \, ds = 0$

3. Face PUTS :

A unit normal $\mathbf{N} = \mathbf{j}$, $y = 1$

$\therefore \quad \mathbf{F} = 4zx\,\mathbf{i} - \mathbf{j} + z\,\mathbf{k} \; (\because y = 1)$

$\mathbf{F} \cdot \mathbf{N} = (4xz\,\mathbf{i} - \mathbf{j} + z\,\mathbf{k}) \cdot \mathbf{j} = -1$

$$ds = \frac{dz\,dx}{|\mathbf{N} \cdot \mathbf{j}|} = \frac{dz\,dx}{|\mathbf{j} \cdot \mathbf{j}|} = dz\,dx$$

$$\therefore \quad \int_{S_3} \mathbf{F} \cdot \mathbf{N} \, ds = \int_0^1 \int_0^1 (-1) \, dz\,dx = -1$$

4. Face ORQV:

A unit normal $\mathbf{N} = -\mathbf{j}$, $y = 0$

$\therefore \quad \mathbf{F} = 4xz\,\mathbf{i} + z\,\mathbf{k}$

$\mathbf{F} \cdot \mathbf{N} = (4xz\,\mathbf{i} + z\,\mathbf{k}) \cdot (-\mathbf{j}) = 0$

$\therefore \quad \int_{S_4} \mathbf{F} \cdot \mathbf{N} \, ds = 0$

5. Face QPUV :

A unit normal $\mathbf{N} = \mathbf{k}$, $z = 1$

$\therefore \quad \mathbf{F} = 4x\,\mathbf{i} - y^2\,\mathbf{j} + y\,\mathbf{k} \quad (\because z = 1)$

$\mathbf{F} \cdot \mathbf{N} = j(4x\mathbf{i} - y^2\mathbf{j} + y\mathbf{k}) = y$

$$ds = \frac{dx\,dy}{|\mathbf{N} \cdot \mathbf{k}|} = \frac{dx\,dy}{|\mathbf{k} \cdot \mathbf{k}|} = dx\,dy$$

$$\therefore \quad \int_{S_5} \mathbf{F} \cdot \mathbf{N} \, ds = \int_0^1 \int_0^1 y \, dx\,dy = \int_0^1 \left(\frac{y^2}{2} \right)_0^1 dx$$

$$= \frac{1}{2}\int_0^1 dx = \frac{1}{2}$$

6. Face ORST :

A unit normal $\mathbf{N} = -\mathbf{k}$, $z = 0$

$$\therefore \quad \mathbf{F} = -y^2\mathbf{j} \quad (\because \ z = 0)$$

$$\mathbf{F} \cdot \mathbf{N} = (-y^2\mathbf{j}) \cdot (-\mathbf{k}) = 0$$

$$\therefore \quad \int_{S_6} \mathbf{F} \cdot \mathbf{N}\, ds = 0$$

$$\therefore \quad \int_S \mathbf{F} \cdot \mathbf{N}\, ds = \int_{S_1} \mathbf{F} \cdot \mathbf{N}\, ds + \int_{S_2} \mathbf{F} \cdot \mathbf{N}\, ds + \int_{S_3} \mathbf{F} \cdot \mathbf{N}\, ds + \int_{S_4} \mathbf{F} \cdot \mathbf{N}\, ds + \int_{S_5} \mathbf{F} \cdot \mathbf{N}\, ds + \int_{S_6} \mathbf{F} \cdot \mathbf{N}\, ds$$

$$= 2 + 0 + (-1) + 0 + \frac{1}{2} + 0 = 3/2$$

Example 21. If $\mathbf{F} = y\mathbf{i} + x(1-2z)\mathbf{j} - xy\mathbf{k}$ evaluate, $\int_S (\nabla \times \mathbf{F}).\mathbf{N}\, ds$ where S is the surface of the sphere $x^2 + y^2 + z^2 = 1$ above the xy plane.

Solution: $\nabla \times \mathbf{F} = \begin{vmatrix} \mathbf{i} & \mathbf{j} & \mathbf{k} \\ \dfrac{\partial}{\partial x} & \dfrac{\partial}{\partial v} & \dfrac{\partial}{\partial z} \\ y & x(1-2z) & -xy \end{vmatrix}$

$$= [-x + 2x]\,\mathbf{i} + [0 - (-y)]\,\mathbf{j} + [1 - 2z + 1]\,\mathbf{k}$$

$$= x\mathbf{i} + y\mathbf{j} - 2z\mathbf{k}$$

A normal to the surface $= \nabla (x^2 + y^2 + z^2)$

$$= 2x\,\mathbf{i} + 2y\,\mathbf{j} + 2z\,\mathbf{k}$$

Unit normal vector $= \mathbf{N} = \dfrac{2x\,\mathbf{i} + 2y\,\mathbf{j} + 2z\,\mathbf{k}}{\sqrt{4(x^2 + y^2 + z^2)}}$

$$= x\mathbf{i} + y\mathbf{j} + z\mathbf{k}$$

$$(\nabla \times \mathbf{F}).\ \mathbf{N} = (x\,\mathbf{i} + y\,\mathbf{j} - 2z\,\mathbf{k}) \cdot (x\,\mathbf{i} + y\,\mathbf{j} + z\,\mathbf{k})$$

$$= x^2 + y^2 - xz^2 = x^2 + y^2 - 2(1 - x^2 - y^2)$$

$$= 3(x^2 + y^2) - 2$$

Now $\int_S (\nabla \times \mathbf{F}) . \mathbf{N}\, ds = \iint_R (\nabla \times \mathbf{F}) . \mathbf{N} \frac{dx\, dy}{|\mathbf{N} . \mathbf{k}|}$

Where **R** is the projection of **S** on xy-plane.

$$\mathbf{N} . \mathbf{k} = z = \sqrt{1 - x^2 - y^2}$$

$$\therefore \quad \int_S (\nabla \times \mathbf{F}) . \mathbf{N}\, ds = \iint_R [3(x^2 + y^2) - 2] \frac{dx\, dy}{\sqrt{1 - x^2 - y^2}} \qquad ...(1)$$

Now **R** is the projection of $x^2 + y^2 + z^2 = 1$ in the xy-plane and is given by $x^2 + y^2 = 1$.

Put $x = r \cos \theta$, $y = r \sin \theta$ in (1)

$\therefore$ $dx\, dy = r\, dr\, d\theta$

So that $0 \le \theta \le 2\pi$ and $0 \le r \le 1$

$$\therefore \quad \int_S (\nabla \times \mathbf{F}) . \mathbf{N}\, ds = \int_0^{2\pi} \int_0^1 (3r^2 - 2) . \frac{r\, dr\, d\theta}{\sqrt{1 - r^2}}$$

$$= 2\pi \int_0^1 \frac{(3r^2 - 2)}{\sqrt{1 - r^2}} r dr$$

Put $1 = r^2 = t \Rightarrow -2rdr = 2tdt$

or $\qquad rdr = tdt$

$$= 2\pi \int_1^0 \frac{1}{t} [3(1 - t^2) - 2] . (-tdt)$$

$$= 2\pi \int_0^1 (1 - 3t^2)\, dt$$

$$= 2\pi \left[t - t^3 \right]_0^1 = 2\pi (1 - 1) = 0.$$

Example 22. If $\phi = xyz$ evaluate $\int_S \phi\, \mathbf{N}\, ds$ where **S** is the surface of cylinder $x^2 + y^2 = 4$ included in the first octant between $z = 0$ and $z = 3$.

Solution: A normal to $x^2 + y^2 = 4$ is along

$$\nabla (x^2 + y^2) = 2x\, \mathbf{i} + 2y\, \mathbf{j}$$

$$\mathbf{N} = \text{Unit normal vector} = \frac{2x\,\mathbf{i} + 2y\,\mathbf{j}}{\sqrt{4(x^2+y^2)}}$$

$$= \frac{1}{2}(x\mathbf{i} + y\mathbf{j})$$

Also $\quad dS = \dfrac{dx\,dz}{|\mathbf{N}\cdot\mathbf{j}|}$

$$\mathbf{N}\cdot\mathbf{j} = \frac{1}{2}(x\mathbf{i} + y\mathbf{j})\cdot\mathbf{j} = \frac{y}{2}$$

$$\int \phi\,\mathbf{N}\,ds = \int_{z=0}^{3}\int_{x=0}^{2} xyz\cdot\frac{1}{2}(x\mathbf{i} + y\mathbf{j})\cdot\frac{2dx\,dz}{y}$$

where $y = \sqrt{4 - x^2}$

$$= \int_{z=0}^{3}\int_{x=0}^{2} (x\mathbf{i} + \sqrt{4 - x^2}\mathbf{j})\,dx\,dz$$

$$= \int_{x=0}^{2} x^2\mathbf{i} + x\sqrt{4 - x^2}\mathbf{j})\,dx\,.\int_{z=0}^{3} z\,dz$$

$$= \left[\frac{x^3}{3}\mathbf{i} - \frac{1}{2}\cdot\frac{2}{3}(4 - x^2)^{3\backslash 2}\mathbf{j}\right]_0^2\left[\frac{z^2}{2}\right]_0^3$$

$$= \frac{9}{2}\left[\frac{8}{3}\mathbf{i} + \frac{8}{3}\mathbf{i}\right] = 12(\mathbf{i} + \mathbf{j})$$

13.10. Volume integrals

Any integral which is to be evaluated over a volume is called a volume integral.

Let **V** be a volume bounded by a surface $\mathbf{r} = f(u,v)$. Let $\mathbf{F}(\mathbf{r})$ be continuous vector point function defined over **V**. Divide **V** into n sub-regions of volumes $\delta v_1, \delta v_2, \delta v_3$, ... δv_i .... δv_n let $\mathbf{P_i}(\mathbf{r_i})$ be a point in δv_i.

Consider $\quad \sum_{1=1}^{n} \mathbf{F}(\mathbf{r_i})\,\delta v_i$

The limit of this sum as $n \to \infty$ in such a way that $\delta v_1 \to 0$, if it exists, is called the volume integral and is written as

$$\int_V \mathbf{F}(\mathbf{r})\, dv \text{ or } \int_V \mathbf{F}\, dv \text{ which mean } \iiint_V \mathbf{F}\, dv$$

Cartesian form :

Let $\mathbf{F}(\mathbf{r}) = \mathbf{F_1 i} + \mathbf{F_2 j} + \mathbf{F_3 k}$

where $\mathbf{F_1} = \mathbf{F_1}\ (x, y, z)$

$\mathbf{F_2} = \mathbf{F_2}\ (x, y, z)$

$\mathbf{F_3} = \mathbf{F_3}\ (x, y, z)$

Also we have $d\mathbf{V} = dx\ dy\ dz$

$$\therefore \quad \int_V \mathbf{F}\, d\mathbf{V} = \int_V (\mathbf{F_i i} + \mathbf{F_2 j} + \mathbf{F_3 k})\, dx\ dy\ dz.$$

Example 23. Evaluate $\int_V$ when $\mathbf{f} = x\mathbf{i} + y\mathbf{j} + z\mathbf{k}$ and $\mathbf{V}$ is the region bounded by $x = 0$, $y = 0$, $y = 6$, $z = 4$ and $z = x^2$.

Solution: Limits of integration $0 \le x \le \sqrt{z}$, $0 \le y \le 6$, $0 \le z \le 4$.

$$\int_V \mathbf{F} \,.\, dv = \left\{\int_V x\ dx\ dy\ dz\right\} \mathbf{i} + \left\{\int_V y\ dx\ dy\ dz\right\} \mathbf{j} + \left\{\int_V x\ dx\ dy\ dz\right\} \mathbf{k}$$

$$= \mathbf{I_1 i} + \mathbf{I_2 j} + \mathbf{I_3 k} \quad \text{say}$$

$$\mathbf{I_1} = \int_V x\ dx\ dy\ dz$$

$$= \int_{z=0}^{4} \int_{x=0}^{6} \int_{x=0}^{\sqrt{z}} x\ dx\ dy\ dz$$

$$= \frac{1}{2} \int_{z=0}^{4} \int_{y=0}^{6} \left\{\frac{x^2}{2}\right\}_0^{\sqrt{z}} dy\ dz$$

$$= \frac{1}{2} \int_{z=0}^{4} \int_{y=0}^{6} z\ dy\ dz$$

$$= \frac{1}{2} \int_{z=0}^{4} z\ dz \int_{y=0}^{6} dy$$

$$= \frac{1}{2} \left(\frac{z^2}{2}\right)_0^4 \cdot (y)_0^6$$

$$= \frac{1}{2} \cdot \frac{16}{2} \cdot 6 = 24$$

$$\mathbf{I_2} = \int_{z=0}^{4} \int_{x=0}^{6} \int_{y=0}^{\sqrt{z}} y \, dx \, dy \, dz$$

$$= \int_{z=0}^{4} \int_{y=0}^{6} \sqrt{z} \, y \, dy \, dz$$

$$= \int_{z=0}^{4} \sqrt{z} \, dz \int_{y=0}^{6} y \, dy$$

$$= \left\{\frac{2}{3} z^{3/2}\right\}_0^4 \left\{\frac{y^2}{2}\right\}_0^6$$

$$= \frac{2}{3} (8)(18) = 96$$

$$\mathbf{I_3} = \int_{z=0}^{4} \int_{x=0}^{6} \int_{y=0}^{\sqrt{z}} z \, dx \, dy \, dz$$

$$= \int_{z=0}^{4} \int_{y=0}^{6} z \cdot \sqrt{z} \, dx \, dy$$

$$= \int_{z=0}^{4} z^{3/2} \, dz \int_{y=0}^{6} dy$$

$$= \left\{\frac{2}{5} z^{5/2}\right\}_0^4 \{y\}_0^6$$

$$= \frac{2}{5} (32)(6) = \frac{384}{5}$$

$$\therefore \int_V \mathbf{F} \, dv = 24\mathbf{i} + 96\mathbf{j} + \frac{384}{5} \mathbf{k}$$

Example 24. If $\mathbf{F} = (2x^2 - 3z)\,\mathbf{i} - 2xy\,\mathbf{j} - 4x\,\mathbf{k}$ then evaluate

(i) $\int_V \nabla \times \mathbf{F} \, dv$,

(ii) $\int_V \nabla \cdot \mathbf{F} \, dv$, where **V** is the closed region bounded by the planes $x = 0$, $y = 0$, $z = 0$ and $2x + 2y + z = 4$.

Solution: Given $\mathbf{F} = (2x^2 - 3z)\,\mathbf{i} - 2xy\mathbf{j} - 4x\,\mathbf{k}$

$$\text{(i) } \nabla \times \mathbf{F} = \begin{vmatrix} \mathbf{i} & \mathbf{j} & \mathbf{k} \\ \dfrac{\partial}{\partial x} & \dfrac{\partial}{\partial y} & \dfrac{\partial}{\partial z} \\ 2x^2 - 3z & -2xy & -4x \end{vmatrix}$$

$$= (0 - 0)\,\mathbf{i} + (-3 + 4)\,\mathbf{j} + (-2y - 0)\,\mathbf{k}$$

$$= (\mathbf{j} - 2y\mathbf{k})$$

$$\therefore \int_V \nabla \times \mathbf{F}\, dv = \iiint (\mathbf{j} - 2y\mathbf{k})\, dx\, dy\, dz$$

$$= \int_{x=0}^{2} \int_{y=0}^{2-x} \int_{z=0}^{4-2x-2y} (\mathbf{j} - 2y\mathbf{k})\, dx\, dy\, dz$$

$$= \int_{x=0}^{2} \int_{y=0}^{2-x} (\mathbf{j} - 2y\,\mathbf{k})(4 - 2x - 2y)\, dx\, dy$$

$$= \int_{x=0}^{2} \left[(4y - 2xy - y^2)\,\mathbf{j} - 2\left(2y^2 - xy^2 - \frac{2y^3}{3}\right)\mathbf{k} \right]_0^{(2-x)} dx$$

$$= \int_{x=0}^{2} [(2 - x)^2\,\mathbf{j} - \frac{2}{3}(2 - x)^2\,\mathbf{k}]\, dx$$

$$= \left[-\frac{(2-x)^3}{3}\,\mathbf{j} + \frac{2}{3} \cdot \frac{(2-x)^4}{4}\,\mathbf{k} \right]_0^2$$

$$= \left[\frac{8}{3}\,\mathbf{j} + \frac{1}{8}(-16)\,\mathbf{k} \right] = \frac{8}{3}(\mathbf{j} - \mathbf{k})$$

$$\text{(ii) } \nabla \cdot \mathbf{F} = \frac{\partial}{\partial x}(2x^2 - 3z) + \frac{\partial}{\partial y}(-2xy) + \frac{\partial}{\partial z}(-4x)$$

$$= 4x - 2x = 2x$$

$$\int_V \nabla \cdot \mathbf{F} dv = \int_{x=0}^{2} \int_{y=0}^{2-x} \int_{z=0}^{4-2x-2y} 2x\, dx\, dy\, dz$$

$$= \int_{x=0}^{2} \int_{y=0}^{2-x} 2x(4 - 2x - 2y)\, dx\, dy$$

$$= \int_{x=0}^{2} (8xy - 4x^2y - 2xy^2)_{y=0}\, dx$$

$$= \int_{x=0}^{2} [8x(2-x) - 4x^2(2-x) - 2x(2-x)^2]\, dx$$

$$= \int_{x=0}^{2} (2x^3 - 8x^2 + 8x)\, dx$$

$$= \left(\frac{x^4}{2} - \frac{8}{3}x^3 + 4x^2 \right)_0^2$$

$$= \left(8 - \frac{64}{3} + 16 \right) = \frac{8}{3}.$$

Example 25. Evaluate $\int_V \phi\, dv$ where $\phi = xyz$ and **V** is the closed region bounded by the planes $x + y + z = 1$, $x = 0$, $y = 0$, $z = 0$.

Solution: The given plane meets the axis of x at (1, 0, 0).

The limits of integration are

$$0 \le z \le 1 - x - y$$
$$0 \le y \le 1 - x$$
$$0 \le x \le 1.$$

$$\int_V \phi\, dv = \int_{x=0}^{1} \int_{y=0}^{1-x} \int_{z=0}^{1-x-y} xyz\, dx\, dy\, dz$$

$$= \int_{x=0}^{1} \int_{y=0}^{1-x} xy \left(\frac{z^2}{2} \right)_0^{1-x-y} dx\, dy$$

$$= \frac{1}{2} \int_{x=0}^{1} \int_{y=0}^{1-x} xy(1-x-y)^2\, dx\, dy$$

$$= \frac{1}{2} \int_{x=0}^{1} \int_{y=0}^{1-x} [xy + x^3y + xy^3 - 2x^2y + 2x^2y^2 - 2xy^2]\, dx\, dy$$

$$= \int_{x=0}^{1} \left[\frac{xy^2}{2} + \frac{x^3y^2}{2} + \frac{xy^4}{4} - \frac{2x^2y^2}{2} + \frac{2x^2y^3}{3} - \frac{2xy^3}{3} \right]_0^{1-x} dx$$

$$= \int_{x=0}^{1} \left[\frac{1}{2}(x + x^3 - 2x^2)(1 - x)^2 + \frac{x}{4}(1 - x)^4 + \frac{2}{3}(x^2 - x)(1 - x)^3 \right] dx$$

$$= \frac{17}{24} \int_{x=0}^{1} (1 - x)^4 \, dx$$

$$= \frac{17}{24} \int_{x=0}^{1} (1 - x)\, x^4 \, dx.$$

$$\left[\therefore \int_0^a x^m (1 - x)^n \, dx = \int_0^a (1 - x)^m x^n \, dx \right]$$

$$= \frac{17}{24} \left(\frac{x^5}{5} - \frac{x^6}{6} \right)_0^1$$

$$= \frac{17}{24} \cdot \frac{1}{30} = \frac{17}{720}.$$

EXERCISE

1. If $\mathbf{F} = x(5y - 6x)\,\mathbf{i} + 2(y - 2x)\,\mathbf{j}$, evaluate $\int_C \mathbf{F} . d\mathbf{r}$ along the curve **C** in the xy-plane, $y = x^2$ from the point (1, 1) to (2, 8).
2. If $\mathbf{F} = 3xy\,\mathbf{i} - y^2\,\mathbf{j}$, $\int_C \mathbf{F} . d\mathbf{r}$ and the are of parabola $y = 2x^2$ from (0, 0) to (1, 2).
3. Evaluate $\int_C \mathbf{F} . d\mathbf{r}$, where $\mathbf{F} = xy\mathbf{i} + yz\mathbf{j} + z\mathbf{k}$ and the curve **C** is $\mathbf{r} = t\mathbf{i} + t^2\mathbf{j} + t^3\mathbf{k}$ varying from -1 to 1.
4. Evaluate $\int_C \mathbf{F} . d\mathbf{r}$, where $\mathbf{F} = z\mathbf{i} + x\mathbf{j} + y\mathbf{k}$ and **C** is the curve $\mathbf{r} = \cos t\mathbf{i} + \sin t\mathbf{j} + t\mathbf{k}$ from $t = 0$ to $t = 2\pi$.
5. Evaluate $\int_C \mathbf{F} . d\mathbf{r}$ where $\mathbf{F} = y^2 - x^2\mathbf{j}$ about the triangle whose vertices are (1, 0), (0, 1) and (−1, 0).

6. Evaluate $\int_C \mathbf{F}\,.\,d\mathbf{r}$ where $\mathbf{F} = (x^2 + y^2)\,\mathbf{i} - 2xy\,\mathbf{j}$, curve **C** is the rectangle in the xy-plane bounded by y = 0, x = a, y = b, x = 0.

7. If $\mathbf{F} = (3x^2 + 6y)\,\mathbf{i} - 14yz\mathbf{j} + 20xz^2\mathbf{k}$, evaluate the line integral $\int_C \mathbf{F}\,.\,d\mathbf{r}$ from (0,0,0) to (1,1,1) along the following paths **C**

(i) $x = t,\ y = t,\ z = t^3$

(ii) Straight line joining (0, 0, 0) to (1, 1, 1)

(iii) The straight line from (0, 0, 0) to (1, 0, 0) then to (1, 1, 0) and then to (1, 1, 1).

8. Find the line integral of the vector $\mathbf{F} = z\mathbf{i} + x\mathbf{j} + y\mathbf{k}$ over the curve **C** given by x = a cos t, y = a sin t, z = t/2π from z = 0 to z = 1.

9. Find the circulation of **F** round the curve **C** where $\mathbf{F} = e^x(\sin y\mathbf{i} + \cos y\mathbf{j})$ and **C** is the rectangle whose vertices are (0, 0), (1, 0) (1, π/2), (0, π/2).

10. The **C** is a simple closed curve in the xy-plane not enclosing the origin, show that

$$\oint_C \mathbf{F}\,.\,d\mathbf{r} = 0, \text{ where } \mathbf{F} = \frac{y\mathbf{i} + x\mathbf{j}}{x^2 + y^2}$$

11. Find the circulation of **F** round the curve **C** where $\mathbf{F} = y\mathbf{i} + z\mathbf{j} + x\mathbf{k}$ and **C** is the circle $x^2 + y^2 = 1,\ z = 0$.

12. Find the circulation of **F** round the curve **C**, where $\mathbf{F} = (x - y)\,\mathbf{i} + (x + y)\,\mathbf{j}$ and **C** is the circle $x^2 + y^2 + 4,\ z = 0$.

13. Find the circulation of **F** round the curve **C** where $\mathbf{F} = (2x + y^2)\,\mathbf{i} + (3y - 4x)\,\mathbf{j}$ and **C** is the curve $y = x^2$ from (0, 0) to (1,1) and the curve $y^2 = x$ from (1, 1) to (0, 0).

14. Find the work done when a force

$$\mathbf{F} = (x^2 - y^2 + x)\,\mathbf{i} - (2xy + y)\,\mathbf{j}$$

moves a particle in the xy-plane from (0, 0) to (1, 1) along the parabola $y^2 = x$. is this work different when the path is the straight line y = x ?

15. Find the total work done in moving a particle in a force field given by $\mathbf{F} = 3xy\mathbf{i} - 5z\mathbf{j} + 10x\mathbf{k}$ along the curve

$$x = t^2 + 1, y = 2t^2, z = t^3, \text{ from } t = 1 \text{ to } t = 2.$$

16. Find the work done in moving a particle in the force field $\mathbf{F} = 3x^2\mathbf{i} + (2xz - y)\,\mathbf{j} + z\mathbf{k}$, along

(i) the straight line from (0, 0, 0) to (2, 1, 3).

(ii) the curve defined by $x^2 = 4y$, $3x^3 = 8z$, from $x = 0$ to $x = 2$.

(iii) the space curve $x = 2t^2$, $y = t$, $z = 4t^2 - t$, from $t = 0$ to $t = 1$.

17. If $\mathbf{f} = xy\mathbf{i} - z\mathbf{j} + x^2\mathbf{k}$ $\phi = 2xyz^2$ and $\mathbf{C}$ is the curve $x = t^2$, $y = 2t$, $x = t^3$ from $t = 0$ to $t = 1$, evaluate the integrals

(i) $\int_C \mathbf{F} \times d\mathbf{r}$ (ii) $\int_C \phi\, d\mathbf{r}$.

18. Evaluate $\int_S \mathbf{F} \,.\, \mathbf{N}\, ds$ where $\mathbf{F} = y\mathbf{i} + 2x\mathbf{j} - z\mathbf{k}$ and $\mathbf{S}$ is the surface of the plane, $2x + y = 6$ in the first octant cut off by the plane $z = 4$.

19. Evaluate $\int_S \mathbf{F} \,.\, \mathbf{N}\, ds$ where $\mathbf{F} = 12x^2y\mathbf{i} - 3yz\mathbf{j} + 2x\mathbf{k}$ and $\mathbf{S}$ is the portion of the plane $x + y + z = 1$ included in the first octant.

20. Evaluate $\int_S \frac{\mathbf{r}}{r^3}\, \mathbf{N}\, ds$ where $\mathbf{S}$ denotes the spheres of radius a with centre at the origin.

21. Evaluate $\int_S \mathbf{F} \,.\, \mathbf{N}\, ds$ where $\mathbf{F} = z\mathbf{i} + x\mathbf{j} - yz\,\mathbf{k}$ and $\mathbf{S}$ is the surface of the cylinder $x^2 + y^2 = 9$ included in the first octant between $z = 0$ and $z = 4$.

22. Evaluate $\int_S \mathbf{F} \,.\, \mathbf{N}\, ds$ where $\mathbf{F} = 2xy\,\mathbf{i} - zy\,\mathbf{j} + x^2\,\mathbf{k}$ over the surface of cube bounded by the coordinate planes and the planes $x = a$, $y = a$, $z = a$.

23. If $\mathbf{F} = 2y\mathbf{i} - 3\mathbf{j} + x^2\mathbf{k}$ and $\mathbf{S}$ is the surface of the parabolic cylinder $y^2 = 8x$ in the first octant bounded by the planes $y = 4$ and $z = 6$ evaluate $\int_S \mathbf{F}\; \mathbf{N}\, ds$.

24. If $\mathbf{F} = \{xye^x + \log(z+1)\sin x\}\,\mathbf{k}$, evaluate $\int_S (\nabla \times \mathbf{F}) \cdot \mathbf{N}\, ds$ over the part of the sphere $x^2 + y^2 + z^2 = 1$ above the xy plane.

25. Evaluate $\int_S \phi\, \mathbf{N}\, ds$, where $\phi = 3/8\, xyz$ and **S** is the surface of the cylinder $x^2 + y^2 = 16$ included in the first octant between $z = 0$ and $z = 5$.

26. If $\mathbf{F} = 2xz\,\mathbf{i} - x\,\mathbf{j} + y^2\,\mathbf{k}$, evaluate $\int_V \mathbf{F} \cdot dv$ where **V** is the region bounded by the surfaces $x = 0, y = 6, z = x^2, z = 4$.

27. Evaluate $\int_V \phi\, dv$ where $\phi = 45x^2y$ and **V** is the closed region bounded by the planes $4x + 2y + z = 8, x = 0, y = 0, z = 0$.

28. Evaluate $\int_V (2x + y)\, dv$ where **V** is closed region bounded bythe cylinder $z = 4 - x^2$ and the planes $x = 0, y = 0, y = 2$ and $z = 0$.

ANSWERS

(1) 35 (2) $-\frac{7}{6}$ (3) $\frac{10}{7}$ (4) 3π

(5) $-\frac{2}{3}$ (6) $-2ab^2$ (7) (i) 5 (ii) $\frac{13}{3}$ (iii) $\frac{23}{3}$

(8) $\pi a^2 + a$ (9) 0 (11) $-\pi$ (12) 8π

(13) $-49/30$ (14) $-2/3$, No, it is the same. (15) 303

(16) (i) 16 (ii) 6 (iii) 14.7

(17) (i) $\frac{9}{10}\mathbf{i} - \frac{2}{3}\mathbf{j} + \frac{7}{5}\mathbf{k}$ (ii) $\frac{8}{11}\mathbf{i} + \frac{4}{5}\mathbf{j} + \mathbf{k}$

(18) 108 (19) 49/120 (20) 4π (21) 42

(22) $\frac{1}{2}a^4$ (23) 132 (24) 0 (25) 100 (**i** +**j**)

(26) $128\mathbf{i} - 24\mathbf{j} + 384\mathbf{k}$ (27) 128 (28) $\frac{80}{3}$.

14

Green's, Gauss's and Stoke's Theorems

14.1. Green's Theorem in Plane

Statement: If M (x, y) and N (x, y) be continuous functions of x and y having continuous partial derivatives $\frac{\partial M}{\partial y}$ and $\frac{\partial N}{\partial x}$ in a region **R** of the xy-plane bounded by closed curve C, then

$$\oint_C M\,dx + N\,dy = \iint_R \left(\frac{\partial N}{\partial x} - \frac{\partial M}{\partial y}\right) dx\,dy$$

where C is traversed in the anticlockwise direction.

Proof: Let us assume that the region **R** is such that any line parallel to either axes meets the boundary curve **C** in at the most two points.

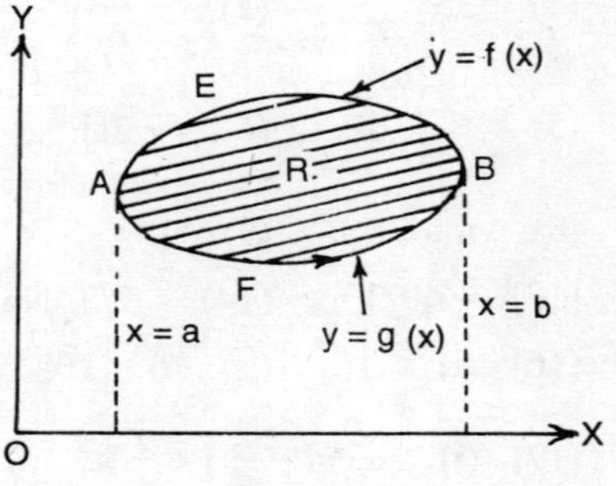

Suppose the region **R** is bounded between the lines x = a and x = b and two arcs AFB and BEA whose equations are y=g(x) and y = f(x) respectively such that f(x) > g(x).

Fig. 14.1

$$\text{Now} \iint_R \frac{\partial M}{\partial y}\,dx\,dy = \int_{x=a}^{b} \left\{ \int_{y=g(x)}^{f(x)} \frac{\partial M}{\partial y}\,dy \right\} dx$$

$$= \int_{x=a}^{b} \left[M(x, y) \right]_{y=g(x)}^{f(x)} dx$$

$$= \int_{x=a}^{b} [m(x, f(x)) - \mathbf{M}(x, g(x))]\, dx$$

$$= \int_{x=b}^{b} \mathbf{M}(x, f(x))dx - \int_{x=a}^{b} \mathbf{M}(x, g(x))\, dx$$

$$= -\int_{x=a}^{b} \mathbf{M}(x, f(x))\, dx - \int_{x=a}^{b} \mathbf{M}(x, g(x))\, dx$$

$$= -\left\{\int_{\mathbf{BEA}} \mathbf{M}(x, y))\, dx + \int_{\mathbf{AFB}} \mathbf{M}(x, y)\, dx\right\}$$

$$= -\int_{\mathbf{C}} \mathbf{M}\, dx$$

or
$$-\iint_{\mathbf{R}} \frac{\partial \mathbf{M}}{\partial y} dx\, dy = \int_{\mathbf{C}} \mathbf{M}\, dx \qquad \text{...(1)}$$

Now suppose once again **R** is the region bounded between the lines y=c and y=d and two arcs AEF and BFA whose equations are $x = g^*(x)$ and $x = f^*(y)$ respectively such that $f^*(y) > g^*(y)$

$$\therefore \iint_{\mathbf{R}} \frac{\partial \mathbf{N}}{\partial x} dx\, dy = \int_{y=c}^{d} \left\{\int_{x=g^*(y)}^{f^*(y)} \frac{\partial \mathbf{N}}{\partial x} dx\right\} dy$$

$$= \int_{y=c}^{d} \{\mathbf{N}(x, y)\}_{x=g^*(y)}^{f^*(y)}\, dy$$

$$= \int_{y=c}^{d} \{\mathbf{N}(f^*(y), y) - \mathbf{N}(g^*(y), y)\}\, dy$$

$$= \int_{y=c}^{d} \mathbf{N}(f^*(y), y)\, dy + \int_{y=c}^{d} \mathbf{N}(g^*(y), y)\, dy$$

$$= \int_{\text{BFA}} \mathbf{N}(x, y)\, dy + \int_{\text{AEB}} \mathbf{N}(x, y)\, dy$$

$$= \oint_{\mathbf{C}} \mathbf{N}(x, y)\, dy$$

$$\therefore \qquad \iint_{\mathbf{R}} \frac{\partial \mathbf{N}}{\partial x} dx\, dy = \oint_{\mathbf{C}} \mathbf{N}(x, y)\, dy. \qquad \text{...(2)}$$

Adding (1) and (2)

$$\iint_{\mathbf{R}} \left(\frac{\partial \mathbf{N}}{\partial x} - \frac{\partial \mathbf{M}}{\partial y}\right) dx\, dy = \int_{\mathbf{C}} \mathbf{M}\, dx + \mathbf{N}\, dy.$$

Hence the theorem.

The result can now be extended to a region **R** which can be subdivided into finitely many sub-regions.

Consider a region **R** which is subdivided into two sub-regions $\mathbf{R_1}$ and $\mathbf{R_2}$ as shown in the following figure [14.3].

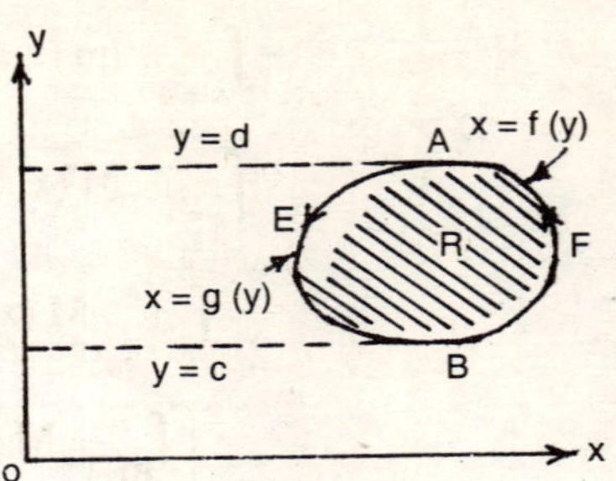

Fig. 14.2

Here the regions $\mathbf{R_1}$ *i.e.* **PSQP** and region $\mathbf{R_2}$ *i.e.* **PQTP** are simply connected and so Green's theorem is valid.

$$i.e. \quad \int_{\mathbf{PSQP\,QTP}} \mathbf{M}\,dx + \mathbf{N}\,dy = \iint_{\mathbf{R_1}} \left(\frac{\partial \mathbf{N}}{\partial x} - \frac{\partial \mathbf{M}}{\partial y}\right) dx\,dy$$

$$+ \iint_{\mathbf{R_2}} \left(\frac{\partial \mathbf{N}}{\partial x} - \frac{\partial \mathbf{M}}{\partial y}\right) dx\,dy \qquad \ldots(1)$$

Now

$$\text{L.H.S} = \int_{\mathbf{PSQP}} + \int_{\mathbf{PQTP}}$$

$$= \int_{\mathbf{PSQ}} + \int_{\mathbf{QP}} + \int_{\mathbf{PQ}} + \int_{\mathbf{QTP}}$$

$$= \int_{\mathbf{PSQ}} + \int_{\mathbf{QP}} - \int_{\mathbf{QP}} + \int_{\mathbf{QTP}}$$

$$= \int_{\mathbf{PSQ}} + \int_{\mathbf{QTP}} + \int_{\mathbf{C}} \qquad \ldots(2)$$

Fig. 14.3

As $\mathbf{R} = \mathbf{R_1} \cup \mathbf{R_2}$

$$\text{R.H.S of (1)} = \iint_{\mathbf{R}} \left(\frac{\partial \mathbf{N}}{\partial x} - \frac{\partial \mathbf{M}}{\partial y}\right) dx\,dy$$

$$\text{Hence } \oint_{\mathbf{C}} \mathbf{M}\,dx + n\,dy = \iint_{\mathbf{R}} \left(\frac{\partial \mathbf{N}}{\partial x} - \frac{\partial \mathbf{M}}{\partial y}\right) dx\,dy.$$

14.2. Extension of Green's theorem in plane to multiply-connected regions

Let the multiply connected region **R** be as in Fig [14.4]. Consider the boundary **C** of **R** consisting of two parts (i) the exterior boundary $\mathbf{C_1}$ and (ii) the interior boundary $\mathbf{C_2}$.

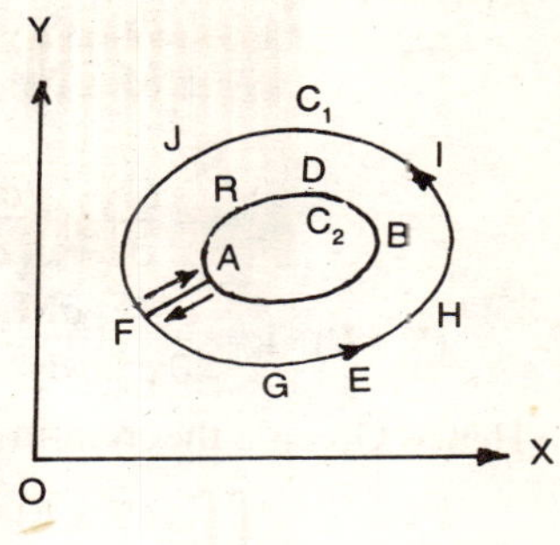

Fig. 14.4

The exterior boundary $\mathbf{C_1}$ is traversed in the anticlockwise direction and the interior boundary $\mathbf{C_2}$ is traversed in the clockwise direction.

Now we construct a cross-cut connecting the exterior and interior boundaries, in order to establish the theorem.

Now the region is bounded by **FADBEFGHIJF** and is simply-connected and so Green's theorem is varied for it.

i.e.
$$\int_{\mathbf{FADBEFGHIJF}} = \int_{\mathbf{FA}} + \int_{\mathbf{ADBEA}} + \int_{\mathbf{AF}} + \int_{\mathbf{FGHIJF}}$$
$$= -\int_{\mathbf{AF}} + \int_{\mathbf{C_2}} + \int_{\mathbf{AF}} + \int_{\mathbf{C_1}}$$
$$= \int_{\mathbf{C_2}} + \int_{\mathbf{C_1}} = \int_{\mathbf{C}}$$

Hence
$$\int_{\mathbf{FADBEFGHIJF}} \mathbf{M}\,dx + \mathbf{N}\,dy = \int_{\mathbf{C}} \mathbf{M}\,dx + \mathbf{N}\,dy$$
$$= \iint_{\mathbf{R}} \left(\frac{\partial N}{\partial x} - \frac{\partial M}{\partial y}\right) dx\,dy$$

14.3. Green's theorem in the plane in vector notation:

We have $\mathbf{r} = x\mathbf{i} + y\mathbf{j}$ so that $dr = dx\,\mathbf{i} + dy\,\mathbf{j}$

Let $\mathbf{F} = \mathbf{M}\,\mathbf{i} + \mathbf{N}\,\mathbf{j}$

$m\,dx + \mathbf{N}\,dy = (\mathbf{M}\,\mathbf{i} + \mathbf{N}\,\mathbf{j}) \cdot (dx\,\mathbf{i} + dy\,\mathbf{j}) = \mathbf{F} \cdot d\mathbf{r}$

Also $\nabla \times \mathbf{F} = \begin{vmatrix} \mathbf{i} & \mathbf{j} & \mathbf{k} \\ \frac{\partial}{\partial x} & \frac{\partial}{\partial y} & \frac{\partial}{\partial z} \\ \mathbf{M} & \mathbf{N} & 0 \end{vmatrix}$

$$= -\frac{\partial \mathbf{N}}{\partial z}\mathbf{i} + \frac{\partial \mathbf{M}}{\partial z}\mathbf{j} + \left(\frac{\partial \mathbf{N}}{\partial x} - \frac{\partial \mathbf{M}}{\partial y}\right)\mathbf{k}$$

$\because \quad (\nabla \times \mathbf{F}) \cdot \mathbf{k} = \frac{\partial \mathbf{N}}{\partial x} - \frac{\partial \mathbf{M}}{\partial y}$

Hence Green's theorem in plane can be written as

$$\iint_{\mathbf{R}} (\nabla \times \mathbf{F}) \cdot \mathrm{k}\, d\mathrm{R} = \int_{\mathbf{C}} \mathbf{F} \cdot \mathbf{dr}$$

where $d\mathbf{R}$ = dx dy and **k** is a unit vector perpendicular to the xy plane.

If s denotes the are length of **C** and **T** is the unit tangent vector to C, then

$$\iint_{\mathbf{R}} (\nabla \times \mathbf{F}) \cdot \mathrm{k}\, d\mathrm{R} = \int_{\mathbf{C}} \mathbf{F} \cdot \mathbf{T}\, ds.$$

Example 1. Verify Green's theorem in plane for

$$\int_{\mathbf{C}} (3x^2 - 8y^2)\, dx + (4y - 6xy)\, dy,$$

where C is the boundary of the region defined by

(i) $y^2 = x, \; y = x^2$

(ii) $x = 0, \; y = 0, \; x + y = 1.$

Solution: The two parabolas intersect at (0,0) and (1,1).

$$\int_{\mathbf{C}} \mathbf{M}\, dx + \mathbf{N}\, dy = \int_{\mathbf{C}_1} \mathbf{M}\, dx + \mathbf{N}\, dy + \int_{\mathbf{C}_2} \mathbf{M}\, dx + \mathbf{N}\, dy \quad ...(1)$$

Along $\mathbf{C}_1 : x^2 = y, \; dy = 2x\, dx, \; 0 \le x \le 1$

$$\therefore \quad \int_{\mathbf{C}_1} \mathbf{M}\, dx + \mathbf{N}\, dy = \int_0^1 (3x^2 - 8x^4)\, dx + (4x^2 - 6x^3)\, 2x\, dx$$

$$= \int_0^1 (3x^2 + 8x^3 - 20x^4)\, dx$$

$$= \left(x^3 + 2x^4 - 4x^5\right)_0^1$$

$$= 1 + 2 - 4 = -1$$

Along $\mathbf{C_2} : y^2 = x,\ 2y\,dy = dx;\ 1 \le y \le 0$

$$\therefore \quad \int_{C_2} \mathbf{M}\,dx + \mathbf{N}\,dy = \int_1^0 (3y^4 - 8y^2)\,2y\,dy + (4y - 6y^3)\,dy$$

$$= \int_1^0 (6y^5 - 22y^5 + 4y)\,dy$$

$$= \left(y^6 - \frac{11}{2}y^4 = 2y^2\right)_1^0$$

$$= -\left(1 - \frac{11}{2} + 2\right) = \frac{5}{2}.$$

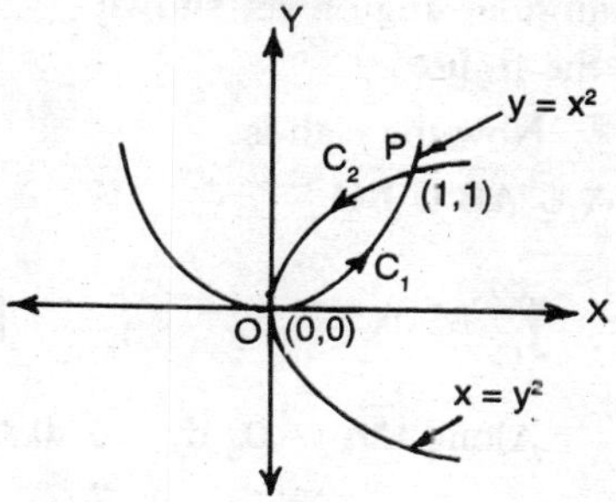

Fig. 14.5

$\therefore$ by substituting in (1), we get

$$\int_C \mathbf{M}\,dx + \mathbf{N}\,dy = -1 + \frac{5}{2} = \frac{3}{2.} \qquad ...(2)$$

We have $\mathbf{M} = 3x^2 - 8y^2 \quad \therefore \quad \frac{\partial \mathbf{M}}{\partial y} = -16y$

$\mathbf{N} = 4y - 6xy \quad \therefore \quad \frac{\partial \mathbf{N}}{\partial x} = -6y$

$$\therefore \quad \left(\frac{\partial \mathbf{N}}{\partial x} - \frac{\partial \mathbf{M}}{\partial y}\right) - 6y + 16y = 10y$$

$$\therefore \quad \iint_R \left(\frac{\partial \mathbf{N}}{\partial x} - \frac{\partial \mathbf{M}}{\partial y}\right) dx\,dy$$

$$= \int_{x=0}^{1} \int_{y=x^2}^{\sqrt{x}} 10y\,dx\,dy = 10 \int_{x=0}^{1} \left(\frac{y^2}{2}\right)_{x^2}^{\sqrt{x}} dx$$

$$= 5 \int_{x=0}^{1} (x - x^4)\,dx = 5\left(\frac{x^2}{2} - \frac{x^5}{5}\right)_0^1$$

$$= 5\left(\frac{1}{2} - \frac{1}{5}\right) = \frac{3}{2} \qquad ...(3)$$

$\therefore$ From (2) and (3), we have

$$\int_C \mathbf{M}\,dx + \mathbf{N}\,dy = \iint_R \left(\frac{\partial \mathbf{N}}{\partial x} - \frac{\partial \mathbf{M}}{\partial y}\right) dx\,dy$$

Hence Green's theorem is verified.

(ii) $x=0, y=0, x+y=1$ is triangular region as shown in the figure

Now the path is

$\overrightarrow{OA} \cup \overrightarrow{AB} \cup \overrightarrow{BO}$

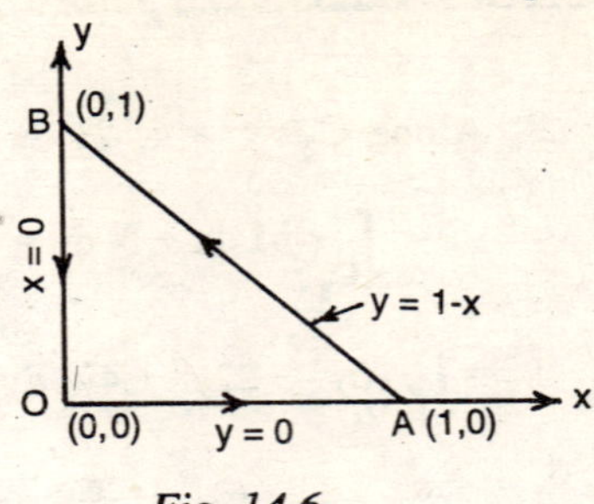

Fig. 14.6

i.e. $\int_C M\,dx + N\,dy = \left\{ \int_{\overrightarrow{OA}} + \int_{\overrightarrow{AB}} + \int_{\overrightarrow{BO}} \right\} M\,dx + N\,dy \quad ...(1)$

Along $\overrightarrow{OA}$ $y = 0,\ dy = 0,\ 0 \le x \le 1.$

$$\therefore \quad \int_{\overrightarrow{OA}} M\,dx + N\,dy = \int_0^1 3x^2\,dx = \left[x^3\right]_0^1 = 1$$

Along $\overrightarrow{AB}$ $y = 1 - x,\ dy = -\,dx$

$$x = -(y-1);\ 0 \le y \le 1$$

$$\therefore \int_{\overrightarrow{AB}} M\,dx + N\,dy = \int_0^1 [3(y-1)^2 - 8y^2](-\,dy) + [4y + 6y(y-1)]\,dy$$

$$= \int_0^1 (-5y^2 - 6y + 3)(-\,dy) + (6y^2 - 2y)\,dy$$

$$\int_0^1 (11y^2 + 4y - 3)\,dy$$

$$= \left(\frac{11y^3}{3} + 2y^2 - 3y\right)_0^1$$

$$= \frac{11}{3} + 2 - 3 = \frac{8}{3}$$

Along $\overrightarrow{BO}$, $x = 0;\ dx = 0;\ 1 \le y \le 0$

$$\therefore \int_{\overrightarrow{BO}} M\,dx + N\,dy = \int_1^0 4y\,dy = \left(2y^2\right)_1^0 = -2.$$

$\therefore$ By substituting in (1), we get

$$\int_C M\,dx + N\,dy = 1\,\frac{8}{3} - 2 = \frac{5}{3} \quad ...(2)$$

We have $\left(\frac{\partial N}{\partial x} - \frac{\partial M}{\partial y}\right) = 10y$

$$\therefore \quad \iint_R \left(\frac{\partial N}{\partial x} - \frac{\partial M}{\partial y}\right) dx\, dy$$

$$= \int_{x=0}^{1} \int_{y=0}^{1-x} 10y\, dx\, dy$$

$$= 5\int_0^1 \left(y^2\right)_0^{1-x} dx$$

$$= 5\int_0^1 (1-x)^2\, dx$$

$$= 5\left[\frac{(1-x)^3}{-3}\right]_0^1 = \frac{5}{3} \quad \text{...(3)}$$

∴ From (2) and (3), we have

$$\int_C M\, dx + N\, dy \iint_R \left(\frac{\partial N}{\partial x} - \frac{\partial M}{\partial y}\right) dx\, dy$$

Hence Green's theorem is verified.

Example 2. Evaluate by Green's theorem

$\int_C$ (y sin x) dx + cos x dy where **C** is the triangle enclosed by x = 0, 2x = π, πy = 2x.

Solution : Here **M** = y - sin x; **N** = cos x.

$$\therefore \quad \frac{\partial M}{\partial y} = 1; \quad \frac{\partial N}{\partial x} = -\sin x.$$

and $\left(\frac{\partial N}{\partial x} - \frac{\partial M}{\partial y}\right) = -(\sin x + 1)$

By Green theorem

$$\int_C M\, dx + N\, dy = \iint_R \left(\frac{\partial N}{\partial x} - \frac{\partial M}{\partial y}\right) dx\, dy$$

$$= -\int_{x=0}^{\pi/2} \int_{y=0}^{2x/\pi} (\sin x + 1)\, dx\, dy$$

$$= -\int_{x=0}^{\pi/2} (\sin x + 1) \frac{2x}{\pi}\, dx$$

$$= -\frac{2}{\pi}\int_{x=0}^{\pi/2} (x \sin x + x)\,dx$$

$$= -\frac{2}{\pi}\left[-x\cos x + \sin x + \frac{x^2}{2}\right]_0^{\pi/2}$$

$$= -\frac{2}{\pi}\left(1 + \frac{\pi^2}{8}\right)$$

$$= -\left(\frac{2}{\pi} + \frac{\pi}{4}\right)$$

Example 3. Verify Green's theorem for

$$\int_C e^{-x}(\sin y\,dx + \cos y\,dy),$$

where **C** is the rectangle with the vertices (0, 0), (π, 0) (π, π/2) and (0, π/2).

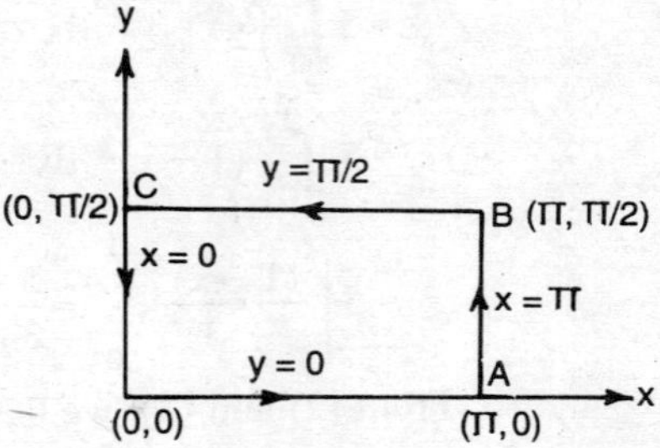

Fig. 14.7

Solution : Here $\mathbf{C} = \overrightarrow{\mathbf{OA}} \cup \overrightarrow{\mathbf{AB}} \cup \overrightarrow{\mathbf{BC}} \cup \overrightarrow{\mathbf{CO}}$

$$\therefore \quad \int_c \mathbf{M}\,dx + \mathbf{N}\,dy$$

$$= \left\{\int_{\overrightarrow{OA}} + \int_{\overrightarrow{AB}} + \int_{\overrightarrow{BC}} + \int_{\overrightarrow{CO}}\right\} \mathbf{M}d\,x + \mathbf{N}\,dy \qquad \ldots(1)$$

Along $\overrightarrow{\mathbf{AB}}$: $y = 0,\ dy = 0,\ 0 \le x \le \pi$

$$\therefore \quad \int_{\overrightarrow{AB}} \mathbf{M}\,dx + \mathbf{N}\,dy = \int_{x=0}^{\pi} e^{-x}\,(0+0) = 0$$

Along $\overrightarrow{\mathbf{AB}}$: $x = \pi,\ dx = 0,\ 0 \le y \le \pi/2.$

$$\int_{\overrightarrow{AB}} \mathbf{M}\,dx + \mathbf{N}\,dy = \int_{y=0}^{\pi/2} e^{-\pi}\,(0 + \cos y)\,dy$$

$$= e^{-\pi}\,(\sin y)_0^{\pi/2} = e^{-\pi}$$

Along $\overrightarrow{\mathbf{BC}}$: $y = \frac{\pi}{2},\ dy = 0,\ \pi \le x \le 0$

$$\int_{\overrightarrow{BC}} \mathbf{M}\,dx + \mathbf{N}\,dy = \int_{x=\pi}^{0} e^{-x} \cdot 1 \cdot dx$$

$$= -\left[e^{-x}\right]_{\pi}^{0} = -[1 - e^{-\pi}]$$

$$= e^{-\pi} - 1.$$

Along $\overrightarrow{CO}$: $x = 0,\ dx = 0,\ \pi/2 \le y \le 0$

$$\therefore \quad \int_{\overrightarrow{CO}} M\,dx + N\,dy = \int_{y=\pi/2}^{0} 1 \cdot \cos y\,dy$$

$$= (\sin y)_{\pi/2}^{0} = -1$$

$\therefore$ By substituting in (1), we get

$$\int_C M\,dx + N\,dy = 0 + e^{-\pi} + e^{-\pi} - 1 - 1$$

$$= 2\left(e^{-\pi} - 1\right) \qquad \text{...(2)}$$

We have $M = e^{x} \sin y \ \therefore\ \dfrac{\partial M}{\partial y} = e^{-x} \cos y$

$$N = e^{-x} \cos y \ \therefore\ \frac{\partial N}{\partial x} = e^{-x} \cos y$$

$$\therefore \quad \frac{\partial N}{\partial x} - \frac{\partial M}{\partial y} = -2e^{-x} \cos y$$

$$\therefore \quad \iint_R \left(\frac{\partial N}{\partial x} - \frac{\partial M}{\partial y}\,dx\,dy\right)$$

$$= \int_{x=0}^{\pi} \int_{y=0}^{\pi/2} -2e^{-x} \cos y\,dx\,dy$$

$$= -2 \int_{x=0}^{\pi} e^{-x}\ (\sin y)_0^{\pi/2}\ dx = -2 \int_{x=0}^{\pi} e^{-x}\,dx$$

$$= +2\left(e^{-x}\right)_0^{\pi} = 2\left(e^{-\pi} - 1\right) \qquad \text{...(3)}$$

From (2) and (3), we have

$$\int_C M\,dx + N\,dy = \iint_R \left(\frac{\partial N}{\partial x} - \frac{\partial M}{\partial y}\right) dx\,dy$$

Hence Green's theorem is verified

Example 4. Evaluate by Green's theorem

$$\int_C (\cos x \sin y - xy)\,dx + \sin x \cos y\,dy$$

where **C** is the circle $x^2 + y^2 = 1$.

Solution : By Green's theorem in plane, we have

$$\int_C \mathbf{M}\,dx + \mathbf{N}\,dy \iint_R \left(\frac{\partial \mathbf{N}}{\partial x} - \frac{\partial \mathbf{M}}{\partial y}\right) dx\,dy$$

Here $\mathbf{M} = \cos x \sin y - xy$

$$\therefore \qquad \frac{\partial \mathbf{M}}{\partial y} = \cos x \cos y - x$$

and $\mathbf{N} = \sin x \cos y$

$$\therefore \qquad \frac{\partial \mathbf{N}}{\partial x} = \cos x \cos y$$

Hence $\left(\frac{\partial \mathbf{N}}{\partial x} - \frac{\partial \mathbf{M}}{\partial y}\right) = x$

$$\therefore \quad \iint_R \left(\frac{\partial \mathbf{N}}{\partial x} - \frac{\partial \mathbf{M}}{\partial y}\right) dx\,dy = \iint x\,dx\,dy = \iint x\,dx\,dy$$

$$= \int_{r=0}^{1} \int_{\theta=0}^{2\pi} r\cos\theta\, r\,dr\,d\theta$$

(By changing into polar co-ordinates)

$$= \int_{\theta=0}^{2\pi} \left(\frac{r^3}{3}\right)_0^1 \cos\theta\,d\theta$$

$$= \frac{1}{2}\int_{\theta=0}^{2\pi} \cos\theta\,d\theta = \frac{1}{3}(\sin\theta)_0^{2\pi} = 0.$$

Example 5. Apply Green's theorem to prove that the area enclosed by a plane curve is $\frac{1}{2}\int_C (x\,dy - y\,dx)$. Hence find (i) the area of an ellipse whose semi major and minor axes are of lengths a and b (ii) the area bounded by one arc of the cycloid $x = a(\theta - \sin\theta)$, $y = a(1 - \cos\theta)$, $a > 0$ and the x-axis.

Solution : Consider $\int_C (x\,dy - y\,dx)$ ...(1)

By Green's theorem

$$\int_C \mathbf{M}\,dx + \mathbf{N}\,dy = \iint_R \left(\frac{\partial \mathbf{N}}{\partial x} - \frac{\partial \mathbf{M}}{\partial y}\right) dx\,dy \qquad ...(2)$$

Here $\mathbf{M} = -y \quad \therefore \quad \frac{\partial \mathbf{M}}{\partial y} = -1$

$\mathbf{N} = -x \quad \therefore \quad \frac{\partial \mathbf{N}}{\partial x} = 1$

Hence $\left(\frac{\partial \mathbf{N}}{\partial x} - \frac{\partial \mathbf{M}}{\partial y}\right) = 2$

$$\therefore \quad \iint_{\mathbf{R}} \left(\frac{\partial \mathbf{N}}{\partial x} - \frac{\partial \mathbf{M}}{\partial y}\right) dx\,dy = \iint_{\mathbf{R}} 2\,dx\,dy$$

$$= 2\iint_{\mathbf{R}} dx\,dy = 2\,A$$

where A is the area of region **R**

$\therefore$ From (1) and (2), $\frac{1}{2}\int_{\mathbf{C}} (x\,dy - y\,dx) = A.$

(i) Now for ellipse $x = a\cos\theta,\ y = b\sin\theta$

$$A = \frac{1}{2}\int_0^{2\pi} \left\{(a\cos\theta), (b\cos\theta) - (b\sin\theta)(-a\sin\theta)\right\} d\theta$$

$$= \frac{1}{2}\,ab\int_0^{2\pi} d\theta = \pi\,ab$$

(ii) $x = a(\theta - \sin\theta),\ y = a(1 - \cos\theta)$

$$A = \frac{1}{2}\int_0^{2\pi} a(\theta - \sin\theta)(a\sin\theta\,d\theta)$$

$$- a(1 - \cos\theta)\,a(1 - \cos\theta)\,d\theta$$

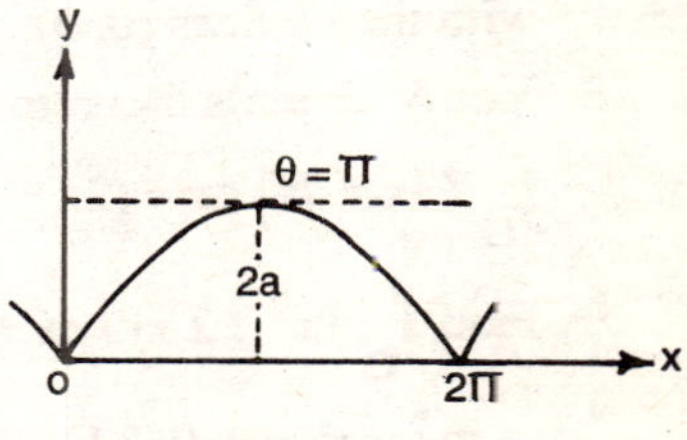

Fig. 14.8

$$= \frac{a^2}{2}\int_0^{2\pi} \{\theta\sin\theta - \sin^2\theta$$

$$- (1 + \cos^2\theta - 2\cos\theta)\}\,d\theta$$

$$= \frac{a^2}{2}\int_0^{2\pi} (\theta\sin\theta - 2 + 2\cos\theta)\,d\theta$$

$$= \frac{a^2}{2}\left[-\theta\cos\theta + \sin\theta - 2\theta + 2\sin\theta\right]_0^{2\pi}$$

$$= \frac{a^2}{2}\left[-2\pi\cos 2\pi - 4\pi\right]$$

$$= -3\pi a^2 \quad (\because \cos 2\pi = 1)$$

$\therefore$ Area $= 3\pi a^2$

EXERCISE

1. Verify Green's theorem in the plane for $\int_C (xy + y^2)\, dx + x^2\, dy$ where C is bounded by $y = x$ and $y = x^2$.
2. Verify Green's theorem in the plane for $\int_C (2xy - x^2)\, dx + (x^2 + y^2)\, dy$, where C is the boundary of the region enclosed by $y = x^2$ and $y^2 = x$.
3. Use Green's theorem in a plane to evaluate the integral $\int_C (2x^2 - y^2)\, dx + (x^2 + y^2)\, dy$, where C is the boundary in the xy plane of the area enclosed by the x-axis and the semi circle $x^2 + y^2 = 1$ in the upper half xy-plane.
4. Evaluate by Green's theorem $\int_C (x^2 - \cos h\, y)\, dx + (y + \sin x)\, dy$, where C is the rectangle with the vertices (0, 0), $(\pi, 0)$ $(\pi, 1)$ and (0, 1).
5. Verify Green's theorem in plane $\int_C (3x + 4y)\, dx + (2x - 3y)\, dy$, where C is circle $x^2 + y^2 = 4$.
6. Find $\int_C (x^2 - 2\, xy)\, dx + (x^2 y + z)\, dy$ around the boundary C of the region defined by $y^2 = 8x$ and $x = 2$, by Green's theorem.
7. Verify Green's theorem in the plane for $\int_C (x^2 - xy^2)\, dx + (y^2 - 2xy)\, dy$, where C is the square with vertices (0,0), (2,0), (2,2), (0,2).
8. Find the area bounded by the hypocycloid $x^{\frac{2}{3}} + y^{\frac{2}{3}} = a^{\frac{2}{3}}$; a = 0 Using Green's thorem.

   [*Hint*: Take $x = a\cos^3\theta$, $y = a\sin^3\theta$]

ANSWERS

3. $\frac{4}{3}$ 4. $\pi(\cosh 1 - 1)$ 6. $\frac{128}{5}$ 8. $\frac{3\pi^2}{8}$

14.6. Gauss's divergence theorem: *(Relation between surface and volume integrals).*

If **F** is a continuously differentiable vector function in the region **V** bounded by a closed surface **S**, then

$\int_S \mathbf{F}.\mathbf{N}\, ds = \int_V \nabla \cdot \mathbf{F}\, dv$; where **N** is the outwards drawn unit normal vector to the surface **S**.

Proof : Let $\mathbf{F} = F_1\mathbf{i} + F_2\mathbf{j} + F_3\mathbf{k}$

Let $\theta_1, \theta_2, \theta_3$ be the angles which the outwards drawn unit normal vector **N** makes with the positive directions of **x, y, z**—axes respectively.

Then $\cos\theta_1$, $\cos\theta_2$, $\cos\theta_3$ are the direction cosines of **N**.

$$\mathbf{N} = \cos\theta_1\,\mathbf{i} + \cos\theta_2\,\mathbf{j} + \cos\theta_3\,\mathbf{k}$$

$$\mathbf{F}\cdot\mathbf{N} = F_1\cos\theta_1 + F_2\cos\theta_2 + F_3\cos\theta_3$$

Now $$\nabla\cdot\mathbf{F} = \frac{\partial F_1}{\partial x} + \frac{\partial F_2}{\partial y} + \frac{\partial F_3}{\partial z}$$

$$dv = dx\, dy\, dz$$

∴ The Cartesian equivalent of divergence theorem is

$$\int_S (F_1\cos\theta_1 + F_2\cos\theta_2 + F_3\cos\theta_3)\, ds$$

$$= \int_V \left(\frac{\partial F_1}{\partial x} + \frac{\partial F_2}{\partial y} + \frac{\partial F_3}{\partial z}\right) dx\, dy\, dz \qquad ...(1)$$

Suppose that **S** is such a closed surface that a line parallel to the co-ordinate axes meets it in two points only. Let S_1 and S_2 denote the lower and upper portions of **S**. With the equations $z = f(x, y)$ and $z = g(x, y)$ respectively.

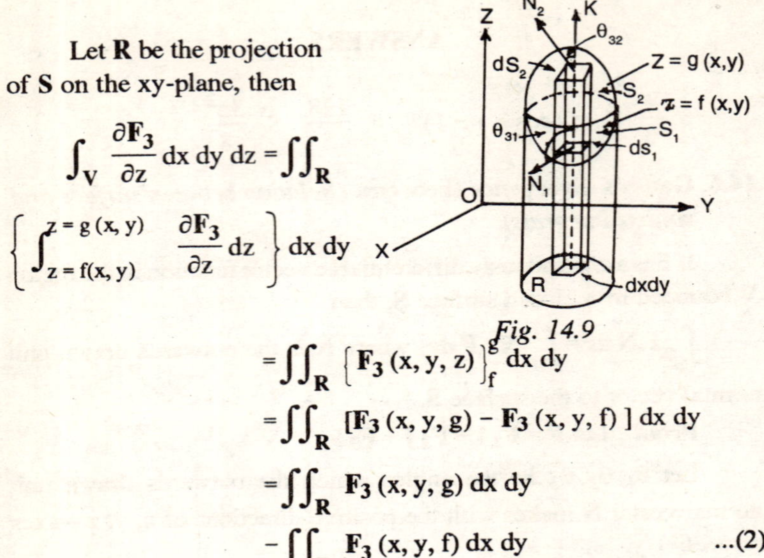

Fig. 14.9

Let **R** be the projection of **S** on the xy-plane, then

$$\int_V \frac{\partial \mathbf{F_3}}{\partial z}\, dx\, dy\, dz = \iint_R \left\{ \int_{z=f(x,y)}^{z=g(x,y)} \frac{\partial \mathbf{F_3}}{\partial z}\, dz \right\} dx\, dy$$

$$= \iint_R \left[\mathbf{F_3}(x, y, z) \right]_f^g dx\, dy$$

$$= \iint_R [\mathbf{F_3}(x, y, g) - \mathbf{F_3}(x, y, f)]\, dx\, dy$$

$$= \iint_R \mathbf{F_3}(x, y, g)\, dx\, dy - \iint_R \mathbf{F_3}(x, y, f)\, dx\, dy \qquad ...(2)$$

Now for the upper portion S_2 of S, the normal $\mathbf{N_2}$ to S_2 makes an acute angle θ_{32} with **k**.

$$\therefore \quad dx\, dy = \cos\theta_{32}\, ds_2 = \mathbf{k} \cdot \mathbf{N_2}\, ds_2$$

For the lower portion S_1 of S, the normal $\mathbf{N_1}$ to S_1 makes an obtuse angle θ_{31} with **k**.

$$\therefore \quad dx\, dy = \cos\theta_{31}\, ds_1 = -\mathbf{k} \cdot \mathbf{N_1}\, ds_1$$

$$\therefore \quad \iint_R \mathbf{F_3}(x, y, g)\, dx\, dy = \int_{S_2} \mathbf{F_3}\, \mathbf{k} \cdot \mathbf{N_2}\, ds_2 \qquad ...(3)$$

and $$\iint_R \mathbf{F_3}(x, y, f)\, dx\, dy = -\int_{S_1} \mathbf{F_3}\, \mathbf{k} \cdot \mathbf{N_1}\, ds_1 \qquad ...(4)$$

Using (3) and (4), (2) be comes

$$\int_V \frac{\partial \mathbf{F_3}}{\partial z}\, dx\, dy\, dz = \int_{S_2} \mathbf{F_3}\, \mathbf{k} \cdot \mathbf{N_1}\, ds_2 + \int_{S_1} \mathbf{F_3}\, \mathbf{k} \cdot \mathbf{N_1}\, ds_1$$

$$= \int_S \mathbf{F_3}\, \mathbf{k} \cdot \mathbf{N}\, ds = \int_S \mathbf{F_3} \cos\theta_3\, ds \qquad ...(5)$$

Similarly, by considering the projection of **S** on yz and zx-planes, we have

$$\int_V \frac{\partial F_1}{\partial x} dx\, dy\, dz = \int_S F_1 \mathbf{i} \cdot N_1 ds$$

$$= \int_S F_1 \cos \theta_1 ds \qquad ...(6)$$

and
$$\int_V \frac{\partial F_2}{\partial y} dx\, dy\, dz = \iint_S F_2 \mathbf{j} \cdot N ds$$

$$= \iint_S F_2 \cos \theta_1 ds \qquad ...(7)$$

Adding (5), (6) and (7), we get (1).

i.e.
$$\iint_S (F_1 \cos \theta_1 + F_2 \cos \theta_2 + F_3 \cos \theta_3) ds$$

$$= \int_V \left(\frac{\partial F_1}{\partial x} + \frac{\partial F_2}{\partial y} + \frac{\partial F_3}{\partial z} \right) dx\, dy\, dz$$

or
$$\int_S \mathbf{F} . \mathbf{N} ds = \int_V \nabla \cdot \mathbf{F} dv$$

Note 1 : In case the region be such that the lines drawn parallel to the co-ordinate axes meet it in more than two points, then we divide the region into various sub-regions each of which is met by a line parallel to any axis in only two points. Applying the theorem to each of these sub-regions and adding the results, we get the volume intergral over the whole region.

Note 2 : The divergence theorem is also applicabe for a regions **R** if it is bounded by two closed surfaces one of which lies with in the other.

14.7. Some deductions from sub divergence theorem :

(1) Green's theorem in symmetrical form : If ϕ and Ψ are two continuous and differentiable scalar point functions over the region **V** enclosed by the surface **S**, then

$$\int_V (\phi \nabla^2 \Psi - \Psi \nabla^2 \phi) dv = \int_S (\phi \nabla \Psi - \Psi \nabla \phi) \cdot \mathbf{N} ds$$

Proof : By divergence theorem, we have

$$\int_V \nabla \cdot \mathbf{F} dv = \int_S \mathbf{F} \cdot \mathbf{N} ds \qquad ...(1)$$

Putting $\mathbf{F} = \phi \nabla \Psi$, we get

$$\nabla \cdot \mathbf{F} = \nabla \cdot (\phi \nabla \Psi) = \phi (\nabla \cdot \nabla \Psi) + (\nabla \phi \cdot \nabla \Psi)$$

$$= \phi \nabla^2 \Psi = (\nabla \phi) : (\nabla \Psi)$$

and $\mathbf{F} \cdot \mathbf{N} = (\phi \nabla \Psi) \cdot \mathbf{N}$

$\therefore$ From (1)

$$\int_V [\phi \nabla^2 \Psi + (\nabla \phi) \cdot (\nabla \Psi)] \, dv$$

$$= \int_S (\phi \nabla \Psi) \cdot \mathbf{N} \, ds \qquad ...(2)$$

This is called *Green's first identity.*

Interchanging ϕ and Ψ in (2), we get

$$\int_V [\Psi \nabla^2 \phi + (\nabla \Psi) \cdot (\nabla \phi)] \, dv$$

$$= \int_S (\Psi \nabla \phi) \cdot \mathbf{N} \, ds \qquad ...(3)$$

By (3)–(1), we get

$$\int_V (\phi \nabla^2 \Psi - \Psi \nabla^2 \phi) \, dv = \int_S (\phi \nabla \Psi - \Psi \nabla \phi) \cdot \mathbf{N} \, ds \qquad ...(4)$$

which is called *Green's second identity.*

If $\frac{\partial}{\partial \mathbf{n}}$ denotes differentiation in the direction of the external normal to the bounding surface **S** enclosing the volume **V**, then $\nabla \Psi \cdot \mathbf{N} = \frac{\partial \Psi}{\partial \mathbf{n}}$ is the directional derivative of Ψ along the external normal at any point of **S**.

Hence we can also write Green's second identity as

$$\int_V (\phi \nabla^2 \Psi - \Psi \nabla^2 \phi) \, dv = \int_S \left(\phi \frac{\partial \Psi}{\partial \mathbf{n}} - \Psi \frac{\partial \phi}{\partial \mathbf{n}} \right) ds.$$

Note : Harmonic function :

If a scalar point function f satisfies Laplace's equation $\nabla^2 f = 0$, then f is called *harmonic function.*

If ϕ and Ψ are harmonic functions, then $\nabla^2 \phi = 0$ and $\nabla^2 \Psi = 0$. Hence sub Green's second identity, we get

$$\int_S (\phi \nabla \Psi - \Psi \nabla \phi) \cdot \mathbf{N}\, ds = 0$$

(2) Prove that $\int_V (\text{grad} \times \mathbf{F}) = \int_S (\mathbf{N} \times \mathbf{F})\, ds$

Proof : Let $\mathbf{f} = \mathbf{A} \times \mathbf{F}$, where A is a constant vector. Applying Gauss's divergence theorem on **f** we have

$$\int_S \mathbf{F} \cdot \mathbf{N}\, ds = \int_V \nabla \mathbf{f}\, dv$$

$$\Rightarrow \quad \int_S (\mathbf{A} \times \mathbf{F}) \cdot \mathbf{N}\, ds = \int_V \nabla \cdot (\mathbf{A} \times \mathbf{F})\, dv$$

$$\Rightarrow \quad \int_S \mathbf{A} \cdot (\mathbf{F} \times \mathbf{N})\, ds = -\int_V \nabla \cdot (\mathbf{F} \times \mathbf{A})\, dv$$

$$\text{or} \quad \mathbf{A} \cdot \int_S (\mathbf{N} \times \mathbf{F})\, ds = -\mathbf{A} \cdot \int_V (\nabla \times \mathbf{F})\, dv$$

$$\text{or} \quad \mathbf{A} \cdot \int_S (\mathbf{N} \times \mathbf{F})\, ds = \mathbf{A} \cdot \int_V (\nabla \times \mathbf{F})\, dv$$

$$\text{or} \quad \mathbf{A} \cdot \left\{ \int_S (\mathbf{N} \times \mathbf{F})\, ds - \int_V (\nabla \times \mathbf{F})\, dv \right\} = 0$$

[$\because$ **A** is a constant vector]

$$\Rightarrow \quad \int_S (\mathbf{N} \times \mathbf{F})\, ds - \int_V (\nabla \times \mathbf{F})\, dv = 0$$

$$\Rightarrow \quad \int_S \mathbf{N} \times \mathbf{F}\, ds = \int_V \nabla \times \mathbf{F}\, dv$$

(3) Prove that $\int_S \mathbf{N}\phi\, ds = \int_V \nabla \phi\, dv$

Proof : Let **A** be a constant vector, Applying Gauss's theorem to $\mathbf{A}\phi$, we have

$$\int_S (\mathbf{A}\phi) \cdot \mathbf{N}\, ds = \int_V \nabla \cdot (\mathbf{A}\phi)\, dv$$

$$\Rightarrow \quad \mathbf{A} \cdot \int_S \phi \mathbf{N} ds = \mathbf{A} \cdot \int_V \nabla \phi\, dv$$

$$\Rightarrow \quad \mathbf{A} \cdot \left\{ \int_S \phi \mathbf{N}\, ds - \int_V \nabla\phi\, dv \right\} = 0$$

[$\because$ **A** is a constant vector]

$$\int_S \phi \mathbf{N}\, ds - \int_V \nabla \phi\, dv = 0$$

or $\int_S \phi \mathbf{N}\, ds = \int_V \nabla\phi\, dv.$

Note : These above results can be systematically written as

$$\int_S \mathbf{N}\cdot\mathbf{F}\, ds = \int_V \nabla\cdot\mathbf{F}\, dv$$

$$\int_S \mathbf{N}\times\mathbf{F}\, ds = \int_V \nabla\times\mathbf{F}\, dv$$

$$\int_S \mathbf{N}\phi\, ds = \int_V \nabla\phi\, dv.$$

Example 6. For any closed surface **S**, prove that

(i) $\int_V \nabla\times\mathbf{F}\cdot\mathbf{N}\, ds = 0$

(ii) $\int_S \mathbf{r}\cdot\mathbf{N}\, ds = 3\mathbf{V}$, where **V** is the volume enclosed by **S.**

(iii) $\int_V \nabla\times\mathbf{N}\, dv = \mathbf{S}$

Solution : (i) By Gauss's divergence theorem, we have

$$\int_S \nabla\times\mathbf{F}\cdot\mathbf{N}\, ds = \int_V (\nabla\cdot\nabla\times\mathbf{F})\, dv$$

where **V** is the volume enclosed by **S**

$$= 0 \quad (\because \nabla\cdot\nabla\times\mathbf{F} = 0)$$

(ii) $$\int_S \mathbf{r}\cdot\mathbf{N}\, ds = \int_V \nabla\cdot\mathbf{r}\, dv$$

$$= \int_V 3\, dv \; (\because \nabla\cdot\mathbf{r} = 3)$$

$$= 3\mathbf{V}$$

(iii) $$\int_V \nabla\times\mathbf{N}\, dv = \int_S \mathbf{N}\cdot\mathbf{N}\, ds$$

$$\int_S ds = \mathbf{S} \quad (\because \mathbf{N}\cdot\mathbf{N}) = 1$$

Example 7. If l, m, n are the direction cosines of the outward drawn normal at a point **P** (x, y, z) of the closed surface **S** enclosing the volume **V**, then prove that

$$\int_S (lu + mv + nw)\, ds = \int_V \left(\frac{\partial u}{\partial x} + \frac{\partial v}{\partial y} + \frac{\partial w}{\partial z}\right) dx\, dy\, dz$$

Solution : From the problem it is clear that

$$\mathbf{N} = l\,\mathbf{i} + m\,\mathbf{J} + n\,\mathbf{k},$$

Let $\mathbf{F} = u\,\mathbf{i} + v\,\mathbf{j} + w\,\mathbf{k}, \therefore\ \mathbf{F} \cdot \mathbf{N} = lu = mv + nw$

But from Gauss's divergence theorem, we know that

$$\int_S \mathbf{F} \cdot \mathbf{N}\, ds = \int_V \nabla \cdot \mathbf{F}\, dv$$

i.e.
$$\int_S (lu + mv + nw) \int_V [\nabla \cdot (u\,\mathbf{i} + v\,\mathbf{j} + w\,\mathbf{k})]\, dv$$

$$= \int_V \left(\frac{\partial u}{\partial x} + \frac{\partial v}{\partial y} + \frac{\partial w}{\partial z}\right) dx\, dy\, dz$$

$(\because dv = dx\, dy\, dz)$

Hence proved.

Example 8. Using Gauss's divergence theorem show that

$$V = \int_S x\, dy\, dz = \int_S y\, dz\, dx = \int_S z\, dx\, dy.$$

$$= \frac{1}{3} \iint (x\, dy\, dz + y\, dz\, dx + dx\, dy$$

where **V** is the volume of a region bounded by surface **S.**

Solution : By Gauss's divergence theorem, we have

$$\int_S x\, dy\, dz = \int_V \frac{\partial}{\partial x}(x)\, dx\, dy\, dz$$

$$= \int_V dv \quad (\because dv = dx\, dy\, dz) = V$$

$$\therefore \qquad \int_S x\, dy\, dz = V \qquad ...(1)$$

Similarly
$$\int_S y\, dz\, dx = V \qquad ...(2)$$

and
$$\int_S z\, dx\, dy = V \qquad ...(3)$$

Adding (1), (2) and (3), we get

$$3V = \int_S x\, dy\, dz + \int_S y\, dz\, dx + \int_S z\, dx\, dy$$

or
$$V = \frac{1}{3}\int_S [x\, dy\, dz + y\, dz\, dx + z\, dx\, dy]$$

Hence the proof.

Example 9. Using divergence theorem, prove that

(i) $\int_S r^n \mathbf{r}\, ds = (n + 3) \int_V r^n\, dv; (n \neq -3)$

(ii) $\int_S \mathbf{F} \cdot \mathbf{N}\, ds = -4\pi \int_V \rho\, dv,$ where $\mathbf{F} = \nabla\phi$

and $\nabla^2 \phi = -4\pi\rho$

Solution : (i) From divergence theorem

$$\int_S r^n \mathbf{r}\, ds = \int_V \nabla \cdot (r^n \mathbf{r})\, dv \quad ...(1)$$

Now
$$\nabla \cdot (r^n \mathbf{r}) = \nabla r^n \mathbf{r} + r^n \nabla \cdot \mathbf{r}$$
$$= (nr^{n-1} \nabla r) \cdot \mathbf{r} + r^n \nabla \cdot \mathbf{r}$$
$$= \left(nr^{n-1} \frac{\mathbf{r}}{r}\right) \cdot \mathbf{r} + 3r^n$$
$$= n\, r^{n-1} \frac{r^2}{r} + 3\, r^n$$
$$= (n + 3)\, r^n \qquad \left(\because \ \nabla r = \frac{\mathbf{r}}{r}; \nabla \cdot \mathbf{r} = 3\right)$$

$\therefore$ From (1)

$$\int_S r^n \mathbf{r}\, ds = (n = 3) \int_V r^n\, dv$$

Hence the proof.

(ii) From divergence theorem, we have

$$\int_S \mathbf{F} \cdot \mathbf{N}\, ds = \int_V \nabla \cdot \mathbf{F}\, dv$$
$$= \int_V (\nabla \cdot \nabla\phi)\, dv \quad (\because \ \mathbf{F} = \nabla\phi)$$
$$= \int_V \nabla^2 \phi\, dv$$
$$= \int_V (-4\,\pi\rho)\, dv \ (\because \ \nabla^2 \phi = -4\pi\rho)$$
$$= -4\pi \int_V \rho\, dv$$

Hence the proof.

Example 10. It **S** be a closed surface and **r** denotes the position vector of any point (x, y, z) measured from the region O, prove that

$$\int_S \frac{\mathbf{N}\cdot\mathbf{r}}{r^3}\,ds = 0, \text{ if O lies outside } \mathbf{S}$$

$$= 4\pi, \qquad \text{if O lies inside } \mathbf{S}.$$

Solution : Case (i) *Origin* **O** *lies outside* **S.**

Here $\mathbf{F} = \frac{\mathbf{r}}{r^3}$ is continuously differentiable throughout the region enclosed by **S,**

Hence we can apply divergence theorem.

$$\therefore \quad \int_S \frac{\mathbf{N}\cdot\mathbf{r}}{r^3}\,ds = \int_V \left(\nabla\cdot\frac{\mathbf{r}}{r^3}\right) dv \qquad ...(1)$$

$$\nabla\cdot\frac{\mathbf{r}}{r^3} = r^{-3}\,\nabla\cdot\mathbf{r} + \mathbf{r}\cdot\nabla r^{-3}$$

$$= 3r^{-3} + \text{atr}\cdot(-3r^{-4}\,\nabla\mathbf{r}) \qquad (\because\ \nabla\cdot\mathbf{r} = 3)$$

$$= 3r^{-3} + \mathbf{r}\cdot(-3r^{-5}\,\mathbf{r}) \qquad \left(\because\ \nabla r = \frac{r}{r}\right.$$

$$= 3r^{-3} - 3r^{-3} = 0 \qquad \left.\mathbf{r}\,.\,\mathbf{r} = r^2\right)$$

$\therefore$ From (1)

$$\int_S \frac{\mathbf{N}\cdot\mathbf{r}}{r^3}\,ds = 0$$

Case (ii) *Origin* **O** *lies inside* **S**

Here we can not apply divergence theorem because $\mathbf{F} = \frac{\mathbf{r}}{r^3}$ has a point of discontinuity at the origin.

$\therefore$ In order to get rid of this difficulty let the origin O be enclosed by a small sphere $\mathbf{S_1}$ of radius a. Now **F** is continuously differentiable for all points with in the region $\mathbf{V_1}$ enclosed between **S** and $\mathbf{S_1}$.

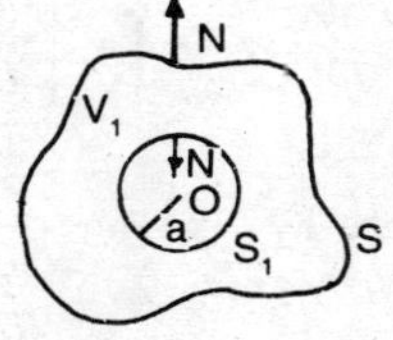

Fig. 14.10

Hence $\int_S \frac{\mathbf{N}\cdot\mathbf{r}}{r^3}\,ds + \int_{S_1} \frac{\mathbf{N}\cdot\mathbf{r}}{r^3}\,ds_1$

$$= \int_{V_1} \left(\nabla \cdot \left(\frac{\mathbf{r}}{r^3} \right) \right) dv_1$$

$$= 0 \quad \text{(as in case (i))}$$

$$\therefore \quad \int_S \frac{\mathbf{N} \cdot \mathbf{r}}{r^3} ds = - \int_{S_1} \frac{\mathbf{N} \cdot \mathbf{r}}{r^3} ds_1 \qquad ...(1)$$

Now from the figure it is evidient that on the shpere $\mathbf{S_1}$ the outward drawn unit normal vector **N** is directed towards the centre **O** of the sphere $\mathbf{S_1}$

$$\therefore \quad \text{on } S_1 \ \mathbf{N} = -\frac{\mathbf{r}}{a}$$

$$\therefore \quad = \int_{S_1} \frac{\mathbf{N} \cdot \mathbf{r}}{r^3} ds_1 = \int_{S_1} \frac{\mathbf{r}}{a} \cdot \frac{\mathbf{r}}{a^3} ds_1$$

$$= \int_{S_1} \frac{r^2}{a^4} ds_1$$

$$= \int_{S_1} \frac{1}{a^2} ds_1 \quad (\because r = a, r^2 = a^2 \text{ on } \mathbf{S_1}$$

$$= \frac{1}{a^2} \mathbf{S_1} = \frac{4\pi a^2}{a^2} = 4\pi$$

$$\therefore \text{ From (1)} \int_S \frac{\mathbf{N} \cdot \mathbf{r}}{r^3} ds = 4\pi$$

Example 11. Evaluate

$\int_S [x^2 \, dy \, dz + y^2 \, dz \, dx + 2z(xy - x - y) \, dx \, dy]$ where S is the surface of the cube $0 \le x \le 1, 0 \le y \le 1, 0 \le z \le 1$.

Solution : From Gauss's divergence theorem, we know that

$$\int_S [F_1 \, dy \, dz + F_2 \, dz \, dx + z \, dx \, dy]$$

$$= \int_V \left(\frac{\partial F_1}{\partial x} + \frac{\partial F_2}{\partial y} + \frac{\partial F_3}{\partial z} \right) dx \, dy \, dz \qquad ...(1)$$

(Cartesian form).

Here $\mathbf{F_1} = x^2 \quad \therefore \quad \frac{\partial F_1}{\partial x} = 2x$

$\mathbf{F_2} = y^2 \quad \therefore \quad \frac{\partial F_2}{\partial y} = 2y$

$$F_3 = 2x\,(xy - x - y) \quad \therefore \quad \frac{\partial F_3}{\partial z} = 2(xy - x - y)$$

$$\therefore \quad \frac{\partial F_1}{\partial x} + \frac{\partial F_2}{\partial y} + \frac{\partial F_3}{\partial z} = 2xy$$

$$\therefore \quad \int_S [x^2\, dy\, dz + y^2\, dz\, dx + 2z\,(xy - x - y)\, dx\, dy]$$

$$= \int_{x=0}^{1} \int_{y=0}^{1} \int_{z=0}^{1} 2\,xy\, dx\, dy\, dz$$

$$= 2\left\{\int_0^1 x\, dy\right\}\left\{\int_0^1 y\, dy\right\}\left\{\int_0^1 dz\right\}$$

$$= 2 \cdot \frac{1}{2} \cdot \frac{1}{2} \cdot 1 = \frac{1}{2}.$$

Example 12. Verify Gauss's divergence theorem for $\mathbf{F} = (x^2 - yz)\,\mathbf{i} + (y^2 - zx)\,\mathbf{j} + (z^2 - xy)\,\mathbf{k}$ taken over the rectangular parallelopiped $0 \le x \le a,\ 0 \le y \le b,\ 0 \le z \le c$.

Solution : Divergence theorem is

$$\int_V \nabla \cdot \mathbf{F}\, dv = \int_S \mathbf{F} \cdot \mathbf{N}\, ds \qquad ...(1)$$

$$\nabla \cdot \mathbf{F} = \frac{\partial}{\partial x}(x^2 - yz) = \frac{\partial}{\partial y}(y^2 - zx) + \frac{\partial}{\partial z}(z^2 - xy)$$

$$= 2\,(x + y + z)$$

$$\int_V \nabla \cdot \mathbf{F}\, dv = 2 \int_0^c \int_0^b \int_0^a (x + y + z)\, dx\, dy\, dz$$

$$= 2 \int_0^c \int_0^b \left(\frac{a^2}{2} + ya + za\right) dy\, dz$$

$$= 2 \int_0^c \left(\frac{a^2 b}{2} + \frac{ab^2}{2} + abz\right) dz$$

$$= 2\left(\frac{a^2bc}{2} + \frac{ab^2c}{2} + \frac{abc^2}{2}\right)$$

$$= abc\,(a + b + c) \qquad ...(2)$$

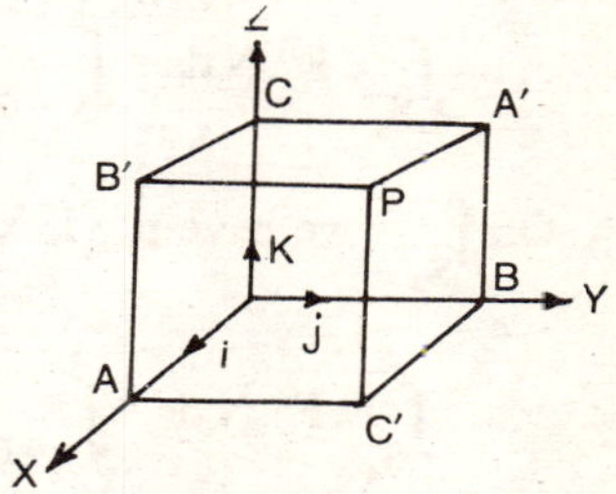

Fig. 14.11

To evaluate the surface integral, divide the closed surface **S** of the rectangular paralellopiped into 6 parts.

i.e. S_1 The face OAC'B ; S_4 : The face AC'PB'

S_2 The face CB'PA' ; S_5 : The face OCB'A

S_3 The face **OBA'C** ; S_6 : The face BA'PC'

$$\int_S \mathbf{F}\cdot\mathbf{N}\,ds = \int_{S_1} \mathbf{F}\cdot\mathbf{N}\,ds + \int_{S_2} \mathbf{F}\cdot\mathbf{N}\,ds + \int_{S_3} \mathbf{F}\cdot\mathbf{N}\,ds$$
$$+ \int_{S_4} \mathbf{F}\cdot\mathbf{N}\,ds \int_{S_5} \mathbf{F}\cdot\mathbf{N}\,ds + \int_{S_6} \mathbf{F}\cdot\mathbf{N}\,ds$$

On **S**, we have N = − **k**, z = 0

$$\mathbf{F} = x^2\,\mathbf{i} + y^2\,\mathbf{j} - xy\,\mathbf{k};\ \mathbf{F}\cdot\mathbf{N} = xy$$

$$\therefore \quad \int_{S_1} \mathbf{F}\cdot\mathbf{N}\,ds = \int_{x=0}^{a}\int_{y=0}^{b} xy\,dx\,dy$$
$$= \int_{x=0}^{a} \frac{xb^2}{2}\,dx = \frac{a^2b^2}{4}$$

On S_2, we have **N** = **k**; z = c ∴ $\mathbf{F}\cdot\mathbf{N} = c^2 - xy$

$$\int_{S_2} \mathbf{F}\cdot\mathbf{N}\,ds = \int_{x=0}^{a}\int_{x=0}^{b} (c^2 - xy)\,dx\,dy$$
$$= \int_{x=0}^{a} \left(c^2\,b - \frac{x\,b^2}{2}\right) dx.$$
$$= c^2\,ba - \frac{a^2\,b^2}{4}$$

On S_3 we have **n** = − **i**, x = 0 ∴ $\mathbf{F}\cdot\mathbf{N} = yz$

$$\therefore \quad \int_{S_3} \mathbf{F}\cdot\mathbf{N}\,ds = \int_{y=0}^{b}\int_{z=0}^{c} yz\,dy\,dz = \frac{b^2\,c^2}{4}$$

On $\int_{S_4}$ we have **n** − **i**, x = a ∴ $\mathbf{F}\cdot\mathbf{N} = (a^2 - yz)$

$$\therefore \quad \int_{S_4} \mathbf{F}\cdot\mathbf{N}\,ds = \int_{y=0}^{b}\int_{z=0}^{c} (a^2 - yz)\,dy\,dz = a^2\,bc - \frac{b^2\,c^2}{4}$$

On S_5, we have N = - **j**, y = 0 ∴ $\mathbf{F}\cdot\mathbf{N} = xz$

$$\therefore \quad \int_{S_5} \mathbf{F}\cdot\mathbf{N}\,ds = \int_{z=0}^{c}\int_{x=0}^{a} xz\,dz\,dz = \frac{a^2\,c^2}{4}$$

On S_6, we have $\mathbf{n} = \mathbf{j}$, $y = b$ $\therefore$ $\mathbf{F} \cdot \mathbf{N} = (b^2 - xz)$

$$\therefore \quad \int_{S_6} \mathbf{F} \cdot \mathbf{N}\, ds = \int_{z=0}^{c} \int_{x=0}^{a} (b^2 - xz)\, dz\, dx$$

$$= b^2 ca - \frac{a^2 c^2}{4}$$

$$\therefore \quad \int_S \mathbf{F} \cdot \mathbf{N}\, ds = \frac{a^2 b^2}{4} + c^2 ab - \frac{a^2 b^2}{4}$$

$$+ \frac{b^2 c^2}{4} + a^2 bc - \frac{b^2 c^2}{4}$$

$$+ \frac{c^2 a^2}{4} + b^2 ca - \frac{c^2 a^2}{4}$$

$$= abc\,(a + b + c) \qquad \text{...(3)}$$

(2) and (3) says that the theorem is verified.

Example 13. Apply divergence theorem to evaluate

$$\int_S (x + z)\, dy\, dz + (y + z)\, dz\, dx + (x + y)\, dx\, dy$$

where S is the surface of the sphere $x^2 + y^2 + z^2 = 4$.

Solution : From divergence theorem, we know that

$$\int_S (F_1\, dy\, dz + F_2\, dz\, dx + z\, dx\, dy)$$

$$= \int_V \left(\frac{\partial F_1}{\partial x} + \frac{\partial F_2}{\partial y} + \frac{\partial F_3}{\partial z}\right) dx\, dy\, dz \qquad \text{...(1)}$$

(Cartesian form).

Here $F_1 = x + z$ $\therefore$ $\dfrac{\partial F_1}{\partial x} = 1$

$F_2 = y + z$ $\therefore$ $\dfrac{\partial F_2}{\partial y} = 1$

$F_3 = x + y$ $\therefore$ $\dfrac{\partial F_3}{\partial z} = 0$

Hence $\dfrac{\partial F_1}{\partial x} + \dfrac{\partial F_2}{\partial y} + \dfrac{\partial F_3}{\partial z} = 1 + 1 + 0 = 2$

$$\therefore \quad \int_S (x + z)\, dy\, dz + (y + z)\, dz\, dx + (x - y)\, dx\, dy$$

$$= \int_V 2\, dx\, dy\, dz$$

$$= 2 \int_V dx\, dy\, dz$$

= 2 (Volume of the given sphere)

$$= 2 \cdot \frac{4}{3}\pi (2)^3 = \frac{64\pi}{3.}$$

Example 14. Evaluate $\int_S (ax^2 + by^2 + cz^2)\, ds$ over the unit sphere.

Solution : By divergence theorem

$$\int_S \mathbf{F} \cdot \mathbf{N}\, ds = \int_V \nabla \cdot \mathbf{F}\, dv \qquad ...(1)$$

Given $\quad \mathbf{F} \cdot \mathbf{N} = ax^2 + by^2 + cz^2 \qquad ...(2)$

Let $\phi = x^2 + y^2 + z^2 - 1 \quad (\because\ x^2 + y^2 + z^2 = 1$ in the unit sphere)

$\therefore$ Unit Normal vector **N** to the surface ϕ is

$$= \frac{\nabla\phi}{|\nabla\phi|} = \frac{2(x\mathbf{i} + y\mathbf{i} + z\mathbf{k})}{2\sqrt{x^2 + y^2 + z^2}}$$

$$= x\mathbf{i} + y\mathbf{j} + z\mathbf{k}$$

But $\mathbf{F} \cdot \mathbf{N} = \mathbf{F} \cdot (x\mathbf{i} + y\mathbf{i} + z\mathbf{k})$

$$\therefore \quad ax^2 + by^2 + cz^2 = \mathbf{F} \cdot (x\mathbf{i} + y\mathbf{i} + z\mathbf{k})$$

$$\Longleftrightarrow \quad \mathbf{F} = ax\,\mathbf{i} + by\,\mathbf{j} + cz\,\mathbf{k}$$

$\therefore$ From (1)

$$\int_S (ax^2 + by^2 + cz^2)\, ds = \int_V \nabla \cdot (ax\mathbf{i} + by\mathbf{j} + cz\mathbf{k})\, dv$$

$$= (a + b + c) \int_V dv$$

$$= (a + b + c)\, \mathbf{V}$$

$$= \frac{4\pi}{3}(a + b + c) \quad \left(\because\ \mathbf{V} = \frac{4\pi}{3}\right)$$

Example 15. Verify divergence theorem for $\mathbf{F} = 4x\mathbf{i} - 2y^2\mathbf{j} + z^2\mathbf{k}$ taken over the region bounded by the cylinder $x^2 + y^2 = 4$, $z = 0$, $z = 3$.

Solution : Gauss's divergence theorem is

$$\int_V \nabla \cdot \mathbf{F}\, dv = \int_S \mathbf{F} \cdot \mathbf{N}\, ds \qquad ...(1)$$

$$\mathbf{F} = 4\,x\mathbf{i} - 2\,y^2\mathbf{j} + z^2\mathbf{k}$$

$$\nabla \cdot \mathbf{F} = \frac{\partial}{\partial x}(4x) - \frac{\partial}{\partial y}(2y^2) + \frac{\partial}{\partial z}(z^2) = 4 - 4y + 2z$$

$\therefore \quad \int_V \nabla \cdot \mathbf{F}\, dv$

$$= \int_{x=-2}^{2} \int_{y=-\sqrt{4-x^2}}^{\sqrt{4-x^2}} \int_{z=0}^{3} (4 - 4y + 2z)\, dx\, dy\, dz$$

$$= \int_{-2}^{2} \int_{-\sqrt{4-x^2}}^{\sqrt{4-x^2}} \left(4z - 4yz + z^2\right)_0^3 dy\, dx$$

$$= \int_{-2}^{2} \int_{-\sqrt{4-x^2}}^{\sqrt{4-x^2}} (21 - 12y)\, dy\, dx \qquad ...(2)$$

Here 12 y is an odd function i.e. if f(y) = 12 y

then $f(-y) = -f(y) \quad \therefore \int_{-\sqrt{4-x^2}}^{\sqrt{4-x^2}} 12y\, dy = 0$

and 21 is even function i.e., if f(y) = 21

$$f(-y) = f(y)$$

$$\therefore \quad \int_{-\sqrt{4-x^2}}^{\sqrt{4-x^2}} 21\, dy = 2 \int_0^{\sqrt{4-x^2}} 21\, dy$$

$\therefore$ From (2)

$$\int_V \nabla \cdot \mathbf{F}\, dv = 2 \int_{-2}^{2} \int_0^{\sqrt{4-x^2}} 21\, dy\, dx$$

$$= 42 \int_{-2}^{2} [y]_0^{\sqrt{4-x^2}} dx$$

$$= 42 \int_{-2}^{2} \sqrt{4-x^2}\, dx = 84 \int_0^2 \sqrt{4-x^2}\, dx$$

$$= 84\, \frac{1}{2} \left\{ x\sqrt{4-x^2} + 4\sin^{-1}\frac{x}{2} \right\}_0^2$$

$$= 42\,(4\sin^{-1} 1) = 168 \cdot \frac{\pi}{2} = 84\,\pi$$

$$\therefore \quad \int_V \nabla \cdot \mathbf{F}\, dv = 84\pi \qquad \text{...(3)}$$

To evaluate the surface integral, divide the closed surface **S** < of the cylinder into 3 parts

$\mathbf{S_1}$: The circular base in the plane z = 0

$\mathbf{S_2}$: The circular top in the plane z = 3

$\mathbf{S_3}$: The curved surface of the cylinder given by $x^2 = y^2 = 4$.

$$\therefore \quad \int_S \mathbf{F}\cdot\mathbf{N}\, ds = \int_{S_1} \mathbf{F}\,.\,\mathbf{N}\, dx + \int_{S_2} \mathbf{F}\cdot\mathbf{N}\, ds + \int_{S_3} \mathbf{F}\cdot\mathbf{N}\, ds \qquad \text{...(4)}$$

On $\mathbf{S_1}$ we have

$\mathbf{N} = -\,\mathbf{k}, \mathbf{F} = 4x\,\mathbf{i} - 2y^2\,\mathbf{j}$

$(\because\ z = 0)$

$$\therefore \quad \mathbf{F}\cdot\mathbf{N} = 0 \Rightarrow \int_{S_1} \mathbf{F}\,.\,\mathbf{N}\, ds = 0 \quad \text{...(5)}$$

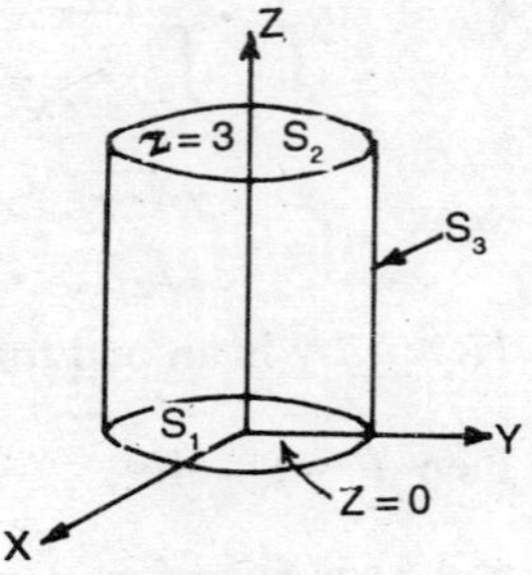

Fig. 14.12

On $\mathbf{S_2}$, we have

$\mathbf{N} = \mathbf{k}, \mathbf{F} = 4x\,\mathbf{i} - 2y^2\,\mathbf{j} + 9\mathbf{k}$

$(\because\ z = 3)$

$\therefore \quad \mathbf{F}\cdot\mathbf{N} = 9$

$$\Rightarrow \quad \int_{S_2} \mathbf{F}\cdot\mathbf{N}\, ds = 9 \int_{S_3} dx\, dy$$

$$= 9\ (\text{area of surface } S_2)$$

$$= 9\,(\pi \cdot 2^2) = 36\pi \qquad \text{...(6)}$$

On $\mathbf{S_3}$:

A unit normal vector to the surface $x^2 + y^2 = 4$ is

$$\mathbf{N} = \frac{\nabla\,(x^2 + y^2)}{|\,\nabla\,(x^2 = y^2)\,|} = \frac{x\mathbf{i} + y\mathbf{j}}{2}$$

$$\mathbf{F}\cdot\mathbf{N} = (4x\,\mathbf{i} - 2y^2\,\mathbf{j} + z^2\,\mathbf{k}) \cdot \left(\frac{x\mathbf{i} = y\,\mathbf{j}}{2}\right)$$

$$= 2x^2 - y^3$$

Also on S_3 :

$$x^2 + y^2 = 4$$

$\therefore$ $x = 2\cos\theta$, $y = 2\sin\theta$ and $ds = 2d\theta\, dz$

And to cover the total curved surface

$$0 \le \theta \le 2\pi,\ 0 \le z \le 3$$

$$\therefore \int_{S_3} \mathbf{F}\cdot\mathbf{N}\,ds = \int_{\theta=0}^{2\pi}\int_{z=0}^{3} \{2(2\cos\theta)^2 - (2\sin\theta)^3\}\, 2\,dz\,d\theta$$

$$= 16\int_{\theta=0}^{2\pi} (\cos^2\theta - \sin^2\theta)\,d\theta \int_{z=0}^{3} dz$$

$$= 16\left\{\int_0^{2\pi} \cos^2\theta\,d\theta - \int_0^{2\pi} \sin^3\theta\,d\theta\right\}(3)$$

$$= 48\left\{4\int_0^{\pi/2} \cos^2\theta\,d\theta - 0\right\}$$

$$= 48\pi \qquad ...(7)$$

$\therefore$ From (4), (5), (6), and (7)

$$\int_S \mathbf{F}\cdot\mathbf{N}\,ds = 0 + 36\pi + 48\pi = 84\pi \qquad ...(8)$$

Hence from (1) and (8), the divergence theorem is proved.

Example 16. Evaluate $\int_S \mathbf{F}\cdot\mathbf{N}\,ds$, where $\mathbf{F} = x\mathbf{i} + y\mathbf{j} + z^2\mathbf{k}$ and S is the surface bounded by cone $x^2 + y^2 = z^2$ in the plane $z = 4$.

Solution : If **V** is the volume enclosed by **S**, then **V** is bounded by the surfaces $z = 0$, $z = 4$, and $z^2 = x^2 + y^2$.

By divergence theorem, we have

$$\int_S \mathbf{F}\cdot\mathbf{N}\,ds = \int_V \nabla\cdot\mathbf{F}\,dv$$

$$\mathbf{F} = x\mathbf{i} = y\mathbf{j} + z^2\mathbf{k},$$

$$\therefore\ \nabla\cdot\mathbf{F} = 2(1 + z)$$

$$\therefore\ \int_V \nabla\cdot\mathbf{F}\,dv$$

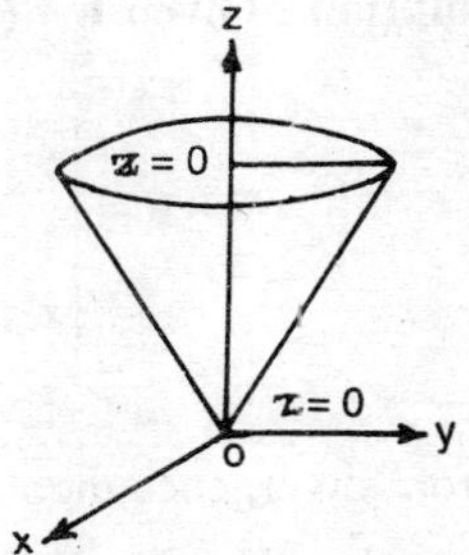

Fig. 14.13

$$= 2\int_V (1 + z)\, dv$$

$$= 2\int_{z=0}^{4} \int_{y=-z}^{z} \int_{x=-\sqrt{z^2-y^2}}^{\sqrt{z^2-y^2}} (1 + z)\, dx\, dy\, dz$$

$$= 4\int_0^4 \int_{-z}^{z} \int_0^{\sqrt{z^2-y^2}} (1 + z)\, dx\, dy\, dz$$

$$= 4\int_0^4 \int_{-z}^{z} (1 + z)\sqrt{z^2-y^2}\, dy\, dz$$

$$= 8\int_0^4 \int_0^{z} (1 + z)\sqrt{z^2-y^2}\, dy\, dz$$

$$= 8\int_0^4 (1 + z) \left\{\frac{y\sqrt{z^2-y^2}}{2} + \frac{z^2}{2}\sin^{-1}\frac{y}{z}\right\}_0^z dz$$

$$= 8\int_0^4 (1 + z)\frac{z^2}{2}\cdot\frac{\pi}{2}\, dz$$

$$= 2\pi\int_0^4 (z^2 + z^3)\, dz = 2\pi \left\{\frac{z^3}{3} + \frac{z^4}{4}\right\}_0^4$$

$$= 2\pi\left\{\frac{64}{3} + 64\right\} = \frac{512\pi}{3}.$$

Example 17. Evaluate $\int_S (\nabla \times \mathbf{F}) \cdot \mathbf{N}\, ds$ where **S** is the surface of the paraboloid $x^2 + y^2 + z = 4$ above the xy-plane and $\mathbf{F} = (x^2 + y)\,\mathbf{i} + 3xy\,\mathbf{j} + (2xz + z^2)\,\mathbf{k}$.

Solution : Given $\mathbf{F} = (x^2 + y - 4)\,\mathbf{i} = 3xy\,\mathbf{j} + (2xz + z^2)\,\mathbf{k}$

$$\nabla \times \mathbf{F} = \begin{vmatrix} \mathbf{i} & \mathbf{j} & \mathbf{k} \\ \frac{\partial}{\partial x} & \frac{\partial}{\partial y} & \frac{\partial}{\partial z} \\ x^2 + y - 4 & 3xy & 2xz + z^2 \end{vmatrix}$$

$$= -2z\,\mathbf{j} + (3y - 1)\,\mathbf{k}$$

From divergence theorem

$$\int_S (\nabla \times \mathbf{F}) \cdot \mathbf{N}\, ds = \int_V \nabla \cdot (\nabla \times \mathbf{F})\, dv \qquad ...(1)$$

$$\therefore \quad \nabla \cdot \nabla \times \mathbf{F} = \frac{\partial}{\partial x}(0) + \frac{\partial}{\partial y}(-2z) + \frac{\partial}{\partial z}(3y - 1)$$

Hence from (1) $\int_S (\nabla \times \mathbf{F})\, \mathbf{N}\, ds = 0.$

Example 18. Evaluate $\int_S (x^2\mathbf{i} + y^2\mathbf{j} + z^2\mathbf{k}) \cdot \mathbf{N}\, ds$, where S denotes the surface of the ellipsoid $\frac{x^2}{a^2} + \frac{y^2}{b^2} + \frac{z^2}{c^2} = 1.$

Solution : $\int_S \mathbf{F} \cdot \mathbf{N}\, ds = \int_V \nabla \cdot \mathbf{F}\, dv$ ...(1)

Here $\mathbf{F} = x^2\mathbf{i} + y^2\mathbf{j} + z^2\mathbf{k}$

$\therefore \quad \nabla \cdot \mathbf{F} = 2(x + y + z).$

$\therefore$ From (1)

$$\int_S \mathbf{F} \cdot \mathbf{N}\, ds = \int_{-a}^{a} \int_{-b}^{b} \int_{-c}^{c} 2(x + y + z)\, dx\, dy\, dz$$

$$= 2\int_{-a}^{a} \int_{-b}^{b} \left\{(x + y)\, z + \frac{z^2}{2}\right\}_{-c}^{c} dx\, dy$$

$$= 4 \int_{-a}^{a} \int_{-b}^{b} (x + y)\, c\, dx\, dy$$

$$= 4c\int_{-a}^{a} \left(xy + \frac{y^2}{2}\right)_{-b}^{b} dx$$

$$= 4c\int_{-a}^{a} 2\, bx\, dx$$

$$= 8\, bc\int_{-a}^{a} x\, dx = 0.$$

Example 19. Evaluate $\int_S (y^2 z^2\, \mathbf{i} + z^2 x^2\, \mathbf{j} + z^2 y^2\, \mathbf{k}) \cdot \mathbf{N}\, ds$ where S is the part of the unit sphere above the xy-plane and bounded by this plane.

Solution :

Let **V** be the volume enclosed by the surface **S**.

Here

$$\mathbf{F} = y^2 z^2 \mathbf{i} + z^2 x^2 \mathbf{j} + z^2 y^2 \mathbf{k}$$

and $\nabla \cdot \mathbf{F} = 2\,zy^2$

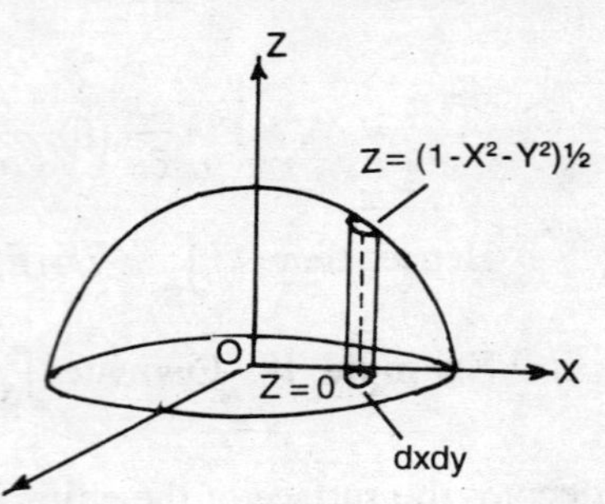

Fig. 14.14(a)

$\therefore$ By Gauss's divergence theorem

$$\int_S \mathbf{F}\cdot\mathbf{N}\,ds = \int_V \nabla\cdot\mathbf{F}\,dv$$

$$= \int_V 2\,zy^2\,dv \qquad ...(1)$$

Here **V** is the volume enclosed by surface **S**, i.e. **V** is the hemisphere $x^2 + y^2 + z^2 = 1$, above the xy plane [Fig. 14.14(a)].

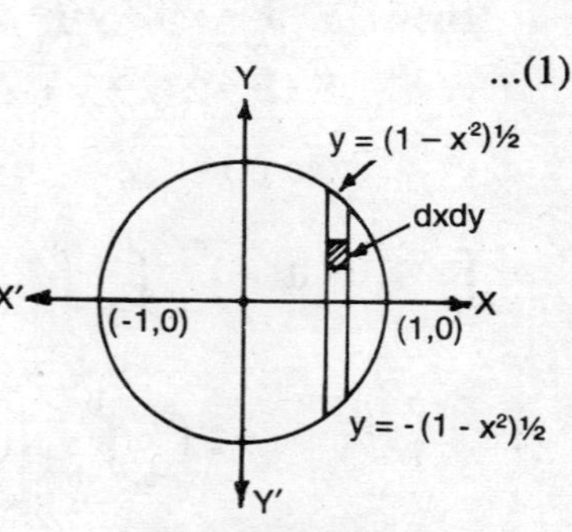

Fig. 14.14(b)

From the figures it is evident that the limits of integration of R.H.S. of (1) are

$$0 \le z \le \sqrt{1 - x^2 - y^2}$$

$$-\sqrt{1 - x^2} \le y \le \sqrt{1 - x^2}$$

$$-1 \le x \le 1$$

Hence R.H.S. of (1)

$$= \int_{x=-1}^{1} \int_{y=-\sqrt{1-x^2}}^{\sqrt{1-x^2}} \int_{z=0}^{z=\sqrt{1-x^2-y^2}} 2\,zy^2\,dx\,dy\,dz$$

$$= \int_{x=-1}^{1} \int_{y=-\sqrt{1-x^2}}^{\sqrt{1-x^2}} (1 - x^2 - y^2)\,y^2\,dx\,dy$$

$$= \int_{x=-1}^{1} \left\{ \frac{1}{3}(1 - x^2)\, y^3 - \frac{1}{5} y^5 \right\}_{-\sqrt{1-x^2}}^{\sqrt{1-x^2}} dx$$

$$= 2\int_{x=-1}^{1} \left\{ \frac{1}{3}(- x^2)^{5/2} - \frac{1}{5}(1 - x^2)^{5/2} \right\} dx$$

$$= \frac{8}{15}\int_0^1 (1 - x^2)^{5/2}\, dx$$

$$= \frac{8}{15}\int_0^{\pi/2} (1 - \sin^2\theta)^{5/2} \cos\theta\, d\theta; \quad (x = \sin\theta)$$

$$= \frac{8}{15}\int_0^{\pi/2} \cos^6\theta\, d\theta = \frac{8}{15}\,\frac{5}{6}\,\frac{3}{4}\,\frac{1}{2}\cdot\frac{\pi}{2} = \frac{\pi}{12}$$

$$\therefore \quad \int_S \mathbf{F}\cdot\mathbf{N}\, ds = \frac{\pi}{12}$$

Another method: From eqn. (1)

$$\int_S \mathbf{F}\cdot\mathbf{N}\, ds = 2\int_V zy^2\, dv \qquad \text{...(1)}$$

Changing R.H.S. to spherical polar co-ordinater by putting $x = r\sin\theta\cos\phi,\ y = r\sin\theta\sin\phi,\ z = r\cos\theta$

and $\quad dy = r^2 \sin\theta\, dr\, d\theta\, d\phi$

so that $\quad 0 \le r \le 1\ ; 0 \le \theta \le \pi/2;\ \ 0 \le \phi \le 2\pi$

$$\therefore \quad 2\int_V zy^2\, dv$$

$$= 2\int_{\phi=0}^{2\pi}\int_{\phi=0}^{\pi/2}\int_{r=0}^{1} (r\cos\theta)(r^2\sin^2\theta\sin^2\phi)\, r^2\sin\theta\, dr\, d\theta\, d\phi$$

$$= 2\int_{\phi=0}^{2\pi}\int_{\phi=0}^{\pi/2}\int_{r=0}^{1} r^5 \sin^3\theta\cos\theta\sin^2\phi\, dr\, d\theta\, d\phi$$

$$= 2\int_{\phi=0}^{2\pi} \sin^2\phi\, d\phi \cdot \int_{\theta=0}^{\pi/2} \sin^3\theta\cos\theta\, d\theta \cdot \int_{r=0}^{1} r^5\, dr$$

$$= 2(\pi)\left(\frac{1}{4}\right)\left(\frac{1}{6}\right) = \frac{\pi}{12}.$$

EXERCISE

1. Show that (i) $\int_S \mathbf{N}\, ds = 0$ (ii) $\int_S \mathbf{r} \times \mathbf{N}\, ds = 0$

(iii) $\int_S \frac{\mathbf{N} \cdot \mathbf{r}}{r^2}\, ds = \int\int_V \frac{dv}{r^2}$

(iv) $\int_S \nabla r^2 \cdot \mathbf{N}\, ds = 6\mathbf{V}$

where **S** is the surface enclosing volume **V.**

2. Show that

(i) $\int_V (u\nabla \cdot \mathbf{V} + \nabla u \cdot \mathbf{V})\, dv = \int_S u\mathbf{V} \cdot \mathbf{N}\, ds$

(ii) $\int_V [\mathbf{f} \cdot \nabla \times \mathbf{F}]\, dv = \int_S \mathbf{F} \times \mathbf{f} \cdot \mathbf{N}\, ds + \int_V \mathbf{F} \cdot \nabla \mathbf{f}\, dv$

3. (i) Evaluate $\int_S \mathbf{F} \cdot \mathbf{N}\, ds$, where $\mathbf{F} = 18z\mathbf{i} - 12\mathbf{j} + 3y\mathbf{k}$ and **S** is the surface of the plane $2x + 3y + 6z = 12$ in the first octant.

(ii) Evaluate $\int_S \mathbf{F}\, dS$, where $\mathbf{F} = (x + y^2)\,\mathbf{i} - 2x\,\mathbf{j} + 2yz\,\mathbf{k}$ and **S** is the surface of the plane $2x + y + 2z = 6$ in the first octant.

4. Verify the divergence theorem for $\mathbf{F} = x^2\mathbf{i} + z\mathbf{j} + y\mathbf{k}$ taken over the surface bounded by $x = 0$, $x = 1$, $y = 0$, $y = 1$, $z = 0$ and $z = 1$

5. Verify the divergence theorem for $\mathbf{F} = y\mathbf{i} + x\mathbf{j} + z^2\mathbf{k}$ over the cylindrical region bounded by $x^2 + y^2 = a^2$, $z = 0$, $z = h$

6. Verify the divergence theorem for $\mathbf{F} = 6z\mathbf{i} + (2x + y)\,\mathbf{j} - x\mathbf{k}$ taken over the region bounded by the surface of the cylinder $x^2 + z^2 = 9$ included between $x = 0$, $y = 0$, $z = 0$ and $y = 8$.

7. Verify the divergence theorem for $\mathbf{F} = 2x^2 y\mathbf{i} - y^2\mathbf{j} + 4xz^2\mathbf{k}$ taken over the region in the first octant bounded by $y^2 + z^2 = 9$ and $x = 2$.

8. Evaluate $\int_S [x^2\, dy\, dz + y^3\, dz\, dx + z^3\, dx\, dy]$ over the surface **S** of a cube bounded by the co-ordinate planes $x = y = z = a$, using divergence theorem.

9. Evaluate $\int_S$ [x dy dz + y dz dx + z dx dy 3 where **S** is the surface $x^2 + y^2 + z^2 = 1$, using divergence theorem.

10. Evaluate using divergence theorem
$\int_S (x^3\, dy\, dz + x^2y\, dz\, dx + x^2z\, dx\, dy)$, where **S** is the closed surface consisting of the cylinder $x^2 + y^2 = a^2$ $(0 \le z \le b)$ and the circular discs z = 0 and z = b.

11. Evaluate $\int_S$ [x^3 – yz) dy dz – 2 x^2y dz dx + z dx dy] over the surface of the cube bounded by the coordinate planes and the planes x = y = z = a, using divergence theorem.

12. Evaluate $\int_S \mathbf{F} \cdot \mathbf{N}\, ds$ using divergence theorem where $\mathbf{F} = (2x + 3z)\,\mathbf{i} - (xz + y)\,\mathbf{j} + (y^2 + 2z)\,\mathbf{k}$ and **S** is the surface of the sphere $(x - 3)^2 + (y + 1)^2 + (z - 2)^2 = 9$.

13. Evaluate $\int_S \mathbf{F} \cdot \mathbf{N}\, ds$, using divergence theorem where $\mathbf{F} = x^3\mathbf{i} + y^3\mathbf{j} + z^3\mathbf{k}$ and **S** is the surface of sphere $x^2 + y^2 + z^2 = a^2$.

14. If $\mathbf{F} = (x^2 + y - 4)\,\mathbf{i} = 3xy\,\mathbf{j} + (2xz + z^2)\,\mathbf{k}$, then by divergence theorem evaluate $\int_S (\nabla \times \mathbf{F}) \cdot \mathbf{N}\, ds$ where **S** is the surface of gthe sphere $x^2 + y^2 + z^2 = 16$ in the xy plane.

15. Evaluate $\int_S \mathbf{F} \cdot \mathbf{N}\, ds$ over the entire surface of the region above the xy plane bounded by the cone $z^2 = x^2 + y^2$ and the plane z = 4, if $\mathbf{F} = 4\,xz\,\mathbf{i} + xyz^2\,\mathbf{j} + 3z\,\mathbf{k}$ using divergence theorem.

16. Evaluate $\int_S (\nabla \times \mathbf{F}) \cdot \mathbf{N}\, ds$ using divergence theorem where $\mathbf{F} = (x - z)\,\mathbf{i} + (x^2 + yz)\,\mathbf{j} - 3xy^2\,\mathbf{k}$ and **S** in the surface of the cone $z = 2 - \sqrt{x^2 + y^2}$ above the xy plane.

17. Verify the divergence theorem for $\mathbf{F} = x\mathbf{i} + y^2\mathbf{j} = z^2\mathbf{k}$ where S is the surface of the solid cut off by the plane x + y + z = a in the first octant.

18. Show that

$$\nabla\phi = \underset{V \to 0}{\text{Lt}} \frac{\int_S N\phi \, ds}{V}$$

$$\nabla \cdot \mathbf{F} = \underset{V \to 0}{\text{Lt}} \frac{\int_S N \cdot \mathbf{F} \, ds}{V}$$

and $$\nabla \times \mathbf{F} = \underset{V \to 0}{\text{Lt}} \frac{\int_S N \times \mathbf{F} \, ds}{V}$$

19. Evaluate $\int_S (lx^2 + my^2 + nz^2) \cdot \mathbf{N} \, ds$ taken over the sphere $(x - a)^2 + (y - b)^2 + (z - c)^2 = r^2$, l, m, n being the direction cosines of external normal to the sphere.

20. (i) Using divergence theorem evaluate

$$\int_S \mathbf{F} \cdot \mathbf{N} \, ds \text{ where } \mathbf{F} = x\mathbf{i} + 2y\mathbf{j} + 3z\mathbf{k}.$$

where **S** in the surface enclosing a volume **V**.

(ii) If $\mathbf{F} = ax\mathbf{i} + by\mathbf{j} + cz\mathbf{k}$; where a, b, c are constants evaluate $\int_S \mathbf{F} \cdot \mathbf{N} \, ds$.

Answers

3. (i) 24 (ii) 180 8. $3a^5$ 9. 3π 10. $\frac{5}{4}\pi a^4 b$

11. $\frac{1}{3}a^3(a^2 + 3)$ 12. 180π 13. $\frac{12}{5}\pi a^5$ 14. -16π

15. 320π 16. 12π 19. $-\frac{8\pi}{3}(a + b + c)r^3$

20. $\frac{4}{3}\pi(a + b + c)$.

14.6. Stoke's theorem : *(Relation between line and surface integrals)*

If **S** be an open surface bounded by a closed curve **C**, and **F** be any vector point function having continuous first order partial derivatives, then

$$\int_C \mathbf{F} \cdot \mathbf{dr} = \int_S (\nabla \times \mathbf{F})\, \mathbf{N} \cdot ds$$

where **N** is a unit external normal at any point of **S** (drawn in the sense in which a right handed screw would advance when rotated in the sense of description of **C**)

Proof : Let $\mathbf{F} = \mathbf{F_1\, i + F_2\, j + F_3\, k}$

Let $\theta_1, \theta_2, \theta_3$ be the angles which the unit outward normal vector **N** makes with the positive directions of x, y, z axes respectively.

$\therefore \quad \mathbf{N} = \cos\theta_1\, \mathbf{i} + \cos\theta_2\, \mathbf{j} + \cos\theta_3\, \mathbf{k}$

Also

$\mathbf{r} = x\mathbf{i} + y\mathbf{j} + z\mathbf{k}, \quad \therefore \quad \mathbf{dr} = dx\, \mathbf{i} + dy\, \mathbf{j} + dz\, \mathbf{k}$

$$\nabla \times \mathbf{F} = \begin{vmatrix} \mathbf{i} & \mathbf{j} & \mathbf{k} \\ \dfrac{\partial}{\partial x} & \dfrac{\partial}{\partial y} & \dfrac{\partial}{\partial z} \\ F_1 & F_2 & F_3 \end{vmatrix}$$

$$= \left(\frac{\partial F_3}{\partial y} - \frac{\partial F_2}{\partial z}\right)\mathbf{i} + \left(\frac{\partial F_1}{\partial z} - \frac{\partial F_3}{\partial x}\mathbf{j}\right) + \left(\frac{\partial F_2}{\partial x} - \frac{\partial F_1}{\partial y}\right)\mathbf{k}$$

$$\therefore \quad (\nabla \times \mathbf{F}) \cdot \mathbf{N} = \left(\frac{\partial F_3}{\partial y} - \frac{\partial F_2}{\partial z}\right)\cos\theta_1$$

$$+ \left(\frac{\partial F_1}{\partial z} - \frac{\partial F_3}{\partial x}\right)\cos\theta_2 + \left(\frac{\partial F_2}{\partial x} - \frac{\partial F_1}{\partial y}\right)\cos\theta_3$$

Also $\mathbf{F} \cdot \mathbf{r} = F_1\, dx + F_2\, dy + F_3\, dz$

Hence the Stoke's theorem can be written as

$$\int_C F_1\, dx + F_2\, dy + F_3\, dz$$

$$= \int_S \left\{ \left(\frac{\partial F_3}{\partial y} - \frac{\partial F_2}{\partial z} \right) \cos \theta_1 + \left(\frac{\partial F_1}{\partial z} - \frac{\partial F_3}{\partial x} \right) \cos \theta_2 + \left(\frac{\partial F_2}{\partial x} - \frac{\partial F_1}{\partial y} \right) \cos \theta_3 \right\} ds \quad \text{...(1)}$$

Let $z = f(x, y)$ be the equation of the surface **S** whose projection on the xy-plane is the region **R**. Then the projection of **C** on the xy-plane is the curve $\mathbf{C}^1$ enclosing **R**.

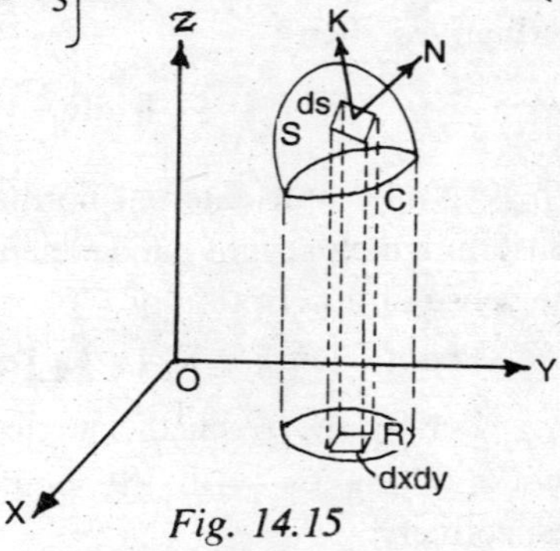

Fig. 14.15

$$\therefore \quad \int_C F_1 (x, y, z)\, dx$$

$$= \int_{C^1} F_1 (x, y, f)\, dx$$

$$= \int_{C^1} [F_1 (x, y, f)\, dx + 0\, dy]$$

$$= -\iint_R \frac{\partial}{\partial y} F_1 (x, y, f)\, dx\, dy$$

[Using Green's theorem is plane for **R**]

$$= -\iint_R \left\{ \frac{\partial F_1}{\partial y} + \frac{\partial F_1}{\partial z} \cdot \frac{\partial f}{\partial y} \right\} dx\, dy \quad \text{...(2)}$$

Now the direction ratios of the normal **N** to the surface $z = f(x, y)$ are

$$\frac{\partial f}{\partial x}; \frac{\partial f}{\partial y}; -1$$

$$\Rightarrow \quad \frac{\cos \theta_1}{\frac{\partial f}{\partial x}} = \frac{\cos \theta_2}{\frac{\partial f}{\partial y}} = \frac{\cos \theta_3}{-1}$$

$$\Rightarrow \quad \frac{\partial f}{\partial y} = -\frac{\cos \theta_2}{\cos \theta_3} \quad \text{...(3)}$$

$\because$ dx dy is the projection of ds on the xy-plane

$\therefore \quad dx\,dy = \cos\theta_3\,ds$ ...(4)

Substituting (3) and (4) in (2), we get

$$F_1(x, y, z)\,dx = -\iint_S \left\{\frac{\partial F_1}{\partial y} - \frac{\partial F_1}{\partial z}\cdot\frac{\cos\theta_2}{\cos\theta_3}\right\}\cos\theta_3\,ds$$

$$= -\iint_S \left\{\frac{\partial F_1}{\partial y}\cos\theta_3 - \frac{\partial F_1}{\partial z}\cos\theta_2\right\}ds$$

$$= \iint_S \left\{\frac{\partial F_1}{\partial z}\cos\theta_2 - \frac{\partial F_1}{\partial y}\cos\theta_3\right\}ds \qquad ...(5)$$

Similarly, we can prove that

$$\oint_C F_2(x, y, z)\,dy = \iint_S \left\{\frac{\partial F_2}{\partial x}\cos\theta_3 - \frac{\partial F_2}{\partial z}\cos\theta_1\right\}ds \qquad ...(6)$$

$$\int_C F_3(x, y, z)\,dz = \iint_S \left\{\frac{\partial F_3}{\partial y}\cos\theta_1 - \frac{\partial F_3}{\partial x}\cos\theta_2\right\}ds \qquad ...(7)$$

Adding (5), (6) and (7) and rearranging the terms (on R.H.S.) we get equation (1).

Hence the theorem is proved.

Note : In words, the Stoke's theorem states that

"The line integral of the tangential component of a vector point function **F** *taken around a closed curve* **C** *is equal to the surface integral of the normal component of* **F** *take over any surface having* **C** *as its boundary".*

14.7. Green's theorem in a plane *(As a deduction from Stoke's theorem)*

Let the surface **S** be in xy-plane, The Z-axis will be taken along the normal **N**. *i.e.*, **N** = **k**.

Let $\mathbf{F} = F_1\mathbf{i} + F_2\mathbf{j};\ \mathbf{r} = x\mathbf{i} + y\mathbf{j}$

So that $d\mathbf{r} = dx\mathbf{i} + dy\mathbf{j}$ and $\mathbf{F}\cdot d\mathbf{r} = F_1\,dx = F_2\,dy$

$$\nabla\times\mathbf{F} = \begin{vmatrix} \mathbf{i} & \mathbf{j} & \mathbf{k} \\ \frac{\partial}{\partial x} & \frac{\partial}{\partial y} & \frac{\partial}{\partial z} \\ F_1 & F_2 & 0 \end{vmatrix}$$

$$= -\frac{\partial F_2}{\partial z}\mathbf{i} + \frac{\partial F_1}{\partial z}\mathbf{j} + \left(\frac{\partial F_2}{\partial x} - \frac{\partial F_1}{\partial y}\right)\mathbf{k}$$

$$\therefore (\nabla \times \mathbf{F}) \cdot \mathbf{N} = (\nabla \times \mathbf{F}) \cdot \mathbf{k} = \frac{\partial F_2}{\partial x} - \frac{\partial F_1}{\partial x}$$

In xy plane ds = dx dy

$$\therefore \qquad \int_C \mathbf{F} \cdot \mathbf{dr} = \int_S (\nabla \times \mathbf{F}) \cdot \mathrm{N\,ds}$$

$$\Rightarrow \qquad \int_C F_1\,dx + F_2\,dy = \int_S \left(\frac{\partial F_2}{\partial x} - \frac{\partial F_1}{\partial y}\right) dx\,dy$$

which is same as *Green's theorem in plane.*

Example 20. Prove that

(i) $\int_C \mathbf{r} \cdot \mathbf{dr} = 0$ (ii) $\int_C \mathbf{r} \times \mathbf{dr} = 2\int_S \mathbf{dS}$

(iii) $\int_S [\nabla \times \nabla\phi] \cdot \mathbf{N}\,ds = 0$, using Stoke's theorem where the symbols have their usual meanings.

Solution : (i) $\int_C \mathbf{r} \cdot \mathbf{dr} = \int_S (\nabla \times \mathbf{r}) \cdot \mathrm{N\,ds} = 0 \quad (\because \nabla \times \mathbf{r} = 0)$

(ii) $\int_C \mathbf{r} \times \mathbf{dr}$

Choose $\mathbf{F} = \mathbf{A} \times \mathbf{r}$ where $\mathbf{A}$ is arbitrary constant vector.

$$\therefore \qquad \nabla \times \mathbf{F} = \nabla \times (\mathbf{A} \times \mathbf{r}) = 2\mathbf{A}$$

$$\therefore \qquad \int_C (\mathbf{A} \times \mathbf{r}) \cdot \mathbf{dr} = \int_S [\nabla \times (\mathbf{A} \times \mathbf{r})] \cdot \mathrm{N\,ds}$$

$$\text{or} \qquad \int_C \mathbf{A} \cdot (\mathbf{r} \times \mathbf{dr}) = 2\mathbf{A} \cdot \int_S \mathrm{N\,ds}$$

$$\Longleftrightarrow \qquad \mathbf{A} \cdot \left\{\int_C \mathbf{r} \times \mathbf{dr} - 2\int_S \mathbf{dS}\right\} = 0$$

$\because$ **A** is arbitrary

$$\Longleftrightarrow \qquad \int_S \mathbf{r} \times \mathbf{dr} = 2\int_S \mathbf{dS} \ (\because \ \mathrm{N\,ds} = \mathbf{dS})$$

(iii) $\int_S [\nabla \times \nabla\phi] \cdot \mathbf{n}\,ds = \int_S \nabla\phi \cdot \mathbf{dr}$

$$= \int_C \left(\frac{\partial \phi}{\partial x} dx + \frac{\partial \phi}{\partial y} dy + \frac{\partial \phi}{\partial z} dz \right)$$

$$= \int_C d\phi = [\phi]_p^p = 0$$

where p is any point on **C**.

$\therefore$ Hence the proofs.

Example 21. Verify stokes theorem for $\mathbf{F} = (2x - y)\,\mathbf{i} - yz^2\,\mathbf{j} - y^2 z\,\mathbf{k}$ over the upper half surface of $x^2 + y^2 + z^2 = 1$, bounded by its projection on the xy-plane.

Solution : Given $\mathbf{F} = (2x - y)\,\mathbf{i} - yz^2\,\mathbf{j} - y^2 z\,\mathbf{k}$.

Let $\mathbf{r} = x\mathbf{i} + y\mathbf{j} + z\mathbf{k};\ d\mathbf{r} = dx\,\mathbf{i} + dy\,\mathbf{j}\ dz\,\mathbf{k}$

$$\mathbf{F} \cdot d\mathbf{r} = (2x - y)\,dx - yz^2\,dy - y^2 z\,dz.$$

The projection of the upper half surface $x^2 + y^2 + z^2 = 1$ on the xy-plane is the boundary **C** of the surface *i.e.*, $x^2 + y^2 = 1$.

$\therefore$ The parametric equations are

$$x = \cos\theta;\ y = \sin\theta;\ 0 \le \theta \le 2\pi$$

$$\therefore \int_C \mathbf{F} \cdot d\mathbf{r} = \int_C (2x - y)\,dx \qquad [\because\ z = 0 \text{ in xy-plane}]$$

$$= \int_0^{2\pi} (2\cos\theta - \sin\theta)\,d(\cos\theta)$$

$$= \int_0^{2\pi} (2\cos\theta - \sin\theta)(-\sin\theta)\,d\theta$$

$$= \int_0^{2\pi} [-\sin 2\theta + \sin^2\theta]\,d\theta = \pi \qquad ...(1)$$

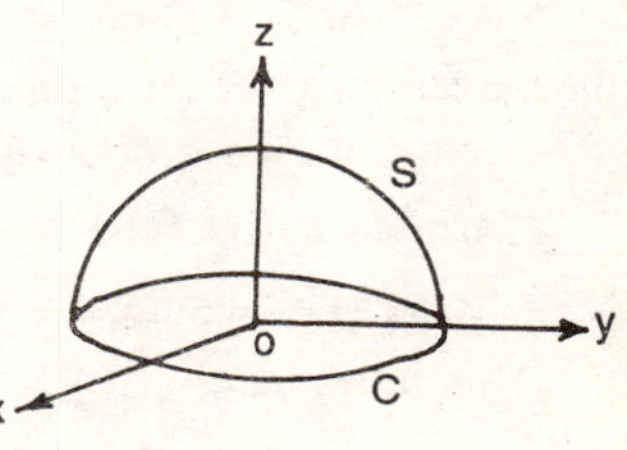

Fig 14.16

Again

$$\nabla \times \mathbf{F} = \begin{vmatrix} \mathbf{i} & \mathbf{j} & \mathbf{k} \\ \frac{\partial}{\partial x} & \frac{\partial}{\partial y} & \frac{\partial}{\partial z} \\ 2x - y & -yz^2 & -y^2 z \end{vmatrix}$$

$$= (-2\,yz + 2yz)\,\mathbf{i} - (0 - 0)\,\mathbf{j} + (0 + 1)\,\mathbf{k}$$
$$= \mathbf{k}$$

$(\nabla \times \mathbf{F}) \cdot \mathbf{N} = \mathbf{k} \cdot \mathbf{N}$

But $\int_{\mathbf{S}} (\nabla \times \mathbf{F}) \cdot \mathbf{N}\, ds = \int_{\mathbf{S}} \mathbf{N} \cdot \mathbf{k}\, ds = \int_{\mathbf{R}} \mathbf{N} \cdot \mathbf{k} \frac{dx\, dy}{\mathbf{N} \cdot \mathbf{k}}$

where **R** is the protection of **S** on xy-plane.

$$= \int_{x=-1}^{1} \int_{y=\sqrt{1-x^2}}^{\sqrt{1-x^2}} dx\, dy$$

$$= 4\int_0^1 \int_0^{\sqrt{1-x^2}} dx\, dy$$

$$= 4\int_0^1 \sqrt{1-x^2}\, dx$$

$$= 4 \cdot \frac{1}{2} \{x\sqrt{1-x^2} + \sin^{-1} x\}_0^1$$

$$= 4 \cdot \frac{1}{2} \cdot \frac{\pi}{2} = \pi \qquad \text{...(2)}$$

$\therefore$ From (1) and (2)

$$\int_{\mathbf{C}} \mathbf{F} \cdot d\mathbf{r} = \int_{\mathbf{S}} (\nabla \times \mathbf{F}) \cdot \mathbf{N}\, ds \qquad \text{...(3)}$$

Hence Stoke's theorem is verified.

Note : The R.H.S. of (3) in above problem can also be evaluated by using direction cosines of a point. **P** on the surface (*i.e.,* in terms of spherical polar co-ordinates).

Methods : Let **P** be any point on the surface so that

$\mathbf{P}(x, y, z) = \mathbf{P}(\sin\theta\cos\phi,\ \sin\theta\sin\phi,\ \cos\theta)$ [Here radius = 1]

Direction cosines of line OP are

$$\sin\theta\cos\phi,\ \sin\theta\sin\phi,\ \cos\theta$$

$\therefore \quad \mathbf{N} = \sin\theta\cos\phi\,\mathbf{i} + \sin\theta\sin\phi\,\mathbf{j} + \cos\theta\,\mathbf{k}.$

$\nabla \times \mathbf{F} = \mathbf{k} \quad \therefore \quad (\nabla \times \mathbf{F}) \cdot \mathbf{N} = \cos\theta$

$\therefore \quad \int_{\mathbf{C}} \mathbf{F} \cdot d\mathbf{r} = \int_{\mathbf{S}} \cos\theta\, ds$

(where $ds = \sin\theta\, d\theta\, d\phi$
and $0 \le \theta \le \pi/2,\ 0 \le \phi \le 2\pi$)

$$= \int_{\theta=0}^{\pi/2} \int_{\phi=0}^{2\pi} \cos\theta \sin\theta \, d\theta \, d\phi$$

$$= \frac{1}{2}\int_{\theta=0}^{\pi/2} \sin 2\theta \, d\theta \int_{\phi=0}^{2\pi} d\phi$$

$$= \frac{1}{2}\left[-\frac{\cos 2\theta}{2}\right]_0^{\pi/2} \cdot 2\pi = \pi(1) = \pi$$

Example 22. Evaluate by Stoke's theorem

(i) $\int_C (e^x\,dx + 2y\,dy - dz$ where **C** is the curve $x^2 + y^2 = 9$, and z=2.

(ii) $\int_C [(x + y)\,dx + (2x - z)\,dy + (y + z)\,dz$, where **C** is the boundary of the triangle with the vertices (0, 0, 0), (1, 0, 0) and (1, 1, 0).

(iii) $\int_C [\sin z\,dx - \cos x\,dy + \sin y\,dz]$, **C** is the boundary of the rectangle $0 \le x \le \pi$, $0 \le y \le 1$, $z = 3$.

Solution : (i) Given $\oint_C e^x\,dx + 2y\,dy - dz)$

$$= \oint_C \mathbf{F} \cdot d\mathbf{r} \text{ (say)}$$

where $\mathbf{F} = e^x\,\mathbf{i} + 2y\,\mathbf{j} - \mathbf{k}$

and $d\mathbf{r} = dx\,\mathbf{i} + dy\,\mathbf{j} + dz\,\mathbf{k}$

$$\nabla \times \mathbf{F} = \begin{vmatrix} \mathbf{i} & \mathbf{j} & \mathbf{k} \\ \frac{\partial}{\partial x} & \frac{\partial}{\partial y} & \frac{\partial}{\partial z} \\ e^x & 2y & -1 \end{vmatrix} = 0$$

By Stoke's theorem

$$\oint_C \mathbf{F} \cdot d\mathbf{r} = \iint_S (\nabla \times \mathbf{F}) \cdot \mathbf{N}\,ds$$

$$= \iint_S (0) \cdot \mathbf{N}\,ds = 0$$

(ii) Given

$$\oint_C [x + y) dx + (2x - z) dy + (y + z) dz]$$

$$= \int_C \mathbf{F} \cdot d\mathbf{r} \text{ (say)}$$

where $\mathbf{F} = (x + y)\,\mathbf{i} + (2x - z)\,\mathbf{j} + (y + z)\,\mathbf{k}$

and $\mathbf{r} = x\,\mathbf{i} + y\,\mathbf{j} + z\,\mathbf{k}$; $d\mathbf{r} = dx\,\mathbf{i} + dy\,\mathbf{j} + dz\,\mathbf{k}$

$$\nabla \times \mathbf{F} = \begin{vmatrix} \mathbf{i} & \mathbf{j} & \mathbf{k} \\ \dfrac{\partial}{\partial x} & \dfrac{\partial}{\partial y} & \dfrac{\partial}{\partial z} \\ x + y & 2x - z & y + z \end{vmatrix} = 2\mathbf{i} + \mathbf{k}$$

And $\mathbf{N} = \mathbf{k}$

The triangle lies in xy-plane

Equation to OB is y=x

$\therefore (\nabla \times \mathbf{F}) \cdot \mathbf{N} = (2\mathbf{i} + \mathbf{k}) \cdot \mathbf{k} = 1$

$$ds = \frac{dx\,dy}{\mathbf{N} \cdot \mathbf{k}} = dx\,dy$$

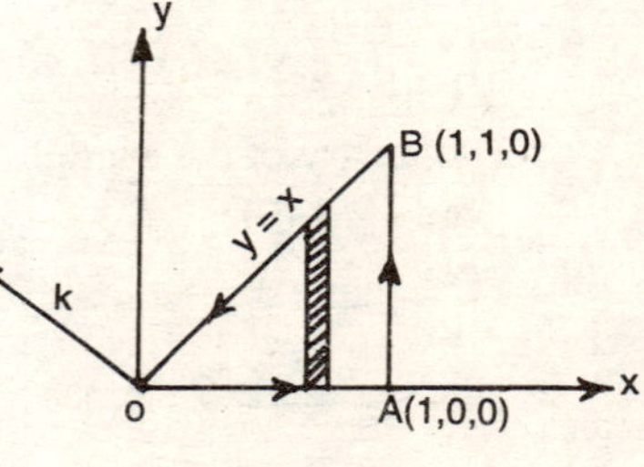

Fig. 14.17

$\therefore$ By Stoke's theorem

$$\oint_C \mathbf{F} \cdot d\mathbf{r} = \int_{x=0}^{1} \int_{y=0}^{x} dx\,dy = \int_{x=0}^{1} x\,dx = \frac{1}{2}$$

(iii) Given $\oint_C (\sin z\,dx - \cos x\,dy + \sin y\,dz)$

$$= \oint_C \mathbf{F} \cdot d\mathbf{r} \text{ (say)}$$

where $\mathbf{F} = \sin z\,\mathbf{i} - \cos x\,\mathbf{j} + \sin y\,\mathbf{k}$

and $\mathbf{r} = x\,\mathbf{i} + y\,\mathbf{j} + z\,\mathbf{k}$

so that $d\mathbf{r} = dx\,\mathbf{i} + dy\,\mathbf{j} + dz\,\mathbf{k}$

$$\nabla \times \mathbf{F} = \begin{vmatrix} \mathbf{i} & \mathbf{j} & \mathbf{k} \\ \dfrac{\partial}{\partial x} & \dfrac{\partial}{\partial y} & \dfrac{\partial}{\partial z} \\ \sin z & -\cos x & \sin y \end{vmatrix}$$

$$= \cos y\,\mathbf{i} + \cos z\,\mathbf{j} + \sin x\,\mathbf{k}$$

$$(\nabla \times \mathbf{F}) \cdot \mathbf{N} = \mathbf{k} \cdot (\cos y\,\mathbf{i} + \cos z\,\mathbf{j} + \sin x\,\mathbf{k}) = \sin x$$

$\therefore$ By Stoke's theorem

$$\oint_C \mathbf{F} \cdot d\mathbf{r} = \int_S (\nabla \times \mathbf{F}) \cdot \mathbf{N}\, ds = \int_{x=0}^{\pi} \int_{y=0}^{1} \sin x\, dx\, dy$$

$$= \int_{x=0}^{\pi} \sin x\, dx \int_{y=0}^{1} dy = (-\cos x)_0^{\pi} (y)_0^1$$

$$= (\cos \pi - \cos 0) = 2,$$

Example 23. Verify Stoke's theorem for the vector field $\mathbf{F} = (x^2 - y^2)\,\mathbf{i} + 2\,xy\,\mathbf{j}$ integrated round the rectangle in the plane $z = 0$ and bounded by the lines $x = 0$, $y = 0$, $x = a$, and $y = b$

Solution : Given $\mathbf{F} = (x^2 - y^2)\,\mathbf{i} + 2\,xy$;

Let $\mathbf{r} = x\mathbf{i} + dy\mathbf{j}$ $\therefore$ $d\mathbf{r} = dx\,\mathbf{i} + dy\,\mathbf{j}$

$$\mathbf{F} \cdot d\mathbf{r} = (x^2 - y^2)\, dx + 2\,xy\, dy$$

$$\therefore \int_C \mathbf{F} \cdot d\mathbf{r} = \int_{\overrightarrow{OB}} \mathbf{F} \cdot d\mathbf{r} + \int_{\overrightarrow{BA}} \mathbf{F} \cdot d\mathbf{r} + \int_{\overrightarrow{AC}} \mathbf{F} \cdot d\mathbf{r} + \int_{\overrightarrow{CO}} \mathbf{F} \cdot d\mathbf{r} \quad ...(1)$$

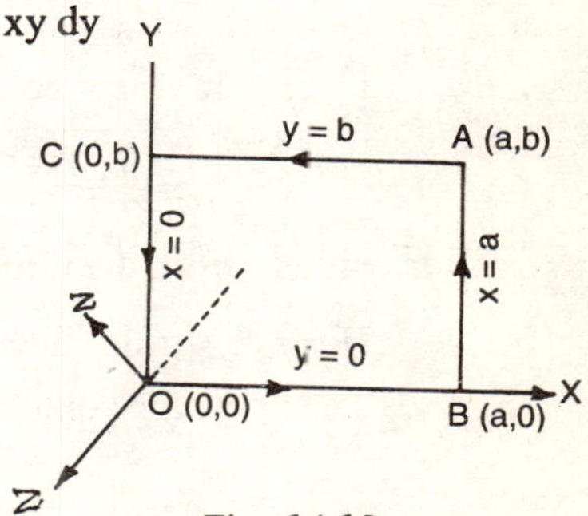

Fig. 14.18

Consider

$$\int_{\overrightarrow{BO}} \mathbf{F} \cdot d\mathbf{r} = \int_{\overrightarrow{OB}} (x^2 - y^2)\, dx + 2\,xy\, dy$$

Along $\overrightarrow{\mathbf{OB}}$: $y = 0$ $\therefore$ $dy = 0$ and $0 \le x \le a$

Hence $\displaystyle\int_{\overrightarrow{OB}} \mathbf{F} \cdot d\mathbf{r} = \int_0^a x^2\, dx = \frac{a^3}{3}$ ...(2)

Consider

$$\int_{\overrightarrow{BA}} \mathbf{F} \cdot d\mathbf{r} = \int_{\overrightarrow{BA}} (x^2 - y^2)\, dx + 2\,xy\, dy$$

Along $\overrightarrow{\mathbf{BA}}$: $x = a$ $\therefore$ $dx = 0$ and $0 \le y \le b$

Hence $\displaystyle\int_{\overrightarrow{AB}} \mathbf{F} \cdot d\mathbf{r} = \int_0^b 2\,ay\, dy = ab^2$ ...(3)

Consider

$$\int_{\overrightarrow{AC}} \mathbf{F} \cdot d\mathbf{r} = \int_{\overrightarrow{AC}} (x^2 - y^2)\, dx + 2\, xy\, dy$$

Along $\overrightarrow{AC}$: $y = b$ $\therefore$ $dy = 0$ and $a \le x \le 0$

Hence $$\int_{\overrightarrow{AC}} \mathbf{F} \cdot d\mathbf{r} = \int_a^0 (x^2 - b^2)\, dx = \left(\frac{a^3}{3} - b^2 a\right) \qquad ...(4)$$

Consider

$$\int_{\overrightarrow{CO}} \mathbf{F} \cdot d\mathbf{r} = \int_{\overrightarrow{CO}} (x^2 - y^2)\, dx + 2\, xy\, dy$$

Along $\overrightarrow{CO}$: $x = 0$, $\therefore$ $dx = 0$ and $b \le y \le 0$

Hence $$\int_{\overrightarrow{CO}} \mathbf{F} \cdot d\mathbf{r} = \int_b^0 2(0)\, y\, dy = 0 \qquad ...(5)$$

$\therefore$ By substituting (2) , (3), (4) and (5) in (1) we get

$$\therefore \quad \int_C \mathbf{F} \cdot d\mathbf{r} = \frac{a^3}{3} + ab^2 - \left(\frac{a^3}{3} - b^2 a\right) + 0$$

$$= 2ab^2 \qquad ...(6)$$

It is evidient from the figure that

$\mathbf{N} = \mathbf{k}$ (the vector field is integrated round the rectangle in the plane z=0)

$$\nabla \times \mathbf{F} = \begin{vmatrix} \mathbf{i} & \mathbf{j} & \mathbf{k} \\ \frac{\partial}{\partial x} & \frac{\partial}{\partial y} & \frac{\partial}{\partial z} \\ x^2 - y^2 & 2\, xy & O \end{vmatrix}$$

$$= (0)\, \mathbf{i} + (0)\, \mathbf{j} + \left\{\frac{\partial}{\partial x}(2\, xy) - \frac{\partial}{\partial y}(x^2 - y^2)\right\} \mathbf{k}$$

$$= 4y\, \mathbf{k}$$

$$\therefore \quad (\nabla \times \mathbf{F}) \cdot \mathbf{N} = 4y\, \mathbf{k} \cdot \mathbf{k} = 4y$$

Also $ds = dx\, dy$

$$\therefore \quad \int_S (\nabla \times \mathbf{F}) \cdot \mathbf{N}\, ds = \int_{x=0}^{a} \int_{y=0}^{b} 4y\, dx\, dy$$

$$= 4\int_0^a \frac{b^2}{2}\, dx$$

$$= 2\, ab^2 \qquad \text{...(7)}$$

$\therefore$ From (6) and (7) $\oint_C \mathbf{F} \cdot d\mathbf{r} = \int_S (\nabla \times \mathbf{F}) \cdot \mathbf{N}\, ds$

Hence the theorem is verified.

Example 24. Evaluate $\int\int_S (\nabla \times \mathbf{V}) \cdot d\mathbf{S}$ over the surface of the paraboloid $z + x^2 + y^2 = 1,\ z \geq 0$, where $\mathbf{V} = y\mathbf{i} + z\mathbf{j} + x\mathbf{k}$.

Solution : The boundary **C** of the surface **S** is the circle $x^2 + y^2 = 1,\ z = 0$

$\therefore$ Parametric equations are

$$x = \cos\theta,\ y = \sin\theta,\ z = 0 \text{ and } 0 \leq \theta \leq 2\pi$$

By stoke's theorem, we have

$$\int_S (\nabla \times \mathbf{V}) \cdot d\mathbf{S} = \int_S (\nabla \times \mathbf{V}) \cdot \mathbf{N}\, ds = \int_C \mathbf{V}\, d\mathbf{r}$$

$$= \int_C (y\mathbf{i} + z\mathbf{j} + x\mathbf{k}) \cdot (dx\mathbf{i} + dy\mathbf{j} + dz\mathbf{k})$$

$$= \int_C y\,dx \quad (z = 0,\ dz = 0 \text{ on } \mathbf{C})$$

$$= \int_{\theta=0}^{2\pi} \sin\theta\, d(\cos\theta)$$

$$= \int_0^{2\pi} \sin^2\theta\, d\theta = -\pi$$

Example 25. Evaluate $\int_C [y\,dx + z\,dy + x\,dz]$ where C is the curve of intersection of $x^2 + y^2 + z^2 = a^2$ and $x + z = a$.

Solution : The curve **C** is evidiently a circle lying in $x + z = a$ plane.

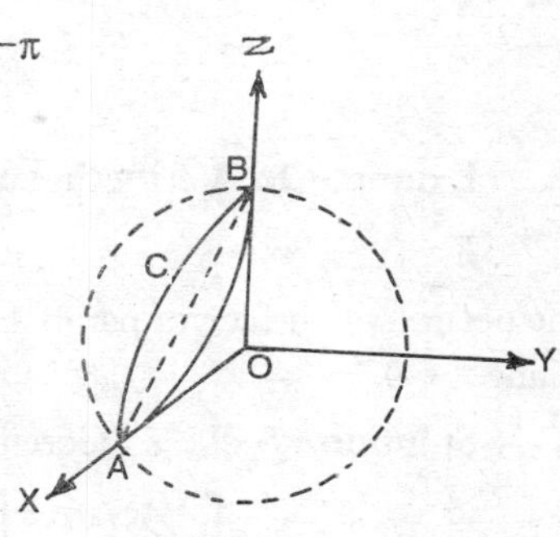

Fig. 14.19

Here **A** = (a, 0, 0), **B** (0, 0, a) are the extremities of diameter **AB**.

$$\therefore \quad \int_C y\,dx + z\,dy + x\,dz$$

$$= \int_C (y\mathbf{i} + z\mathbf{j} + x\mathbf{k}) \cdot (dx\,\mathbf{i} + dy\,\mathbf{j} + dz\,\mathbf{k})$$

$$= \int_C \mathbf{F} \cdot d\mathbf{r}$$

where $\mathbf{F} = y\mathbf{i} + z\mathbf{j} + x\mathbf{k}$ and $\mathbf{r} = x\mathbf{i} + y\mathbf{j} + z\mathbf{k}$

so that $d\mathbf{r} = dx\,\mathbf{i} + dy\,\mathbf{j} + dz\,\mathbf{k}$

$$\nabla \times \mathbf{F} = \begin{vmatrix} \mathbf{i} & \mathbf{j} & \mathbf{k} \\ \dfrac{\partial}{\partial x} & \dfrac{\partial}{\partial y} & \dfrac{\partial}{\partial z} \\ y & z & x \end{vmatrix}$$

$$= -(\mathbf{i} + \mathbf{j} + \mathbf{k}).$$

Here $N = \frac{1}{\sqrt{2}}\mathbf{i} + \frac{1}{\sqrt{2}}\mathbf{k}$ (see figure)

$$\therefore \quad (\nabla + \mathbf{F}) \cdot \mathbf{N} = -\left(\frac{1}{\sqrt{2}} + \frac{1}{\sqrt{2}}\right) = -\sqrt{2}$$

We have by Stoke's theorem

$$\int_C \mathbf{F} \cdot d\mathbf{r} = \int_S (\nabla \times \mathbf{F}) \cdot \mathbf{N}\,ds$$

$$= \int_S (-\sqrt{2})\,ds - \sqrt{2}\int_S ds$$

$$= -\sqrt{2}\,\pi\left(\frac{a}{\sqrt{2}}\right)^2 = \frac{-\pi a^2}{\sqrt{2}}.$$

Example 26. Verify Stoke's theorem for

$\mathbf{F} = (y^2 + x^2 - x^2)\,\mathbf{i} + (z^2 + x^2 - y^2)\,\mathbf{j} + (x^2 + y^2 - z^2)\,\mathbf{k}$ < over the portion of surface of paraboloid $x^2 + y^2 - 2\,ax + az = 0$ above the plane $z = 0$.

Solution : Stoke's theorem is

$$\int_C \mathbf{F} \cdot d\mathbf{r} = \int_S (\nabla \times \mathbf{F}) \cdot \mathbf{N}\,ds$$

$$\nabla \times \mathbf{F} = \begin{vmatrix} \mathbf{i} & \mathbf{j} & \mathbf{k} \\ \dfrac{\partial}{\partial x} & \dfrac{\partial}{\partial y} & \dfrac{\partial}{\partial z} \\ y^2 + z^2 - x^2 & z^2 + x^2 - y^2 & x^2 + y^2 - z^2 \end{vmatrix}$$

$$= 2(y - z)\,\mathbf{i} + 2(z - x)\,\mathbf{j} + 2(x - y)\,\mathbf{k}$$

The given surface is $x^2 + y^2 - 2ax + az = 0$, can be written as $(x - a)^2 + y^2 = -a(z - a)$, which is a paraboloid.

It meets the plane $z = 0$

$\therefore \quad (x - a)^2 + y^2 = a^2$ which is a circle, whose centre is at O′.

Any element of the circle is $r\,d\theta\,dr$ referred to O′ as origin

$\therefore \quad x = a + r\cos\theta, \; y = r\sin\theta$

$\therefore \quad ds = r\,d\theta\,dr; \; \mathbf{N} = \mathbf{k}$

$\therefore \quad (\nabla \times \mathbf{F}) \cdot \mathbf{N} = 2(x - y) = 2[a + r\cos\theta - r\sin\theta]$

$$\therefore \quad \int_S (\nabla \times \mathbf{F}) \cdot \mathbf{N}\,ds = 2\int_{r=0}^{a}\int_{\theta=0}^{2\pi} (a + r\cos\theta - r\sin\theta)\,r\,d\theta\,dr$$

$$= 2\int_0^a a\,r\,dr \cdot \int_0^{2\pi} d\theta$$

$$= 4\pi a\left[\frac{r^2}{2}\right]_0^a = 2\pi a^3.$$

Now $\oint_C \mathbf{F} \cdot = \oint_C F_1\,dx + F_2\,dy + F_3\,dz$

$$= \int_C \left\{(y^2 + z^2 - x^2)\,dx + (z^2 + x^2 - y^2)\,dy + (x^2 + y^2 - z^2)\,dz\right\}$$

$$= \int_C (y^2 - x^2)(dx - dy) \qquad (\because \; z = 0, \; dz = 0 \text{ on } \mathbf{C})$$

$$= \int_{\theta=0}^{2\pi}\left[a^2\sin^2\theta - (a + a\cos\theta)^2\right]$$

$[-a\sin\theta - a\cos\theta]\,d\theta$

{Putting

$x = a + a\cos\theta, \; y = a\sin\theta$}

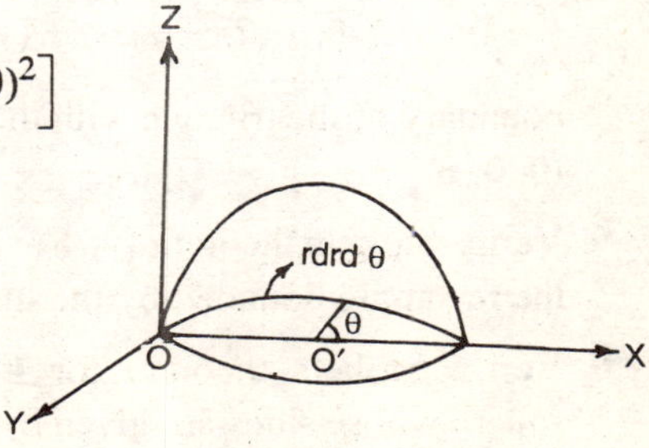

Fig. 14.20

$$= -a^3 \int_0^{2\pi} 2\cos\theta\,(1+\cos\theta)(\sin\theta+\cos\theta)\,d\theta$$

$$= 2a^3 \int_0^{2\pi} \cos\theta\,(1+\cos\theta)\sin\theta\,d\theta$$

$$+ 2a^3 \int_0^{2\pi} \cos^2\theta\,(1+\cos\theta)\,d\theta$$

$$= 2a^3 \int_0^{2\pi} \cos^2\theta\,d\theta + 2a^3 \int_0^{2\pi} \cos^3\theta\,d\theta$$

(The value of first integral being zero)

$$= a^3 \int_0^{2\pi} (1+\cos 2\theta)\,d\theta + \frac{a^3}{2}\int_0^{2\pi} (\cos 3\theta + 3\cos\theta)\,d\theta$$

$$= a^3 \left[\theta + \frac{\sin 2\theta}{2}\right]_0^{2\pi} + \frac{a^3}{2}\left[\frac{\sin 3\theta}{3} + 3\sin\theta\right]_0^{2\pi}$$

$$= 2\pi a^3$$

Hence $\oint_C \mathbf{F}\cdot d\mathbf{r} = \int_S (\nabla\times\mathbf{F})\cdot\mathbf{N}\,ds$ i.e. stoke's theorem is verified.

EXERCISE

1. Apply Stokes theorem to evaluate

   $\mathbf{F} = y^2\,\mathbf{i} + x^2\,\mathbf{j} - (x+z)\,\mathbf{k}$, and **C** is the boundary of triangle with the vertices (0, 0, 0), (1, 0, 0) and (1, 1, 0).

2. Apply Stoke's theorem to evaluate

   $\int_C (x+y)\,dx + (2x-z)\,dy + (y+z)\,dz$ where **C** is the boundary of the triangle with the vertices (2, 0, 0), (0, 3, 0) and (0, 0, 6).

3. Verify Stoke's theorem for $\mathbf{F} = (x^2+y^2)\,\mathbf{i} - 2xy\,\mathbf{j}$ taken round the rectangle bounded by the lines $x = \pm a$, $y = 0$, $y = b$.

4. Verify Stoke's theorem for $\mathbf{F} = x^2\,\mathbf{i} + xy\,\mathbf{j}$ taken round the square whose sides are given by $x = 0, y = 0, x = a, y = a$ in the plane $z = 0$.

5. Evaluate $\int_C (xy\,dx + xy^2\,dy)$ taken round the positively oriented curve **C** with vertices (1, 0) (–1, 0) (0, 1) and (0, –1) by using Stoke's theorem.

6. Verify Stoke's theorem when $\mathbf{F} = y\,\mathbf{i} + z\,\mathbf{j} + x\,\mathbf{k}$ and surface **S** is the part of the sphere $x^2 + y^2 + z^2 = 1$ above the xy-plane.

7. Show that $\oint_C (yz\,dx + zx\,dy + xy\,dz) = 0$ where **C** is the curve $x^2 + y^2 = 1,\ z = y^2$.

8. Evaluate $\int_S (\nabla \times \mathbf{F}) \cdot \mathbf{N}\,ds$ where $\mathbf{F} = y\,\mathbf{i} + (x - 2xz)\,\mathbf{j} - xy\,\mathbf{k}$ and **S** is the surface of sphere of radius 'a' above the xy-plane.

9. Evaluate $\int_S (\nabla \times \mathbf{F}) \cdot \mathbf{N}\,ds$ where

   $\mathbf{F} = (x^2 + y - 4)\,\mathbf{i} + 3\,xy\,\mathbf{j} + (2\,xz + z^2)\,\mathbf{k}$ and **S** is the surface of the paraboloid $z = 4 - (x^2 + y^2)$ above the xy-plane.

10. Verify Stoke's theorem for $\mathbf{F} = y\,\mathbf{i} + z\,\mathbf{j} + x\,\mathbf{k}$ and **S** is the portion of the surface $x^2 + y^2 - 2\,ax + az = 0$ above the plane $z = 0$.

11. Apply Stoke's theorem to prove that

    $\int_C (y\,dx + z\,dy + x\,dz) = -2\sqrt{2}\pi a^2$, where **C** is the curve given by $x^2 + y^2 + z^2 - 2ax - 2ay = 0;\ x + y = 2a$ and begins at the point (2a, 0, 0) and goes at first below the xy- plane.

    $\left\{\text{Hint: Here } \mathbf{N} = \frac{1}{a\sqrt{2}}[x - a)\,\mathbf{i} + (y - a)\,\mathbf{j} + z\,\mathbf{k}.]\right\}$

ANSWERS

(1) $\frac{1}{3}$ (2) 21 (5) $\frac{4}{3}$ (8) 0 (9) -4π

14.8. Irrotational Vector function :

If **F** is a vector point function then it is called irrotational vector function (or field),

if $$\nabla \times \mathbf{F} = 0 \qquad ...(1)$$

But from Stoke's theorem

$$\int_C \mathbf{F} \cdot d\mathbf{r} = \int_S (\nabla \times \mathbf{F}) \cdot \mathbf{N}\, ds = 0$$

i.e.,
$$\int_C \mathbf{F} \cdot d\mathbf{r} = 0 \qquad ...(2)$$

i.e., the circulation of **F** around a closed curve **C** is zero.

We also know that

$\nabla \times \nabla \phi = 0$, where ϕ is any scalar function.

$\therefore$ In an irrotational field for which $\nabla \times \mathbf{F} = 0$, the vector **F** can always be expressed as the gradient of a scalar function ϕ.

Thus
$$\mathbf{F} = \nabla\phi \qquad ...(3)$$

Here ϕ is called *potential.*

$\therefore$ an irrotational vector function (field) **F** is characterised by any one of the following conditions :

(i) $\nabla \times \mathbf{F} = 0$ (ii) $\oint_C \mathbf{F} \cdot d\mathbf{r} = 0$ (iii) $\mathbf{F} = \nabla\phi$.

Note 1 : In rotational fields $\mathbf{F} \neq \nabla\phi$.

Note 2 : In an irrotational field, the line integral $\int_a^b \mathbf{F} \cdot d\mathbf{r}$ is independent of the path of integration and is equal to the potential difference between the points.

i.e.,
$$\int_a^b \mathbf{F} \cdot d\mathbf{r} = \phi(b) - \phi(a) \qquad (\because\ \mathbf{F} = \nabla\phi)$$

Note 3 : If $\int_C \mathbf{F} \cdot d\mathbf{r} = 0$ then the field is said to be conservative. Thus every irrotational field is conservative.

Example 27. Show that $r^k\mathbf{r}$ is an irrotational vector for any value k, where $\mathbf{r} = x\mathbf{i} + y\mathbf{j} + z\mathbf{k}$ and $r = |\mathbf{r}|$.

Solution : Given $\mathbf{r} = x\mathbf{i} + y\mathbf{j} + z\mathbf{k}$; $r = |\mathbf{r}|$

then
$$\frac{\partial r}{\partial x} = \frac{x}{r}, \frac{\partial r}{\partial y} = \frac{y}{r}, \frac{\partial r}{\partial z} = \frac{z}{r}$$

and
$$1^k\mathbf{r} = r^k x\,\mathbf{i} + r^k r\,\mathbf{j} + r^k\,\mathbf{k}$$

$$\nabla \times (r^k \mathbf{r}) = \begin{vmatrix} \mathbf{i} & \mathbf{j} & \mathbf{k} \\ \frac{\partial}{\partial x} & \frac{\partial}{\partial y} & \frac{\partial}{\partial z} \\ r^k x & r^k y & r^k z \end{vmatrix}$$

$$= \sum \left\{ \frac{\partial}{\partial y}(r^k z) - \frac{\partial}{\partial z}(r^k y) \right\} \mathbf{i}$$

$$= \sum \left\{ kr^{k-1} \frac{\partial r}{\partial y} z - kr^{k-1} \frac{\partial r}{\partial z} y \right\} \mathbf{i}$$

$$= \sum \left\{ kr^{k-1} \frac{yz}{\mathbf{r}} - kr^{k-1} \frac{z}{\mathbf{r}} y \right\} \mathbf{i} = \mathbf{0}$$

$\therefore$ $\nabla \times (r^k \mathbf{r}) = 0$; $r^k \mathbf{r}$ is an irrotational vector.

Example 28. A vector field i s given by

$\mathbf{F} = (x^2 + xy^2)\,\mathbf{i} + (y^2 + x^2 y)\,\mathbf{j}$. Show that the field is irrotational and find the scalar potential.

Solution : Given $\mathbf{F} = (x^2 + xy^2)\,\mathbf{i} + (y^2 + x^2 y)\,\mathbf{j}$

$$\nabla \times \mathbf{F} = \begin{vmatrix} \mathbf{i} & \mathbf{j} & \mathbf{k} \\ \frac{\partial}{\partial x} & \frac{\partial}{\partial y} & \frac{\partial}{\partial z} \\ x^2 + xy^2 & y^2 + x^2 y & 0 \end{vmatrix}$$

$$= (0 - 0)\,\mathbf{i} + (0 - 0)\,\mathbf{j} + (2xy - 2xy)\,\mathbf{k} = 0$$

$\because$ $\nabla \times \mathbf{F} = \mathbf{O}$, field $\mathbf{F}$ is irrotational.

If ϕ is the scalar potential, then we can write $\mathbf{F} = \nabla\phi$

i.e., $$(x^2 + xy^2)\,\mathbf{i} + (y^2 + x^2 y)\,\mathbf{j} = \frac{\partial \phi}{\partial x}\mathbf{i} + \frac{\partial \phi}{\partial y}\mathbf{j} + \frac{\partial \phi}{\partial z}\mathbf{k}$$

$$\Longleftrightarrow \quad \frac{\partial \phi}{\partial x} = x^2 + xy^2 \qquad \text{...(1)}$$

$$\frac{\partial \phi}{\partial y} = y^2 + x^2 y \qquad \text{...(2)}$$

$$\frac{\partial \phi}{\partial z} = 0 \qquad \text{...(3)}$$

Integrating (1) partially with respect to x

[Here y and z are to be treated as constants]

$$\phi = \frac{x^3}{3} + \frac{x^2y^2}{2} + \text{a function independent of x}$$

i.e. $$\phi = \frac{x^3}{3} + \frac{x^2y^2}{2} + f(y, z) \quad ...(4)$$

Similarly from (2) and (3), we get

$$\phi = \frac{y^3}{3} + \frac{x^2y^2}{2} + g(z, x) \quad ...(5)$$

$$\phi = h(x, y) \quad ...(6)$$

Comparing the (4), (5), (6), we have

$f(y, z)$ = terms in ϕ independent of $x = y^3/3$

$g(z, x)$ = terms in ϕ independent of $y = x^3/3$

$h(x, y)$ = terms in ϕ independent of z

$$= \frac{x^3}{3} + \frac{y^3}{3} + \frac{x^2y^2}{2}$$

Hence $$\phi(x, y, z) = \frac{x^3}{3} + \frac{y^3}{3} + \frac{x^2y^2}{2} + \text{constant}$$

Another Method

We also know that

$$d\phi = \frac{\partial\phi}{\partial x}dx + \frac{\partial\phi}{\partial y}dy + \frac{\partial\phi}{\partial z}dz$$

$$= (x^2 + xy^2)\,dx + (y^2 + x^2y)\,dy$$

$$= (x^2dx + y^2\,dy) + (xy^2\,dx + x^2y\,dy)$$

$$= \frac{1}{3}d\,(x^2 + y^3) + \frac{1}{2}d\,(x^2y^2)$$

$$\phi(x, y, z) = \frac{1}{3}(x^3 + y^3) + \frac{1}{2}x^2y^2 + \text{constant.}$$

EXERCISE

1. Prove that a necessary and sufficient condition that

$\oint_C$ **F** · dr = 0 for every closed curve **C** is that $\nabla \times \mathbf{F} = 0$

2. Show that the vector field $\mathbf{F} = x\mathbf{i} - y\mathbf{j} - z\mathbf{k}$ is irrotational, and find a scalar function ϕ such that $\mathbf{F} = \nabla\phi$.
3. Show that the vector field $\mathbf{F} = yz\,\mathbf{i} + zx\,\mathbf{j} + xy\,\mathbf{k}$ is irrotational, and find a scalar function ϕ such that $\mathbf{F} = -\nabla\phi$.
4. Show that the vector field

   $\mathbf{F} = (x^2 - yz)\,\mathbf{i} + (y^2 - zx)\,\mathbf{j} + (z^2 - xy)\,\mathbf{k}$ is irrotational, and find a scalar function ϕ such that $\mathbf{F} = \mathbf{k}\nabla\phi$ where $\mathbf{k}$ is constant.
5. If $\mathbf{F} = (\sin y + z\cos x)\mathbf{i} + (\sin z + x\cos y)\mathbf{j} + (\sin x + y\cos z)\mathbf{k}$ is irrotational, find a scalar function ϕ such that $\mathbf{F} = \nabla o$.

ANSWERS

2. $\phi(x, y, z) = \frac{1}{2}(x^2 - y^2 - z^2) + \text{constant}.$

3. $\phi(x, y, z) = -xyz + \text{constant}.$

4. $\phi(x, y, z) = \frac{k}{3}(x^2 + y^2 + z^2) - k\,xyz + \text{constant}.$

5. $\phi(x, y, z) = x\sin y + y\sin z + z\sin x + \text{constant}.$

15

Curvilinear Co-ordinates

15.1. Definitions

The position of a point **P** (x, y, z) in a cartesian coordinate system is determined by the intersection of three mutually perpendicular planes

$$x = \mathbf{C^*}_1 \text{ (constant)}$$

$$y = \mathbf{C^*}_2 \text{ (constant)}$$

$$z = \mathbf{C^*}_3 \text{ (constant)}$$

Let the cartesian coordinates (x, y, z) of any point be expressed as functions of three new quantities u_1, u_2, u_3, so that

$$\left.\begin{aligned} x &= x\,(u_1, u_2, u_3) \\ y &= y\,(u_1, u_2, u_3) \\ z &= z\,(u_1, u_2, u_3) \end{aligned}\right\} \qquad ...(1)$$

Suppose that the equations (1) are solvable for u_1, u_2, u_3, interms of x, y, z,

i.e.,
$$\left.\begin{aligned} u_1 &= u_1\,(x, y, z) \\ u_2 &= u_2\,(x, y, z) \\ u_3 &= u_3\,(x, y, z) \end{aligned}\right\} \qquad ...(2)$$

The functions in (1) and (2) are assumed to be single valued and to have continuous derivatives so that the correspondence between (x, y, z) and (u_1, u_2, u_3) is unique.

(**Note:** The point where this assumption is not applied there a special consideration is required]

$\therefore$ To each point **P** (x, y, z) in the region there corresponds a unique set of values (u_1, u_2, u_3) called *curvilinear co-ordinates* of **P** (x, y, z).

The set of equations (1) and (2) define a *transformation of coordinates.*

For a given points **P** (x, y, z) the set of equations (2) become

$$\left.\begin{aligned} u_1 &= c_1 \text{ (constant)} \\ u_2 &= c_2 \text{ (constant)} \end{aligned}\right\} \qquad \text{...(3)}$$

$u_3 = c_3$ (constant), and are called *co-ordinates surface.* Each pair of these surface intersect in curves called *coordinate curves or lines* (see figure).

The coordinate axes are determined by the tangents to the coordinate curves at the intersections of the surfaces. They are not in general fixed directions in space.

If the coordinate surfaces (3) intersect at right angles at every point **P** then the curvilinear coordinate system is called *orthogonal*, and the coordinates of **P** are called *orthogonal curvilinear coordinates.*

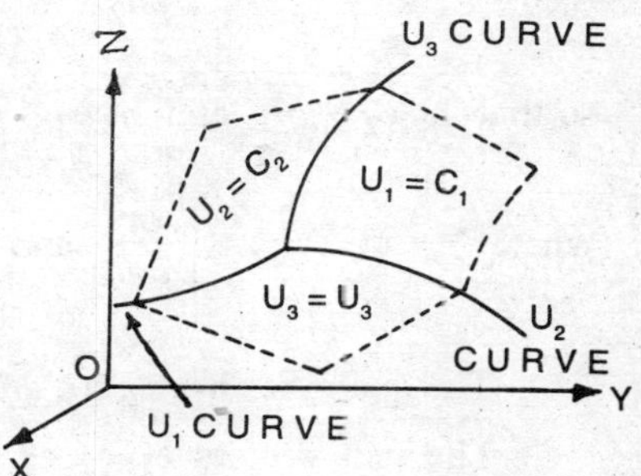

Fig. 15.1

15.2. Unit vectors in curvilinear systems

Let **r** =x**i** + y**j** +z**k** be the position vector **P**.

Then the set of equations (1) can be written as

$$\mathbf{r} = \mathbf{r}(u_1, u_2, u_3) \qquad \text{...(4)}$$

A tangent vector to the u_1 curve at **P** (for which u_2 and u_3 are constants) $\dfrac{\partial \mathbf{r}}{\partial u_1}$.

Then a unit tangent vector in this direction is

$$e_1 = \frac{\frac{\partial r}{\partial u_1}}{\left|\frac{\partial r}{\partial u_1}\right|}$$

so that $h_1 e_1 = \frac{\partial r}{\partial u_1}$, where $h_1 = \left|\frac{\partial r}{\partial u_1}\right|$

Similarly if e_2 and e_3 are unit tangent vectors to the u_2 and u_3 curves at **P** respectively, then

$$e_2 = \frac{\frac{\partial r}{\partial u_2}}{\left|\frac{\partial r}{\partial u_2}\right|} \quad \text{and} \quad e_3 = \frac{\frac{\partial r}{\partial u_3}}{\left|\frac{\partial r}{\partial u_3}\right|}$$

so that $h_2 e_2 = \frac{\partial r}{\partial u_2}$ and $h_3 e_3 = \frac{\partial r}{\partial u_3}$

where $h_2 = \left|\frac{\partial r}{\partial u_2}\right|$ and $h_3 = \left|\frac{\partial r}{\partial u_3}\right|$

The quantities h_1, h_2, h_3, are called *scale factors*

The unit vectors e_1, e_2, e_3, are in the directions of increasing u_1, u_2, u_3, respectively.

$\therefore$ ∇u_1 is a vector at **P** normal to the surface $u_1 = c_1$ a unit vector in this direction is given by

$$E_1 = \frac{\nabla u_1}{|\nabla u_1|}$$

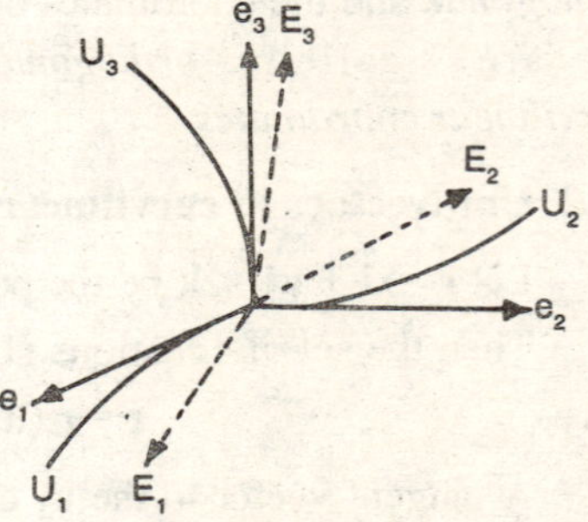

Fig. 15.2

Similarly

$E_2 = \frac{\nabla u_2}{|\nabla u_2|}$ and $E_3 = \frac{\nabla u_3}{|\nabla u_3|}$ at **P** are normal to the surfaces $u_2 = c_2$ and $u_3 = c_3$ respectively.

∴ In a curvilinear system, at each point **P** there exist in general, two sets of unit vectors ($\mathbf{e}_1, \mathbf{e}_2, \mathbf{e}_3$) and ($\mathbf{E}_1, \mathbf{E}_2, \mathbf{E}_3$), former tangent to the coordinate curves and later normal to the coordinate surfaces. If the coordinate system in orthogonal then these sets become identical. Both the sets are analogous to the **i, j, k** unit vectors, in rectangular co-ordinates. They may change in directions from one point to another.

It can also be shown that the sets $\frac{\partial \mathbf{r}}{\partial u_1}, \frac{\partial \mathbf{r}}{\partial u_2}, \frac{\partial \mathbf{r}}{\partial u_3}$ and ∇u_1, ∇u_2, ∇u_3 constitute *reciprocal system of vectors*

i.e $$\frac{\partial \mathbf{r}}{\partial u_1} \cdot \nabla u_1 = \frac{\partial \mathbf{r}}{\partial u_2} \cdot \nabla u_2 = \frac{\partial \mathbf{r}}{\partial u_3} \cdot \nabla u_3 = 1$$

and $$\frac{\partial \mathbf{r}}{\partial u_2} \cdot \nabla u_1 = \frac{\partial \mathbf{r}}{\partial u_3} \cdot \nabla u_1 = 0$$

$$\frac{\partial \mathbf{r}}{\partial u_1} \cdot \nabla u_2 = \frac{\partial \mathbf{r}}{\partial u_3} \cdot \nabla u_2 = 0$$

$$\frac{\partial \mathbf{r}}{\partial u_1} \cdot \nabla u_3 = \frac{\partial \mathbf{r}}{\partial u_2} \cdot \nabla u_3 = 0$$

A vector **F** can be represented in term of the unit base vectors $\mathbf{e}_1, \mathbf{e}_2, \mathbf{e}_3$ and $\mathbf{E}_1, \mathbf{E}_2, \mathbf{E}_3$ in the form

$$\mathbf{F} = \mathbf{F}_1 \mathbf{e}_i + \mathbf{F}_2 \mathbf{e}_2 + \mathbf{F}_3 \mathbf{e}_3$$

$$= f_1 \mathbf{E}_1 = f_2 \mathbf{E}_2 + f_3 \mathbf{E}_3,$$

where $\mathbf{F}_1, \mathbf{F}_2, \mathbf{F}_3$ andf_1, f_2, f_3 are components of **F** in each system.

The base vectors $\frac{\partial \mathbf{r}}{\partial u_1}, \frac{\partial \mathbf{r}}{\partial u_2}, \frac{\partial \mathbf{r}}{\partial u_3}$ or $\nabla u_1, \nabla u_2, \nabla u_3$ are called *unitary base vectors* but are not unit vectors in general.

In this case we represent **F** in the form

$$\mathbf{F} = C_i \frac{\partial \mathbf{r}}{\partial u_1} + C_2 \frac{\partial \mathbf{r}}{\partial u_2} + C_3 \frac{\partial \mathbf{r}}{\partial u_3}$$

$$= C^*_1 \nabla u_1 + C^*_2 \nabla u_2 + C^*_3 \nabla u_3$$

where C_1, C_2, C_3 are called *contravariant components* of **F** and C^*_1, C^*_2, C^*_3 are called covariant components of F.

15.3. Arc length and volume element

The curve u_1 is the intersection of the surfaces $u_2 = c_2$ and $u_3 = c_3$.

$\therefore$ Along the u_1 curve, u_2, u_3 are constants and u_1 is variable.

Thus along each coordinate curve arc length is a function of a single variable. Let s_1, s_2, s_3 be the are lengths measured along the coordinate curve in the positive directions of u_1, u_2, u_3 respectively.

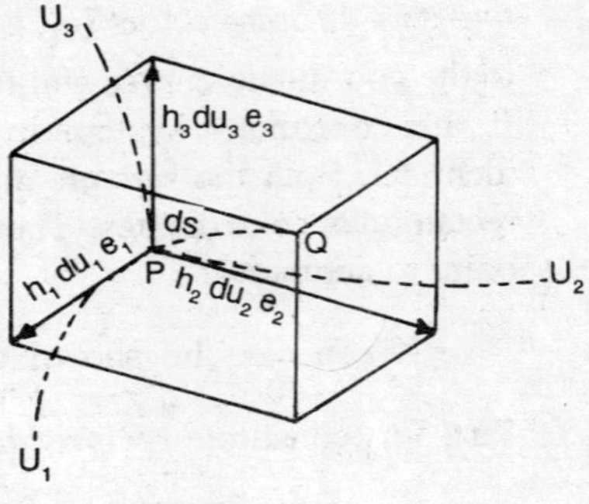

Fig. 15.3

Let the derivatives taken at **P** be given by

$$\frac{ds_1}{du_1} = h_1; \frac{ds_2}{du_2} = h_2; \frac{ds_3}{du_3} = h_3$$

Hence $\quad ds_1 = h_1, du_1, ds_2 = h_2 \, du_2, ds_3 = h_3 \, du_3$

The rectangular parallelopiped with edges $ds_1 = h_1 \, du_1$, $ds_2 = h_2 \, du_2$, $ds_3 = h_3 = du_3$ extending along **i, j, k** is called *volume element.*

Let PQ = ds = diagonal of rectangular parallelopiped.

$$ds_2 = ds \, . \, ds = h_1^2 \, du_1^2 + h_2^2 \, du_2^2 + h_3^2 \, du_3^2 \quad ...(1)$$

this is the expression for ds, the element of arc along any curve tangent to **PQ** at **P**. h_i, h_2, h_3 and called *scale factors*.

The *differential volume element* is given by

$$dv = h_1 \, h_2 \, h_3 \, du_1 \, du_2 \, du_3 \quad ...(2)$$

Note: It is not necessary that u must have the dimensions of length, but the product h du must have the dimensions of length.

15.4. Important relations:

(I) Prove that

(i) $| \nabla u_i | = \frac{1}{h_i}; i = 1, 2, 3$

(ii) $\mathbf{E_i} = \mathbf{e_i}$; i = 1, 2, 3.

Proof: (i) Let $\phi = u_i$ and also let $\nabla \phi = f_i \mathbf{e_i}$ ...(1)

$$\therefore \quad \mathbf{dr} = \frac{\partial \phi}{\partial u_i} du_i = h_i \mathbf{e_i} du_i$$

We have $d\phi = \nabla \phi \cdot dr = h_i f_i du_i$

$$= \frac{\partial \phi}{\partial u_i} du_i$$

$$\Leftrightarrow \quad \frac{\partial \phi}{\partial u_i} = h_i f_i$$

or $$\frac{\partial u_i}{\partial u_i} = 1 = h_i f_i \qquad (\because \ \phi = u_i)$$

or $$f_i = \frac{1}{h_i}$$

Substituting in (1) we get

$$\nabla u_i = f_i \mathbf{e_i} \qquad (\phi = u_i)$$

$$| \nabla u_i | = f_i | \mathbf{e_i} |$$

$$= f_i \qquad (\because | \mathbf{e_i} | = 1)$$

$$= \frac{1}{h_i}$$

It is true for i = 1, 2, 3.

$$\therefore \quad \nabla i_1 = \frac{\mathbf{e_1}}{h_1}, \nabla u_2 = \frac{\mathbf{e_1}}{\mathbf{h_1}}, \nabla u_3 = \frac{\mathbf{e_3}}{\mathbf{h_3}}$$

and $$| \nabla u_1 | = \frac{1}{h_1}, | \nabla u_2 | = \frac{1}{h_2}, | \nabla u_3 | = \frac{1}{h_3}$$

Hence the proof.

(ii) We can write $\mathbf{E_i} = \dfrac{\nabla u_i}{| \nabla u_i |}$

But $| \nabla u_i | = \dfrac{1}{h_i}$ [from (i) of 1 in (7.3)]

$$\therefore \quad \mathbf{E_i} = h_i \nabla u_i = h_i \frac{\mathbf{e_i}}{h_i} = \mathbf{e_i}$$

Hence $\mathbf{E_i} = \mathbf{e_i}$ for i = 1, 2, 3

ie $\mathbf{e_i} = \mathbf{E_i}, \mathbf{e_2} = \mathbf{E_2}, \mathbf{e_3} = \mathbf{E_3}$

(ii) Prove that $\mathbf{e_2} = h_2 h_3 \nabla u_2 \times \nabla u_3$ $\mathbf{e_2} = h_3 h_1 \nabla u_3 \times \nabla u_1$ and $\mathbf{e_3} = h_1 h_2 \nabla u_1 \times \nabla u_2$

Proof: From (i) of 1 in (7.3), we have

$$\nabla u_1 = \frac{\mathbf{e_1}}{h_1}, \nabla u_2 = \frac{\mathbf{e_2}}{h_2}, \nabla u_3 = \frac{\mathbf{e_3}}{h_3}$$

Then $$\nabla u_2 \times \nabla u_3 = \frac{\mathbf{e_2} \times \mathbf{e_3}}{h_2 h_3} = \frac{\mathbf{e_1}}{h_2 h_3} \qquad (\because \ \mathbf{e_2} \times \mathbf{e_3} = \mathbf{e_1})$$

$\therefore$ $h_2 h_3 \nabla u_2 \times \nabla u_3 = \mathbf{e_1}$. Hence the proof.

Similarly we can prove that

$$\mathbf{e_2} = h_3 h_1 \nabla u_3 \times \nabla u_1$$

$$\mathbf{e_3} = h_1 h_2 \nabla u_1 \times \nabla u_2$$

Example 1. Prove that

$$\left\{\frac{\partial r}{\partial u_1} \cdot \frac{\partial r}{\partial u_2} \times \frac{\partial r}{\partial u_3}\right\} \left\{ \nabla u_1 \cdot \nabla u_2 \times \nabla u_3 \right\} = 1$$

Solution: We have

$$dr = \frac{\partial r}{\partial u_1} du_1 + \frac{\partial r}{\partial u_2} du_2 + \frac{\partial r}{\partial u_3} du_3 \qquad ...(1)$$

Multiplying (1) by ∇u_1. on both sides.

$$\nabla u_i \cdot dr = du_1 = \left(\nabla u_1 \cdot \frac{\partial r}{\partial u_1}\right) du_1 + \left(\nabla u_1 \cdot \frac{\partial r}{\partial u_2}\right) du_2 + \left(\nabla u_1 \cdot \frac{\partial r}{\partial u_3}\right) du_3$$

$$\Leftrightarrow \quad 1 = \nabla u_i \cdot \frac{\partial r}{\partial u_1}; 0 = \nabla u_1 \frac{\partial r}{\partial u_2}$$

and $$0 = \nabla u_1 \cdot \frac{\partial r}{\partial u_3}$$

Similarly multiplying (1) by ∇u_2 . and ∇u_3. we get

$$0 = \nabla u_2 \cdot \frac{\partial r}{\partial u_1}; 1 = \nabla u_2 \cdot \frac{\partial r}{\partial u_2}; 0 = \nabla u_2 \cdot \frac{\partial r}{\partial u_3}$$

$$0 = \nabla u_3 \cdot \frac{\partial r}{\partial u_1}; 0 = \nabla u_3 \cdot \frac{\partial r}{\partial u_2}; 1 = \nabla u_3 \cdot \frac{\partial r}{\partial u_3}$$

$$\therefore \quad \frac{\partial r}{\partial u_1}, \frac{\partial r}{\partial u_2}, \frac{\partial r}{\partial u_3} \text{ and } \nabla u_1, \nabla u_2, \nabla u_3$$

reciprocal system of vectors.

$$\therefore \quad \text{If} \quad \left\{\frac{\partial r}{\partial u_1} \cdot \frac{\partial r}{\partial u_2} \times \frac{\partial r}{\partial u_3}\right\} = V$$

$$\text{Then} \qquad \nabla u_1 = \frac{\dfrac{\partial r}{\partial u_2} \times \dfrac{\partial r}{\partial u_3}}{V}$$

$$\nabla u_2 = \frac{\dfrac{\partial r}{\partial u_3} \times \dfrac{\partial r}{\partial u_1}}{V}$$

$$\text{and} \qquad \nabla u_3 = \frac{\dfrac{\partial r}{\partial u_1} \times \dfrac{\partial r}{\partial u_2}}{V}$$

$$\therefore \quad \nabla u_1 \cdot (\nabla u_2 \times \nabla u_3)$$

$$= \frac{1}{V^3}\left[\left(\frac{\partial r}{\partial u_2} \times \frac{\partial r}{\partial u_3}\right) \cdot \left\{\left(\frac{\partial r}{\partial u_3} \times \frac{\partial r}{\partial u_1}\right) \times \left(\frac{\partial r}{\partial u_1} \times \frac{\partial r}{\partial u_2}\right)\right\}\right]$$

$$= \frac{1}{V^3}\left[\frac{\partial r}{\partial u_1} \cdot \left(\frac{\partial r}{\partial u_2} \times \frac{\partial r}{\partial u_3}\right)\right]^2 = \frac{V_2}{V_3} = \frac{1}{V}$$

or $$V \cdot \nabla u_i \cdot (\nabla u_2 \times \nabla u_{3)} = 1$$

or $$\left\{\frac{\partial r}{\partial u_1} \cdot \left(\frac{\partial r}{\partial u_2} \times \frac{\partial r}{\partial u_3}\right)\right\} [\nabla u_1 \cdot (\nabla u_2 \times \nabla u_3) = 1.$$

Hence the proof.

15.5. Calculation of gradient in orthogonal curvilinear co-ordinates

Let $\phi\,(u_i, u_2, u_3)$ be a scalar point function given in terms of orthogonal curvilinear co-ordinates.

Let $$\nabla\phi = f_1\, e_1 + f_2\, e_2 + f_3\, e_3 \qquad ...(1)$$

where f_1, f_2, f_3, are to be determined.

$$\because \quad d\mathbf{r} = \frac{\partial \mathbf{r}}{\partial u_1} du_1 + \frac{\partial \mathbf{r}}{\partial u_2} du_2 + \frac{\partial \mathbf{r}}{\partial u_3} du$$

$$= h_i \mathbf{e_i} \; du_i + h_2 \mathbf{e_2} \, du_2 + h_2 \mathbf{e_3} \, du_3$$

$$\therefore \quad \nabla\phi \, . \, d\mathbf{r} = h_1 f_1 \, du_1 + h_2 f_2 \, du_2 + h_3 f_3 \, du_3 = d\phi \quad ...(2)$$

But $$d\phi = \frac{\partial \phi}{\partial u_1} du_1 + \frac{\partial \phi}{\partial u_2} du_2 + \frac{\partial \phi}{\partial u_3} du_3 \quad ...(3)$$

Equating (1) and (2)

$$h_1 f_1 = \frac{\partial \phi}{\partial u_1} \text{ or } f_1 = \frac{1}{h_1}\frac{\partial \phi}{\partial u_1}$$

$$f_2 = \frac{1}{h_2}\frac{\partial \phi}{\partial u_2} \text{ and } f_3 = \frac{1}{h_3}\frac{\partial \phi}{\partial u_3}$$

Then $$\nabla\phi = \frac{1}{h_1}\frac{\partial \phi}{\partial u_1}\mathbf{e_1} + \frac{1}{h_2}\frac{\partial \phi}{\partial u_2}\mathbf{e_2} + \frac{1}{h_3}\frac{\partial \phi}{\partial u_3}\mathbf{e_3} \quad ...(4)$$

$\therefore$ The general expression for ∇ (grad) in terms of orthogonal curvilinear co-ordinates is

$$\nabla = \frac{\mathbf{e_1}}{h_1}\frac{\partial}{\partial u_1} + \frac{\mathbf{e_2}}{h_2}\frac{\partial}{\partial u_2} + \frac{\mathbf{e_3}}{h_3}\frac{\partial}{\partial u_3}$$

15.6. Calculation of divergence in orthogonal curvilinear co-ordinates

Let $F(u_1, u_2, u_3)$ be a vector point function given in terms of orthogonal curvilinear coordinates.

Let $$\mathbf{F} = F_1 \mathbf{e_1} + F_2 \mathbf{e_2} + F_3 \mathbf{e_3}$$

$$\nabla \, . \, \mathbf{F} = \nabla \, . \, (F_1 \mathbf{e_1} + F_2 \mathbf{e_2} + F_3 \mathbf{e_3})$$

$$= \nabla \, . \, (F_1 \mathbf{e_1}) + \nabla \, . \, (f_2 \mathbf{e_2}) + \nabla \, . \, (F_3 \, . \mathbf{e_3}) \quad ...(1)$$

Consider $\nabla \, . \, (F_1 \mathbf{e_1})$

$$= \nabla \, . \, (F_1 h_2 h_3 \nabla u_2 \times \nabla u_3) \quad (\because \; \mathbf{e_1} = h_2 h_3 \nabla u_2 \times \nabla u_3)$$

$$= \nabla (F_1 h_2 h_3) \nabla u_2 \times \nabla u_3 + f_1 h_2 h_3 \nabla \, . \, (\nabla u_2 \times \nabla u_3)$$

$$= \nabla (F_1 h_2 h_3) \, . \, \frac{\mathbf{e_3}}{h_2} \times \frac{\mathbf{e_3}}{h_3} + 0$$

$$= \nabla (F_1 h_2 h_3) \, . \, \frac{\mathbf{e_1}}{h_2 h_3} \quad (\therefore \; \mathbf{e_1} = \mathbf{e_2} \times \mathbf{e_3})$$

$$= \left(\frac{\mathbf{e_1}}{h_1} \frac{\partial}{\partial u_1} + \frac{\mathbf{e_2}}{h_2} \frac{\partial}{\partial u_2} + \frac{\mathbf{e_3}}{h_3} \frac{\partial}{\partial u_3} \right) (F_1\, h_2 h_3) \, . \, \frac{\mathbf{e_1}}{h_2 h_3}$$

$$= \frac{\mathbf{e_1}}{h_2 h_3} \left\{ \frac{\mathbf{e_1}}{h_1} \frac{\partial}{\partial u_1} (F_1\, h_2 h_3) + \frac{\mathbf{e_2}}{h_2} \frac{\partial}{\partial u_2} (F_2\, h_2 h_3) \right.$$

$$\left. + \frac{\mathbf{e_3}}{h_3} \frac{\partial}{\partial u_3} (F_2\, h_2 h_3) \right\}$$

$$\therefore \quad \nabla \, . \, (F_1\, \mathbf{e_1}) = \frac{1}{h_1 h_2 h_3} \frac{\partial}{\partial u_1} (F_1\, h_2 h_3) \quad \text{...(2)}$$

Similarly $\nabla \, . \, (F_2\, \mathbf{e_2}) = \dfrac{1}{h_1 h_2 h_3} \dfrac{\partial}{\partial u_2} (F_2\, h_3 h_1)$ ...(3)

and $\nabla \, . \, (F_3\, \mathbf{e_3}) = \dfrac{1}{h_1 h_2 h_3} \dfrac{\partial}{\partial u_3} (F_3\, h_1 h_2)$ ...(4)

Substituting (2), (3) and (4) in (1)

$$\nabla \, . \, \mathbf{F} = \frac{1}{h_1 h_2 h_3} \frac{\partial}{\partial u_1} (F_1\, h_2 h_3)$$

$$+ \frac{1}{h_1 h_2 h_3} \frac{\partial}{\partial u_2} (F_2\, h_3 h_1) + \frac{1}{h_1 h_2 h_3} \frac{\partial}{\partial u_3} (F_3\, h_1 h_2)$$

$$= \frac{1}{h_1 h_2 h_3} \left\{ \frac{\partial}{\partial u_1} (F_1\, h_2 h_3) + \frac{\partial}{\partial u_1} (F_2\, h_3 h_1) \right.$$

$$\left. + \frac{\partial}{\partial u_3} (F_3\, h_1 h_2) \right\} \quad \text{...(5)}$$

15.7. Calculation of curl in orthogonal curvilinear co-ordinates

Let **F** (u_1, u_2, u_3) be a vector point function given interms of orthogonal curvilinear co- ordinates.

Let $\mathbf{F} = F_1\, \mathbf{e_1} + F_2\, \mathbf{e_2} + F_3\, \mathbf{e_3}$

Then $\nabla \times \mathbf{F} = \nabla \times (F_1\, \mathbf{e_1} + F_2\, \mathbf{E_2} + F_3\, \mathbf{e_3})$

$$= \nabla \times (F_1\, \mathbf{e_1}) + \nabla \times (F_2\, \mathbf{e_2}) + \nabla \times (F_3\, \mathbf{e_3}) \quad \text{...(1)}$$

Consider $\nabla \times (F_1\, \mathbf{e_1}) = \nabla \times (F_1\, h_1\, \nabla u_1)$ $(\therefore \mathbf{e_1} = h_1\, \nabla u_1)$

$$= \nabla (F_1 h_1) \times \nabla u_1 + F_1 h_1 \nabla \times \nabla u_1$$

$$= \nabla (F_1 h_1) \times \frac{\mathbf{e_1}}{h_1} + 0 \qquad (\because \nabla \times \nabla u_1 = 0)$$

$$= \left\{ \frac{\mathbf{e_1}}{h_1} \frac{\partial}{\partial u_1}(F_1 h_1) + \frac{\mathbf{e_2}}{h_2} \frac{\partial}{\partial u_2}(F_1 h_1) + \frac{\mathbf{e_3}}{h_3} \frac{\partial}{\partial u_3}(F_1 h_1) \right\} \times \frac{\mathbf{e_1}}{h_1}$$

$$= \left\{ \frac{\mathbf{e_2}}{h_3 h_1} \frac{\partial}{\partial u_3}(F_1 h_1) - \frac{\mathbf{e_3}}{h_1 h_2} \frac{\partial}{\partial u_2}(F_1 h_1) \right\} \qquad ...(2)$$

Similarly

$$\nabla \times (F_2 \mathbf{e_2}) = \left\{ \frac{\mathbf{e_3}}{h_1 h_2} \frac{\partial}{\partial u_1}(F_2 h_2) - \frac{\mathbf{e_2}}{h_2 h_3} \frac{\partial}{\partial u_3}(F_2 h_2) \right\} \qquad ...(3)$$

and $$\nabla \times (F_3 \mathbf{e_3}) = \left\{ \frac{\mathbf{e_1}}{h_2 h_3} \frac{\partial}{\partial u_2}(F_3 h_3) - \frac{\mathbf{e_2}}{h_3 h_1} \frac{\partial}{\partial u_1}(F_3 h_3) \right\} \qquad ...(4)$$

Substituting (2), (3) and (4) in (1), after adjusting the terms, we get

$$\nabla \times \mathbf{F} = \frac{\mathbf{e_1}}{h_2 h_3} \left\{ \frac{\partial}{\partial u_2}(F_3 h_3) - \frac{\partial}{\partial u_2}(F_2 h_2) \right\}$$

$$+ \frac{\mathbf{e_2}}{h_3 h_1} \left\{ \frac{\partial}{\partial u_3}(F_1 h_1) - \frac{\partial}{\partial u_1}(F_3 h_3) \right\}$$

$$+ \frac{\mathbf{e_3}}{h_1 h_3} \left\{ \frac{\partial}{\partial u_1}(F_2 h_2) - \frac{\partial}{\partial u_2}(F_1 h_1) \right\}$$

Which also can be written as

$$\nabla \times F = \frac{1}{h_1 h_2 h_3} \begin{vmatrix} h_1 e_1 & h_2 e_2 & h_3 e_3 \\ \frac{\partial}{\partial u_1} & \frac{\partial}{\partial u_2} & \frac{\partial}{\partial u_3} \\ F_1 h_1 & F_2 h_2 & F_3 h_3 \end{vmatrix} \qquad ...(5)$$

15.8. Calculation of Laplacian in orthogonal curvilinear co-ordinates

Let $\phi(u_1, u_2, u_3)$ be a scalar point function given in terms of orthogonal curvilinear co-ordinates.

From (4) of art (7.4)

$$\nabla\phi = e_1 h_1 \frac{\partial\phi}{\partial u_1} + \frac{e_2}{h_2}\frac{\partial\phi}{\partial u_2} + \frac{e_3}{h_3}\frac{\partial\phi}{\partial u_3}$$

If $\quad f = \nabla\phi$ then $\nabla . f = \nabla . \nabla\phi - \nabla^2\phi.$

If $\quad f = f_1 e_1 + f_2 e_2 + f_3 e_3$

then $\quad f_1 = \frac{1}{h_1}\frac{\partial\phi}{\partial u_1};\ f_2 = \frac{1}{h_2}\frac{\partial\,\phi}{\partial\, u_3};\ f_3 = \frac{1}{h_3}\frac{\partial\phi}{\partial u_3}$

$$\therefore \quad \nabla^2\phi = \nabla\cdot f = \frac{1}{h_1 h_2 h_3}\left\{\frac{\partial}{\partial u_1}(f_1 h_2 h_3) + \frac{\partial}{\partial u^2}(f_2 h_3 h_1) + \frac{\partial}{\partial u_3}(f_3 h_1 h_2)\right\}$$

$$= \frac{1}{h_1 h_2 h_3}\left\{\frac{\partial}{\partial u_1}\left(\frac{h_2 h_3}{h_1}\frac{\partial\phi}{\partial u_1}\right) + \frac{\partial}{\partial u_2}\left(\frac{h_3 h_1}{h_2}\frac{\partial\phi}{\partial u_2}\right) + \frac{\partial}{\partial u_3}\left(\frac{h_1 h_2}{h_3}\frac{\partial\phi}{\partial u_3}\right)\right\} \quad ...(1)$$

15.9. Special curvilinear co-ordinate systems

(1) Cartesian co-ordinate system

Here $u_1 = x,\ u_2 = y,\ u_3 = z$

$$\therefore \quad ds^2 = h_1^2\, du_1^2 + h_2^2\, du_2^2 + h_3^2\, du_3^2$$

$$= h_1^2\, dx^2 + h_2^2\, dy^2 + h_3^2\, dz^2$$

But $ds^2 = dx^2 + dy^2 + dz^2 \quad \therefore \quad h_1 = h_2 = h_2 = 1$

Now if we replace (e_1, e_2, e_3) by **(i, j, k)**,

Now from (4) of art (7.4)

$$\nabla\phi = \frac{\partial\phi}{\partial x}\mathbf{i} + \frac{\partial\phi}{\partial y}\mathbf{j} + \frac{\partial\phi}{\partial z}\mathbf{k}$$

From (5) of art (7.5)

$$\nabla \cdot \mathbf{F} = \frac{\partial}{\partial x}\mathbf{F_1} + \frac{\partial}{\partial y}\mathbf{F_2} + \frac{\partial}{\partial z}\mathbf{F_3}$$

From (5) of art (7.6)

$$\nabla \times \mathbf{F} = \begin{vmatrix} \mathbf{i} & \mathbf{j} & \mathbf{k} \\ \frac{\partial}{\partial x} & \frac{\partial}{\partial y} & \frac{\partial}{\partial z} \\ F_1 & F_2 & F_3 \end{vmatrix}$$

And from (1) of art (7.7)

$$\nabla^2 \phi = \frac{\partial^2 \phi}{\partial x^2} + \frac{\partial^2 \phi}{\partial y^2} + \frac{\partial^2 \phi}{\partial z^2}$$

which are identical to those expressions, we used in earlier

2 Cylindrical co-ordinate system

Let P(x, y, z) be a point and PN perpendicular to the plane XOY

Let ON = ρ, ∠XON = θ, PN = z

Then (ρ, θ, z) are called the cylindrical co-ordinates of **P.**

The co-ordinate surfaces are

(i) ρ = constant : It gives a family of coaxial cylinders.

(ii) θ = constant: It gives a family of planes intersecting along the z-axis.

(iii) z=constant : it gives a family of planes normal to the z-axis.

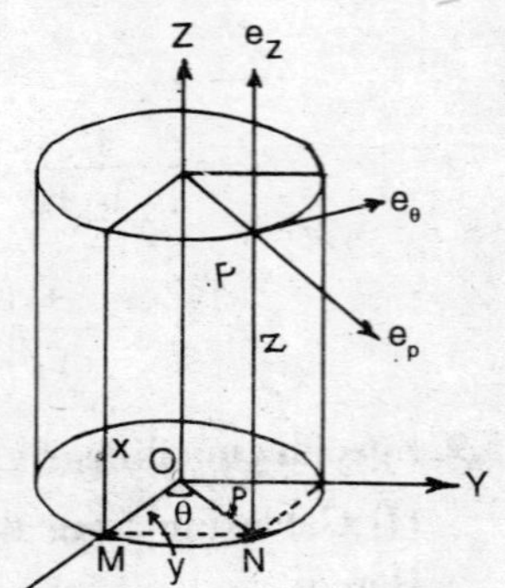

Fig. 15.4

∴ From fig 15.4 $x = OM = ON \cos\theta = \rho\cos\theta$

$$y = NM = ON \sin\theta = \rho \sin\theta$$

$$z = NP = z$$

$$(\rho \geq 0,\ 0 \leq \theta \leq 2\pi,\ -\infty < z < \infty)$$

∴ $x = \rho\cos\theta$, $y = \rho\sin\theta$, $z = z$ *are transformation equations between* (x, y, z) *and* (ρ, θ, z)

Now $$dx = \frac{\partial x}{\partial \rho} d\rho + \frac{\partial x}{\partial \theta} d\theta + \frac{\partial x}{\partial z} dz$$
$$= \cos\theta\, d\rho - \rho \sin\theta\, d\theta$$

Similarly $dy = \sin\theta\, d\rho + \rho\cos\theta\, d\theta$ and $dz = dz$

We know that $ds^2 = dx^2 + dy^2 + dz^2$

On substituting the values of dx, dy, dz from above, we get in cylindrical co-ordinate system the square of arc length as

$$ds^2 = d\rho^2 + \rho^2 d\theta + dz^2 \qquad \text{...(1)}$$

If $u_1 = \rho,\ u_2 = \theta,\ u_3 = z$

Then $$ds^2 = du_1^2 + \rho^2\, du_2^2 + du_3^2$$
$$= h_1^2\, du_1^2 + h_2^2\, du_2^2 + h_3^2\, du_3^2$$

$\therefore$ $h_1 = 1, h_2 = \rho,\ h_3 = 1.$

Let e_ρ, e_θ, e_z be the unit vectors in the directions of ρ, θ and z directions.

$\therefore$ We replace $e_1, e_2\ e_3$ by e_ρ, e_θ, e_z

Then volume element

$$dv = \rho\, d\rho\, d\theta\, dz$$

Now from (4) of art 15.6

$$\nabla\Psi = \frac{\partial \Psi}{\partial \rho} e_\rho + \frac{1}{\rho}\frac{\partial \Psi}{\partial \theta} e_\theta + \frac{\partial \Psi}{\partial z} e_z$$

[Here Ψ is a scalar function].

From (5) of art (15.6)

$$\nabla \cdot \mathbf{F} = \frac{1}{\rho}\left\{\frac{\partial f}{\partial \rho}(\rho F_\rho) + \frac{\partial}{\partial \theta}(F_\theta) + \frac{\partial}{\partial z}(\rho F_z)\right\}$$

where $F = F_\rho\ e_\rho + F_\theta\ e_\theta + F_\phi\ e_\phi$

From (5) of art 15.7

$$\nabla \times \mathbf{F} = \frac{1}{\rho}\begin{vmatrix} e_\rho & e_\theta & e_z \\ \frac{\partial}{\partial \rho} & \frac{\partial}{\partial \theta} & \frac{\partial}{\partial z} \\ F_\rho & F_\theta & F_z \end{vmatrix}$$

and from (2) of art 15.8

$$\nabla^2\Psi = \frac{1}{\rho}\frac{\partial}{\partial\rho}\left(\rho\frac{\partial\Psi}{\partial\rho}\right) + \frac{1}{\rho^2}\frac{\partial^2\Psi}{\partial\theta^2} + \frac{\partial^2\Psi}{\partial z^2}$$

Example 2. (i) Find the unit vectors e_ρ, e_θ, e_z of a cylindrical co-ordinate system in terms of **i, j, k,** (ii) Solve for **i, j, k,** interms of e_ρ, e_θ, e_z. (iii) Represent the vector $\mathbf{V} = zx\,\mathbf{i} + xy\,\mathbf{j} + yz\,\mathbf{k}$ in cylindrical co-ordinates and hence determine $\mathbf{V}_\rho$, $\mathbf{V}_\theta$, $\mathbf{V}_z$.

Solution: (i) The position vector of any point in cylindrical co-ordinates is

$$\mathbf{r} = x\,\mathbf{i} + y\,\mathbf{j} + z\,\mathbf{k} = \rho\cos\theta\,\mathbf{i} + \rho\sin\theta\,\mathbf{j} + z\,\mathbf{k}$$

Then $\frac{\partial \mathbf{r}}{\partial \rho}, \frac{\partial \mathbf{r}}{\partial \theta}, \frac{\partial \mathbf{r}}{\partial z}$ are tangent vectors to the ρ_1, θ and z curves.

$$\therefore \quad \frac{\partial \mathbf{r}}{\partial \rho} = \cos\theta\,\mathbf{i} + \sin\theta\,\mathbf{j}$$

$$\frac{\partial \mathbf{r}}{\partial \theta} = \rho\sin\theta\,\mathbf{i} + \rho\cos\theta\,\mathbf{j}$$

$$\frac{\partial \mathbf{r}}{\partial z} = \mathbf{k}$$

The unit vectors in these directions are

$$e_\rho = \frac{\frac{\partial \mathbf{r}}{\partial \rho}}{\left|\frac{\partial \mathbf{r}}{\partial \rho}\right|} = \frac{\cos\theta\,\mathbf{i} + \sin\theta\,\mathbf{j}}{\sqrt{\cos^2\theta + \sin^2\theta}} = \cos\theta\,\mathbf{i} + \sin\theta\,\mathbf{j} \qquad \text{...(1)}$$

$$e_\theta = \frac{\frac{\partial \mathbf{r}}{\partial \theta}}{\left|\frac{\partial \mathbf{r}}{\partial \theta}\right|} = \frac{-\rho\sin\theta\,\mathbf{i} + \rho\cos\theta\,\mathbf{j}}{\rho\sqrt{\cos^2\theta + \sin^2\theta}} = -\sin\theta\,\mathbf{i} + \cos\theta\,\mathbf{j} \qquad \text{...(2)}$$

$$e_z = \frac{\frac{\partial \mathbf{r}}{\partial z}}{\left|\frac{\partial \mathbf{r}}{\partial z}\right|} = \frac{k}{\sqrt{1}} = k. \qquad \text{...(3)}$$

(ii) Solving (1) and (2) simultaneously, we get

$$\mathbf{i} = \cos\theta\, e_\rho = \sin\theta\, e_\theta \qquad \text{...(4)}$$

$$\mathbf{j} = \sin\theta\, e_\rho + \cos\theta\, e_\theta \qquad ...(5)$$

$$\mathbf{k} = e_z \qquad ...(6)$$

(iii) $\mathbf{V} = zx\,\mathbf{i} + xy\,\mathbf{j} + yz\,\mathbf{k}$

$$= z(\rho\cos\theta)(\cos\theta\, e_\rho - \sin\theta\, e_\theta)$$

$$+ (\rho\cos\theta)(\rho\sin\theta)(\sin\theta\, e_\rho + \cos\theta\, e_\theta)$$

$$+ (\rho\sin\theta)(z)\, e_z$$

$$= (\rho z\cos^2\theta + \rho^2\cos\theta\sin^2\theta)\, e_\rho$$

$$+ (-\rho z\cos\theta\sin\theta + \rho^2\cos^2\theta\sin\theta)\, e_\theta$$

$$+ (z\rho\sin\theta)\, e_z$$

$\therefore$ If $V = V_\rho e_\rho + V_\theta e_\theta + V_z e_z$

Then $V_\rho = \rho z\cos^2\theta + \rho^2\cos\theta\sin^2\theta$

$$= \rho\cos\theta(z\cos\theta + \rho\sin^2\theta)$$

$$V_\theta = -\rho\cos\theta\sin\theta + \rho^2\cos^2\theta\sin\theta$$

$$= \rho\cos\theta\sin\theta(-z + \rho\cos\theta)$$

$$V_z = z\rho\sin\theta$$

3. Spherical co-ordinate system:

Let P (x, y, z) be a point and PN perpendicular to the plane XOY.
Let OP = r, $\lfloor$ ZOP = θ, $\lfloor$ XON = φ

The spherical co-ordinates of P are (r, θ, φ).

The co-ordinate surfaces are

(ii) θ = constant: It gives a system of cones having the vertex at the origin and axis along the z-axis.

(iii) φ = constant: It gives a family of planes intersecting along the z-axis.

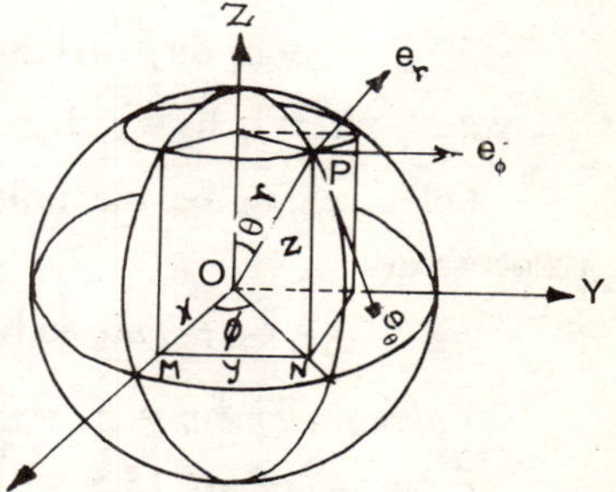

Fig. 15.5

Now $\mathbf{X} = \mathbf{OM} = ON\cos\phi = \mathbf{OP}\sin\theta\cos\phi$

$\therefore \quad x = r \sin\theta \cos\phi$

$y = MN = ON \sin\phi = OP \sin\theta \sin\phi$

$\therefore \quad y = r \sin\theta \sin\phi.$

$z = \mathbf{NP} = \mathbf{OP} \cos\theta = r\cos\theta$

Hence $x = r \sin\theta \cos\phi$

$y = r \sin\theta \sin\phi$

$z = r \sin\theta$

are the *transformation equations between* (x, y, z) *and* $(\mathbf{r}, \theta, \phi)$

$$\text{Now} dx = \frac{\partial x}{\partial r} dr + \frac{\partial x}{\partial \theta} d\theta + \frac{\partial x}{\partial \phi} d\phi$$

$$= \sin\theta \cos\phi \, dr + r \cos\theta \cos\phi \, d\theta + r \sin\theta \sin\phi \, d\phi.$$

Similarly

$$dy = \sin\theta \sin\phi \, dr + r\cos\theta \sin\phi \, d\theta + r \sin\theta \cos\phi \, d\phi$$

$$dz = \cos\theta \, dr - r\sin\theta \, d\theta$$

We know that $ds^2 = dx^2 + dy^2 + dz^2$

On substituting the values of dx, dy, dz after simplification, we get

$$ds^2 = dr^2 + r^2 d\theta^2 + r^2 \sin^2\theta \, d\phi^2 \quad \ldots(1)$$

If $\quad u_1 = r, \; u_2 = \theta, u_3 = \phi$

$$ds^2 = du_1^2 + r^2 du_2^2 + r^2 \sin^2\theta \, du_3^2$$

$$= h_1^2 du_1^2 + h_2^2 du_2^2 + h_3^2 du_3^2$$

$\therefore \quad h_1 = 1, \; h_2 = r, \; h_3 = r\sin\theta.$

Let e_r, e_θ, e_ϕ be the unit vectors in the directions of r, θ, ϕ respectively

ie. $\quad (e_1, e_2, e_3)$ are to be replaced by (e_r, e_θ, e_ϕ).

Volume element = $dv = r^2 \sin^2\theta \, dr \, d\theta \, d\phi$.

From (4) of **art 15.5**

$$\nabla\Psi = \frac{\partial \Psi}{\partial r} e_r + \frac{1}{r}\frac{\partial \Psi}{\partial \theta} e_\theta + \frac{1}{r\sin\theta}\frac{\partial \Psi}{\partial \phi} e_\phi$$

[Here Ψ is a scalar function]

From (5) of art **15.6**

$$\nabla.\mathbf{F} = \frac{1}{r^2}\frac{\partial}{\partial r}(F, r^2) + \frac{1}{r \sin\theta}\frac{\partial}{\partial \theta}(F_\theta \sin\theta) + \frac{1}{r\sin\theta}\ \frac{\partial F_\phi}{\partial \phi}$$

From (5) of **art 15.7**

$$\nabla \times \mathbf{F} = \frac{1}{r^2 \sin\theta}\begin{vmatrix} e_r & re_\theta & r\sin\theta\, e_\phi \\ \frac{\partial}{\partial r} & \frac{\partial}{\partial \theta} & \frac{\partial}{\partial \phi} \\ F_r & r\,F_\theta & r\sin\theta \mathbf{F}_\phi \end{vmatrix}$$

where $F = F_r\, e_r + F_\theta\, e_\theta + F_\phi\, e_\phi$

From (1) of **art 15.8**

$$\nabla^2\Psi = \frac{1}{r^2}\frac{\partial}{\partial r}\left(r^2\ \frac{\partial \Psi}{\partial r}\right) + \frac{1}{r^2\sin\theta}\frac{\partial}{\partial\theta}\left(\sin\theta\ \frac{\partial\Psi}{\partial\theta}\right)$$

$$+ \frac{1}{r^2\sin^2\theta}\frac{\partial^2\Psi}{\partial\phi^2}$$

Example 3. (i) Find the unit vectors e_r, e_θ, e_ϕ of a spherical co-ordinate system interms of **i, j, k** (ii) Solve for **i, j, k** interms of e_r, e_θ, e_ϕ (iiI) Represent $\mathbf{F} = 2\,y\mathbf{i} - z\mathbf{j} + 3x\mathbf{k}$ in spherical co-ordinates and hence determine F_r, F_θ, F_ϕ.

Solution: (i) The position vector of any point in spherical co-ordinates is

$$\mathbf{R} = x\mathbf{i} + y\mathbf{j} + z\mathbf{k}$$

$$= r\sin\theta\cos\phi\,\mathbf{i} + r\sin\theta\sin\phi\,\mathbf{j} + r\cos\theta\,\mathbf{k}$$

Then $\frac{\partial \mathbf{R}}{\partial r} = \sin\theta\cos\phi\,\mathbf{i} + \sin\theta\sin\phi\,\mathbf{j} + \cos\theta\,\mathbf{k}$

$$\frac{\partial \mathbf{R}}{\partial \theta} = r\cos\theta\cos\phi\,\mathbf{i} + r\cos\theta\sin\phi\,\mathbf{j} - r\sin\theta\,\mathbf{k}$$

$$\frac{\partial \mathbf{R}}{\partial \phi} = -\,r\sin\theta\sin\phi\,\mathbf{i} + r\sin\theta\cos\phi\,\mathbf{j}$$

Then unit vectors in these directions are e_r, e_θ, e_ϕ.

$$\left|\frac{\partial \mathbf{R}}{\partial r}\right| = \sqrt{(\sin\theta\cos\phi)^2 + (\cos\theta\sin\phi)^2 + (\cos\theta)^2}$$

$$= \sqrt{\sin^2\theta\,(\cos^2\phi + \sin^2\phi) + \cos^2\theta}$$

$$= \sqrt{\sin^2\theta + \cos^2\theta = 1}$$

$$\left|\frac{\partial \mathbf{R}}{\partial \theta}\right| = \sqrt{(r\cos\theta\cos\phi)^2 + (r\cos\theta\sin\phi)^2 + (-r\sin\theta)^2}$$

$$= \sqrt{r^2\cos^2\theta\,(\cos^2\theta + \sin^2\phi) + r^2\sin^2\theta}$$

$$= \sqrt{r^2(\cos^2\theta + \sin^2\theta)} = \mathbf{r}$$

$$\left|\frac{\partial \mathbf{R}}{\partial \phi}\right| = \sqrt{(-r\sin\theta\sin\phi)^2 + (r\sin\theta\cos\phi)^2}$$

$$= \sqrt{r^2\sin^2\theta\,(\sin^2\phi + \cos^2\phi)} = r\sin\theta$$

$$\therefore\ e_r = \frac{\dfrac{\partial \mathbf{r}}{\partial r}}{\left|\dfrac{\partial \mathbf{R}}{\partial r}\right|} = \sin\theta\cos\phi\mathbf{i} + \sin\theta\sin\phi\mathbf{j} + \cos\theta\,\mathbf{k} \qquad ...(1)$$

$$e_\theta = \frac{\dfrac{\partial \mathbf{R}}{\partial \theta}}{\left|\dfrac{\partial \mathbf{R}}{\partial \theta}\right|} = \cos\theta\cos\phi\mathbf{i} + \cos\theta\sin\phi\mathbf{j} - \sin\theta\,\mathbf{k} \qquad ...(2)$$

$$e_\phi = \frac{\dfrac{\partial \mathbf{R}}{\partial \phi}}{\left|\dfrac{\partial \mathbf{R}}{\partial \phi}\right|} = -\sin\phi\,\mathbf{i} + \cos\phi\,\mathbf{j} \qquad ...(3)$$

(ii) Writing equations (1), (2), (3) in the matrix form, we get

$$\begin{bmatrix} e_r \\ e_\theta \\ e_\phi \end{bmatrix} = \begin{bmatrix} \sin\theta\cos\phi & \sin\theta\sin\theta & \cos\theta \\ \cos\theta\cos\phi & \cos\theta\sin\phi & -\sin\theta \\ -\sin\phi & \cos\phi & 0 \end{bmatrix} \begin{bmatrix} \mathbf{i} \\ \mathbf{j} \\ \mathbf{k} \end{bmatrix}$$

or

$$\begin{bmatrix} \mathbf{i} \\ \mathbf{j} \\ \mathbf{k} \end{bmatrix} = \begin{bmatrix} \sin\theta\cos\theta & \sin\theta\sin\phi & \cos\theta \\ \cos\theta\cos\phi & \cos\theta\sin\phi & -\sin\theta \\ -\sin\phi & \cos\phi & 0 \end{bmatrix}^{-1} \begin{bmatrix} e_r \\ e_\theta \\ e_\phi \end{bmatrix}$$